Heterocycles via Cross Dehydrogenative Coupling

Ananya Srivastava · Chandan K. Jana
Editors

Heterocycles via Cross Dehydrogenative Coupling

Synthesis and Functionalization

Springer

Editors
Ananya Srivastava
National Institute of Pharmaceutical Education and Research
Guwahati, Assam, India

Chandan K. Jana
Indian Institute of Technology Guwahati
Guwahati, Assam, India

ISBN 978-981-13-9143-9 ISBN 978-981-13-9144-6 (eBook)
https://doi.org/10.1007/978-981-13-9144-6

This Springer imprint is published by the registered company Springer Nature Singapore Pte Ltd.
The registered company address is: 152 Beach Road, #21-01/04 Gateway East, Singapore 189721, Singapore

Preface

Carbon–carbon and carbon–heteroatom bond formation plays a central role in many chemical syntheses, and revolutions in these reactions will intensely improve the overall synthetic efficiency. This book describes the studies of past several years for the synthesis and functionalization of different heterocycles by making carbon–carbon (C-C) and carbon–heteroatom (C-X) bonds via cross-dehydrogenative coupling (CDC) of C-H and C-H or X-H. CDC, representing a new state of the art in the field of organic chemistry, expedites the synthesis and functionalization of heterocycles to provide architecturally more complex heterocyclic molecules which are the important constituents of most of the biologically active molecules, agrochemicals, etc. These methods offer an alternative to the separate steps of prefunctionalization and defunctionalization that have traditionally been part of the synthetic design. Therefore, in view of step and atom economies, direct utilization of C-H and X-H bonds without preactivation has gained enormous attention over the past decade as an effective and straightforward method in synthetic chemistry.

This book containing 13 chapters systematically describes the advances in the utilization of CDC reactions for the synthesis and functionalization of different types of nitrogen, oxygen, and sulfur heterocycles. Different mechanistic pathways have also been discussed to help readers understand the CDC reactions in-depth. In this book, a detailed account of metal-free CDC reactions for the synthesis and functionalization of heterocycles has been provided to promote the green chemistry aspect which has attracted much attention in both academic and industrial researches.

The development of CDC reactions necessitates a rethinking of some of the most fundamental concepts and theories of chemical reactivity. We hope that this book will provide a comprehensive overview of this topic, shed light on new perspectives, and inspire chemists to work toward further improvement and expansion of the application of CDC reactions.

The experts of the field contributed all the chapters so that it will be beneficial to the students and researchers in the field of synthetic chemistry. Informative and straightforward description makes this book well suited for the students at undergraduate, postgraduate, and research levels.

We wish to extend our deep appreciation to all the authors who have contributed to this book and made this project possible. We also thank our families for giving us the time, space, and freedom required to accomplish such an enormous project.

Guwahati, Assam, India

Ananya Srivastava
Chandan K. Jana

Contents

About the Editors

Dr. Ananya Srivastava completed her Ph.D. in Synthetic Organic Chemistry at the Department of Chemistry, Indian Institute of Technology Delhi, India. Prior to this, she earned her Master of Science in Chemistry with a gold medal from D.D.U. Gorakhpur University, India. She is currently a scientific officer at the National Institute of Pharmaceutical Education and Research, Guwahati, India. Her highly interdisciplinary research spans a wide range of areas in synthetic organic chemistry, nano-catalysis, material characterization, and nanobiosensors. She has published her work in leading journals and at various conferences and symposiums, including the 2nd International Symposium on C-H Bond Activation, Rennes, France. She has received numerous prestigious awards and fellowships, including JRF and SRF from the Council of Scientific & Industrial Research (CSIR), Government of India, and was an invited visiting scholar at Pusan National University, South Korea under an NRF fellowship. She has also worked with Professor Yoon Bo Shim as a visiting scientist at the Department of Chemistry, PNU, South Korea. She has completed the Patent Law course at the Nalsar University of Law, Hyderabad, India.

Dr. Chandan K. Jana graduated in Chemistry from Calcutta University in 2002, and received his M.S. in Chemical Science from the Indian Institute of Science Bangalore in 2005. He then completed his Ph.D. in Synthetic Organic Chemistry at the University of Muenster, Germany in 2008. He pursued postdoctoral research at EPEL and the University of Basel, Switzerland, and is currently an Associate Professor at the Indian Institute of Technology, Guwahati, India. His research and teaching interests focus on organic chemistry.

Functionalization of Aromatic *N*-Heterocycles with C(sp^3)–H Sources via CDC Reactions

Ramil F. Fatykhov, Igor A. Khalymbadzha and Oleg N. Chupakhin

Abstract Nitrogen-containing heterocycles are a ubiquitous nature and synthetic compounds having wide spectrum of activities, which has found applications in various industrial fields. Among a variety of synthetic approaches toward substituted *N*-heterocycles, C(sp^2)–H functionalization represents the most rapid and convenient transformation. In this review, we concentrated attention on the methods of construction of new C–C bonds via direct coupling of *N*-heterocyclic C(sp^2)–H with C(sp^3)–H derivatives, which is called cross-dehydrogenative coupling and satisfied the requirements of "atom economy" and "green chemistry." Alkanes, ethers, amines and amides, methylarenes, etc., were involved in the oxidative process with *N*-heterocyclic compound.

Keywords CDC reactions · Nitrogen heterocycles · Functionalization

Abbreviation

Ar	Aryl
BHP	2,6-Di-*tert*-butyl-4-methylphenol
Binap	2,2′-Bis(diphenylphosphino)-1,1′-binaphthyl
BPO	Benzoyl peroxide
CCHE	Cross-coupling hydrogen evolution
CDC	Cross-dehydrogenative coupling
DCDC	Double cross-dehydrogenative coupling
CFL	Compact fluorescent lamp
DCE	1,2-Dichloroethane
DCP	Dicumyl peroxide
DDQ	2,3-Dichloro-5,6-dicyano-1,4-benzoquinone
DG	Directing group

R. F. Fatykhov · I. A. Khalymbadzha (✉) · O. N. Chupakhin
Ural Federal University, Mira 19, 620002 Ekaterinburg, Russian Federation
e-mail: i.a.khalymbadzha@urfu.ru

I. A. Khalymbadzha · O. N. Chupakhin
I. Postovsky Institute of Organic Synthesis, Ural Branch of the Russian Academy of Sciences, S. Kovalevskaya/Akademicheskaya, 22/20, 620219 Ekaterinburg, Russian Federation

A. Srivastava and C. K. Jana (eds.), *Heterocycles via Cross Dehydrogenative Coupling*, https://doi.org/10.1007/978-981-13-9144-6_1

DMA	*N,N*-Dimethylacetamide
DMF	*N,N*-Dimethylformamide
DTBP	Di-*tert*-butyl peroxide
HAT	Hydrogen atom transfer
LED	Light-emitting diode
PG	Protecting group
Ph	Phenyl
PIFA	Phenyliodine bis-(trifluoroacetate)
PMP	*p*-Methoxyphenyl
Py-GC-MS	Pyrolysis–gas chromatography–mass spectrometry
RT	Room temperature
SET	Single-electron transfer
TBAB	Tetra-*n*-butylammonium bromide
TBAI	Tetra-*n*-butylammonium iodide
TBPB	*Tert*-butyl perbenzoate
TEMPO	(2,2,6,6-Tetramethylpiperidin-1-yl)oxidanyl
TFA	Trifluoroacetic acid
THF	Tetrahydrofuran
THIQ	Tetrahydroisoquinoline
TMU	*N,N,N′,N′*-tetramethylurea
Tol	Tolyl
TsOH	*p*-Toluenesulfonic acid

1 Introduction

The nitrogen heterocyclic derivatives are ubiquitous in the nature. It is present in DNA/RNA bases, amino acids, alkaloids, and many other biologically important compounds. Nonetheless, the majority of valuable heterocyclic compounds are substituted heterocycles. Among the substituents, the carbon ones, especially sp^3-carbon, provide the variability that determines the variety of properties of heterocyclic derivatives.

In this connection, carbon–carbon bond-forming reactions always attract an attention and are of primary interest for an organic chemist. Such methods developed should ideally meet with four important requirements: (1) The method should be single step; (2) environmentally benign reagents and catalysts should be employed; (3) readily available compounds should be used as starting materials; and (4) the reaction should be performed under mild and operationally simple conditions.

The oxidative heteroaromatic C–H/C–H CDC reactions satisfy most of these requirements, since it provide a convenient, one-step method for modifying readily available heterocycles with simple C–H derivatives. Moreover, the oxidative coupling corresponds to the principles of green chemistry, since water is the only by-product formed in the oxidative coupling. In addition, the reaction is carried out in one operationally simple stage using the available starting materials.

CDC reactions have a number of undoubted advantages over the "usual" coupling reactions. While the traditional method involves the metal-catalyzed cross-coupling between two prefunctionalyzated substrates, an organometallic species and a hetaryl halide or triflate, C–H/C–H CDC does not require the activation of substrates by halogens and organometallic groups.

Thus, the oxidative coupling is one of the hottest topics of modern organic synthetic chemistry.

This review systematizes the data on cross-dehydrogenative C–C coupling reactions of nitrogen-containing aromatic heterocycles with C(sp^3)–H derivatives for the last 10 years.

2 Historical Aspects of the CDC Methodology

One of the first examples of the coupling of C(sp^3)–H with C(sp^2)–H of *N*-heterocycles, which can be called "cross-dehydrogenative coupling reaction", was the works of Italian chemist Francesco Minisci and his colleagues, published in 1970–1980s and dedicated coupling of nucleophilic carbon radicals with protonated *N*-heterocyclic bases [1–5]. Since, the reaction has evolved into a very effective method for the functionalization of electron-deficient heteroaromatics. For this reason, homolytic substitutions of protonated hetarenes are often called "Minisci reaction" in the chemical literature [6]. So, Punta and Minisci have described the radical-based methods as "a Friedel–Crafts-type process with opposite reactivity and selectivity" [4]. In general, the Minisci reaction is shown in Scheme 1. The nucleophilic radical added to the cationic substrate to afford an *aminyl* radical cation which is subsequently oxidized to the corresponding alkylated compound under the reaction conditions. Due to the nucleophilic nature of the radical, the reaction proceeds exclusively at the electrophilic centers of the heterocyclic system at the α- or γ-positions with high regioselectivity.

Using this methodology, important substitution reactions, such as alkylation [3], hydroxymethylation [3], benzylation [7], carbamoylation [8], acylation [3], α-amidoalkylation [8], were developed utilizing an easily available reagent as source of nucleophilic radical. Minisci et al. used quite simple and available compounds such as cyclohexane, methanol, tetrahydrofuran (THF), dioxane, toluene, *N,N*-dimethylformamide (DMF), formamide and benzaldehyde, that acted as the source of the nucleophilic carbon-centered radical. In the course of CDC reaction, the active radical formed was generated by Fe^{II}-catalyzed reaction in the presence of an oxidant (Scheme 1) [3, 7, 8]. However, nucleophilic radical was produced from the reaction of silver salts with carboxylic acid derivatives [2, 3] in the "classical" Minisci reaction [6].

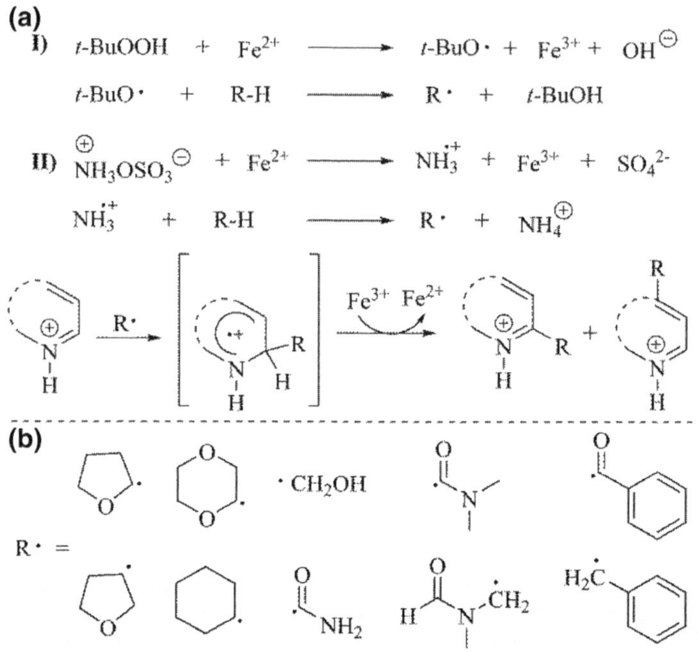

Scheme 1 General mechanism of the Minisci reaction with initial generation of radical (**a**) and diversity of nucleophilic radicals (**b**)

3 CDC *N*-Heterocycles with Alkanes

In 2009, Li et al. published several papers of the direct alkylation of electron-deficient *N*-heterocycles and their N-oxides by reaction with cycloalkanes in the presence of the di-*tert*-butyl peroxide (DTBP) under catalysis by transition metals [9] or their absence [10], respectively. Both processes require a large excess of alkanes at high temperature that initiates homolytic decomposition of DTBP. Cycloalkanes from C_5 to C_8 included norbornane were used in these coupling. Pyridine core was alkylated on both C2 and C6 positions in the presence of catalytic amount of $Sc(OTf)_3$ Lewis acid (Scheme 2a) [9]. The alkylation of pyridine N-oxide in transition-metal-free conditions also gave rise to the mixture products: 2,6-disubstituted- and 2,4,6-trisubstituted-pyridines (Scheme 2b) [10]. When dioxane was involved in the coupling with pyridine N-oxide, a mixture of di- and trisubstituted products of oxyalkylation was observed in good yield [10]. Interestingly, a complex mixture of products in overall yields 70% was produced, when heptane was used as substrate [9].

Antonchick et al. reported the alkylation of various azaheterocycles, i.e., quinoline, 1,10-phenantroine, caffeine, quinazoline, benzimidazoles etc. with alkanes under catalysis by phenyliodine bis-(trifluoroacetate) (PIFA) in the presence of NaN_3 as additive [11]. Results of this coupling are depicted in Scheme 3.

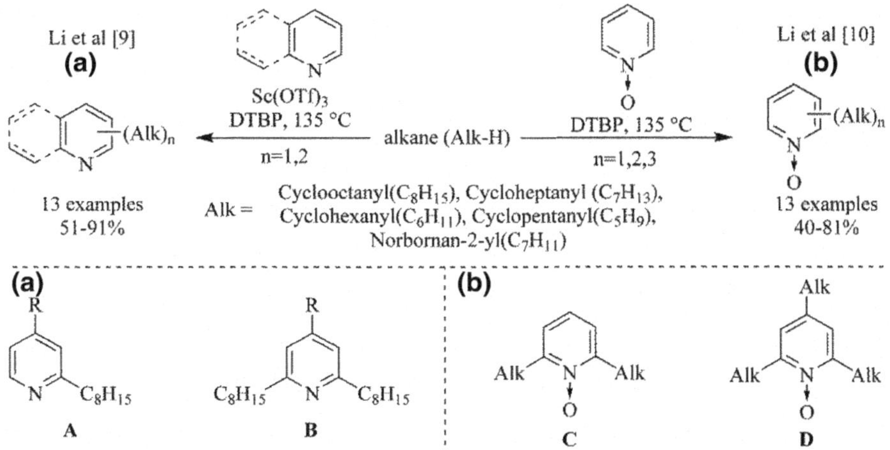

R	yields	**A**:**B**
H	64%	1:1.3
Me	68%	1:1
Ph	68%	1:1.5
Co_2Me	52%	only **A**

Akl	yields	**C**:**D**
C_8H_{15}	71%	6:1
C_7H_{13}	61%	4:1
C_6H_{11}	70%	4.8:1
C_5H_{19}	40%	trace **D**

Scheme 2 Alkylation of hetarenes under Sc-catalyzed (**a**) and transition-metal-free (**b**) conditions

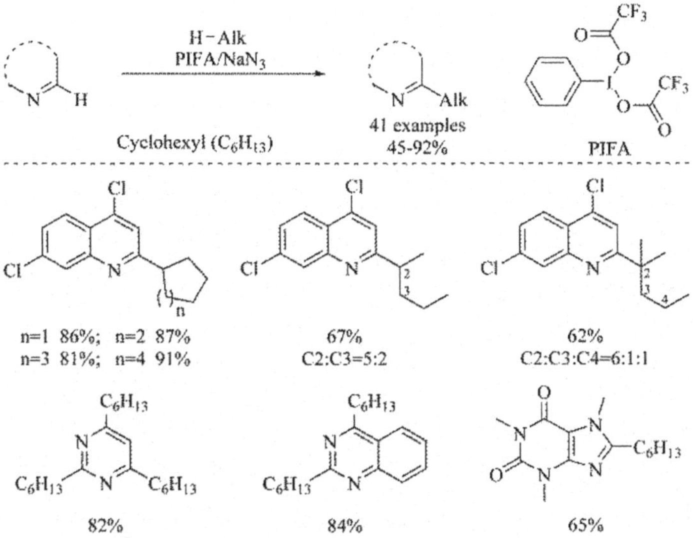

Scheme 3 Alkylation of *N*-heterocycles by PIFA/NaN_3

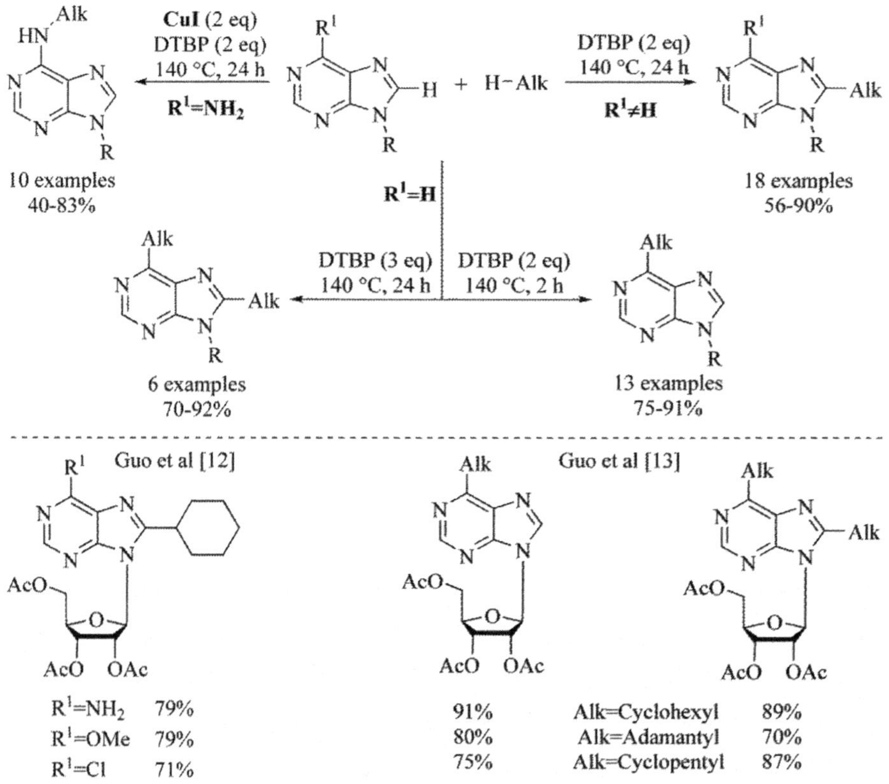

Scheme 4 DTBP-mediated alkylation of the purines

Products of alkylation with cycloalkanes were formed with excellent yields. It is interesting to note that for linear acyclic alkanes, the binding site is located exclusively at the secondary C–H bond. However, for the branched acyclic alkanes, a clear preference for hetarylation at the 3° site over the 2° site was observed with good yields.

Guo et al. described the DTBP-mediated alkylation of the purine core at the C6 and C8 positions with various cycloalkanes in the absence of transition metal (Scheme 4) [12, 13]. When the C6 position of purines was unsubstituted (Scheme 4, R^1=H), the C6-alkylated product was formed in the presence of DTBP (2 equiv) at 140 °C for 2 h in 75–91% yields. By extending the reaction time to 24 h and the amount of peroxide to 3 equiv, the 6,8-dialkylated products were produced [13]. Under the same conditions, alkylation of the 6-substituted purines gave rise to 8-alkylpurine derivatives (Scheme 4, $R^1 \neq$H) [12]. Interestingly, the addition of CuI to the CDC reaction conditions, the reaction of 6-aminopurine (adenine) with cycloalkane led to exclusively *N*-alkyladenines (Scheme 4, R^1=NH_2), as suggested by the authors, due to the generation of alkyl cation instead of carbon radical [12].

4 CDC *N*-Heterocycles with (Thio)Ethers and Alcohols

Recently, Phillips and Pombeiro [14] and Lakshman and Vuram [15] summarized a lot of the information of CDC reactions of *N*-heteroaromatics with (thio)ethers and alcohols. Various azoles, including indole, pyrrole, thia- and oxazole and their benzannulated analogs, *N*-heterocyclic bases and their N-oxides, were involved in CDC process with alcohols and ethers. The basic scope of oxy- and thio-alkylation of *N*-heterocyclic compounds is shown in Scheme 5. It should be noted that these couplings mainly proceed similar to the radical Minisci-type reaction.

After, Zhang et al. reported the direct access to the oxyalkyl-substituted pyrazine N-oxides by CDC reaction of the pyrazine N-oxides and ethers in the presence of TBHP (Scheme 6) [16]. α-Oxyalkylated products were produced with not only pyrazine N-oxides, but also quinoline and pyridine N-oxides in excellent regioselectivity. Interestingly, using pyrolysis–gas chromatography–mass spectrometry (Py-GC-MS) experiment was shown that 1,4-(dioxan-2-yl)-3,6-dimethylpyrazine N-oxide is a potential flavor precursor that can release several characteristic flavor components upon heating.

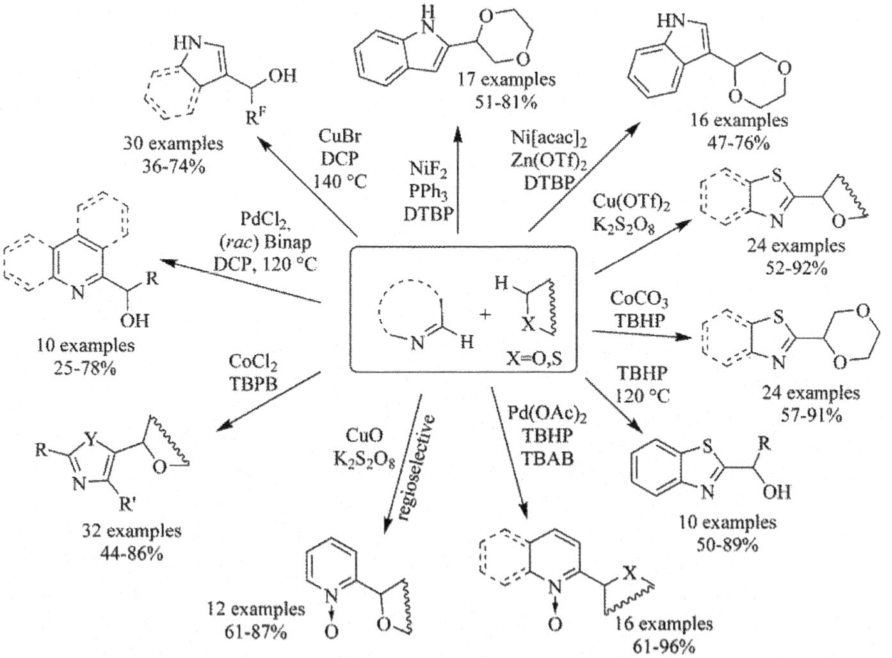

Scheme 5 Scope of CDC reaction of *N*-heteroaromatic compounds with (thio)ethers and alcohols reviewed by Faisca Phillips and Pombeiro [14] and Lakshman and Vuram [15] (DCP = dicumyl peroxide, TBPB = *tert*-butyl peroxybenzoate, TBAB = tetra-*n*-butylammonium bromide, Binap = 2,2′-bis(diphenylphosphino)-1,1′-binaphthyl)

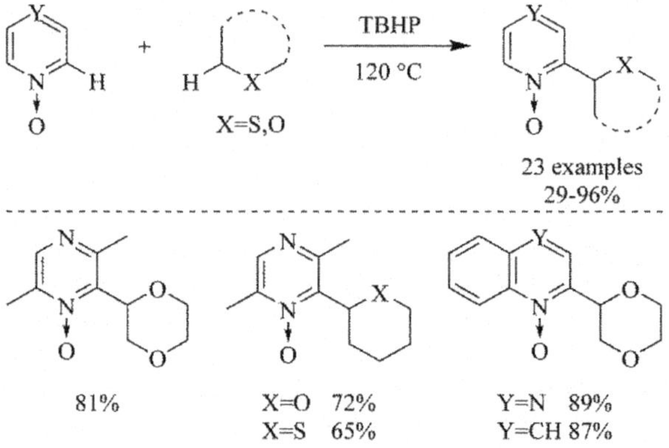

Scheme 6 Alkylation of *N*-heterocycles with ethers

Wang group published aerobic co-catalyzed CDC reaction of 2-arylimidazo-[1,2-*a*]pyridines with isochroman using molecular oxygen as oxidant (Scheme 7) [17]. A variety of 3-(isochroman-1-yl)-imidazo[1,2-*a*]pyridines were prepared in moderate to good yields. It should be noted, other ethers, i.e., isothiochroman,

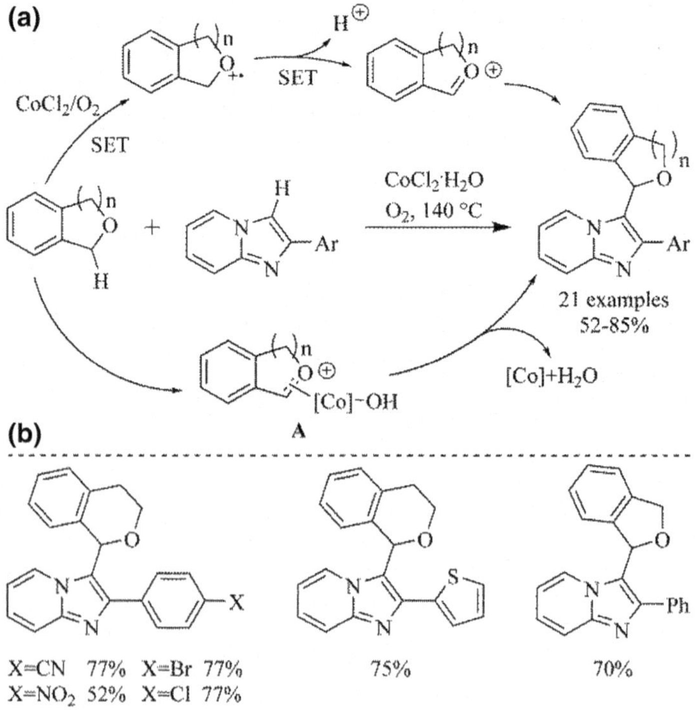

Scheme 7 Co-catalyzed CDC imidazopyridine with isochroman

1,3-benzodioxole, 1,4-dioxane, and benzyl methyl ether, were inferior to the reaction, and little products were detected by GC analysis except 1,3-dihydroisobenzofuran, which was found to be compatible and the product was prepared with 70% yield. Interestingly, CDC product was obtained in 70% yield in the present radical trapping (2,2,6,6-tetramethylpiperidin-1-yl)oxidanyl (TEMPO). On the basic of this observation, two possible mechanisms of reaction were proposed by the authors. The first pathway proceeded through single-electron transfer (SET) to generate oxonium ion. The oxonium ion presumably was formed by another SET and deprotonation (Scheme 7a). The another mechanism was that the reaction of isochroman with O_2 and $CoCl_2$ generated co-centered oxonium ion; the reactive nucleophile trapped the oxonium ion **A**, giving the coupling product and water (Scheme 7b).

Zhu et al. researched Au-catalyzed interaction of hetarenes with alcohols [18]. Six gold catalysts shown in Scheme 8 were evaluated in combination with TBHP. While all the catalysts gave product, $HAuCl_4 \cdot 4H_2O$ proved to be most promising. Using optimized conditions ($HAuCl_4 \cdot 4H_2O$, TBHP, 120 °C, 24 h),

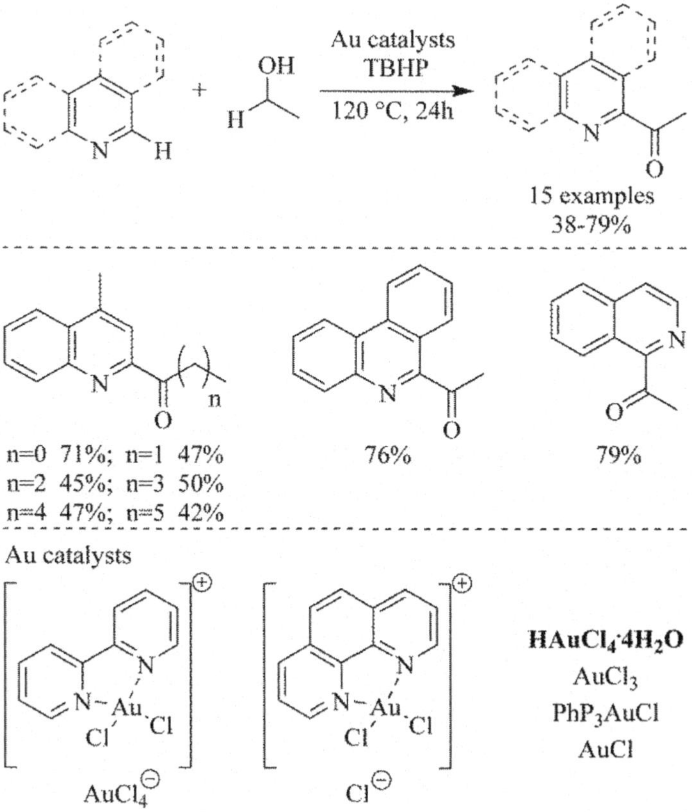

Scheme 8 Au-catalyzed acylation of benzazines with alcohols

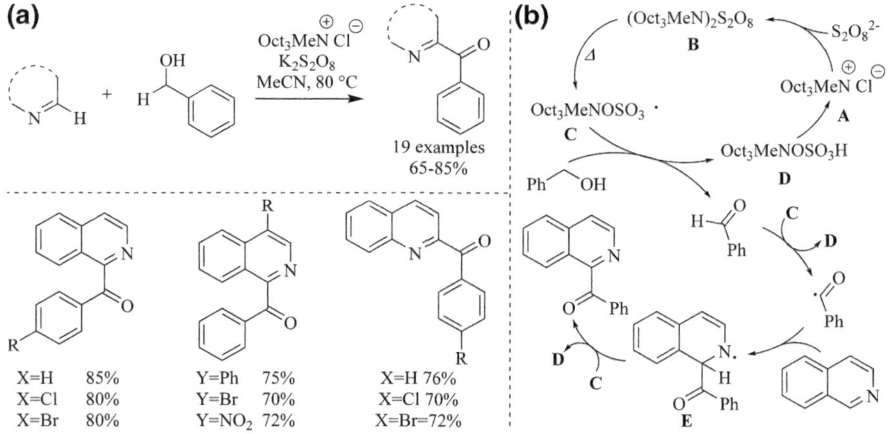

Scheme 9 Benzoylation reaction of (iso)quinolines with benzylic alcohols (**a**) and plausible mechanism (**b**) of this process

several *N*-heterocycles were acylated with different alcohols. The reaction was fully suppressed in the presence of radical inhibitors, indicating the radical nature of the process. Interestingly, no reaction occurred in the absence of Au. The authors suggested coordination of Au to the heterocycle group, as well a role of Au in the final oxidation.

Later, transition metal-free acylation of benzazines with benzyl alcohols was proposed [19]. A variety of the α-benzoyl-quinolines and isoquinolines were prepared in the presence of $K_2S_2O_8$ on combination with methyltrioctylammonium chloride (Oct_3N^+ $MeCl^-$) as a phase transfer agent in MeCN at 80 °C. Results of this process and plausible mechanism are depicted in Scheme 9. It was likely that initially benzaldehyde was formed by hydrogen atom transfer (HAT) from benzylic alcohol to methyltrioctylammonium sulfate radical **C**. Second HAT converted benzaldehyde into benzoyl radical, which attacked the azine core at the α-position to form acyl amine radical **E**. Finally, HAT and rearomatization gave the product. No product was formed in the presence of the radical trap TEMPO under the CDC conditions that radical intermediates were involved in the CDC reaction.

5 CDC *N*-Heterocycles with Methylarenes

In 2015, Liu et al. published CDC reaction between isoquinoline and methylarenes (Scheme 10) [20]. Several Lewis acids (trifluoroacetic acid (TFA) and triflate salts of transition metal such as Cu, La, Sc, Y, In) and oxidants (H_2O_2, DTBP, TBHP, DCP, $K_2S_2O_8$/TBAB, etc.) were tested for the optimized conditions of coupling of isoquinoline with toluene derivatives. While combination of DTBP with Lewis acids led to a mixture of products of the benzylation and the benzoylation reactions,

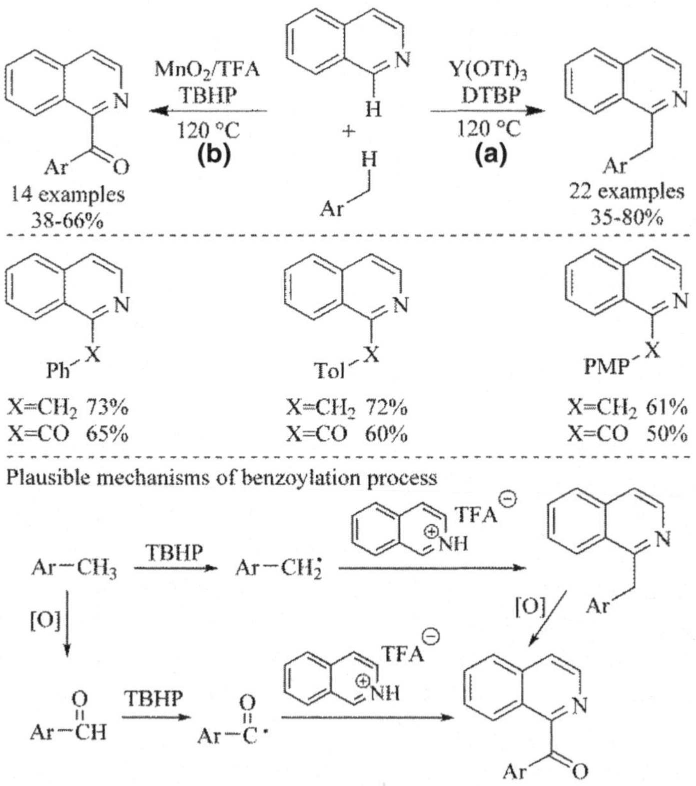

Scheme 10 α-Benzylation (**a**) and α-benzoylation (**b**) of isoquinolines (PMP = *p*-metoxyphenyl)

catalytic system DTBP/$Y(OTf)_3$ at high temperature predominantly produced α-benzylisoquinoline and α-benzoylisoquinoline was observed in trace amount. On the other hand, combination of TFA with TBHP in the presence of additive of MnO_2 predominantly gave the benzoylation product in good yield. Series of α-benzyl- and α-benzoylisoquinoline derivatives were prepared using these catalytic systems. The authors proposed that the benzylation proceeded via mechanism of the Minisci reaction, but benzoylation process may be realized by means of two possible mechanisms, that are shown in Scheme 11 [20].

In the following, Yang et al. modified the protocol of benzylation reaction by Liu et al. [20] using molecular iodine instead of $Y(OTf)_3$ under the same condition (Scheme 11) [21]. It should be noted that *N*-benzylated isoquinolin-1-one product was formed in trace amount under the optimized condition. However, using TBHP instead of DTBP exclusively led to the *N*-benzylated product in good yield.

Patel research group has expanded the process of α-benzoylation on the other heterocyclic bases such as quinoxaline and quinoline [22]. $AlCl_3$ as catalyst was used in the presence of TBHP for preparation of the α-benzoylated azines.

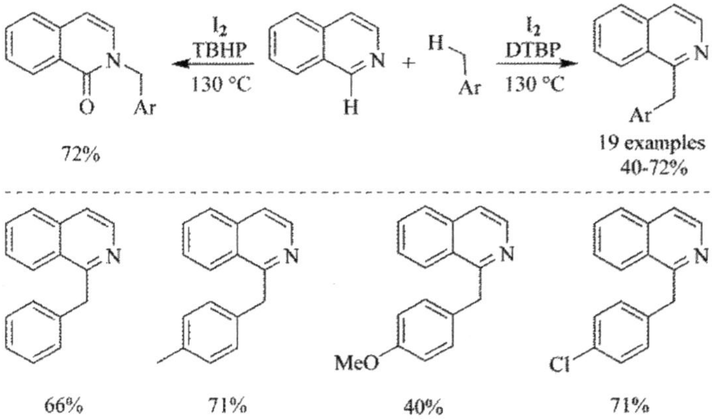

Scheme 11 I_2-catalyzed α-benzylation of isoquinolines

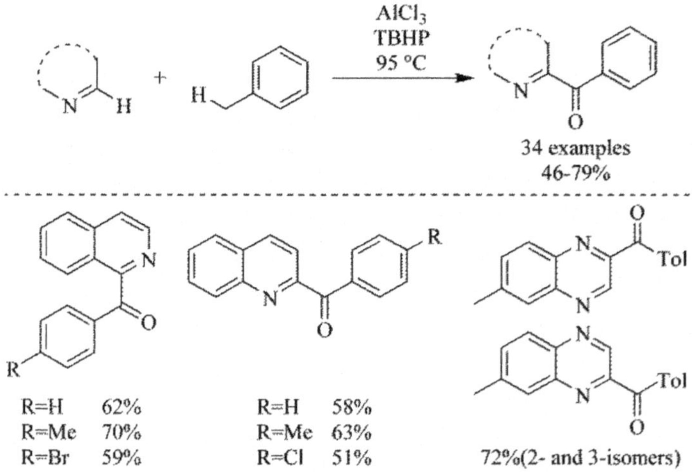

Scheme 12 $AlCl_3$-catalyzed benzoylation of benzazines with toluene

Results of this process are shown in Scheme 12. When the asymmetric quinoxaline was applied, a mixture of the isomers was obtained in the yield of 72%.

At the same time, Kianmehr et al. developed the Pd-catalyzed benzylation of pyridine N-oxides in the presence of the oxidant $K_2S_2O_8$ at high temperature (Scheme 13) [23]. Later, they proposed visible light promoted conditions for preparation of the 2-benzylpyridine N-oxides. The reaction was carried out in the presence of catalytic amount of CuO under irradiation with blue light-emitting diode (LED) at room temperature (RT) using the same oxidant (Scheme 13) [24]. Visible light-mediated process gave comparable results in comparison with the Pd-catalyzed coupling. Interestingly, when 2-ethylpyridine N-oxide was underwent

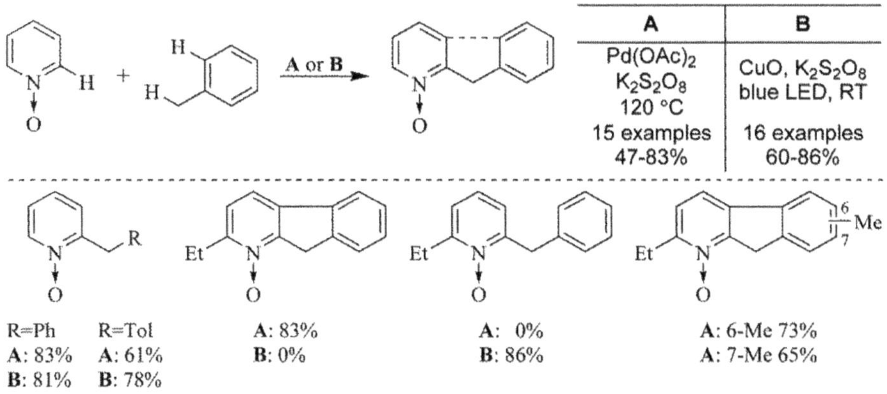

Scheme 13 Pd^{II}- and Cu^{II}-catalyzed benzylation of pyridine N-oxides

thermal conditions, the formation of the indeno[2,1-*b*]pyridine N-oxide derivative was observed. On the other hand, no indene ring was formed under irradiation media with 1-ethylpyridine N-oxide and only benzylation product was yielded.

Also coupling of the heterocyclic N-oxides with methylarene derivatives may take place in the absence of catalyst. For example, Guo et al. found that using DCP as oxidant in toluene solvent at 110 °C gave rise to the benzylation product of azine N-oxides in moderate to good yields (Scheme 14) [25]. Similar to protocol of Liu et al. [20], side product of benzoylation was observed in trace amount. Interestingly, using the 4-phenyl pyridine N-oxide as substrate led to the deoxygenated product in moderate yield. The reaction of methylation was observed when 8-hydroxy by or 8-ethoxy quinoline N-oxides are underwent standard condition. The authors proposed that DCP is the source of methyl fragment what was earlier shown via direct methylation of hetarenes with DCP [26, 27].

Subsequently, Kianmehr et al. showed that high temperature is required for the benzoylation reaction in the Cu-catalyzed coupling of pyridine N-oxide with toluene (Scheme 15) [24], since, as shown in Scheme 13, 2-benzylpyridine derivatives were obtained, when $S_2O_8^{2-}$ was underwent homolysis under irradiation at RT. In benzoylation case, authors believed that initially toluene was oxidized to form benzaldehyde at high temperature by the sulfate radical anion that was produced from $K_2S_2O_8$. Authors proposed radical mechanism of this reaction, so no product was observed in the presence of radical scavenger. Plausible mechanism is illustrated in Scheme 15.

The CDC reaction of N-protected indoles with methylarenes was also developed. For example, *N*-carbamoyl indoles react with benzylic C(sp^3)–H of diphenylmethanes under Fe^{II} catalysis in the presence of 2,3-dichloro-5,6-dicyano-1,4-benzoquinone (DDQ) to give 3-(diphenyl)methyl-indoles (Scheme 16) [28]. No benzylation product was observed, when indoles with the other N-protected groups (PGs) were underwent CDC conditions except *N*-acetyl group. In this case, the product was isolated in 32% yield. The authors assumed that reaction proceeded via formation of

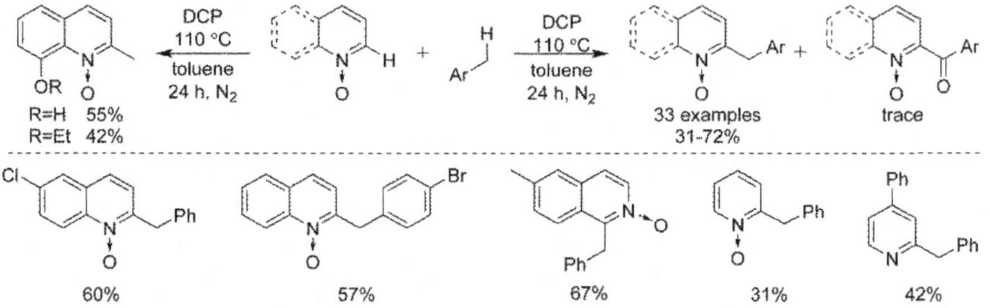

Scheme 14 Transition-metal-free benzylation of azine N-oxides

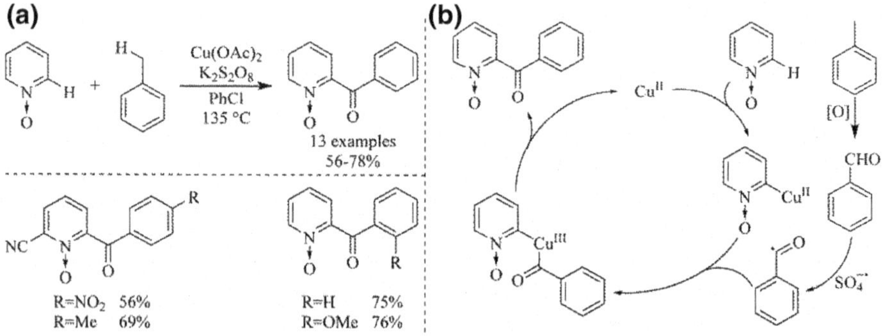

Scheme 15 Cu-catalyzed benzoylation of pyridine N-oxides (**a**) and plausible mechanism (**b**)

the benzyl cation by HAT and SET processes from diphenylmethane to DDQ. Subsequently, a nucleophilic reaction between *N*,*N*-dimethyl-1*H*-indole-1-carboxamide and the benzyl cation furnished the final product.

The replacement of *N*-carbamoyl fragment with *N*-pyrimidinyl allows to conduct metal-catalyzed benzylation at the position C2 of the indole [29]. Set of *N*-pyrimidin-2-yl 2-benzylindoles was prepared under Cu^{II}/DTBP catalytic system at 120 °C in the presence of benzaldehyde as an additive (Scheme 17). The pyrimidin-2-yl fragment acts as coordinating moiety (an "internal ligand"), which directed a metal catalyst into the proximity of a certain C–H bond in the molecule, leading to its selective cleavage and subsequent functionalization, so-called directing group (DG) [30, 31]. It should be noted, no product was observed, when *N*-phenylindole was underwent standard conditions. 3-Bromo-*N*-pyrimidylindole alone underwent homocoupling in DCE to give 2,2′-bisindole in 59% yield. The CDC product was isolated in 11% yield in the presence of radical scavenger. Based on these observations, the authors suggested the reaction mechanism, which is depicted in Scheme 17. Firstly, coordination-directed C–H cupration of *N*-(2-pyrimidyl)-indoles afforded the metallacycle intermediate **A**. The benzyl radical that arises from the reaction between toluene and tert-butoxy radical interacted with intermediate **A** to form the desired benzylation product [29].

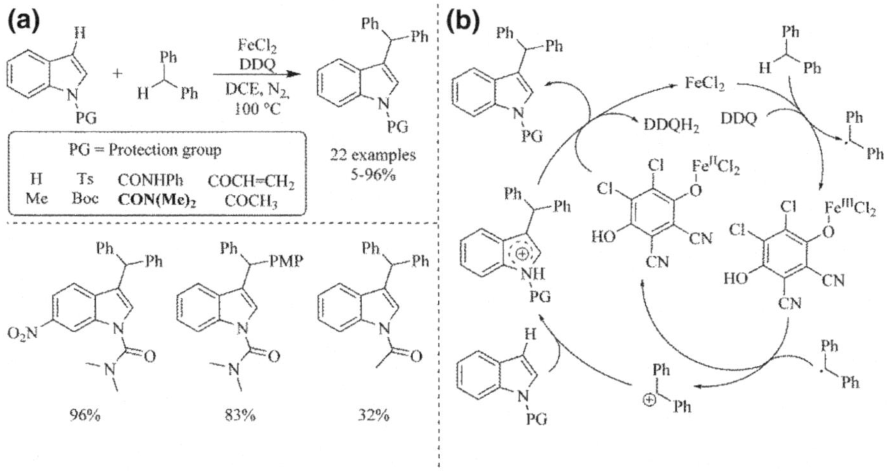

Scheme 16 Fe-catalyzed benzylation of *N*-carbamoylindoles (**a**) and plausible mechanism (**b**)

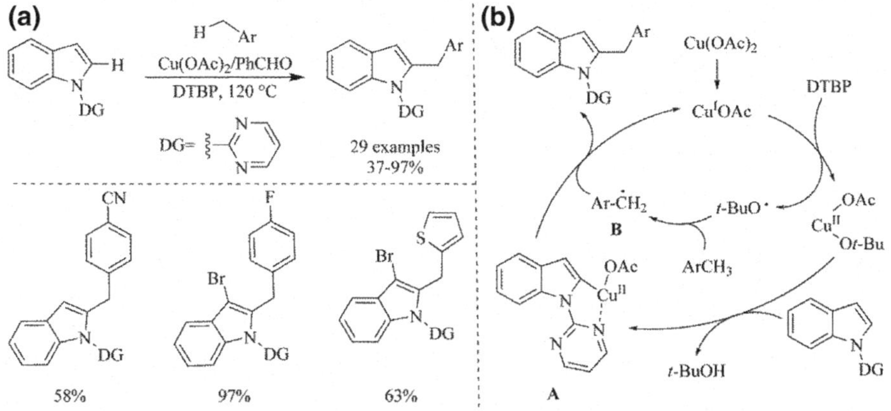

Scheme 17 Cu-catalyzed C2-benzylation of indoles (**a**) and tentative mechanism (**b**)

6 CDC *N*-Heterocycles with Alkylamines and Alkylamides

As shown in Scheme 1, the Minisci reaction between *N*-alkylamides and protonated heterocycles bases proceeds under metal catalysis. Recently, Togo et al. researched transition-metal-free conditions for the amidoalkylation reaction. Various oxidants, i.e., H_2O_2, TBHP, DTBP, $K_2S_2O_8$, and benzoyl peroxide (BPO), in the presence of TFA were examined under irradiation with a Hg lamp [32]. While all oxidant promoted the Minisci reaction, system BPO/TFA proved to be most promising. Various amidoalkyl-substituted benzenes were prepared under optimized conditions in moderate to good yields (Scheme 18). It should be noted that amidoalkylation of unsubstituted quinoline proceeded at the α-/γ-positions simultaneously in 84% yield.

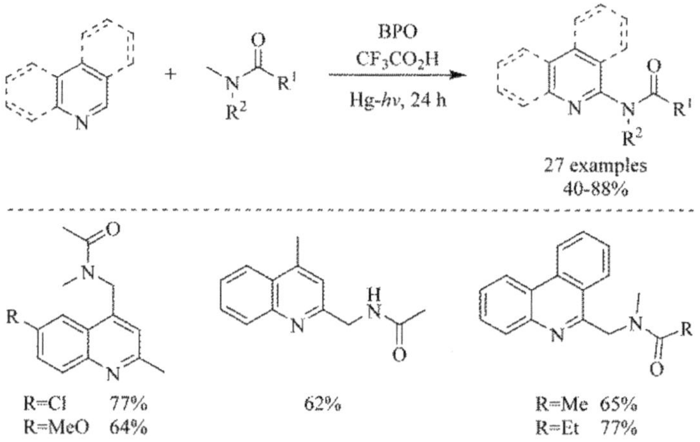

Scheme 18 Transition-metal-free Minisci-type reaction between azines and *N*-alkyl amides

In 2005, Li et al. developed coupling of $C(sp^2)$–H indole with $C(sp^3)$–H tertiary amines (Scheme 19) [33]. Coupling *N*-aryl tetrahydroisoquinoline (THIQ) with unprotected indoles was successfully performed in the presence of a catalytic amount of CuBr and TBHP. It is worth noting that the reaction was not sensitive to moisture or air and that the products were obtained in good yields even when the

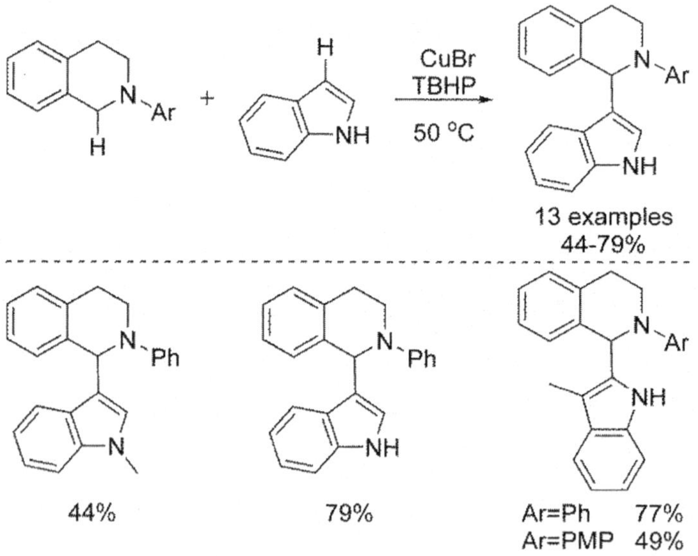

Scheme 19 Cu-catalyzed CDC reaction of indoles with THIQ

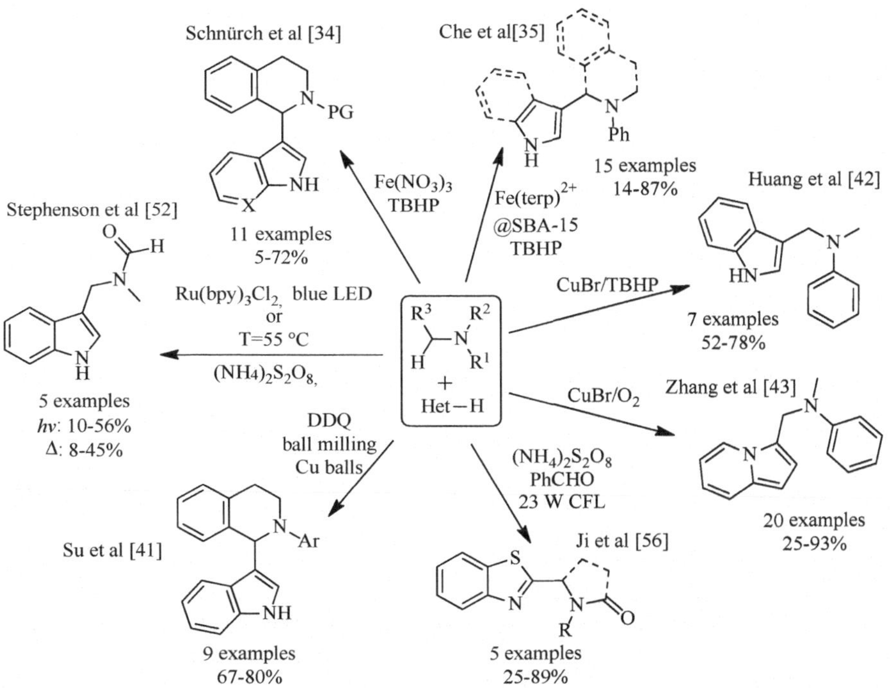

Scheme 20 Scope of CDC reaction of azoles with tertiary amine (CFL = compact fluorescent lamp)

reaction was carried out in water under an atmosphere of air. The reactions occurred selectively at the C3 position of indoles, if both C2 and C3 positions of indoles are unoccupied. When the C3 position of indoles is substituted, the C2-substituted products were obtained.

In the following years, the scope of CDC azoles with tertiary amines was significantly extended by the development of highly efficient catalyst systems with transition metal such as iron [34–37], copper [38–44], manganese [45], cobalt [46, 47], antimony [48], vanadium [49, 50], ruthenium [51, 52], platinum [53, 54], or without transition metal [55, 56]. The main results of this coupling are depicted in Scheme 20.

In the contrast of the free radical Minisci reaction, coupling of tertiary amines with pronucleophiles, in particular, electron-donating heterocycles, proceeds via C–H activation of amine to form iminium cation, which added nucleophile to produce CDC product [57]. In the last few years, mechanical researches by Doyle [58] and Klussmann [59–62] have shed new light on the CDC reaction of amines. Hypothetical iminium intermediate was isolated in the aerobic CDC reaction of *N*-phenyl THIQ catalyzed by copper salts [59, 60, 63]. The key feature of this reaction under catalytic system $CuCl_2/O_2$ is that the amine is oxidized by two molecules $Cu^{II}Cl_2$ to give the iminium dichlorocuprate и $Cu^{I}Cl$ (Scheme 21a).

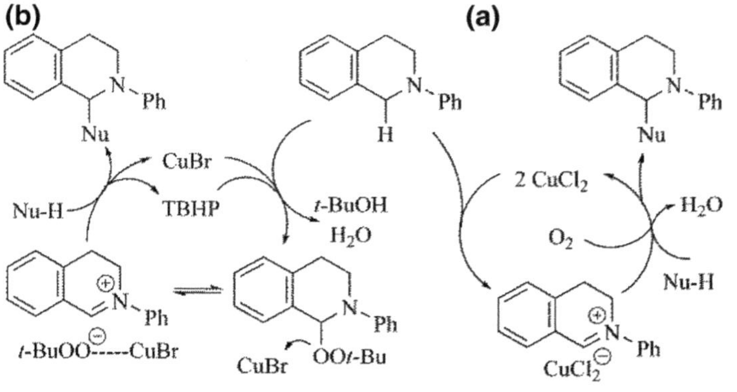

Scheme 21 Mechanism of the oxidative coupling of the tertiary amine with pronucleophile using the $CuCl_2/O_2$ system (**a**) and the CuBr/TBHP system (**b**)

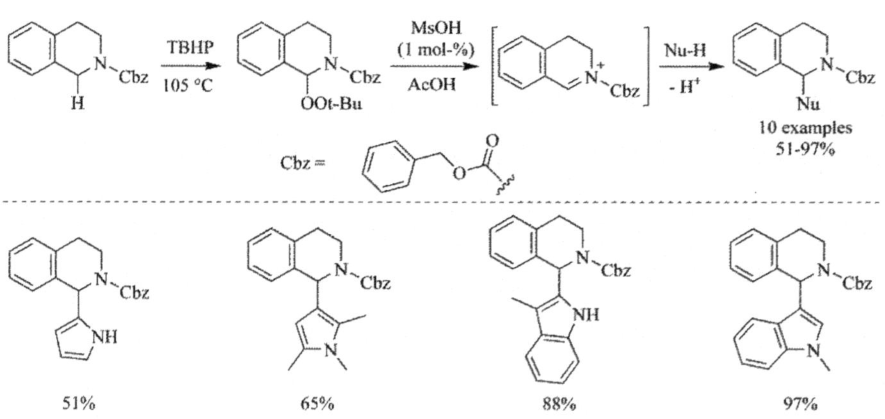

Scheme 22 CDC indoles and pyrroles with THIQs via intermediate peroxides of amine

The role of oxygen is the reoxidation of Cu^I to Cu^{II}. A recent computational study by Wu and co-workers supports this proposed mechanism [64].

However, in the process catalyzed by system CuBr/TBHP the key intermediate is α-amino peroxide, which was formed from amine and TBHP by means of the Cu-catalyzed radical reaction and acts as iminium ion precursor (Scheme 21b). The role of CuBr is to catalyze the decomposition of the oxidant TBHP into $^tBuO^·$ and $^tBuOO^·$ [58, 60, 62]. To confirm this theory, Klussmann and co-workers developed the CDC reaction of N-protected THIQs with pronucleophiles via intermediate peroxides, which were synthesized from TBHP under metal-free thermal conditions and were converted into the final products by Brønsted acid catalyzed substitution (Scheme 22) [65].

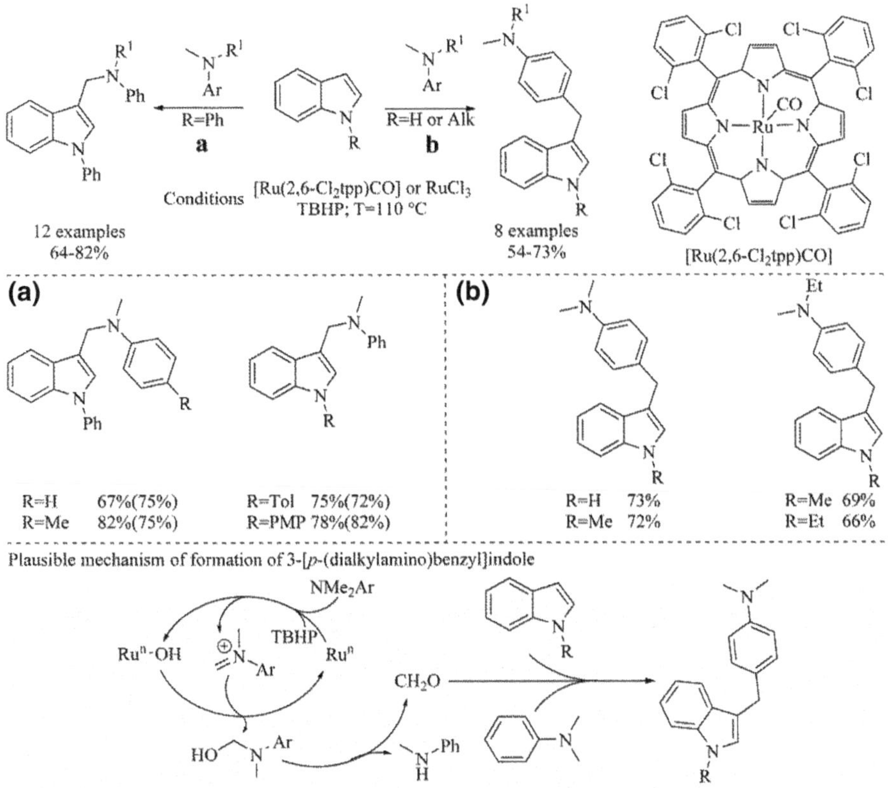

Scheme 23 CDC reaction of indoles with *N*-alkylated anilines (yields of $RuCl_3$-catalyzed coupling are shown in parentheses)

In 2010, Che group reported the Ru-catalyzed aminomethylation of indoles with *N*, *N*-dialkylanilines in the presence of TBHP as oxidant at high temperature [51]. $RuCl_3$ and ruthenium porphyrins (particularly [Ru(2,6-Cl_2tpp)CO]; tpp = tetraphenylporphinato) were utilized in this process. The aminomethylation of *N*-phenyl indoles gave rise to 3-{[(*N*-aryl-*N*-alkyl)amino]methyl}indole (Scheme 23a). However, coupling of *N*-alkyl or N-H indoles with *N*-methyl, *N*-alkylanilines led to 3-[*p*-(dialkylamino)benzyl]indole (Scheme 23b). Using *N*,*N*-di(trideuteromethyl)-aniline the authors showed that amine acted as methylene source (MA). Oxidizing of *N*,*N*-dimethylaniline in the presence of Ru catalyst led to *N*-methylaniline in moderate yield. Based on these observations, the authors suggested that the formaldehyde was formed from *N*,*N*-dimethylaniline, which reacted with indole to form the 3-alkylidene-3H-indolium cation intermediate. Nucleophilic conjugate addition of the second aniline molecule to alkylidene intermediate gave rise 3-[*p*-(dialkylamino) benzyl]indole.

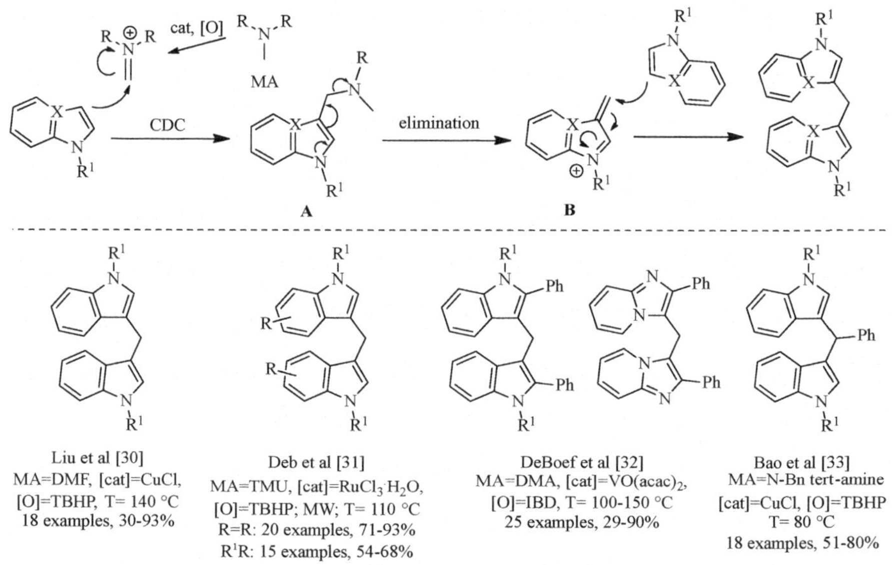

Scheme 24 Synthesis of bis(hetaryl)methanes

Subsequently, several protocols included CDC process were proposed for the preparation of the 3,3′-bis(indolyl)methanes and bis(imidazo[1,2-*a*]pyridin-3-yl) methanes by transition metal salts catalyzed reactions at high temperature. The scope of this process is depicted in Scheme 24. DMF [66], *N,N,N′,N′*-tetramethylurea (TMU) [67], *N,N*-dimethylacetamide (DMA) [68], and dialkylbenzylamine [69] acted as MA. It should be noted that in case of the dialkylbenzylamine, more reactive methylene center of the benzyl fragment was involved in the reaction. In contrast to mechanistic assumptions of Che et al. [51], another mechanism was proposed (Scheme 24). The key feature is the cleavage C–N bond of intermediate **A** to form intermediate **B** via the elimination reaction [66, 68].

Nakamura et al. reported an intramolecular hydroarylation-redox CDC of N-substituted *N*-propargyl anilines with the excess of indoles proceeded in the presence of $Zn(OAc)_2$ catalyst under high temperature to give 2-indolyltetrahydroquinolines (Scheme 25) [70]. When R is not Ph in the aniline, then hydroarylation-redox CDC product **A** was formed. When R is Ph, the reaction of *N*-propargyl aniline with 4 equivalents of indole proceeded to produce the bis (indolyl)methane. Interestingly, the reaction of diphenylpropargylaniline with 2 equivalents of indole gave rise only product **A**, which can be transformed into compound **B** by the addition of excess of indole. In the following DFT studies, Zhang and co-workers suggested the mechanism of this hydroarylation-redox CDC reaction [71]. The catalytic cycle consists of two subcycles, through which three C–H bonds (two sp^2 and one sp^3) are activated sequentially. One subcycle carries out the transformation of *N*-propargylaniline to the intermediate product

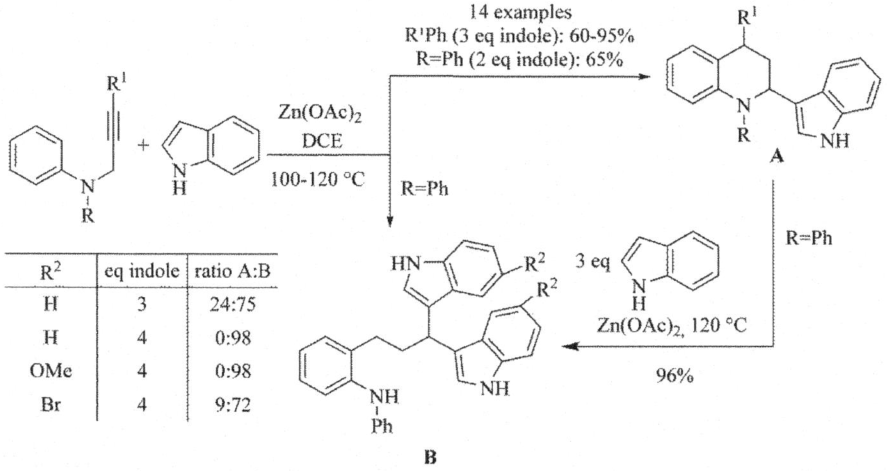

R^2	eq indole	ratio A:B
H	3	24:75
H	4	0:98
OMe	4	0:98
Br	4	9:72

Scheme 25 $Zn(OAc)_2$-catalyzed coupling of indoles with *N*-propargyl aniline (DCE = 1,2-dichloroethane)

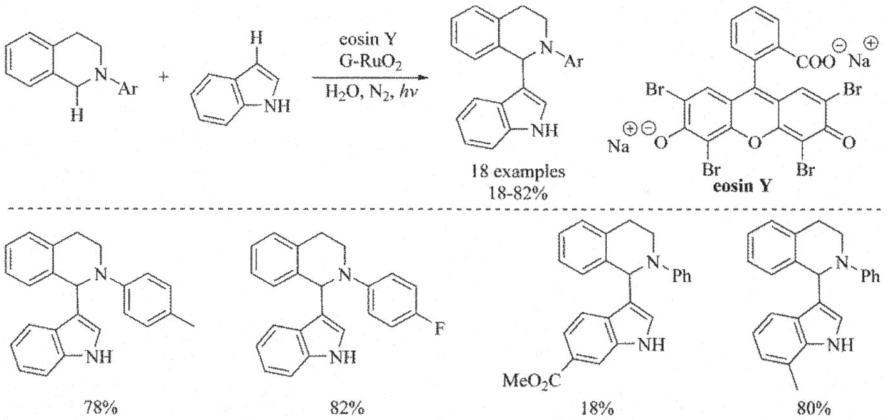

Scheme 26 CCHE reaction of indoles with THIQs

1,4-dihydroquinoline and another realizes the transformation of *N*-benzyl-1,4-dihydroquinoline with indole into the final product.

In 2013, Wu et al. reported a novel type of CDC reaction of *N*-aryl THIQs with undoles, which they called cross-coupling hydrogen evolution (CCHE) [72]. This method does not require any external oxidant, and only hydrogen is released as side product. Combining of eosin Y and graphene-supported RuO_2 nanocomposite (G-RuO_2) as photosensitizer and catalyst, authors produced coupling products of indoles with *N*-aryl THIQs under visible light irradiation at RT (Scheme 26).

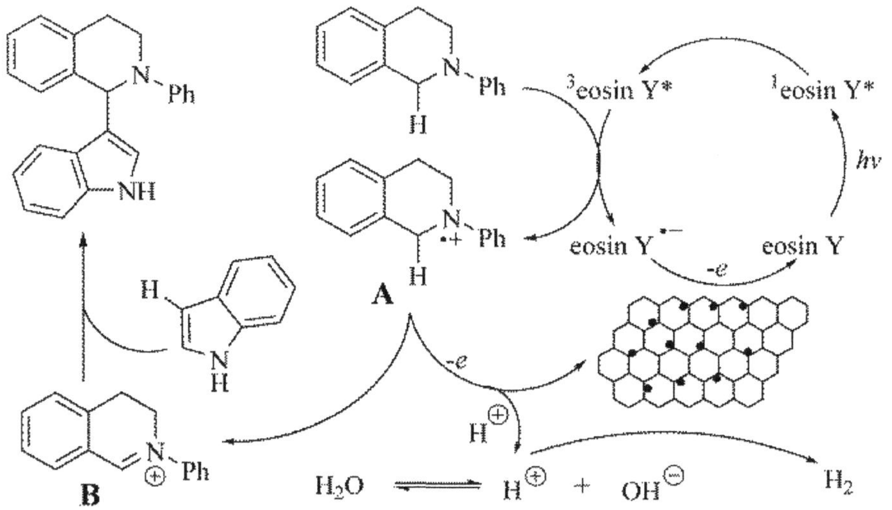

Scheme 27 Proposed mechanism of CCHE reaction by Wu et al. [72]

Proposed mechanism of CCHE reaction is depicted in Scheme 27. Upon visible light irradiation, eosin Y was turned to its singlet excited state ^{1}[eosin Y]* that quickly transferred to its triplet state ^{3}[eosin Y]*. With the addition of N-protected THIQ into the solution, an SET from THIQ to ^{3}[eosin Y]* state resulted in the formation of cation radical **A** and eosin Y radical anion [eosin Y]$^{\cdot-}$. The generated intermediate **A** further released a proton into water and then was oxidized to afford an iminium ion intermediate **B**. Subsequently, the nucleophilic addition of indole to the iminium ion **B** gave rise to the final product [72].

Recently, Segundo and Correa published the review [73] dedicated functionalization of α-aminocarbonyl compounds included the peptides with a pronucleophiles included various *N*-heterocycles by means of the CDC process. The reactions also proceed through primary activation of the aminocompound **A** to form the iminium cation **B**, to which is added with nucleophile to produce the CDC product **C** (Scheme 28a). It should be noted that the compound **C** can be transferred to the appropriate α-imino- **D** or keto-compound **E**. Interestingly, ^{18}O-labeling experiments confirmed that the newly formed carbonyl group upon hydrolysis was derived from water.

Scheme 28b displayed the scope of the CDC reaction of indoles with α-aminocarbonyl compounds that is reported by Correa et al. in their review [73]. An interesting coupling reaction including benzoxazone-2 and indole as substrates was discussed. This coupling was successfully applied to the gram-scale synthesis of the natural product *cephalandole A*.

Other than indoles, a number of *N*-heteroarenes including pyrrole [73, 74], indolizine [73], azaindole [73], imidazopyrimidine [73], imidazothiazole [73],

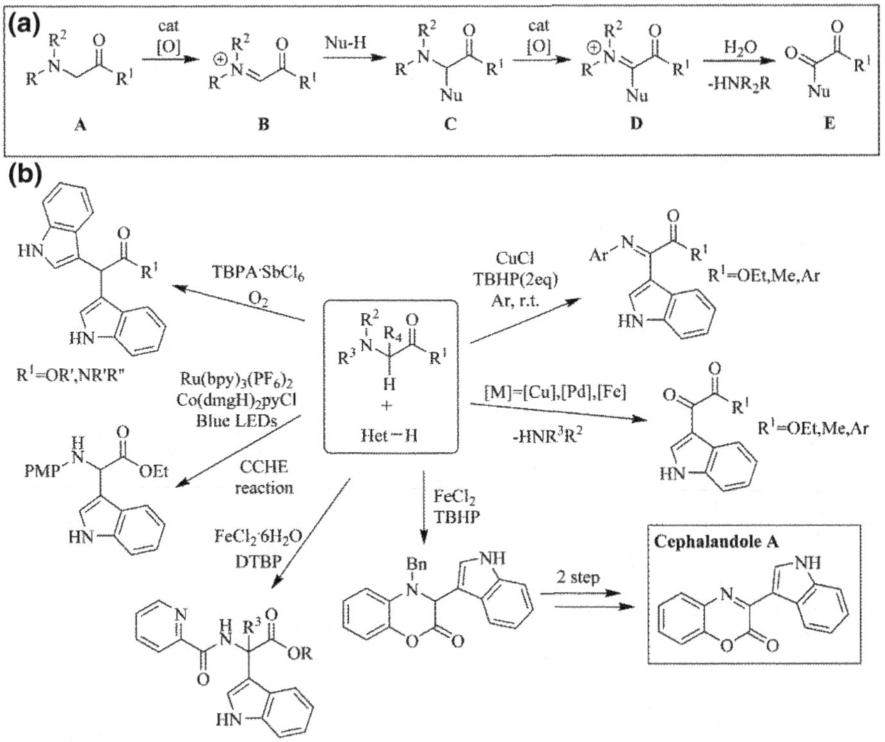

Scheme 28 CDC reaction of indoles with α-aminocarbonyl compounds: **a** general mechanism; **b** examples of catalytic cross-dehydrogenative coupling reaction from Correa review

imidazopyridine [73, 75, 76] were explored as powerful pronucleophiles in these oxidative coupling reactions with α-aminocarbonyl compounds.

For example, a protocol of visible-light-promoted CDC reaction between α-aminocarbonyl compounds and imidazo[1,2-*a*]pyridines was recently developed by the group of Zhu [75]. A catalytic amount of the $Cu(OTf)_2$ allowed the hetarylation of *N*-arylglycine esters with a set of imidazopyridines. The reaction proceeded smoothly at RT under irradiation with blue light. Interestingly, the arylation also occurred under the same conditions at temperatures above 40 °C [76]. In direct comparison, the visible-light-promoted protocol proved to be more efficient than the thermal variant, and higher yields were observed (Scheme 29).

Liu group developed the site-specific method for modifying proline-based peptides by Cu-catalyzed oxidative coupling of tertiary amine N-oxides with indoles in the presence of *p*-toluenesulfonic acid (TsOH) (Scheme 30) [77]. Feature of this protocol is that the reaction center is C5 position of N-substituted proline in the peptide system with high regio- and diastereoselectivity at ambient temperature. It should be noted that this coupling can proceed *one pot* through two-step process under the same conditions.

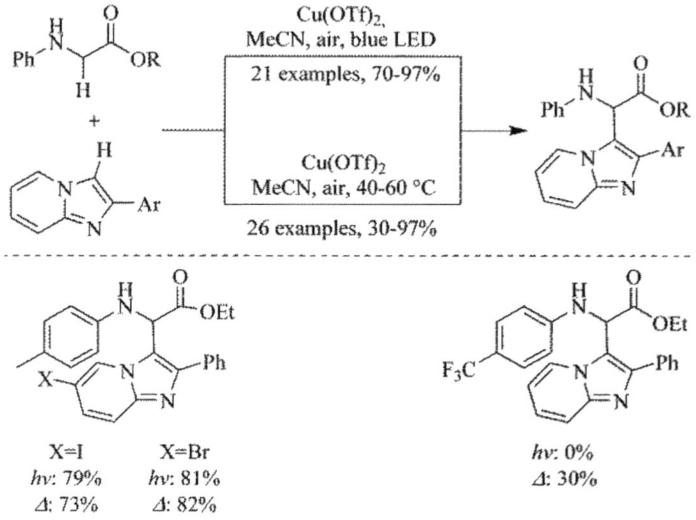

Scheme 29 Visible-light or thermal-promoted coupling of imidazo[1,2-*a*]pyridines with *N*-arylglycine esters

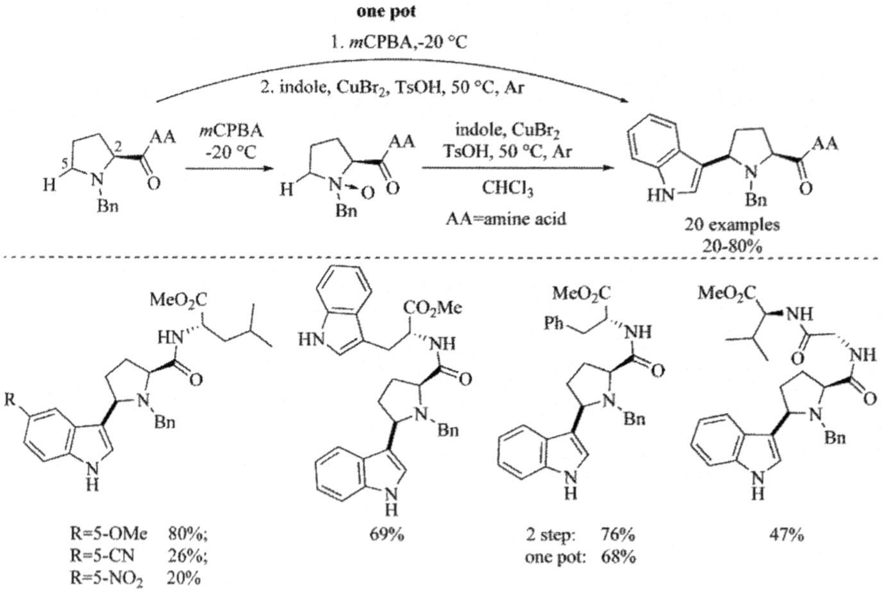

Scheme 30 Indolation of N-protected proline-based peptides

7 CDC *N*-Heterocycles with Other C(sp^3)–H Sources

Li et al. developed the site-specific CDC reaction of azoles with nitriles catalyzed with Cu^I salts in the presence of DCP (Scheme 31) [78]. The cyanalkylation reaction proceeded at the C2 position of pyrroles and indoles with high regioselectivity. Nevertheless, using 1*H*-pyrrolo[2,3-*b*]pyridine as substrate gave rise products mixture with ratio C2:C3=7:1 with overall yield of 33%. Mechanistically, this process is plausibly a radical reaction because the formation of product was suppressed in the presence of radical scavenger. Catalytic system CuCl/DCP generated the α-cyano methylenyl radical, which added to the heterocycle led to the 2-cyanomethyl-2,3-dihydroindole radical **A**. Next, direct HAT and SET from the heterocycle radical **A** gave the final product (Scheme 31). The authors applied their strategy for the preparation of *tolmetin*, a non-steroidal anti-inflammatory drug.

Pihko group published Pd-catalyzed oxidative coupling of indoles with α-substituted *β*-keto esters, which proceeded in the presence of slight excess of TBPB at room temperature with high regioselectivity (C3-selective for the indole partner and *β*′-selective for the *β*-ketoester) and good yields [79]. Introduction of the third

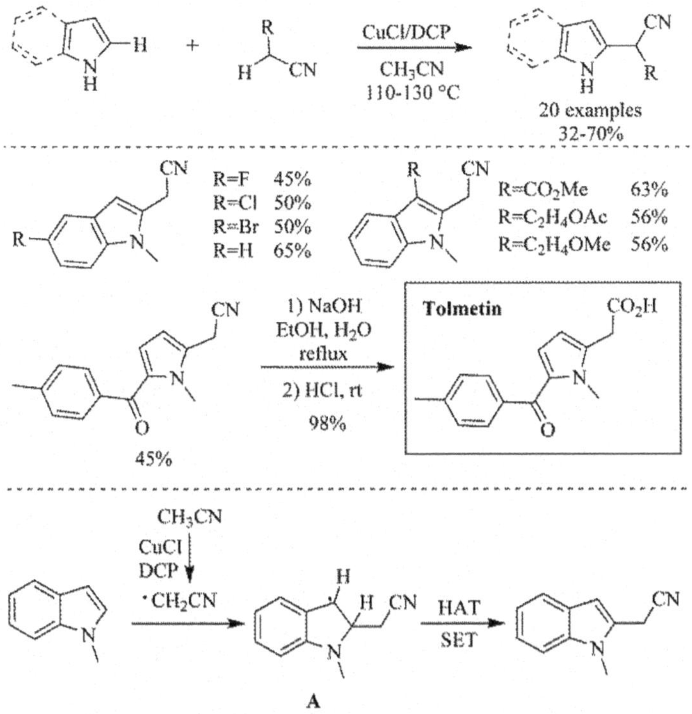

Scheme 31 Cyanalkylation of indoles and pyrroles

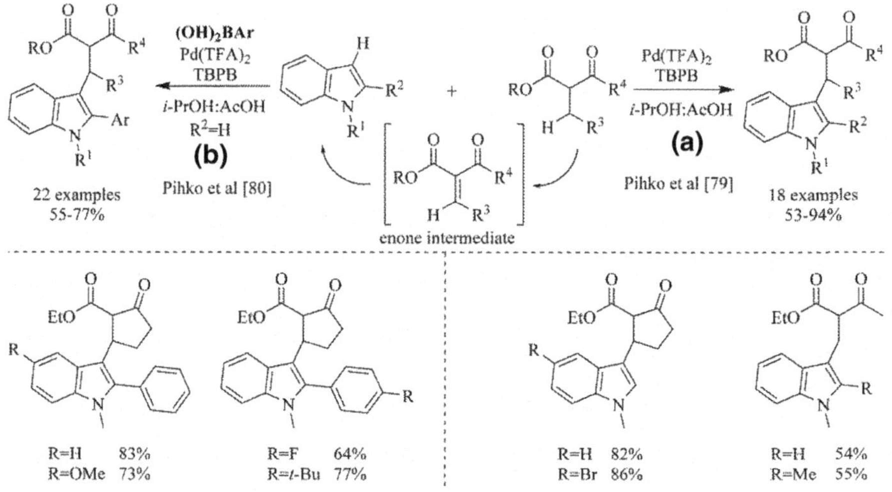

Scheme 32 Pd-catalyzed two-component CDC indoles with *β*-ketoesters (**a**) and three-component reaction between indole, *β*-ketoesters and arylboronic acid (**b**)

component, arylboronic acid, as an additional reagent in coupling of 2,3-unsubstituted indoles with *β*-keto esters under the same condition gave rise to 2-aryl-3-alkylindoles in 55–77% yields with excellent regioselectivity [80]. Results of both processes are shown in Scheme 32. Authors established that an enone intermediate was involved into the mechanism of coupling of indoles and *β*-keto esters [79]. Subsequently, using a computational studies Pihko et al. showed that the reaction proceeds via two catalytic cycles: (1) *the dehydrogenation cycle* produced the enone intermediate and (2) *the C–C coupling cycle* gave rise the final product, wherein the indole plays a role of ligand and substrate, respectively [81].

Wu et al. reported double cross-dehydrogenative coupling (DCDC) of indoles with acetophenones by using molecular iodine as an oxidant in the absence of metal catalyst (Scheme 33) [82]. Three mechanisms–different reactions (iodination, Kornblum oxidation, and Friedel–Crafts reaction) take place step by step in one pot to give a variety of 2,2-bisindolyl-1-arylethanones. Plausible mechanism of reaction process is depicted in Scheme 33.

Mo and Bao published the protocol for dehydrogenative coupling of indoles with 1,3-diphenylpropenes [83]. Set of 3-allylindoles was prepared in the presence of catalytic system $PdCl_2$/DDQ in moderate to excellent yields (Scheme 34). Asymmetric di-*para*-substituted 1,3-diphenylpropene reacted with indole to form only one isomer, while *α*- and *γ*-indolated isomers were detected when monosubstituted or asymmetric diphenylpropene was used as substrate. Interestingly, when the C3 position of indole was substituted, the C2-alkylation product was also obtained in 55% yield. Plausible mechanism is illustrated in Scheme 34. First, DDQ abstracted a hydride from 1,3-diphenylpropene to form the conjugated allylic intermediate **A**, which then underwent carbometallation with palladium chloride

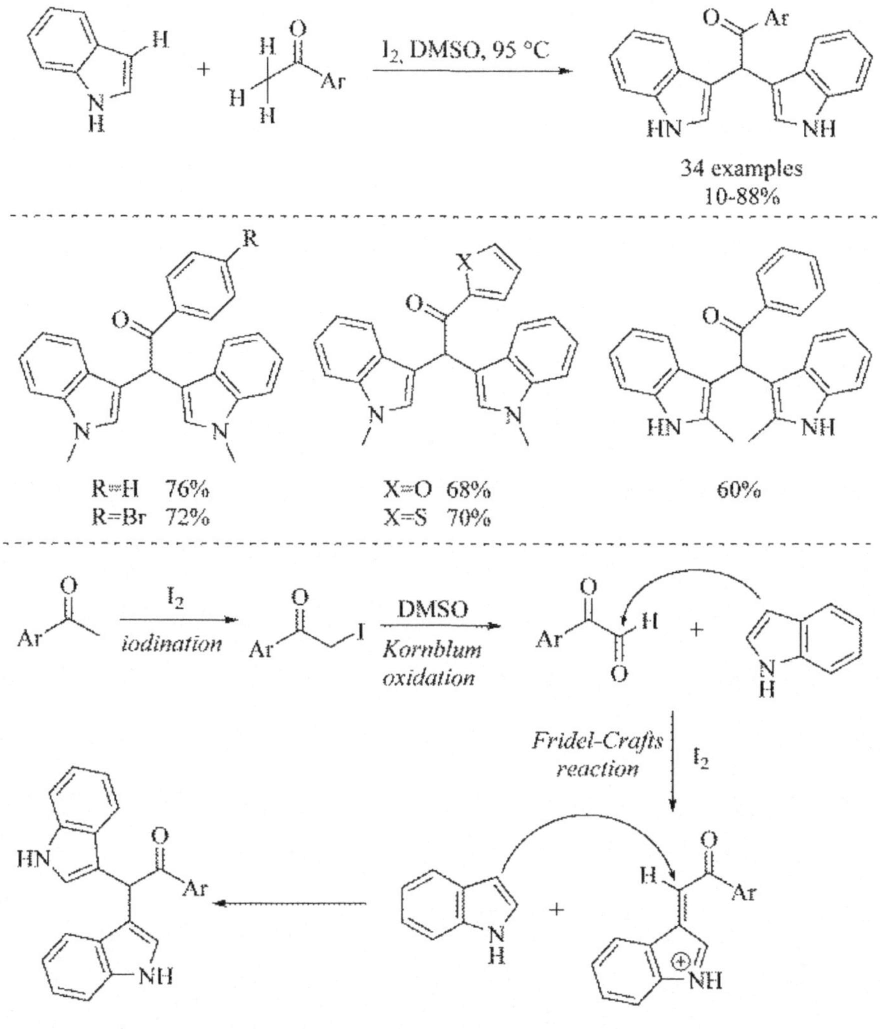

Scheme 33 CDDC reaction of indoles with arylmethylketones

and indole to form the allyl palladium intermediate **B**. Subsequent elimination of $PdCl_2$ and hydrogen abstraction afforded the product.

Recently, Boens and co-workers reported general and facile synthesis of a set of α-alkylated boron dipyrromethenes (BODIPYs) via tetra-*n*-butylammonium iodide (TBAI)/TBHP catalyzed CDC alkylation reaction of BODIPY by allylic alkenes (Scheme 35) [84]. The reaction was carried out at 90 °C for 12 h to give products in moderate to good yields. Extending the reaction time to 24 h, the corresponding diallylic alkylation products were produced in 30–40% yields. The formation of product was suppressed in the presence of 2,6-di-tert-butyl-4-methylphenol (BHP) radical scavenger that evidenced the radical mechanism involved.

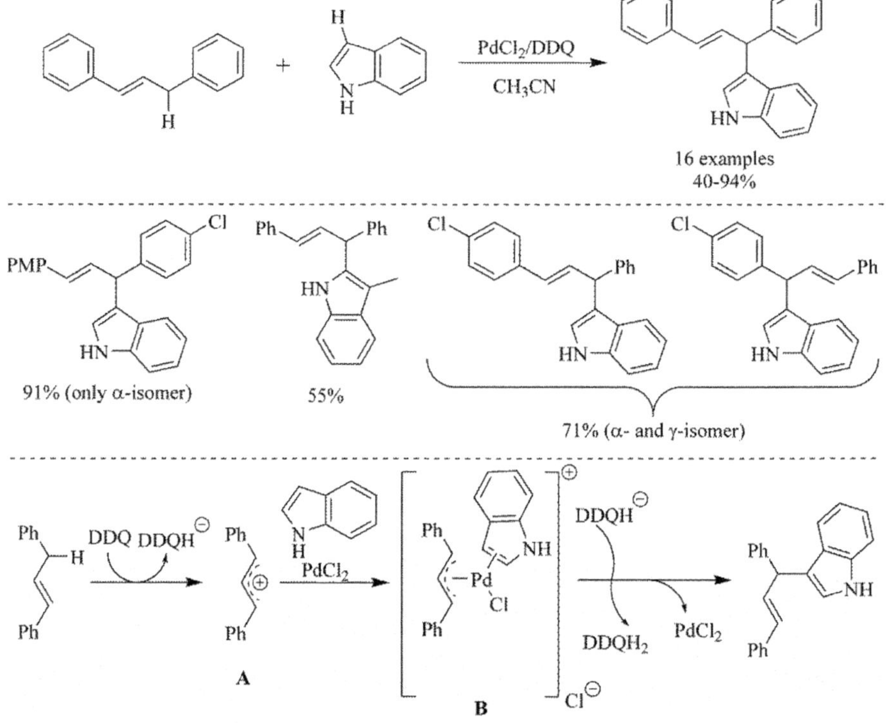

Scheme 34 Pd-catalyzed allylation of indoles

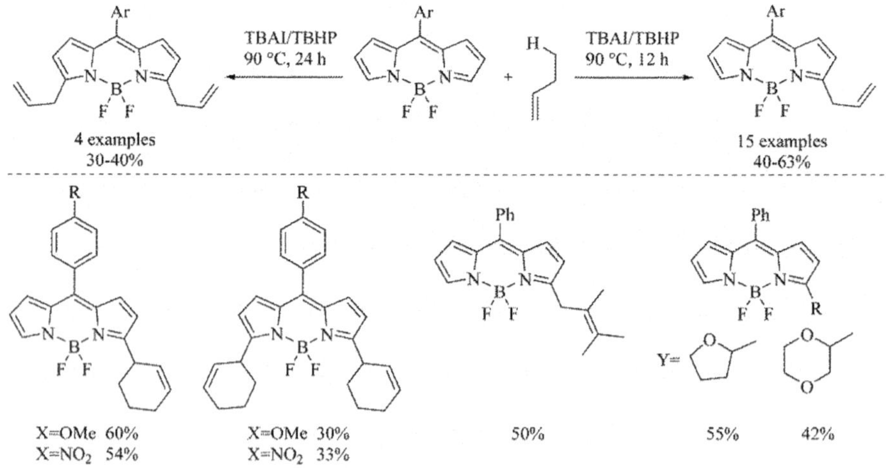

Scheme 35 Transition-metal-free allylation of BODIPYs

The authors also extended protocol to dioxane and THF that led to obtain the coupling product with a yield of 42–55%.

8 Conclusions

CDC reactions are an extremely convenient synthetic tool that allows introducing into the structure of a heterocycle such fragments as alkanes, ethers, thioesters, methylarenes, and methylamines. From the point of view of fine organic synthesis, the value of CDC reactions lies in the fact that it is possible to combine various fragments, achieving high molecular complexity in few steps.

Moreover, not functionalized alkane derivatives can be used as starting materials, and there is no need to use a halogenated or organometallic derivative. The simplest C–H alkane derivatives are active in CDC reactions.

As a result, the C–H bond is now considered not as something inert, but as a *functional group* that is capable of substitution.

Acknowledgments This work was supported by the Russian Science Foundation (project #18-13-00365).

References

1. Minisci F (1976) Recent aspects of homolytic aromatic substitutions. Top Curr Chem 62:1–48
2. Minisci F, Vismara E, Fontana F (1989) Recent developments of free-radical substitutions of heteroaromatic bases. Heterocycles 28:489–519
3. Minisci F, Fontana F, Vismara E (1990) Substitutions by nucleophilic free radicals: a new general reaction of heteroaromatic bases. J Heterocycl Chem 27:79–96
4. Punta C, Minisci F (2008) Minisci reaction: a Friedel-Crafts type process with opposite reactivity and selectivity. Selective homolytic alkylation, acylation, carboxylation and carbamoylation of heterocyclic aromatic bases. Trends Heterocycl Chem 13:1–68
5. Duncton MAJ (2011) Minisci reactions: versatile CH-functionalizations for medicinal chemists. Med Chem Commun 2:1135–1161
6. Li JJ (2014) Minisci reaction. In: Name reactions, 5th edn. Springer International Publishing, pp 361–362
7. Minisci F, Vismara E, Morini G, Fontana F, Levi S, Serravalle M, Giordano C (1986) Polar effects in free-radical reactions. selectivity and reversibility in the homolytic benzylation of protonated heteroaromatic bases. J Org Chem 51:476–479
8. Citterio A, Gentile A, Minisci F, Serravalle M, Ventura S (1984) Polar effects in free-radical reactions. Carbamoylation and α-*N*-amidoalkylation of heteroaromatic bases by amides and hydroxylamine-*O*-sulfonic acid. J Org Chem 49:3364–3367
9. Deng G, Li C-J (2009) $Sc(OTf)_3$-catalyzed direct alkylation of quinolines and pyridines with alkanes. Org Lett 11:1171–1174
10. Deng G, Ueda K, Yanagisawa S, Itami K, Li C-J (2009) Coupling of nitrogen heteroaromatics and alkanes without transition metals: a new oxidative cross-coupling at C–H/C–H bonds. Chem Eur J 15:333–337

11. Antonchick AP, Burgmann L (2013) Direct selective oxidative cross-coupling of simple alkanes with heteroarenes. Angew Chem Int Ed 52:3267–3271
12. Xia R, Niu H-Y, Qu G-R, Guo H-M (2012) CuI controlled C–C and C–N bond formation of heteroaromatics through C(sp3)–H activation. Org Lett 14:5546–5549
13. Wang D-C, Xia R, Xie M-S, Qu G-R, Guo H-M (2016) Synthesis of cycloalkyl substituted purine nucleosides via a metal-free radical route. Org Biomol Chem 14:4189–4193
14. Faisca Phillips AM, Pombeiro AJL (2018) Recent developments in transition metal-catalyzed cross-dehydrogenative coupling reactions of ethers and thioethers. ChemCatChem 10:3354–3383
15. Lakshman MK, Vuram PK (2017) Cross-dehydrogenative coupling and oxidative-amination reactions of ethers and alcohols with aromatics and heteroaromatics. Chem Sci 8:5845–5888
16. Lai M, Li Y, Wu Z, Zhao M, Ji X, Liu P, Zhang X (2018) Synthesis of alkyl-substituted pyrazine *N*-oxides by transition-metal-free oxidative cross-coupling reactions. Asian J Org Chem 7:1118–1123
17. Yang Q, Li S, Wang J (Joelle) (2018) Cobalt-catalyzed cross-dehydrogenative coupling of imidazo[1,2-a]pyridines with isochroman using molecular oxygen as the oxidant. Org Chem Front 5:577–581
18. Jiang H, Xie J, Lin A, Cheng Y, Zhu C (2012) The Au(III)-catalyzed coupling reactions between alcohols and *N*-heterocycles via C–H bond activation. RSC Adv 2:10496–10498
19. Adib M, Pashazadeh R, Rajai-Daryasarei S, Kabiri R, Gohari SJA (2016) Transition-metal-free acylation of quinolines and isoquinolines with arylmethanols via oxidative cross-dehydrogenative coupling reactions. Synlett 27:2241–2245
20. Wan M, Lou H, Liu L (2015) C1-benzyl and benzoyl isoquinoline synthesis through direct oxidative cross-dehydrogenative coupling with methyl arenes. Chem Commun 51:13953–13956
21. Shi X, Zhang F, Luo W-K, Yang L (2017) Oxidant-triggered C1-benzylation of isoquinoline by iodine–catalyzed cross-dehydrogenative-coupling with methylarenes. Synlett 13:494–498
22. Ali W, Behera A, Guin S, Patel BK (2015) Regiospecific benzoylation of electron-deficient N-heterocycles with methylbenzenes via a Minisci-type reaction. J Org Chem 80:5625–5632
23. Kianmehr E, Faghih N, Khan KM (2015) Palladium-catalyzed regioselective benzylation–annulation of pyridine *N*-oxides with toluene derivatives via multiple C-H bond activations: benzylation versus arylation. Org Lett 17:414–417
24. Kianmehr E, Gholamhosseyni M (2018) Visible-light-promoted copper-catalyzed regioselective benzylation of pyridine *N*-oxides versus thermal acylation reaction with toluene derivatives. Eur J Org Chem 2018:1559–1566
25. Wan L, Qiao K, Sun XN, Di ZC, Fang Z, Li ZJ, Guo K (2016) Benzylation of heterocyclic *N*-oxides via direct oxidative cross-dehydrogenative coupling with toluene derivatives. New J Chem 40:10227–10232
26. Zhang Y, Feng J, Li C-J (2008) Palladium-catalyzed methylation of aryl C–H bond by using peroxides. J Am Chem Soc 130:2900–2901
27. Li G, Yang S, Lv B, Han Q, Ma X, Sun K, Wang Z, Zhao F, Lv Y, Wu H (2015) Metal-free methylation of a pyridine *N*-oxide C–H bond by using peroxides. Org Biomol Chem 13:11184–11188
28. Guo S, Li Y, Wang Y, Guo X, Meng X, Chen B (2015) Iron-catalyzed cross dehydrogenative coupling (CDC) of indoles and benzylic C–H bonds. Adv Synth Catal 357:950–954
29. Zhang H-J, Su F, Wen T-B (2015) Copper-catalyzed direct C2-benzylation of indoles with alkylarenes. J Org Chem 80:11322–11329
30. Sambiagio C, Schönbauer D, Blieck R, Dao-Huy T, Pototschnig G, Schaaf P, Wiesinger T, Zia MF, Wencel-Delord J, Besset T, Maes BUW, Schnürch M (2018) A comprehensive overview of directing groups applied in metal-catalysed C–H functionalisation chemistry. Chem Soc Rev 47:6603–6743
31. Rasheed OK, Sun B (2018) Advances in development of C–H activation/functionalization using a catalytic directing group. ChemistrySelect 3:5689–5708

32. Okugawa N, Moriyama K, Togo H (2017) Introduction of quinolines and isoquinolines onto nonactivated α-C–H bond of tertiary amides through a radical pathway. J Org Chem 82:170–178
33. Li Z, Li C-J (2005) CuBr-catalyzed direct indolation of tetrahydroisoquinolines via cross-dehydrogenative coupling between sp^3 C–H and sp^2 C–H bonds. J Am Chem Soc 127:6968–6969
34. Ghobrial M, Harhammer K, Mihovilovic MD, Schnürch M (2009) Facile, solvent and ligand free iron catalyzed direct functionalization of N-protected tetrahydroisoquinolines and isochroman. Chem Commun 46:8836–8838
35. Liu P, Zhou C-Y, Xiang S, Che C-M (2010) Highly efficient oxidative carbon–carbon coupling with SBA-15-support iron terpyridine catalyst. Chem Commun 46:2739–2741
36. Ohta M, Quick MP, Yamaguchi J, Wünsch B, Itami K (2009) Fe-catalyzed oxidative coupling of heteroarenes and methylamines. Chem Asian J 4:1416–1419
37. Shirakawa E, Yoneda T, Moriya K, Ota K, Uchiyama N, Nishikawa R, Hayashi T (2011) Iron-catalyzed oxidative coupling of alkylamines with arenes, nitroalkanes, and 1,3-dicarbonyl compounds. Chem Lett 40:1041–1043
38. Marset X, Pérez JM, Ramón DJ (2016) Cross-dehydrogenative coupling reaction using copper oxide impregnated on magnetite in deep eutectic solvents. Green Chem 18:826–833
39. Ho HE, Ishikawa Y, Asao N, Yamamoto Y, Jin T (2015) Highly efficient heterogeneous aerobic cross-dehydrogenative coupling via C–H functionalization of tertiary amines using a nanoporous gold skeleton catalyst. Chem Commun 51:12764–12767
40. Liu Y, Wang C, Xue D, Xiao M, Li C, Xiao J (2017) Reactions catalysed by a binuclear copper complex: aerobic cross dehydrogenative coupling of *N*-aryl tetrahydroisoquinolines. Chem Eur J 23:3051–3061
41. Su W, Yu J, Li Z, Jiang Z (2011) Solvent-free cross-dehydrogenative coupling reactions under high speed ball-milling conditions applied to the synthesis of functionalized tetrahydroisoquinolines. J Org Chem 76:9144–9150
42. Yang F, Li J, Xie J, Huang Z-Z (2010) Copper-catalyzed cross dehydrogenative coupling reactions of tertiary amines with ketones or indoles. Org Lett 12:5214–5217
43. Huang L, Niu T, Wu J, Zhang Y (2011) Copper-catalyzed oxidative cross-coupling of *N,N*-dimethylanilines with heteroarenes under molecular oxygen. J Org Chem 76:1759–1766
44. Romo-Pérez A, Miranda LD, García A (2015) Synthesis of *N*-methyl-5,6-dihydrobenzo[c] phenanthridine and its sp^3 C(6)–H bond functionalization via oxidative cross-dehydrogenative coupling reactions. Tetrahedron Lett 56:6669–6673
45. Dutta B, Sharma V, Sassu N, Dang Y, Weerakkody C, Macharia J, Miao R, Howell AR, Suib SL (2017) Cross dehydrogenative coupling of *N*-aryltetrahydroisoquinolines (sp3 C–H) with indoles (sp2 C–H) using a heterogeneous mesoporous manganese oxide catalyst. Green Chem 19:5350–5355
46. Patil MR, Dedhia NP, Kapdi AR, Kumar AV (2018) Cobalt(II)/*N*-hydroxyphthalimide-catalyzed cross-dehydrogenative coupling reaction at room temperature under aerobic condition. J Org Chem 83:4477–4490
47. Wu C-J, Zhong J-J, Meng Q-Y, Lei T, Gao X-W, Tung C-H, Wu L-Z (2015) Cobalt-catalyzed cross-dehydrogenative coupling reaction in water by visible light. Org Lett 17:884–887
48. Tanoue A, Yoo W-J, Kobayashi S (2013) Antimony/*N*-hydroxyphthalimide as a catalyst system for cross-dehydrogenative coupling reactions under aerobic conditions. Adv Synth Catal 355:269–273
49. Alagiri K, Kumara GSR, Prabhu KR (2011) An oxidative cross-dehydrogenative-coupling reaction in water using molecular oxygen as the oxidant: vanadium catalyzed indolation of tetrahydroisoquinolines. Chem Commun 47:11787–11789
50. Jones KM, Karier P, Klussmann M (2012) C1-substituted *N*-alkyl tetrahydroisoquinoline derivatives through V-catalyzed oxidative coupling. ChemCatChem 4:51–54
51. Wang M-Z, Zhou C-Y, Wong M-K, Che C-M (2010) Ruthenium-catalyzed alkylation of indoles with tertiary amines by oxidation of a sp^3 C–H bond and Lewis acid catalysis. Chem Eur J 16:5723–5735

52. Dai C, Meschini F, Narayanam JMR, Stephenson CRJ (2012) Friedel-Crafts amidoalkylation via thermolysis and axidative photocatalysis. J Org Chem 77:4425–4431
53. Shelar DP, Li T-T, Chen Y, Fu W-F (2015) Platinum(II) Schiff base complexes as photocatalysts for visible-light-induced cross-dehydrogenative coupling reactions. ChemPlusChem 80:1541–1546
54. Zhong J-J, Meng Q-Y, Wang G-X, Liu Q, Chen B, Feng K, Tung C-H, Wu L-Z (2013) A highly efficient and selective aerobic cross-dehydrogenative-coupling reaction photocatalyzed by a platinum(II) terpyridyl complex. Chem Eur J 19:6443–6450
55. Chen W, Zheng H, Pan X, Xie Z, Zan X, Sun B, Liu L, Lou H (2014) A metal-free cross-dehydrogenative coupling of *N*-carbamoyl tetrahydroisoquinoline by sodium persulfate. Tetrahedron Lett 55:2879–2882
56. Zhang Y, Teuscher KB, Ji H (2016) Direct α-heteroarylation of amides (α to nitrogen) and ethers through a benzaldehyde-mediated photoredox reaction. Chem Sci 7:2111–2118
57. Jones KM, Klussmann M (2012) Oxidative coupling of tertiary amines: scope, mechanism and challenges. Synlett 2012:159–162
58. Ratnikov MO, Doyle MP (2013) Mechanistic investigation of oxidative Mannich reaction with tert-butyl hydroperoxide. The role of transition metal salt. J Am Chem Soc 135:1549–1557
59. Boess E, Sureshkumar D, Sud A, Wirtz C, Farès C, Klussmann M (2011) Mechanistic studies on a Cu-catalyzed aerobic oxidative coupling reaction with *N*-phenyl tetrahydroisoquinoline: structure of intermediates and the role of methanol as a solvent. J Am Chem Soc 133:8106–8109
60. Boess E, Schmitz C, Klussmann M (2012) A comparative mechanistic study of Cu-catalyzed oxidative coupling reactions with *N*-phenyltetrahydroisoquinoline. J Am Chem Soc 134:5317–5325
61. Scott M, Sud A, Boess E, Klussmann M (2014) Reaction progress kinetic analysis of a copper-catalyzed aerobic oxidative coupling reaction with *N*-phenyl tetrahydroisoquinoline. J Org Chem 79:12033–12040
62. Boess E, Wolf LM, Malakar S, Salamone M, Bietti M, Thiel W, Klussmann M (2016) Competitive hydrogen atom transfer to oxyl- and peroxyl radicals in the Cu-catalyzed oxidative coupling of *N*-aryl tetrahydroisoquinolines using tert-butyl hydroperoxide. ACS Catal 6:3253–3261
63. Tsang AS-K, Jensen P, Hook JM, Hashmi ASK, Todd MH (2011) An oxidative carbon–carbon bond-forming reaction proceeds via an isolable iminium ion. Pure Appl Chem 83:655–665
64. Cheng G-J, Song L-J, Yang Y-F, Zhang X, Wiest O, Wu Y-D (2013) Computational studies on the mechanism of the copper-catalyzed sp^3-C–H cross-dehydrogenative coupling reaction. ChemPlusChem 78:943–951
65. Schweitzer-Chaput B, Klussmann M (2013) Brønsted acid catalyzed C–H functionalization of N-protected tetrahydroisoquinolines via intermediate peroxides. Eur J Org Chem 2013:666–671
66. Pu F, Li Y, Song Y-H, Xiao J, Liu Z-W, Wang C, Liu Z-T, Chen J-G, Lu J (2016) Copper-catalyzed coupling of indoles with dimethylformamide as a methylenating reagent. Adv Synth Catal 358:539–542
67. Deb ML, Borpatra PJ, Saikia PJ, Baruah PK (2017) Introducing tetramethylurea as a new methylene precursor: a microwave-assisted $RuCl_3$-catalyzed cross dehydrogenative coupling approach to bis(indolyl)methanes. Org Biomol Chem 15:1435–1443
68. Kaswan P, Nandwana NK, DeBoef B, Kumar A (2016) Vanadyl acetylacetonate catalyzed methylenation of imidazo[1,2-*a*]pyridines by using dimethylacetamide as a methylene source: direct access to bis(imidazo[1,2-*a*]pyridin-3-yl)methanes. Adv Synth Catal 358:2108–2115
69. Yang J, Wang Z, Pan F, Li Y, Bao W (2010) CuBr-catalyzed selective oxidation of *N*-azomethine: highly efficient synthesis of methine-bridged bis-indole compounds. Org Biomol Chem 8:2975–2978

70. Li G, Nakamura H (2016) Synthesis of 2-indolyltetrahydroquinolines by zinc(II)-catalyzed intramolecular hydroarylation-redox cross-dehydrogenative coupling of *N*-propargylanilines with indoles. Angew Chem Int Ed 55:6758–6761
71. Wang H, Dong M, Liu C, Zhang D (2018) Theoretical insight into the zinc(II)-catalyzed synthesis of 2-indolyltetrahydroquinolines from *N*-propargylanilines and indoles: cross-dehydrogenative coupling with temporally separated catalytic activity. Catal Sci Technol 8:1997–2007
72. Meng Q-Y, Zhong J-J, Liu Q, Gao X-W, Zhang H-H, Lei T, Li Z-J, Feng K, Chen B, Tung C-H, Wu L-Z (2013) A cascade cross-coupling hydrogen evolution reaction by visible light catalysis. J Am Chem Soc 135:19052–19055
73. Segundo MS, Correa A (2018) Cross-dehydrogenative coupling reactions for the functionalization of α-amino acid derivatives and peptides. Synthesis 50:2853–2866
74. Sonobe T, Oisaki K, Kanai M (2012) Catalytic aerobic production of imines en route to mild, green, and concise derivatizations of amines. Chem Sci 3:3249–3255
75. Zhu Z-Q, Xiao L-J, Zhou C-C, Song H-L, Xie Z-B, Le Z-G (2018) A visible-light-promoted cross-dehydrogenative-coupling reaction of *N*-arylglycine esters with imidazo[1,2-*a*]pyridines. Tetrahedron Lett 59:3326–3331
76. Zhu Z-Q, Xiao L-J, Chen Y, Xie Z-B, Zhu H-B, Le Z-G (2018) A highly efficient copper(II)-catalyzed cross-dehydrogenative-coupling reaction of *N*-arylglycine esters with 2-arylimidazo [1,2-*a*]pyridines. Synthesis 50:2775–2783
77. Wu X, Zhang D, Zhou S, Gao F, Liu H (2015) Site-specific indolation of proline-based peptides via copper(II)-catalyzed oxidative coupling of tertiary amine *N*-oxides. Chem Commun 51:12571–12573
78. Liu Z-Q, Li Z (2016) Radical-promoted site-specific cross dehydrogenative coupling of heterocycles with nitriles. Chem Commun 52:14278–14281
79. Leskinen MV, Yip K-T, Valkonen A, Pihko PM (2012) Palladium-catalyzed dehydrogenative *β′*-functionalization of *β*-keto esters with indoles at room temperature. J Am Chem Soc 134:5750–5753
80. Nimje RY, Leskinen MV, Pihko PM (2013) A three-component palladium-catalyzed oxidative C–C coupling reaction: a domino process in two dimensions. Angew Chem Int Ed 52:4818–4822
81. Leskinen MV, Madarász Á, Yip K-T, Vuorinen A, Pápai I, Neuvonen AJ, Pihko PM (2014) Cross-dehydrogenative couplings between indoles and *β*-keto esters: ligand-assisted ligand tautomerization and dehydrogenation via a proton-assisted electron transfer to Pd(II). J Am Chem Soc 136:6453–6462
82. Zhu Y, Liu M, Jia F, Yuan J, Gao Q, Lian M, Wu A (2012) Metal-free sp3 C–H bond dual-(het)arylation: I2-promoted domino process to construct 2,2-bisindolyl-1-arylethanones. Org Lett 14:3392–3395
83. Mo H, Bao W (2009) Efficient palladium-catalyzed oxidative indolation of allylic compounds with DDQ via sp^3 C–H bond activation and carbon-carbon bond formation under mild conditions. Adv Synth Catal 351:2845–2849
84. Yu Y, Jiao L, Wang J, Wang H, Yu C, Hao E, Boens N (2017) $Bu_4NI/^tBuOOH$ catalyzed, α-regioselective cross-dehydrogenative coupling of BODIPY with allylic alkenes and ethers. Chem Commun 53:581–584

Functionalization of Aromatic *N*-Heterocycles via C(sp^2)–H/C(sp^2)–H CDC Reactions

Igor A. Khalymbadzha, Ramil F. Fatykhov and Oleg N. Chupakhin

Abstract This review is devoted to dehydrogenative C(sp^2)–C(sp^2) cross-coupling reactions. In the review we discussed the reactions of aromatic *N*-heterocycles with such reagents as amides and aldehydes, alkenes, aromatic and heterocyclic compounds with the formation of the corresponding C–H/C–H coupling products. The review is focused on mechanism as well as scope and limitations of dehydrogenative cross-coupling reactions.

Keywords *N*-heterocycles · Dehydrogenative cross-coupling · C(sp^2)–C(sp^2) bond formation

1 Introduction

Formation of new C(sp^2)–C(sp^2) bond was and is still a straightforward goal in the field of organic chemistry since sp^2 carbon atom comprises carbonyl, olefinic, aromatic and heteroaryl carbon.

The first cross-coupling reaction is, probably, Ullmann reaction [1], known for more than 100 years. This reaction involves the coupling of two aryl halides in the presence of copper to form a carbon–carbon bond with the elimination of copper halide. The drastic conditions of the Ullmann reaction (temperature is often higher than 200 °C) impose significant restrictions on the choice of substrates.

The next step in the development of C(sp^2)–C(sp^2) bond formation methodology was taken in the 1970s and associated with Kumada, Heck, Sonogashira, Negishi,

I. A. Khalymbadzha (✉) · R. F. Fatykhov · O. N. Chupakhin
Ural Federal University, Mira, 19, 620002 Ekaterinburg, Russian Federation
e-mail: i.a.khalymbadzha@urfu.ru

I. A. Khalymbadzha · O. N. Chupakhin
I. Postovsky Institute of Organic Synthesis, Ural Branch of the Russian Academy of Sciences, S. Kovalevskaya/Akademicheskaya, 22/20, 620219 Ekaterinburg, Russian Federation

A. Srivastava and C. K. Jana (eds.), *Heterocycles via Cross Dehydrogenative Coupling*, https://doi.org/10.1007/978-981-13-9144-6_2

Stille and Suzuki reactions [2]. These transformations allow significantly milder conditions, and the scope of the corresponding was sufficiently extended.

Nonetheless, the traditional metal-catalysed cross-coupling requires preliminary functionalization and activation of the partners. One of them should be provided as organometallic-boosted nucleophile species (magnesium, tin, boron) and other as heteroaryl halide or triflate. Furthermore, if halogenide and organometallic nucleophile were introduced into the starting materials, a waste comprising these elements should appear. Thus, traditional transition metal-catalysed cross-coupling cannot be considered as atom economy and green method.

To address these environmental issues, oxidative CDC reactions were developed. Interestingly, both oxidative and halogen-based reactions were discovered approximately simultaneously. However, the oxidation variants of the coupling have developed especially rapidly in recent times in context of green chemistry.

Thus, nowadays functionalization of aromatic *N*-heterocycles via $C(sp^2)$–H/C(sp^2)–H CDC reactions is one of the hottest topics in organic chemistry.

The present review has the following structure: the first part is devoted to coupling reactions of aromatic *N*-heterocycles with non-aromatic $C(sp^2)$ substrates, such as amides, aldehydes, alkenes; the second part is directed to coupling between aromatic *N*-heterocycles and aromatic and heteroaromatic $C(sp^2)$ substrates.

2 Functionalization of Aromatic *N*-Heterocycles with Non-aromatic Substrates

2.1 Minisci-Type Reaction

Heterocyclic carboxamides are a ubiquitous class of biologically active compounds having wide spectrum activity exemplified such well-known compounds as nicotinic acid, ribavirin, nevirapine and LSD (Fig. 1).

On the other hand, obtaining of heterocyclic carboxamides through a halogenation reaction, substitution of the halogen by a cyano-group and subsequent hydrolysis of the cyano-group are cumbersome multistep procedures. An ideal

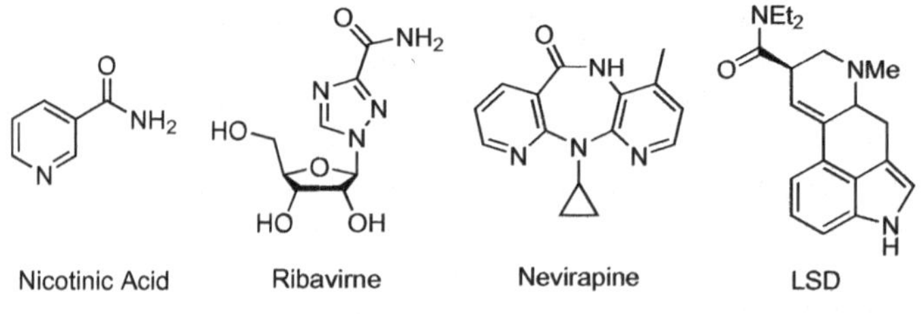

Fig. 1 Well-known heterocyclic carboxamides

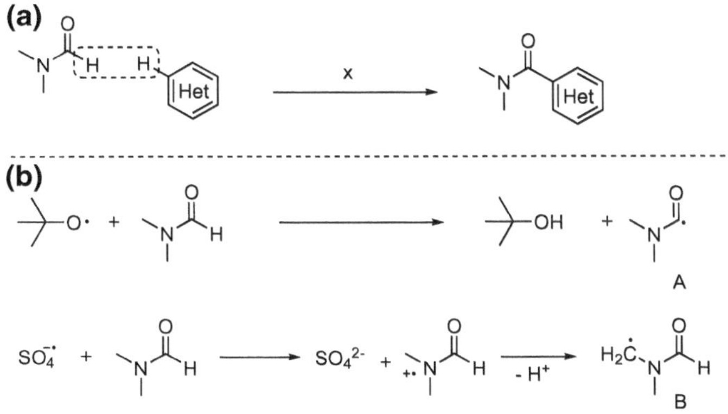

Radical initiator	A,%	B,%
TBHP/Fe^{2+}	97	3
S_2O_8/Fe^{2+}	2	98
H_2O_2/Fe^{2+}	85	15
DTBP	33	67
DBP	40	60

Scheme 1 Formation of nucleophilic radical: carbamoyl radical (**a**) and amido-alkyl-radical (**b**)

reaction should include experimentally simple one-step procedure that allows to introduce a carboxamide group instead of an aromatic hydrogen.

The Minisci reaction between protonated heterocyclic bases and formamide under Fe^{II} catalyst in the presence of oxidant to yield carbamoylation product may be considered as such ideal reaction (Scheme 1a). However, utilizing an amide, for example, DMF as radical source leads to the formation of two radical: carbamoyl radical **A** and amido-alkyl-radical **B** (Scheme 1b) [3–5].

Minisci et al. showed that formation of radicals A and B depended on the nature of the radical initiator. Two radical sources showed a high selectivity: peroxodisulphate, which leads to α-*N*-amido-alkyl radicals, and the redox system TBHP/Fe^{2+}, which mainly leads to carbamoyl radicals (Scheme 1b). This dramatic difference of selectivity in similar experimental conditions was explained by assuming two different mechanisms; with $S_2O_8^{2-}$, the primary process would be an electron transfer; with TBHP, an actual hydrogen abstraction would occur. The hydrogen abstraction is less selective with other radical sources, and both types of radicals are formed in significant amounts [3].

In order to improve the selectivity of carbamoylation reaction of isoquinoline N-oxide with N-substituted formamides, Pd-catalysed protocol was proposed [6]. Library of isoquinoline-1-carbamides was prepared under the optimized conditions ($Pd(MeCN)_2Cl_2$, n-Bu_4NOAc, Yb_2O_3, 120 °C) (Scheme 2a). Plausible mechanism of this process is depicted in Scheme 2b.

When isoquinoline without N-oxide group was underwent coupling conditions, no product was obtained. This observation showed that N-oxide group is required for this coupling.

Han et al. reported Ag-catalysed acid-free and highly selective method for the C2-carbamoylation of pyridines with unprotected formamide and *N*-methyl formamide (Scheme 3). The reaction proceeded similar to the Minisci reaction. When 2-methoxy-6-methylpyridine was subjected to the typical conditions ($AgNO_3$, $K_2S_2O_8$, HCO_2Na, 80 °C), only the C4-selective carbamoylated product was

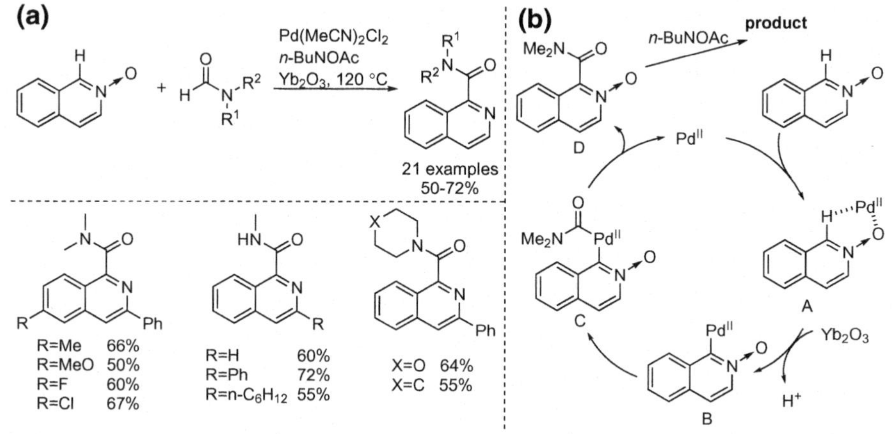

Scheme 2 Pd-catalysed carbamoylation of isoquinoxaline N-oxide (**a**) and plausible mechanism (**b**)

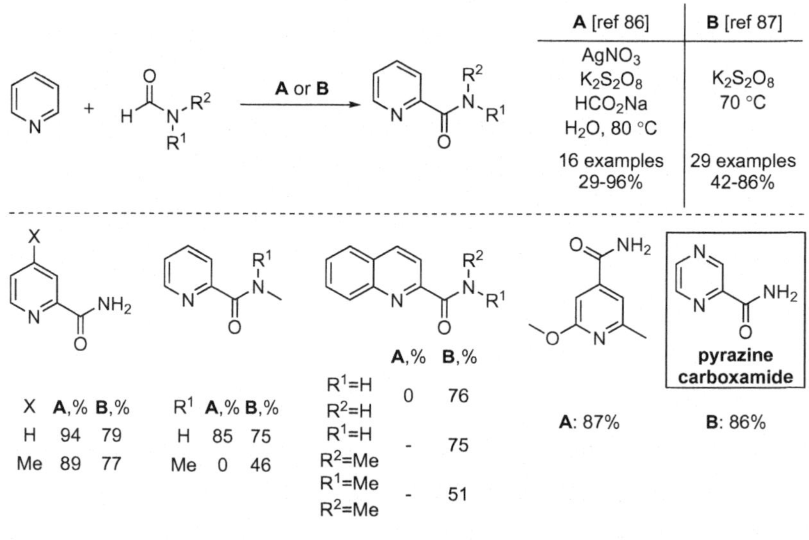

Scheme 3 Carbamoylation of pyridine, quinoxaline and pyrazine

detected, what demonstrated the strong preference for pyridines without C2 positions to react at the C4 position in this transformation [7].

Later, Bhat et al. proposed cross-dehydrogenative coupling (CDC) of heterocycles bases, i.e. pyridine, quinoline and pyrazine, with formamide, *N*-methyl formamide and DMF under TM-free conditions, using $K_2S_2O_8$ as oxidant [8]. In contrast to Ag-catalysed protocol [7], the TM-free method [8] allows to produce 2-(*N*,*N*-dimethylcarbamoyl)azines in moderate yields (Scheme 3). Interestingly, the therapeutically useful antitubercular agent pyrazinecarboxamide was successfully synthesized by employing the TM-free protocol [8].

Group of Harwood published [9] visible-light mediated synthesis of 1,10-phenanthroline-derived ligand systems via cross-coupling of formamide with 1,10-phenanthroline in the TM-free conditions. Di-, tri- and tetra-carbamoyl 1,10-phenanthroline derivatives were obtained in moderate to good yields by irradiation with a 23-W compact fluorescent lamp (CFL) in the presence of $(NH_4)_2S_2O_8$, benzaldehyde and formamide. This protocol provides a streamlined, scalable approach to the synthesis of perspective bis-triazinyl-1,10-phenanthroline (BTPhen) ligands. $CyMe_4$-BTPhen ligands were used for separating actinides from lanthanides (Scheme 4) [10, 11].

Wang et al. developed a method for the direct CDC reaction between azoles and different formamides in the presence of *tert*-butyl perbenzoate (TBPB) as oxidant under metal- and base-free conditions yielded C2 carbamoylation products in good yields [12]. Authors believed that radical initiator TBPB underwent a homolytic cleavage to generate a carboxyl radical and an alkoxyl radical. Subsequently, they abstracted hydrogen atoms from benzothiazole and DMF, forming the corresponding free radicals, which reacted with each other to form the corresponding cross-coupling product through termination of two radicals. The homocoupling

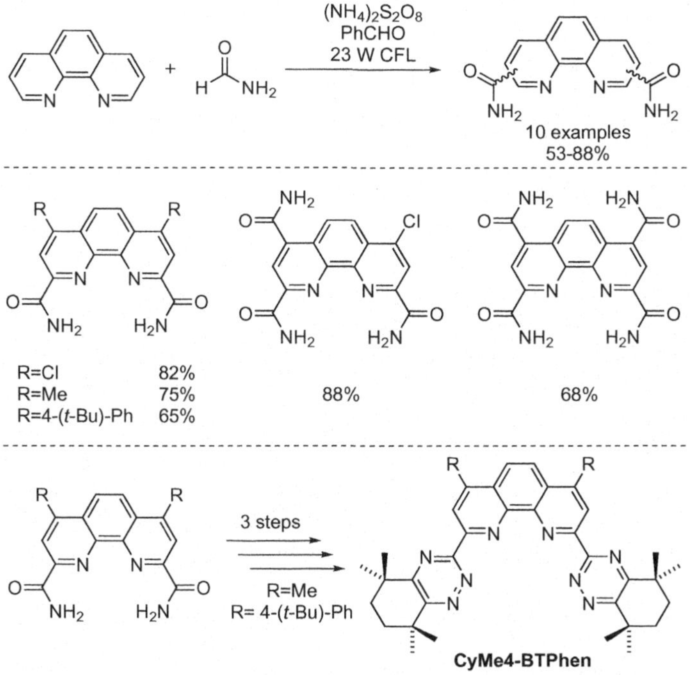

Scheme 4 Carbamoylation of 1,10-phenanthroline

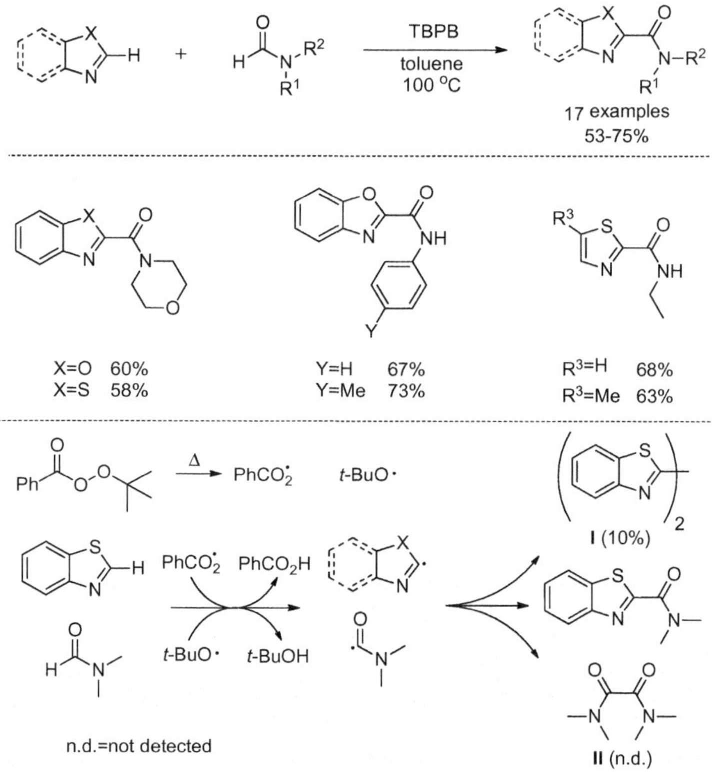

Scheme 5 TM-free carbamoylation of azoles

product (**I**) was also obtained in 10% yield, and no (**II**) was observed in the reaction (Scheme 5).

Subsequently, Ji group has suggested benzaldehyde-mediated photoredox carbamoylation of azoles with amides. This reaction proceeded smoothly with household 23-W CFL bulbs as the energy source under TM-free conditions in the presence of $(NH_4)_2S_2O_8$ as oxidant and *p*-toluenesulfonic acid hydrate via Minisci-type mechanism. It should be noted, when DMF was used as substrate under optimized conditions, reaction of amidoalkylation proceeded in 80% yield (Scheme 6) [13].

Antonchick et al. developed an efficient, mild and scalable method for the direct CDC of α-/γ position of *N*-heterocyclic bases with aldehydes (Scheme 7) [14]. Acylation reaction was carried out under TM-free conditions in the presence of phenyliodine bis(trifluoroacetate) (PIFA) as oxidant and trimethylsilylazide ($TMSN_3$) as additive. Isoquinoline, quinoline, acridine, pyridine, phthalazine, quinoxaline and purine were applied as substrate. Using α- or γ-unsubstituted quinoline or phthalazine in the excess of reagents gave rise to diacylated derivatives in good yields. However, when 3-bromoquinoline underwent typical conditions,

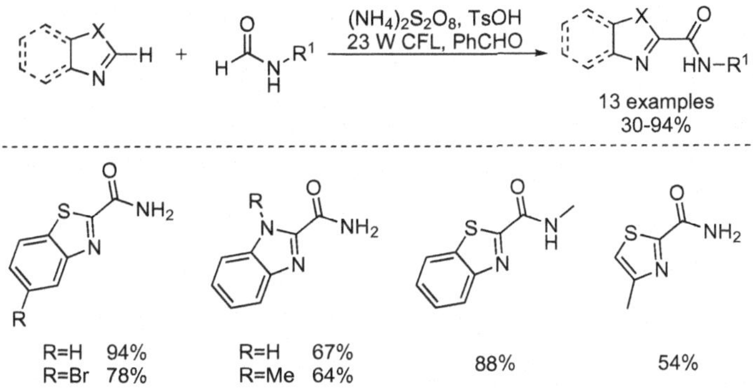

Scheme 6 Benzaldehyde-mediated photoredox carbamoylation of azoles

Scheme 7 Acylation of *N*-heterocycles in the presence of PIFA

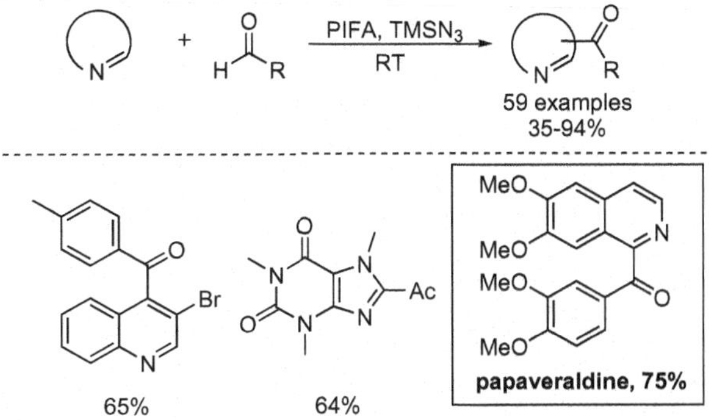

only 4-acylated quinoline was formed. It should be noted that authors showed the applicability of the proposed methodology for the synthesis of natural alkaloids including papaveraldine, pulcheotine and thalimicrinone by one-step cross-coupling between isoquinolines and corresponding aldehyde.

In the following, $K_2S_2O_8$/TBAB [15, 16], TBHP/TFA [17] and TBHP/NSC [18] catalytic systems were proposed for catalysed CDC reaction between heterocyclic bases and aldehydes. The scope of these processes is demonstrated in Scheme 8.

In 2014, Kianmehr reported a novel approach to the synthesis of 3-benzoylindoles via Pd-catalysed CDC reaction of indoles with aldehydes [19]. The reaction was carried out in the presence of combination of catalytic amount of $Pd(OAc)_2$ with TBHP in the chlorobenzene at 140 °C to give products in good yields. The data from these transformations are shown in Scheme 9. The authors believed that transformation began with the 3-palladation of indole with $Pd(OAc)_2$. At the same time, the aldehyde was transferred to an acyl radical by TBHP. The Pd^{II} intermediate **A** reacted with the acyl radical to afford the Pd^{IV} intermediate **D**, which then underwent reductive elimination to produce the desired product and regenerate the Pd^{II} species.

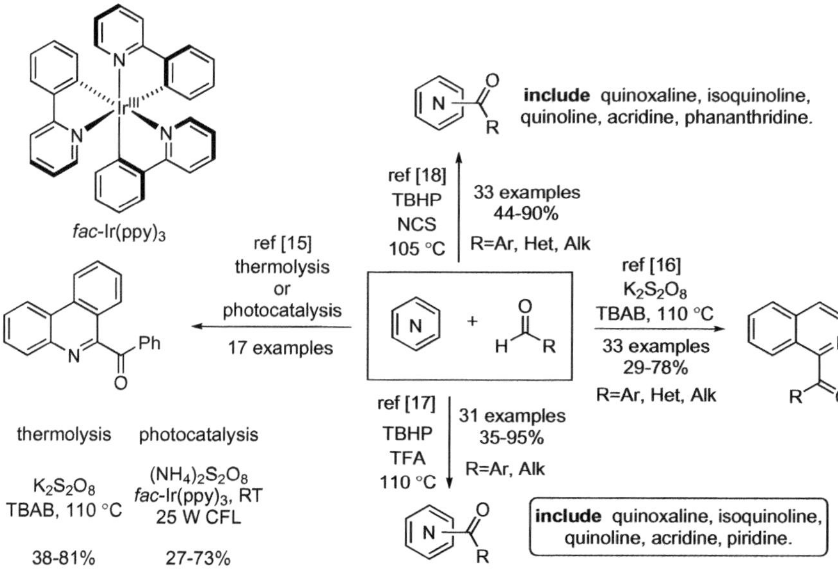

Scheme 8 Basic scope of CDC of azines with aldehydes

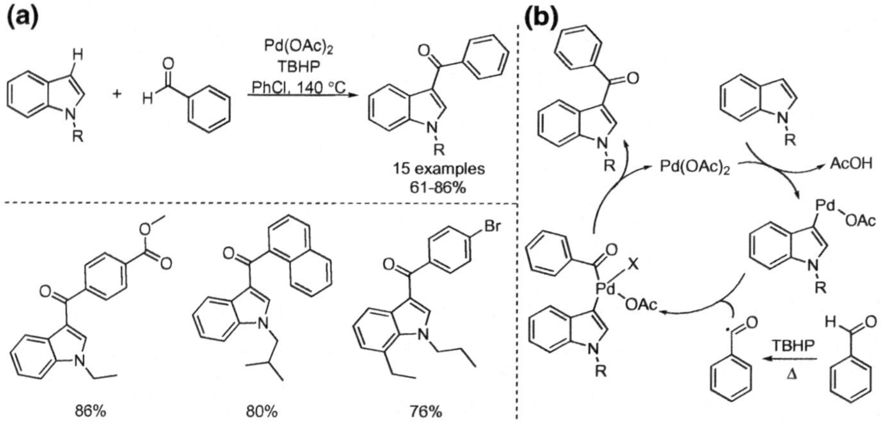

Scheme 9 Pd-catalysed acylation of indoles (**a**) and plausible mechanism (**b**)

Group of Sakhuja developed [20] the direct aerobic oxidative CDC aryl acetaldehydes at the C3 position of imidazoheterocycles through Cu^{I}-catalysed reaction. α,β-Dicarbonylated compounds were formed in the presence of CuBr as catalyst and 2,2′-bipyridine as ligand in 55–85% yields (Scheme 10a). Interestingly, when acetaldehyde or propionaldehyde was used, no α,β-dicarbonylated product was observed. Formation of product under optimized conditions in the presence of

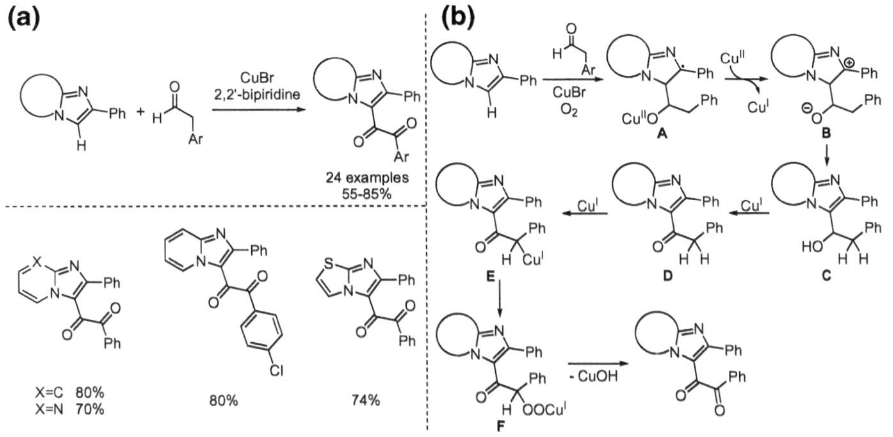

Scheme 10 Double-oxidative CDC of imidazoheterocycles with aryl acetaldehyde (**a**) and plausible mechanism (**b**)

radical scavenger (2,2,6,6-Tetramethylpiperidin-1-yl)oxyl (TEMPO) was suppressed. This indicates that the reactions are likely radical processes. Tentative mechanism is depicted in Scheme 10b.

Later, Hajra et al. also reported synthesis of the C3 dicarbonylatedimidazoheterocycles via Fe-catalysed CDC of the imidazoheterocycles with aryl 2-oxo-acetaldehydes under ambient air in good yields (Scheme 11). Indolizine substitution at position 1 with electron-withdrawing groups also reacted with arylglyoxal at C3 position to produce CDC product. The reaction take place in equal ease in the presence of radical scavengers confirming that the reaction is

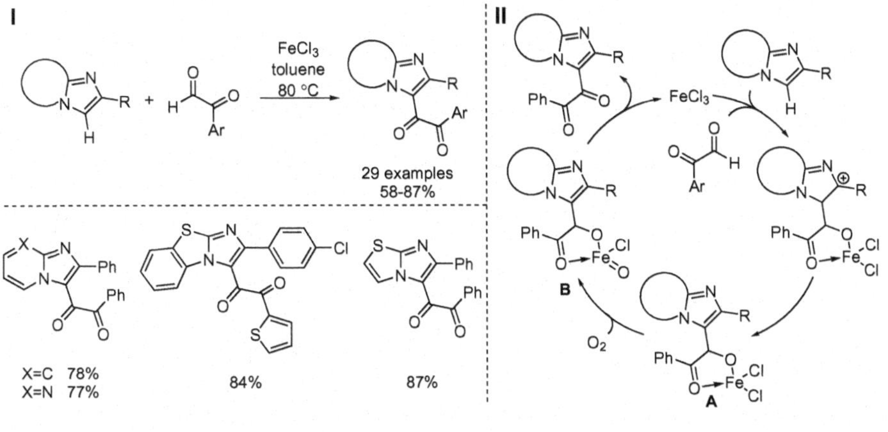

Scheme 11 Fe-catalysed CDC of imidazoheterocycles with 2-oxoaldehydes (**a**) and tentative mechanism (**b**)

likely nonradical process. Authors suggested that $FeCl_3$-catalysed addition of phenylglyoxal to imidazoheterocycles occurred at the C3 position to produce the Fe^{III}-chelated intermediate **A**. Then, Fe^{III} was converted to Fe^{IV}-oxo complex **B** via air oxidation. Finally, product was obtained via reductive elimination and Fe^{II} was oxidized to Fe^{III} in the presence of oxygen to complete the catalytic cycle [21].

2.2 Oxidative Heck-Type Reaction

The oxidative Heck reaction, also called Fujiwara–Moritani reaction (for recent review oxidative Heck reaction, see [22]), is an interesting example of CDC process discovered in 1967 [23]. In the classical interpretation, the Heck reaction is a Pd-catalysed reaction of olefins with prefunctionalized alkenes or aryls [24]. Halide and triflate groups are used as the functional groups in the Heck reaction. In contrast, the Fujiwara–Moritani reaction represents the reaction of alkenylation of non prefunctionalized arenes via C–H activation bond in the presence of palladium (II) salts [23]. This reaction does not yield halogen by-products and, in general, is more environmentally preferable over classic Heck reaction (Scheme 12).

Several examples of coupling alkenes with hetarenes and application for synthesis of natural compounds, for example, ergot alkaloid *clavicipitic acid*, through the oxidative Heck coupling are depicted in Scheme 13. It should be noted, that not only palladium was used as catalyst, but also bismuth and zinc salts were utilized for catalysis of the coupling of indole with alkenes.

Alkenylation of indolizines [25] and imidazo[1,2-*a*]pyridines [26] with acrylic acid derivatives also catalysed by palladium (II) acetate with copper (II) acetate as oxidant as displayed in Scheme 14. The reaction proceeded with high regioselectivity at C3 position in both cases. Interestingly, when ethyl methacrylate was utilized, ethyl 2-(indolizin-3-ylmethyl)acrylate was obtained in moderate yields (Scheme 14) [25].

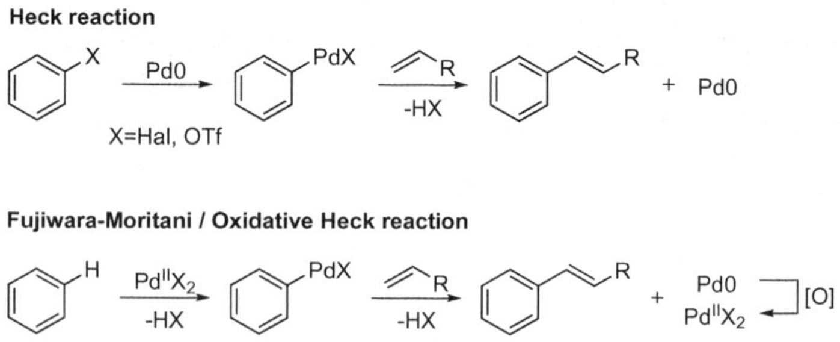

Scheme 12 Heck and the Fujiwara–Moritani reactions [22]

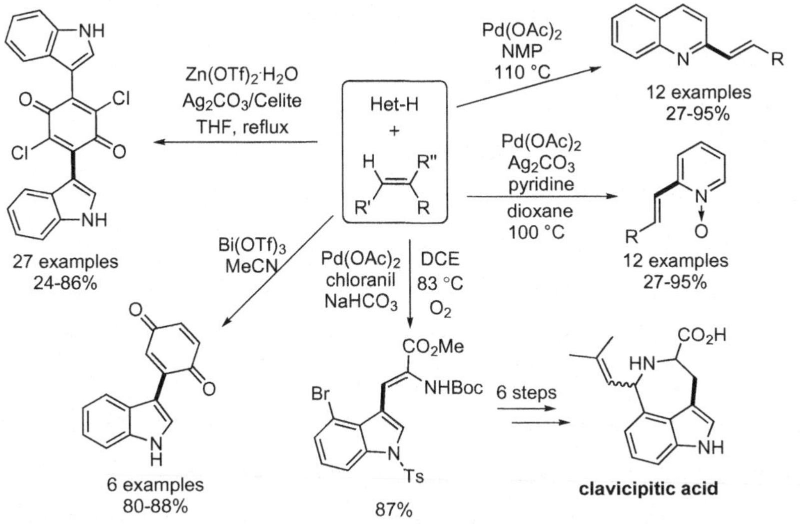

Scheme 13 Examples of oxidative Heck reaction

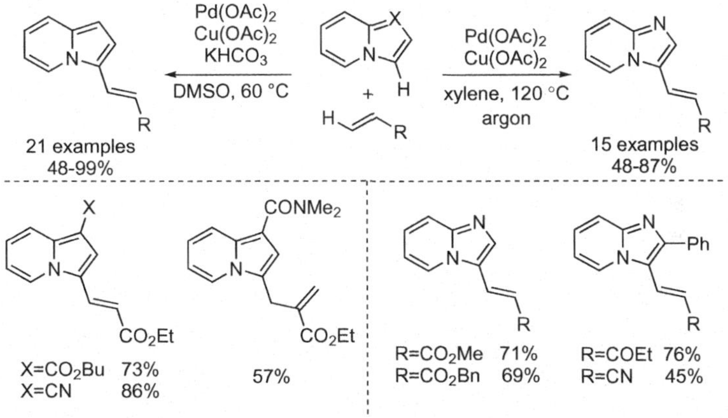

Scheme 14 Alkenylation of indolizines and imidazo[1,2-*a*]pyridines

Later, Ru-catalysed coupling of imidazo[1,2-*a*]pyridines with acrylates was proposed [27]. 3-Regioselective alkenylated compounds were formed in 50–76% yields with catalytic amount of [$RuCl_2$(*p*-cymene)]$_2$, $Cu(OAc)_2$ as oxidant and silver hexafluoroantimonate(V) ($AgSbF_6$) as additive in 1,2-dichloroethane at 120 °C.

Interestingly, using styrene as alkene in the coupling with imidazo[1,2-*a*] pyridines under conditions, shown in Scheme 13, gave rise the mixture of isomeric 3-(α-(phenyl)vinyl)- and 3-(β-(phenyl)vinyl)-imidazo[1,2-*a*]pyridines in 28 and 39% yields, respectively [26]. Later protocol for preparation of 3-(α-(phenyl)vinyl)-

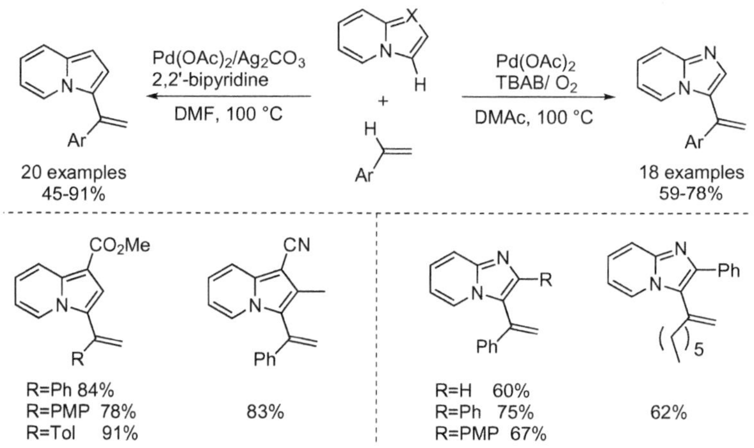

Scheme 15 Regioselective α-hetarylation of styrenes

imidazo[1,2-*a*]pyridines [28] and indolizines [29] with high regioselectivity via Pd-catalysed alkenylation was published. Results of these works are shown in Scheme 15. Using octene-1 instead of styrene under standard condition led to 3-(oct-1-en-2-yl)imidazo[1,2-*a*]pyridine in 62% yield (Scheme 15) [28].

On the other hand, Ong et al. proposed protocol for selective β-hetarylation of styrenes with azoles (Scheme 16) [30]. In this method, palladium(II) trifluoroacetate (Pd(TFA)$_2$) was used instead of $PdCl_2$ or Pd(OAc)$_2$. Oxazole, thiazole, imidazole and their benzo-annulated derivatives were coupled with styrenes in the presence of Pd(TFA)$_2$, 1,10-phenanthroline as ligand and AgTFA as oxidant in toluene at 130 °C. 1*H*-Indene, acrylates, cycloocta-1,5-diene were underwent standard condition yielding the corresponding coupling products. In addition, 1-phenyl-1,3-butadiene reacted exclusively at its terminal double bond. Naturally occurring *Annuloline* was synthesized in good yields [30] using this method.

Later Ma et al. published synthesis of 5,8-distyrylquinoxaline via the Fujiwara–Moritani reaction of 2,3-disubstituted-6,7-difluoroquinoxalines as initial substrate with styrenes (Scheme 16). The reaction was carried out under Pd(TFA)$_2$ catalyst in the presence of argentum salt and 1,10-phenantroline at high temperature to give products in good yields [31].

Wang et al. utilized thiazolo[3,2-*b*]-1,2,4-triazoles in the oxidative Heck reaction with various alkenes using Pd(OAc)$_2$/Cu(OAc)$_2$/O_2 catalytic system (Scheme 17) [32]. When the C2- and the C5 positions of thiazolo[3,2-*b*]-1,2,4-triazoles were not occupied, the reaction of alkenylation proceeded exclusive at the C5 position led to 5-alkenyl-thiazolo[3,2-*b*]-1,2,4-triazoles in good yields, so the C5 position is more nucleophilic than the C2 position; therefore, the C5 is most likely first attacked by the Pd(II) species.

In 2006, Gaunt et al. suggested method of alkenylation of pyrroles with control of regioselectivity (Scheme 18) [33]. Authors used a steric and electronic control

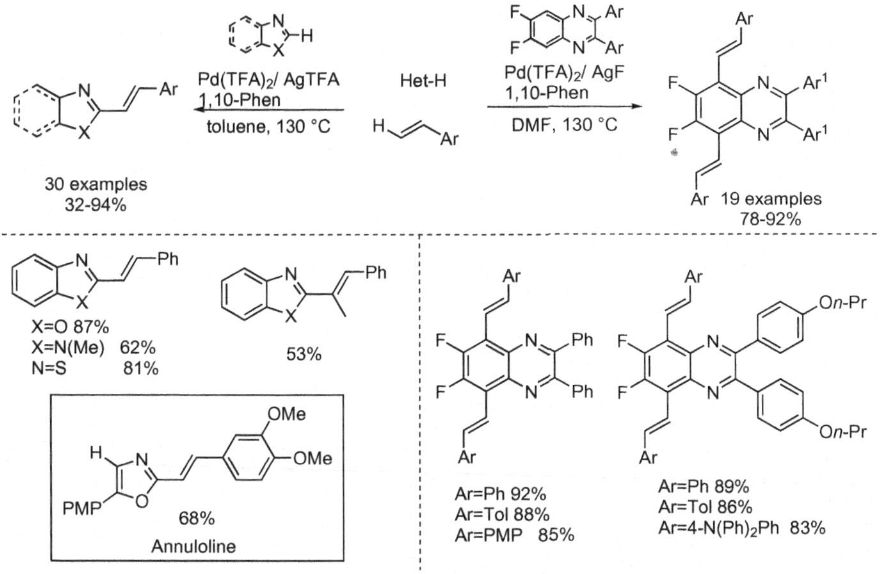

Scheme 16 Pd-catalysed coupling of hetaryls with styrenes

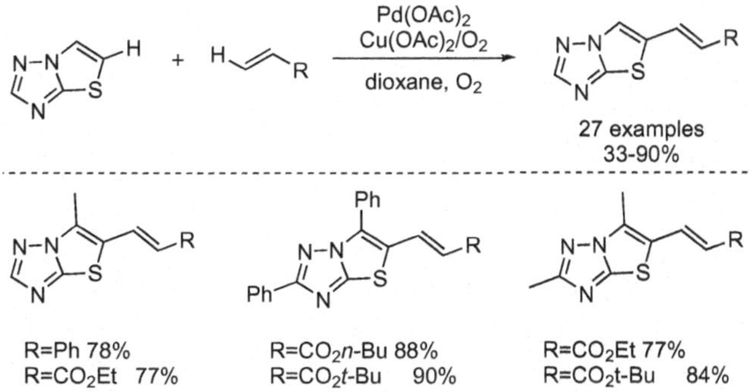

Scheme 17 Alkenylation of thiazolo[3,2-*b*]-1,2,4-triazoles

strategy to influence positional control in the C–H bond functionalization process that resulted in either the C2 or C3 alkenylated products. *N*-acetyl, *N*-tosyl (Ts) and *N*-*tert*-butyloxycarbonyl (BOC) pyrrole afforded only the C2 product. N-BOC pyrrole gave higher yield in comparison with other protecting groups. In contrast, *N*-triisopropylsilyl (TIPS) pyrrole gave only the C3 product. Set of alkenylated pyrroles was prepared in good yields in the presence of $Pd(OAc)_2$ in a dioxane-AcOH-DMSO solvent system and oxidant. When cyclohex-2-en-1-one was supposed standard condition, no product was observed with N-BOC pyrroles.

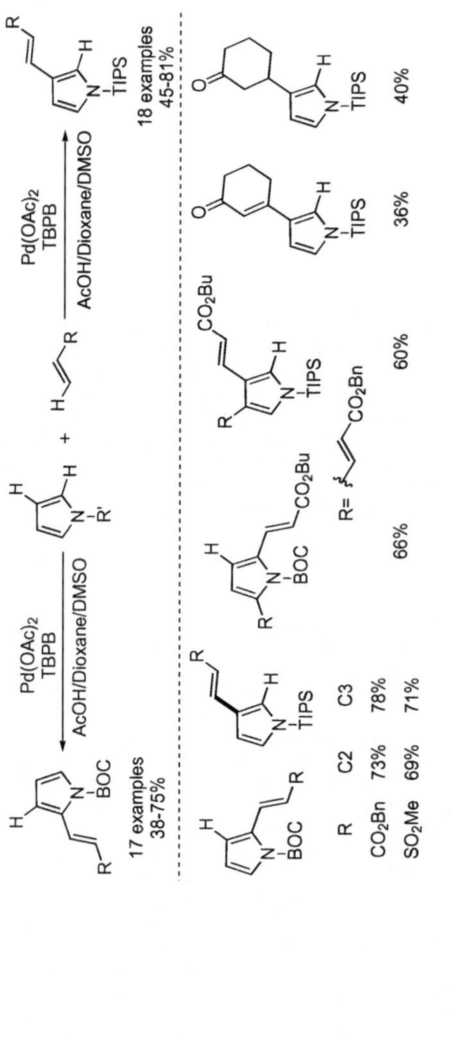

Scheme 18 C2 and C3 alkenylation of pyrroles

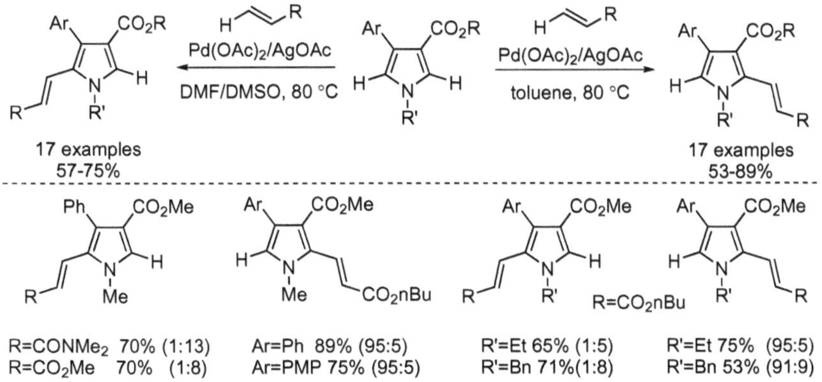

Scheme 19 Regioselective alkenylation of 4-aryl-1*H*-pyrrole-3-carboxylates (ratio C2:C5 shown in parentheses)

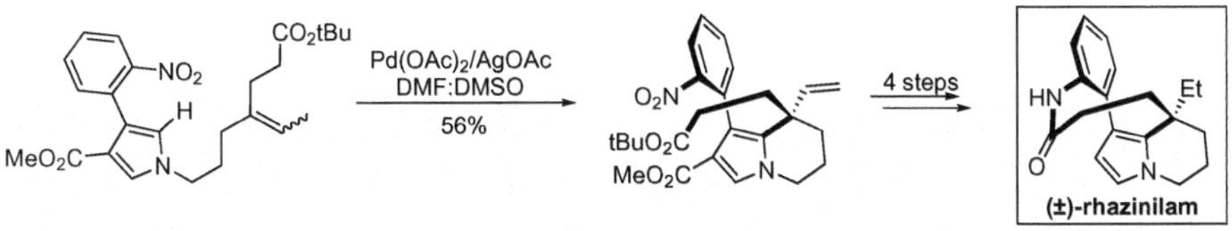

Scheme 20 Total synthesis of (±)-Rhazinilam via a Pd-catalysed intramolecular oxidative Heck-type reaction

At the same time, mixture of saturated and unsaturated product in 76% overall yield was produced with N-TIPS pyrrole.

Subsequently, Yao et al. developed solvent-controlled selectivity C2- and C5-alkenylated of 4-aryl-1*H*-pyrrole-3-carboxylates through the oxidative Heck reaction under $Pd(OAc)_2/AgOAc$ catalytic system (Scheme 19). The C2-alkenylated pyrroles were the major product with ratio C2:C5 up to 95:5 when toluene was used as solvent. On the other hand, using DMF-DMSO (v/v = 4:1) solvent system led to C5-alkenylated product with ratio C2:C5 up to 1:13. Authors believed that in the toluene palladation preferentially occurred at the C2 position via a carboxylate-assisted chelation. On the contrary, DMSO as a strong coordinating solvent overrode the chelation effect of the carboxylate group, promoting the palladation at the more electron-rich C5 position [34].

The intramolecular oxidative Heck reaction is well developed within *N*-heterocycles, and examples of intramolecular type reaction are shown in detail review [22].

As another example of oxidative Heck-type reaction should be mentioned total synthesis of naturally occurring (±)-Rhazinilam (Scheme 20) [34].

3 Functionalization of Aromatic *N*-Heterocycles with Aromatic Substrates

Heteroaromatic biaryls are important framework with wide scope of applications from advanced materials and liquid crystals to bioactive compounds and natural products. Due to this fact, the development of new approaches towards the synthesis of heteroaromatic biaryls framework remains an ongoing challenge for organic chemists. The traditional method involves the TM-catalysed cross-coupling between two prefunctionalyzated substrates, an organometallic species and a heteroaryl halide or pseudohalide. Over the last decades (for recent reviews on CDC reactions, see [35–38]), C–H/C–H CDC has emerged as an active field in organic synthesis which allows for atom- and step-economic transformations [39]. This novel approach is even more suitable for heteroaromatic substrates since their organometallic derivatives are often unstable or troublesome to prepare.

Nonetheless, CDC reactions have some drawbacks. One of the main disadvantages is the low reactivity of substrates with one heteroatom which requires high temperatures; in addition, some reactions require expensive palladium or rhodium catalysts.

3.1 TM-Catalysed Coupling

3.1.1 Azole–Azole Coupling

Five-membered aromatic π-excessive heteroarenes are reluctant, of course, to react with nucleophilic reagents in common conditions. Contrary to that, metal activated π-excessive systems readily undergo the coupling. Often the metal of the highest degree of oxidation acts as both a catalyst and an oxidant. As a rule, a copper, palladium, silver or rhodium salts are used as such metal salts.

The probable mechanism of TM-catalysed CDC reaction on the example of the $M(OAc)_2$-mediated dehydrogenative cross-coupling between the azole derivatives is outlined in Scheme 21. The mechanism involves twofold aromatic C−H bond metalation via M(II) to M(0) route by convergent disproportionation of the M(I) mediator. The mechanism is supported by metal, for example, copper [40] mirror formation during the reaction. Kinetic isotope effect study clearly indicates that the C−H bond metalation of azoles was not involved in the rate-determining step in the present reaction [41].

This mechanism, apparently, *mutatis mutandis*, is applicable to all metal-catalysed CDC reactions.

Scope of such CDC reactions was extended to variety of heterocyclic compounds, including pyrroles, imidazoles, furans, oxazoles, thiophenes, thiazoles as well as their benzo-annulated analogues. An increase in the number of heteroatoms facilitates the metalation of the heterocycle and facilitates the reaction. Thus,

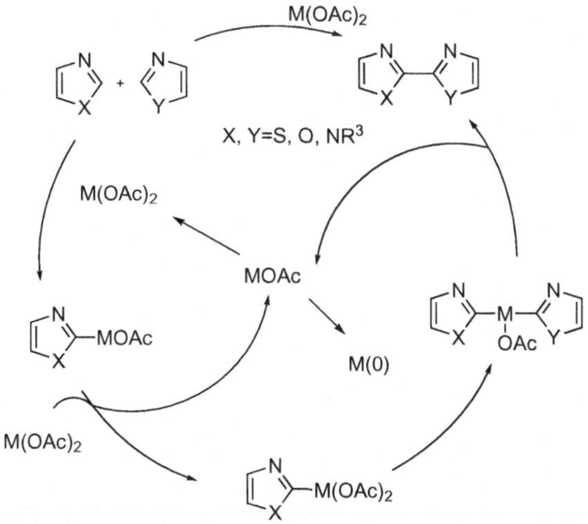

Scheme 21 Plausible mechanism of TM-catalysed coupling

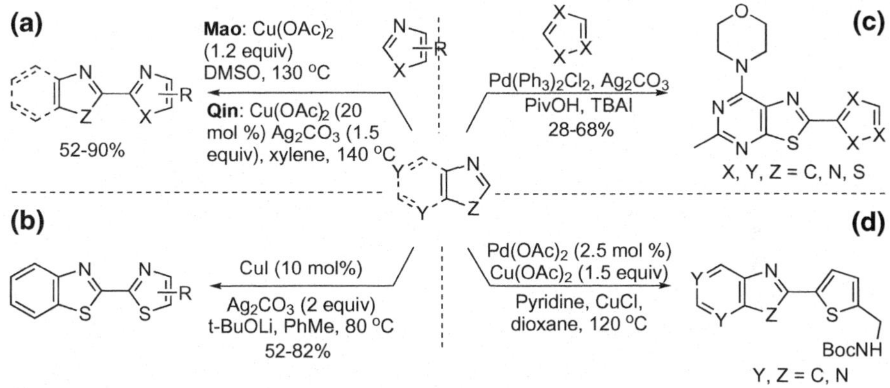

Scheme 22 Scope of TM-catalysed coupling of azoles

oxazoles, imidazoles, thiazoles and triazoles are most studies substrates in this type reaction (Scheme 22).

In their reports, Qin et al. [41] and Mao [40] described the $Cu(OAc)_2$-catalysed CDC reactions between two electron-rich five-membered heterocycles (thiazoles and oxazoles, imidazoles, along with their benzo-annulated analogues, Scheme 22a).

Fan et al. [42] used CuI as catalyst in combination with t-BuOLi as base to achieve coupling between benzothiazoles and thiazoles (Scheme 19b). In these reactions, as an oxidant used excess of $Cu(OAc)_2$ [41] or Ag_2CO_3 [40, 42].

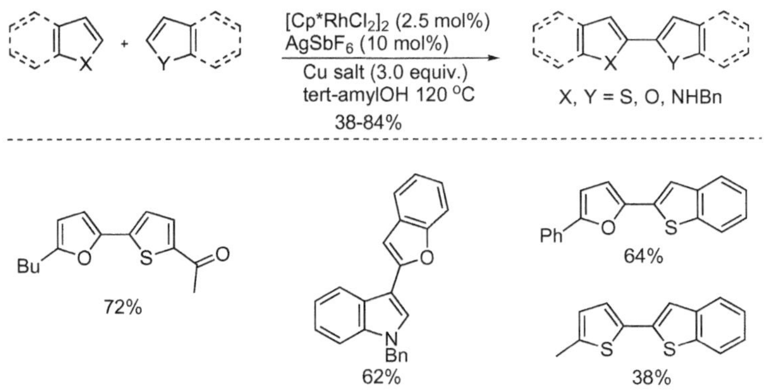

Scheme 23 Rhodium(III) catalysed coupling of benzothiophenes and thiophenes

Another examples of Pd-catalysed coupling of azolels are provided in Scheme 22c, d [43].

For cross-coupling of two heterocycles with one heteroatom each and that do not contain activating (chelating) groups, the most active metal catalysts are required. For example, CDC reaction between furans and benzothiophenes or thiophenes yielding the corresponding 2,2′-bi(heteroaryl) compounds may be achieved in the presence of rhodium(III) catalysis (Scheme 23) [44]. Additionally, the reaction conditions could also be applied to other substrate combinations including indoles and pyrroles, i.e. X = NBn (Scheme 23) [44].

If there is no suitable coupling partner, a dimerization of heterocycle (i.e. homocoupling) takes place. Indoles [45, 46], indolizines [47] and imidazopyridines [48] may be easy dimerized in Pd- or Cu-catalysed conditions in high yields (Scheme 24). When two indolizine moiety are already linked through linker (R+R = –COO(CH_2CH_2O)$_3$CO–), synthesis of corresponding cyclophanes may be achieved via such intramolecular double C–H coupling [47].

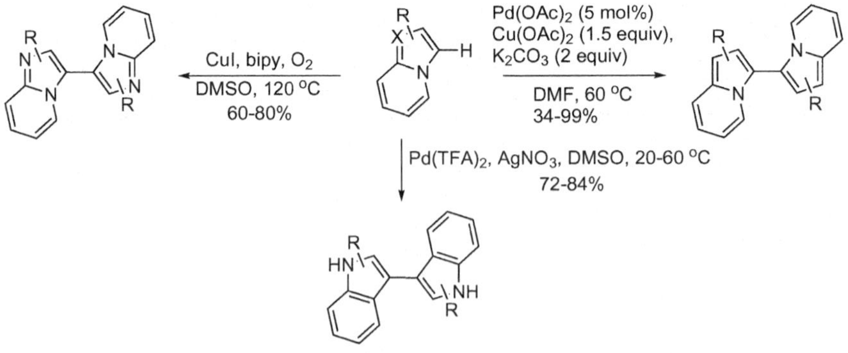

Scheme 24 Dimerization of indoles, indolizines and imidazopyrimidines

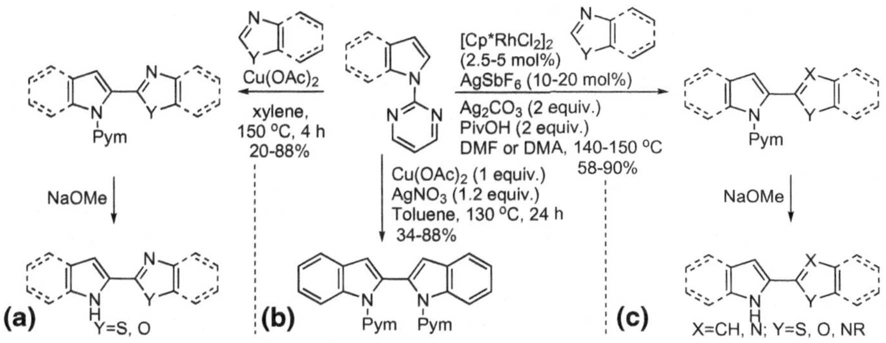

Scheme 25 TM-catalysed CDC reactions of activated indoles with azole derivatives

Chelation in CDC Reactions of Azoles

The chelation-assisted strategy is widely used in metal-catalysed CDC reactions to involve less active heterocycles with one heteroatom. The matter is that the chelation stabilizes a bond between the metal and the aromatic substrate, thereby allowing the metalation step to be carried out more easily, even for slightly reactive substrates.

An excellent example of the chelation-assisted CDC is a copper-mediated intermolecular cross-coupling between 1,3-azoles (thiazoles and oxazoles) and activated indoles and pyrroles with the chelation assistance of a 2-pyrimidyl group (Scheme 25a) published by Nishino group [49]. The reaction was achieved by using atmospheric oxygen as the sole oxidant and can be regarded as one of the most environmentally benign cross-couplings with the liberation of H_2O as the sole by-product. The N-pyrimidyl group exerted C2 regiocontrol via selective C–H activation; the pyrimidyl group may be smoothly removed with sodium methylate after the coupling.

The authors have analysed isotopic exchange and concluded that first cupration takes place in 1,3-azole ring rather than in indole ring even activated by chelating group [49].

A 2,2′-dimerization of indole derivatives being yet another example of Cu-mediated chelation-assisted CDC coupling (Scheme 25b) was reported by Le [50]. The dehydrogenative homocoupling of indoles was effective in the presence of $AgNO_3$ as the oxidant.

Rhodium catalysts are necessary for coupling of less active substrates containing single heteroatom. Rh(III)-catalysed dehydrogenative coupling between indoles/pyrroles activated with 2-pyrimidinyl group and heteroarenes was described (Scheme 25c) by Qin [51] and Reddy [52]. The activity of the rhodium catalyst is so high that a coupling of two heterocycles, the heterocycles contain only one heteroatom, can be carried out. Another important consequence of the high stability of rhodium complexes is isolation [51] and characterization using XRD such rhodium complex (Fig. 2), the formation of which was only postulated earlier for copper-containing one.

Intramolecular CDC Reactions

As mentioned above, chelation plays an important role in the metal-catalysed CDC reactions. At the same time, the presence of a chelating group necessitates the subsequent removal of such a group. A much more attractive option is when the chelating group is part of the starting molecule which should remain in the product, and even better to participate as a coupling partner. This concept was implemented in the intramolecular CDC reactions. Presumably, the presence of a linker between two azole rings, both of which must be metallated, provides favourable steric effects for the formation of chelate [53].

A series of works on intramolecular Cu- [54] and Pd- [53] catalysed cross-coupling of imidazoles with pyrroles as well as their benzo-annulated analogues was published by Tripathi group (Scheme 26). The simplest example of such a reaction includes the intramolecular coupling of 2-(pyrrol-1-yl)ethylbenzimidazoles (Scheme 26a). There is competition between intramolecular cyclization and intermolecular coupling in this reaction and intramolecular/intermolecular product ratio depends on the amount of $Cu(OAc)_2$, selected ligand, base, etc., and varies from 1:20 to 7:1 [53].

Another work [55] of Tripathi group relates to a coupling between indole and imidazole or triazole moieties yielding corresponding intramolecular product (Scheme 26c); the coupling proceeds as formation of 2,2′-bond. Interestingly, the

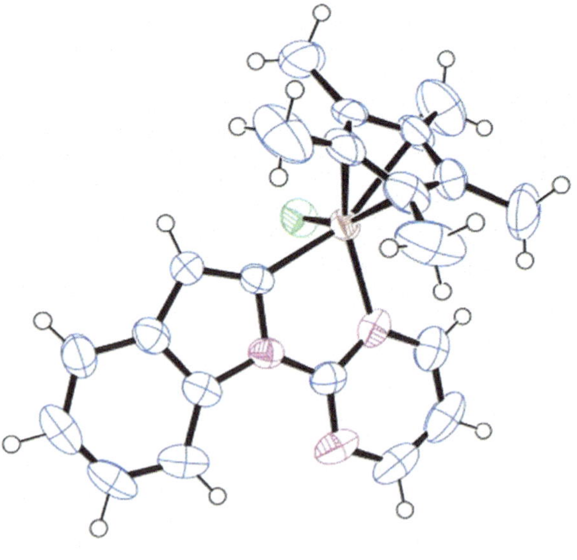

Fig. 2 X-ray crystallographic data for isolated coupling reaction intermediate

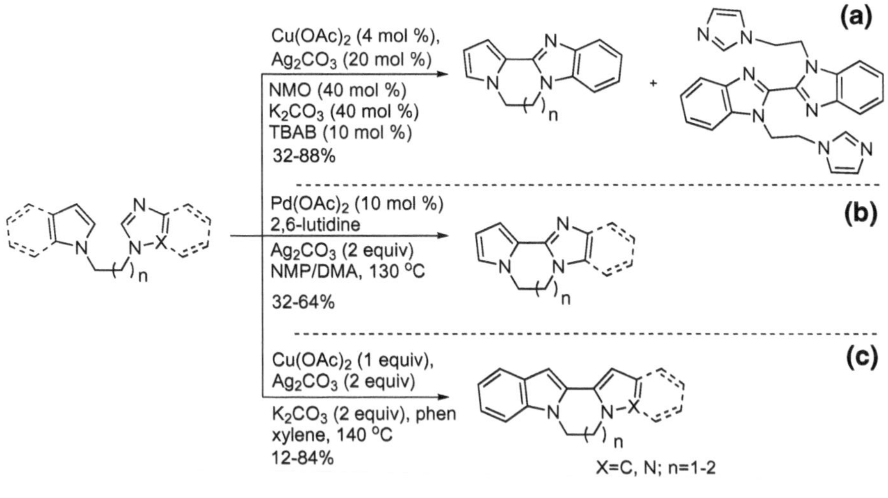

Scheme 26 Intramolecular CDC reactions in azole derivatives

authors claim no intermolecular coupling products; however, low yields in some cases suggest formation of such products.

An intramolecular Pd(II)-catalysed CDC of indolyl-3-alkylazoles (Scheme 27) described by Mantenuto et al. demonstrated that it is not necessary for the azole moieties to be linked through nitrogen atom [56]. The suggested method may be useful for synthesis of indole-derived natural products and pharmaceuticals due to wide distribution of indolylalkylamines in the nature and as medicines.

Two azole rings connected through o-phenylene moiety provide favourable steric effect for intramolecular coupling. Such reactions are often carried out as domino- or one-pot processes involving the formation (or addition) of an azole ring and the subsequent intramolecular coupling.

One example of such one-pot process is reaction between phenylpropionicacid and o-(indol-1-yl)phenylazide [57]. The authors have suggested mechanism comprising decarboxylative Cu-catalysed click reaction followed by reaction between benzimidazole and 1,2,3-triazole thus formed (Scheme 28). The mechanism was supported by the isolation of 1,2,3-triazole intermediate from reaction mixture.

Scheme 27 Intramolecular Pd-catalysed coupling in indolyl alkyl derivatives

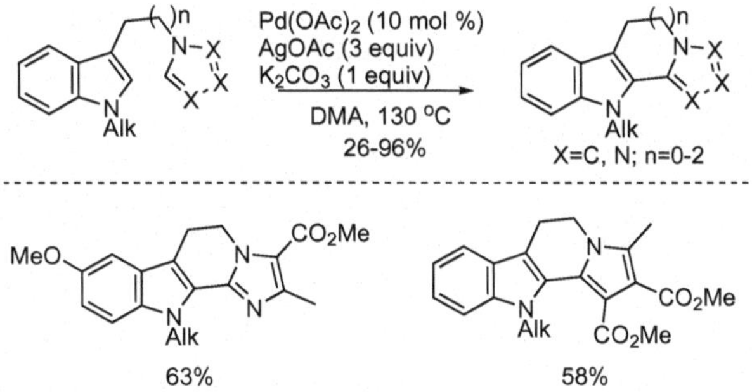

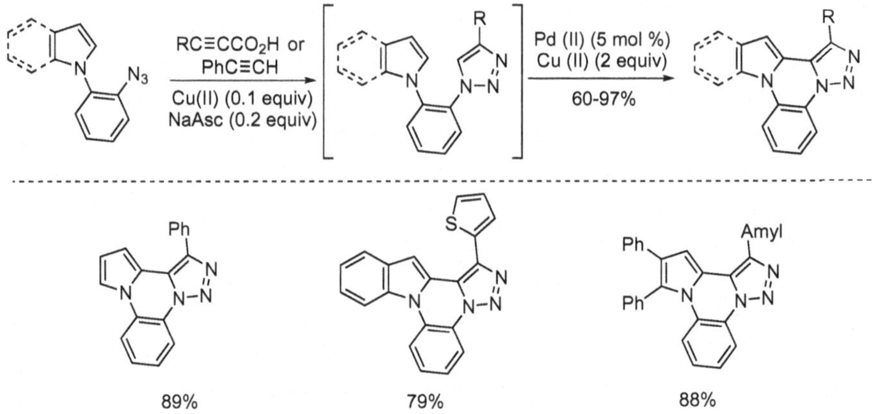

Scheme 28 Click reaction CDC domino reactions' sequences

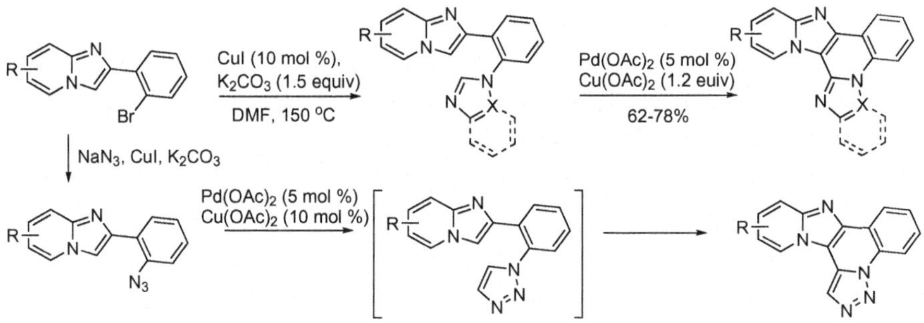

Scheme 29 CHal/NH intermolecular and CDC CH/CH intramolecular coupling domino reactions

Similar approach including substitution of halogen with azide followed by 2+3 cycloaddition and CDC reaction was developed by Wang. In contrast to previous work, the cycloaddition and CDC reactions were performed as single step [58].

Another one-pot protocol comprises sequential CHal/NH intermolecular and CDC CH/CH intramolecular coupling (Scheme 29) [59]. Intermediate o-phenylene di-azole was also isolated supporting proposed mechanism.

3.1.2 Azole–Benzene Couplings

Examples of CDC reaction of not activated benzene ring are outlined in Scheme 30. The reactions were carried out using simple not activated benzene derivatives such as halobenzene, methylbenzene or methoxybenzene and oxazole or thiazole derivatives (Scheme 30a) [60], caffeine [61], thiazolo[5,4-d]pyrimidine [43] or thiazole [62] derivatives.

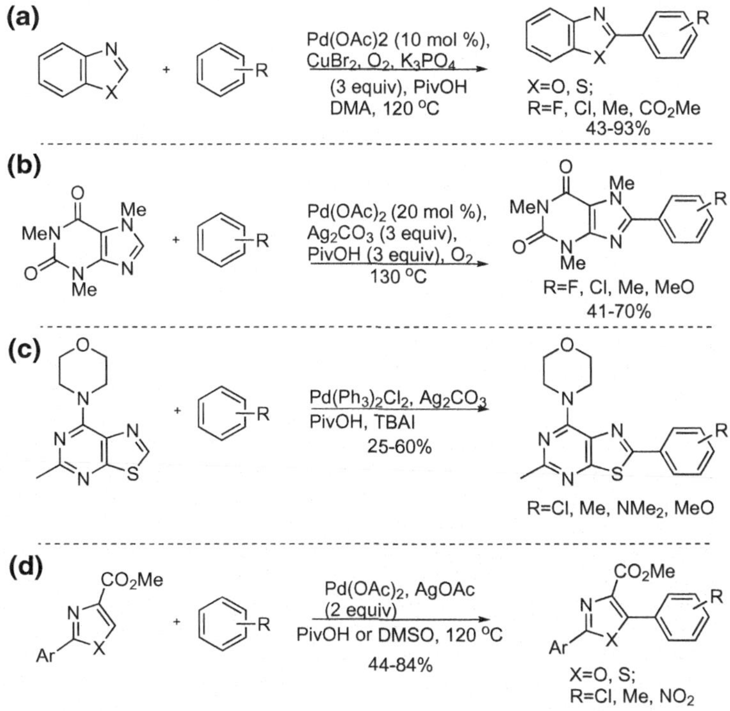

Scheme 30 Pd-catalysed coupling of azoles with not activated simple benzenes

Pd-catalysed CDC reaction between annulated pyrazoles and phenyl rings (Scheme 28) was described for imidazo[1,2-*a*]pyridines (Scheme 31a) as inter- [63] and intramolecular [64] reactions (Scheme 31b) and for benzimidazole substrates [65].

Since the benzene cycle does not contain heteroatoms, the way to improve the activity and regioorientation in such substrates is introduction of chelating (directing) groups that contain atoms with lonely electron pair. Furthermore, coupling of activated benzene derivatives more than azole derivatives requires the most active metals, such as rhodium and ruthenium.

One of the most studied processes in this field is metal-catalysed CDC of 2-phenylpyridine framework with π-excessive 5-membered heterocycles (Scheme 32). The reaction was carried out as rhodium- [52, 66, 67] or copper- [68] mediated process at 140 or 170 °C, respectively. Labelled atoms study showed that the deuterium exchange occurs more easily in activated benzene than in the 5-membered heterocycle [66]. The scope of this reaction was extended from 2-phenopyridine to azaphenanthrene [52, 66, 68], 2-phenylquinoline [66], 2-naphthylpyridine [52, 66], 2-naphthylquinoline [66], 2-phenylpyrimidine [66, 68] and 2-phenylquinaxoline [66].

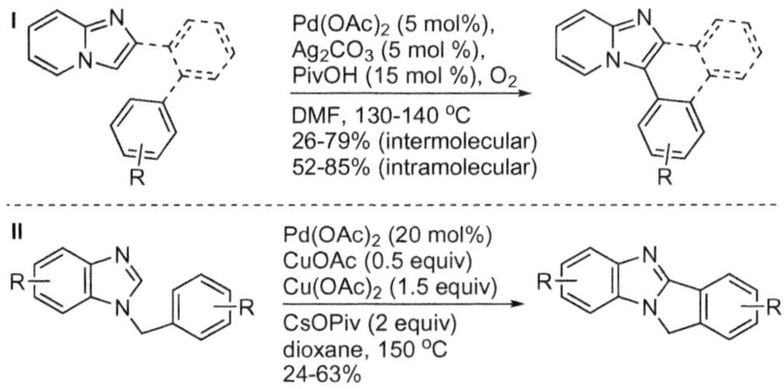

Scheme 31 Pd-catalysed cyclization of imidazole derivatives

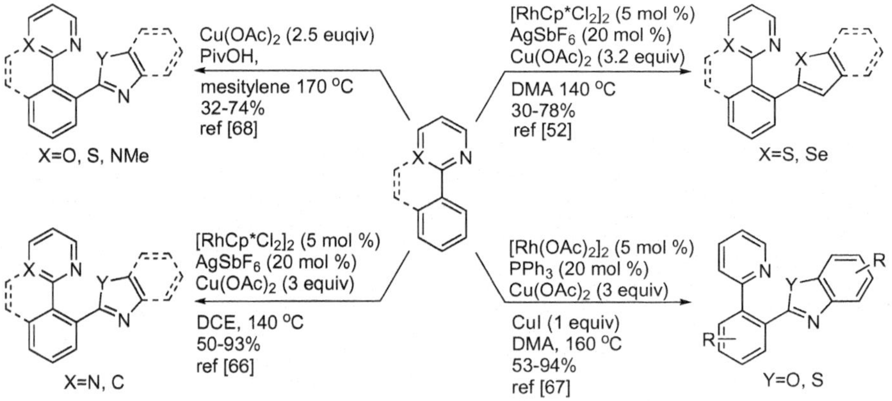

Scheme 32 Hetarylation of 2-phenylpyridine framework

As mentioned above, high temperatures and strong directing group are required to achieve the coupling using copper catalyst. An one example of such directing group is *N,N*-bidentate coordination system based on the picolinamide (Scheme 33). The reaction proceeds smoothly in the presence of $Cu(OAc)_2$ catalysts and provides heterobiaryls [69]. In this case, the formation of an intermediate aryl cuprate is facilitated by the presence of a strongly directing group located at peri-position of naphthalene ring. The directing picolinamide group is easily removed after the coupling using Boc_2O and NaOEt subsequent treatment, and free amine is recovered.

Similar example of double N,N-coordination strategy is copper-mediated CDC reaction of benzoic acid derivatives and 1,3-azoles with the aid of 8-aminoquinoline framework (Scheme 34). Such a bidentate coordination system broadens the substrate scope of both nonacidic arenes and 1,3-azoles in the copper-promoted direct heterobiaryl coupling. The directing 8-aminoquinoline group is easy removed as described for picolinamide one [70].

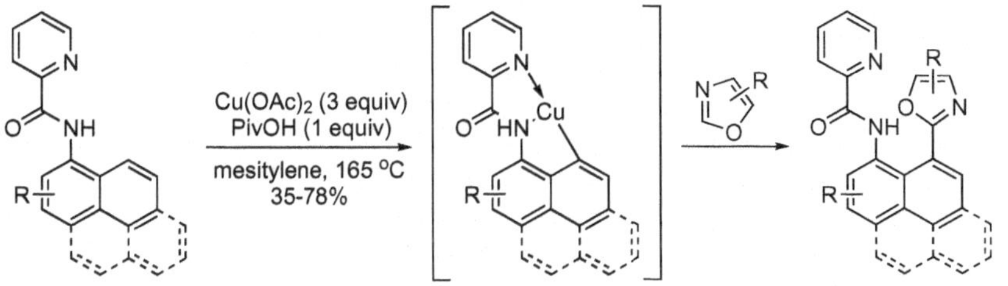

Scheme 33 Cu-mediated picolinamide-directed CDC reaction of oxazioles with naphthalenes

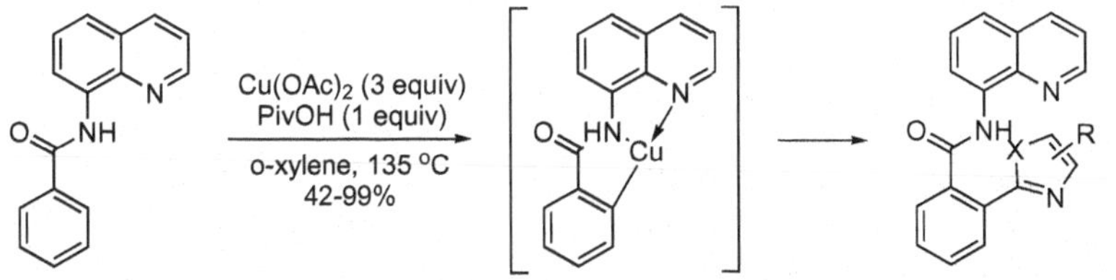

Scheme 34 Cu-mediated 8-aminoquinoline-directed CDC reaction of azioles with benzamides

One of the drawbacks of the copper-catalysed reactions is the requirement of a high temperature, about 170 °C, and the corresponding high-boiling solvent which makes it impossible to involve labile substrates in the reaction.

At the same time, the use of a ruthenium catalyst can reduce the temperature to 140 °C and use a wider range of solvents. Such rhodium-catalysed reaction was used for coupling 9-pyrimidinyl activated carbazoles with thio- and selenophene derivatives (Scheme 35a) via double CH bond using $Cu(OAc)_2/AgSbF_6$ as an oxidant [52].

This catalytic methodology was developed for the synthesis of heteroaryled pyridines [71] and azabenzenes [72] with a good substrate scope with a broad range of functional group tolerance and high regioselectivity of the pyridyl C–H activation (Scheme 35b, c).

Less active acetylamino group requires palladium catalyst and phenantranyl ligand to achieve the coupling (Scheme 36) [73]. At the same time, the ready availability of acylated derivatives in contrast to picolinic acid derivatives repays the disadvantages related to the use of palladium catalyst and silver nitrate oxidant.

An unusual example of a double C–H coupling was described by Bhunia et al. According to author's rationalization (Scheme 37), after the electrophilic C3 palladation of indole, an unexpected "through-space" 1,4-palladium migration to the 2-aryl moiety, by remote C–H bond activation followed by C–H arylation with diaryliodonium salt, and an unprecedented 1,2-aryl shift take place. Finally, an intramolecular CDC at the C2 position affords dibenzo[*a*,*c*]carbazoles in good to high yields [74].

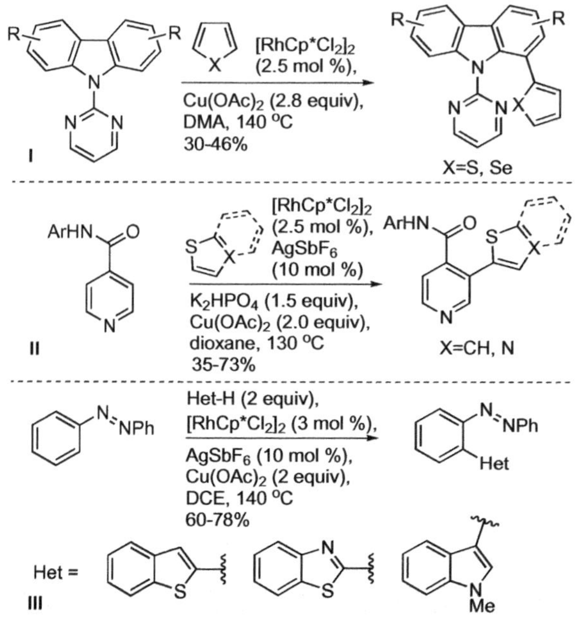

Scheme 35 Rh-catalysed directing group-assisted CDC reactions of azoles with aromatics

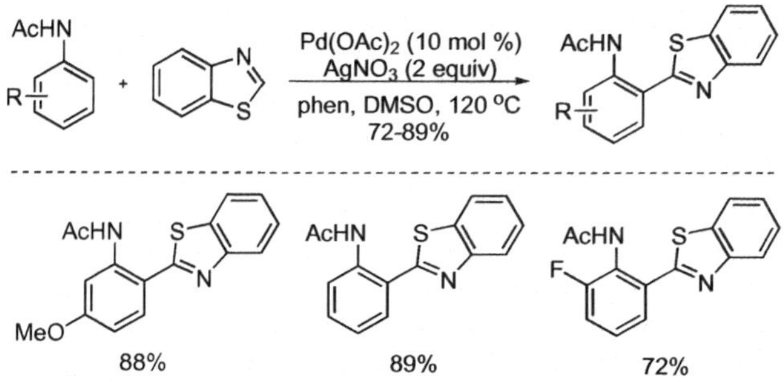

Scheme 36 Pd-catalysed CDC between acetylanilines and benzothiazole

An example of copper-mediated coupling of pentafluorobenzene with benzothiazoles [42] and oxadiazoles [75] was described (Scheme 38a). Particularly noteworthy, the fluorine atom does not leave aromatic ring as in the case of aromatic nucleophilic substitution but acts as a kind of protective group that directs the reaction at a single hydrogen-containing position.

Due to low reactivity of pyridine itself, N-oxides may be considered as activated analogues thereof. The reactions of N-oxides of pyridine and related azines with azoles are presented in Scheme 39. The scope of the reaction was shown for

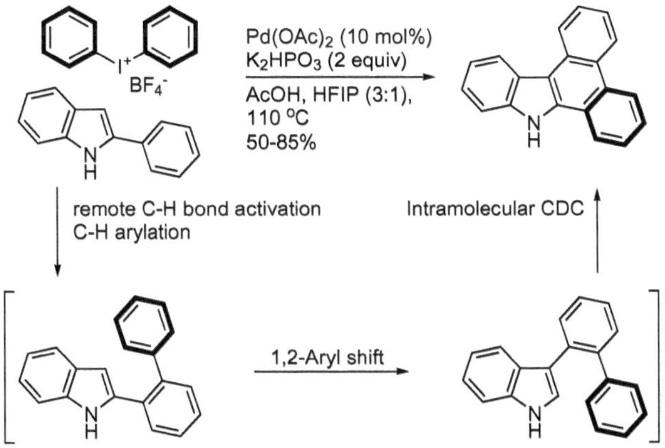

Scheme 37 Double C–H coupling of 2-phenylindoles with phenyl iodonium salts

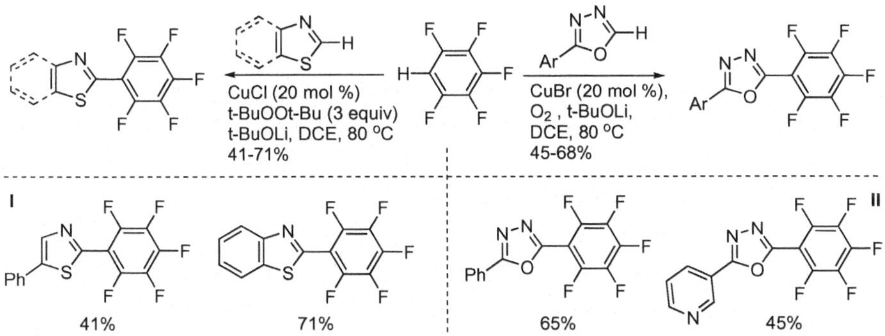

Scheme 38 Fluorine atom as protecting group in benzene ring

pyrroles [74, 76], thiophenes [77], furans [77], indoles [77, 78], thiazoles [79, 80] and benzothiazoles [80].

Pyridine-N-oxides may react both with heteroaromatic and aromatic compounds.

A palladium-catalysed CDC reaction of pyridine-N-oxides with toluenes under mild conditions has been developed by Kianmehr group. 2-Benzylpyridines can be obtained directly by this method via a CDC reaction between not activated toluenes and pyridine-N-oxides (Scheme 40). In addition, azafluorene N-oxides can be obtained successfully by this procedure via four tandem C–H bond activations [81].

A protocol for the regioselective CDC pyridine-N-oxides with *N*,*N*-dialkyluracils via twofold C–H activation was reported by Kianmehr in 2014 [82]. The obtained uracils substituted with pyridine-N-oxide structural motif may be readily deoxygenated, making the present procedure a highly attractive route for the synthesis of 5-pyridine substituted uracils (Scheme 41).

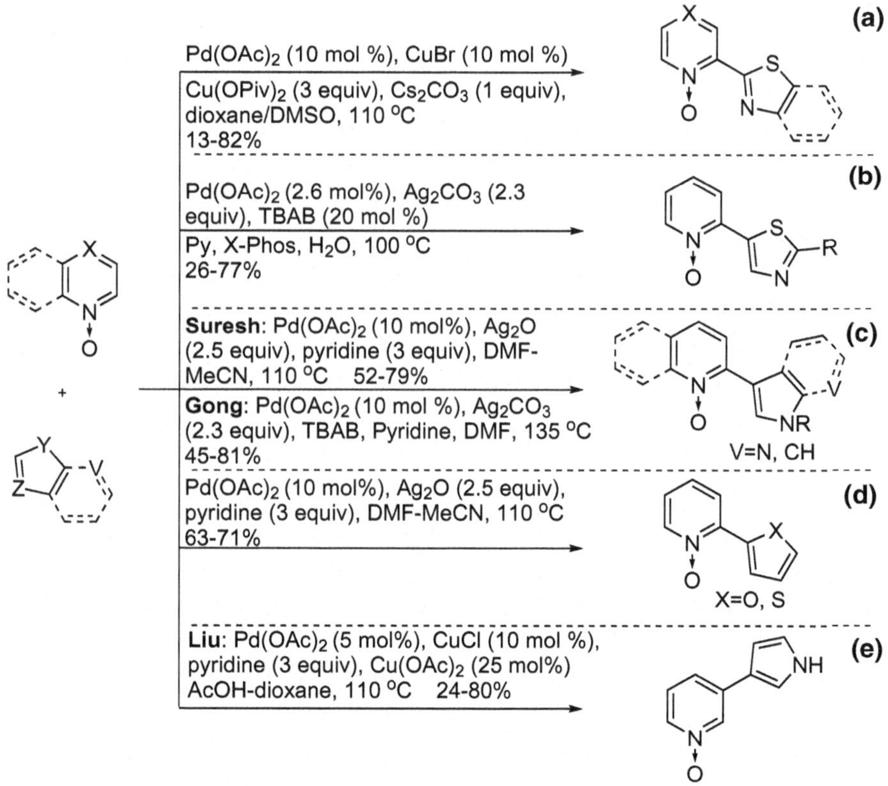

Scheme 39 Reactions of pyridine-N-oxides and related azines with azoles

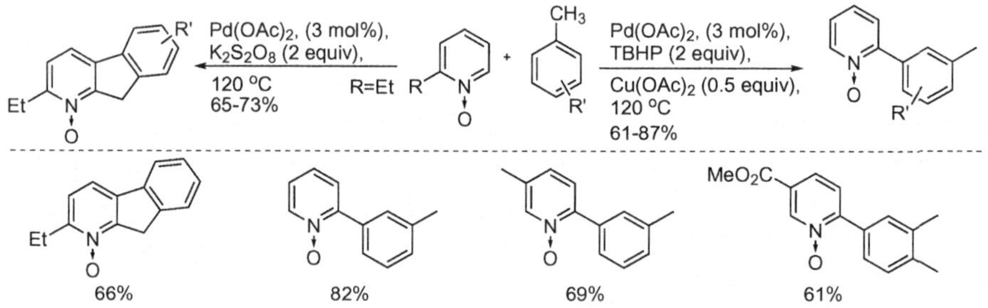

Scheme 40 Reaction of pyridine-N-oxides with toluene

This reaction also may be considered as an example of the Fujiwara–Moritani (the oxidative Heck reaction) coupling due to low aromaticity of uracil ring.

Oldani group reported an excellent example of chelation-assisted CDC coupling reaction of 2-pyridone with pyridine-N-oxide (Scheme 42). It was found that

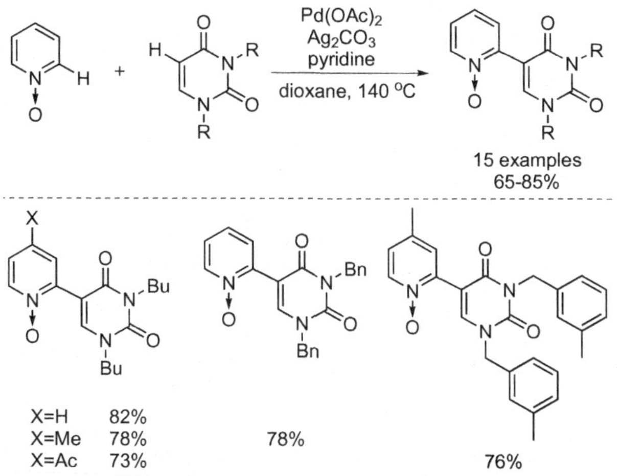

Scheme 41 Pd-catalysed synthesis of uracil–pyridine hybrids

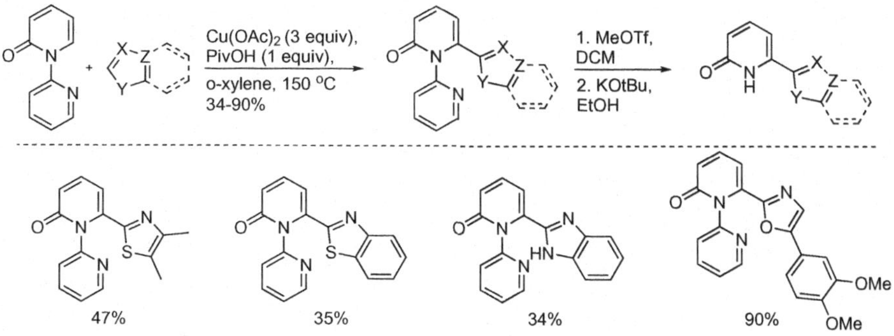

Scheme 42 Pyrimininyl as directing group in arylation of pyridin-2-one

pyridine-2-yl-pyridone-2 reacts with azole derivatives in the presence of $Cu(OAc)_2$. According to Oldani's rationalization, reaction proceeds as cupration of azole followed by copper complex formation and elimination of copper [83].

Palladium-catalysed CDC reaction of pyridine with benzoxazole was described by Itami group [84]. The regioselectivity depends on organic halide used as an oxidant. Thus, then triisopropylbromo benzene was used, and the reaction takes place at the C3 position of pyridine; benzyl bromide directs the reaction at the C2 position of pyridine (Scheme 43) [84]. The effect of aromatic bromides additive may be rationalized from point of view steric hindrances of the corresponding aromatic groups.

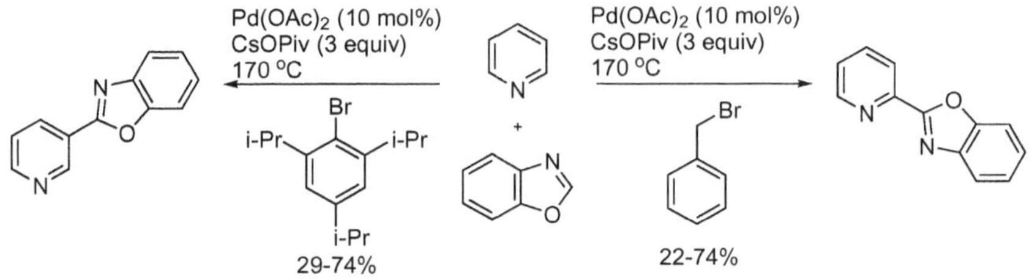

Scheme 43 Regiocontrolled reaction of pyridine with benzodioxole

3.1.3 Azine–Benzene Coupling

Coupling azines with a benzene derivative are usually carried out in the presence of palladium acetate as catalyst and an oxidant being an argentum salt, air oxygen or iodonium salt.

Two related works directed to synthesis of indolo- and benzofuranoquinolines were published by groups of Chen and Cheng (Scheme 44).

The synthesis of the benzofuranoquinolines includes one-pot, two-step cyclization reaction of iodoarenes with 4-hydroxyquinoline or 4-hydroxypyridine. The reaction sequence comprises iodonium salt formation, followed by O-arylation and CDC reaction (Scheme 44a) [85]. The iodonium salt formation step may be omitted and isolated quinolinyl-aniline may be subjected to oxidative cyclization (Scheme 44b) [86].

Ren and co-workers reported Pd-catalysed regioselective C-2 arylation of quinolines, isoquinolines and quinoxalines with benzene derivatives (Scheme 45) [87].

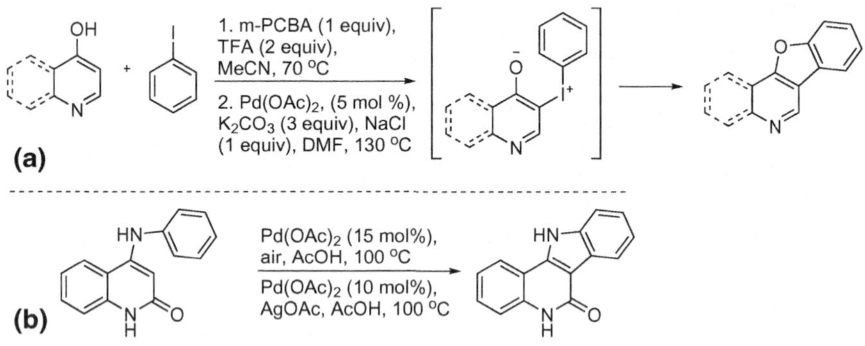

Scheme 44 Synthesis of indoloquinolines and benzofuranoquinolines

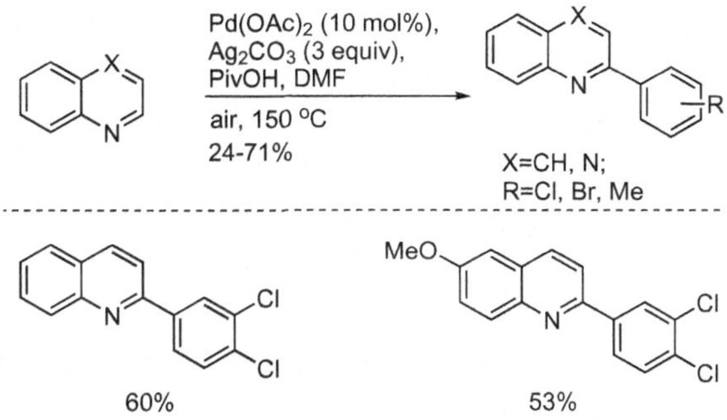

Scheme 45 Pd-catalysed arylation of quinolines and quinoxalines

3.2 *TM-Free Coupling Reactions*

As mentioned above, C–H/C–H CDC reactions provide extremely wide opportunities for the construction of heterocyclic frameworks. However, these reactions cannot avoid a criticism, since TM (usually Pd or Ru) is rather expensive and toxic, and even traces of TM are not allowed to be impurities in drugs, organic solar cells and other high-tech materials [39]. Furthermore, such reactions usually require high temperatures (100–150 °C) and may be not applicable for labile substrates.

Metal-free coupling reactions have not so wide scope as metal-catalysed reaction; however, they have some advantages, since there is no requirement of expensive catalyst, and metal-free reactions are usually carried out at ambient or slightly elevated temperature. Thus, metal-free and metal-catalysed reactions are complementing each other.

The common mechanism of metal-free coupling reaction involves the addition of a nucleophile at a double C=N bond followed by oxidation of the dihydroadduct (σ^H-adduct) thus formed [39]. Depending on the stability of the dihydroadduct, the reaction may stop at the stage of its formation or rearomatization occurred with the restoration of aromaticity. Oxidation often occurs under the action of atmospheric oxygen or oxidizers can be used (including anode oxidation) (Scheme 46). Such reactions can be considered as nucleophilic substitution of hydrogen (S_N^H reactions).

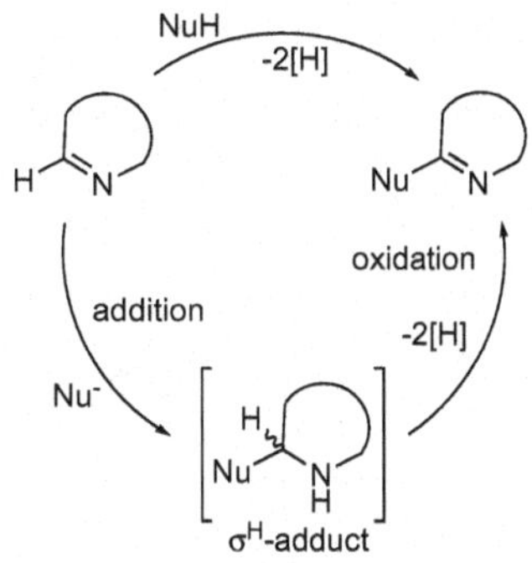

Scheme 46 Mechanism of TM-free CDC reactions

Scheme 47 TM-free coupling of acridine with aromatics and hetaromatics

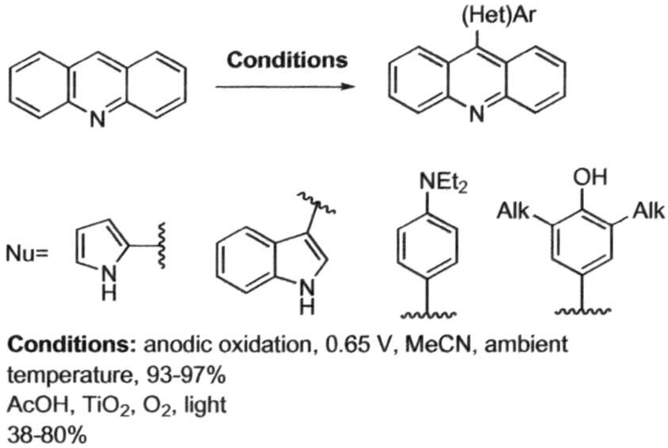

The activity of azine in metal-free CDC reactions is determined by its ability for addition of a nucleophile, and, obviously, related to the aromaticity of the azine component. Most active azines, such as pyrimidine, triazines, quinazoline, quinoxaline, and acridine, may be used for metal-free coupling reactions.

The coupling with acridines (Scheme 47) was carried out by electrochemical oxidation [88] or by oxidation on activated titanium dioxide [89]. Reactions are characterized by high yields and, moreover, correspond to the principles of PASE.

Pyrimidines and quinazolines, their benzo-annulated analogues, comprise additional heteroatom and characterized by increased reactivity in contrast to acridine. TM-free CDC reactions are outlined in Scheme 48.

The reaction of nucleophilic arenes and hetarenes with quinoxalones giving the corresponding C–H/C–H coupling products is a well-studied process (Scheme 49). Derivatives of indole [89, 90], pyrrole [91], pyrazole [92] and benzene activated by electron donor groups [89] were used as nucleophilic substrates.

To promote the reaction, such catalysts as I_2 [91], TiO_2 [89] and TfOH [90] were used. Furthermore, *p*-toluenesulphonyl chloride may be oxidant in this reaction [93].

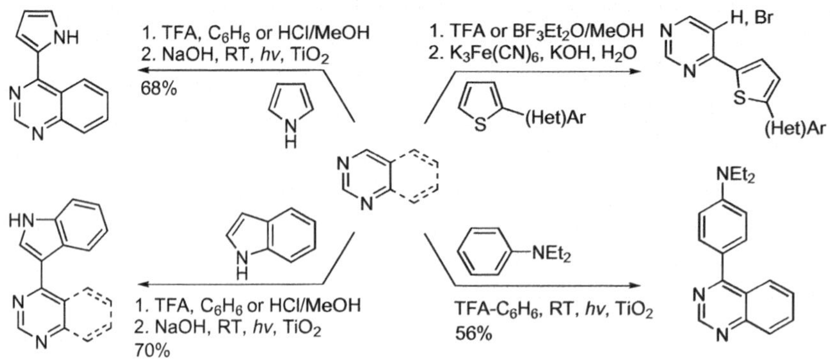

Scheme 48 TM-free coupling reactions of pyrimidines

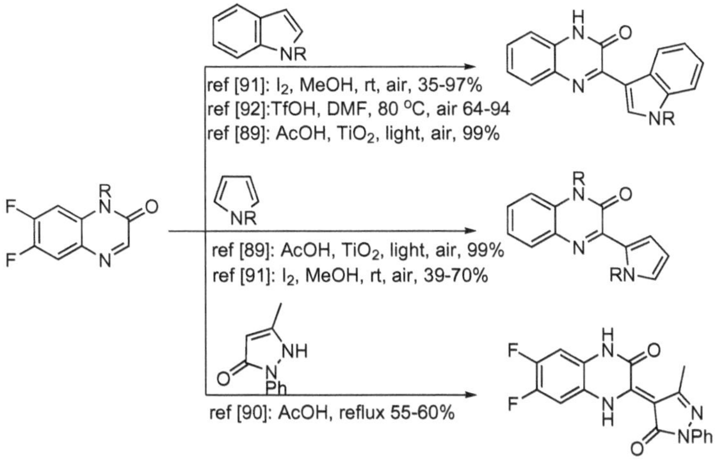

Scheme 49 Scope of coupling reactions of quinoxalone with electron-donating heterocyles

Examples of the coupling of activated benzene derivatives, such as *N*,*N*-diethylaniline [89] and resorcinol [92], with quinoxalones are shown in Scheme 49. 5,7-Dihydroxycoumarins react at the benzene ring [94] and can also be considered as annelated derivatives of resorcinol. Most of the reactions take place under mild conditions, which makes it possible to introduce rather complex substrates into the reaction. For example, mild conditions allow modification of natural occurring 5,7-dihydroxychromone genistein [94] (Scheme 50).

Regiochemistry in this reaction corresponds to the usual electrophilic reactions and proceeds to positions 2 and 3 for pyrrole and indole, respectively. The structure of 3-substituted pyrroles attributable to the products of the coupling pyrrole with quinoxalone is probably erroneous, since the spectra of these products coincide with the spectra for 2-substituted pyrroles [93].

The reactions of indoles and pyrroles with strongly electron-deficient azines (Scheme 51) proceed in the presence of air oxygen/TiO_2 as a photocatalytic oxidative system in a similar way. Proposed approach can be applied for functionalisation of such azaaromatic systems as mono-, di- and triazines, substituted and unsubstituted azines and their benzo-annulated analogues [89].

Borovlev et al. have described reaction of 1,3,7-triazapyrene with both inactivated arenes such as benzene, toluene, xylene [95] and π-excessive arenes such as phenol, anisole, phenetol and aniline [96] in polyphosphoric acid (PPA) or concentrated hydrochloric acid, respectively (Scheme 52). The reaction yielded 6-aryl-triazapyrenes or 6,8-biaryl- double arylation product.

Pteridinone derivatives have attracted attention as substrates for nucleophilic modification due to structural similarity with biogenic pterins, including folic acid. The reactions were carried out using both electron-rich heterocycles, such as indole and pyrazolidone, and resorcinol derivatives and take place at position 7 of

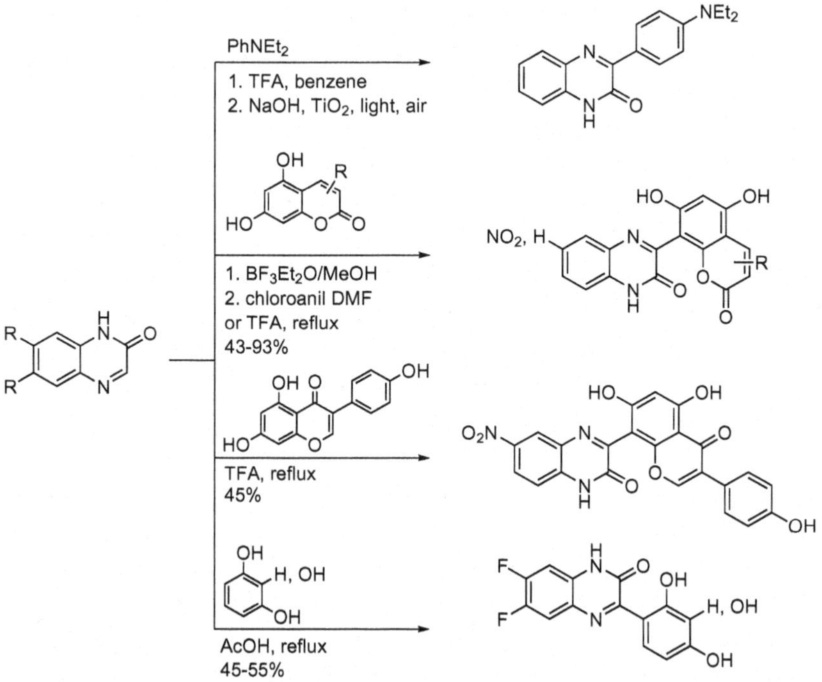

Scheme 50 Scope of coupling reactions of quinoxalone with electron-donating aromatics

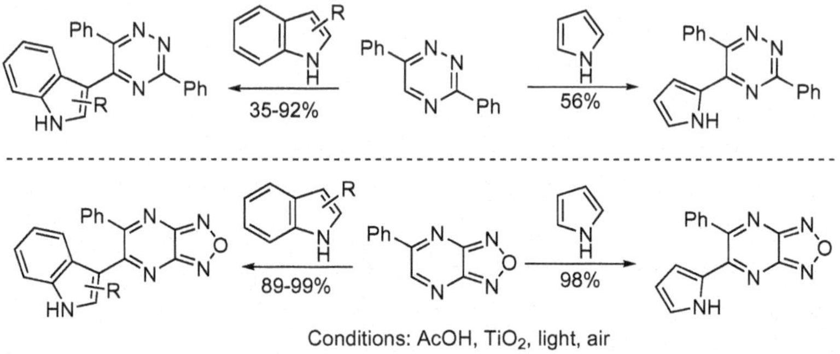

Scheme 51 Reaction of high electron-withdrawing azines with indole and pyrrole

pteridinone scaffold to yield the corresponding coupling product (Scheme 53). The reactions with indole and pyrazolidone proceed under rather harsh conditions, probably because of this their yield is low [97]. At the same time, using trifluoroacetic acid allows to decrease the temperature and increase the yield of the desired products [94].

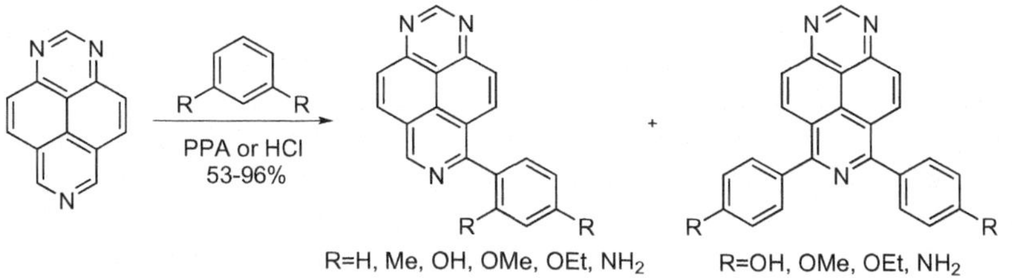

Scheme 52 Reaction of 1,3,7-triazapyrene with arenes

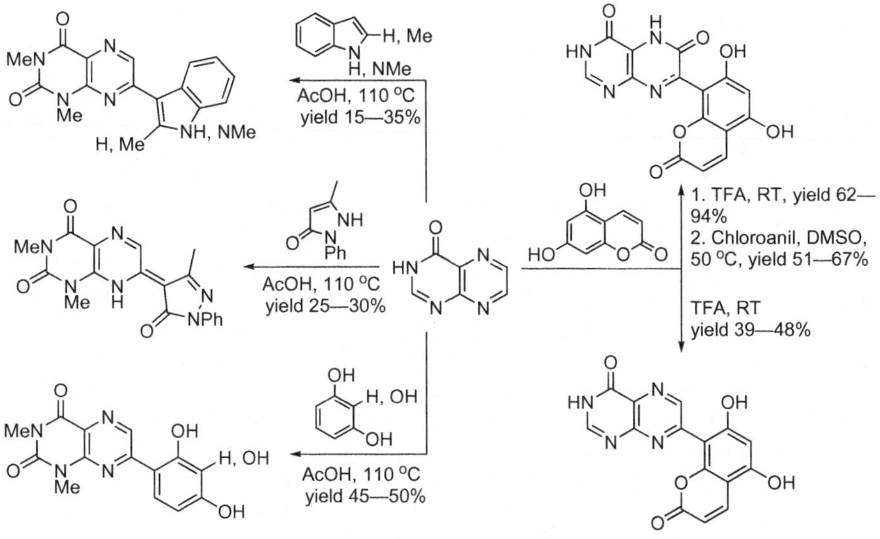

Scheme 53 Scope of coupling reaction of pteridin-4-ones with electron-donating heterocycles and aromatics

4 Conclusion

Oxidative cross-coupling reactions are important tool for the formation of new C(sp^2)–C(sp^2) bond due to simple starting reagents and high complexity of products formed in the reaction. Another aspect of CDC reactions is their consistency with the principles of pot–atom–step economy (PASE). These advantages determine the relevance of this direction in the modern organic chemistry.

The term "CDC reactions" involves a variety of types of reactions, such as free radical, transition metal-catalysed, and TM-free ionic-type reactions. The vast majority of examples of CDC reactions are transition metal-catalysed transformations (e.g. oxidative Heck-type reaction). Reactions of such type are characterized by the formation of an intermediate metal complex, in which corresponding (het) aryls are contained as ligands and a TM (usually copper, rhodium or palladium) as a core. Example of free radical CDC is Minisci-type reaction. TM-free CDC is another direction in cross-coupling reactions avoiding expensive and toxic transition metals and allowing mild reaction conditions.

Currently, all the common types of heterocycles, carbocyclic aromatics, olefins and carbonyl compounds have been involved in these cross-coupling reactions.

CDC reactions were used in the synthesis of biologically active and photoactive compounds, as well in total synthesis of natural compounds (rhazinilam, papaveraldine, thalimicrinone, etc.).

Acknowledgements This work was supported by the Russian Science Foundation (project #19-73-10144).

References

1. Fanta PE (1974) The Ullmann synthesis of biaryls. Synthesis 1974:9–21
2. Colacot TJ (2015) New trends in cross-coupling: theory and applications. The Royal Society of Chemistry, Cambridge
3. Minisci F (1976) Recent aspects of homolytic aromatic substitutions. In: Minisci F, Hendrickson JB, Wentrup C (eds) Synthetic and mechanistic organic chemistry. Springer, Berlin, pp 1–48
4. Citterio A, Gentile A, Minisci F, Serravalle M, Ventura S (1984) Polar effects in free-radical reactions. Carbamoylation and α-N-amidoalkylation of heteroaromatic bases by amides and hydroxylamine-O-sulfonic acid. J Org Chem 49:3364–3367
5. Citterio A, Gentile A, Minisci F, Serravalle M, Ventura S (1983) Redox-chain decomposition of hydroxylamine-O-sulphonic acid. A novel general source of nucleophilic radicals for the functionalization of heteroaromatic bases. J Chem Soc Chem Commun (16):916–917
6. Yao B, Deng C-L, Liu Y, Tang R-Y, Zhang X-G, Li J-H (2015) Palladium-catalyzed oxidative carbamoylation of isoquinoline N-oxides with formylamides by means of dual C-H oxidative coupling. Chem Commun 51:4097–4100
7. Han W, Jin F, Zhao Q, Du H, Yao L (2016) Acid-free silver-catalyzed cross-dehydrogenative carbamoylation of pyridines with formamides. Synlett 27:1854–1859
8. Mete TB, Singh A, Bhat RG (2017) Transition-metal-free synthesis of primary to tertiary carboxamides: a quick access to prodrug-pyrazinecarboxamide. Tetrahedron Lett 58:4709–4712
9. Edwards AC, Geist A, Müllich U, Sharrad CA, Pritchard RG, Whitehead RC, Harwood LM (2017) Transition metal-free, visible-light mediated synthesis of 1,10-phenanthroline derived ligand systems. Chem Commun 53:8160–8163
10. Lewis FW, Harwood LM, Hudson MJ, Drew MGB, Hubscher-Bruder V, Videva V, Arnaud-Neu F, Stamberg K, Vyas S (2013) BTBPs versus BTPhens: some reasons for their differences in properties concerning the partitioning of minor actinides and the advantages of BTPhens. Inorg Chem 52:4993–5005
11. Higginson MA, Kyle ND, Marsden OJ, Thompson P, Livens FR, Heath SL (2015) Synthesis of functionalised BTPhen derivatives—effects on solubility and americium extraction. Dalton Trans 44:16547–16552
12. He T, Li H, Li P, Wang L (2011) Direct amidation of azoles with formamides via metal-free C-H activation in the presence of tert-butyl perbenzoate. Chem Commun 47:8946–8948

13. Zhang Y, Teuscher KB, Ji H (2016) Direct α-heteroarylation of amides (α to nitrogen) and ethers through a benzaldehyde-mediated photoredox reaction. Chem Sci 7:2111–2118
14. Matcha K, Antonchick AP (2013) Metal-free cross-dehydrogenative coupling of heterocycles with aldehydes. Angew Chem Int Ed 52:2082–2086
15. Cheng P, Qing Z, Liu S, Liu W, Xie H, Zeng J (2014) Regiospecific Minisci acylation of phenanthridine via thermolysis or photolysis. Tetrahedron Lett 55:6647–6651
16. Siddaraju Y, Lamani M, Prabhu KR (2014) A transition metal-free Minisci reaction: acylation of isoquinolines, quinolines, and quinoxaline. J Org Chem 79:3856–3865
17. Chen J, Wan M, Hua J, Sun Y, Lv Z, Li W, Liu L (2015) TBHP/TFA mediated oxidative cross-dehydrogenative coupling of N-heterocycles with aldehydes. Org Biomol Chem 13:11561–11566
18. Siddaraju Y, Prabhu KR (2016) Transition metal-free Minisci reaction promoted by NCS, and TBHP: acylation of heteroarenes. Tetrahedron 72:959–967
19. Kianmehr E, Kazemi S, Foroumadi A (2014) Palladium-catalyzed oxidative C-H bond coupling of indoles and benzaldehydes: a new approach to the synthesis of 3-benzoylindoles. Tetrahedron 70:349–354
20. Shakoor SMA, Agarwal DS, Kumar A, Sakhuja R (2016) Copper catalyzed direct aerobic double-oxidative cross-dehydrogenative coupling of imidazoheterocycles with aryl acetaldehydes: an articulate approach for dicarbonylation at C-3 position. Tetrahedron 72:645–652
21. Samanta S, Mondal S, Santra S, Kibriya G, Hajra A (2016) FeCl3-catalyzed cross-dehydrogenative coupling between imidazoheterocycles and oxoaldehydes. J Org Chem 81:10088–10093
22. Yeung CS, Dong VM (2011) Catalytic dehydrogenative cross-coupling: forming carbon −carbon bonds by oxidizing two carbon−hydrogen bonds. Chem Rev 111:1215–1292
23. Ferreira EM, Zhang H, Stoltz BM (2009) Oxidative heck-type reactions (Fujiwara–Moritani reactions). In: The Mizoroki–Heck reaction. Wiley-Blackwell, Hoboken, pp 345–382
24. Li JJ (2009) Heck reaction. In: Li JJ (ed) Name reactions: a collection of detailed mechanisms and synthetic applications. Springer, Berlin, pp 277–280
25. Hu H, Liu Y, Zhong H, Zhu Y, Wang C, Ji M (2012) Heck-type cross-dehydrogenative coupling reactions of indolizines at the 3-position with electron-deficient alkenes through palladium-catalyzed C–H activation. Chem Asian J 7:884–888
26. Koubachi J, Berteina-Raboin S, Mouaddib A, Guillaumet G (2009) Pd/Cu-catalyzed oxidative C-H alkenylation of imidazo[1,2-a]pyridines. Synthesis 2009:271–276
27. Zhan H, Zhao L, Li N, Chen L, Liu J, Liao J, Cao H (2014) Ruthenium-catalyzed direct C-3 oxidative olefination of imidazo[1,2-a]pyridines. RSC Adv 4:32013–32016
28. Ghosh M, Naskar A, Mitra S, Hajra A (2015) Palladium-catalyzed α-selective alkenylation of imidazo[1,2-a]pyridines through aerobic cross-dehydrogenative coupling reaction. Eur J Org Chem 2015:715–718
29. Yang Y, Cheng K, Zhang Y (2009) Highly regioselective palladium-catalyzed oxidative coupling of indolizines and vinylarenes via C−H bond cleavage. Org Lett 11:5606–5609
30. Lee W-C, Wang T-H, Ong T-G (2014) Ligand promoted Pd-catalyzed dehydrogenative alkenylation of hetereoarenes. Chem Commun 50:3671–3673
31. Zhang Z, Zheng Y, Sun Z, Dai Z, Tang Z, Ma J, Ma C (2017) Direct olefination of fluorinated quinoxalines via cross-dehydrogenative coupling reactions: a new near-infrared probe for mitochondria. Adv Synth Catal 359:2259–2268
32. Liu W, Wang S, Zhan H, Lin J, He P, Jiang Y (2014) Highly regioselective palladium-catalyzed direct alkenylation of thiazolo[3,2-b]-1,2,4-triazoles via CH activation. Tetrahedron Lett 55:3549–3552
33. Beck EM, Grimster NP, Hatley R, Gaunt MJ (2006) Mild aerobic oxidative palladium (II) catalyzed C−H bond functionalization: regioselective and switchable C−H alkenylation and annulation of pyrroles. J Am Chem Soc 128:2528–2529
34. Su Y, Zhou H, Chen J, Xu J, Wu X, Lin A, Yao H (2014) Solvent-controlled switchable C-H alkenylation of 4-aryl-1H-pyrrole-3-carboxylates: application to the total synthesis of (±)-rhazinilam. Org Lett 16:4884–4887

35. Bugaut X, Glorius F (2011) Palladium-catalyzed selective dehydrogenative cross-couplings of heteroarenes. Angew Chem Int Ed 50:7479–7481
36. Hirano K, Miura M (2015) Copper-mediated intermolecular C–H/C–H and C–H/N–H couplings via aromatic C–H cleavage. C-H bond activation and catalytic functionalization II. Springer, Cham, pp 47–65
37. Varun BV, Dhineshkumar J, Bettadapur KR, Siddaraju Y, Alagiri K, Prabhu KR (2017) Recent advancements in dehydrogenative cross coupling reactions for CC bond formation. Tetrahedron Lett 58:803–824
38. Yang Y, Lan J, You J (2017) Oxidative C-H/C–H coupling reactions between two (hetero) arenes. Chem Rev 117:8787–8863
39. Charushin V, Chupakhin O (2018) Metal-free C–H functionalization of aromatic compounds through the action of nucleophilic reagents. SpringerLink. https://link.springer.com/chapter/10.1007/7081_2013_119. Accessed 21 Sept 2018
40. Mao Z, Wang Z, Xu Z, Huang F, Yu Z, Wang R (2012) Copper(II)-mediated dehydrogenative cross-coupling of heteroarenes. Org Lett 14:3854–3857
41. Qin X, Feng B, Dong J, Li X, Xue Y, Lan J, You J (2012) Copper(II)-catalyzed dehydrogenative cross-coupling between two azoles. J Org Chem 77:7677–7683
42. Fan S, Chen Z, Zhang X (2012) Copper-catalyzed dehydrogenative cross-coupling of benzothiazoles with thiazoles and polyfluoroarene. Org Lett 14:4950–4953
43. Yang S-W, Su Y-X, Sun L-P (2014) Palladium-catalyzed oxidative C-H/C–H cross-couplings of thiazolo[5,4-d]pyrimidine with aromatic (hetero)cycles. Tetrahedron 70:3730–3734
44. Kuhl N, Hopkinson MN, Glorius F (2012) Selective rhodium(III)-catalyzed cross-dehydrogenative coupling of furan and thiophene derivatives. Angew Chem Int Ed 51:8230–8234
45. Li Y, Wang W-H, Yang S-D, Li B-J, Feng C, Shi Z-J (2010) Oxidative dimerization of N-protected and free indole derivatives toward 3,3′-biindoles via Pd-catalyzed direct C-H transformations. Chem Commun 46:4553–4555
46. Total synthesis of 5,5′,6,6′-tetrahydroxy-3,3′-biindolyl, the proposed structure of a potent antioxidant found in beetroot (Beta vulgaris)—ScienceDirect. https://www.sciencedirect.com/science/article/pii/S0040402004002996. Accessed 9 Nov 2018
47. Xia J-B, Wang X-Q, You S-L (2009) Synthesis of biindolizines through highly regioselective palladium-catalyzed C−H functionalization. J Org Chem 74:456–458
48. Lei S, Cao H, Chen L, Liu J, Cai H, Tan J (2015) Regioselective oxidative homocoupling reaction: an efficient copper-catalyzed synthesis of biimidazo[1,2-a]pyridines. Adv Synth Catal 357:3109–3114
49. Nishino M, Hirano K, Satoh T, Miura M (2012) Copper-mediated and copper-catalyzed cross-coupling of indoles and 1,3-azoles: double C-H activation. Angew Chem Int Ed 51:6993–6997
50. Le J, Gao Y, Ding Y, Jiang C (2016) Cu-mediated C2-dehydrogenative homocoupling of indoles via C-H activation assisted by a removable N-pyrimidyl group. Tetrahedron Lett 57:1728–1731
51. Qin X, Liu H, Qin D, Wu Q, You J, Zhao D, Guo Q, Huang X, Lan J (2013) Chelation-assisted Rh(III)-catalyzed C2-selective oxidative C-H/C–H cross-coupling of indoles/pyrroles with heteroarenes. Chem Sci 4:1964–1969
52. Reddy VP, Qiu R, Iwasaki T, Kambe N (2013) Rhodium-catalyzed intermolecular oxidative cross-coupling of (hetero)arenes with chalcogenophenes. Org Lett 15:1290–1293
53. Tripathi KN, Ray D, Singh RP (2017) Pd-Catalyzed regioselective intramolecular dehydrogenative C-5 cross coupling in an N-substituted pyrrole-azole system. Org Biomol Chem 15:10082–10086
54. Tripathi KN, Ray D, Singh RP (2017) Synthesis of pyrrole-annulated heterocycles through copper-catalyzed site-selective dehydrogenative cross-coupling. Eur J Org Chem 2017:5809–5813

55. Ray D, Manikandan T, Roy A, Tripathi KN, Singh RP (2015) Ligand-promoted intramolecular dehydrogenative cross-coupling using a Cu catalyst: direct access to polycyclic heteroarenes. Chem Commun 51:7065–7068
56. Mantenuto S, Ciccolini C, Lucarini S, Piersanti G, Favi G, Mantellini F (2017) Palladium(II)-catalyzed intramolecular oxidative C-H/C–H cross-coupling reaction of C3, N-linked biheterocycles: rapid access to polycyclic nitrogen heterocycles. Org Lett 19:608–611
57. Bharathimohan K, Ponpandian T, Ahamed AJ, Bhuvanesh N (2014) Sequential decarboxylative azide–alkyne cycloaddition and dehydrogenative coupling reactions: one-pot synthesis of polycyclic fused triazoles. Beilstein J Org Chem 10:3031–3037
58. Wang Z, Li B, Zhang X, Fan X (2016) One-pot cascade reactions leading to pyrido[2′,1′:2,3] imidazo[4,5-c][1,2,3]triazolo[1,5-a]quinolines under bimetallic relay catalysis with air as the oxidant. J Org Chem 81:6357–6363
59. Pericherla K, Khedar P, Khungar B, Kumar A (2013) One-pot sequential C-N coupling and cross dehydrogenative couplings: synthesis of novel azole fused imidazo[1,2-a]pyridines. Chem Commun 49:2924–2926
60. Wu G, Zhou J, Zhang M, Hu P, Su W (2012) Palladium-catalyzed direct arylation of benzoxazoles with unactivated simple arenes. Chem Commun 48:8964–8966
61. Malakar CC, Schmidt D, Conrad J, Beifuss U (2011) Double C–H activation: the palladium-catalyzed direct C-arylation of xanthines with arenes. Org Lett 13:1378–1381
62. Li Z, Ma L, Xu J, Kong L, Wu X, Yao H (2012) Pd(II)-catalyzed direct C5-arylation of azole-4-carboxylates through double C-H bond cleavage. Chem Commun 48:3763–3765
63. Wang S, Liu W, Cen J, Liao J, Huang J, Zhan H (2014) Pd-catalyzed oxidative cross-coupling of imidazo[1,2-a]pyridine with arenes. Tetrahedron Lett 55:1589–1592
64. Banerji B, Chatterjee S, Chandrasekhar K, Bera S, Majumder L, Prodhan C, Chaudhuri K (2017) Expedient synthesis of a phenanthro-imidazo-pyridine fused heteropolynuclear framework via CDC coupling: a new class of luminophores. Org Biomol Chem 15:4130–4134
65. Pereira KC, Porter AL, DeBoef B (2014) Intramolecular arylation of benzimidazoles via Pd (II)/Cu(I) catalyzed cross-dehydrogenative coupling. Tetrahedron Lett 55:1729–1732
66. Dong J, Long Z, Song F, Wu N, Guo Q, Lan J, You J (2013) Rhodium or ruthenium-catalyzed oxidative C-H/C-H cross-coupling: direct access to extended π-conjugated systems. Angew Chem Int Ed 52:580–584
67. Liu H, Xu H, Yuan Y (2014) Rhodium-catalyzed direct oxidative cross-coupling of 2-aryl pyridine with benzothiazoles. Tetrahedron 70:6474–6481
68. Kitahara M, Umeda N, Hirano K, Satoh T, Miura M (2011) copper-mediated intermolecular direct biaryl coupling. J Am Chem Soc 133:2160–2162
69. Odani R, Hirano K, Satoh T, Miura M (2013) Copper-mediated dehydrogenative biaryl coupling of naphthylamines and 1,3-azoles. J Org Chem 78:11045–11052
70. Nishino M, Hirano K, Satoh T, Miura M (2013) Copper-mediated C-H/C-H biaryl coupling of benzoic acid derivatives and 1,3-azoles. Angew Chem Int Ed 52:4457–4461
71. Shang Y, Jie X, Zhao H, Hu P, Su W (2014) Rh(III)-catalyzed amide-directed cross-dehydrogenative heteroarylation of pyridines. Org Lett 16:416–419
72. Deng H, Li H, Wang L (2016) Ortho-heteroarylation of azobenzenes by Rh-catalyzed cross-dehydrogenative coupling: an approach to conjugated biaryls. Org Lett 18:3110–3113
73. Kianmehr E, Fardpour M, Kharat AN (2017) Palladium-catalyzed chemo- and regioselective oxidative cross-dehydrogenative coupling of acetanilides with benzothiazole. Eur J Org Chem 2017:3017–3021
74. Bhunia SK, Polley A, Natarajan R, Jana R (2015) Through-space 1,4-palladium migration and 1,2-aryl shift: direct access to dibenzo[a,c]carbazoles through a triple C–H functionalization cascade. Chem Eur J 21:16786–16791
75. Zou L-H, Mottweiler J, Priebbenow DL, Wang J, Stubenrauch JA, Bolm C (2013) Mild copper-mediated direct oxidative cross-coupling of 1,3,4-oxadiazoles with polyfluoroarenes by using dioxygen as oxidant. Chem Eur J 19:3302–3305

76. Liu S, Tzschucke CC (2016) Palladium-catalyzed regioselective dehydrogenative C-H/C–H cross-coupling of pyrroles and pyridine N-oxides. Eur J Org Chem 2016:3509–3513
77. Suresh R, Muthusubramanian S, Senthilkumaran R (2014) Regioselective cross-dehydrogenative coupling of indole-2-carboxylic acids and pyridine-N-oxides followed by protodecarboxylation. Synlett 25:2064–2066
78. Gong X, Song G, Zhang H, Li X (2011) Palladium-catalyzed oxidative cross-coupling between pyridine N-oxides and indoles. Org Lett 13:1766–1769
79. Willis NJ, Smith JM (2014) An operationally simple, palladium catalysed dehydrogenative cross-coupling reaction of pyridine N-oxides and thiazoles "on water". RSC Adv 4:11059–11063
80. Fu X-P, Xuan Q-Q, Liu L, Wang D, Chen Y-J, Li C-J (2013) Dual C-H activations of electron-deficient heteroarenes: palladium-catalyzed oxidative cross coupling of thiazoles with azine N-oxides. Tetrahedron 69:4436–4444
81. Kianmehr E, Faghih N, Khan KM (2015) Palladium-catalyzed regioselective benzylation-annulation of pyridine N-oxides with toluene derivatives via multiple C-H bond activations: benzylation versus arylation. Org Lett 17:414–417
82. Kianmehr E, Rezaeefard M, Khalkhali MR, Khan KM (2014) Pd-catalyzed dehydrogenative cross-coupling of pyridine-N-oxides with uracils. RSC Adv 4:13764–13767
83. Odani R, Hirano K, Satoh T, Miura M (2014) Copper-mediated C6-selective dehydrogenative heteroarylation of 2-pyridones with 1,3-azoles. Angew Chem Int Ed 53:10784–10788
84. Yamada S, Murakami K, Itami K (2016) Regiodivergent cross-dehydrogenative coupling of pyridines and benzoxazoles: discovery of organic halides as regio-switching oxidants. Org Lett 18:2415–2418
85. Hong F, Chen Y, Lu B, Cheng J (2016) One-pot assembly of fused heterocycles via oxidative palladium-catalyzed cyclization of arylols and iodoarenes. Adv Synth Catal 358:353–357
86. Cheng C, Chen W-W, Xu B, Xu M-H (2016) Access to indole-fused polyheterocycles via Pd-catalyzed base-free intramolecular cross dehydrogenative coupling. J Org Chem 81:11501–11507
87. Ren X, Wen P, Shi X, Wang Y, Li J, Yang S, Yan H, Huang G (2013) Palladium-catalyzed C-2 selective arylation of quinolines. Org Lett 15:5194–5197
88. Chupakhin ON, Shchepochkin AV, Charushin VN (2017) Atom- and step-economical nucleophilic arylation of azaaromatics via electrochemical oxidative cross C-C coupling reactions. Green Chem 19:2931–2935
89. Utepova IA, Trestsova MA, Chupakhin ON, Charushin VN, Rempel AA (2015) Aerobic oxidative C-H/C–H coupling of azaaromatics with indoles and pyrroles in the presence of TiO2 as a photocatalyst. Green Chem 17:4401–4410
90. Han Y-Y, Wu Z-J, Zhang X-M, Yuan W-C (2010) An efficient synthesis of 3-(indol-3-yl) quinoxaline-2-ones with TfOH-catalyzed Friedel-Crafts type coupling reaction in air. Tetrahedron Lett 51:2023–2028
91. Noikham M, Kittikool T, Yotphan S (2018) Iodine-catalyzed oxidative cross-dehydrogenative coupling of quinoxalinones and indoles: synthesis of 3-(indol-2-yl)quinoxalin-2-one under mild and ambient conditions. Synthesis 50:2337–2346
92. Azev YA, Ermakova OS, Berseneva VS, Bakulev VA, Ezhikova MA, Kodess MI (2017) Synthesis of fluoroquinoxalin-2(1H)-one derivatives containing substituents in the pyrazine and benzene fragments. Russ J Org Chem 53:90–95
93. Chupakhin ON, Egorov IN, Rusinov VL, Slepukhin PA (2010) Asymmetric induction in the reactions of azinones with C-nucleophiles. Russ Chem Bull 59:991–1001
94. Khalymbadzha IA, Fatykhov RF, Chupakhin ON, Charushin VN, Tseitler TA, Sharapov AD, Inytina AK, Kartsev VG (2018) Transition-metal-free C-C coupling of 5,7-dihydroxybenzopyrones with quinoxalones and pteridinones. Synthesis 50:2423–2431
95. Borovlev IV, Demidov OP, Borovlev II, Saigakova NA (2013) 1,3,7-triazapyrene: the first case of hetarylation of benzene and its analogs. Chem Heterocycl Compd 49:952–954

96. Borovlev IV, Demidov OP, Saigakova NA (2013) SNH arylation of 1,3,7-triazapyrenes in acidic aqueous solution. Chem Heterocycl Compd 49:618–623
97. Khalymbadzha IA, Chupakhin ON, Fatykhov RF, Charushin VN, Schepochkin AV, Kartsev VG (2016) Transition-metal-free cross-dehydrogenative coupling of triazines with 5,7-dihydroxycoumarins. Synlett 27:2606–2610

The C-1 Functionalization of Tetrahydroisoquinolines via Cross-Dehydrogenative Coupling Reactions

Marlyn C. Ortiz Villamizar, Carlos E. Puerto Galvis and Vladimir V. Kouznetsov

Abstract As a fast-growing research field in modern organic chemistry, the cross-dehydrogenative coupling (CDC) has shown its power in the synthesis of diverse potentially bioactive functionalized tetrahydroisoquinolines. To date, several strategies have been developed to provide a viable, predictable and selective protocol for the C-1 functionalization of tetrahydroisoquinolines, making active an inert C–H bond. Thereby, this chapter is aimed to highlight the recent and impressive progress in the field of CDC reactions where reactive intermediates are formed in situ through different methodologies (organometallic, transition-metal-free, enantioselective and photocatalyst approaches) in order to activate C–H bonds, given a comprehensive overview on existing procedures and reported methodologies.

Keywords Cross-dehydrogenative coupling · C(sp^3)–H bond activation · Tetrahydroisoquinolines · *N*-phenyl-tetrahydroisoquinoline derivatives · Functionalization · Direct coupling

1 Introduction

The direct coupling of two C–H bonds or C–H and C–X bonds to enable the formation of a new C–C(X) bond has become an attractive and powerful strategy to build simple to complex organic frameworks. Since Li's and Murahashi's groups developed this approach in 2000 known as a cross-dehydrogenative coupling (CDC) reactions [35, 56], this concept has evolved from the use of heavy

M. C. Ortiz Villamizar · C. E. P. Galvis · V. V. Kouznetsov (✉)
Laboratorio de Química Orgánica y Biomolecular, CMN, Universidad Industrial de Santander, Parque Tecnológico Guatiguará Km 2 Vía Refugio, A.A. 681011 Bucaramanga, Colombia
e-mail: vkuznechnik@gmail.com; kouznet@uis.edu.co

C. E. P. Galvis
Laboratorio de Química Orgánica Aplicada, Universidad Manuela Beltrán, Cl. 33 # 26-34, 680002 Bucaramanga, Colombia

A. Srivastava and C. K. Jana (eds.), *Heterocycles via Cross Dehydrogenative Coupling*, https://doi.org/10.1007/978-981-13-9144-6_3

transition-metals or organometallic catalysts, to photocatalytic C–H activation, and has gained a great interest of the organic and medicinal chemistry when it is combined with enantio- and regioselective approaches.

The reaction conditions of this atom's efficient transformation usually involve a metal catalyst (Fe, Cu, Pd, Ru, Ir, etc.) along with a suitable oxidant (oxygen or peroxides) [68], but nowadays, in some examples, CDC reactions have been conducted under metal-free conditions, water as a solvent and photochemically or mechanically assisted, enhancing the environmental friendliness [61]. However, beyond establishing other reaction parameters, the real challenge nowadays is to find the adequate and novel substrates to couple a sp^3 C–H bond with another sp^3 C–H bond, a sp^2 C–H bond or various nucleophiles (Nu–H). Although the most common substrates in the CDC reactions are aliphatic and simple unfunctionalized precursors as starting materials, being the major advantage of this strategy, issues well summarized in recent well-designed reviews [35–37, 47, 61, 68, 80], it is more interesting the use of substrates that contain a heteroatom like nitrogen, phosphor, oxygen or sulfur at the α-position of an active sp^3 C–H bond. This heteroatom not only generates an electrophilic center at the C–H carbon, but also acts as a directing group to promote the coupling with the nucleophilic partner and form the new C–C or C–X (X=N, O, S, P) bond [25].

Among the vast number of substrates employed in CDC reactions, Seidel and his group found that pyrrolidines generate an iminium ion that could be coupled with terminal alkynes, furnishing α-propargylamines in what can be called the first CDC with secondary amines [13]. This finding encourages other groups to explore different substrates until Yu and co-workers used 1,2,3,4-tetrahydroisoquinolines (THIQs) as starting materials in a CDC reaction [107] and thus, a new research topic was uncovered in the field of organic chemistry.

Definitely, the THIQ core, which is present in many natural and synthetic products, represents an interesting scaffold for pharmaceutical purposes since it exhibited a wide spectrum of biological activities [69, 71]. Nevertheless, those THIQs derivatized through the CDC reaction, mainly the corresponding α-alkynyl THIQs, displayed a remarkable affinity to D_2 and D_3 dopamine receptors [8], a sirtuin protein inhibition [67] and antitumor properties against HeLa cell lines [94], that clearly stands the CDC reaction as a powerful transformation in organic chemistry which allows the synthesis of small molecules of great value.

Whereas most pioneering work regarding CDC reactions of THIQs has been focused in the use of a variety of metal catalysts, the aim of this chapter is to cover the recent progress in CDC reactions as a C-1 functionalization strategy for THIQs, which have been reported in the last six years (2013–2018), specifically in those protocols where the reactivity of the THIQs is studied besides the transition-metal approach, including metal-free conditions, photochemically or mechanically assisted and stereoselective controlled methodologies. This chapter is not intended to be comprehensive, but we hope that this overview will encourage the further investigation of this trending area.

2 The Transition-Metal-Catalyzed C-1 Activation in THIQs

Cross-dehydrogenative coupling is an ideal atom-economical protocol for the formation of new C–C bonds. Transition metals (TM) have proved to be very useful catalysts coupling reactions.

Copper is one of the transition metal mostly used in CDC reactions, due to its low toxicity and cost. However, CDC reactions are not only limited to the use of copper (I and II) salts, but also copper complexes supported on zeolite [2] and porous materials composed by metal–organic frameworks [93] such as binuclear copper-salicylate complex have been employed [45]. In the aforementioned cases, the catalytic activity of copper enhanced via α-functionalization of the *N*-phenyl tetrahydroisoquinoline **1** using different pronucleophiles (Nu–H) through the formation of the iminium ion (Table 1). Copper oxide on magnetite in deep eutectic solvents (DES) has also shown to be a good catalytic system that can be reused up to ten times without losing its effectiveness [51]. On the other hand, the oxidant has an important role in the mechanism facilitating the re-oxidation of Cu(I) to Cu(II) by an amine oxidation after the iminium ion formation. For these reasons, different oxidizing agents such as TEMPO, TBHP, O_2 are typically tested in the CDC reaction performance.

In addition, nanoparticles have also shown an outstanding efficiency in the oxidation of amines [72]. This fact boosted the study of new methodologies to deposit gold nanoparticles on porous materials (Scheme 1a) [28] and polymers such as the poly(2-methoxyaniline-5-sulfonic acid) (Scheme 1b) [27]. Previous studies have reported the functionalization of the *N*-substituted tetrahydroisoquinoline **14** with nitroalkanes **2**, indoles **7**, phosphites **8** and malonates **11** exhibiting good yields and employing “greener” methodologies.

TM catalysts such as ruthenium (Ru) (Scheme 2a) [42], iron (Fe) (Scheme 2b) [9], cobalt (Co) (Scheme 2c) [62], iridium (Ir) (Scheme 2d) [76, 77], cerium (Ce) (Scheme 2e) [70] and manganese (Mn) (Scheme 2f) [18] have been employed in CDC reactions to form new bonds through an inter or intramolecular coupling of the THIQs **1**. Without any doubt, transition metals have favored the development of this methodology and also have allowed to access new systems with excellent biological and pharmacological properties.

3 The C-1 Functionalization of THIQs Through Metal-Free Approaches

Transition-metal-catalyzed CDC reactions have undoubtedly made a significant progress in organic chemistry. However, this approach is still limited in some applications and confronts the same challenges and the owing drawbacks of the catalytic systems [78]. Thus, the TM-free C-1 functionalization of THIQs has been

Table 1 Copper complex catalyzed in CDC reaction

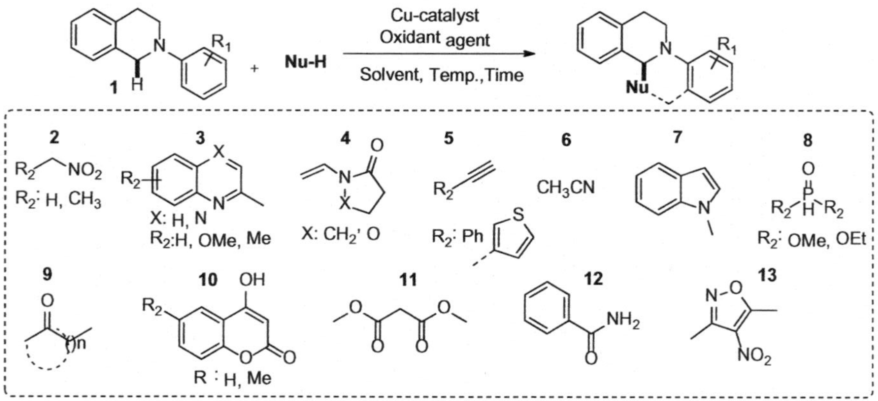

Entry	Cu-catalyst (mol%)	Additive	Nu–H	[O]	Solvent	Temperature (°C)	Time (h)	Yield (%)	Refs.
1	$[Cu_4(Ni\text{-}OCPP)(H_2O)_4] \cdot 10DMF \cdot 11H_2O$	–	2	TBHP	–	rt	6	11–90	Yang et al. [93]
2	$Cu(OTf)_2$	PivOH	3	O_2	DCE	60	24	50–90	Wang et al. [83]
3	CuBr	–	4	TBHP	DCE	60	24–36	26–68	Min et al. [55]
4	CuNPS/Zeolite Y	–	5	TBHP	–	70	15–20	62	Alonso et al. [2]
5	$CuCl_2$	Cs_2CO_3	6	TEMPO	CH_3CN	120	20	37–87	Zhang et al. [99, 105]
6	$CuO–Fe_3O_4$	–	2,4,6,7,9	–	ChCl: $(CH_2OH)_2$	50	3 d	24–95	Marset et al. [51]
7	$[Cu(Sal)_2(NCMe)]_2$	TBAC	2,7,10,11,12	O_2	CH_3CN	30	4–24	50–96	Liu et al. [45]

(continued)

Table 1 (continued)

Entry	Cu-catalyst (mol%)	Additive	Nu–H	[O]	Solvent	Temperature (°C)	Time (h)	Yield (%)	Refs.
8	CuBr	–	13	Air	H_2O	50–80	17–40	41–71	Zhang et al. [100, 104]
9	CuCl	–	5	Air	H_2O	50	24	71–95	Gröll et al. [22]

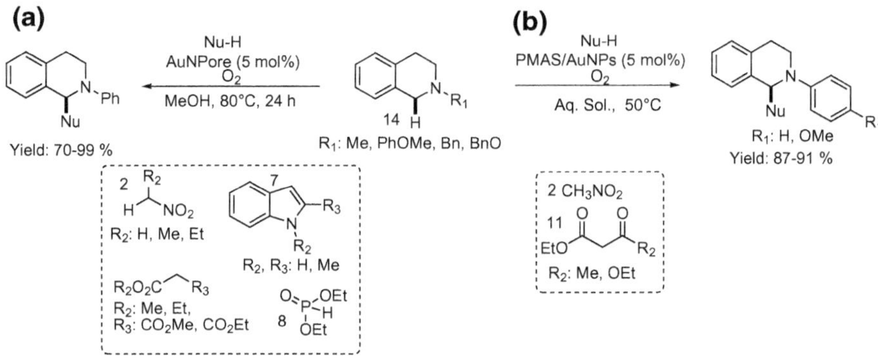

Scheme 1 TM-based nanomaterials as catalysts in the CDC reaction

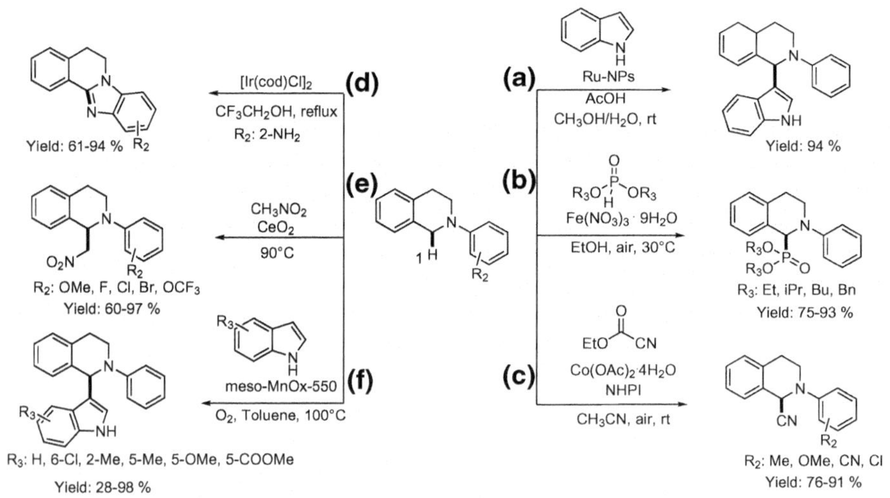

Scheme 2 Different catalytic approach for the α-functionalization of *N*-phenyl THIQs

successfully addressed during the last couple of years based on the formation of a reactive intermediate, an iminium ion, which is formed through the hydride abstraction of one proton at C-1 by a strong and selective oxidant.

The 2,3-dichloro-5,6-dicyano-1,4-benzoquinone (DDQ) is a well-known oxidant agent with a high reduction potential that commonly mediates hydride transfer reactions [85]. Taking this into consideration, Prabhu and co-workers developed a method for the C-1 functionalization of *N*-aryl THIQs **1** with phosphites **8**, ketones **9** and coumarins **10** as pronuclephiles (Nu–H). This CDC reaction formed new C–C and C–P bond mediated by a catalytic amount of DDQ and azobisisobutyronitrile (AIBN) as an additive to regenerate the DDQ [1] (Scheme 3).

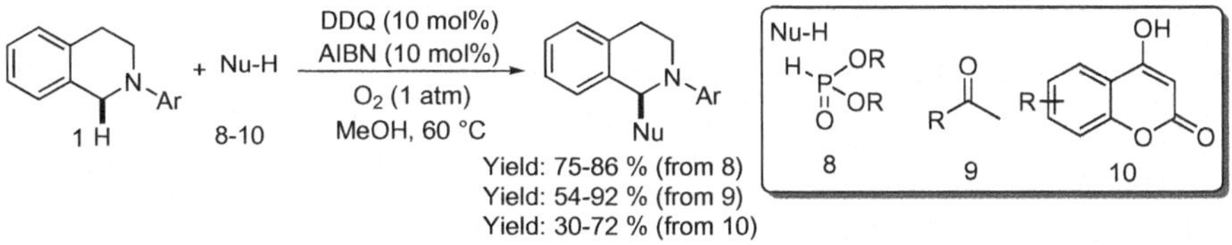

Scheme 3 DDQ-catalyzed CDC reactions of *N*-aryl THIQs **1** with different pronucleophiles **8–10**

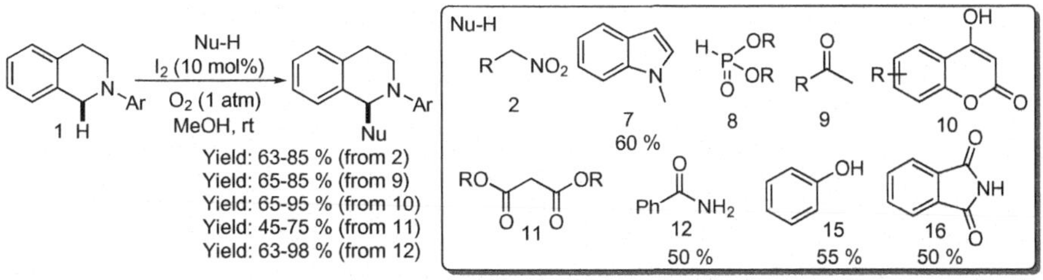

Scheme 4 I_2-catalyzed CDC reactions of *N*-aryl THIQs **1** with different pronucleophiles **Nu–H**

In virtue of its low toxicity, commercial availability, stability, easy handling and versatile reactivity, iodine has been used as a catalyst in several organic reactions [20, 61, 98]. Prabhu and his research group achieved the C–H functionalization of *N*-aryl THIQs **1** catalyzed by iodine under aerobic conditions with several Nu–H, including nitroalkanes **2**, indoles **7** and phosphites **8**, ketones **9**, coumarins **10** and amides **12**, phenols **15** and imides **16** [14] (Scheme 4).

Another mild and efficient oxidants like the triarylaminium salt, tris (4-bromophenyl)aminium hexachloro-antimonate ($TBPA^{+\cdot}SbCl_6^-$) and sodium persulfate ($Na_2S_2O_8$) could catalyze CDC reactions within different THIQs. Luo and co-workers demonstrated that the stable radical cation $TBPA^{+\cdot}SbCl_6^-$ promotes the aerobic oxidative C–H functionalization of *N*-aryl THIQs **1** with phosphites **8** and indoles **7** with excellent yields, in comparison with DDQ and I_2 as a catalyst, and in the absence of any other additives and only required atmosphere air as a friendly co-oxidant (Scheme 5a), while *N*-carbamoyl THIQs **17** reacted with ketones **9** and indoles **7** in the presence of the inexpensive reagent $Na_2S_2O_8$ without any other additive, leading to the C–C bond formation in moderate yield but with an easy and attractive workup reaction protocol (Scheme 5b) [12]. On the other hand, and despite of its high toxicity, sulfuryl chloride (SO_2Cl_2) also have showed that can promote CDC reactions between *N*-aryl THIQs **1** and phosphites **8**, nitroalkanes **2** and malonates **18** [79]. On the basis of the nature of SO_2Cl_2, the authors proposed a mechanism in which a radical initiator is formed through an ammonium cation species to furnish the reactive iminium ion intermediate. Finally, the nucleophilic attack takes place and the desired products are obtained in good to excellent yields (Scheme 5c).

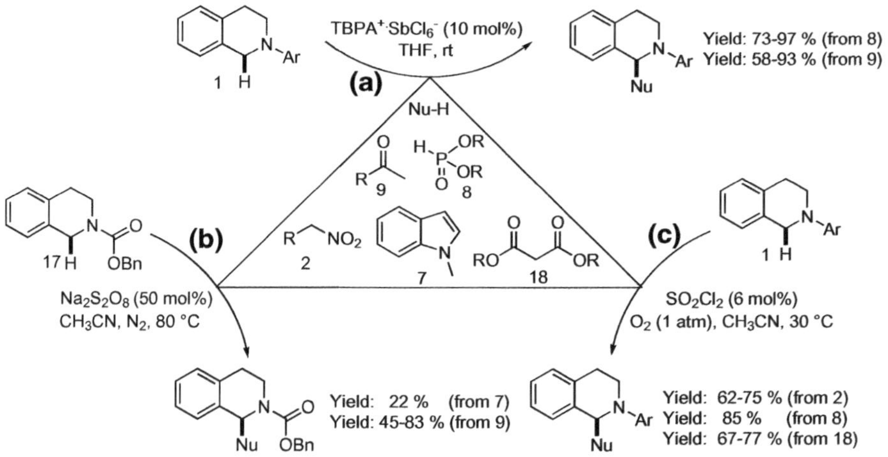

Scheme 5 CDC reactions of *N*-aryl THIQs **1** and *N*-carbamoyl THIQs **17** with different Nu–H promoted by the triarylaminium salt $TBPA^{+\cdot}SbCl_6^-$, $Na_2S_2O_8$ and SO_2Cl_2

Certainly, one of the major benefits of the CDC reactions is the forming new C–C(X) bond between two different C–H bonds without using prefunctionalized starting materials, but until 2017, this goal was achieved by various laboratories using various TM catalysts and/or with strong and selective oxidants. However, aware of the biological importance of α-amino phosphonates, Prabhu's research team discovered that *N*-aryl THIQs **1** could be coupled with phosphites **8** under open atmosphere (using oxygen as the only oxidant) and in the absence of any catalysts in dichloroethane (DCE) at 80 °C for 48 h, affording the corresponding α-amino phosphonates **19–23** in moderate yields (Scheme 6). Unfortunately, control experiments revealed that these reaction conditions could not be applied to other Nu–H like indoles and coumarins [15].

The tropylium ion is a cycloheptatrienyl system with six electrons and a positive charge delocalized over seven carbon atoms that confers to him a great combination

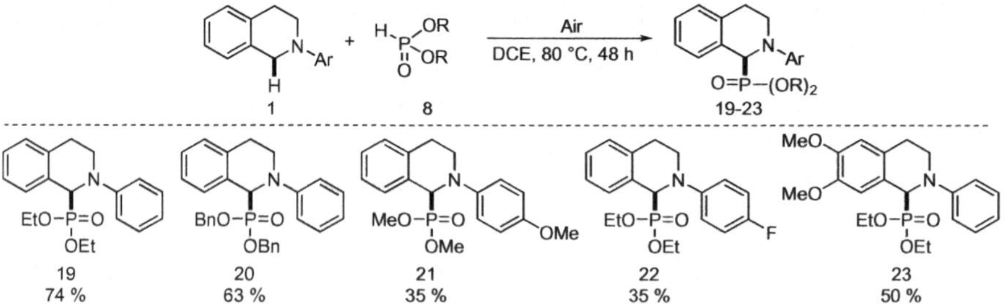

Scheme 6 Catalyst-free CDC reaction for the synthesis of α amino phosphonates **19–23**

of stability and reactivity as an electrophile and a Lewis acid [48, 58]. But more interestingly, this species can act as a hydride acceptor with a similar electrophilicity parameter to the common exhibit by the DDQ as an oxidant [23]. Based on this fact, Nguyen and his research group explored and reported the use of different tropylium (Trop) salts in the C-1 functionalization of *N*-substituted THIQs, through the generation of iminium intermediates with diverse types of Nu–H [59]. First, the authors examined the reactivity of various tropylium salts (BF_4^-, Br^- BPh_4^-, $Ph_3C^+BF_4^-$ and $NO^+BF_4^-$), and it was established that tropylium tetrafluoroborate ($Trop^+BF_4^-$) oxidized *N*-substituted THIQs where the reactivity against the group bonded to the heterocyclic nitrogen follows the order Bn > alkyl > Me > Ph. Thus, a series of *N*-benzyl THIQs **24** were subjected to the CDC reaction with different Nu–H such as nitroalkanes **2**, Grignard reagents **25** and silyl enol ethers **29** as novel substrates (Scheme 7).

Finally, one of the most growing and interesting oxidants in organic chemistry is the *N*-oxoammonium salts derived from the radical 2,2,6,6-tetramethylpiperidinyloxy (TEMPO) [5], and since 2014, the reactivity and robustness of the 2,2,6,6-tetramethylpiperidine-1-oxoammonium tetrafluoroborate ($TEMPO^+BF_4^-$) as an oxidant have been studied in the α-oxidation of THIQs to furnish the corresponding *N*-acyliminium species as intermediates. The pioneer work of Wang's group [91], where the direct α-C–H cyanation of THIQs *N*-substituted with carbamates, amides and sulfamides was reported, would reveal that one of the most reactive substrates in this CDC transformation is the corresponding *N*-carbobenzyloxy THIQ (*N*-Cbz THIQ) 33, which rapidly oxidized by $TEMPO^+BF_4^-$ **34** to the stable iminium cation **35** susceptible to the nucleophilic attack of species, such as trimethylsilyl cyanide (TMSCN) [91], allyltrimethylsilane [92], alkynyl potassium trifluoroborates ($RC{\equiv}CBF_3K$) [11] and trimethylsilyl azide ($TMSN_3$), protocol that allowed the subsequent formation of aminals, a functional group that has two amine groups attached to the same carbon atom, through the rapid reduction and transformation of the azide group [17] (Scheme 8).

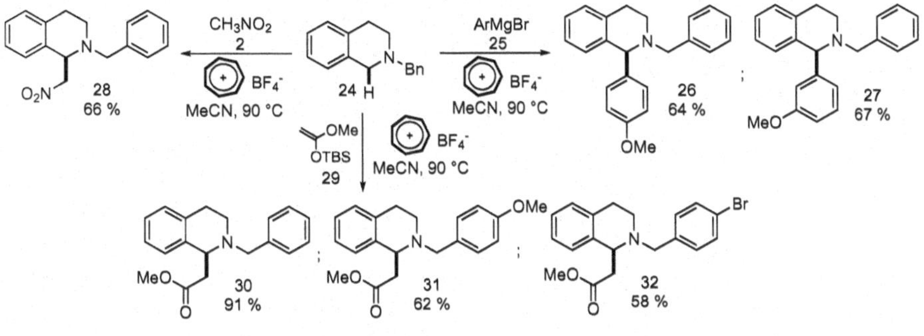

Scheme 7 α-alkylation through a CDC reaction of *N*-benzyl THIQs promoted by $Trop^+BF_4^-$

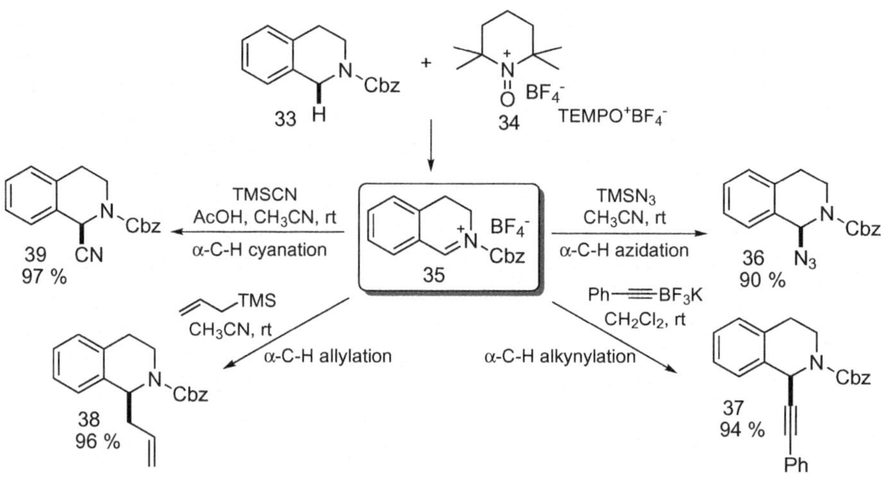

Scheme 8 Different C-1 functionalization reactions of *N*-Cbz THIQ **33** promoted by TEMPO$^+$BF$_4^-$

Even though the reactivity of *N*-aryl THIQs **1** has been less studied in CDC reactions promoted by *N*-oxoammonium salts, these substrates can be transformed under these reaction conditions with some modifications. The first study reported the coupling between THIQs **40** and 2-methyl quinolines **3** using a more powerful oxoammonium salt than the TEMPO$^+$BF$_4^-$ oxidant, the 4-acetylamino-2,2,6,6-tetramethylpiperidine-*N*-oxoammonium tetrafluoroborate **41**, known as Bobbitt's salt [54], that besides of favored the hydride transfer, can be easily recovered and recyclable [19] (Scheme 9a). While for the oxidative difluoromethylation of different *N*-substituted THIQs, employing the tetrafluoroborate **34** as a catalyst, the reaction required a coupled of additives like cesium fluoride (excess) and acetic acid (stochiometric) to obtain the respective C1-difluoro(phenylsulfanyl)methylated THIQ derivatives **47–64** in good yields [64] (Scheme 9b).

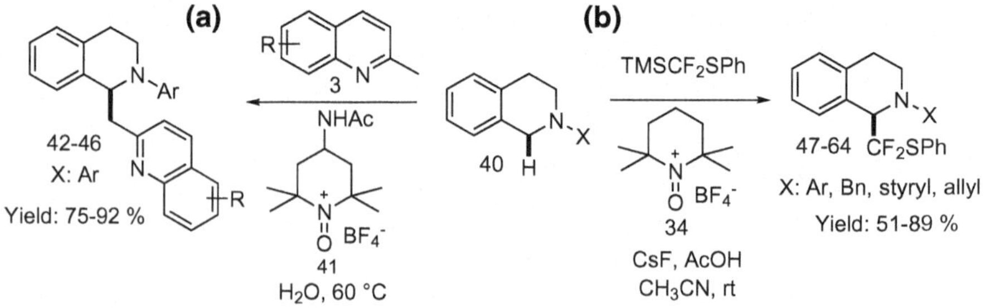

Scheme 9 Cross-dehydrogenative coupling of unreactive THIQs promoted by *N*-oxoammonium salts

4 Stereoselective Protocols for the C-1 Functionalization of THIQs

Certainly, one of the biggest challenges in the preparation of pharmaceutical agents and for the total synthesis of natural products through the CDC reaction is the simultaneously control of the regioselectivity and enantioselectivity of this transformation, mostly due to the tough reaction conditions described so far that affect the interaction of chiral ligands with the intermediates formed, leading to unsuccessful attempts [60, 65, 75, 90]. By 2012, two pioneer enantioselective CDC reactions of THIQs were reported, the $Cu(OTf)_2$-catalyzed C–H-olefination of *N*-(*p*-methoxy)phenyl THIQs with acrolein derivatives, using quinine as a chiral ligand, that reach ee% up to 85% [102, 103], and the $CuBr_2$-catalyzed oxidative coupling of *N*-aryl THIQs and propionaldehydes in the presence of proline-based chiral ligands, achieving ee% between 72 and 92% [102, 103]. In perspective, the current efforts during the developing of new strategies for the CDC reaction are focused in incrementing the stability of the iminium ion, to increase the regioselectivity, and in choosing the adequate chiral ligand that interacts with this ion that can improve the enantioselectivity of the CDC reaction [44]. Otherwise, enantioselective intramolecular CDC reactions are also being part of the actual topics of study [73].

Based on that the iminium ion intermediate could form a π-complex with metal catalysts, the major efforts in enantioselective CDC reactions is to find a suitable ligand that interact with this species and form a chiral ion pair that direct the respective nucleophilic addition [6].

Inspired by this concept, Wang and co-workers reported that *N*-aryl THIQs **1** undergo an oxidative coupling with cyclic ketones **65** in the presence of copper(II) triflate ($Cu(OTf)_2$), a chiral secondary amine derived from glycine **66** and under oxidizing conditions such as DDQ to yield a series of asymmetric C1-alkylated THIQs **67** with a good to excellent diastereo- (13:1) and enantioselectivities (94 ee %) [101] (Scheme 10).

Based under the same approach, but improving in the reaction conditions by carrying out the CDC transformation without the use of any transition-metal catalyst. A series of *N*-aryl THIQs **1** was coupled with α,β-unsaturated γ-butyrolactams **68**, generating some kind of Morita–Baylis–Hillman-type products **70**, when a thiourea derived from quinine **69** was used as a sole chiral ligand [49]

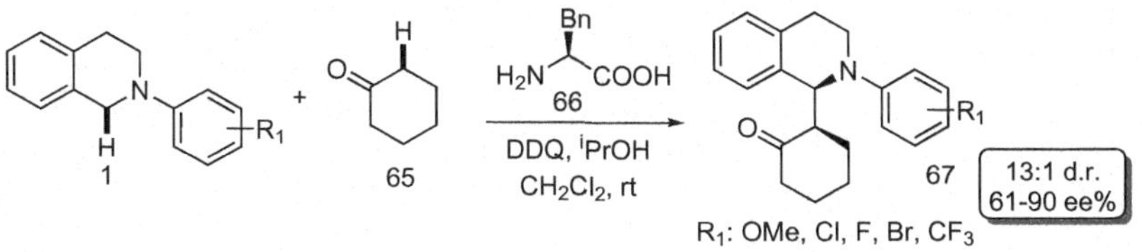

Scheme 10 Synthesis of asymmetric C1-alkylated THIQ derivatives **67** using a secondary amine as an auxiliary ligand

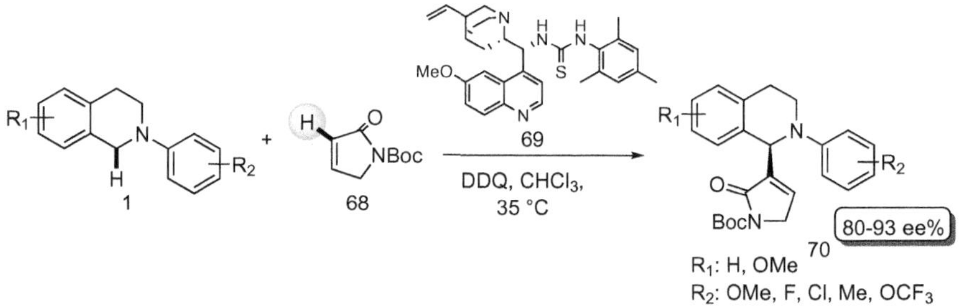

Scheme 11 Organocatalytic CDC reaction of THIQs **1** via bifunctional thiourea catalysis

(Scheme 11). The authors demonstrated that metal catalysts ($M(OTf)_2$, M=Mg, Cu) decrease dramatically the enantioselectivity of this CDC reaction, no matter that the racemic products were isolated in excellent yields, and that a simple oxidant conditions (DDQ) afforded the desired THIQs scaffolds **70** in moderate yield and excellent enantioselectivity.

One of the most widely PNu studied for the functionalization of THIQs in the traditional CDC reactions is phenylacetylene derivatives, and clearly, these substrates have been employed as successful model reactants in the enantioselective approach. The first attempt was performed between *N*-Cbz THIQs **33** and terminal alkynes **71** in a heterobimetallic cooperative catalytic system (CuCl/$Yb(OTf)_3$) and $TEMPO^+BF_4^-$ as an oxidant. Under these "racemic" CDC reaction conditions, the product was obtained in 59% yield, but adding the bis(oxazoline) **72** as a chiral ligand, the desired products **73** were obtained in moderate yields and with an ee% between 85 and 95% [76, 77] (Scheme 12a). Phenylacetylenes are relatively weak nucleophiles, so this explains the use of two metal catalysts to stabilize the iminium ion intermediate and to activate the triple C≡C–H bond. This fact was recently corroborated by Feng's research group, which followed the reaction through electroparamagnetic resonance (EPR) and reported the asymmetric α-alkynylation of *N*-aryl THIQs **1** using a $Zn(NTf)_2/Fe(OTf)_2$ (ratio 20/3) system under aerobic oxidative conditions with tetrakisphenyl borate as an additive [31]. Although in terms of the reaction yields these protocols are comparable, the chiral ligand added on the Zn(II)/Fe(II) methodology, the proline-based *N,N′*-dioxide **74**, increased the enantioselectivity of the desired products **75** to 85–99% (Scheme 12b).

Surprisingly, another enantioselective C-1 functionalization of THIQs was reported without the use of any transition-metal catalyst. This organocatalytic approach was studied for the vinylation and arylation of *N*-acyl THIQs **33** and **76** when they are coupled with the corresponding boronates in the presence of a Brønsted acid. To achieve this goal, *N*-substituted derivatives of tartaric acid were tested in this CDC reaction by Liu's group which established that those THIQs protected with the Cbz group favored the C–H arylation at 70 °C with aryl boronates **77** just in the presence of the chiral ligand *N,N*-diphenyl **78** and DDQ, furnishing the corresponding 1-phenyl THIQs **79** in moderate yields and enantioselectivity (Scheme 13a).

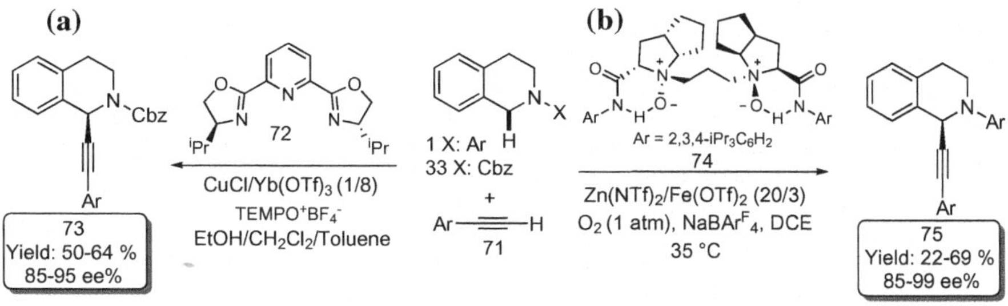

Scheme 12 Enantioselective α-alkynylation of *N*-aryl THIQs via the heterobimetallic cooperative catalytic system

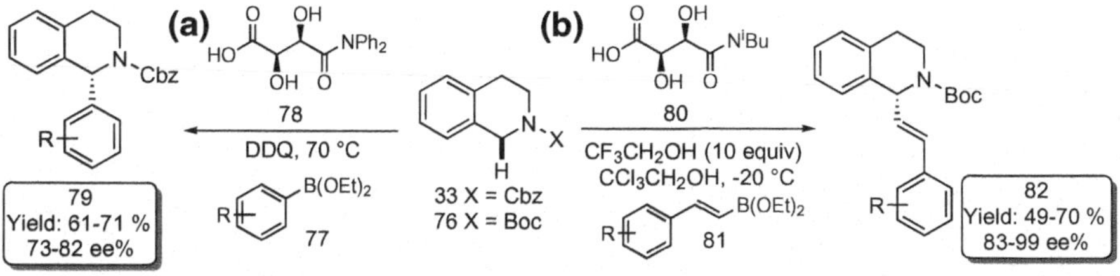

Scheme 13 Enantioselective C–H vinylation and arylation of *N*-acyl THIQs promoted by chiral tartaric acid derivatives

On the other hand, much better results were obtained when *N*-Boc THIQs **76** were subjected to the CDC reaction with vinyl boronates **81** and the *N,N*-diisobutyl **80** ligand; however, this reaction required the use of 10 equiv of 2,2,2-trichloroethanol (CCl_3CH_2OH) as an additive and trifluoroethanol (CF_3CH_2OH) as a solvent, resulting beneficial for the reaction enantiocontrol, affording the 1-styryl THIQs **82** with a protecting group that can be easily cleaved under mild conditions, obtaining the corresponding enantiopure secondary amines with an ee% up to 83% [45] (Scheme 13b).

Finally, an interesting three-step strategy based on an asymmetric CDC reaction was reported by Zhou and Tong for the concise synthesis of methyltetrahydroprotoberberine (MTP) alkaloids like **87**. After having studied the previously reported conditions by Ma [41, 43], the authors selected the chiral compound (*S,R*)-*N*-PINAP as the optimal ligand for the multicomponent redox-A^3 reaction between *N*H-THIQs **83**, 2-bromobenzaldehyde derivatives **84** and the simple alkyne **85**, being the major novelty of this protocol [109]. Thus, the first step, which can be indicated as a crucial step in this strategy, allowed the synthesis of a series of C-1 functionalized THIQs **86** in good yields and excellent enantioselectivity (up to 94%). The routinely and subsequent steps take advantage of the terminal alkyne function present at C-1 of **86** to perform the reductive Heck cyclization, followed by the asymmetric reduction to allow the rapid and efficient access to a 12-membered library of MTP alkaloids **87** (Scheme 14).

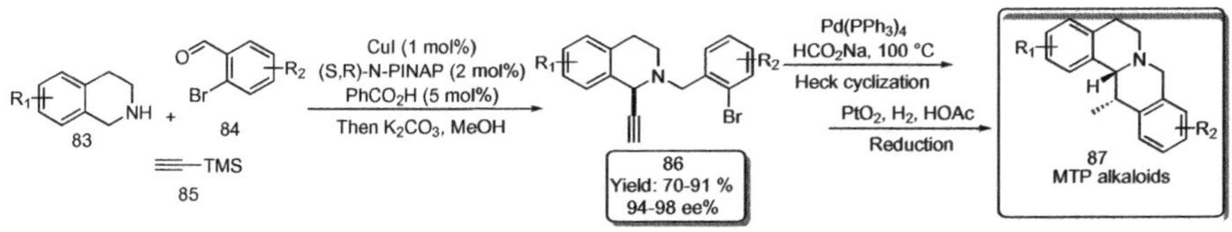

Scheme 14 Catalytic asymmetric total synthesis of MTP alkaloids based on the CDC reaction

5 Photocatalytic CDC Reactions Applied to THIQs

The selective functionalization of C–H bonds through direct oxidation reactions in order to generate reactive species is a difficult task. Cross-dehydrogenative coupling is an interesting path to the formation of C–C bonds by direct coupling from two C–H bonds. This reaction employs C–H bonds adjacent to heteroatoms such as nitrogen due to its good oxidant properties [10].

The use of tertiary amines in photocatalysis has a broad spectrum of applications. Typically, they have been employed as electron donors to reduce the photoexcited state of photocatalyst while oxidize to amine radical cations [30]. The plausible mechanism of reactive pathways of the amine radical cation and the influence of oxygen, and the light source employed discussed in detail by Bartling and co-workers in an exquisite work [3]. In this section, new perspectives about tertiary amines as substrates in photochemistry light-assisted CDC reactions will be presented.

5.1 Transition Metal Complexes Employed in CDC Reactions

In 2014, Yoo and Kobayashi reported the synthesis of a polymer with iridium immobilized and used as heterogeneous photocatalyst for aerobic phosphonylation of the *N*-phenyl-1,2,3,4-tetrahydroisoquinoline **1** through a cross-dehydrogenative coupling reaction mediated by light. The iridium complex was prepared by the immobilization of $Ir(ppy)_2(dtbbpy)PF_6$ (ppy: 2-phenylpyridyl and dtbbpy: 4,4′-ditert-butyl-2,2′-dipyridyl) using a polymerization protocol [95]. The recovery and reusing of the PS-Ir catalyst were evaluated at least four times without a noticeable loss of catalytic activity and iridium leaching (Scheme 15).

Copper is an abundant metal on the earth and therefore, it is a low-cost catalyst in many important organic reactions. The photoactive properties of copper complexes have been widely studied [7, 50, 52]. For instance, the preparation of a zwitterionic copper (I) complex with long-lived triplet excited state as a photocatalyst in cross-dehydrogenative coupling has been recognized as a novel and economic strategy that replace conventional photocatalyst of iridium and ruthenium [84].

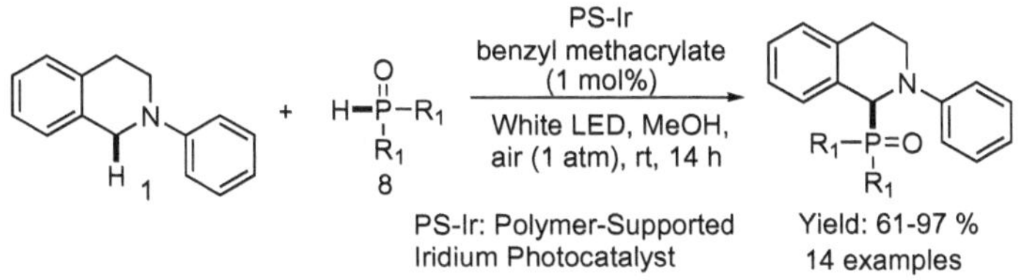

Scheme 15 Aerobic oxidative phosphonylation of *N*-phenyl THIQ **1** mediated by visible light

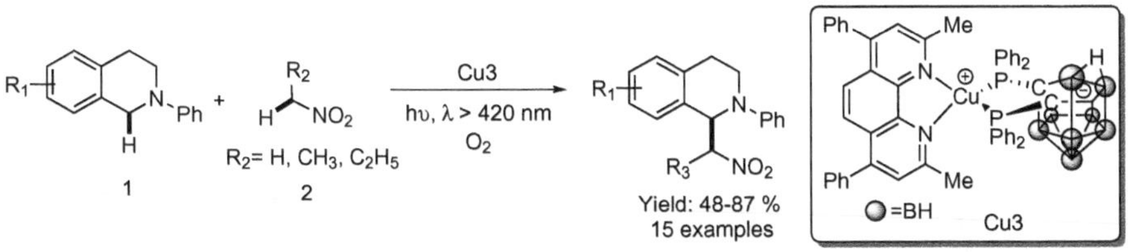

Scheme 16 CDC reaction mediated by zwitterionic copper (I) complex

Wang et al. reported in their work four zwitterionic copper (I) catalysts, named Cu_1–Cu_4, which have a mixed of phenanthroline and nido-carborane-diphosphine ligands. The study showed that complexes Cu_2 and Cu_3 had an intense and relatively long-lived emission which is a crucial factor in achieving high conversions toward the desired product. It was also established that, Cu_3 had the best catalytic activity and light irradiation along with molecular oxygen were essential to carry out the reaction (Scheme 16).

The enantioselective CDC reactions are striking for organic chemist so the mixture of photoredox and asymmetric methodologies introduces a new approach. Stephenson's group described a dual-catalyst strategy for the enantioselective oxidative alkylation of *N*-phenyl tetrahydroisoquinolines **1** with silyl ketene acetals **88** [4]. In this reaction, a ruthenium complex $[Ru(bpy)_3]Cl_2$, used as a photocatalyst and carbon tetrachloride (CCl_4) serves as an oxidant agent in acetonitrile (MeCN), favored the formation of a racemic Mannich product (Scheme 17a). In order to enhance the photocatalytic activity and enantioselectivity, the MeCN solvent is changed by the methyl tert-butyl ether (MTBE) to improve the thiourea-catalyzed alkylation reaction. Therefore, the conversion toward the product was achieved with good yields and a moderate enantiomeric excess. This ruthenium complex has also been used in photocatalytic oxidative coupling reaction of the *N*-phenyl tetrahydroisoquinoline **1** and *β*-keto-acids **89** [34] (Scheme 17b).

Perepichka and co-workers reported the asymmetric *α*-alkynylation of *N*-phenyl-tetrahidroisoquinolines employing the complex $Ir(ppy)_2(dtbbpy)]PF_6$ and CuBr as a dual catalytic system. The catalytic synergy proved an efficient asymmetry CDC protocol for *N*-phenyl THIQs **1** and terminal alkynes **5** where both catalysts improved the formation of the desire product with yields up to 90 ee% [63] (Scheme 18).

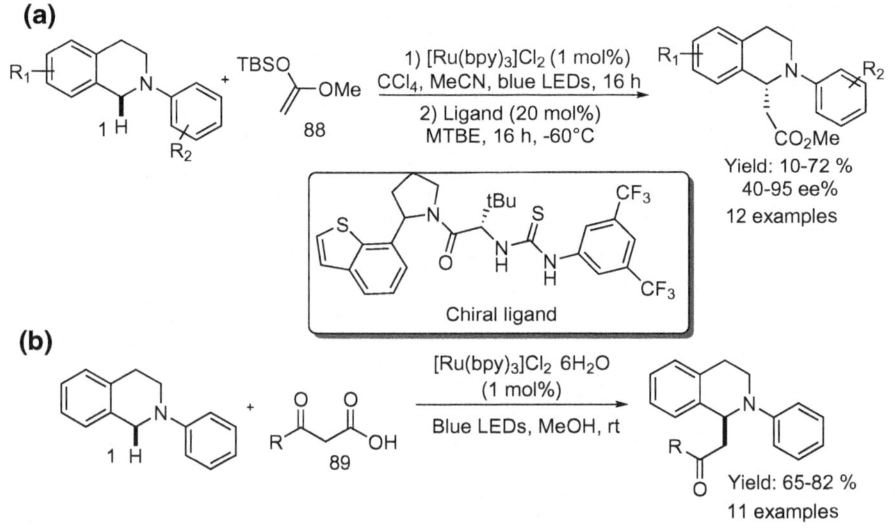

Scheme 17 Ruthenium complex has been employed as a photocatalyst in CDC reactions

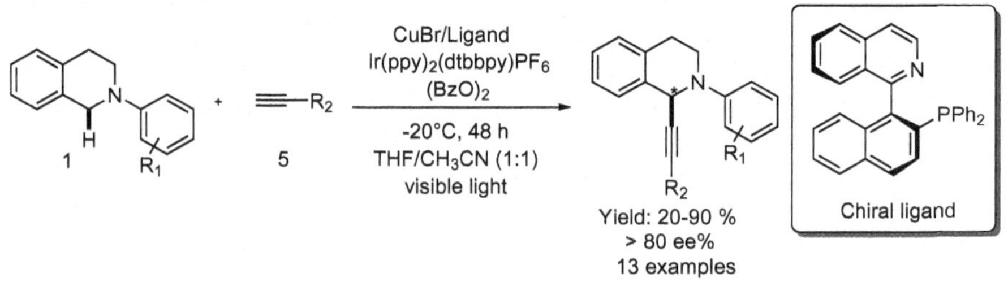

Scheme 18 Asymmetric coupling between THIQs with terminal alkynes in the presence of a photocatalyst and visible light

In 2013, Wu's group reported two novel methodologies described as effective and clean. It was found that the external oxidant agent was not required in cross-coupling hydrogen evolution reaction [53]. The eosin Y and $Co(dmgH)_2Cl_2$ as dual catalytic system improved the *N*-phenyl THIQs **1** and indole **7** coupling and H_2 production (Scheme 19a). This research work provided information about the formation of the THIQ radical cation and reduction of the Co^{II} and Co^{III} intermediate species [108].

In the same year, Wu's group had also reported an interesting example of a visible-light-driven CDC reaction where direct coupling between the same components was promoted in an aqueous phase [86]. In this case, the cobaloxime complex was generated in situ from cobalt salts ($CoCl_2$) and dimethylglyoxime (dmgH) acted as a photocatalyst. Likewise, the role of active species of oxygen was

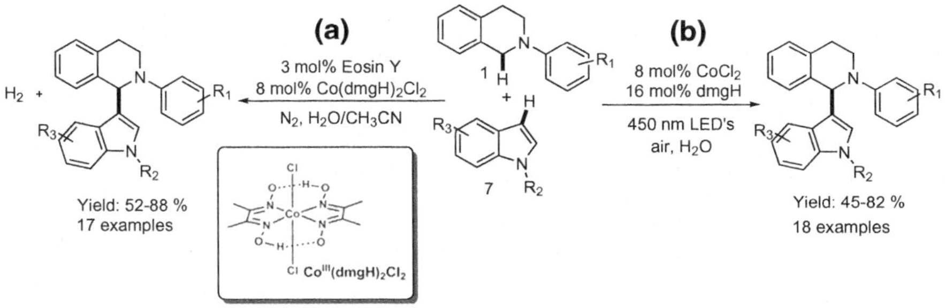

Scheme 19 Photocatalytic direct C–H sp^3-sp^2 coupling of *N*-phenyl THIQs **1** and indole **7**

studied to understand the mechanism. It was established that the molecular oxygen (O_2) helps to the formation of the photocatalyst ($Co^{III}(dmgH)_2Cl_2$) and reacts with the Co^{II} intermediate to generate the superoxide radical anion, which abstracts one proton from the amine cation radical to produce an amine radical and hydrogen peroxide (H_2O_2) (Scheme 19b).

Metal–organic frameworks (MOF) have exhibited promising applications in a wide variety of areas [87]. However, MOF-based photocatalysts for organic reactions are still challenging due to the complicated synthesis of porphyrins and the leaching of toxic elements caused by the decomposition of noble metal complexes [57]. Despite this, two CDC reactions have been recently carried out using MOF as photocatalysts. The first one described the use of isostructural porphyrin MOF named UNLPF-12 (UNLPF: University of Nebraska–Lincoln porous framework) with Sn (IV) cation that can act as chromophores [82]. Hence, photoactive porphyrin MOF could be used in photocatalytic oxidation. Particularly, UNLPF-12 showed high photocatalytic activity in the CDC reaction of THIQs **1** and acetone **9**, allowing a preparation of the corresponding THIQs in good to excellent yields (87–98%) [32] (Scheme 20a). It was also found that air and light were crucial for the accomplishment for this reaction. The recyclability was also good during four cycles where UNLPF-12 did not lose its activity and crystallinity.

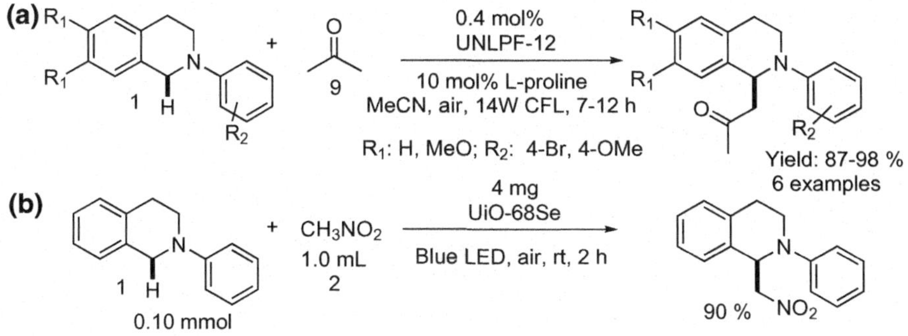

Scheme 20 Metal–organic frameworks as photocatalyst in CDC reactions

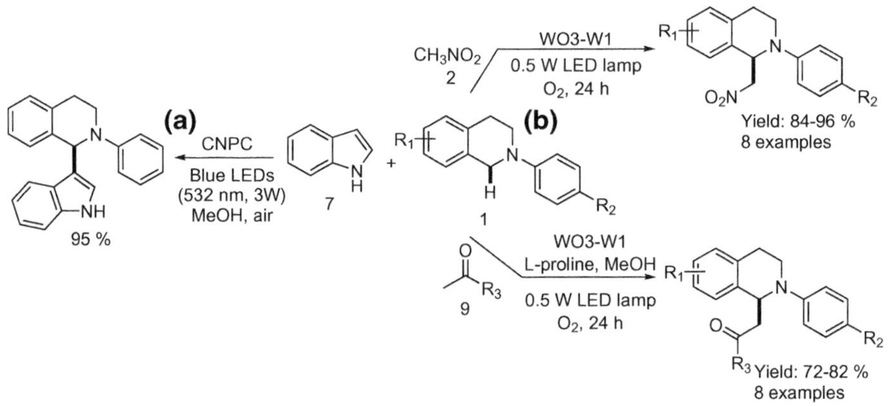

Scheme 21 Nanoparticles of Cu_2O and WO_3 with high photocatalytic activity for visible-light-induced CDC reaction

On the other hand, the second report consisted of the use of zirconium(IV)-based UiO-topological MOF containing benzoselenadiazole (UiO–68Se). This approach revealed a high efficient and recyclable photocatalytic activity for aerobic CDC reaction of *N*-phenyl THIQ **1** and nitroethane **2** under visible-light irradiation (blue LEDs, λ_{max} = 450 nm, 3 W). [46, 105, 99] (Scheme 20b).

In 2018, Chen's group published the effectiveness of anchoring Cu_2O nanoparticles in situ on nitrogen-doped porous carbon yolk-shell cuboctahedral framework (CNPC) which was successfully carried out by direct derivatization from a benzimidazole-modified Cu-btc (btc = benzene-1,3,5-tricarboxylate) MOF. This photocatalyst showed a prolonged lifetime of photogenerated electrons and excellent stability and recyclability. Its photocatalytic properties facilitate the coupling of *N*-phenyl THIQ **1** and indole **7** [24] (Scheme 21a).

On the other hand, the heterogeneous catalysis with nanoparticles of tungsten oxide WO_3 with hollow sphere morphology (WO_3–W1) was used to promote aza-Henry and Mannich reactions of *N*-THIQ **1** and pronucleophiles **2** and **9** in high yields by irradiating this mixture to visible light under aerobic conditions [38] (Scheme 21b).

5.2 Metal-Free Photocatalytic CDC Reactions

Metal-free photocatalyst reactions are attractive due to their novel and sustainable conditions. In this section, we will describe in different approaches where the employment of metal-free visible-light photoredox chemistry induced oxidative coupling of tertiary amines.

Xiao and co-workers reported new metal-free visible-light conditions to carry out CDC reactions of the *N*-phenyl tetrahydroisoquinoline **1** and diazo compounds

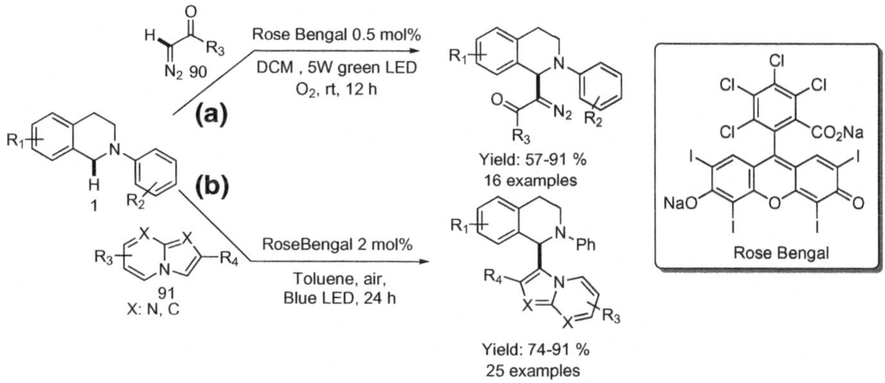

Scheme 22 Metal-free visible-light-mediated CDC reaction with Rose Bengal as photosensitizer

90 [89]. This reaction proceeds under mild conditions by using O_2 as oxidant and Rose Bengal as a photosensitizer employing green light as an energy source at room temperature. The expected products were obtained in a range of 57–91% of yield (Scheme 22a). This research set a precedent for new pathways in the use of photosensitizers such as Rose Bengal for the coupling of *N*-phenyl tetrahydroisoquinolines **1** and imidazopyridine **91**. The desired products were obtained with high yields (74–91%) under an atmosphere of air and toluene as a solvent with blue LED light [33] (Scheme 22b).

In 2015, Rusch et al. developed an interesting work related to the transformation of the *N*-phenyl THIQs **1** into the 12-nitroindoloisoquinoline **92** through cascade reactions (cross-dehydrogenative coupling/dehydrogenation/6π-cyclization/oxidation) mediated by a photocatalytic visible-light system [66]. It was reported that 1,5-diaminoanthraquinone used as a photocatalyst under an atmosphere of air and potassium phosphate (K_3PO_4) is the best conditions to obtain 12-nitroindoloisoquinolines **92** with yield in a range of 24–69% (Scheme 23).

Due to importance of the iminium ions in cross-dehydrogenative coupling, Franz and co-workers studied the oxidative α-CH functionalization of *N*-phenyl THIQs **1** with diverse nucleophiles using polyhalomethanes ($CBrCl_3$) [21]. The visible light

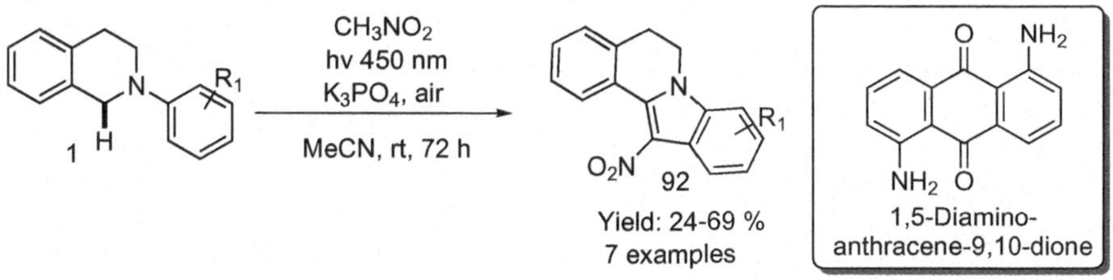

Scheme 23 A visible-light photocatalytic cross-dehydrogenative coupling/dehydrogenation/6π-cyclization/oxidation cascade

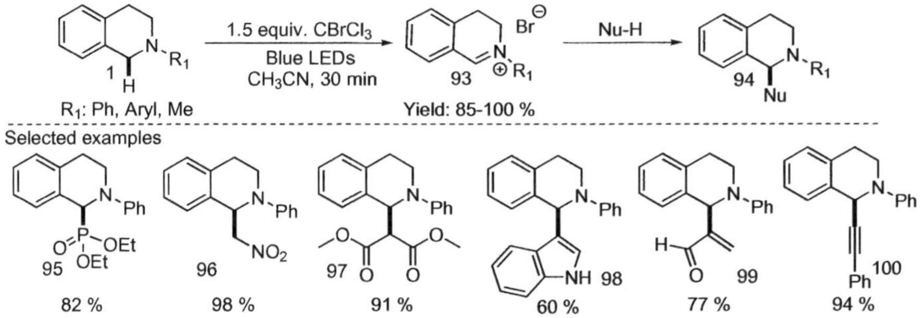

Scheme 24 Visible-light-assisted formation of iminium ions **93** for later functionalization

triggered activation of polyhalomethane offering mild conditions. The intermediate iminium ions **93** were obtained with yield in a range of 85–100% and final products **94–100** with yields up 60% (Scheme 24).

Although, photochemistry is considered an ecofriendly methodology, the photocatalysts are evolving from metal complexes to organic dyes; however, more research is required to get more efficient and environmentally benign protocols. In this way, recently new synthesis strategies have been developed. Wang and co-workers used successfully Eosin Y dye-based porous organic polymer, EY-POP-1 in the CDC reaction of THIQs **1** and nitroalkenes **2** under green conditions (visible light, −14 W fluorescent lamp and air) that gave the desired products in good to excellent yield (90–99%) [81] (Scheme 25a).

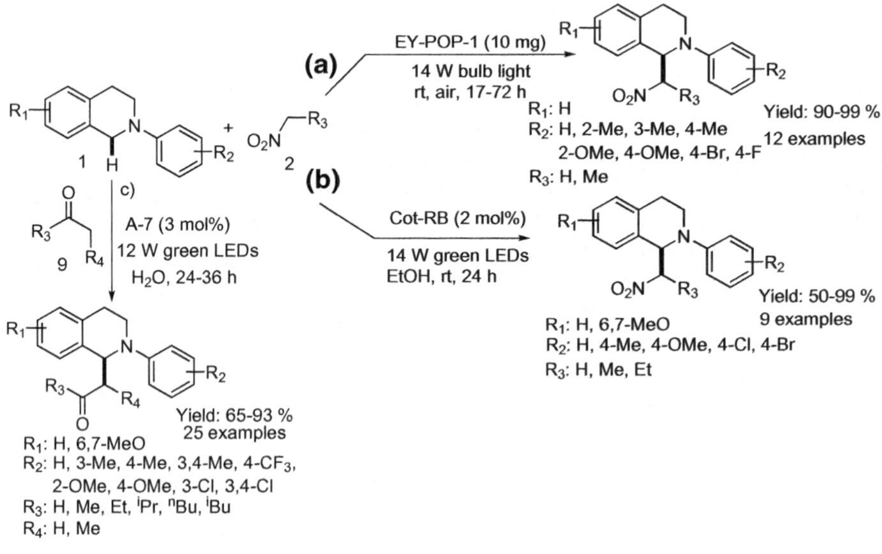

Scheme 25 Organic dyes supported on polymers, sponge and cotton as photosensitizer

Analogous C-1 substituted THIQs were obtained from similar CDC reaction using cotton supported Rose Bengal photocatalyst (Cot-RB). The desired THIQ products are in moderate to excellent yield (50–99%) [88] (Scheme 25b). This catalyst can be easily recovered by simply grabbing out the modified cotton from the reaction system. Meanwhile, Zang and his group developed a novel bifunctional organic sponge photocatalyst containing supported Rose Bengal onto a poly-dimethylsiloxane (A-7) that catalyzes efficiently the coupling reaction of THIQs **1** and ketones **9** in water under visible-light irradiation [100, 104] (Scheme 25c). This bifunctional photocatalyst works also in asymmetric transformation with good enantioselectivities. All these approaches put a new perspective for future research related to the CDC reaction using visible light: their advantages are facile catalyst preparation, environmental friendliness, convenient separation and easy catalyst recycling and workup.

6 Mechanochemistry Applied to the Functionalization of THIQs

Mechanochemistry has emerged as a sustainable and environmentally benign support in organic chemistry for many approaches that seek the construction of diverse synthetic and natural products by eliminating the use of organic solvents and lowering the energy consumption [16, 29]. As it was expected, mechanochemistry would offer a rapid access to a variety of metal-catalyzed and organocatalytic transformations [26] as well as the C–H functionalization in the CDC reactions [106].

One of the pioneer works in this field was reported by Su's research group in 2011, which mechanically promoted the coupling of *N*-aryl THIQs **1** with nitroalkanes **2**, indoles **7** and alkynes **5** under ball-milling conditions using DDQ (BM-DDQ) as an oxidant and copper balls as a gridding material [74] (Scheme 26).

The remarkable aspect of this report is that the authors demonstrated that the CDC reaction took place on the surface/boundary between these recyclable copper balls, supporting the theory that this kind of transformation is benefited by just traces of cooper [39, 40]. Thus, the coupling products **101**, **102** and **103** were obtained in short reaction times (less than 40 min) and with excellent yields.

With the aim of combining the mechanochemistry methodology with the asymmetric approach to develop an enantioselective CDC reaction assisted by ball milling, in 2013 Su and co-workers reported C-1 alkynylation of *N*-aryl THIQs **1** using copper balls as a grinding material and as a catalyst, DDQ as an oxidant and silica gel as a grinding aid. After having tested a few chiral compounds among oxazolines, phosphoric acids and BINOL derivatives, the bis(oxazoline) pyridine PyBox-1 **104** resulted to be the best ligand to furnish a series of 1-alkynyl THIQs derivatives **105** in moderate yields and enantioselectivity; however, this is so far the

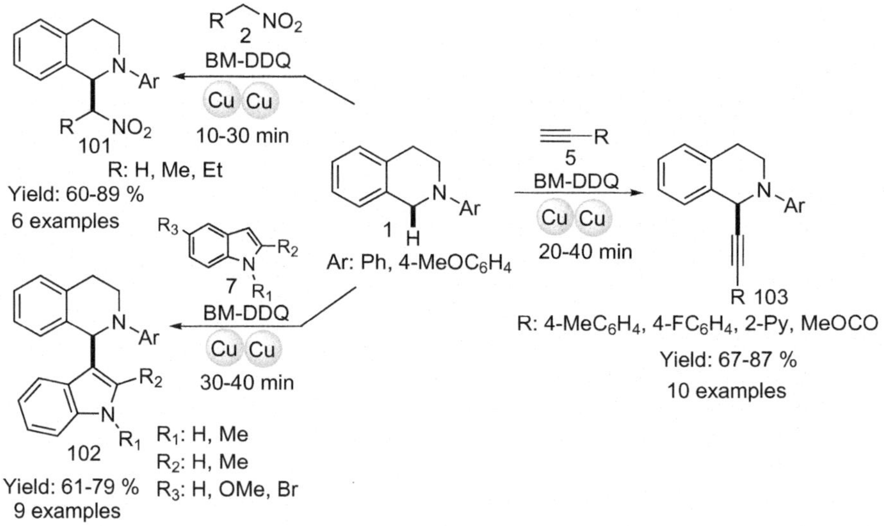

Scheme 26 CDC reaction of THIQs **1** with nitroalkanes **2**, indoles **7** and alkynes **5** via ball milling

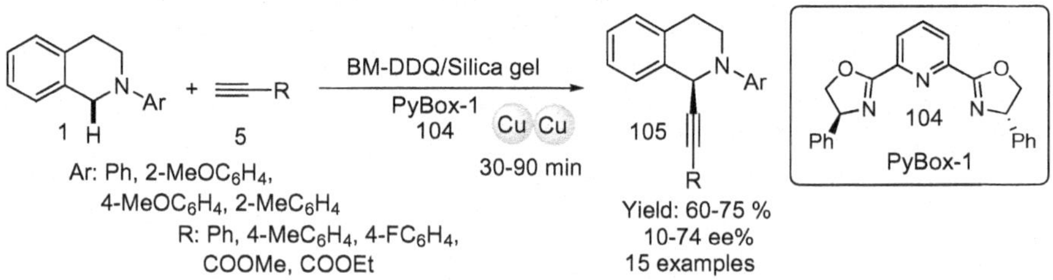

Scheme 27 Enantioselective CDC reaction of THIQs with alkynes **5** via ball milling

unique study for the enantioselective CDC reaction assisted by mechanical energy [96] (Scheme 27).

Finally, Su's group extended their work to promote the coupling of *N*-Cbz THIQ **33** with acyclic and cyclic ketones **9** under the same reaction conditions described above [97]. In this study, the authors evaluated the chemical and mechanical parameters for this CDC reaction, describing the role of silica as grinding auxiliaries which provide a microenvironment to favor the reactants contact, the size of the balls and the milling time and frequency (Scheme 28).

Thereby, a series of Mannich-type products **106** were obtained with acyclic and cyclic ketones **9** in good to excellent yields after the reaction was milled for 45 min at 20 Hz and with two cooper balls (14 mm, $\Phi_{MB} = 0.057$).

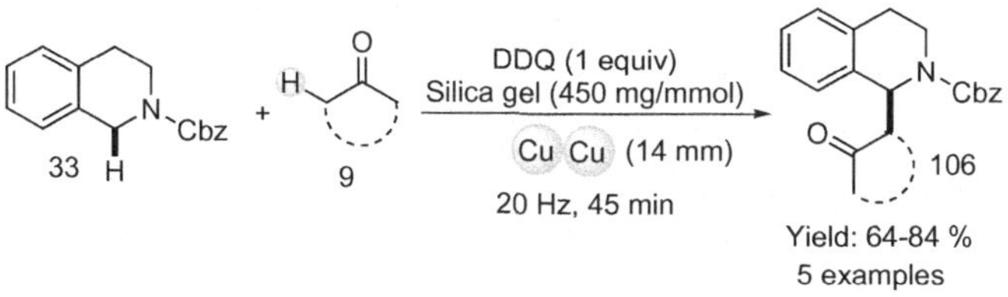

Scheme 28 Oxidative Mannich reaction of THIQs with ketones **9** via ball milling

7 Conclusions and Perspectives

Cross-coupling reactions have proven very useful tools for heterocyclic chemists allowing the selective coupling of C–C or C–X bonds of biologically important heterocycles. 1,2,3,4-Tetrahydroisoquinoline is one of the most important "privileged scaffolds" present in natural products. Due to the importance of these privileged structures, numerous synthetic methods for making THIQ derivatives, especially C-1 substituted tetrahydroisoquinolines, which demonstrate a wide range of bioactivities. However, these last five years have witnessed great advances in the C-1 substituted THIQs syntheses through the cross-dehydrogenative coupling reactions that opened up new avenues for the chemistry of functionalized THIQ molecules. As an ideal atom-economical approach for the formation of new C–C(X) bonds, the CDC methodology makes quickly, easily and environmentally friendly the synthesis of pharmacologically active THIQ derivatives and alkaloid-like products.

Actually, there are numerous attractive methods, protocol and variants in the CDC methodology that allow to select a suitable approach to construct new C-1 functionalized THIQs. Among them, CDC reactions without metals for bioactive THIQ molecules, i.e., organocatalysis and metal-free photocatalysis as well as mechanically assisted solvent-free CDC reactions stand out as very perspective areas of heterocyclic chemistry. These approaches will be more developed in the near future; all these contribute considerably in the development of green heterocyclic chemistry. Despite this exciting progress, some troublesome challenges remain unsolved, among which are highly efficient construction of asymmetric 1,2-disubstituted THIQs via the multicomponent redox-A3 reactions. Moreover, multicomponent redox-A3 reactions under metal-free visible-light conditions neither mechanical activation are practically not developed. Nevertheless, the rapid development in this field continues to provide the impetus for advancement in both heterocyclic chemistry and medicinal chemistry. Authors hope that this chapter will attract attention from synthetic organic chemists toward overcoming the above-mentioned challenges in the near future.

Acknowledgements MCOV and CEPG acknowledge the fellowship given by the doctoral program COLCIENCIAS- Conv. 727 and Conv. 617, respectively. VVK is grateful to the Colombian Institute for Science and Research (COLCIENCIAS) for financial support (project No. RC-007-2017, Cod. 110274558597).

References

1. Alagiri K, Devadig P, Prabhu KR (2012) CDC reactions of *N*-aryl tetrahydroisoquinolines using catalytic amounts of DDQ: C–H activation under aerobic conditions. Chem Eur J 18:5160–5164
2. Alonso F, Arroyo A, Martín-García I, Moglie Y (2015) Cross-dehydrogenative coupling of tertiary amines and terminal alkynes catalyzed by copper nanoparticles on zeolite. Adv Synth Catal 357:3549–3561
3. Bartling H, Eisenhofer A, König B, Gschwind RM (2016) The photocatalyzed Aza-Henry reaction of *N*-aryltetrahydroisoquinolines: comprehensive mechanism, H·- versus H^+-abstraction, and background reactions. J Am Chem Soc 138:11860–11871
4. Bergonzini G, Schindler CS, Wallentin CJ, Jacobsen EN, Stephenson CR (2014) Photoredox activation and anion binding catalysis in the dual catalytic enantioselective synthesis of β-amino esters. Chem Sci 5:112–116
5. Bobbitt JM, Bartelson AL, Bailey WF, Hamlin TA, Kelly CB (2014) Oxoammonium salt oxidations of alcohols in the presence of pyridine bases. J Org Chem 79:1055–1067
6. Brak K, Jacobsen EN (2013) Asymmetric ion-pairing catalysis. Angew Chem Int Ed 52:534–561
7. Buckner M, McMillin DR (1978) Photoluminescence from copper(I) complexes with low-lying metal-to-ligand charge transfer excited states. J Chem Soc, Chemical Communications 759–761
8. Byvatov E, Sasse B, Stark H, Schneider G (2005) From virtual to real screening for D3 dopamine receptor ligands. ChemBioChem 6:997–999
9. Cai J, Liu Y, Jiang Y, Yang Y (2017) Iron-catalyzed aerobic oxidative phosphonation of N-aryl tetrahydroisoquinolines. Phosphorus Sulfur Silicon Relat Elements 192:1068–1073
10. Campos K (2007) Direct Sp3 C–H bond activation adjacent to nitrogen in heterocycles. Chem Soc Rev 36:1069–1084
11. Chen L, Sun C, Feng G, Cao M, Zhao S-l, Yan J, Wan R-Z, Liu L (2018) Direct oxidative C-H alkynylation of N-carbamoyl tetrahydroisoquinolines and dihydroisoquinolines. Org Biomol Chem 16:2792–2799
12. Chen W, Zheng H, Pan X, Xie Z, Zan X, Sun B, Liu L, Lou H (2014) A metal-free cross-dehydrogenative coupling of *N*-carbamoyl tetrahydroisoquinoline by sodium persulfate. Tetrahedron Lett 55:2879–2882
13. Das D, Sun AX, Seidel D (2013) Redox-neutral copper(II) carboxylate catalyzed α-alkynylation of amines. Angew Chem Int Ed 52:3765–3769
14. Dhineshkumar J, Lamani M, Alagiri K, Prabhu KR (2013) A versatile C–H functionalization of tetrahydroisoquinolines catalyzed by iodine at aerobic conditions. Org Lett 15:1092–1095
15. Dhineshkumar J, Samaddar P, Prabhu KR (2017) Catalyst-free cross-dehydrogenative coupling strategy using air as an oxidant: synthesis of α-aminophosphonates. ACS Omega 2:4885–4893
16. Do JL, Friščič T (2017) Mechanochemistry: a force of synthesis. ACS Central Sci 3:13–19
17. Dong J, Xia Q, Yan C, Song H, Liu Y, Wang Q (2018) C(sp^3)–H azidation reaction: a protocol for preparation of aminals. J Org Chem 83:4516–4524
18. Dutta B, Sharma V, Sassu N, Dang Y, Weerakkody C, Macharia J, Miao R, Howella AR, Suib SL (2017) Cross dehydrogenative coupling of N-aryltetrahydroisoquinolines (Sp^3 C–H) with Indoles (Sp^2 C–H) using a heterogeneous mesoporous manganese oxide catalyst. Green Chem 19:5350–5355
19. Fang L, Li Z, Jiang Z, Tan Z, Xie Y (2016) A metal-free oxidative cross-dehydrogenative coupling of *N*-aryl tetrahydroisoquinolines and 2-methylazaarenes using a recyclable oxoammonium salt as oxidant in aqueous media. Eur J Org Chem 21:3559–3567
20. Flores A, Cots E, Bergès J, Muñiz K (2019) Enantioselective iodine(I/III) catalysis in organic synthesis. Adv Synth Catal 361:2–25

21. Franz JF, Kraus WB, Zeitler K (2015) No photocatalyst required—versatile, visible light mediated transformations with polyhalomethanes. Chem Commun 51:8280–8283
22. Gröll B, Schaaf P, Schnürch M (2017) Improved simplicity and practicability in copper-catalyzed alkynylation of tetrahydroisoquinoline. MonShefte Für Chem Chem Mon 148:91–104
23. Guo X, Zipse H, Mayr H (2014) Mechanisms of hydride abstractions by quinones. J Am Chem Soc 136:13863–13873
24. Han X, He X, Sun L, Han X, Zhan W, Xu J, Wang X, Chen J (2018) Increasing effectiveness of photogenerated carriers by in situ anchoring of Cu_2O nanoparticles on a nitrogen-doped porous carbon yolk-shell cuboctahedral framework. ACS Catal 8:3348–3356
25. Heinrich ACJ, Thiedemann B, Gates PJ, Staubitz A (2013) Dual selectivity: electrophile and nucleophile selective cross-coupling reactions on a single aromatic substrate. Org Lett 15:4666–4669
26. Hernández JG, Friščić T (2015) Metal-catalyzed organic reactions using mechanochemistry. Tetrahedron Lett 56:4253–4265
27. Hirao T, Amaya T, Ito T (2012) Selective cross-dehydrogenative coupling of *N*-phenyltetrahydroisoquinolines in aqueous media using poly(aniline sulfonic acid)/gold nanoparticles. Heterocycles 86:927
28. Ho HE, Ishikawa Y, Asao N, Yamamoto Y, Jin T (2015) Highly efficient heterogeneous aerobic cross-dehydrogenative coupling via C–H functionalization of tertiary amines using a nanoporous gold skeleton catalyst. Chem Commun 51:12764–12767
29. Howard JL, Cao Q, Browne DL (2018) Mechanochemistry as an emerging tool for molecular synthesis: what can it offer? Chem Sci 9:3080–3094
30. Hu J, Wang J, Nguyen TH, Zheng N (2013) The chemistry of amine radical cations produced by visible light photoredox catalysis. Beilstein J Org Chem 9:1977–2001
31. Huang T, Liu X, Lang J, Xu J, Lin L, Feng X (2017) Asymmetric aerobic oxidative cross-coupling of tetrahydroisoquinolines with alkynes. ACS Catal 7:5654–5660
32. Johnson JA, Luo J, Zhang X, Chen YS, Morton MD, Echeverría E, Torres FE, Zhang J (2015) Porphyrin-metalation-mediated tuning of photoredox catalytic properties in metal-organic frameworks. ACS Catal 5:5283–5291
33. Kibriya G, Bagdi AK, Hajra A (2018) Visible-light-promoted $C(Sp^3)$–$C(Sp^2)$ cross-dehydrogenative coupling of tertiary amine with imidazopyridine. J Org Chem 83:10619–10626
34. Kwon SJ, Gil MG, Kim DY (2017) Visible light mediated photocatalytic oxidative coupling reaction of *N*-phenyl tetrahydroisoquinoline with β-keto acids. Tetrahedron Lett 58:1592–1594
35. Li CJ (2009) Cross-dehydrogenative coupling (CDC): exploring C–C bond formations beyond functional group transformations. Acc Chem Res 42:335–344
36. Li CJ, Li Z (2006) Green chemistry: The development of cross-dehydrogenative coupling (CDC) for chemical synthesis. Pure Appl Chem 78:935–945
37. Li Z, Scott Bohle D, Li CJ (2006) Cu-catalyzed cross-dehydrogenative coupling: a versatile strategy for C-C bond formations via the oxidative activation of sp3 C–H bonds. Proc Natl Acad Sci USA 103:8928–8933
38. Li S, Shelar DP, Hou CC, Chen QQ, Deng P, Chen Y (2018) WO_3 nanospheres with improved catalytic activity for visible light induced cross dehydrogenative coupling reactions. J Photochem Photobiol A 363:44–50
39. Li Z, Li CJ (2005a) Highly efficient copper-catalyzed nitro-Mannich type reaction: cross-dehydrogenative-coupling between sp^3 C–H bond and sp^3 C–H bond. J Am Chem Soc 127:3672–3673
40. Li Z, Li CJ (2005b) Highly efficient CuBr-catalyzed cross-dehydrogenative coupling (CDC) between tetrahydroisoquinolines and activated methylene compounds. Eur J Org Chem 3173–3176

41. Lin W, Cao T, Fan W, Han Y, Kuang J, Luo H, Miao B, Tang X, Yu Q, Yuan W, Zhang J, Zhu C, Ma S (2014) Enantioselective double manipulation of tetrahydroisoquinolines with terminal alkynes and aldehydes under copper(I) catalysis. Angew Chem Int Ed 53:277–281
42. Lin M, Dai LX, Gu J, Kanga LQ, Wanga YH, Si R, Zhaoa ZQ, Liua CW, Fua X, Sun LD, Zhang YW, Yan CH (2017) Moderate oxidation levels of Ru nanoparticles enhance molecular oxygen activation for cross-dehydrogenative-coupling reactions via single electron transfer. RSC Adv 7:33078–33085
43. Lin W, Ma S (2014) Highly efficient and general enantioselective synthesis of naturally occurring isoquinoline alkaloids. Org Chem Front 1:338–346
44. Liu W, Liu S, Jin R, Guo H, Zhao J (2015) Novel strategies for catalytic asymmetric synthesis of C1-chiral 1,2,3,4-tetrahydroisoquinolines and 3,4-dihydrotetrahydroisoquinolines. Org Chem Front 2:288–299
45. Liu Y, Wang C, Xue D, Xiao M, Li C, Xiao J (2017) Reactions catalysed by a binuclear copper complex: aerobic cross dehydrogenative coupling of *N*-aryl tetrahydroisoquinolines. Chem Eur J 23:3051–3061
46. Liu W, Wu S, Guo B, Ju P, Li G, Wu Q (2019) Difluoroborate-based conjugated organic polymer: a high-performance heterogeneous photocatalyst for oxidative coupling reactions. J Mater Sci 54:1205–1212
47. Lv L, Li Z (2016) Fe-catalyzed cross-dehydrogenative coupling reactions. Top Curr Chem 374(38):1–39
48. Lyons DJM, Crocker RD, Enders D, Nguyen TV (2017) Tropylium salts as efficient organic Lewis acid catalysts for acetalization and transacetalization reactions in batch and flow. Green Chem 19:3993–3996
49. Ma Y, Zhang G, Zhang J, Yang D, Wang R (2014) Organocatalyzed asymmetric oxidative coupling of α-Csp3-H of tertiary amines to α, β unsaturated γ-butyrolactam: synthesis of MBH-Type products. Org Lett 16:5358–5361
50. Mara M, Jackson NE, Huang J, Stickrath A, Zhang X, Gothard NA, Ratner MA, Chen LX (2013) Effects of electronic and nuclear interactions on the excited-state properties and structural dynamics of copper(I) diimine complexes. J Phys Chem B 117:1921–1931
51. Marset X, Pérez J, Ramón D (2016) Cross-dehydrogenative coupling reaction using copper oxide impregnated on magnetite in deep eutectic solvents. Green Chem 18:826–833
52. McCusker CE, Castellano FN (2013) Design of a long-lifetime, earth-abundant, aqueous compatible Cu(I) photosensitizer using cooperative steric effects. Inorg Chem 52:8114–8120
53. Meng QY, Zhong JJ, Liu Q, Gao XW, Zhang HH, Lei T, Li ZJ, Feng K, Chen B, Tung CH, Wu LZ (2013) A cascade cross-coupling hydrogen evolution reaction by visible light catalysis. J Am Chem Soc 135:19052–19055
54. Mercadante MA, Kelly CB, Bobbitt JM, Tilley LJ, Leadbeater NE (2013) Synthesis of 4-acetamido-2,2,6,6-tetramethylpiperidine-1-oxoammonium tetrafluoroborate and 4-acetamido-(2,2,6,6-tetramethyl-piperidin-1-yl)oxyl and their use in oxidative reactions. Nat Protoc 8:666–676
55. Min C, Sanchawala A, Seidel D (2014) Dual C–H functionalization of N-aryl amines: synthesis of polycyclic amines via an oxidative Povarov approach. Org Lett 16:2756–2759
56. Murahashi SI, Komiya N, Terai H, Nakae T (2003) Aerobic ruthenium-catalyzed oxidative cyanation of tertiary amines with sodium cyanide. J Am Chem Soc 125:15312–15313
57. Nasalevich MA, van der Veen M, Kapteijna F, Gascon J (2014) Metal-organic frameworks as heterogeneous photocatalysts: advantages and challenges. CrystEngComm 16:4919–4926
58. Nguyen TV, Lyons DJM (2015) A novel aromatic carbocation-based coupling reagent for esterification and amidation reactions. Chem Commun 51:3131–3134
59. Oss G, de Vos SD, Luc KNH, Harper JB, Nguyen TV (2018) Tropylium-promoted oxidative functionalization of tetrahydroisoquinolines. J Org Chem 83:1000–1010
60. Pan Y, Kee CW, Chen L, Tan CH (2011) Dehydrogenative coupling reactions catalysed by Rose Bengal using visible light irradiation. Green Chem 13:2682–2685

61. Parvatkar PT, Manetsch R, Banik BK (2019) Metal-free cross-dehydrogenative coupling (CDC): molecular iodine as a versatile catalyst/reagent for CDC reactions. Chem Asian J 14: 6–30
62. Patil MR, Dedhia NP, Kapdi AR, Kumar AV (2018) Cobalt(II)/*N*-hydroxyphthalimide-catalyzed cross-dehydrogenative coupling reaction at room temperature under aerobic condition. J Org Chem 83:4477–4490
63. Perepichka I, Kundu S, Hearne Z, Li CJ (2015) Efficient merging of copper and photoredox catalysis for the asymmetric cross-dehydrogenative-coupling of alkynes and tetrahydroisoquinolines. Org Biomol Chem 13:447–451
64. Punirun T, Soorukram D, Kuhakarn C, Reutrakul V, Pohmakotr M (2018) Oxidative difluoromethylation of tetrahydroisoquinolines using TMSCF2SPh: synthesis of fluorinated pyrrolo[2,1-a]isoquinolines and benzo[a]quinolizidines. J Org Chem 32:765–782
65. Rueping M, Vila C, Koenigs RM, Poscharny K, Fabry DC (2011) Dual catalysis: combining photoredox and Lewis base catalysis for direct Mannich reactions. Chem Commun 47:2360–2362
66. Rusch F, Unkel L, Alpers D, Hoffmann F, Brasholz M (2015) A visible light photocatalytic cross-dehydrogenative coupling/dehydrogenation/6π-cyclization/oxidation cascade: synthesis of 12-nitroindoloisoquinolines from 2-aryltetrahydroisoquinolines. Chem Eur J 21:8336–8340
67. Sanders B, Jackson B, Brent M, Taylor A, Dang W, Berger S, Schreiber S, Howitz K, Marmorstein R (2009) Identification and characterization of novel sirtuin inhibitor scaffolds. Bioorg Med Chem 17:7031–7041
68. Scheuermann CJ (2010) Beyond traditional cross couplings: the scope of the cross dehydrogenative coupling reaction. Chem Asian J 5:436–451
69. Scott JD, Williams RM (2002) Chemistry and biology of the tetrahydroisoquinoline antitumor antibiotics. Chem Rev 102:1669–1730
70. Sharma K, Borah A, Neog A, Gogoi DP (2016) CeO_2-catalyzed C–H functionalization of *N*-aryltetrahydroisoquinolines: an aerobic cross-dehydrogenative coupling reaction between two Sp^3 C–H bonds. ChemistrySelect 1:4620–4623
71. Singh IP, Shah P (2017) Tetrahydroisoquinolines in therapeutics: a patent review (2010–2015). Expert Opin Ther Pat 1:17–36
72. So MH, Liu Y, Ho CM, Che CM (2009) Graphite-supported gold nanoparticles as efficient catalyst for aerobic oxidation of benzylic amines to imines and *N*-substituted 1,2,3,4-tetrahydroisoquinolines to amides: synthetic applications and mechanistic study. Chem Asian J 4:1551–1561
73. Stockerl S, Gutiérrez D, Mancheño OG (2017) Click-binol-phosphoric acid catalysts in intramolecular enantioselective oxidative C–H-bond functionalization. J Mol Catal A Chem 426:572–585
74. Su WK, Yu JB, Li ZH, Jiang Z (2011) Solvent-free cross-dehydrogenative coupling reactions under high speed ball-milling conditions applied to the synthesis of functionalized tetrahydroisoquinolines. J Org Chem 76:9144–9150
75. Sud A, Sureshkumar D, Klussmann M (2009) Oxidative coupling of amines and ketones by combined vanadium- and organocatalysis. Chem Commun 3169–3171
76. Sun S, Li C, Floreancig PE, Lou H, Liu L (2015) Highly enantioselective catalytic cross-dehydrogenative coupling of N-carbamoyl tetrahydroisoquinolines and terminal alkynes. Org Lett 17:1684–1687
77. Sun X, Lv XH, Ye LM, Hu Y, Chen YY, Zhang XJ, Yan M (2015) Synthesis of benzimidazoles via iridium-catalyzed acceptorless dehydrogenative coupling. Org Biomol Chem 13:7381–7383
78. Sun CL, Shi ZJ (2014) Transition-metal-free coupling reactions. Chem Rev 114:9219–9280
79. Tanoue A, Yoo WJ, Kobayashi S (2014) Sulfuryl chloride as an efficient initiator for the metal-free aerobic cross-dehydrogenative coupling reaction of tertiary amines. Org Lett 16:2346–2349

80. Varun BV, Dhineshkumar J, Bettadapur KR, Siddaraju Y, Alagiri K, Prabhu KR (2017) Recent advancements in dehydrogenative cross coupling reactions for C–C bond formation. Tetrahedron Lett 58:803–824
81. Wang CA, Li YW, Cheng XL, Zhanga JP, Han YF (2017) Eosin Y dye-based porous organic polymers for highly efficient heterogeneous photocatalytic dehydrogenative coupling reaction. RSC Adv 7:408–414
82. Wang C, Liu D, Lin W (2013) Metal-organic frameworks as a tunable platform for designing functional molecular materials. J Am Chem Soc 135:13222–13234
83. Wang F, Luo C, Deng G, Yang L (2014) C(Sp3)–C(Sp3) bond formation via copper/ brønsted acid Co-catalyzed C(Sp^3)–H bond oxidative cross-dehydrogenative-coupling (CDC) of azaarenes. Green Chem 16:2428–2431
84. Wang B, Shelar DP, Han X, Li T, Guan X, Lu W, Liu K, Chen Y, Fu W, Che C (2015) Long-lived excited states of zwitterionic copper(I) complexes for photoinduced cross-dehydrogenative coupling reactions. Chem Eur J 21:184–190
85. Wendlandt AE, Stahl SS (2015) Quinone-catalyzed selective oxidation of organic molecules. Angew Chem Int Ed 54:14638–14658
86. Wu CJ, Zhong JJ, Meng QY, Lei T, Gao XW, Tung CH, Wu LZ (2015) Cobalt-catalyzed cross-dehydrogenative coupling reaction in water by visible light. Org Lett 17:884–887
87. Xia W, Mahmood A, Zou R, Xu Q (2015) Metal-organic frameworks and their derived nanostructures for electrochemical energy storage and conversion. Energy Environ Sci 8:1837–1866
88. Xiao L, Huang Y, Luo Y, Yang B, Liu Y, Zhou X, Zhang J (2018) Organic cotton photocatalysis. ACS Sustain Chem Eng 6:14759–14766
89. Xiao T, Li L, Lin G, Mao ZW, Zhou L (2014) Metal-free visible-light induced cross-dehydrogenative coupling of tertiary amines with diazo compounds. Org Lett 16:4232–4235
90. Xie J, Huang ZZ (2010) Cross-dehydrogenative coupling reactions by transition-metal and aminocatalysis for the synthesis of amino acid derivatives. Angew Chem Int Ed 49:10181–10185
91. Yan C, Liu Y, Wang Q (2014) Mild and highly efficient metal-free oxidative α-cyanation of N-acyl/sulfonyl tetrahydroisoquinolines. RSC Adv 4:60075–60078
92. Yan C, Liu Y, Wang Q (2015) Direct C–H allylation of N-acyl/sulfonyl tetrahydroisoquinolines and analogues. Org Lett 17:5714–5717
93. Yang X, Zou C, He Y, Zhao M, Chen B, Xiang S, O'Keeffe M, Wu C (2014) A stable microporous mixed-metal metal-organic framework with highly active Cu^{2+} sites for efficient cross-dehydrogenative coupling reactions. Chem Eur J 20:1447–1452
94. Ye K, Ke Y, Keshava N, Shanks J, Kapp J, Tekmal R, Petros J, Joshi H (1998) Opium alkaloid noscapine is an antitumor agent that arrests metaphase and induces apoptosis in dividing cells. Proc Natl Acad Sci USA 95:1601–1606
95. Yoo WJ, Kobayashi S (2014) Efficient visible light-mediated cross-dehydrogenative coupling reactions of tertiary amines catalyzed by a polymer-immobilized iridium-based photocatalyst. Green Chem 16:2438–2442
96. Yu JB, Li ZH, Jia K, Jiang ZJ, Liu M, Su WK (2013) Fast, solvent-free asymmetric alkynylation of prochiral sp3 C–H bonds in a ball mill for the preparation of optically active tetrahydroisoquinoline derivatives. Tetrahedron Lett 54:2006–2009
97. Yu JB, Peng G, Jiang ZJ, Hong ZK, Su WK (2016) Mechanochemical oxidative Mannich reaction: evaluation of chemical and mechanical parameters for the mild and chemoselective coupling of N-tert-butoxycarbonyltetrahydroquinolines and ketones. Eur J Org Chem 2016:5340–5344
98. Yusubov MS, Zhdankin VV (2015) Iodine catalysis: a green alternative to transition metals in organic chemistry and technology. Resour Effic Technol 1:49–67
99. Zhang WQ, Li QY, Zhang Q, Lu Y, Lu H, Wang W, Zhao X, Wang XJ (2016) Robust metal-organic framework containing benzoselenadiazole for highly efficient aerobic cross-dehydrogenative coupling reactions under visible light. Inorg Chem 55:1005–1007

100. Zhang T, Liang W, Huang Y, Li X, Liu Y, Yang B, He C, Zhou X, Zhang J (2017) Bifunctional organic sponge photocatalyst for efficient cross-dehydrogenative coupling of tertiary amines to ketones. Chem Commun 53:12536–12539
101. Zhang G, Ma Y, Wang S, Kong W, Wang R (2013) Chiral organic contact ion pairs in metal-free catalytic enantioselective oxidative cross-dehydrogenative coupling of tertiary amines to ketones. Chem Sci 4:2645–2651
102. Zhang G, Ma Y, Wang S, Zhang Y, Wang R (2012) Enantioselective metal/organo-catalyzed aerobic oxidative sp3 C–H olefination of tertiary amines using molecular oxygen as the sole oxidant. J Am Chem Soc 134:12334–12337
103. Zhang J, Tiwari B, Xing C, Chen X, Chi YR (2012) Enantioselective oxidative cross-dehydrogenative coupling of tertiary amines to aldehydes. Angew Chem Int Ed 51:3649–3652
104. Zhang Y, Wei BW, Wang WX, Deng LL, Nie LJ, Luo HQ, Fan XL (2017) Direct vinylogous oxidative cross-dehydrogenative coupling of 4-nitroisoxazoles with N-Aryl tetrahydroisoquinolines in water under air conditions. RSC Adv 7:1229–1232
105. Zhang W, Yang S, Shen Z (2016) Copper-catalyzed cyanomethylation of substituted tetrahydroisoquinolines with acetonitrile. Adv Synth Catal 358:2392–2397
106. Zhao S, Li Y, Liu C, Zhao Y (2018) Recent advances in mechanochemical C–H functionalization reactions. Tetrahedron Lett 59:317–324
107. Zheng Q, Meng W, Jiang G, Yu Z (2013) CuI-catalyzed C1-alkynylation of tetrahydroisoquinolines (THIQs) by a reaction with tunable iminium ions. Org Lett 15:5928–5931
108. Zhong JJ, Meng QY, Liu B, Li XB, Gao XW, Lei T, Wu CJ, Li ZJ, Tung CH, Wu LZ (2014) Cross-coupling hydrogen evolution reaction in homogeneous solution without noble metals. Org Lett 16:1988–1991
109. Zhou S, Tong R (2017) Three-step catalytic asymmetric total syntheses of 13 methyltetrahydroprotoberberine alkaloids. Org Lett 19:1594–1597

Cross-Dehydrogenative Coupling in the Synthesis and Functionalization of Fused Imidazoheterocycles

Avik Kumar Bagdi and Alakananda Hajra

Abstract Imidazoheterocycles, containing a fused imidazole ring, are recognized as privileged scaffolds due to their wide range of applications in the field of biological activities, pharmaceuticals, optoelectronics, material science, etc. As a consequence, the development of synthetic strategies for the construction of functionalized imidazoheterocycles is an important activity of the synthetic chemists over the last few decades. Cross-dehydrogenative coupling (CDC) has been emerged as an important synthetic strategy for the direct synthesis of bioactive molecules and is carried out employing different catalytic systems like transition-metal catalysis (Pd, Cu, Ag, etc.), iodine catalysis, organophotocatalysis, etc. This chapter will offer an overview of the synthesis and functionalization of various biologically relevant imidazoheterocycle derivatives like imidazopyridine, imidazothiazole, imidazopyrimidine, etc. via cross-dehydrogenative coupling reactions. The importance of this strategy over traditional pathway has been also emphasized in this chapter.

Keywords Cross-dehydrogenative coupling · Imidazoheterocycle · C–C bond · C–heteroatom bond

1 Introduction

Fused imidazoheterocycles are the class of molecules which bear fused imidazole ring, e.g., imidazo[1,2-*a*]pyridine, imidazo[1,2-*a*]pyrimidine, benzo[*d*]imidazo[2,1-*b*]thiazole, imidazo[2,1-*b*]thiazole, etc. (Scheme 1). These scaffolds exhibit diverse biological and pharmaceutical activities and are important building blocks in synthetic organic chemistry [1]. There are several drugs available in the market bearing

A. K. Bagdi
Department of Chemistry, University of Kalyani, Kalyani, Nadia 741235, India

A. Hajra (✉)
Department of Chemistry, Visva-Bharati (a Central University), Santiniketan 731235, India
e-mail: alakananda.hajra@visva-bharati.ac.in

A. Srivastava and C. K. Jana (eds.), *Heterocycles via Cross Dehydrogenative Coupling*, https://doi.org/10.1007/978-981-13-9144-6_4

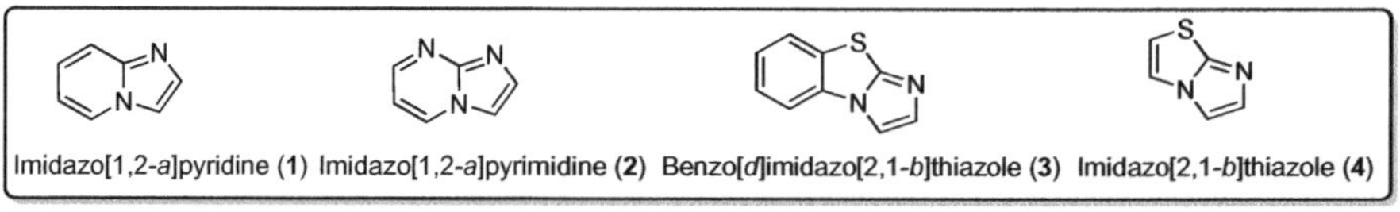

Scheme 1 Structure of fused imidazoheterocycles

these moieties [2]. These are also important in the field of material science [3] and agrochemicals [4]. Some abnormal *N*-heterocyclic carbenes based on imidazo[1,2-*a*] pyridines have also been synthesized [5]. Consequently, a continuous effort has been paid on the synthesis and functionalization of imidazoheterocycles [6]. Different strategies have been employed for these purposes. Among these cross-dehydrogenative coupling (CDC) is the most appealing from the aspect of green chemistry as this is a straightforward and more efficient process. In this chapter, we will discuss the synthesis and functionalization of imidazoheterocycles employing cross-dehydrogenative coupling. Mechanism and advantages of these strategies will be also discussed.

2 C–C Bond Forming Cross-Dehydrogenative Coupling Reaction

*2.1 C–H Alkenylation of Imidazo[1,2-*a*]Pyridines*

Alkenylation of heterocycles is an important strategy for the introduction of functionality in the heteroarenes. So, various methodologies have been developed for the selective alkenylation of heterocycles employing transition-metal (Pd, Ru, Rh, Ir, etc.) catalyzed coupling reaction. Among the various strategies, the oxidative Heck-type reaction is most appealing one. As a consequence, methodologies have been developed by different groups for the alkenylation of imidazoheterocycles. Roboin et al. employed cross-dehydrogenative coupling reaction for the functionalization of imidazo[1,2-*a*]pyridines in the first time (Scheme 2) [7]. They demonstrated that cross-dehydrogenative coupling could be an efficient strategy for the direct C–H functionalization of imidazoheterocycles avoiding pre-functionalization of the scaffolds. Direct coupling of imidazo[1,2-*a*]pyridines with electron-deficient alkenes in the presence of Pd-catalyst and $Cu(OAc)_2$ as oxidant in xylene under argon atmosphere afforded 3-alkenylated product with ease. The reaction is highly regioselective; in the case of unsubstituted imidazo[1,2-*a*]pyridine, only 3-alkenylated product was obtained. Although selectively β-branched product was obtained from electron-deficient alkenes like α, β-unsaturated esters, nitrile, etc., a mixture of both α- and β-products was produced in case of styrene.

Cao et al. modified this alkenylation of imidazo[1,2-*a*]pyridines employing Ru-catalyst (Scheme 3) in 2014 [8]. They demonstrated that 3 mol%

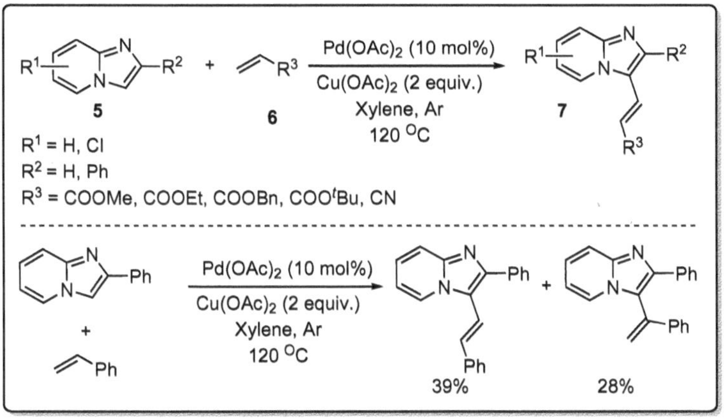

Scheme 2 C–H alkenylation of imidazo[1,2-*a*]pyridines

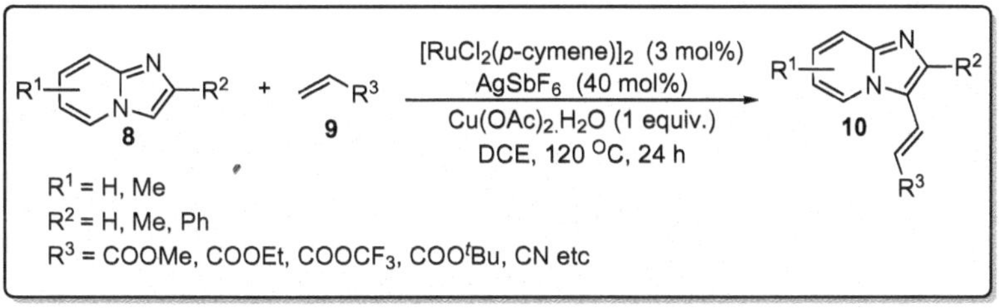

Scheme 3 Ru-catalyzed alkenylation of imidazo[1,2-*a*]pyridines

$[RuCl_2(p\text{-cymene})]_2$ can act as an efficient catalyst in the presence of $Cu(OAc)_2{\cdot}H_2O$ (1 equiv.) as oxidant and 40 mol% of $AgSbF_6$ in DCE for the cross-dehydrogenative coupling between imidazo[1,2-*a*]pyridines and electron-deficient alkenes. This methodology is also regioselective and alkenylation occurred at C-3 position only.

The probable mechanism of this oxidative olefination reaction is represented in Scheme 4. At first, Ru-complex **12** is formed by the reaction of Ru-catalyst with imidazopyridine moiety. This Ru-complex **12** reacts with the unsaturated alkene to form the intermediate **13**. Product is formed from the intermediate **13** through *β*-hydride elimination along with the generation of Ru-hydride (**14**) formation which reacts with $Cu(OAc)_2$ to regenerate the active Ru-catalyst.

Later, this group developed a Pd-catalyzed methodology for the alkenylation of imidazo[1,2-*a*]pyridines (Scheme 5) [9]. They demonstrated that the best result was obtained using $Pd(OAc)_2$ as catalyst, Ag_2CO_3 (5 mol%) and O_2 (with oxygen balloon) as co-oxidant in the presence of AcOH (1 equiv.), and Ac_2O (1 equiv.) as additives in dioxane at 120 °C. The reaction is highly regioselective and only C-3 alkenylation of imidazopyridines took place. In addition, only *α*-product is obtained in case of styrene, whereas only *β*-product is produced in case of acrylates. A wide

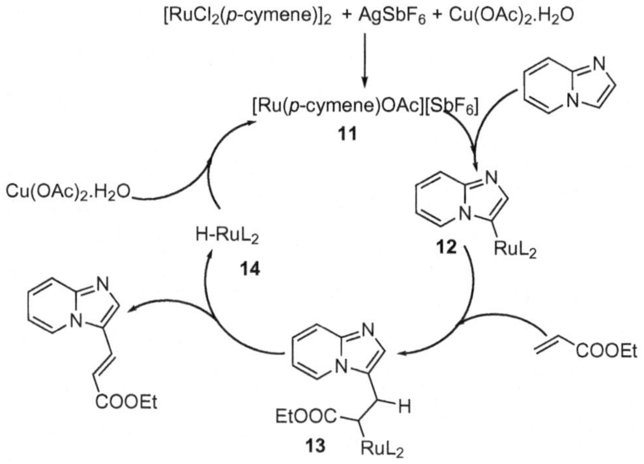

Scheme 4 Probable mechanism of Ru-catalyzed oxidative olefination of imidazo[1,2-*a*]pyridines

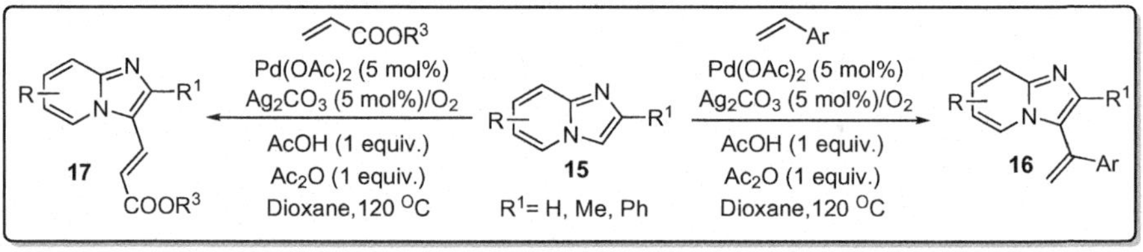

Scheme 5 Pd-catalyzed alkenylation of imidazo[1,2-*a*]pyridines

array of styrenes, acrylates, and imidazopyridines with various groups are susceptible under the reaction conditions.

In 2015, Hajra et al. explored a cross-dehydrogenative coupling reaction between imidazo[1,2-*a*]pyridine and vinylarene for the synthesis of alkenylated imidazo[1,2-*a*]pyridines (Scheme 6) [10]. They developed a new Pd-catalyzed and ligand-free methodology employing oxygen as an oxidant. This Pd-catalyzed reaction is highly regioselective and afforded only α-product. The role of additive is crucial in this cross-dehydrogenative coupling reaction and TBAB is the best one to improve the yield up to 75%. Imidazopyridines with different functionalities like

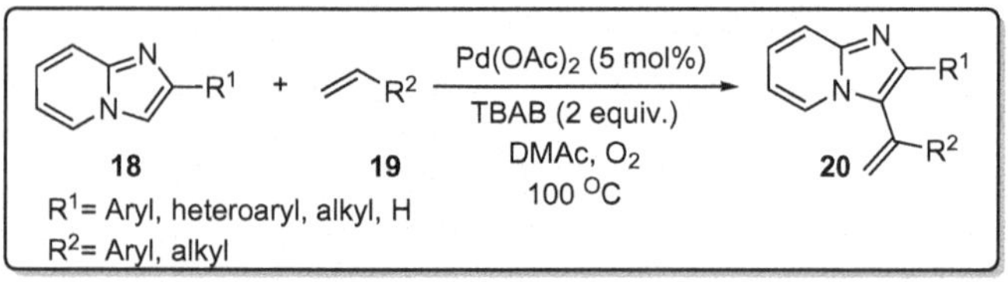

Scheme 6 Vinylation of imidazopyridines

Scheme 7 Probable mechanism

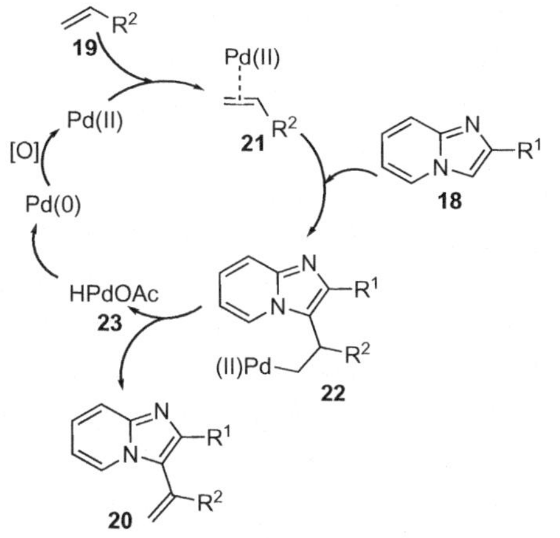

methyl, methoxy, halides, cyanide, etc., as well as hetroaromatic and aliphatic substituent are well tolerant under the optimized conditions. Most important achievement of this methodology is the applicability toward aliphatic alkenes along with various styrenes.

The probable mechanistic pathway is represented in Scheme 7. Initially, the alkene is activated by the Pd-salt to form the complex **21** which on sequential attack by the imidazo[1,2-*a*]pyridine to form the intermediate **22**. The intermediate **22** upon β-hydride elimination afforded α-product (**20**) and HPdOAc (**23**). Through reductive elimination, this HPdOAc (**23**) forms Pd(0) which regenerates the active Pd-catalyst by aerial oxidation.

*2.2 Cross-Dehydrogenative Coupling of Imidazo[1,2-*a*] Pyridine with Arenes*

Biaryls are very important as they are found in the various natural products and functional materials. So, continuous efforts have been devoted for the synthesis of biaryls. Among the different strategies, synthesis of biaryls via direct cross-dehydrogenative coupling is an important approach. A new methodology has been developed for the cross-dehydrogenative coupling between imidazo[1,2-*a*] pyridines and arenes by Zhan et al. (Scheme 8) [11]. The combination of 5 mol% Ag_2CO_3/O_2 (Balloon) acts as the best oxidant for this Pd-catalyzed reaction. The role of additive is crucial for this transformation and PivOH (15 mol%) is proven to be the best one as an additive. This methodology is applicable to different imidazo [1,2-*a*]pyridines and arenes such as benzene, toluene, *o*-xylene, *m*-xylene, etc. It regioselectively afforded 3-substituted product; no homocoupling product was formed. In case of toluene, fluorobenzene, and anisole, a mixture of *ortho* and *para*

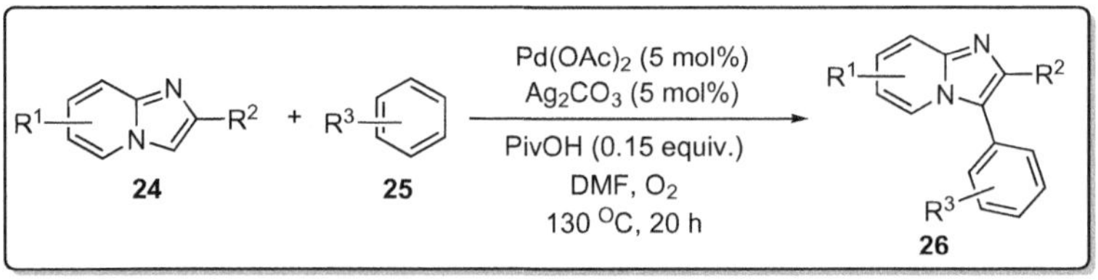

Scheme 8 Cross-dehydrogenative coupling of imidazo[1,2-*a*]pyridine with arenes

Scheme 9 Probable mechanism

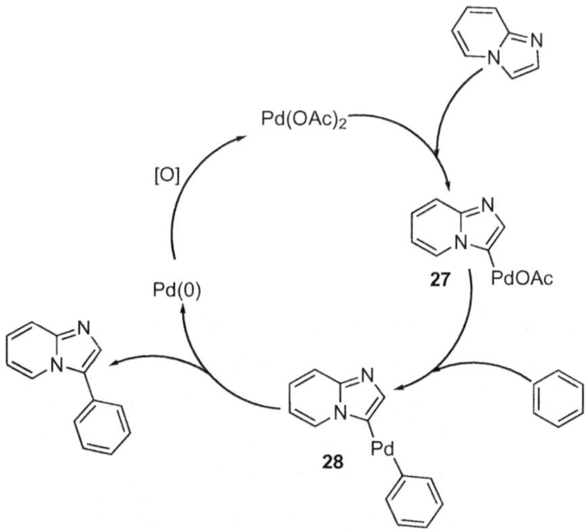

products was formed. The reaction is highly sensitive to steric hindrance and reacts with less hindered site in case of *o*-xylene and *m*-xylene.

The probable mechanism of this reaction is represented in Scheme 9. First step of this reaction is the metallation step in which $Pd(OAc)_2$ reacts with the imidazo [1,2-*a*]pyridine moiety to form the intermediate **27**. In the next step, this intermediate **27** reacts with arene to form the intermediate **28** through concerted metallation deprotonation (CMD) process. Finally, the product is obtained from the intermediate **28** via reductive elimination along with the formation of Pd(0) which gets oxidized to Pd(II) by the oxidant O_2 to complete the catalytic cycle.

2.3 *Synthesis of Biimidazo[1,2-*a*]Pyridines Through Oxidative Homocoupling Reaction*

Synthesis of biheteroaryls is very much important due their wide occurrences in natural products, pharmaceuticals, and agrochemicals. Cao et al. reported a new methodology for the synthesis of biimidazo[1,2-*a*]pyridines through a Cu(I)-

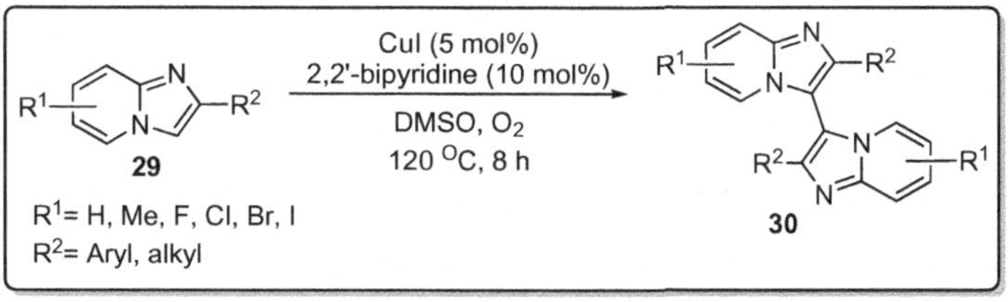

Scheme 10 Synthesis of biimidazo[1,2-*a*]pyridines via oxidative homocoupling reaction

catalyzed homocoupling of imidazo[1,2-*a*]pyridines (Scheme 10) [12]. In this CuI-catalyzed reaction, 2,2′-bipyridine acts as the best ligand and O_2 as the best oxidant for this transformation. This methodology is highly regioselective as well as chemoselective. The reaction of unsubstituted imidazo[1,2-*a*]pyridines afforded the homocoupling product through 3-position and the imidazo[1,2-*a*]pyridines moieties bearing halogens such as −F, −Cl, −Br, −I selectively produced the homocoupling product.

Probable mechanism of this Cu(I)-catalyzed reaction is represented in Scheme 11. The yield of the reaction did not hamper by the presence of the radical scavengers (e.g., TEMPO and galvinoxyl), signifying the non-radical pathway. Initially, CuI reacts with imidazo[1,2-*a*]pyridine to form the intermediate **31** which reacts with another imidazo[1,2-*a*]pyridine moiety to form the intermediate **32**. Finally, the product is formed from the intermediate **32** through reductive elimination in the presence of O_2.

Although the previously discussed methodology has a wide applicability toward various imidazopyridines, employment of Cu-catalyst, ligand, and high temperature is not desirable from the aspect of green chemistry. So, the transition-metal-free approach and room temperature approach are always welcome. Sakhuja et al. modified the synthesis of biimidazo[1,2-*a*]pyridines through a transition-metal-free approach. They demonstrated that the PIDA can catalyze the homocoupling of 2-aryl imidazopyridines in the presence of $BF_3{\cdot}OEt_2$ in DCM at room temperature (Scheme 12) [13]. Although this methodology is highly useful for the homocoupling of imidazopyridines bearing methyl and methoxy substituent, it is not suitable for the imidazopyridines having halogens and nitro functionalities. In addition, aryl substituent at 2-position of imidazopyridines is necessary for this transformation.

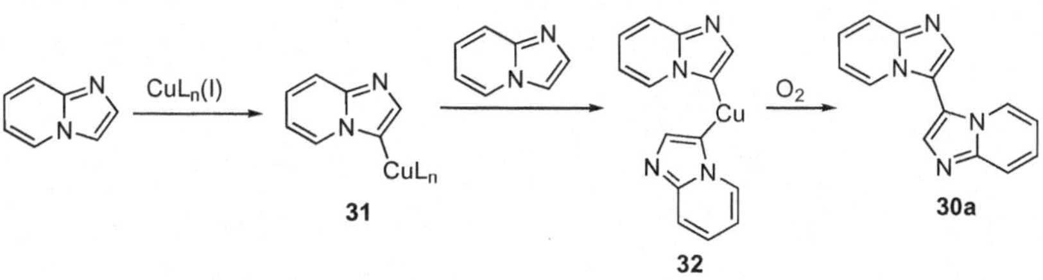

Scheme 11 Probable mechanism

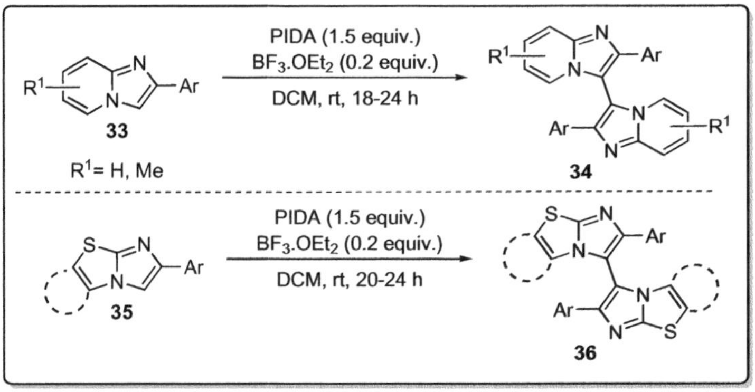

Scheme 12 PIDA-mediated homocoupling of imidazoheterocycles

They also extended this PIDA-mediated strategy for the homocoupling of other imidazoheterocycles such as 2-arylimidazo[2,1-*b*]thiazole and 2-arylbenzo[*d*]imidazo[2,1-*b*]thiazole, and demonstrated the applicability toward gram-scale synthesis of biimidazohetercycles. An organocatalyzed approach of the homocoupling of imidazopyridines was developed by them. In this approach, the use of a catalytic amount of iodobenzene and 3 equiv. of *m*-CPBA as oxidant and 5 equiv. of AcOH in DCM is sufficient to catalyze this transformation. Probably, this reaction proceeds through in situ generation of PIDA by the reaction of iodobenzene and AcOH in the presence of *m*-CPBA. They demonstrated the efficiency of this approach employing different imidazoheterocycles (Scheme 13).

This PIDA-mediated reaction proceeded through a non-radical pathway. The proposed mechanism is represented in Scheme 14. In the first step, imidazopyridine moiety reacts with PIDA in the presence of $BF_3{\cdot}OEt_2$ to form the imidazolenium intermediate **37** and elimination of AcOH affords the intermediate **38**. This intermediate **38** reacts with another molecule of imidazopyridne moiety to form the

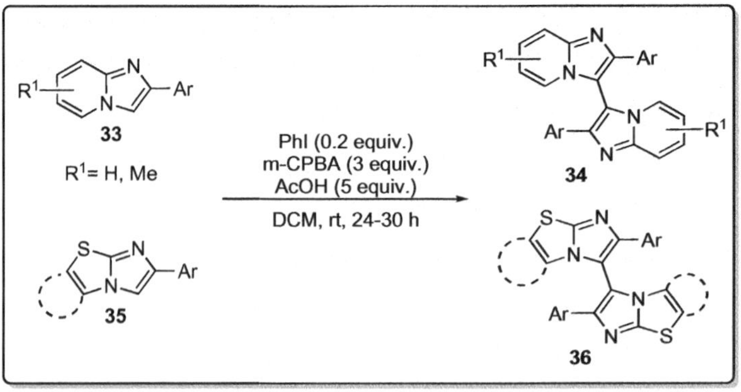

Scheme 13 Organocatalytic homocoupling of imidazoheterocycles

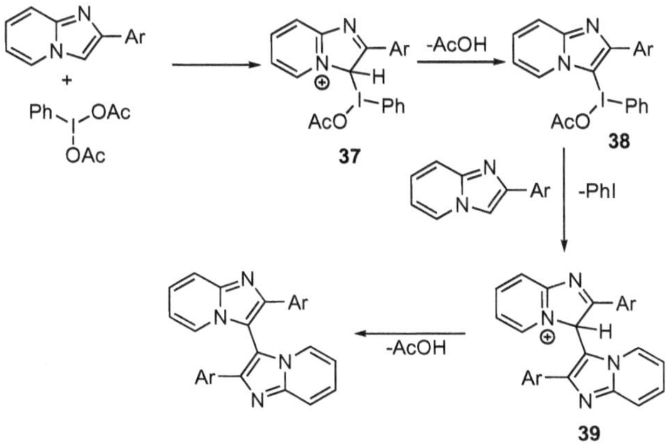

Scheme 14 Probable mechanism of the PIDA-mediated homocoupling of imidazopyridines

intermediate **39** along with the elimination of PhI. Finally, the product is formed from the intermediate **39** via elimination of AcOH.

Electrochemical organic transformation is recently emerged as one of the greener strategies as it is based on only the employment of electrons as the promoters [14]. In addition, generally, there is no need to use catalyst and external oxidant in these transformations. It avoids the formation of waste by-products and it is carried out under room temperature. So, modification of the reported methodologies through an electrochemical pathway is an efficient development from the view of sustainable environment. Han et al. reported the synthesis of biimidazoheterocycles through electrochemical C–C dehydrogenative coupling of imidazoheterocycles (Scheme 15) [15]. They demonstrated that the employment of graphite plate as the anode and platinum plate as the cathode and n-Bu_4NPF_6 as electrolyte in CH_3CN/CF_3CH_2OH (2:1) afforded the best result. Under these conditions, a wide range of

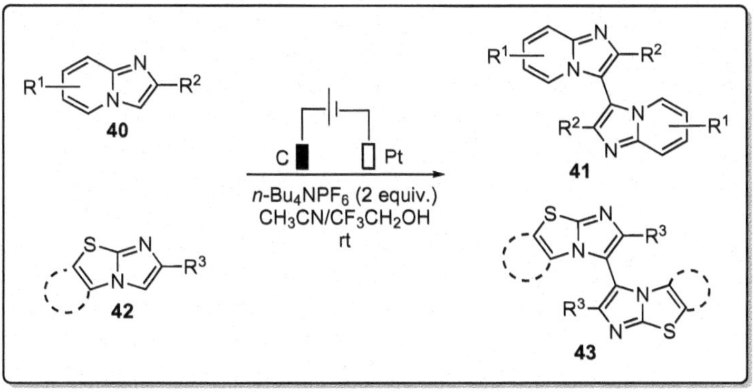

Scheme 15 Electrochemical homocoupling of imidazoheterocycles

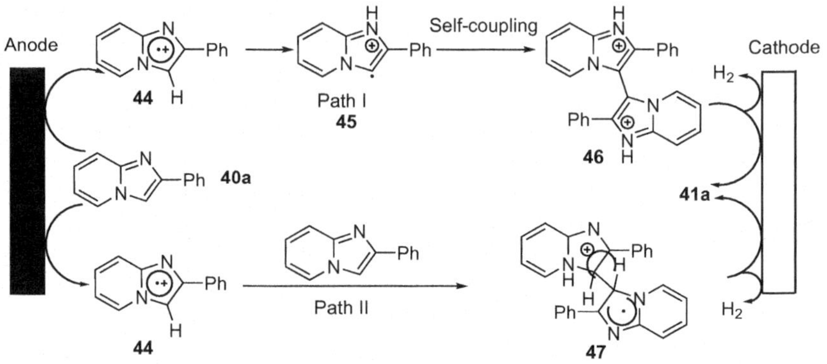

Scheme 16 Probable mechanism of the electrochemical dehydrogenative homocoupling of imidazopyridines

imidazopyridines with aryl/heteroaryl substituent at 2-position reacted efficiently to form the homocoupling product. In addition, the unsubstituted and alkyl-substituted imidazopyridines also produced the products albeit much lower yields were obtained. This electrochemical approach is well suited for other imidazoheterocycles such as imidazo[1,2-*a*]pyrazine, benzo[*d*]imidazo[2,1-*b*]thiazole, and imidazo [2,1-*b*]thiazole.

It proceeds through either of the two pathways as represented in the Scheme 16. First one is the formation of the radical cation intermediate **44** from the imidazopyridine moiety via oxidation at the anode (Path I). This intermediate **44** is transformed into the intermediate **45** which on self-coupling generates the intermediate **46**. Finally, the product is formed from the intermediate **46** via the generation of H_2 at the cathode. The another pathway is the initial formation of radical cation intermediate **44** through SET which reacts with second imidazopyridine molecule to form the intermediate **47** (Path II). Product is obtained from the intermediate **47** via SET at cathode along with generation H_2.

2.4 *Double-Oxidative Cross-Dehydrogenative Coupling of Imidazoheterocycles with Aryl Acetaldehydes*

The direct cross-dehydrogenative coupling reaction between imidazoheterocycles and aryl acetaldehydes was developed by Sakhuja et al. (Scheme 17) [16]. They demonstrated that the reaction of imidazoheterocycles and aryl acetaldehyde in the presence of CuBr (as catalyst), 2,2-bipyridine (as ligand) in toluene afforded imidazoheterocyclic diketones through a cumulative oxidative cross-dehydrogenative coupling and oxidation of methylene next to phenyl group of aryl acetaldehydes. The methodology is applicable to various imidazo[1,2-*a*]pyridines and aryl acetaldehydes with different functionalities. In addition, this strategy is applicable to

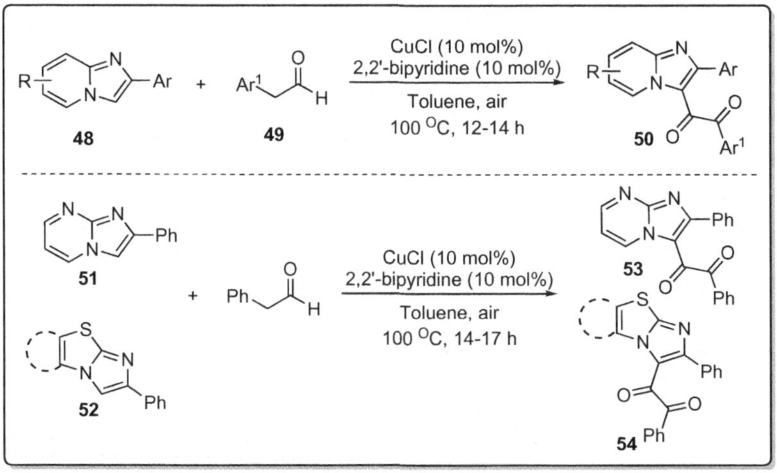

Scheme 17 CDC reaction between imidazoheterocycles and aryl acetaldehydes

other imidazoheterocylces such as imidazo[1,2-*a*]pyrimidine, imidazo[2,1-*b*]thiazole, and benzo[*d*]imidazo[2,1-*b*]thiazole albeit longer reaction time is required.

Relying on the control experiments the plausible mechanism of this reaction is depicted in Scheme 18. The reaction proceeds through a radical pathway. Initially, the intermediate **55** is formed by the reaction of imidazo[1,2-*a*]pyridine and phenyl acetaldehyde through SET pathway. Intermediate **55** gets oxidized to the intermediate **56** and subsequently, it transformed into the intermediate **58**. The intermediate **58** in the presence of aerial oxygen is converted into intermediate **60** which finally afforded the product along with the generation of CuOH. CuBr is regenerated from CuOH by the reaction with HBr.

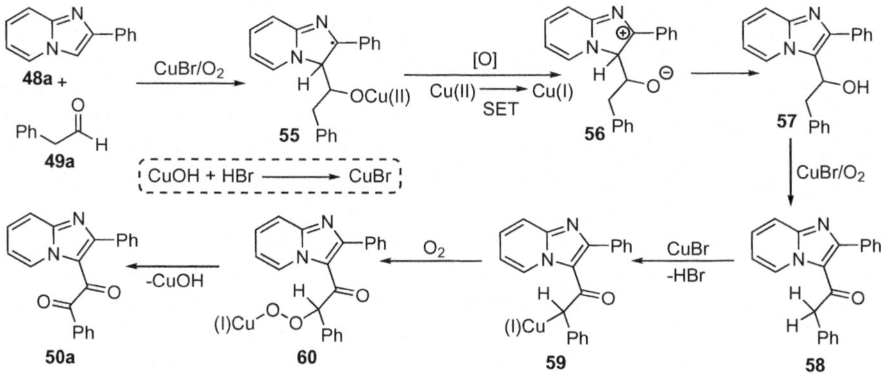

Scheme 18 Probable mechanism

2.5 1,2-Dicarbonylation of Imidazoheterocycles

Cross-dehydrogenative coupling reactions are explored employing different transition-metal catalysts like Pd, Cu, Ag, Rh, and Ru salts, etc. due to their high efficiency. However, because of their expensiveness, the exploration of low price, ready available, and sustainable metal catalysts is highly desirable. In this context, Fe salts have emerged as efficient catalysts in cross-dehydrogenative coupling reactions. Hajra et al. demonstrated the efficiency of iron salt in the cross-dehydrogenative coupling between imidazoheterocycles and oxoaldehydes (Scheme 19) [17]. No additional oxidant is required; the reactions have been carried out under ambient air. Dicarbonylation of different imidazoheterocycles such as benzo[*d*]imidazo[2,1-*b*]thiazole, imidazo[2,1-*b*]thiazole, imidazo[1,2-*a*]pyridine, and imidazo[1,2-*a*]pyrimidine is carried out employing different aryl/heteroaryl oxoaldehyde. The reaction is tolerant with wide range of functional groups such as methyl, methoxy, halides, cyanide, trifluoromethyl, nitro, etc. However, this methodology is not suitable for the dicarbonylation of unsubstituted and aliphatic-substituted imidazoheterocycles. In the case of simple aldehydes, the bisimidazo[1,2-*a*]pyridine is formed.

Probable mechanism of this reaction is represented in Scheme 20. Yield of the reaction is not affected in the presence of radical scavengers like TEMPO, BHT, and BQ. Thus, the reaction is proceeded through a non-radical pathway. The reaction of oxoaldehyde with benzo[*d*]imidazo[2,1-*b*]thiazole in the presence of $FeCl_3$ afforded the intermediate **71**. In the next step, intermediate **71** is transformed into the intermediate **72**. This intermediate **72** via aerial oxidation is converted into Fe(IV) intermediate **73**. From this intermediate **73**, product is formed along with the formation of Fe(II) which regenerates the active catalyst Fe(III) in the presence of air.

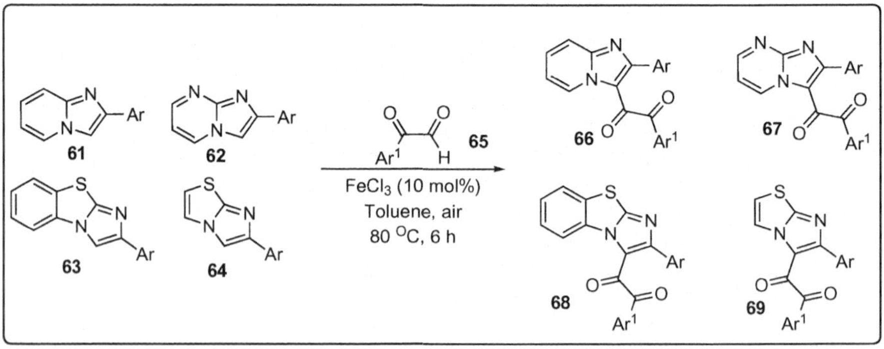

Scheme 19 1,2-dicarbonylation of imidazoheterocycles

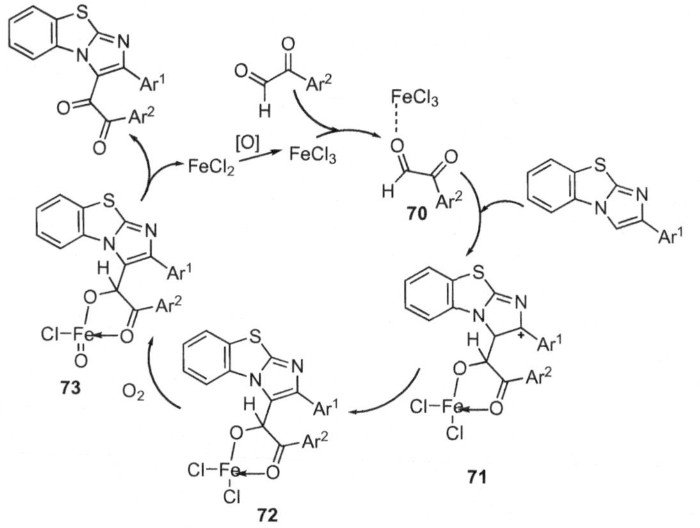

Scheme 20 Plausible mechanism

2.6 *Cross-Dehydrogenative Coupling of Imidazoheterocycle with* N-*Aryl Glycine Esters*

Tang and his co-worker reported a Cu(I)-catalyzed cross-dehydrogenative coupling reaction between imidazopheterocycles and *N*-aryl glycine esters under aerobic atmosphere (Scheme 21) [18]. CuCl is the best catalyst among Cu(I) catalysts, whereas Cu(II) is totally ineffective. Aerial oxygen has been proven to be an effective oxidant for this transformation. This methodology is suitable for different imidazo[1,2-*a*]pyridines and imidazo[2,1-*b*]thiazoles. Although benzo[*d*]imidazo

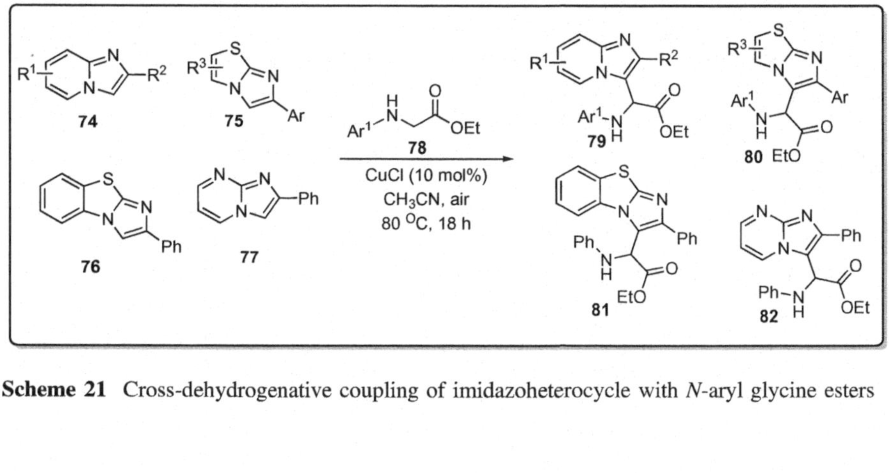

Scheme 21 Cross-dehydrogenative coupling of imidazoheterocycle with *N*-aryl glycine esters

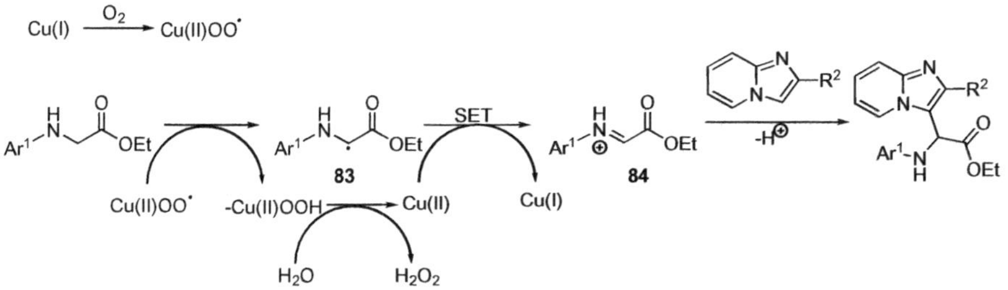

Scheme 22 Probable mechanism

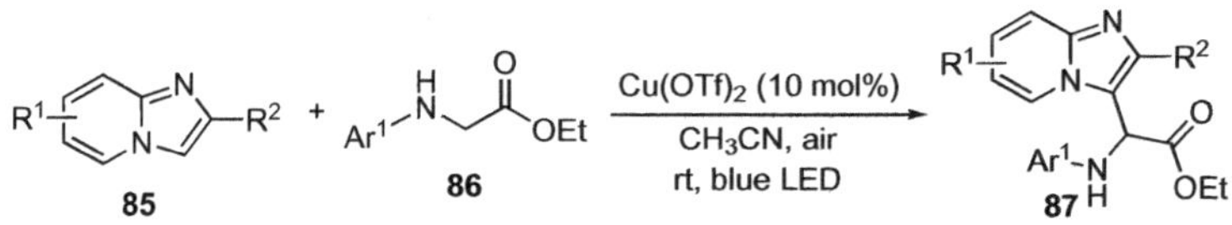

Scheme 23 CDC of imidazo[1,2-*a*]pyridines with *N*-arylglycine esters

[2,1-*b*]thiazole and imidazo[1,2-*a*]pyrimidine worked well, *N*-Boc, *N*-Bz, *N*-methyl protected or unprotected glycine esters did not react under the optimized reaction conditions.

The reaction proceeded through a radical pathway. The reaction is initiated by the formation of Cu(II) peroxide radical by the reaction between Cu(I) and oxygen (Scheme 22). Then Cu(II) peroxide radicals abstract a hydrogen atom to form Cu (II)-OOH and *N*-aryl glycine ester radical **83** which subsequently transformed into the imine cation intermediate **84**. Finally, the product is formed from the intermediate **84** through nucleophilic attack by the imidazo[1,2-*a*]pyridine moiety followed by the aromatization.

Later, Zhu et al. carried out the cross-dehydrogenative coupling of imidazopyridines with *N*-aryl glycine esters using $Cu(OTf)_2$ as the catalyst under the irradiation of 15 W blue LED (Scheme 23) under ambient air [19]. $Cu(OTf)_2$ is the best catalyst for this transformation among the catalysts such as $Ru(bpy)_3Cl_2{\cdot}6H_2O$, Ir $(ppy)_3$, Eosin Y, Acr^+-Mes-ClO_4^-, $Ni(OTf)_2$, $Zn(OTf)_2$, $CuBr_2$, $CuCl_2$ and Cu $(OTf)_2$. Synthetic utility of this strategy is explored employing different 2-aryl/alkyl-substituted imidazopyridines and *N*-aryl glycine esters.

2.7 *Cross-Dehydrogenative Coupling of Tertiary Amine with Imidazoheterocycle*

Recently, visible light has emerged as an energy source for the organic transformations [20]. Employment of this visible light has advantages like (i) operational simplicity, (ii) use of "almost infinitely" energy source, (iii) mild reaction conditions, and (iv) greater selectivity, etc. As a consequence, it becomes powerful strategy to enable versatile functionalization of inert C–H bonds. Chemists have successfully employed the visible light as energy source for the cross-dehydrogenative coupling reaction along with other strategies using visible light photoredox catalysts. Generally, transition-metal catalysts like Ru, Ir, Cu salts are employed as visible light photocatalysts [21]. Organic dyes such as rose bengal (RB), eosin Y, acridines, methylene blue, etc., have been also successfully explored as visible light organophotoredox catalysts [22]. Hajra et al. developed an efficient methodology for the cross-dehydrogenative coupling between imidazoheterocycles and tertiary amines employing rose bengal as photocatalysts (Scheme 24) [23]. Rose bengal exhibits higher catalytic activity compared to the transition-metal catalysts for this C(sp^3)-C(sp^2) cross-dehydrogenative coupling reaction. Only 2% rose bengal in toluene is required for this photo-induced reaction. Imidazo[1,2-*a*]pyridines and imidazo[1,2-*a*]pyrimidine successfully coupled with various *N*-aryl tetrahydroisoquinoline/*N*,*N*-dimethyl aniline derivatives under the irradiation of 34 W blue LED. This methodology is also applicable to other heterocycles such as indolizine, indole, and pyrrole. This efficient methodology has broad functional group compatibility and is amenable to gram-scale synthesis.

From the control experiments, it is confirmed the reaction proceeded through a singlet oxygen generation and the radical formation step is not the key step. The plausible mechanism of this reaction is depicted in Scheme 25. Rose bengal (RB) gets excited (RB*) in the presence of visible light. This excited RB* gets back its ground state by transferring its energy to the triplet oxygen which is transformed into singlet oxygen. This singlet oxygen converts amine to amine radical cation intermediate **93** and forms superoxide radical anion ($O_2^{\cdot-}$). The intermediate **93** gives up a hydrogen atom to generate the iminium intermediate **94** along with the formation of HO_2^-. In the next step, imidazo[1,2-*a*]pyridine attacks the intermediate **95** to form imidazolenium ion which upon deprotonation affords the product.

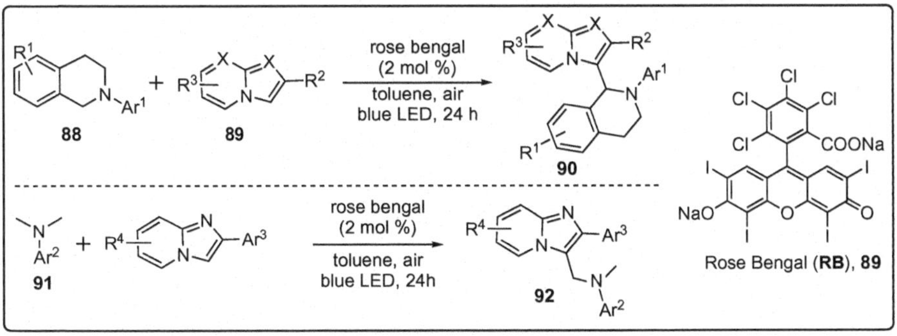

Scheme 24 C(sp^3)-C(sp^2) Cross-dehydrogenative coupling reaction

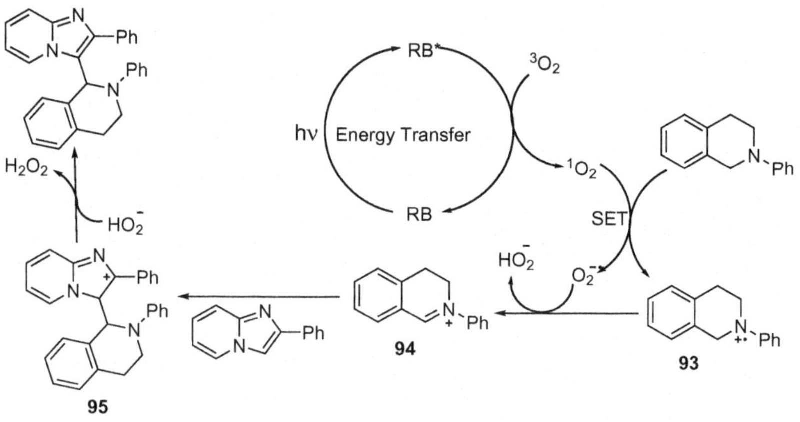

Scheme 25 Probable mechanism

2.8 *Cross-Dehydrogenative Coupling of Imidazo[1,2-*a*] Pyridines with Isochroman*

Wang et al. developed a Co-catalyzed cross-dehydrogenative coupling reaction between imidazo[1,2-*a*]pyridine and isochroman using molecular oxygen as oxidant (Scheme 26) [24]. The presence of aryl substituent at 2-position of imidazo [1,2-*a*]pyridine is essential for this transformation. The methodology exhibits a wide range of functional group tolerances and is applicable to gram-scale synthesis. However, isothiochroman, 1,3-benzodioxole, 1,4-dioxane, benzyl methyl ether, and tetrahydroisoquinoline are not suitable as coupling partners with imidazo[1,2-*a*] pyridine under the optimized reaction conditions.

They proposed two mechanistic pathways depending on the control experiments and literature reports on Co-chemistry (Scheme 27). Initially, the isochroman in the presence of Co-catalyst and oxygen is converted into oxonium ion intermediate **99** which readily reacts with the active nucleophile imidazo[1,2-*a*]pyridine to form the product along with H_2O. Alternative pathway is relying on the single electron

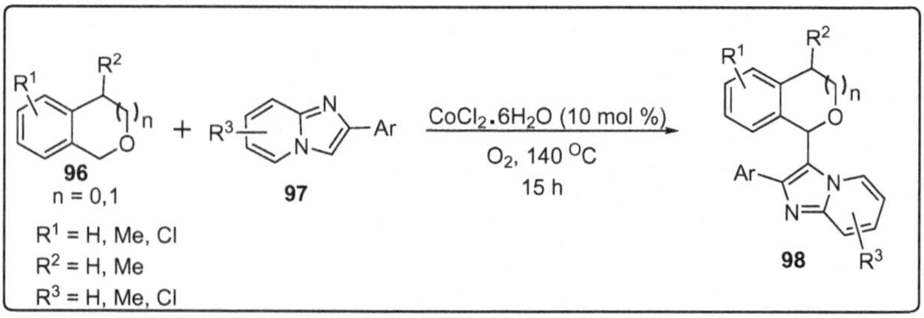

Scheme 26 CDC reaction between imidazo[1,2-*a*]pyridine and isochroman

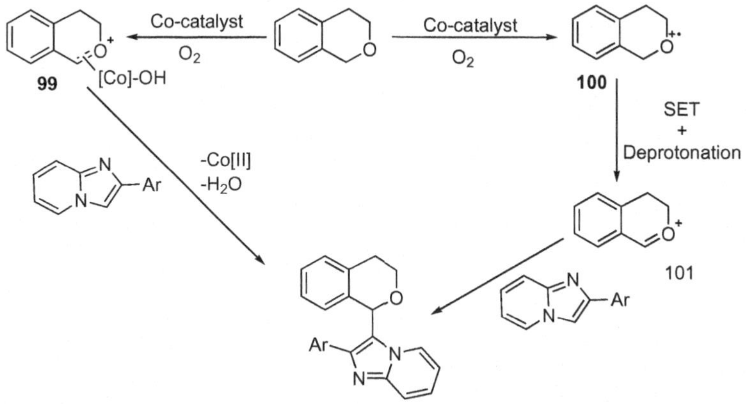

Scheme 27 Probable mechanistic pathways

transfer (SET). At first, the isochroman is converted into the intermediate **100** in the presence of Co-catalyst and O_2. In the next step, this intermediate **101** is transformed into the oxonium ion intermediate X through a SET and deprotonation. Finally, the product is obtained by sequential nuclophilic attack and deprotonation.

3 C–N Bond Forming Cross-Dehydrogenative Coupling Reaction

3.1 C–H Amination of Imidazo[1,2-a]Pyridines

Hajra et al. developed a transition-metal-free environmentally benign methodology for the direct coupling between cyclic secondary amines and imidazo[1,2-*a*] pyridines (Scheme 28) [25]. A mixture of cyclic secondary amines and imidazo [1,2-*a*]pyridines in 1,4-dioxane in the presence of 2 equiv. of PIDA afforded the coupling product at room temperature within 30 min. This methodology is tolerant with various functionalities such as methyl, methoxy, halides, trifluoromethyl, etc., and applicable to cyclic amines like morpholine, thiomorpholine, and piperidine.

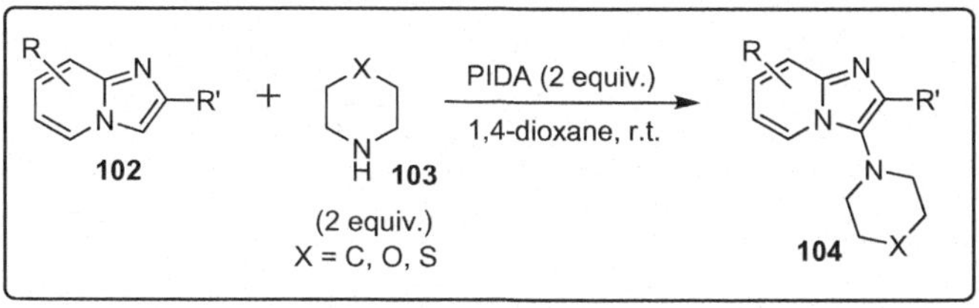

Scheme 28 C–H amination of imidazo[1,2-*a*]pyridines

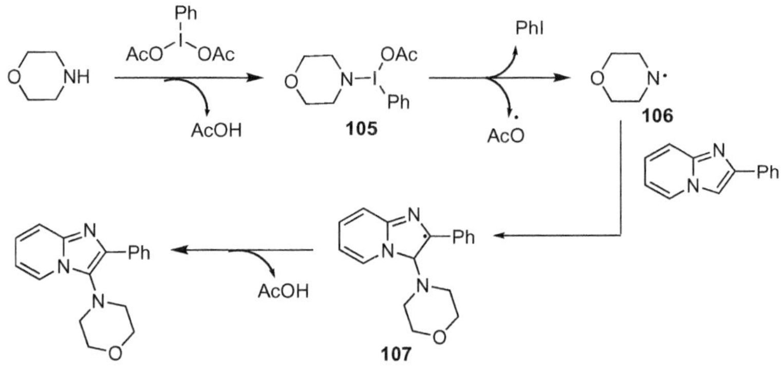

Scheme 29 Probable mechanism

No formation of product was observed in the presence of radical scavengers such as TEMPO, BHT, BQ, and DDQ. From these results, it is confirmed that the reaction proceeds through a radical pathway. The probable mechanism is depicted in Scheme 29. The reaction of a secondary amine with PIDA generates amine radical **106** through the initial formation of *N*-iodoamido species **105**. Then the amine radical **106** reacts with the imidazo[1,2-*a*]pyridines to form the radical intermediate **107**. This intermediate **107** afforded the product via elimination of H˙ along with the generation of AcOH.

3.2 C3–Sulfonamidation of Imidazopyridines with Sulfamides

Sun et al. developed a new and efficient methodology for the C-3 sulfonamidation of imidazopyridines through a visible light induced cross-dehydrogenative coupling reaction between imidazo[1,2-*a*]pyridines and sulfamides (Scheme 30) [26]. Optimum yield is obtained employing Ir{[(dF)CF_3ppy]$_2$(dtbbpy)}PF_6 as photocatalyst and NaClO as oxidant in 1,4-dioxane under irradiation by 5 W white LED under ambient air. Imidazo[1,2-*a*]pyridines with various functionalities such as

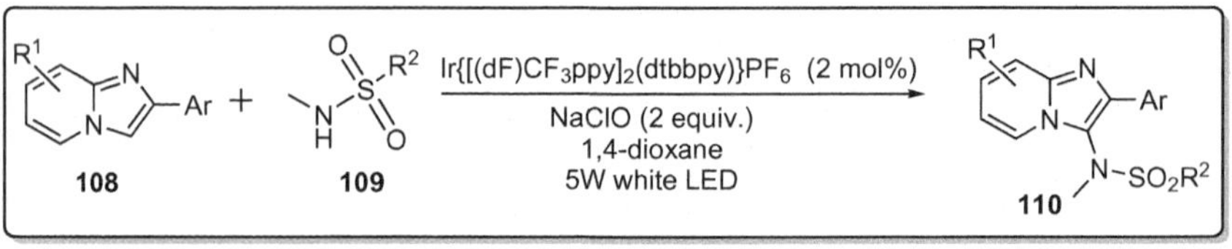

Scheme 30 C3-sulfonamidation of imidazopyridines

methyl, methoxy, *t*-butyl, halides, cyanide, etc. and other imidazoheterocycles, e.g., imidazo[1,2-*a*]pyrazine, imidazo[1,2-*a*]pyrimidine, benzo[*d*]imidazo[2,1-*b*]thiazole, and imidazo[2,1-*b*]thiazole underwent successfully to afford the product. This methodology is highly regioselective and sulfonamidation took place only at C-3 position of imidazo[1,2-*a*]pyridine.

3.3 Cross-Dehydrogenative Coupling of Imidazoheterocycles with Azoles

Lei et al. very recently developed a cross-dehydrogenative coupling between imidazoheterocycles and azoles through photoredox catalysis (Scheme 31) [27]. The combination of photoredox catalyst Acr^+-$MesClO_4^-$ and hydrogen evolution catalyst $Co(dmgH)(dmgH_2)Cl_2$ in DCE under 3 W blue LED acts as the best cooperative catalysts for this coupling reaction. This reaction is carried out under nitrogen atmosphere and no external oxidant is required and only hydrogen is produced as a by-product. This methodology is an application to imidazoheterocycles like imidazo[1,2-*a*]pyridines and benzo[*d*]imidazo[2,1-*b*]thiazoles and azoles such as pyrazole, imidazole, 1,2,3-triazole, and 1,2,4-triazole.

3.4 Transition-Metal-Free CDC of Imidazo[1,2-a]Pyridines with Azoles

Adimurthy et al. reported a transition-metal-free cross-dehydrogenative coupling between imidazo[1,2-*a*]pyridines and azoles under visible light photocatalysis (Scheme 32) [28]. Only 5 mol% of 9-mesityl-10-methylacridinium perchlorate (Acr^+-Mes) is sufficient as photocatalyst in the presence of $K_2S_2O_8$ (oxidant, 2 equiv.) under irradiation of a 12 W blue LED strips to carry out this transformation.

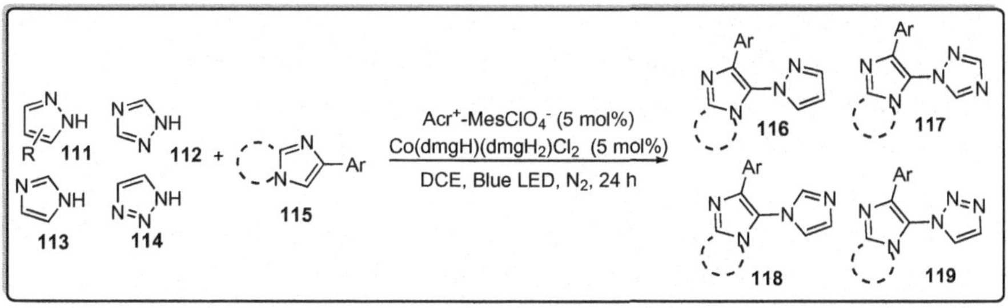

Scheme 31 CDC reaction between imidazoheterocycles and azoles

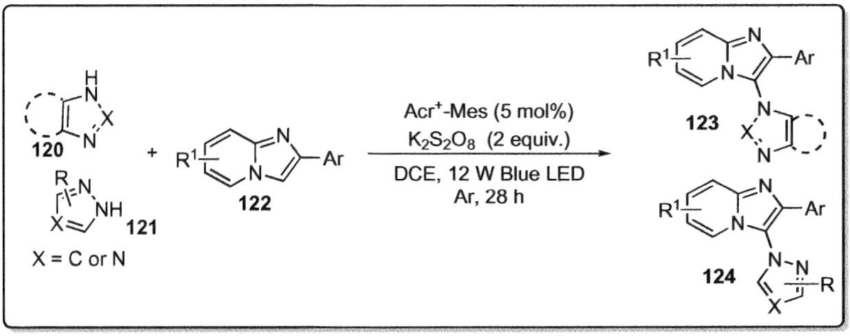

Scheme 32 Transition-metal-free CDC of imidazo[1,2-*a*]pyridines with azoles

This methodology is applicable to different azoles like benzotriazole, pyrazole, imidazole, 1*H*-1,2,4-triazole, and 1*H*-benzo[*d*]imidazole and in all the cases, corresponding C-3 amination products are obtained regioselectively.

*3.5 C-3 Amidation of Imidazo[1,2-*a*]Pyridines*

The amidation of imidazo[1,2-*a*]pyridines with saccharin through cross-dehydrogenative coupling was carried out by Sun et al. (Scheme 33) [29]. This Cu(II)-catalyzed reaction has been developed using selectfluor (2 equiv.) as an oxidant and Na_2CO_3 (2 equiv.) as a base. Various imidazo[1,2-*a*]pyridines bearing halogen, trifluoromethyl, and heteroaryl functionalities are well tolerated under the reaction conditions. The reaction proceeded through a radical pathway. Initially, the imidyl radical is formed from the saccharin in the presence of copper salt and selectfluor. In the next step, this radical reacts with the imidazo[1,2-*a*]pyridine to form the radical intermediate **128**. Finally, the product is obtained from the intermediate **128** via oxidative aromatization.

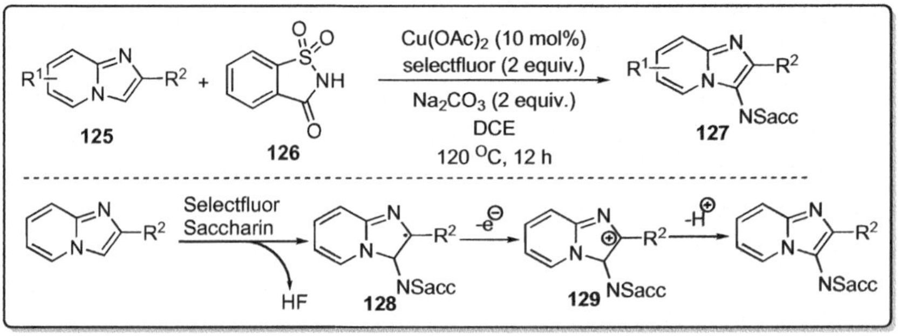

Scheme 33 C3-amidation of imidazo[1,2-*a*]pyridines

4 C–O Bond Forming Cross-Dehydrogenative Coupling Reaction

A continuous effort has been paid on the C–O bond formation as they are widely present in natural products, bioactive molecules, and functional materials. The main strategies for the C–O bond formation include Williamson synthesis, i.e., coupling of alkoxides and alkyl halides, acid-catalyzed condensation of alcohols, alkoxymercuration/demercuration of alkenes, etc. Recently, direct C–O bond formation via C–H functionalization has gained much attention as it is straightforward and environmentally benign approach.

4.1 Alkoxylation of Imidazoheterocycles

Imidazo[1,2-*a*]pyridines bearing alkoxyl group at C-3 position are used as effective inhibitors of mycobacterial adenosine triphosphate (ATP) synthesis. It is also used for the measurement of the luciferase activity in living cells [30]. Hajra et al. reported an efficient methodology for the C-3 alkoxylation of imidazo[1,2-*a*] pyridines (Scheme 34) [31]. They employed visible light photocatalysis in the cross-dehydrogenative coupling reaction between imidazo[1,2-*a*]pyridines and alcohols using rose bengal as an organic photoredox catalyst. The transformation requires only 2 mol% of the photocatalyst without using any external oxidant. A wide range of alcohols such as alkyl, allyl, propargyl, benzyl alcohol, etc., successfully afforded the product under the irradiation of 20 W LED bulb. The reaction is highly selective; only alkoxylation takes place in case of 2-bromo ethanol, and in the case of ethylene glycol, only mono alkoxylation takes place. In addition, the alkoxylation of benzo[*d*]imidazo[2,1-*b*]thiazoles has been carried out employing this methodology.

The probable mechanism of the reaction is represented in Scheme 35. From the control experiment, it is confirmed that reaction proceeded through a radical pathway. Rose bengal gets excited under visible light irradiation. The excited rose

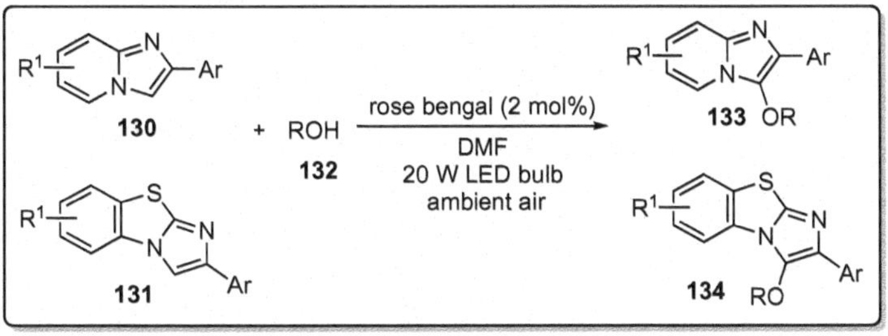

Scheme 34 Etherification of imidazoheterocycles

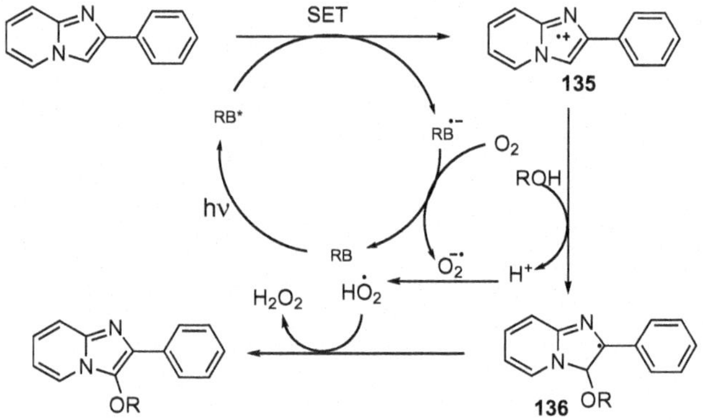

Scheme 35 Probable mechanism

bengal RB* accepts an electron from imidazopyridine to form the intermediate **135** and rose bengal radical anion ($RB^{\cdot-}$). The intermediate **135** upon nuclophilic attack by the alcohol forms the intermediate **136**. Finally, the product is obtained from the intermediate **136** though hydrogen abstraction by $HO_2^{\cdot}$.

5 C–S Cross-Dehydrogenative Coupling Reaction

Direct C–S bond formation is very much important due to the importance of organosulfur compounds in biological applications and pharmaceuticals. Heteroaryl sulfides such as 3-sulfenyl indoles, 3-sulfenyl pyrroles, and 3-sulfenyl imidazopyridines exhibit considerable therapeutic value against a variety of diseases [32]. Thioether containing imidazo[1,2-*a*]pyridines are highly active against human cytomegalovirus and varicella-zoster virus [33].

5.1 Sulfenylation of Imidazoheterocycles with Thiophenols

Various sulfenylating agents such as disulfides, arylsulfonyl chlorides, sodium arylsulfinates, sulfinic acids, arylsulfonyl hydrazides, and thiols have been employed for the sulfenylation of imidazoheterocycles over the years [34]. Among these, the direct coupling of imidazoheterocycles with thiols is more appealing as no pre-functionalization of the thiols is required. Hamdouchi et al. for the first time developed a methodology for the sulfenylation of 2-amino imidazo[1,2-*a*]pyridine derivatives employing thiol as sulfenylating agent (Scheme 36) [35]. Kita's reagent, phenyliodine(III) bis(trifluoroacetate) (PIFA) in 1,1,1,3,3,3-hexafluoropropan-2-ol (HFIP), can efficiently catalyze the reaction between various 2-amino imidazo[1,2-*a*]

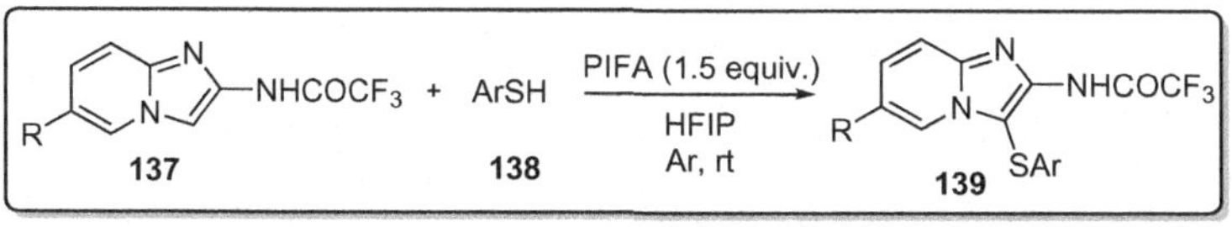

Scheme 36 PIFA-mediated sulfenylation of imidazo[1,2-*a*]pyridines

pyridines and aryl thiols. The reaction is highly chemo- and regioselective; sulfenylation only takes place at 3-position.

Adimurthy et al. reported a direct sulfenylation of imidazoheterocycles with thiols through cross-dehydrogenative coupling reaction (Scheme 37) [36]. They demonstrated that *N*-chloroscuccinamide (NCS) in DCM can efficiently couple imidazoheterocycles and aryl thiols at room temperature. *N*-Bromoscuccinamide (NBS) or *N*-iodoscuccinamide (NIS) is not suitable for this transformation as diaryl disulfide is formed as the by-product in these cases. This is a highly regioselective methodology and only 3-sulfenylated product is obtained from the unsubstituted imidazo[1,2-*a*]pyridine, whereas no product is formed from 3-substituted imidazo [1,2-*a*]pyridine. This metal-free methodology is applicable to wide range of imidazo[1,2-*a*]pyridines bearing aryl, heteroaryl, alkenyl, and alkyl group at 2-position. Imidazo[2,1-*b*]thiazoles and 2-phenylbenzo[*d*]imidazo[2,1-*b*]thiazole derivatives also reacted well under the reaction conditions. In addition, aliphatic secondary thiol reacted well with imidazo[1,2-*a*]pyridine and imidazo[2,1-*b*]thiazole albeit moderate yield is obtained.

The probable mechanism of this NCS-mediated reaction is outlined in Scheme 38. At first, thiol reacts with the NCS to form the sulfenyl chloride intermediate **146** which on subsequent nucleophilic attack by the imidazopyridine moiety forms the imidazolenium intermediate **147**. The sulfenylated imidazopyridine is obtained from the intermediate **147** via dehydrochlorination.

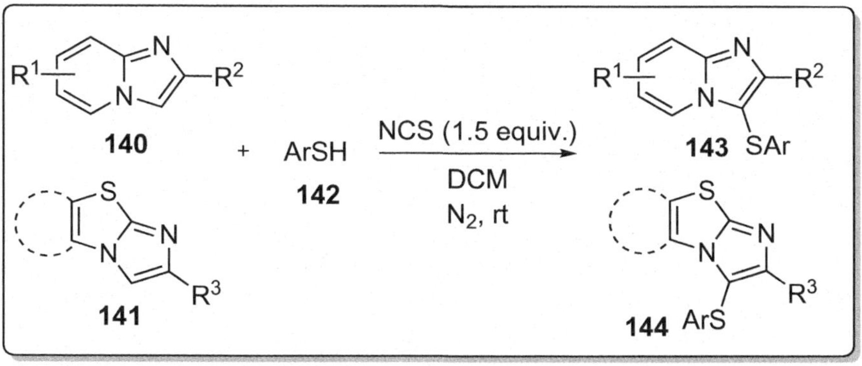

Scheme 37 NCS-mediated sulfenylation of imidazoheterocycles

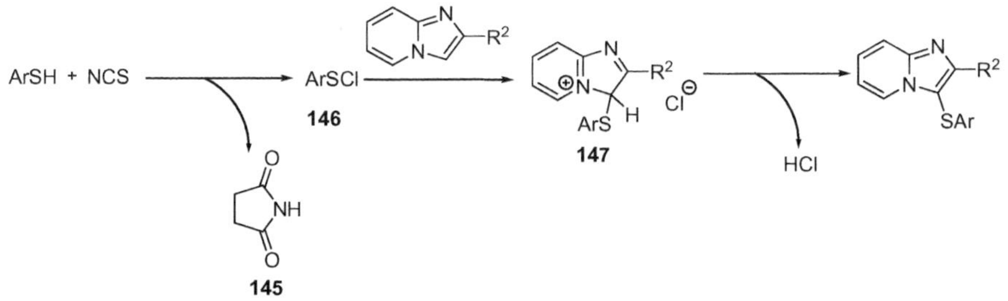

Scheme 38 Probable mechanism of NCS-mediated sulfenylation of imidazoheterocycles

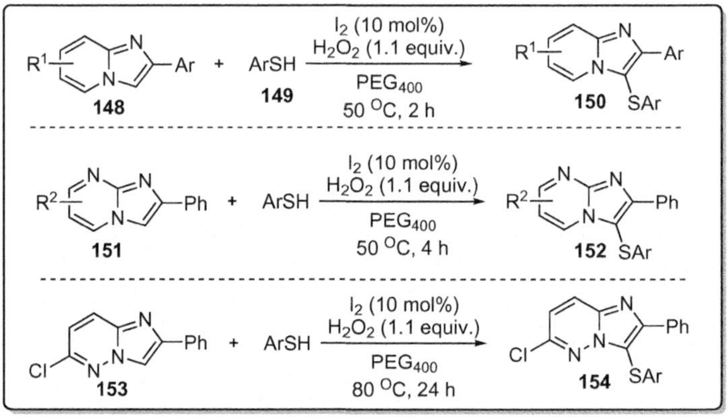

Scheme 39 Iodine-catalyzed regioselective sulfenylation of imidazoheterocycles

Later Hiebel et al. modified this sulfenylation strategy employing iodine catalysis (Scheme 39) [37]. They demonstrated that the combination of 10 mol% of I_2 can act as an efficient catalyst for this transformation. H_2O_2 in PEG_{400} has been employed as an oxidant. Only water is produced as the by-product in this transformation. The employment of green solvent (PEG_{400}) makes this approach a green methodology. Various imidazopyridines and aryl thiols reacted well under the reaction conditions to produce the product with excellent yields. Other imidazoheterocycles such as imidazo[1,2-*b*]pyridazines and imidazopyrimidine also afforded the corresponding sulfenylated products although higher temperature and longer reaction time are required in these cases.

The possible mechanism of this iodine-catalyzed reaction is represented in Scheme 40. This reaction is initiated by the formation of aryl disulfide **155** from the thiol by the reaction with I_2/H_2O_2. This disulfide (**155**) on reaction with I_2 forms the intermediate **156**. In the next step, imidazopyridine moiety gets attacked by the electrophilic intermediate **156** to form the imidazolenium intermediate **157**. Finally, product is obtained from the imidazolenium intermediate **157** through the release of HI which reacts with H_2O_2 to regenerate I_2 and complete the catalytic cycle.

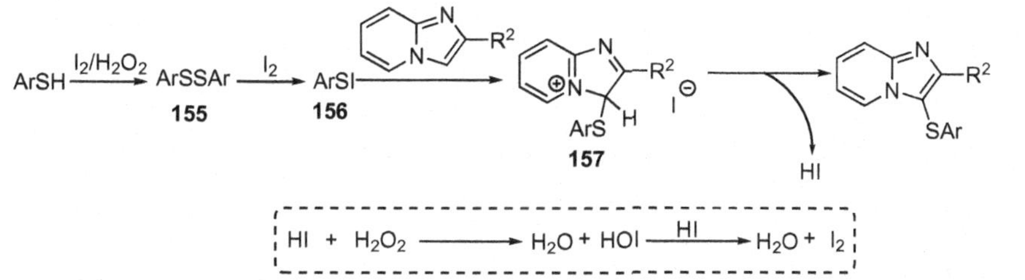

Scheme 40 Probable mechanism of the iodine-catalyzed sulfenylation of imidazoheterocycles

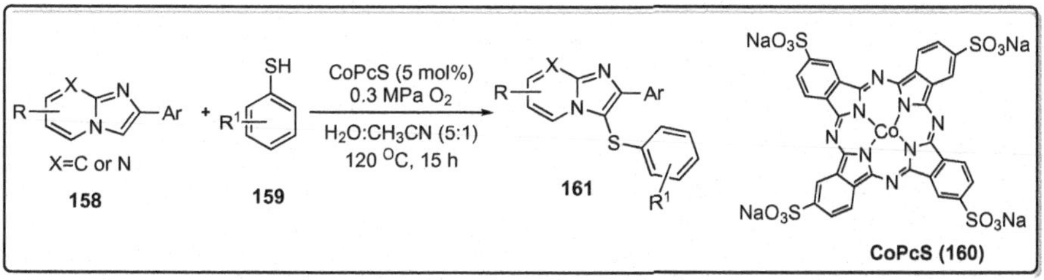

Scheme 41 CoPcS-catalyzed sulfenylation of imidazoheterocylces with thiophenols

The employment of water as solvent is highly desirable from the viewpoint of economic as well as environmental issue. So, the development of cross-dehydrogenative coupling in water is an important achievement from the aspect of green chemistry. From the view point of this, Yuan et al. developed a cross-dehydrogenative coupling reaction for the C–S bond formation in aqueous medium (Scheme 41) [38]. They reported that Cobalt(II) phthalocyanine tetrasodium sulfonate (CoPcS, **160**) can act as efficient catalyst under oxygen atmosphere in water for the direct coupling of aryl C–H and S–H for the synthesis of aryl sufide. They employed this strategy for the direct C–H functionalization of imidazo [1,2-*a*]pyridines with thiophenols under aqueous medium. Thiophenols with electron-donating as well as electron-withdrawing groups efficiently reacted with the imidazo[1,2-*a*]pyridines to afford sulfenylated product with excellent yield. This methodology is also applicable toward the sulfenylation of imidazo[1,2-*a*]pyrimidine with thiophenol. The reaction proceeded through a radical pathway and formation of disulfide as intermediate.

5.2 *Sulfenylation of Imidazoheterocycles with Aliphatic Thiols*

Although the previously described methodologies are highly useful for the sulfenylation of imidazoheterocylces using thiophenols, these are not useful for the

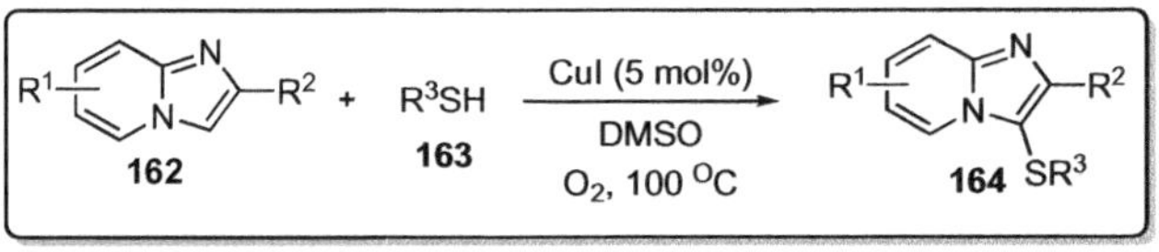

Scheme 42 Cu(I)-catalyzed sulfenylation of imidazoheterocylces using aliphatic thiols

direct coupling between imidazoheterocylces and aliphatic thiols. NCS-mediated methodology is only applicable for aliphatic secondary thio-2-propanol, and however, moderate yield is obtained. Cao et al. developed Cu(I)-catalyzed efficient methodology for the direct coupling between imidazo[1,2-*a*]pyridines and aliphatic thiols (Scheme 42) [39]. They demonstrated that CuI can efficiently catalyze this transformation under oxygen atmosphere in DMSO at 100 °C. This methodology is applicable for different imidazo[1,2-*a*]pyridines and primary and secondary thiols such as dodecane-1-thiol, butane-1-thiol, propane-1-thiol, ethanethiol, phenylmethane thiol, propane-2-thiol, and cyclohexane thiol. However, tertiary thiol is not an effective substrate for this coupling reaction probably due to steric hindrance.

Probable mechanism of this direct sulfenylation is represented in Scheme 43. Initially, the CuI reacts with the thiol RSH to form the species Cu(I)SR (**165**). This Cu(I)SR reacts with imidazo[1,2-*a*]pyridine moiety to generate the intermediate **166** through a concerted metalation–deprotonation. Finally, the product is obtained from the intermediate **166** via reductive elimination under oxygen atmosphere.

They also extend their methodology for the sulfenylation of imidazo[2,1-*b*]thiazoles and benzo[*d*]imidazo[2,1-*b*]thiazoles (Scheme 44) [40]. 20 mol% CuI and 20 mol% phenanthroline in DMF under oxygen atmosphere efficiently catalyzed

Scheme 43 Probable mechanism of Cu(I)-catalyzed sulfenylation of imidazopyridines

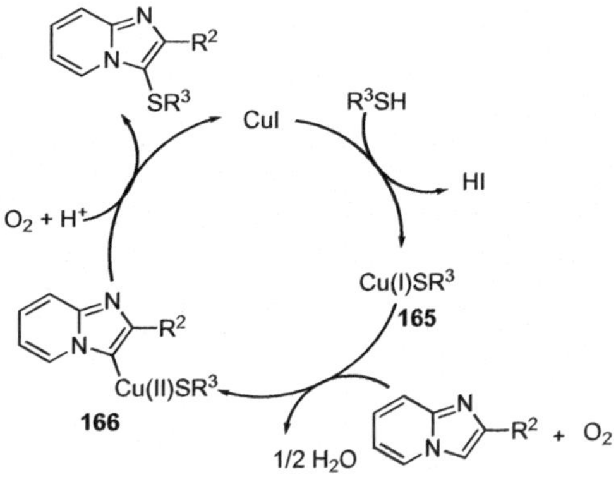

Scheme 44 C-5 Sulfenylation of imidazo[2,1-*b*]thiazoles

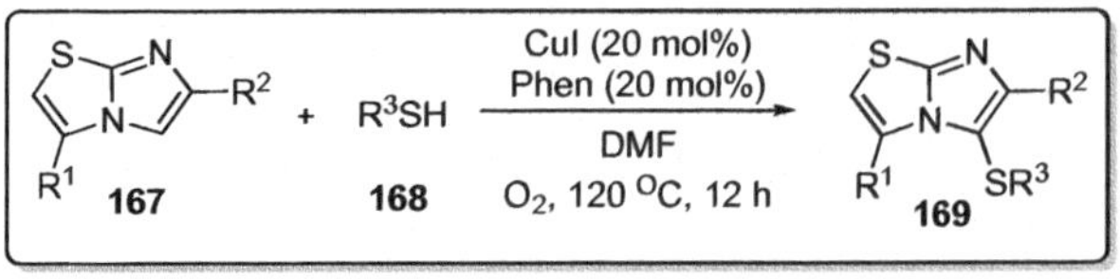

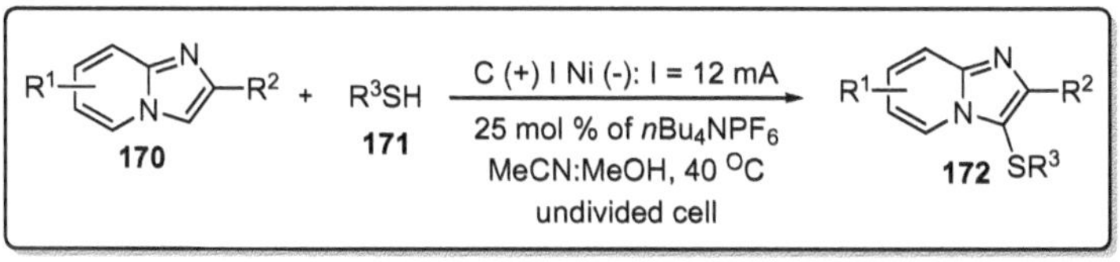

Scheme 45 Electrochemical sulfenylation of imidazo[1,2-*a*]pyridines

the direct coupling between imidazo[2,1-*b*]thiazoles and thiols. In this reaction, ligand plays a crucial role to increase the yield and among the ligands such as DMEDA, TMEDA, phenanthroline, pyridine, and bipyridine, phenanthroline is proven to be the best one. This strategy is applicable to a wide range of thiols including alkyl, aryl, and heteroaryl thiols. Interestingly, ethane-1,2-dithiol reacts with both ends to form 1,2-bis(imidazo[2,1-*b*]thiazol-5-ylthio)ethane. This methodology is highly regioselective for the synthesis of C-5 sulfenated imidazo [2,1-*b*]thiazoles.

Lei et al. modified the sulfenylation of imidazopyridines employing electrochemical oxidative coupling between imidazopyridines and thiols (Scheme 45) [41]. The best result is obtained utilizing 12 mA constant current with carbon rod anode and nickel plate cathode in an undivided cell and 25 mol% of n-Bu_4NPF_6 as electrolyte in MeCN:MeOH at 40 °C. Methodology is applicable to a wide range of aryl/alkyl thiols and imidazopyridines with aryl functionalities at 2-position. Unsubstituted imidazopyridines also undergo this sulfenylation albeit affording low yield. Only hydrogen is produced as the by-product in this electrochemical transformation.

5.3 *Sulfenylation of Imidazoheterocycles with Heteroaryl Thiols*

Zheng et al. developed a Cu-catalyzed efficient strategy for the direct coupling between imidazopyridines and thiols (Scheme 46) [42]. They reported that the best result is obtained carrying out the reaction in the presence of 5 mol% $Cu(OAc)_2$ in dimethyl acetamide (DMA) under oxygen atmosphere at 80 °C. This copper-catalyzed strategy is susceptible to various heteroaryl thiols such as benzo [*d*]thiazole-2-thiol, benzo[*d*]oxazole-2-thiol, 1-methyl-1*H*-imidazole-2-thiol, and pyrimidine-2-thiol. Additionally, they extended this methodology for the sulfenylation of imidazopyridines using thiophenols. Alphatic thiols like ethanethiol, propane-1-thiol, and butane-1-thiol, etc., efficiently react with unsubstituted imidazopyridine to afford regioselectively 3-sulfenylated imidazopyridines. This strategy is also applicable for C-5 sulfenylation of imidazo[2,1-*b*]thiazole albeit moderate yield is obtained.

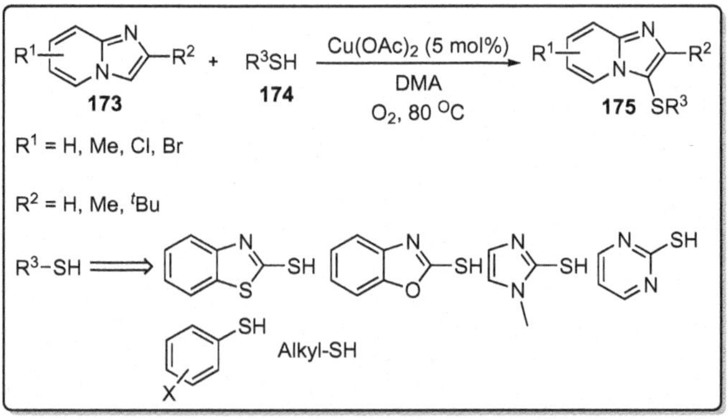

Scheme 46 Cu-catalyzed sulfenylation of imidazopyridines using thiols

Prabhu et al. demonstrated that the direct coupling of heteroaryl thiols and imidazopyridines can be carried out under transition-metal-free conditions (Scheme 47) [43]. Only 20 mol% of I_2 in DMSO at 80 °C temperature efficiently catalyzed this reaction. Under the optimized reaction conditions, various imidazopyridines with 2-aryl/heteroaryl substituents react well to afford the diheteroaryl sulfide. Different heteroaryl thiols/thiones such as 1*H*-tetrazole-5-thiol and 1,3,4-thiadiazole-2-thiol are susceptible to produce the products with ease. In addition, this methodology is also applicable to pyridine-2-thiol, pyrimidine-2-thiol, and thiones derivatives like benzo[*d*]-thiazole-2(3*H*)-thione, 4-methylthiazole-2 (3*H*)-thione, and benzo[*d*]oxazole-2(3*H*)-thione. However, one equivalent of I_2 is necessary in these cases. This methodology is also extendable toward the sulfenylation of imidazo[2,1-*b*]thiazoles and imidazo[1,2-*a*]pyrimidine.

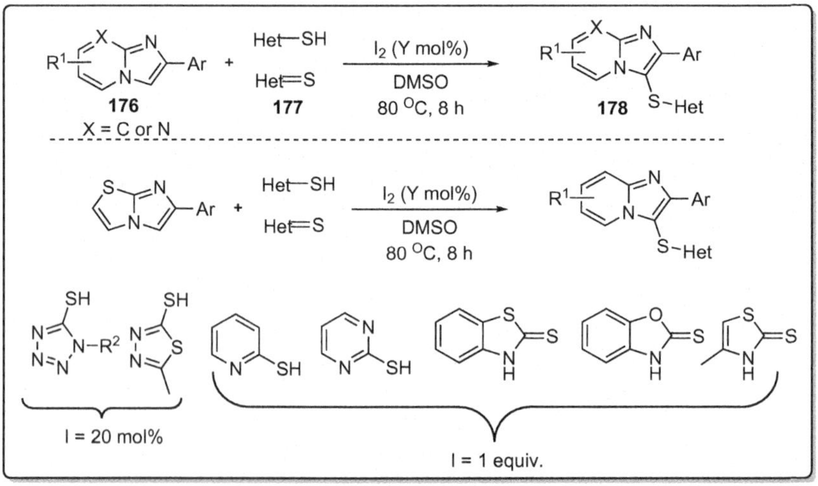

Scheme 47 I_2-catalyzed sulfenylation of imidazopyridines using heteroaryl thiols/thiones

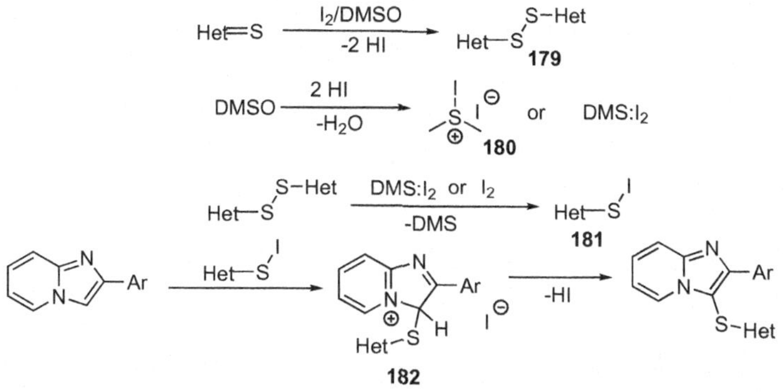

Scheme 48 Plausible mechanism of I_2-catalyzed sulfenylation of imidazopyridines using heteroaryl thiols/thiones

The probable mechanism of this I_2-catalyzed reaction is represented on the basis of control experiments (Scheme 48). This reaction proceeds through a non-radical pathway and DMSO acts as oxidant. From the NMR studies, it is confirmed that disulfide acts as the active intermediate in this reaction. Initially, disulfide intermediate **179** is formed from the thiol by the reaction with I_2/DMSO. This disulfide intermediate **179** is transformed into the intermediate **181** by the reaction with I_2 or DMS:I_2 (**180**, in situ generated by the reaction DMSO with I_2). In the next step, nucleophilic imidazopyridine moiety attacks the highly electrophilic intermediate **181** to form the imidazolenium intermediate **182** which upon elimination of HI affords the final product.

6 C–P Cross-Dehydrogenative Coupling Reaction

Cross-dehydrogenative coupling has also been employed for the C–P bond formation due to the importance of phosphonated heteroaryls in pharmaceutical. Singh et al. demonstrated a Mn(III)-mediated coupling between imidazo[1,2-*a*]pyridines and dialkyl phosphites (Scheme 49) [44]. The best result was obtained by carrying

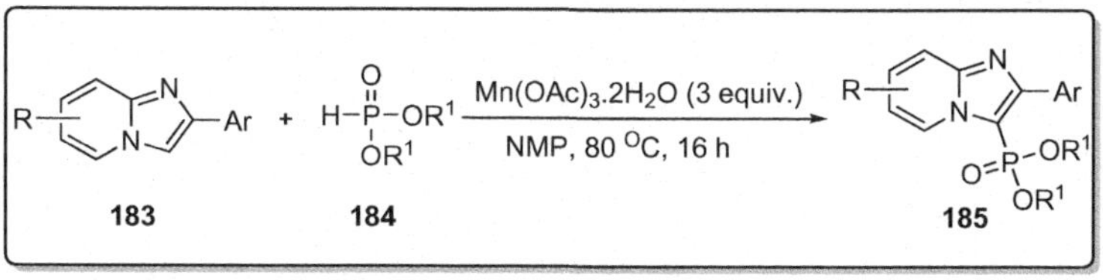

Scheme 49 Cross-dehydrogenative C-P bond formation

out the reaction employing $Mn(OAc)_3 \cdot 2H_2O$ in NMP at 80 °C. This methodology is suitable to different imidazo[1,2-*a*]pyridines and alkyl phosphate like *iso*-propyl, ethyl, methyl, *t*-butyl, benzyl, etc. Reaction proceeded through a radical pathway.

7 Miscellaneous

Apart from the direct functionalization of imidazoheterocycles, the cross-dehydrogenative coupling has been employed for the synthesis of polycyclic heterocycles bearing imidazoheterocycles as the core structure.

*7.1 Synthesis of Azole-Fused Imidazo[1,2-*a*]Pyridines*

Kumar et al. developed an efficient methodology for the synthesis of azole-fused imidazopyridines through a one-pot sequential Ulmann coupling followed by intramolecular cross-dehydrogenative coupling (Scheme 50) [45]. They demonstrated that the reaction of 2-(2-bromophenyl)imidazo[1,2-*a*]pyridine moiety with azoles in the presence of 10 mol% CuI and 1.5 equiv. K_2CO_3 in DMF at 150 °C affords the C–N coupling product and finally in the presence of 5 mol% $Pd(OAc)_2$ as catalyst and 1.2 equiv. $Cu(OAc)_2$ as oxidant in DMF at 150 °C it produces the azole-fused imidazopyridine derivatives. This methodology is susceptible to different azoles such as imidazoles, benzimidazole, 1,2,4-triazoles, and 1*H*-imidazo [4,5-*b*]pyridine etc. and imidazo[1,2-*a*]pyridine derivatives with functionalities like methyl, chloro, fluoro.

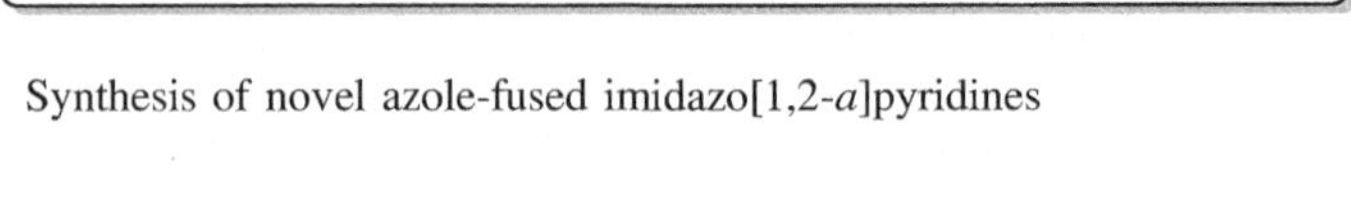

Scheme 50 Synthesis of novel azole-fused imidazo[1,2-*a*]pyridines

7.2 Synthesis of Polycyclic Heterocycles via Dehydrogenative Coupling

Gryko et al. reported the synthesis of fused polycyclic aza-heterocycle from imidazo[1,2-*a*]pyridines and imidazo[1,5-*a*]pyridines through an anion–radical coupling (Scheme 51) [46]. This is a two-step synthesis; in the first step, they synthesized various arylated imidazopyridines from imidazopyridines through Pd-catalyzed arylation. Final step is the intramolecular dehydrogenative coupling of the synthesized arylated imidazopyridines to form the polycylic aza-heterocycle through a K-mediated reaction. They also studied the optical properties of the synthesized polycyclic molecules.

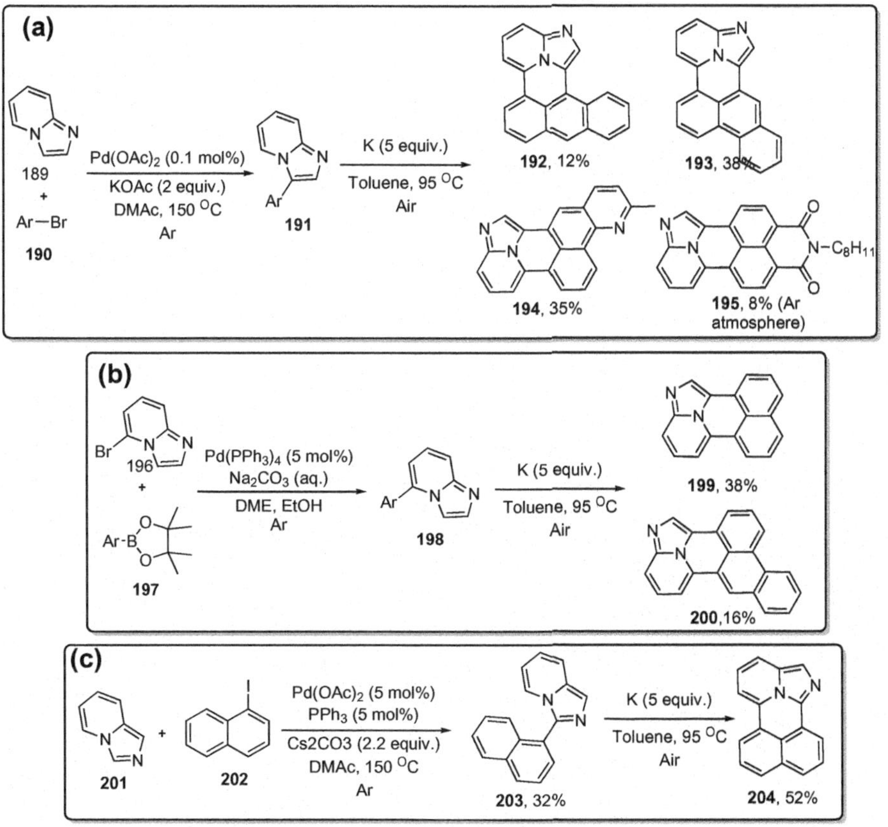

Scheme 51 Synthesis of fused polycyclic aza-heterocycle

8 Concluding Remarks

In this chapter, we have discussed the various methodologies for the functionalization of imidazoheterocycle derivatives through cross-dehydrogenative coupling reaction. Various C–C, C–N, C–O, C–S, and C–P bond forming reactions have been developed via different strategies. These strategies include transition-metal-catalyzed reactions, iodine catalysis, visible light photocatalysis, electrochemical synthesis, etc. Continuous efforts have been paid on the development of cross-dehydrogenative coupling reaction toward the green synthesis. We believe that this discussion will help the chemist for the development of new methodologies employing cross-dehydrogenative coupling reaction toward the synthesis and functionalization of biologically relevant heterocycles.

References

1. (a) Enguehard-Gueiffier C, Gueiffier A (2007) Recent progress in the pharmacology of imidazo[1,2-*a*]pyridines. Mini-Rev Med Chem 7:888–899. (b) Andreani A, Burnelli S, Granaiola M, Leoni A, Locatelli A, Morigi R, Rambaldi M, Varoli L, Calonghi N, Cappadone C, Farruggia G, Zini M, Stefanelli C, Masotti L, Radin NS, Shoemaker RH (2008) New antitumor imidazo[2,1-*b*]thiazole guanylhydrazones and analogues. J Med Chem 51:809–816
2. (a) Harrison TS, Keating GM (2005) Zolpidem: a review of its use in the management of insomnia. CNS Drugs 19:65–89. (b) Hsu N, Jha SK, Coleman T, Frank MG (2009) Paradoxical effects of the hypnotic zolpidem in the neonatal ferret. Behav Brain Res 201:233–236
3. (a) Shao N, Pang G-X, Yan C-X, Shi G-F, Cheng Y (2011) Reaction of β-lactam carbenes with 2-pyridyl isonitriles: a one-pot synthesis of 2-carbonyl-3-(pyridylamino)imidazo[1,2-*a*] pyridines useful as fluorescent probes for mercury ion. J Org Chem 76:7458–7465. (b) Stasyuk AJ, Banasiewicz M, Cyrański MK, Gryko DT (2012) Imidazo[1,2-*a*]pyridines susceptible to excited state intramolecular proton transfer: one-pot synthesis *via* an ortoleva-king reaction. J Org Chem 77:5552–5558
4. Egner U, Gerbling KP, Hoyer G-A, Krüger G, Wegner P (1996) Design of inhibitors of photosystem II using a model of the D1 protein. Pestic Sci 47:145–158
5. John A, Shaikh MM, Ghosh P (2009) Palladium complexes of abnormal *N*-heterocyclic carbenes as precatalysts for the much preferred Cu-free and amine-free sonogashira coupling in air in a mixed-aqueous medium. Dalton Trans 10581–10591
6. (a) Bagdi AK, Santra S, Monir K, Hajra A (2015) Synthesis of imidazo[1,2-*a*]pyridines: a decade update. Chem Commun 51:1555–1575. (b) Bagdi AK, Hajra A (2016) Design, synthesis, and functionalization of imidazoheterocycles. Chem. Rec. 16:1868–1885. (c) Pericherla K, Kaswan P, Pandey K, Kumar A (2015) Recent developments in the synthesis of imidazo[1,2-*a*]pyridines. Synthesis 47:887–912. (d) Koubachi J, Kazzouli SE, Bousmina M, Guillaumet G (2014) Functionalization of imidazo[1,2-*a*]pyridines by means of metal-catalyzed cross-coupling reactions. Eur J Org Chem 5119–5138
7. Koubachi J, Berteina-Raboin S, Mouaddib A, Guillaumet G (2009) Pd/Cu-catalyzed oxidative C–H alkenylation of imidazo[1,2-*a*]pyridines. Synthesis 2:0271–0276
8. Zhan H, Zhao L, Li N, Chen L, Liu J, Liao J, Cao H (2014) Ruthenium-catalyzed direct C-3 oxidative olefination of imidazo[1,2-*a*]pyridines. RSC Adv 4:32013–32016

9. Cao H, Lei S, Liao J, Huang J, Qiu H, Chen Q, Qiu S, Chen Y (2014) Palladium(II)-catalyzed intermolecular oxidative C-3 alkenylations of imidazo[1,2-*a*] pyridines by substrate-contolled regioselective C–H functionalization. RSC Adv. 4:50137–50140
10. Ghosh M, Naskar A, Mitra S, Hajra A (2015) Palladium-catalyzed α-selective alkenylation of imidazo[1,2-*a*]pyridines through aerobic cross-dehydrogenative coupling reaction. Eur J Org Chem 715–718
11. Wang S, Liu W, Cen J, Liao J, Huang J, Zhan H (2014) Pd-catalyzed oxidative cross-coupling of imidazo[1,2-*a*]pyridine with arenes. Tetrahedron Lett 55:1589–1592
12. Lei S, Cao H, Chen L, Liu J, Cai H, Tana J (2015) Regioselective oxidative homocoupling reaction: an efficient copper-catalyzed synthesis of biimidazo[1,2-*a*]pyridines. Adv Synth Catal 357:3109–3114
13. Shakoor SMA, Mandal SK, Sakhuja R (2017) An articulate oxidative transition-metal-free homocoupling of imidazo heterocycles through C(sp^2)-C(sp^2) bond formation. Eur J Org Chem 2596–2602
14. (a) Jiang Y, Xu K, Zeng C (2018) Use of electrochemistry in the synthesis of heterocyclic structures. Chem Rev 118:4485–4540. (b) Waldvogel SR, Lips S, Selt M, Riehl B, Kamp CJ (2018) Electrochemical arylation reaction. Chem Rev 118:6706–6765
15. Gao Y, Wang Y, Zhou J, Mei H, Han J (2018) An electrochemical oxidative homocoupling reaction of imidazopyridine heterocycles to biheteroaryls. Green Chem 20:583–587
16. Shakoor SMA, Agarwal DS, Kumar A, Sakhuja R (2016) Copper catalyzed direct aerobic double-oxidative cross-dehydrogenative coupling of imidazoheterocycles with aryl acetaldehydes: an articulate approach for dicarbonylation at C-3 position. Tetrahedron 72:645–652
17. Samanta S, Mondal S, Santra S, Kibriya G, Hajra A (2016) $FeCl_3$-catalyzed cross-dehydrogenative coupling between imidazoheterocycles and oxoaldehydes. J Org Chem 81:10088–10093
18. Jiao J, Zhang J-R, Liao Y-Y, Xu L, Hu M, Tang R-Y (2017) CuCl/air-mediated oxidative coupling reaction of imidazoheterocycles with *N*-Aryl glycine esters. RSC Adv 7:30152–30159
19. Zhu Z-Q, Xiao L-J, Zhou C-C, Song H-L, Xie Z-B, Le Z-G (2018) A visible-light-promoted cross-dehydrogenativecoupling reaction of *N*-arylglycine esters with imidazo[1,2-*a*]pyridines. Tetrahedron Lett 59:3326–3331
20. (a) Yoon TP, Ischay MA, Du J (2010) Visible light photocatalysis as a greener approach to photochemical synthesis. Nat Chem 2:527–532. (b) Narayanam JMR, Stephenson CRJ (2011) Visible light photoredox catalysis: applications in organic synthesis. Chem Soc Rev 40:102–113
21. (a) Skubi KL, Blum TR, Yoon TP (2016) Dual catalysis strategies in photochemical synthesis. Chem Rev 116:10035–10074. (b) Ghosh I, Marzo L, Das A, Shaikh R, König B (2016) Visible light mediated photoredox catalytic arylation reactions. Acc Chem Res 49:1566–1577
22. Romero NA, Nicewicz DA (2016) Organic photoredox catalysis. Chem Rev 116:10075–10166
23. Kibriya G, Bagdi AK, Hajra A (2018) Visible-light-promoted C(sp3)-C(sp2) cross-dehydrogenative coupling of tertiary amine with imidazopyridine. J Org Chem 83:10619–10626
24. Yang Q, Li S, Wang J(J) (2018) Cobalt-catalyzed cross-dehydrogenative coupling of imidazo [1,2-*a*]pyridines with isochroman using molecular oxygen as the oxidant. Org Chem Front 5:577–581
25. Mondal S, Samanta S, Singsardar M, Hajra A (2017) Aminomethylation of imidazoheterocycles with morpholine. Org Lett 19:3751–3754
26. Gao Y, Chen S, Lu W, Gu W, Liu P, Sun P (2017) Visible light-induced C3-sulfonamidation of imidazopyridines with sulfamides. Org Biomol Chem 15:8102–8109
27. Chen H, Yi H, Tang Z, Bian C, Zhang H, Lei A (2018) External oxidant-free regioselective cross-dehydrogenative coupling of 2-arylimidazoheterocycles and azoles with H_2 evolution *via* photoredox catalysis. Adv Synth Catal 360:3220–3227

28. Samanta S, Ravi C, Rao SN, Joshi A, Adimurthy S (2017) Visible-light-promoted selective C–H amination of heteroarenes with heteroaromatic amines under metal-free conditions. Org Biomol Chem 15:9590–9594
29. Sun K, Mu S, Liu Z, Feng R, Li Y, Pang K, Zhang B (2018) Copper-catalyzed C–N bond formation with imidazo[1,2-*a*]pyridines. Org Biomol Chem 16:6655–6658
30. Tantry SJ, Markad SD, Shinde V, Bhat J, Balakrishnan G, Gupta AK, Ambady A, Raichurkar AV, Kedari C, Sharma S, Mudugal NV, Narayan A, Kumar CNN, Nanduri R, Bharath S, Reddy J, Panduga V, Prabhakar KR, Kandaswamy K, Saralaya R, Kaur P, Dinesh N, Guptha S, Rich K, Murray D, Plant H, Preston M, Ashton H, Plant D, Walsh J, Alcock P, Naylor K, Collier M, Whiteaker J, McLaughlin RE, Mallya M, Panda M, Rudrapatna S, Ramachandran V, Shandil RK, Sambandamurthy VK, Mdluli K, Cooper CB, Rubin H, Yano T, Iyer PS, Narayanan S, Kavanagh S, Mukherjee K, Balasubramanian V, Hosagrahara VP, Solapure S, Ravishankar S, Shahul HP (2017) Discovery of imidazo[1,2-*a*] pyridine ethers and squaramides as selective and potent inhibitors of mycobacterial adenosine triphosphate (ATP) synthesis. J Med Chem 60:1379–1399
31. Kibriya G, Samanta S, Jana S, Mondal S, Hajra A (2017) Visible light organic photoredox-catalyzed C–H alkoxylation of imidazopyridine with alcohol. J Org Chem 82:13722–13727
32. Dong D-Q, Hao S-H, Yang D-S, Li L-X, Wang Z-L (2017) Sulfenylation of C–H bonds for C–S bond formation under metal-free conditions. Eur J Org Chem 6576–6592
33. (a) Gueiffier A, Mavel S, Lhassani M, Elhakmoui A, Snoeck R, Andrei G, Chavignon O, Teulade J-C, Witvrouw M, Balzarini J, De Clercq E, Chapat J-P (1998) Synthesis of imidazo [1,2-*a*]pyridines as antiviral agents. J Med Chem 41:5108–5112. (b) Véron J-B, Allouchi H, Enguehard-Gueiffier C, Snoeck R, Andrei G, De Clercq E, Gueiffier A (2008) Influence of 6- or 8-substitution on the antiviral activity of 3-arylalkylthiomethylimidazo[1,2-*a*]pyridine against human cytomegalovirus (CMV) and Varicella-Zoster Virus (VZV): Part II. Bioorg Med Chem 16:9536–9545
34. Ravi C, Adimurthy S (2017) Synthesis of imidazo[1,2-*a*]pyridines: C–H functionalization in the direction of C–S bond formation. Chem Rec 17:1019–1038
35. Hamdouchi C, Sanchez C, Ezquerra J (1998) Chemoselective arylsulfenylation of 2-aminoimidazo[1,2-*a*]pyridines by phenyliodine(III) bis(trifluoroacetate) (PIFA). Synthesis 867–872
36. Ravi C, Mohan DC, Adimurthy S (2014) *N*-chlorosuccinimide-promoted regioselective sulfenylation of imidazoheterocycles at room temperature. Org Lett 16:2978–2981
37. Hiebel M-A, Berteina-Raboin S (2015) Iodine-catalyzed regioselective sulfenylation of imidazoheterocycles in PEG_{400}. Green Chem 17:937–944
38. Huang X, Chen Y, Zhen S, Song L, Gao M, Zhang P, Li H, Yuan B, Yang G (2018) Cobalt-catalyzed aerobic cross-dehydrogenative coupling of C–H and thiols in water for C–S formation. J Org Chem 83:7331–7340
39. Cao H, Chen L, Liu J, Cai H, Deng H, Chen G, Yan C, Chen Y (2015) Regioselective copper-catalyzed thiolation of imidazo[1,2-*a*]pyridines: an efficient C–H functionalization strategy for C–S bond formation. RSC Adv. 5:22356–22360
40. Liu W, Wang S, Jiang Y, He P, Zhang Q, Cao H (2015) Copper-catalyzed regioselective C5-sulfenylation of imidazo[2,1-*b*]thiazoles with thiols. Asian J Org Chem 4:312–315
41. Yuan Y, Cao Y, Qiao J, Lin Y, Jiang X, Weng Y, Tang S, Lei A (2018) Electrochemical oxidative C–H sulfenylation of imidazopyridines with hydrogen evolution. J Chem, Chinese. https://doi.org/10.1002/cjoc.201800405
42. Zheng Z, Qi D, Shi L (2015) Copper-catalyzed thiolation of imidazo[1,2-*a*]pyridines with (Hetero)aryl thiols using molecular oxygen. Catalysis Commun. 66:83–86
43. Siddaraju Y, Prabhu KR (2016) Iodine-catalyzed cross-dehydrogenative coupling reaction: a regioselective sulfenylation of imidazoheterocycles using dimethyl sulfoxide as an oxidant. J Org Chem 81:7838–7846

44. Yadav M, Dara S, Saikam V, Kumar M Aithagani SK, Paul S, Vishwakarma RA, Singh PP (2015) Regioselective oxidative C–H phosphonation of imidazo[1,2-*a*]pyridines and related heteroarenes mediated by manganese(III) acetate. Eur J Org Chem 6526–6533
45. Pericherla K, Khedar P, Khungar B, Kumar A (2013) One-pot sequential C–N coupling and cross-dehydrogenative couplings: synthesis of novel azole fused imidazo[1,2-*a*]pyridines. Chem Commun 49:2924–2926
46. Firmansyah D, Banasiewicz M, Gryko DT (2015) Vertically-expanded imidazo[1,2-*a*] pyridines and imidazo[1,5-*a*]pyridine *via* dehydrogenative coupling. Org Biomol Chem 13:1367–1374

Synthesis and Functionalization of *N*-Heterocycles Using Transition Metal-Free Cross-Dehydrogenative Coupling (CDC) Approaches

V. Murugesh, Apurba Ranjan Sahoo, Mathieu Achard and Surisetti Suresh

Abstract The creation of carbon–carbon bonds for the construction or the post-functionalization of various *N*-heterocycles is one of the most active research areas in organic chemistry. Among those approaches, cross-dehydrogenative coupling (CDC) processes in the front line because of its simple, practical applicability in organic synthesis. Historically discovered in the presence of transition metal catalysts, this decade witnessed great progress in the metal-free CDC approaches, thus improving the sustainability of this transformation. This chapter aims at providing an update on selected examples of the metal-free approaches toward the access to various *N*-heterocycles. The first part deals with the synthesis of *N*-heterocycles by metal-free CDC approaches in the presence of stoichiometric or catalytic amount of oxidants, whereas the second part is focused on the last stage C–H bond functionalization of *N*-heterocycles via CDC process.

Keywords Functionalization of amines · Oxidant · Metal-free approaches

V. Murugesh (✉) · S. Suresh
Department of Organic Synthesis and Process Chemistry,
CSIR—Indian Institute of Chemical Technology (CSIR-IICT), Hyderabad 500007, India
e-mail: murugeshprasad@gmail.com

Academy of Scientific and Innovative Research (AcSIR), Ghaziabad 201002, India

S. Suresh
e-mail: surisetti@iict.res.in

A. R. Sahoo · M. Achard
Univ Rennes, UMR 6226 (Institut Des Sciences Chimiques de Rennes), 35000 Rennes, France
e-mail: apurbar.sahoo@gmail.com

M. Achard
e-mail: mathieu.achard@univ-rennes1.fr

A. Srivastava and C. K. Jana (eds.), *Heterocycles via Cross Dehydrogenative Coupling*, https://doi.org/10.1007/978-981-13-9144-6_5

Abbreviations

[bpy]I	1-Butylpyridinium iodide
2-*tert*-Bu-AQN	2-*tert*-Butylanthraquinone
BPO	Benzoyl peroxide
DCP	Dicumyl peroxide
DDQ	2,3-Dichloro-5,6-dicyano-1,4-benzoquinone
DTBP	Di-*tert*-butyl peroxide
IL	Ionic liquids
KHMDS	Potassium hexamethyldisilazide
*m*CPBA	*meta*-Chloroperoxybenzoic acid
NBS	*N*-Bromosuccinimide
NHS	*N*-Hydroxysuccinimide
NIS	*N*-Iodosuccinimide
PIDA	(Diacetoxyiodo)benzene
PIFA	[Bis(trifluoroacetoxy)iodo]benzene
TBAB	Tetrabutylammonium bromide
TBAI	Tetrabutylammonium iodide
TBHP	*tert*-Butyl hydroperoxide
TEMPO	(2,2,6,6-Tetramethylpiperidin-1-yl)oxidanyl
TFA	Trifluoroacetic acid

1 Introduction

From synthetic organic chemistry standpoint, the construction of a new chemical bond is a preliminary element. In the past five to six decades, transition metal catalysis has emerged as an important tool toward this end and synthetic chemists have discovered, explored, and exploited several excellent new reactions and methods that largely served for the synthesis of essential molecules including heterocycles for life-saving drugs and lifestyle commodity materials [1]. The synthesis of pharmaceutical, medicinally valuable products, and natural products relies on transition-metal-catalyzed coupling reactions. Transition metal catalysis [2] raises several issues lying on toxicity, metal abundancy, catalyst/product separation, and requirement of prefunctionalized starting material and ligands. To address these drawbacks, scientific community showed increasing interest in the development of transition metal-free approaches for constructing new chemical bonds by using organocatalysts, oxidants, organophotocatalysts, and acid-/base-mediated approaches. Nowadays, scientific community intends to develop metal-free synthesis of chemical bonds.

Construction of new chemical bond through a metal-free reaction conditions is an advisable challenge for chemists. There is a significant progress achieved in this direction by the scientific community. Transition metal-free reactions like organocatalytic reactions, Friedel–Crafts reactions, multicomponent reactions,

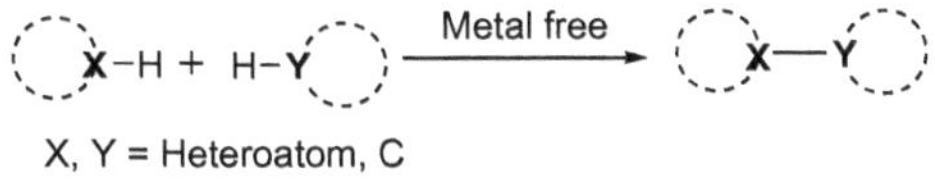

Fig. 1 General representation of CDC reaction

domino reactions, and Diels–Alder reaction play a crucial role in synthetic chemistry development. For making chemical bonds, number of methods available including strong base-mediated reactions to photo-induced reactions. Among those, metal-free cross-dehydrogenative coupling (CDC) process [1e, f, 3] is in the front line because of its simple, practical applicability in organic synthesis. Less by-products and vast applications, atom economy, not sensitive to air and moisture, no need of prefunctionalization of starting material, convergent, greener, and sustainable procedures make CDC as a suitable method for the synthesis of complex molecules by making new chemical bonds.

This decade witnessed great progress in the metal-free CDC approaches in developing methods for synthesizing valuable compounds. Although good progress has been achieved in transition metal-free CDC approach development, it is still needed to achieve greener and improved methods for constructing C–C, C–heteroatom bonds. In this section, we discuss the metal-free CDC approaches for constructing new C–C, C–heteroatom bonds by using oxidants, base, or acid. Photo-induced reactions will be briefly described. This section is divided into two parts: Part one deals with the synthesis of *N*-heterocycles by metal-free CDC approaches, and part two describes the metal-free C–H bond functionalization of *N*-heterocycles via CDC process. In this section, we covered the CDC transformations, which form a new chemical bond (C–C, C–heteroatom, heteroatom–heteroatom) by direct oxidative coupling or at least oxidative coupling at intermediate stage by the (formal) elimination of H_2 by the action of metal-free reagents or conditions (Fig. 1).

2 Synthesis of *N*-Heterocycles via Metal-Free CDC Approaches

N-Heterocyclic compounds present in number of drugs, vitamins, natural products, and biologically active compounds are one of the most important classes of compounds in pharmaceutical industry. According to FDA records, 84% of the approved small-molecule drugs are containing at least one nitrogen, and 59% of the drugs are having *N*-heterocyclic core structure [4]. The synthesis of such a highly valuable molecule with simple CDC under metal-free conditions is desirable and essential for the synthetic chemistry. The metal-free synthesis of *N*-heterocycles by CDC deals with the applications of oxidants, organocatalysts, photocatalysts, sensitizers, etc. Oxidants including halogen reagents and peracids have remarkable applications in synthesis of *N*-heterocycles, and also photo-inducers and organocatalysts are useful in the synthesis of *N*-heterocycles.

2.1 Halogen-Reagent-Mediated CDC

The stereoselective, iodine-mediated asymmetric route to 3,3-oxindoles via CDC process was reported by Petersen and co-workers (Scheme 1) [5a]. The reaction proceeds through the formation of transient iodinated intermediate but not via the usual radical intermediate process [5b]. According to their investigations, stable tertiary-iodide intermediate formation **A** takes place through single-electron oxidation by iodine catalysis; the iodinated intermediate transformed to α-carbonyl radical **B** intermediate, which undergoes cyclization to form the desired oxindoles (Fig. 2).

Li and co-workers synthesized 2,3-disubstituted indole derivatives via oxidative intermolecular CDC process. The treatment of either electron-rich and electron-poor *N*-arylated enamines with catalytic amount of iodine along with a stoichiometric amount of *N*-bromosuccinimide (NBS) led to the 2,3-disubstituted indoles via oxidative cyclizations (Scheme 2) [6]. After excluding the involvement of iodonium intermediates, the authors proposed the formation of *N*-iodo intermediates followed by intramolecular radical cyclization to furnish 2,3-disubstituted indoles.

Tilve and co-workers developed a metal-free, iodine-mediated reaction in combination with TBHP as an efficient method for constructing quinoline derivatives via CDC process (Scheme 3) [7]. The reaction proceeds through the annulation followed by aromatization. The transformation was mediated by iodine/TBHP catalytic system to access indole-fused quinolinium derivatives. Mechanical studies revealed that iodine forms a complex **A** with tertiary amine substituent, which leads to the formation of iminium ion **C**. The desired product is formed by the intramolecular addition of iminium ion followed by the oxidation (Fig. 3).

He, Huang, and co-workers developed a method for the synthesis of aroylated *N*-heterocycles from aldehydes and isocyanides. This methodology has been extended for the construction of various types of isoquinoline and phenanthridine derivatives. Similar to the Minisci reaction, the use of hypervalent iodine PIDA along with azidotrimethylsilane was proposed to generate the key acyl radical intermediate from aldehydes and azide radical which play the role of acylating reagent to deliver arylated *N*-heterocycles by forming two new C–C bonds (Scheme 4) [8]. The reaction is initiated by the formation of radical intermediate **B**. The radical

Scheme 1 CDC for the synthesis of 3,3-disubstituted oxindoles

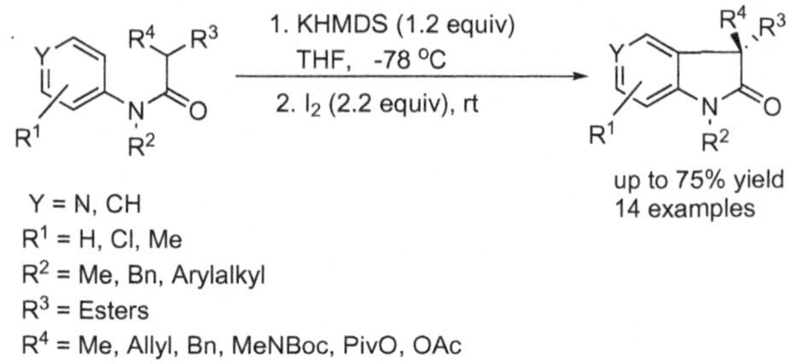

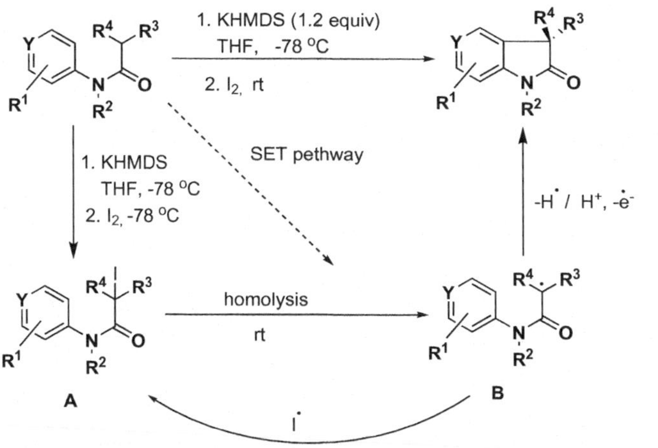

Fig. 2 Mechanism of metal-free cross-dehydrogenative synthesis of 3,3-disubstituted oxindoles

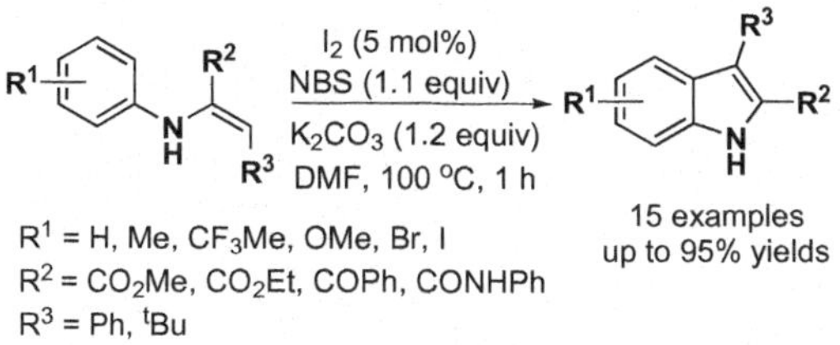

Scheme 2 Iodine-mediated synthesis of indole via oxidative cyclization

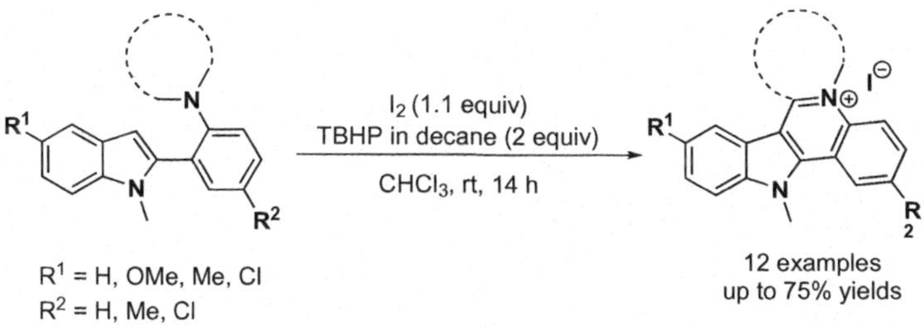

Scheme 3 Iodine-catalyzed synthesis of *N*-alkylindolo[3,2-*c*]quinolinium iodides

aldehyde intermediate **C** generated by the reaction of aldehyde with azide radical, reacts with isocyanide derivative to form the coupled product **D**. Further intramolecular oxidative cyclization leads to the formation of phenanthridine derivatives (Fig. 4).

The transient formation of acyl radicals was recently employed by Zhao and co-workers to develop an intramolecular CDC method to synthesize acridine derivatives. They treated 2-(*N*-arylamino)aldehydes with $PhI(OAc)_2$ in the presence

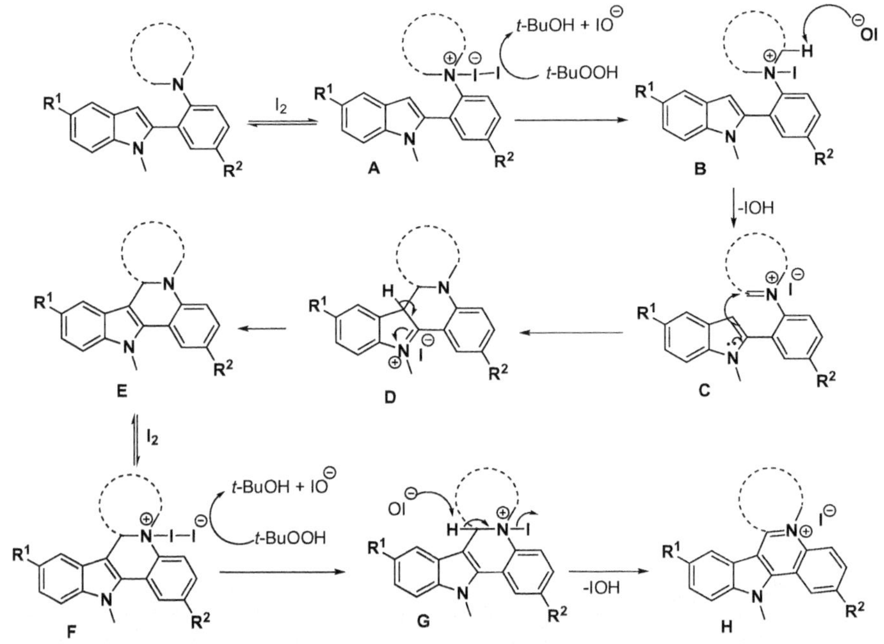

Fig. 3 Mechanism of iodine-catalyzed intramolecular dehydrogenative coupling

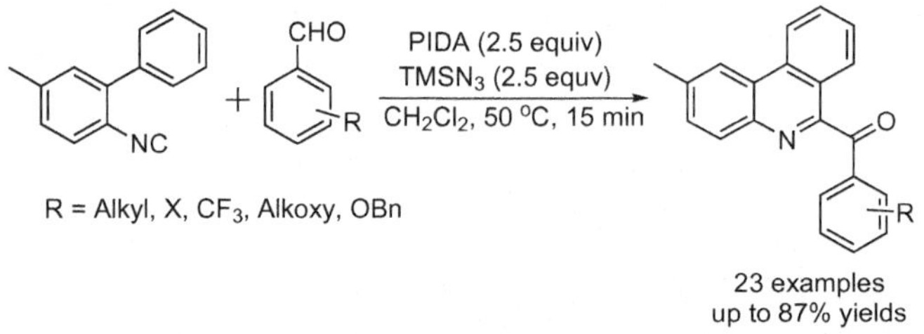

Scheme 4 Metal free synthesis of aroylated *N*-heterocycles

of catalytic amount of benzoyl peroxide (BPO) to furnish the acridine derivatives (Scheme 5) [9]. The authors proposed radical reaction pathway. Initially, aldehyde radical **D** will be formed by the action of iodine reagent followed by the intramolecular cyclization; oxidation leads to the formation of coupled product (Fig. 5).

Following the seminal report of Bhuyan and co-workers [10a] on iodine-mediated synthesis of 6,12-disubstituted 5,7-dihydroindolo[2,3-b]carbazoles from aldehydes and indoles, Xue et al. synthesized indol[2,3-*b*]carbazoles through iodination/Kornblum oxidation/hydrolysis/decarboxylation/deacylation/aromatization cascade sequence. The treatment of indoles with β-ketoesters (or) 1,3-diketones in the

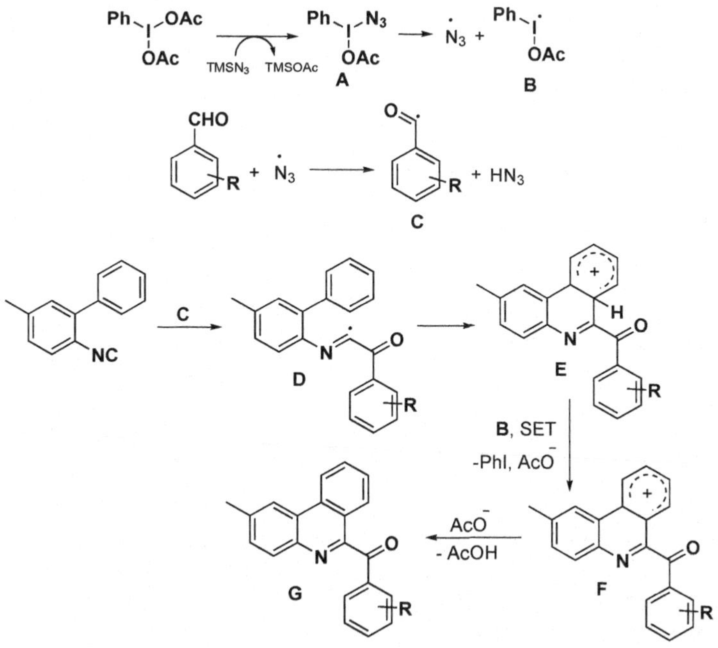

Fig. 4 Plausible mechanism for oxidative insertion of isocyanides with aldehydes

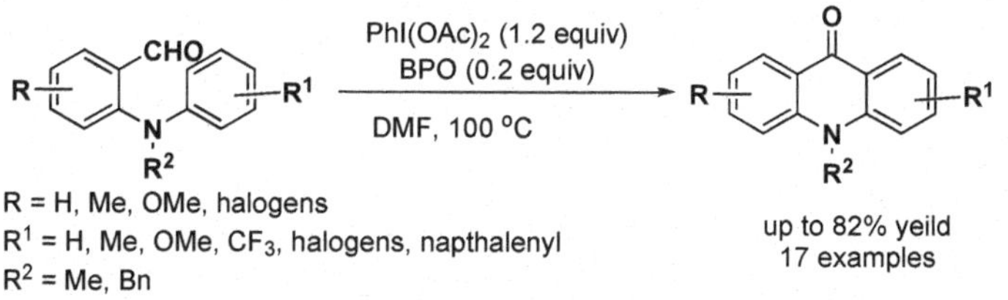

Scheme 5 Iodine catalysed intramolecular C(sp^2)–C(sp^2) bond formation

presence of iodine affords carbazole derivatives with a broad functional group tolerance and good yields (Scheme 6) [10b].

Recently, an intramolecular oxidative cyclo-aromatization methodology to synthesize the 1,10-phenanthrenol derivatives was reported by Wang and co-workers. The reaction operated with electron withdrawing group substituted 1-biphenyl-2-ylethanones with NBS along with TBHP to give 1,10-phenanthrenol (Scheme 7) [11]. The authors proposed the in situ generation of bromine cation as the real initiator for the C–H bond activation. The keto–enol tautomerism facilitates the bromination at the α position followed by homolysis to generate the key radical intermediate which undergoes homolytic aromatic substitution (HAS). The reaction

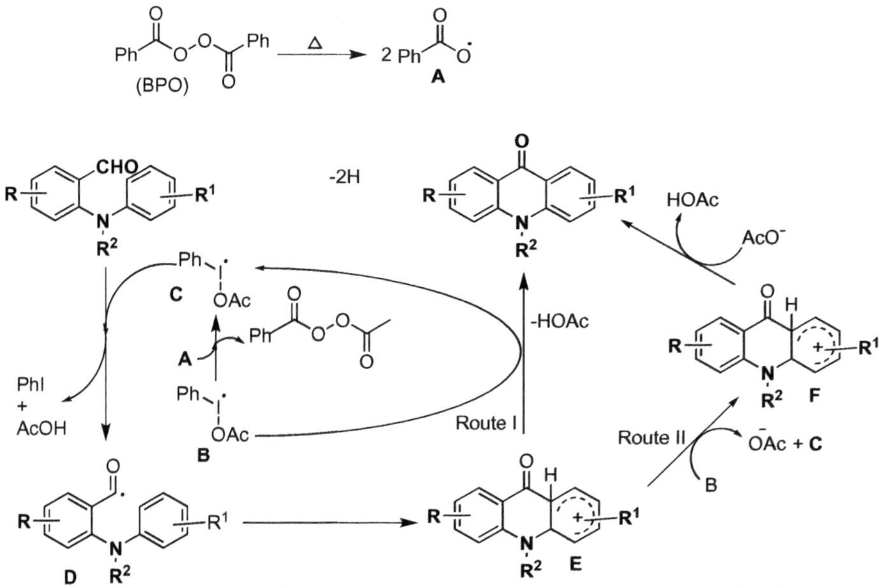

Fig. 5 Plausible mechanism for the synthesis of acridones

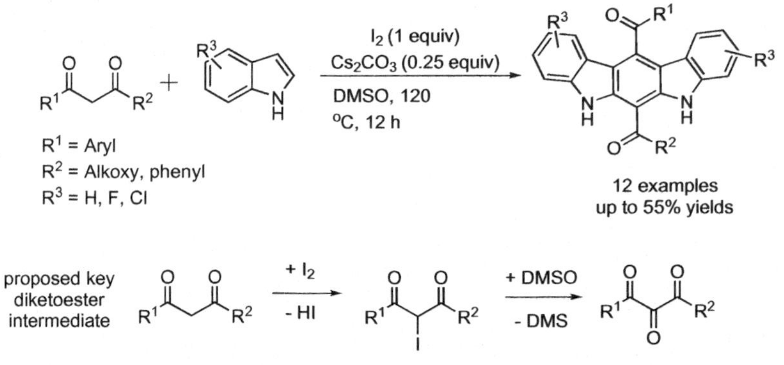

Scheme 6 Oxidative cross-coupling method for the synthesis of indolo[2,3-*b*]carbazoles

proceeds through the radical reaction pathway. From the precursor A′, α-carbonyl radical **B** will be formed that would undergo the intramolecular radical cyclization to lead to the formation of new C–C bond and then the oxidation takes place to form the 1,10-phenanthrenol derivatives (Fig. 6).

Zhao and co-workers developed a method for the synthesis of dibenzodihydro 1,3-diazepin-2-ones by intramolecular CDC and rearrangement in a tandem fashion. The treatment of 2-substituted *N*-phenylbenzamides with hypervalent iodine PIDA gives the corresponding dibenzodihydro 1,3-diazepin-2-ones (Scheme 8) [12].

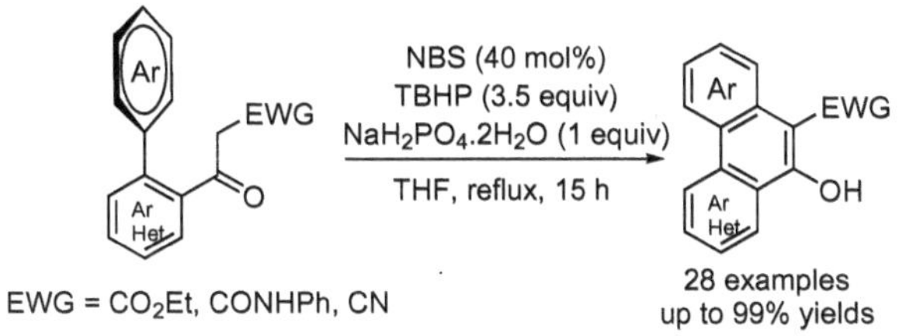

Scheme 7 *N*-Bromosuccinimide-mediated intramolecular cycloaromatization

After isolating the key spirocyclic intermediate, the authors proposed a rapid ring opening followed by lactamization to rationalize the reaction mechanism. The authors proposed a plausible reaction mechanism, which proceeds through iodo complex formation **A** with starting substrate that can cyclize to give spirocyclic intermediate **B**. The spirocyclic intermediate will undergo ring opening followed by

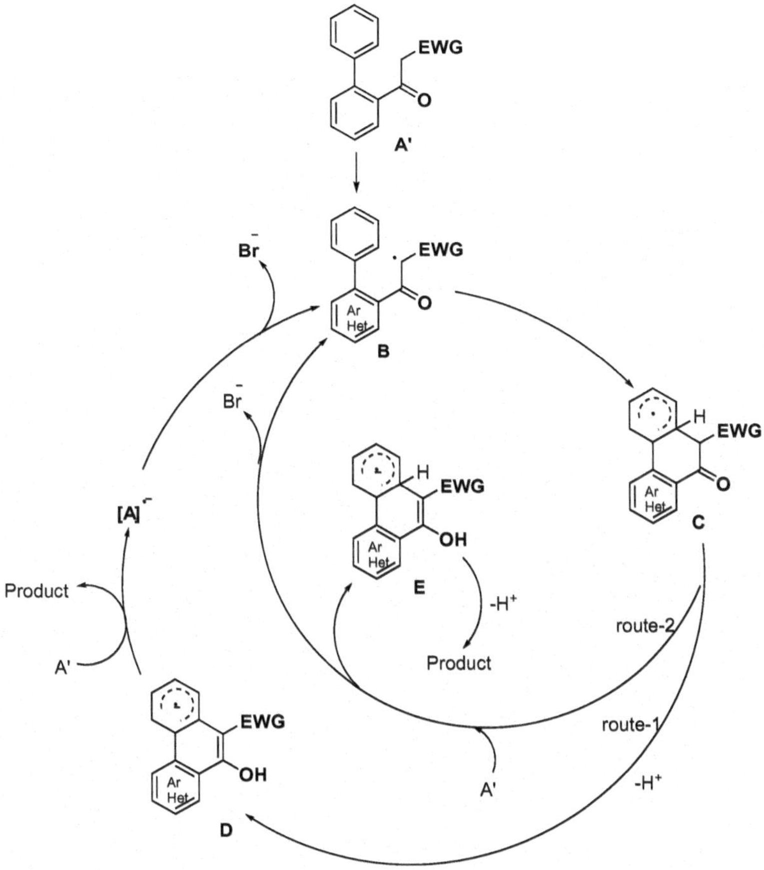

Fig. 6 Proposed mechanism for the synthesis of 10-phenanthrenol

nucleophilic attack to lead to the formation of dibenzodihydro 1,3-diazepin-2-ones derivatives (Fig. 7).

The preparation of various carbazoles arising from an intermolecular CDC of unactivated arenes with anilines in the presence of hypervalent iodines was reported by Mal and co-workers. The use of PIDA or a combination of PhI along with *m*CPBA enabled the transformation (Scheme 9) [13]. In this transformation, three C(sp^2)–H bonds and one N–H bond undergo functionalization to form one C–C and one C–N bonds.

Maiti and Mal developed a CDC process for synthesizing carbazoles involving C–C and C–N bond formations followed by an unprecedented alkyl migration. They treated 1,3,5-trialkylbenzenes with anilides in the presence of I(III) to give the carbazole derivatives (Scheme 10) [14]. The reaction proceeds through the formation of nitrenium ion or equivalent carbenium ion. A regioselective fusion of arenes to anilides was presented.

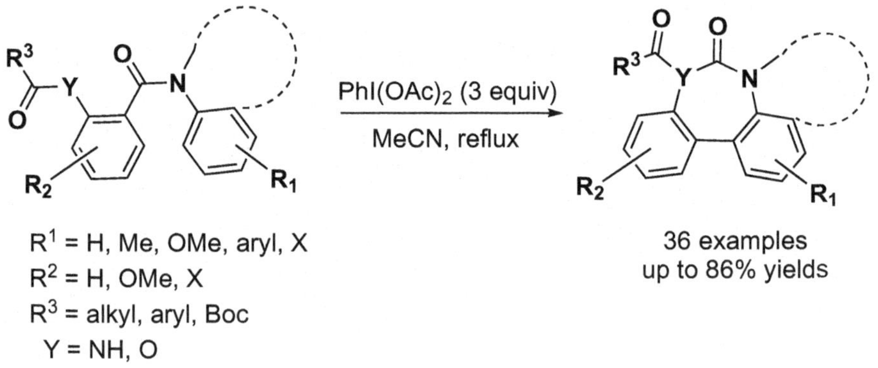

Scheme 8 Hypervalent iodine-mediated oxidative aryl-aryl coupling reaction

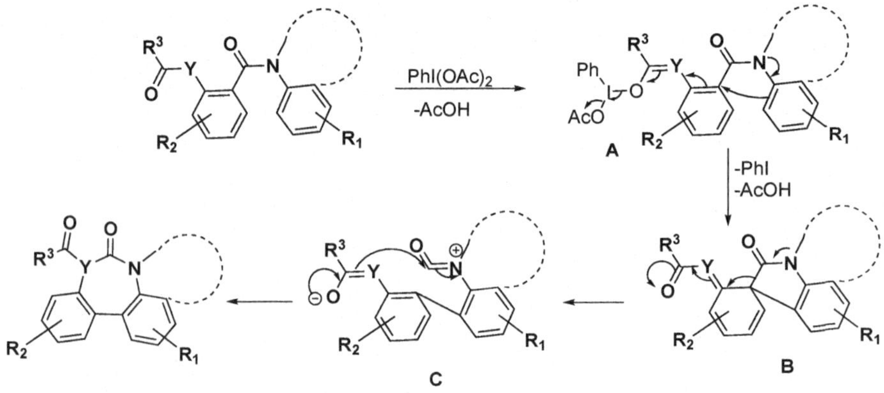

Fig. 7 Mechanism of an unusual rearrangement of 2-substituted *N*-phenylbenzamides

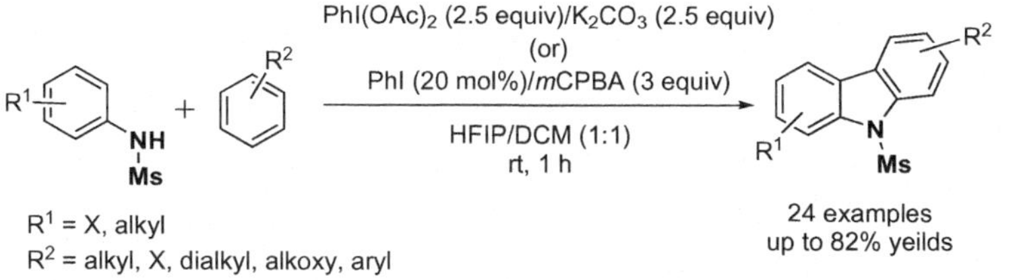

Scheme 9 An organic intermolecular dehydrogenative annulation reaction

Yin and co-workers developed an intramolecular amination protocol for the synthesis of indoline derivatives. A site-selective, iodine-mediated intramolecular amination reaction for the functionalization of unactivated C(sp^3)–H bonds was reported. They treated suitable *ortho*-substituted anilide derivatives with iodine to form indoline derivatives via CDC process by forming C–C bond between C(sp^3)–H and N–H bonds (Scheme 11) [15]. Even though C(sp^2)–H bonds were available, selective amination took place at C(sp^3)–H bonds only. The authors proposed the formation of a *N*-centered radical followed by a 1,5-H shift on the sp^3 C–H bond. Further iodination and cyclization account for product formation.

Satish et al. developed a method for constructing isatins with different substitutions on benzene ring and nitrogen atom. They treated *ortho*-ethynyl/ethenylanilines in the presence of iodine catalytic system to give isatins (Scheme 12) [16].

Li et al. developed a synthetic transformation for the synthesis of polysubstituted-2,3-dihydropyrroles. A CDC process proceeds through Michael addition/intramolecular cyclization. They treated chalcones with β-enamine ketones

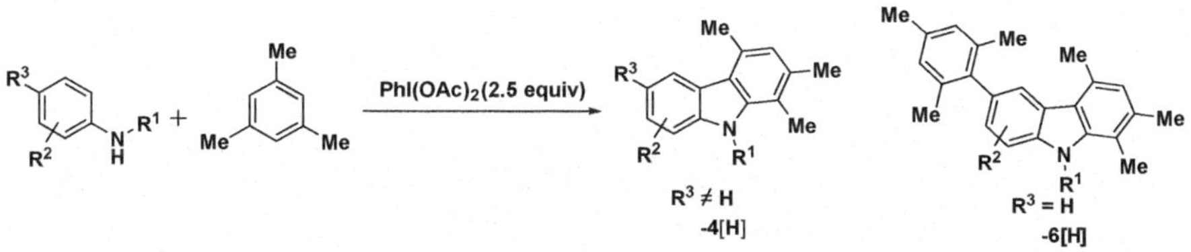

Scheme 10 Dehydrogenative synthesis of carbazoles by ring fusion

Scheme 11 Iodine-mediated synthesis of indulines by amination

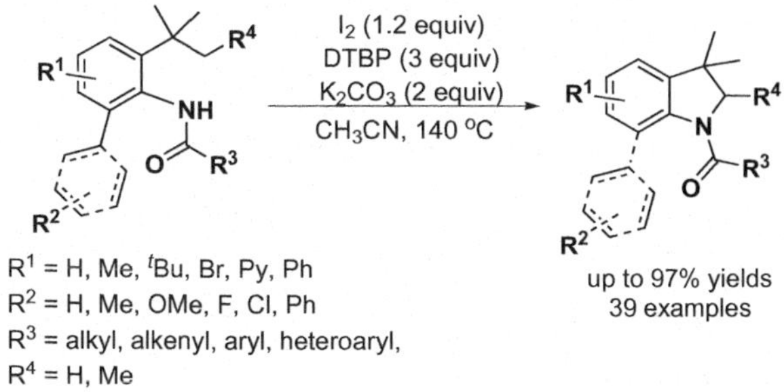

Scheme 12 Iodine-mediated domino protocol for the synthesis of isatins

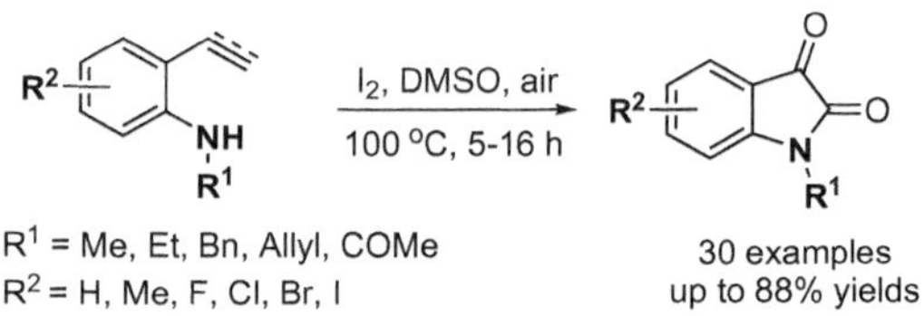

or esters in the presence of iodine and K_2CO_3 to furnish 2,3-dihydropyrrole derivatives (Scheme 13) [17].

In 1970, Ohno and co-workers synthesized 2,3-dihydropyrrolo[2,3-*b*]indole derivatives in the presence of stoichiometric amount of NBS or *t*-BuOCl. The authors proposed halogenation at the 3-position of the indole nucleus generating imine intermediates followed by a cyclization and the elimination of HCl or HBr (Scheme 14) [18].

Wang and co-workers reported intermolecular amination method for the synthesis of pyrrolo[2,3-*b*]indoles from tryptophan esters. They employed the tryptophan derivatives with substoichiometric amount of iodine along with a stoichiometric amount of TBHP to form pyrrole-fused indole derivatives by forming C–N bond via CDC process (Scheme 15) [19]. The combination of iodine/TBHP/Na_2HPO_4 enabled the gram-scale synthesis of pyrrole-fused indole synthesis. In this case, a key cyclic iodonium ion followed the attack of the amino group; then, elimination of HI was proposed.

Gao et al. developed a C(sp^3)–H bond functionalization method to synthesize 3-(methylthio)-4-aryl-1*H*-pyrrole-2,5-diones from methyl ketones and acetamides by iodine-mediated CDC (Scheme 16) [20].

Jiang et al. developed a method for the synthesis of polysubstituted imidazoles. They treated aryl methyl ketones with benzylamines in the presence of iodine catalytic system to furnish highly substituted imidazoles (Scheme 17) [21]. This was a one-pot oxidative cyclization method for the synthesis of imidazoles.

Chen, Ren, and co-workers synthesized highly substituted 1,2,4-triazoles under aerobic oxidative conditions. Hydrazones were reacted with benzylamines in the presence of I_2/TBHP system to obtain substituted 1,2,4-triazoles (Scheme 18) [22].

Du and co-workers developed an intramolecular metal-free CDC transformation to synthesize isoquinolinoquinazolinone derivatives through iminium generation in the presence of hypervalent iodine. Direct formation of new C–C or C–N bond from various tetrahydroisoquinoline derivatives was achieved (Scheme 19) [23].

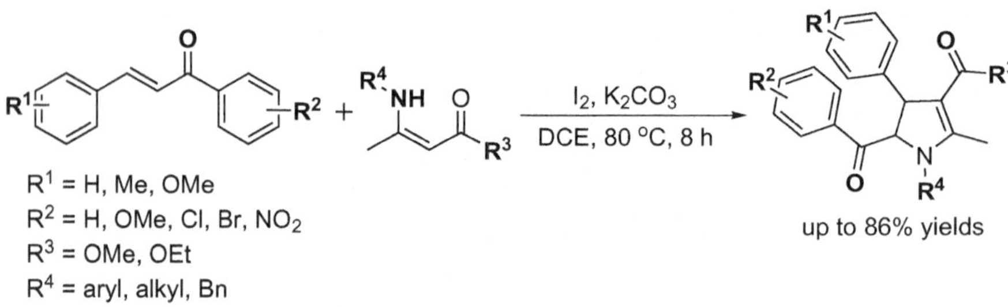

Scheme 13 Iodine-mediated synthesis of polysubstituted 2,3-dihydropyrroles

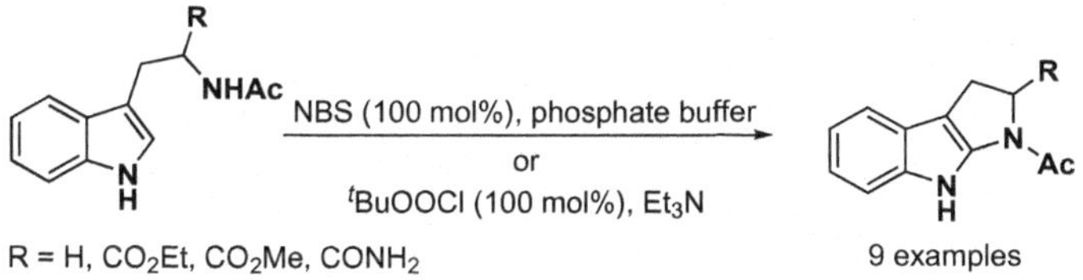

Scheme 14 Synthesis of 2,3-dihydropyrrolo [2,3-*b*]indole from tryptophan and tryptamine derivatives

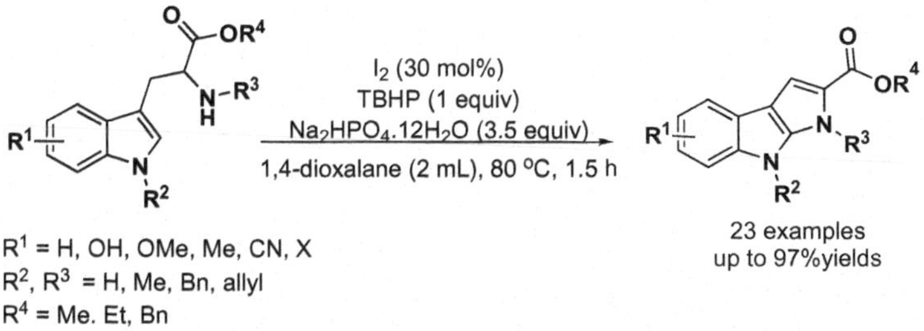

Scheme 15 Synthesis of pyrrolo[2,3-*b*]indoles by intramolecular amination of tryptophan esters

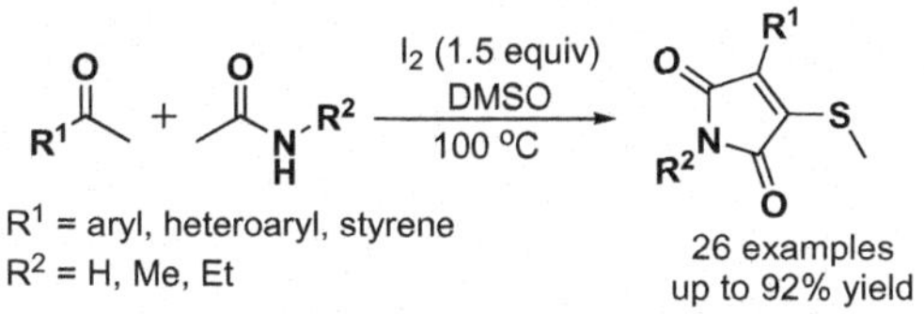

Scheme 16 Iodine-catalysed synthesis of 3-methylthio-4-arylmaleimides via annulations

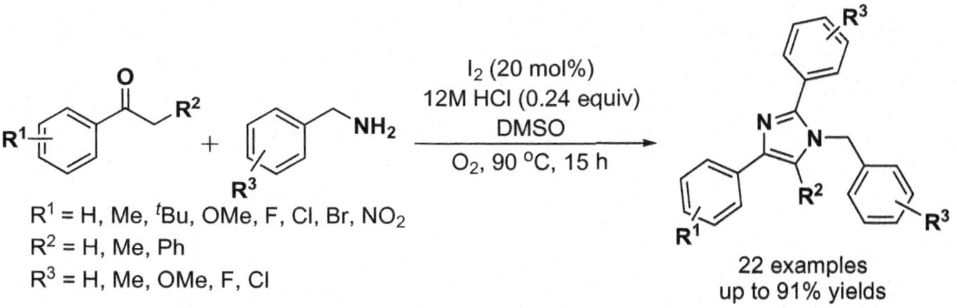

Scheme 17 Iodine-mediated aerobic oxidative cyclization for the synthesis of polysubstituted imidazoles

Scheme 18 Iodine assisted coupling reaction for the synthesis of 1,3,5-trisubstituted 1,2,4-triazoles

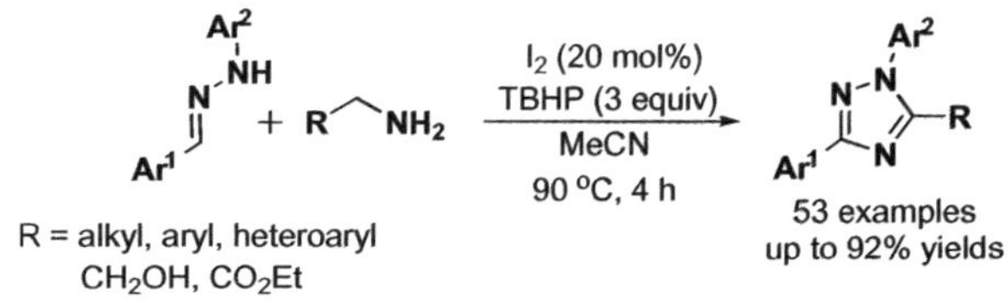

Wang et al. developed a synthetic method for the construction of quinazolines. They treated 2-aminobenzophenones with *N*-alkylamides or ethers or alcohols along with ammonia in the presence of iodine catalytic system to produce quinazolines (Scheme 20) [24]. This domino process is initiated by the generation of iminium ion from acetamide derivatives in the presence of *N*-iodosuccinimide/ TBHP system.

Yu and co-workers developed a method for the synthesis of 3-acylbenzothiadiazine 1,1-dioxide derivatives via CDC by forming new C–N bonds. The transient formation of α-iodinated ketones followed by Kornblum oxidation generates phenylglyoxal derivatives. Then, condensation with 2-aminobenzenesulfonamides and a succession of addition and oxidation account for the product formation (Scheme 21) [25a]. When similar reaction was performed in the presence of catalytic amount of iodine under an O_2 atmosphere, Bharate, Vishwakarma, and Mohammed observed the selective formation of 2-arylquinazolin-4(3H)-ones arising from the elimination of formaldehyde [25b].

Yang and co-workers extended the catalytic system developed by Bharate and Vishwakarma to synthesize 2-heteroaryl quinolinones from azaarenes (Scheme 22)

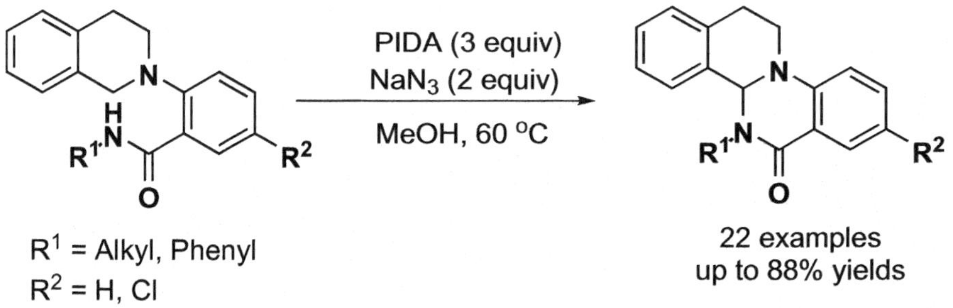

Scheme 19 Intramolecular functionalization of benzylic methylene in tetrahydroisoquinoline derivatives

Scheme 20 Iodine-catalyzed synthesis of quinazolines from *ortho*-carbonyl-substituted anilines

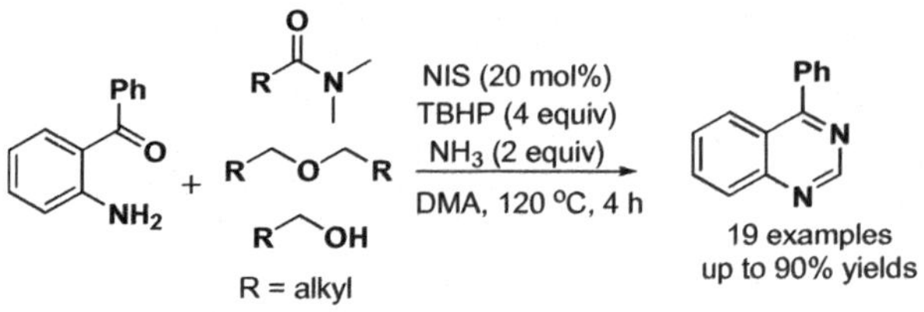

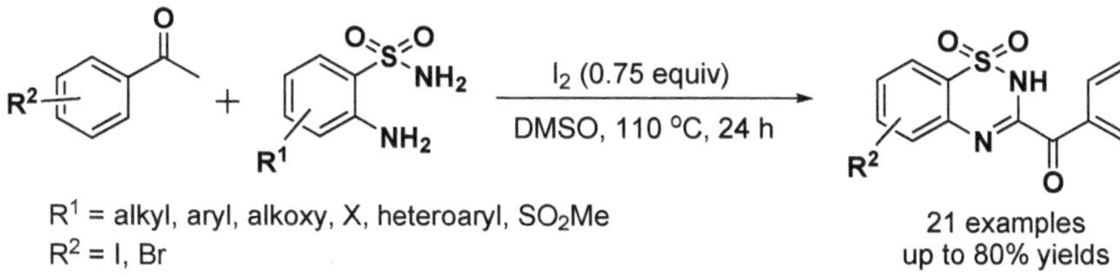

Scheme 21 Iodine-mediated synthesis of 3-acylbenzothiadiazine 1,1-dioxides

[26]. They treated 2-aminobenzamide derivatives with methylheteroarenes in the presence of catalytic amount of iodine and oxygen. Oxygen acted as a terminal oxidant. New C–C bond formed between unactivated methyl and amine groups.

Xicun et al. developed a method for the synthesis of pyrido[2,3-*d*]-pyrimidin-4-ols. They treated 2-amino-4-aryl-6-aryl nicotinamide with benzylamines in the presence of I_2/TBHP system under microwave irradiation to furnish pyrido[2,3-*d*]-pyrimidin-4-ols (Scheme 23) [27].

Ghorbani-Vaghei and Malaekehpoor developed a method for synthesizing polycyclic indol[2,3-*b*]quinoline derivatives. They treated anilines with indole-3-carbaldehydes in the presence of NBS to produce polycyclic indol[2,3-*b*] quinoline derivatives with good yields (Scheme 24) [28].

Scheme 22 Iodine-catalyzed synthesis of quinazolin-4(3*H*)-ones by amination of azaarenes

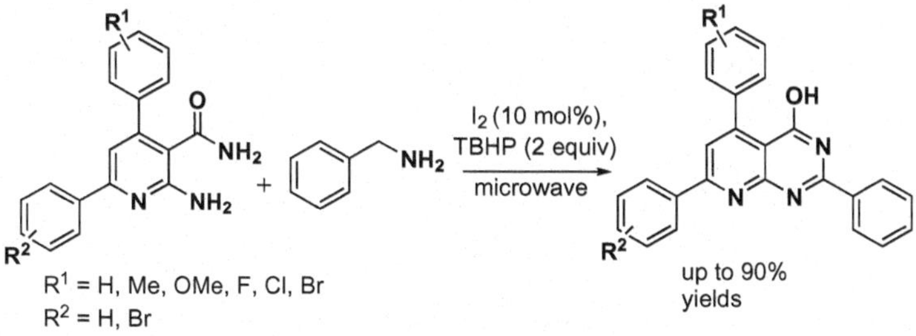

Scheme 23 Iodine-catalyzed synthesis of pyrido[2,3-*d*]pyrimidin-4-ols

Al-Mourabit and co-workers developed a method to functionalize the α-, β-C (sp^3)–H bonds of alicyclic ketones. They treated cyclohexanone with 2-aminopyrimidine and 2-aminopyridine in the presence of catalytic amount of iodine under an oxygen atmosphere in acidic medium to construct imidazole-fused products by forming two new C–C bonds (Scheme 25) [29].

Qu and co-workers developed an oxidative transformation of pyrrolidines to *N*, *O*-acetals catalyzed by iodine. A regio- and diastereoselective functionalization of C–H bonds of pyrrolidines by CDC furnished *N*,*O*-acetals (Scheme 26) [30].

The metal-free formation of C–O bonds has been illustrated by Baruah and co-workers for synthesizing 1,3-oxazines from alicyclic amines. This intermolecular oxidative C–O bond forming reaction is mediated by the combination of iodine and H_2O_2 system at ambient conditions (Scheme 27) [31].

Du and co-workers developed a method for forming C–O bond via CDC process. Intramolecular oxidative C–C bond formation reaction for the synthesis of 1,2-dihydro-(4*H*)-3,1-benzoxazin-4-one derivatives was reported (Scheme 28) [32]. An α-functionalization of tertiary amines mediated by *N*-iodosuccinimide (NIS),

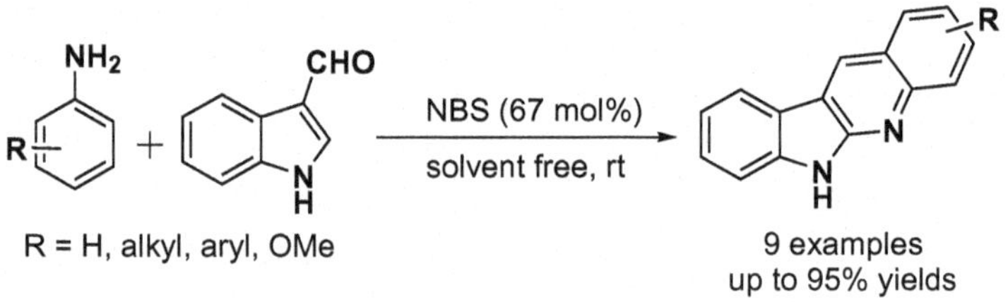

Scheme 24 *N*-Bromosuccinimide-mediated synthesis of indolo[2,3-*b*]quinolines

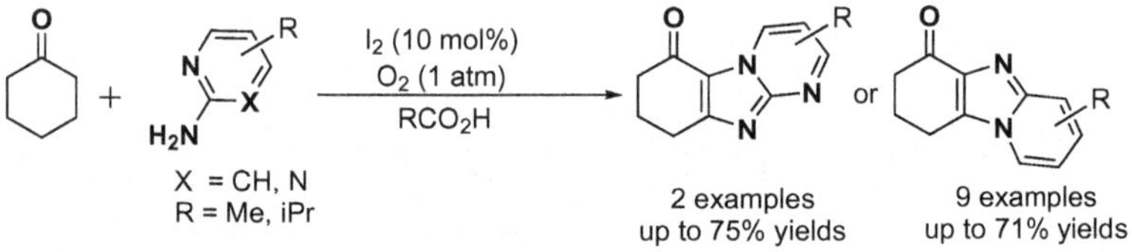

Scheme 25 Iodine-catalyzed synthesis of imidazole-fused heterocycles

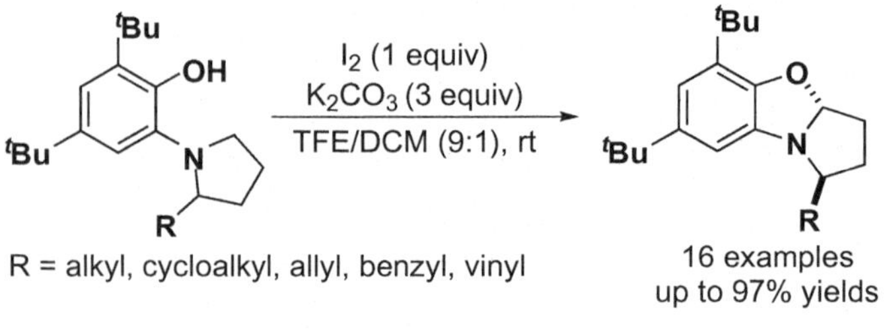

Scheme 26 Iodine-mediated synthesis of *N*,O-acetals from pyrrolidines

Scheme 27 Iodine/hydrogen peroxide promoted synthesis of 1,3-oxazines

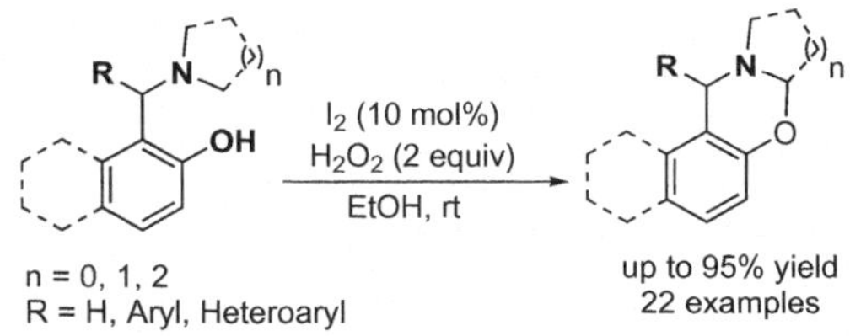

which served as both oxidant and iodine source, was developed for a regioselective synthesis of iodinated or non-iodinated 1,2-dihydro-(4*H*)-3,1-benzoxazin-4-ones.

Karade and co-workers synthesized 1,3-napthoxazines by intermolecular CDC process by treating 1-(α-aminoalkyl)-2-naphthols with hypervalent iodine (III) reagent via the formation of new C–O bond (Scheme 29) [33]. The starting 1-(α-aminoalkyl)-2-naphthols were synthesized by three-component reaction of β-naphthol, aldehyde, and secondary amines.

Jiang and co-workers synthesized 2-aminobenzothiazoles by treating cyclohexanone with thiourea in the presence of I_2/O_2 catalytic system to give 2-aminobenzothiazoles (Scheme 30) [34]. They have reported an oxidative coupling reaction to synthesize divergent 2-aminobenzothiazoles, 2-aminonaphtho[2,1-*d*]thiazoles, and 2-aminonaphtho[1,2-*d*]thiazole derivatives by employing various cyclohexanones, 2-tetralones, and 1-tetralones with different thiourea derivatives.

Fan and co-workers developed a CDC process for constructing C–S bond to synthesize 2-aminobenzoxazole in the presence of catalytic amount of iodine under an O_2 atmosphere. They treated isothiocyanatobenzene with amines in the presence

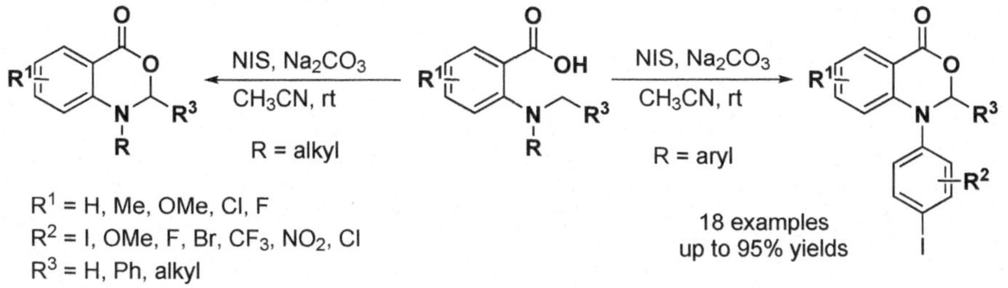

Scheme 28 Synthesis of 1,2-dihydro-(4*H*)-3,1-benzoxazin-4-one derivatives from tertiary amines

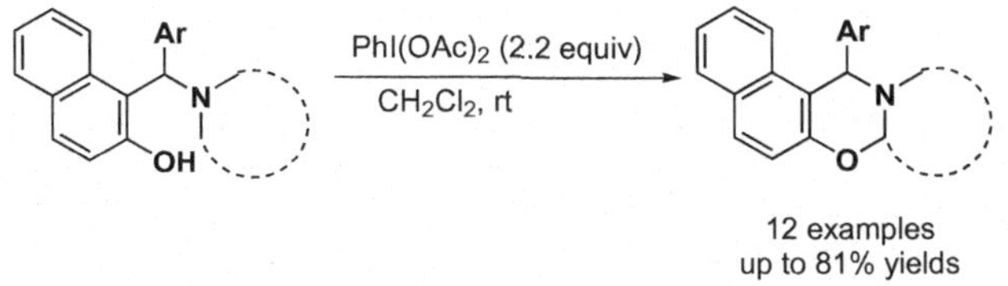

Scheme 29 Hypervalent iodine-mediated synthesis of 1,3-naphthoxazines from 1-(α-aminoalkyl)-2-naphthols

Scheme 30 Synthesis of 2-aminobenzothiazoles by oxidative dehydrogenation of cyclohexanones and thioureas

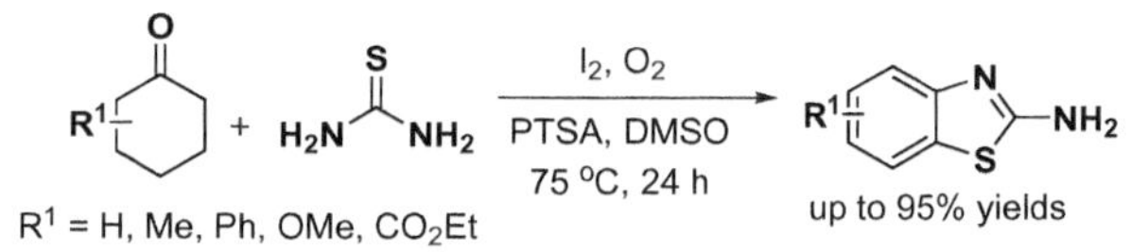

of iodine and oxygen and formed 2-aminobenzothiazole by cascade reaction (Scheme 31) [35]. 2-Aminobenzothiazole formation took place by the intramolecular CDC of C(sp^2)–H and S–H bond. The reaction is initiated by formation of thiourea intermediate **A** from isothionate derivatives and amines. The intermediate **A** reacts with iodine reagent to form iodo-substituted intermediate **B**, which undergoes cyclization followed by elimination of HI to afford 2-aminobenzoxazole (Fig. 8).

Wu and collaborators synthesized diheteroaryl thioethers or thiazole derivatives. They treated aromatic methyl ketones with thiourea in the presence of iodine catalyst to give diheteroarylthioether derivatives (Scheme 32) [36]. Reaction proceeds through iodination, condensation/cyclization/dehydration domino reaction.

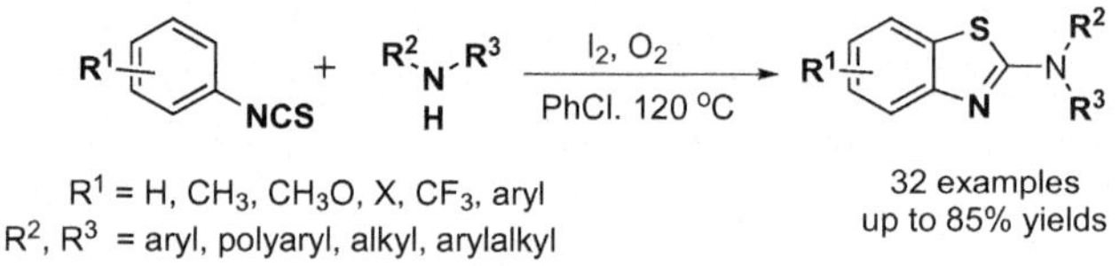

Scheme 31 Iodine-mediated synthesis of 2-aminobenzothiazoles from isothiocyanatobenzenes and amines

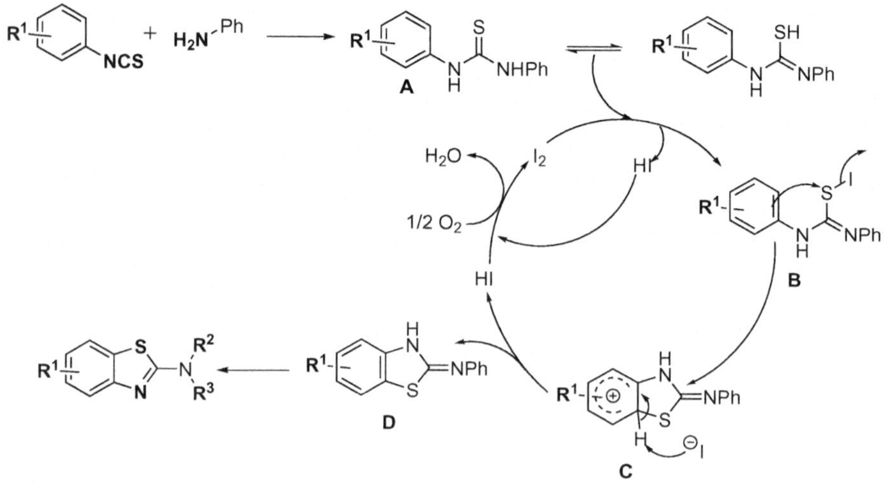

Fig. 8 Mechanism of iodine-catalyzed and oxygen-promoted cascade synthesis of 2-aminobenzothiazoles

Yu and co-workers developed a method for the synthesis of imidazolidin-4-ones by dehydrogenative cyclization under mild conditions. They treated *N*-acyl dipeptide in the presence of KI/TBHP to produce imidazolidin-4-ones by forming peroxide intermediate (Scheme 33) [37].

Kumar and co-workers developed an intramolecular oxidative coupling reaction for the synthesis of *N*-aryl-isoindolinones by functionalizing selective C–H bonds via C–N bond formation. The treatment of 2-alkyl benzamides in the presence of iodine and DTBP catalytic system generates a *N*-centered radical which undergoes a 1,5-H shift followed by iodination and then substitution to give *N*-aryl-isoindolinones (Scheme 34) [38]. The new C–N bond is formed between unactivated C(sp^3)–H and N–H via oxidative coupling. They synthesized indoprofen and DWP205190 drugs by using this method.

Scheme 32 Iodine catalysed synthesis of diheteroaryl thioethers

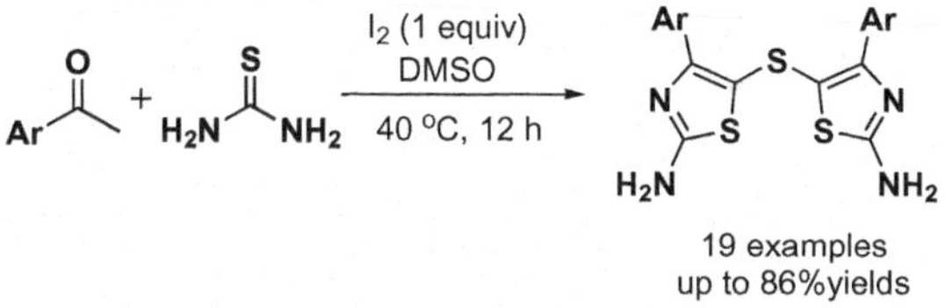

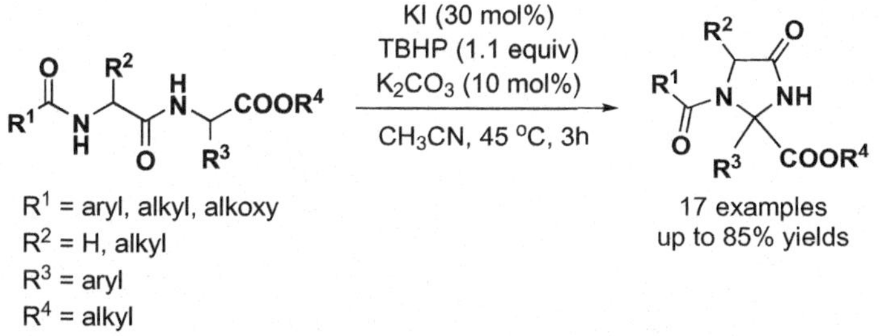

Scheme 33 Synthesis of imidazolidin-4-ones from *N*-acyl dipeptide esters

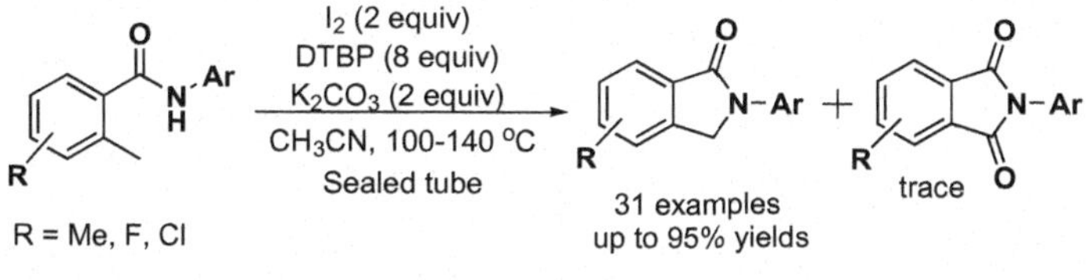

Scheme 34 Synthesis of *N*-aryl-isoindolinones via oxidative C(sp3)–N coupling from 2-alkylbenzamides

2.2 Peroxides Mediated CDC

He and co-workers reported on a DTBP-promoted, radical transformation for the synthesis of 2,5-diaryl 1,3,4-oxadiazoles through *N*-acylation of aryl tetrazoles with aldehydes followed by thermal rearrangement. This one-pot, metal-free, and base-free process was applicable to a wide range of tetrazoles and aryl aldehydes (Scheme 35) [39]. The authors proposed radical reaction pathway. Initially, on heating, formed *tert*-butoxy radical reacts with aldehyde to generate the acyl radical **A**. The acyl radical is further transformed to the corresponding acyl cation **B** in the presence of *tert*-butoxy radical, and then, nucleophilic addition of tetrazole followed by thermal decomposition leads to the formation of oxadiazoles derivatives (Fig. 9).

Li and co-workers described a simple route to access quinazolines by dual amination of methyl arenes and 2-aminobenzamides. The transient formation of *t*-BuOBn arising from the *tert*-butoxy and benzyl radicals is proposed as key intermediate for quinazoline formation (Scheme 36) [40].

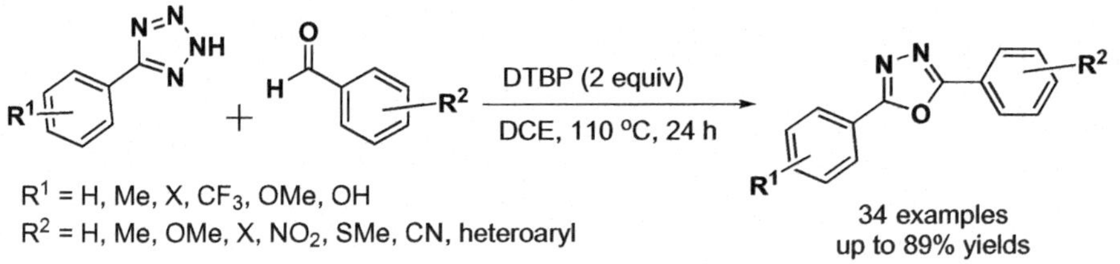

Scheme 35 di-*tert*-Butyl peroxide-catalyzed synthesis of 2,5-diaryl 1,3,4-oxadiazoles

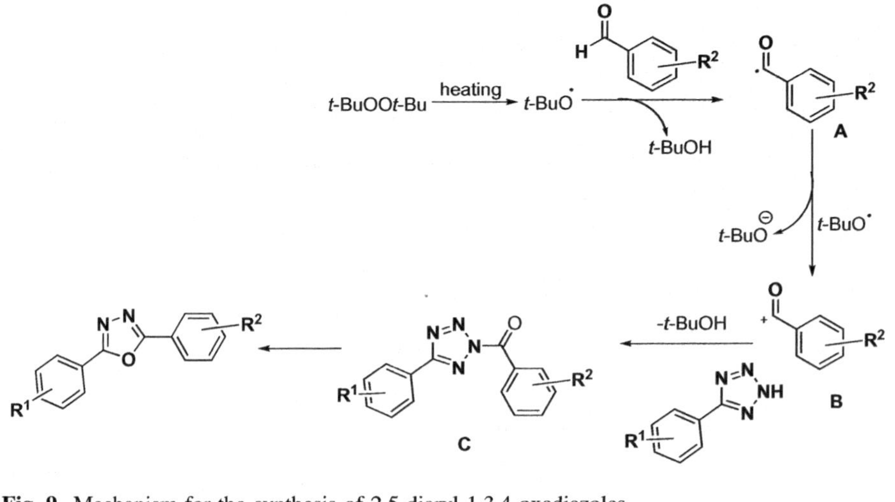

Fig. 9 Mechanism for the synthesis of 2,5-diaryl 1,3,4-oxadiazoles

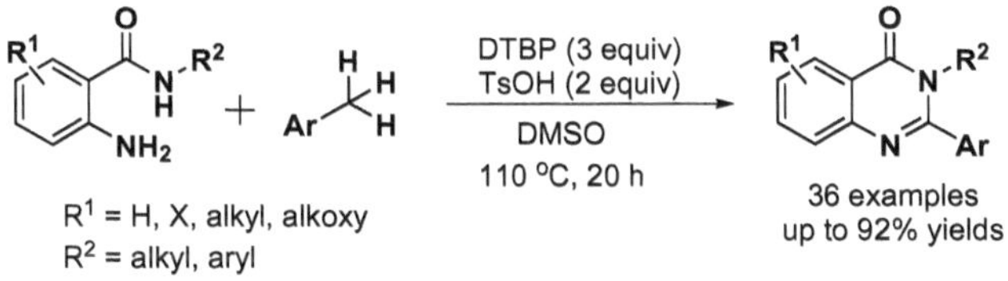

Scheme 36 Synthesis of quinazolinones via oxidative amination

2.3 CDC by Organocatalysts

Wu and co-workers developed a method for the synthesis of polycyclic amines consisting of a tandem CDC process and intramolecular hydroarylation reaction. They treated *N*-aryl tetrahydroisoquinolines with crotonaldehydes in the presence of amine catalyst and acid catalyst to give polycyclic amines (Scheme 37) [41]. This process involves π-conjugated enamine intermediates which act as transient nucleophiles on the in situ generated iminium arising from the oxidation of the tetrahydroisoquinoline derivatives. The authors proposed diamine-initiated reaction pathway. The reaction proceeds through the formation dienamine intermediate **A**. The intermediate **A** will couple with iminium ion intermediate **B**, which is generated from tetrahydroisoquinoline derivatives by the action of DDQ, to give intermediate **C**. The intramolecular hydroarylation of **C** followed by hydrolysis leads to the formation of polycyclic amine derivatives (Fig. 10).

The preparation of various 2-aryl-4-quinolones via intramolecular oxidative C–C coupling reaction has been reported by Long and co-workers. The use of TEMPO enabled a SET process followed by HAT on the starting *N*-aryl methyl-2-aminophenyl ketones. An intramolecular Knoevenagel condensation followed by oxidation accounts for quinolone product formation (Scheme 38) [42].

Liu and co-workers developed a method for the synthesis of chromeno[2,3-*c*]pyrazol-4(1*H*)-ones by using ionic liquid as a catalyst to construct new C–C bond between two C(sp^2)–H bonds. They treated 5-(aryloxy)-1*H*-pyrazole-4-carbaldehydes in the presence of ionic liquids to furnish the chromene-fused pyrazoles via direct annulation (Scheme 39) [43]. 1,3-Dibutyl-1*H*-benzo[*d*][1,2,3]-triazole-3-ium bromide was used as an ionic liquid in this transformation. It was

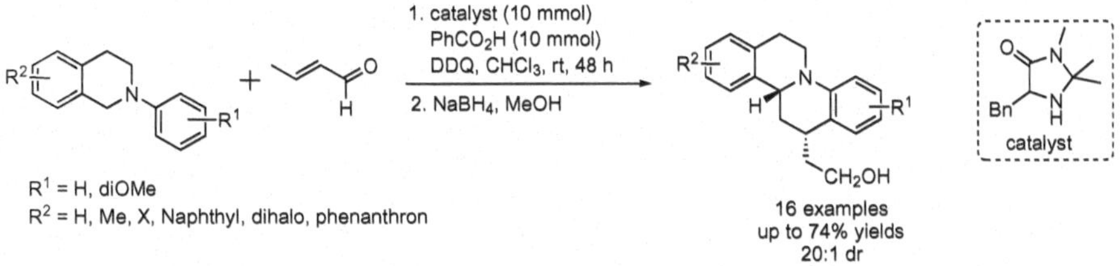

Scheme 37 Tandem method for the synthesis of polycyclic amines from crotonaldehyde

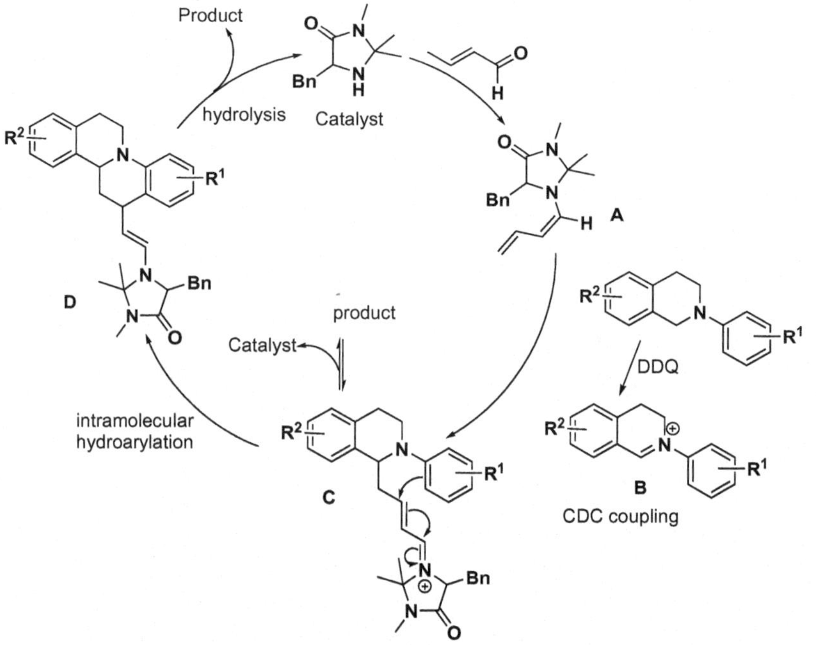

Fig. 10 Mechanism for the synthesis of polycyclic amines

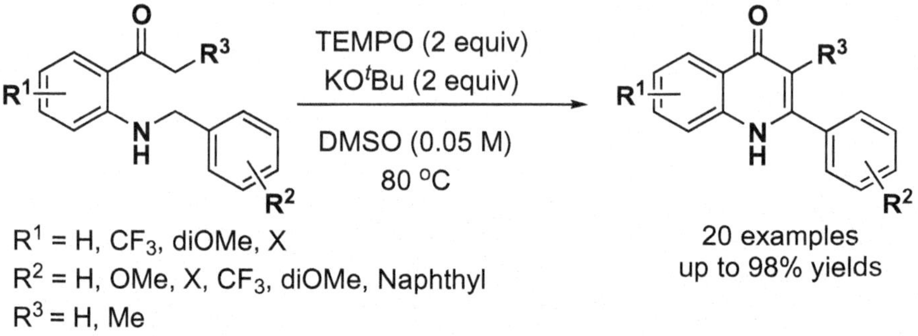

Scheme 38 Metal free synthesis of 2-aryl-4-quinolones via intramolecular oxidative C(sp^3)–H/C(sp^3)–H coupling

shown that the catalyst can be recycled and can be reused up to five reactive cycles without loss of activity.

Xue and co-workers reported a one-pot, sequential reaction of 2-substituted phenanthridinones via CDC process by forming intramolecular C–N bond. They treated *N*-methoxybenzamides with TBAB catalytic system to synthesize 2-substituted phenanthridinones (Scheme 40) [44]. Reaction proceeds through an amidation of arenes followed by regioselective halogenation reaction and

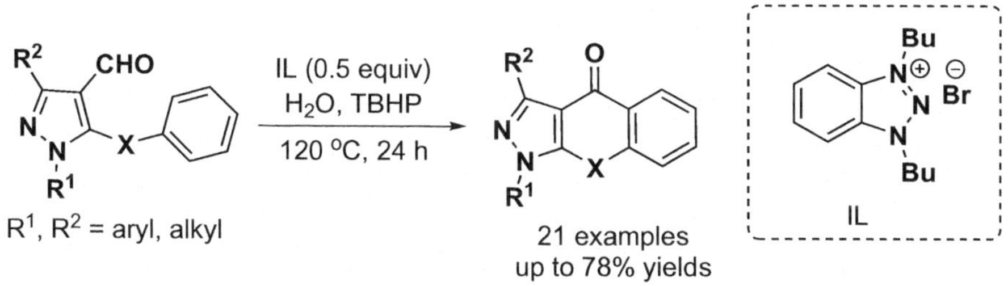

Scheme 39 Ionic liquid promoted synthesis of chromeno[2,3-*c*]pyrazol-4(1*H*)-ones

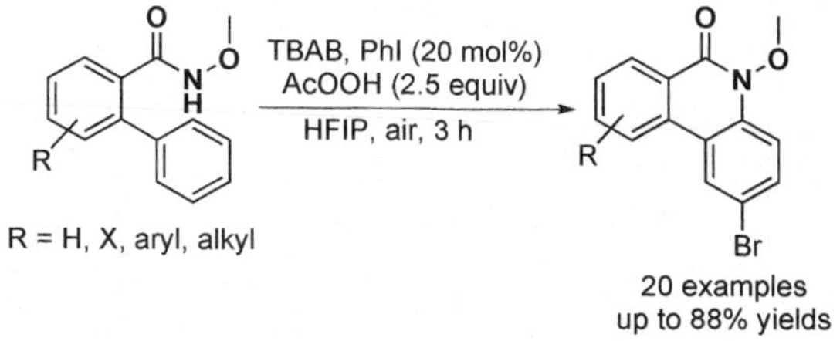

Scheme 40 One-pot sequential synthesis of 2-substitutedphenanthridinones from *N*-methoxybenzamides

eventually forms 2-halophenanthridinones derivatives. The natural product PJ34 was synthesized by using this strategy as a key step. PhI was used as a catalyst and AcOOH acted as oxidant, whereas TBAB served as bromide source. TBAC was used when required to obtain chlorosubstitution on the benzene ring.

Toste and co-workers developed an asymmetric CDC process for constructing C–N bond on tetrahydroisoquinoline derivatives through chiral counter anion catalysis involving CDC. They designed and developed a chiral triazole containing phosphoric acids for constructing C–N bond enantioselectively (Scheme 41) [45].

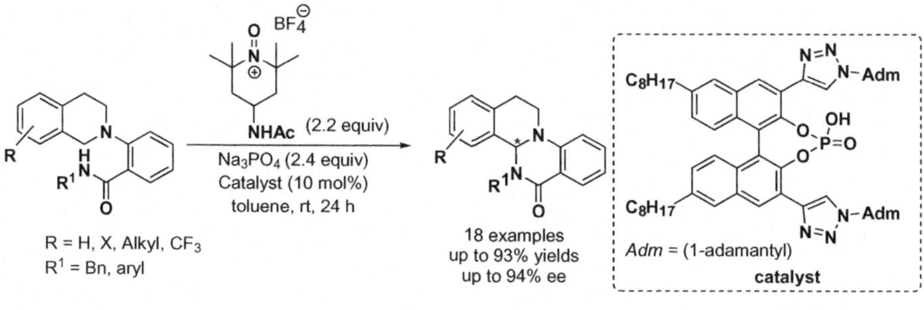

Scheme 41 Construction of C−N bond via Asymmetric CDC

2.4 CDC via Photochemical Reactions

Liu and co-workers synthesized a zwitterionic hydroquinone-fused [1,4]oxazinium derivatives by a photo-induced intramolecular dehydrogenation coupling reaction (Scheme 42) [46]. The new C–C bond was formed between C(sp^3)–H and C(sp^2)–H bonds. They treated semiquinone radicals in the presence of light to afford hydroquinone-fused [1,4]oxazinium derivatives. The crucial semiquinone radicals are stable and prepared from DDQ, and the daylight drives this transformation.

Muñiz and co-workers developed a method for synthesizing piperidine derivatives via intramolecular C(sp^3)–H amination under visible-light-promoted iodine reaction. The reaction proceeds through 1,6-H abstraction from nitrogen-centered radicals. They treated alkylamines in the presence of light and iodine catalytic system to synthesize 2-aryl substituted piperidines in good yields (Scheme 43) [47]. The protocol enlarges the scope of Hofmann–Loffler synthetic method of piperidine synthesis by visible light promotion.

Following the seminal independent reports of König and others on the use of organic dyes as photoredox catalysts to achieve CDC on amines and tetrahydroisoquinoline derivatives [48], Baruah, Deb, and co-workers developed a method for the synthesis of 1,3-oxazines by a visible-light-promoted CDC process. They subjected 1-aminoalkyl-2-napthols to light source, to give 1,3-oxazines under photochemical conditions (Scheme 44) [48]. The reaction proceeds through iminium ion intermediate formation step.

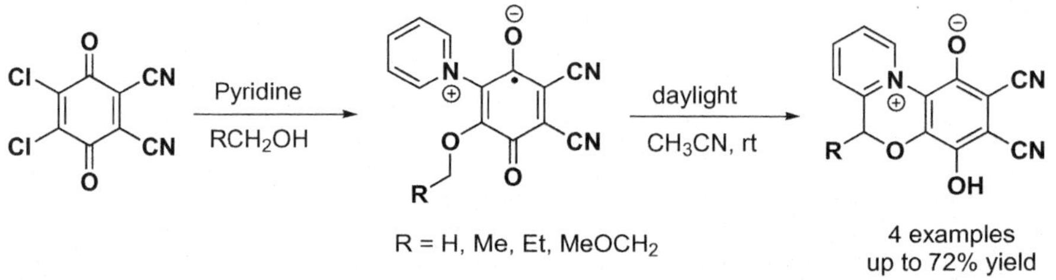

Scheme 42 Photoinduced synthesis of zwitterionic hydroquinone-fused [1,4]oxazinium derivatives

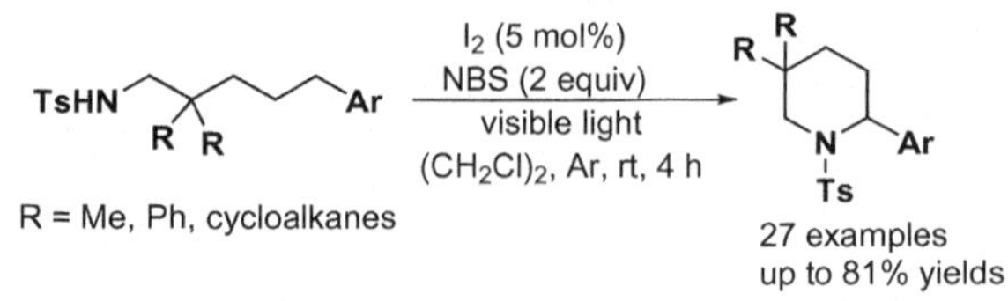

Scheme 43 Iodine-catalyzed synthesis of piperidine via visible light promoted C(sp^3)–H amination

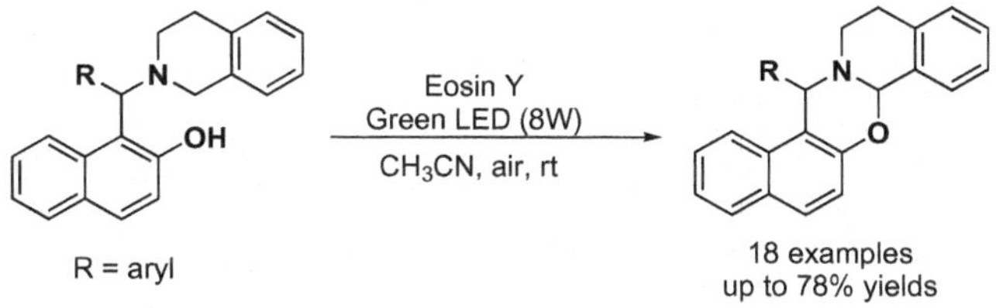

Scheme 44 Eosin Y-catalyzed synthesis of 1,3-oxazines

2.5 Base-mediated CDC

Xia and co-workers developed a method for the synthesis of tetracyclic-3-spiro-oxindole from 3-(2-bromoethyl)indoline-2-one via intramolecular CDC process (Scheme 45) [49]. The reaction proceeds through in situ formation of pyridinium intermediate followed by condensation. Natural products belonging to corynoxine family were synthesized by using this transformation as a key step.

Bisai and co-workers developed a method for the synthesis of oxindoles featuring a quaternary carbon center via intramolecular CDC process. They prepared 2-oxindoles by treating β-*N*-arylamido esters with alkyl halides in the presence of KO^{t}Bu (Scheme 46) [50]. The reaction proceeds through C-alkylation followed by dehydrogenative coupling; the new C–C bond is formed between C(sp^2)–H and C (sp^3)–H via radical pathway [5]. In the presence of base and iodine, α-carbonyl radical intermediate **A** is formed followed by radical cyclization to give

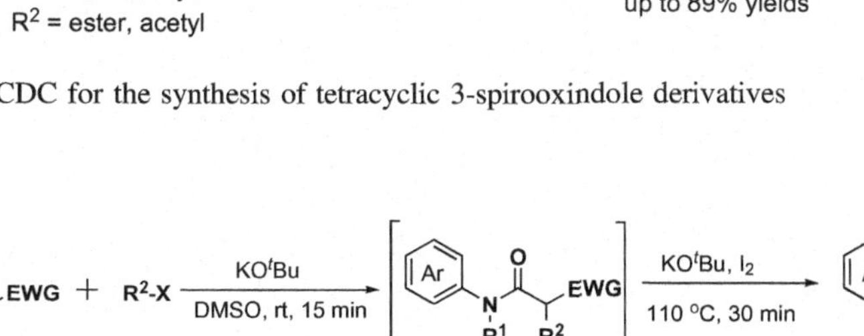

Scheme 45 CDC for the synthesis of tetracyclic 3-spirooxindole derivatives

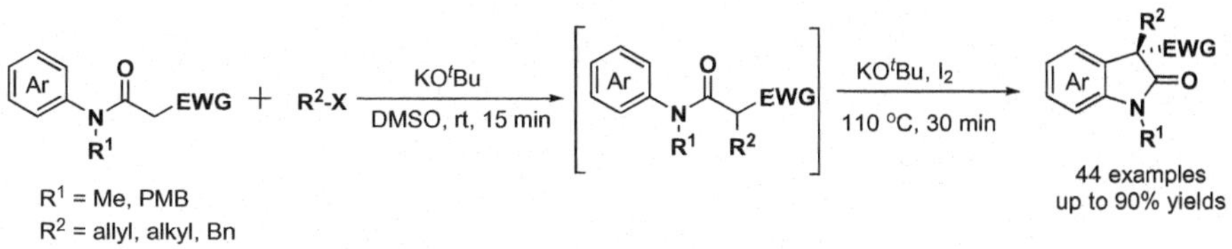

Scheme 46 Intramolecular dehydrogenative coupling for the synthesis of 2-oxindoles

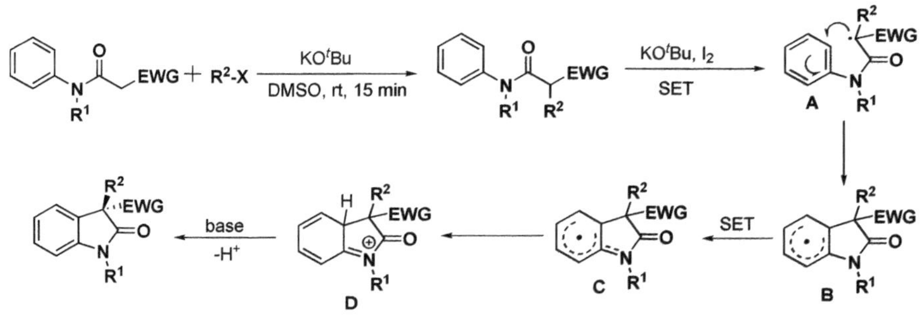

Fig. 11 Mechanism of intramolecular dehydrogenative coupling of C(sp^2)–H and C(sp^3)–H bonds

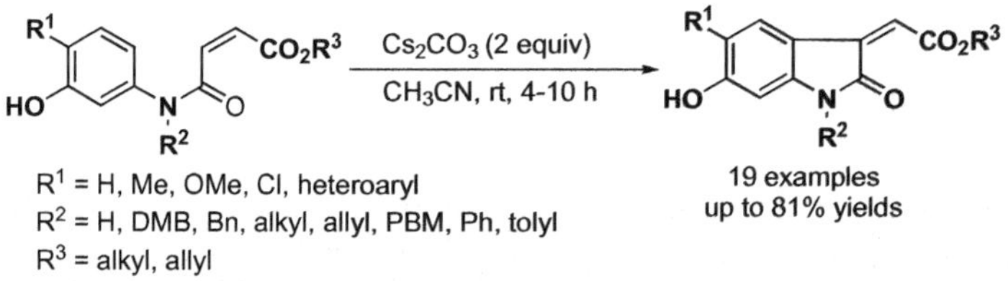

Scheme 47 Synthesis of 6-hydroxy-2-oxindole derivatives by phenoxide cyclization

intermediate **B**. Iminium intermediate **D** will be formed via SET process. The oxindole derivatives arise from aromatization of the corresponding iminium intermediate **D** (Fig. 11).

Ramana and co-workers reported an intramolecular CDC process for making new C–C bond between two C(sp^2)–H bonds to synthesize 6-hydroxy-2-oxindole derivatives via a base-promoted phenoxide cyclization followed by aerobic oxidation. They treated *N*-(3-hydroxyl)monoanilides of maleic esters in the presence of base to give 2-(6-hydroxy-2-oxoindoline-3-ylidene)-acetate derivatives (Scheme 47) [51]. The proposed mechanism was supported by the isolation of intermediate cyclized product, which was subjected to dehydrogenation with base to form oxindole derivatives. The reaction proceeds through intermolecular phenoxide Michael addition followed by aerobic oxidation.

2.6 Miscellaneous Reactions Involving CDC

Tian and co-workers developed a method for synthesizing quinoline derivatives by constructing C–C bond between aliphatic and aromatic C–H bonds. They have developed an intramolecular regioselective coupling reaction for the synthesis of quinoline derivatives by using anilide derivatives and $K_2S_2O_8$ (Scheme 48) [52].

Scheme 48 Metal free synthesis of quinoline derivatives by regioselective couplings of aliphatic and aromatic C–H bonds

Singh and co-workers synthesized a library of 3,4,5-trisubstituted 1*H*-pyrazoles by sulfur-mediated CDC reaction. They treated acetophenone hydrazones with aldehydes in the presence of sulfur to give highly substituted pyrazole derivatives by forming new C–C bond between C(sp^3)–H and C(sp^2)–H bonds (Scheme 49) [53]. Sulfur was used as a elemental catalyst and also as hydrogen acceptor. The sulfur plays a dual role as promoter and also oxidant. This was the first sulfur-mediated CDC process. The reaction initiated by the formation of diimine derivatives **A** formed from the reaction of hydrazine and aldehyde derivatives. The diimine derivatives react with sulfur to form sulfonated intermediate **C** followed by cyclization that leads to the formation of pyrazole derivatives via the intermediates **E** and **F** (Fig. 12).

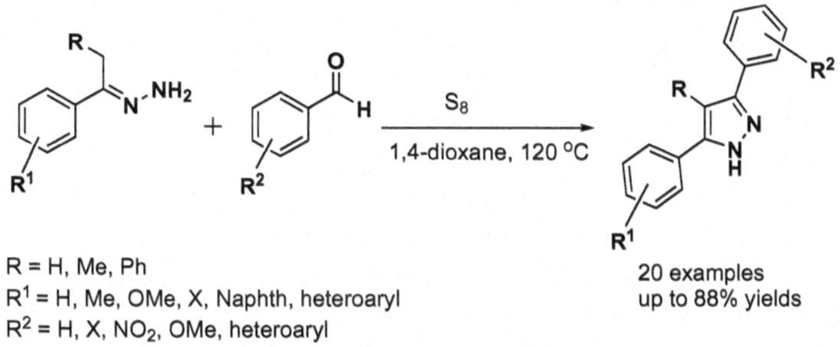

Scheme 49 Sulphur-mediated synthesis of 3,4,5-trisubstituted 1H-pyrazoles

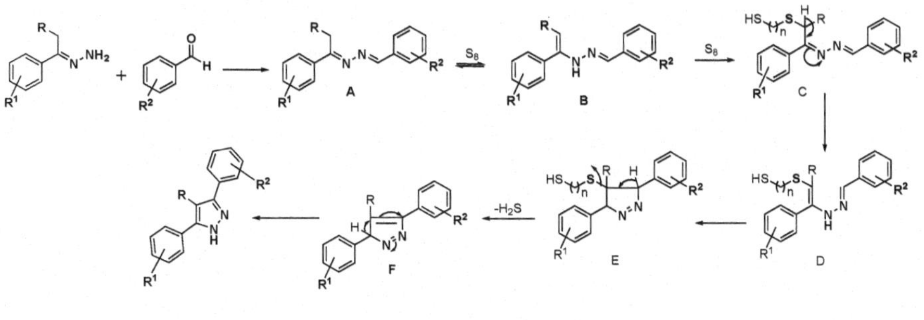

Fig. 12 Mechanism of sulphur-mediated cross-dehydrogenative cyclization of acetophenone hydrazones with aldehydes

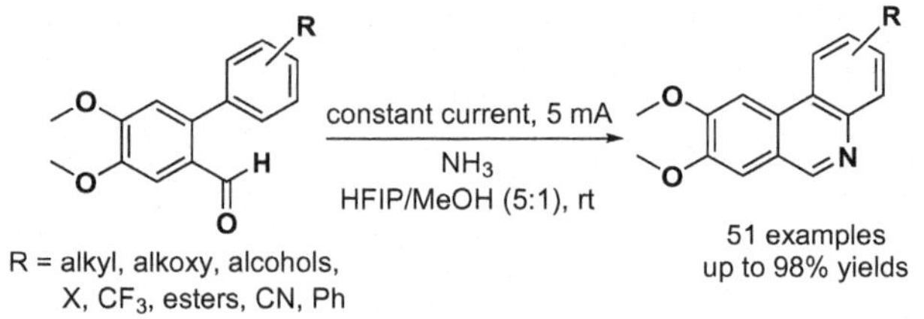

Scheme 50 Regioselective synthesis of *N*-heteroaromatics from biaryl aldehydes and NH_3

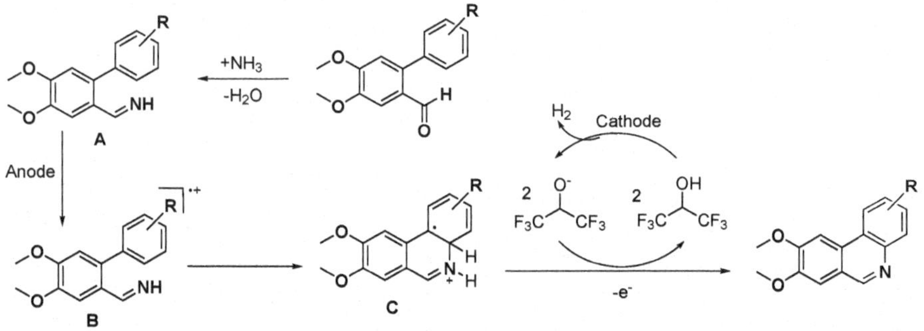

Fig. 13 Mechanism of the C–H/N–H cross-coupling

Since the first report on anodic cyanation of *N*-alkyl anilines at a platinum electrode in 1969, CDC through electrosynthesis has attracted a lot of attention [54a]. Xu and co-workers developed a reagent-free method for constructing C–N bond via CDC process. Anodic condensation of biaryl aldehydes and NH_3 affords pyridine-fused polycyclic *N*-heteroaromatic compounds (Scheme 50) [54b]. The reaction proceeds through the formation of aldimine intermediate **A**, from the aldehyde and ammonia. At the anode, imine intermediate forms radical cation intermediate **C** by loosing electron, and then followed by cyclization. At the cathode, formation of hydrogen from HFIP occured (Fig. 13).

3 Functionalization of *N*-Heterocycles via Metal-Free CDC Approaches

The complexity of *N*-heterocycles directly relates to the potentiality of the biological activities of the molecules. Functionalization of chemical bonds in *N*-heterocycles is one of the best ways of increasing complexity of molecule. CDC methods have emerged as better choices for the functionalization of different C–H and heteroatom–H bonds in heterocycles. Metal-free CDC approaches for the functionalization of *N*-heterocyclic bonds are attractive because these have wide range of applications to green, sustainable chemistry. Metal-free, CDC approaches are scattered in the literature as the applications of oxidants, organocatalysts, and photocatalysts.

3.1 Functionalization of N*-Heterocycles via CDC Using Halogen Reagents*

Wang and co-workers developed a method for stereoselective construction of C–C bond by treating *N*-aryl tetrahydroisoquinolines and 5*H*-oxazole-4-one by using copper catalysis or metal-free strategy. In metal-free approach, $PhI(OAc)_2$ was used as an oxidant (Scheme 51) [55]. The *N*-aryl tetrahydroisoquinolines derivatives react with $PhI(OAc)_2$ to form iminium ion intermediate **A**, which is trapped with heterocyclic enophile to provide the coupled product (Fig. 14).

Itoh and co-workers developed a photo-oxidative C–C bond forming reaction for the construction of new C–C bond via CDC process between tertiary amines and

Scheme 51 Metal-free oxidative CDC of *N*-aryl tetrahydroisoquinolines and 5*H*-oxazol-4-ones

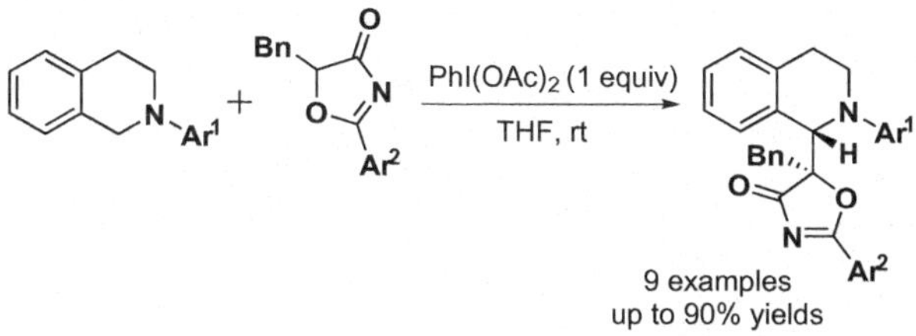

Fig. 14 Mechanism of oxidative CDC of *N*-aryl tetrahydroisoquinolines and 5*H*-oxazol-4-ones

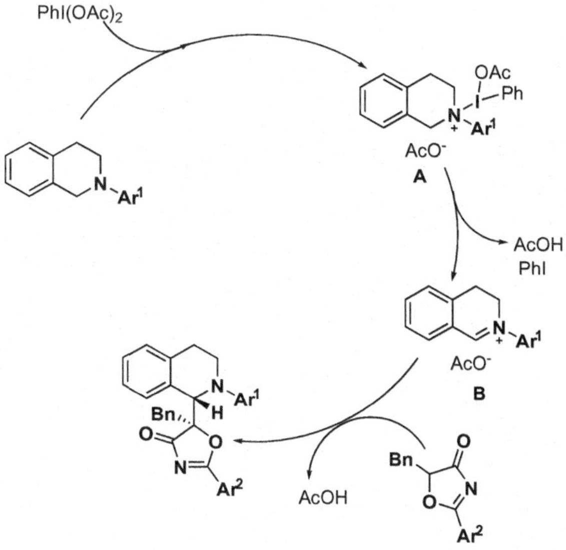

carbon nucleophiles by simply using iodine in catalytic amount under visible light irradiation (Scheme 52) [56].

Itoh and co-workers developed a CDC process for the construction of C–C bond between two C(sp^3)–H bonds. Various THIQs with nucleophiles in the presence of iodine along with H_2O_2 were employed to afford the corresponding coupled products (Scheme 53) [57].

Li and co-workers developed a CDC method for direct functionalization of C (sp^3)–H bond adjacent to nitrogen in tetrahydroisoquinolines by constructing new chemical bond. They treated tetrahydroisoquinolines with different kinds of nucleophiles in the presence of iodine to form new C–C and C–N bonds (Scheme 54) [58]. Direct allylation, alkylation, and amidation were achieved on tetrahydroisoquinolines at C1 position with the use of various Grignard reagents.

Lou and co-workers developed a method for synthesizing structurally divergent α-substituted *C*-benzyl and *C*-allyl heterocycles. They treated *N*-protected

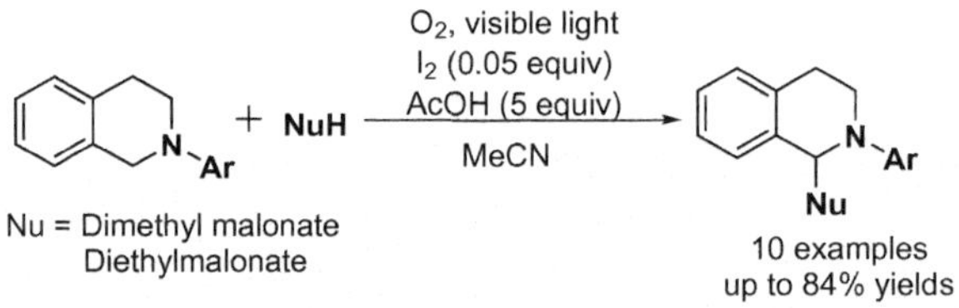

Scheme 52 Photooxidative C–C bond formation between tertiary amines and carbon nucleophiles

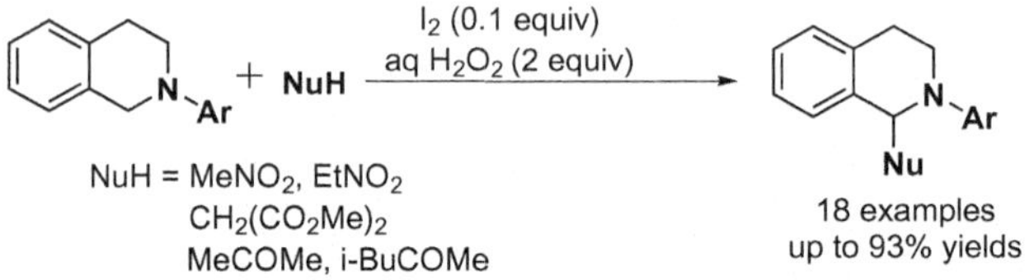

Scheme 53 Iodine-catalyzed CDC for C–C bond formation

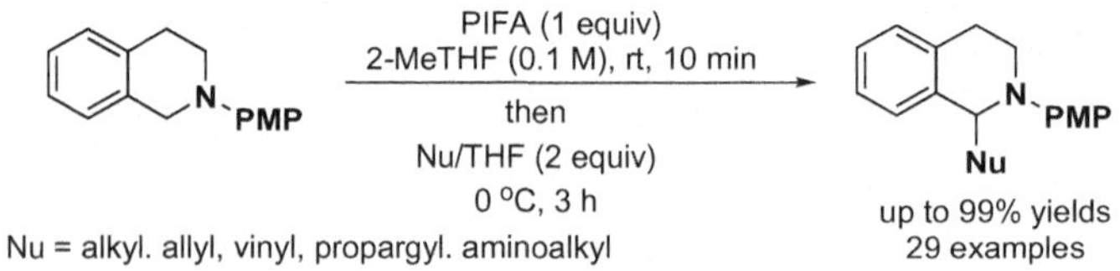

Scheme 54 Iodine-catalyzed C(sp^3)–H bond functionalization

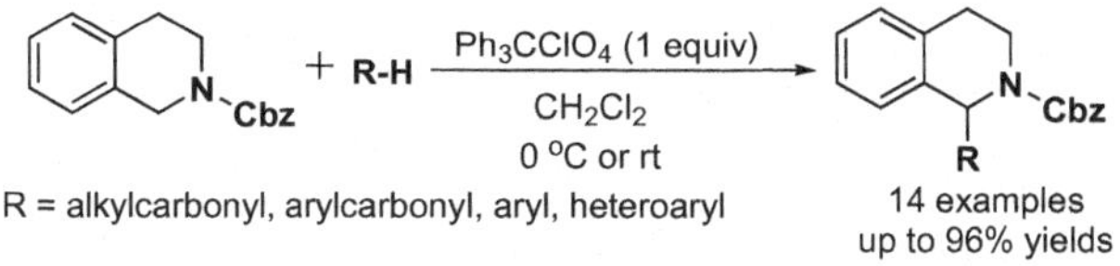

Scheme 55 Metal-free C(sp^3)–H functionalization for α-substituted tetrahydroisoquinolines

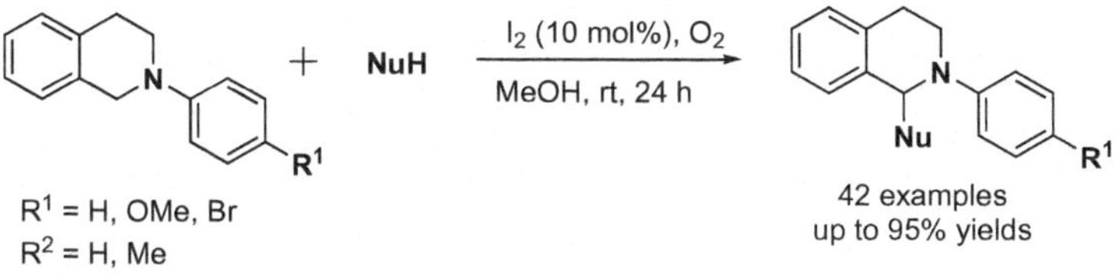

Scheme 56 Iodine-catalyzed C–H functionalization of tetrahydroisoquinolines

carbamates with variety of C–H nucleophiles in the presence of Ph_3CClO_4 to furnish the coupled *N*-heterocycles (Scheme 55) [59].

Dhineshkumar et al. developed an iodine-catalyzed coupling reaction for constructing C–C bond between tetrahydroisoquinolines and carbon nucleophiles. They treated *N*-substituted tetrahydroisoquinolines with nucleophiles in the presence of iodine to give coupled products via CDC process (Scheme 56) [60]. Variety of nucleophiles like 4-hydroxy coumarins, nitroalkanes, phosphites, phenols, amides, imides, organic peroxides, and *N*-methylindoles were employed as coupling partners.

Antonchick and co-workers pioneered the metal-free Minisci-type acylation of *N*-heteroarenes. They treated *N*-heteroarene derivatives with aldehydes in the presence of iodine catalytic system to give acylated *N*-heteroarenes by constructing new C–C bond by CDC process (Scheme 57) [61]. The nucleophilic attack of the in situ generated acyl radical to the protonated *N*-heteroarene derivatives is proposed to rationalize product formation.

Following the first metal-free cross-coupling of quinoline N-oxides with boronic acids to 2-substituted quinolines by Antonchick and co-workers [62a], Wu and co-workers reported a C-alkenylation method for forming C–C bond by CDC process. They treated azaheterocyclic *N*-oxide derivatives with alkenes in the presence of iodine catalyst to give C-alkenylated azaheterocyclic compounds (Scheme 58) [62b]. This is a regioselective alkenylation for the synthesis of *E*-2-styrylazaheterocycles. *N*-Oxide group was found to be crucial for this transformation, acting as directing group for alkenylation and also as internal oxidant during the transformation. Iodine forms iminium ion intermediate **A** with quinoline *N*-oxide by coordinating with oxygen atom. The iminium ion further transformed into radical cation intermediate **B**. The nucleophilic addition of radical cation on styrene derivatives leads to the formation of desired product (Fig. 15).

Wang and co-workers developed an iodine-catalyzed CDC process for coupling pyrazolones with alkenes by forming C–C bond and all-carbon quaternary center.

Scheme 57 CDC of heterocycles with aldehydes

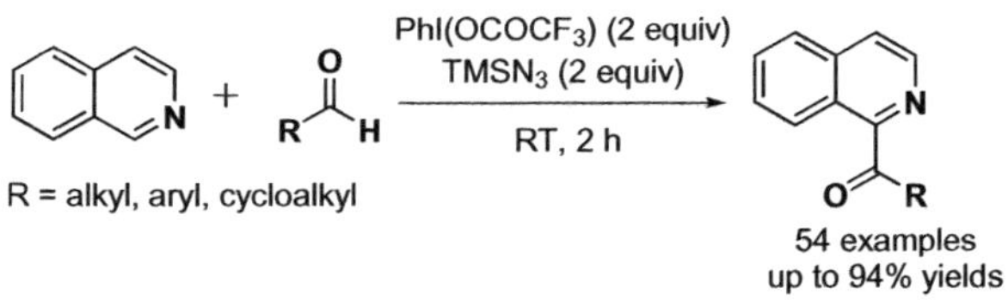

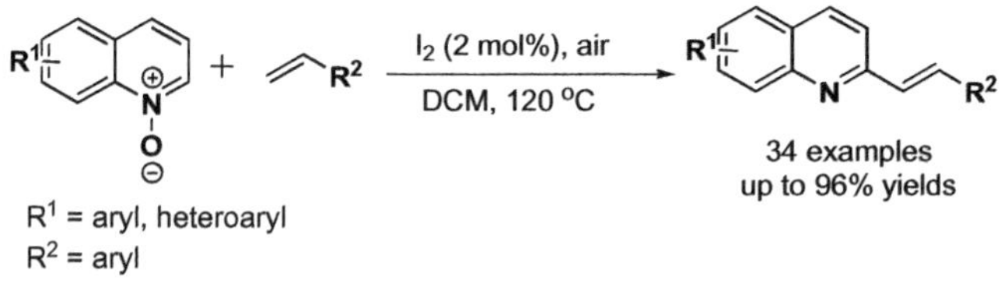

Scheme 58 Iodine-catalyzed C–H alkenylation of azaheterocycle *N*-oxides with alkenes

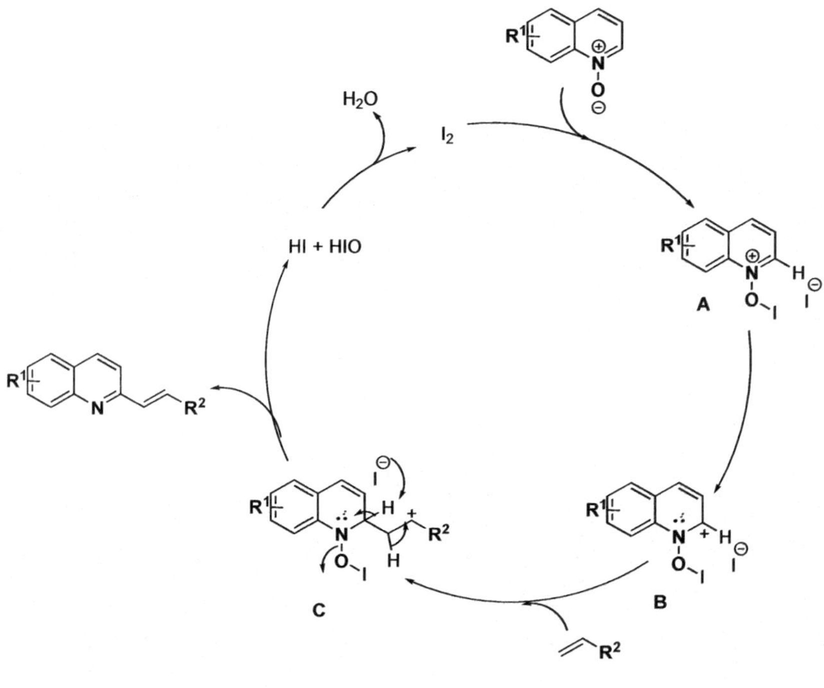

Fig. 15 Mechanism of iodine-catalyzed direct C–H alkenylation

This method enabled the synthesis of alkenyl derivatives of pyrazolones with good substrate scope, moderate-to-excellent yields (Scheme 59) [63].

Liu and co-workers developed a CDC method for olefination of 2-oxindoles at C3 position with alkenes in the presence of iodine catalyst. This is an intermolecular olefination transformation by forming new C–C bond between C(sp^3)–H and C(sp^2)–H bonds by using catalytic amount of iodine and atmospheric oxygen as an oxidant (Scheme 60) [64].

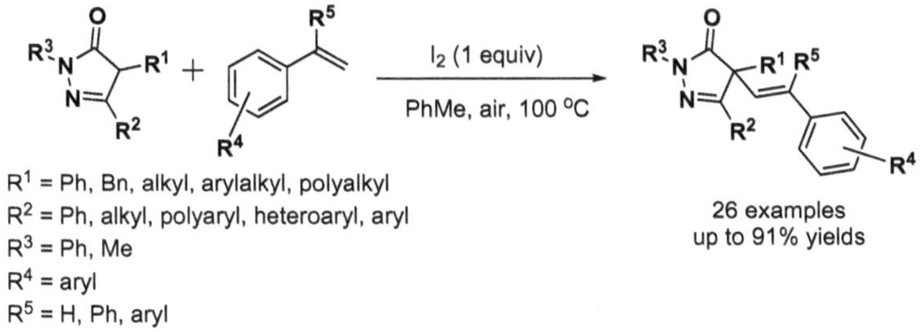

Scheme 59 Iodine-catalyzed CDC of pyrazolones and alkenes

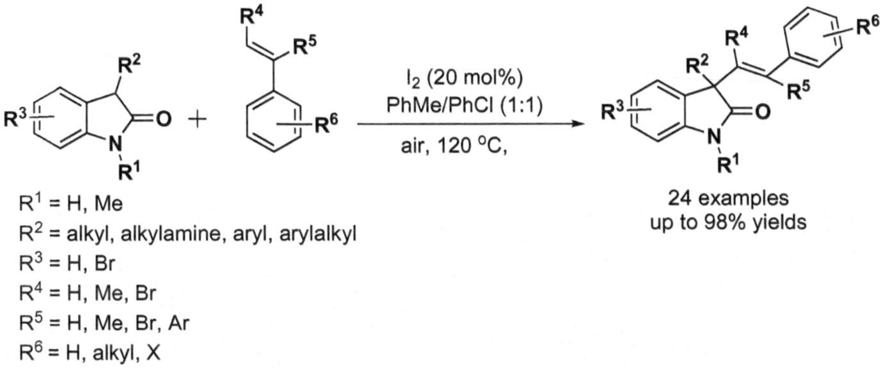

Scheme 60 Iodine-catalyzed CDC of 2-oxindoles and alkenes

Yavari et al. developed a diastereoselective method for cyclopropanation of arylidine malononitriles. They treated 2,6-dimethylquinolines with arylidine malononitriles in the presence of iodine to furnish cyclopropanated arylidine-malanonitriles with good yields. The reaction proceeds through the formation of pyridinium ylide intermediate followed by cyclopropanation (Scheme 61) [65].

Bharate and co-workers developed a CDC process for constructing C–C bonds between C(sp^3)–H and C(sp^2)–H bonds by treating benzylamines with various *N*-heterocycles in the presence of different oxidants. The reaction involved an oxidant control CDC process for acylation and alkenylation at C1 position of various *N*-heterocycles. Ionic liquids promoted the alkenylation with benzylamines,

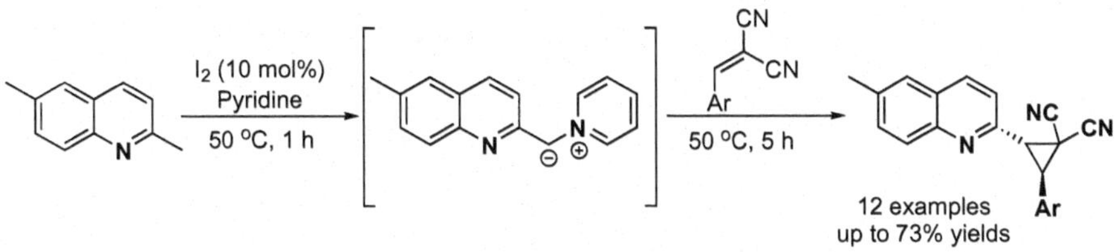

Scheme 61 Iodine-catalyzed diastereoselective cyclopropanation

ionic liquids along with $(NH_4)_2S_2O_8$ promoted the C-benzoylation, and iodine along with HCl promoted the alkenylation of *N*-heterocycles. Benzylation proceeds through the formation of benzoyl radical formation, and alkenylation proceeds through deamination of benzylamines followed by C(sp^3)–H bond activation (Scheme 62) [66].

Patureau and co-workers developed a periodate-catalyzed C–N bond formation reaction for the synthesis of aminated phenols from phenol and amines under mild conditions (Scheme 63) [67].

Kita and co-workers developed a C–C bond formation reaction for the synthesis of *N*-pyrrolyl azoles via oxidative CDC process by using a recyclable iodine reagent as a catalyst. They treated pyrrole with 1,2,3-triazoles in the presence of

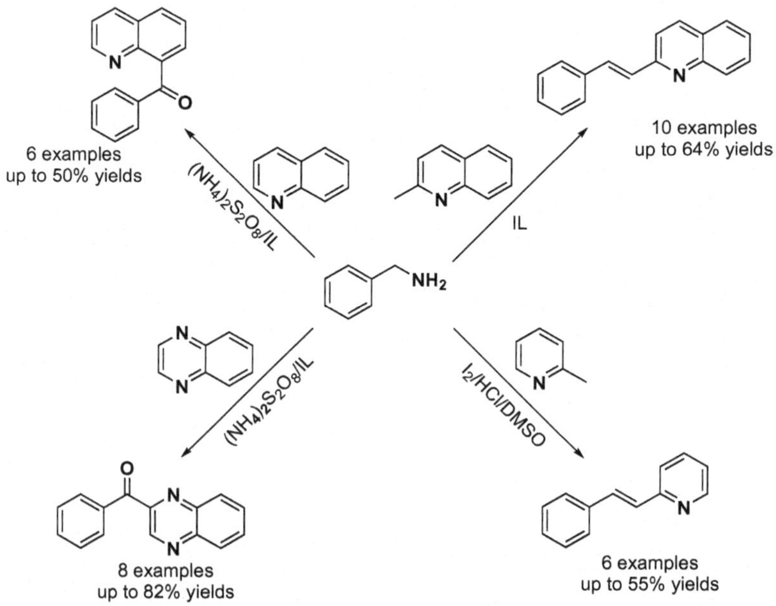

Scheme 62 Oxidant-controlled CDC of *N*-heterocycles with benzylamines

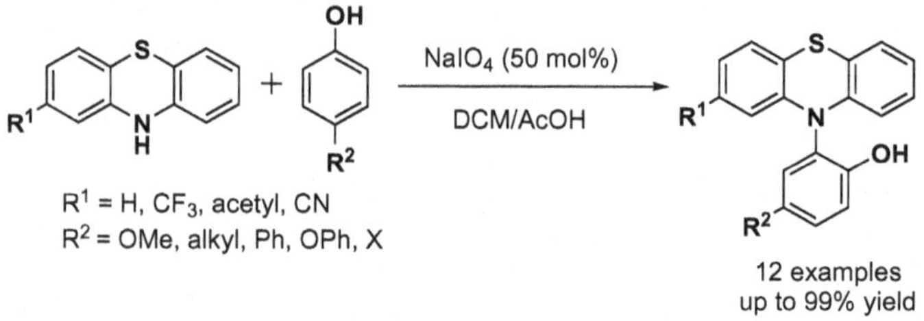

Scheme 63 Periodate-mediated dehydrogenative C−N bond formation

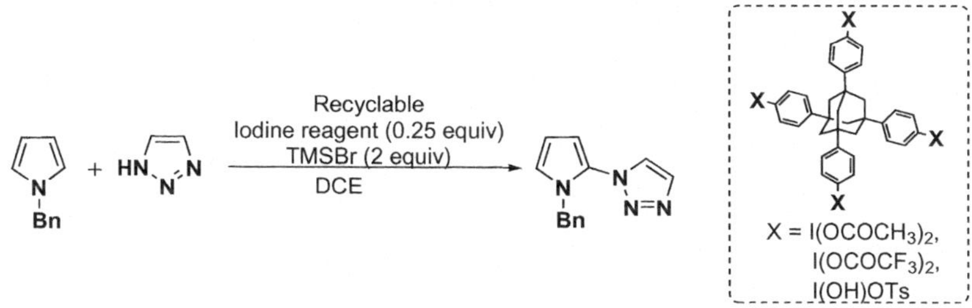

Scheme 64 Iodine-catalyzed oxidative cross-coupling of *N*-pyrrolyl azoles

hypervalent iodine catalyst to furnish *N*-pyrrolyl azoles. They showed that catalyst can be recycled and the catalyst recycling process involved easy methods like solid–liquid separation (Scheme 64) [68].

Yang and co-workers developed an oxidant-controlled CDC protocol for the benzylation of isoquinolines. They treated isoquinolines with methylarenes in the presence of iodine/DTBP or TBHP catalytic system to synthesize benzylated isoquinoline derivatives. DTBP promoted the C-benzylation, while TBHP promoted the *N*-benzylation (Scheme 65) [69]. It was proposed that initially, benzyl radical **A** was generated by the reaction of TBHP and toluene derivatives. The iminium ion intermediate **B** was formed in the presence of iodine, which coupled with benzyl radical in the presence of TBHP/iodine leads to the formation of C-benzylated product (Fig. 16).

Jain and co-workers developed a CDC process for constructing C–N bond between quinoxalinones and amines. They treated quinoxalinones and amines in the presence of iodine to give 3-aminoquinoxalinones via CDC process. The new C–N bond was formed between C(sp^2)–H and N–H bonds, this protocol was applicable for wide range of amines including primary and secondary amines (Scheme 66) [70]. The reaction is initiated by the formation of *N*-iodo-amine **A**, which is formed from the amine and iodine. *N*-iodo-amine reacts with quinoxalinone formed the iodine-coordinated complex **B**, which undergoes nucleophilic addition by amine that leads to the formation of amine-substituted intermediate **C**. By the oxidation, HI was generated along with aminated quinoxalinone derivatives. Iodine was regenerated with the assistance of TBHP from HI (Fig. 17).

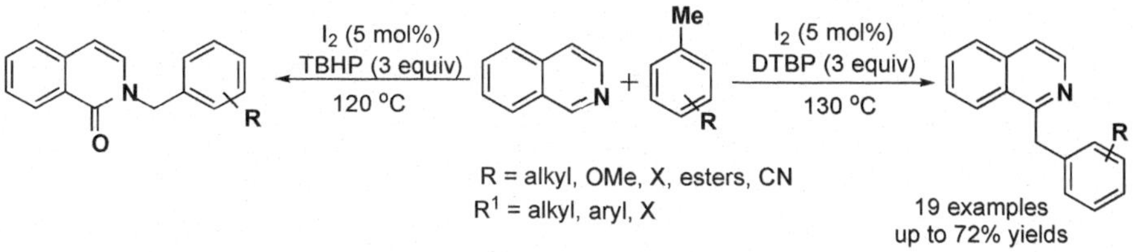

Scheme 65 Iodine-catalyzed CDC for benzylation of isoquinolines

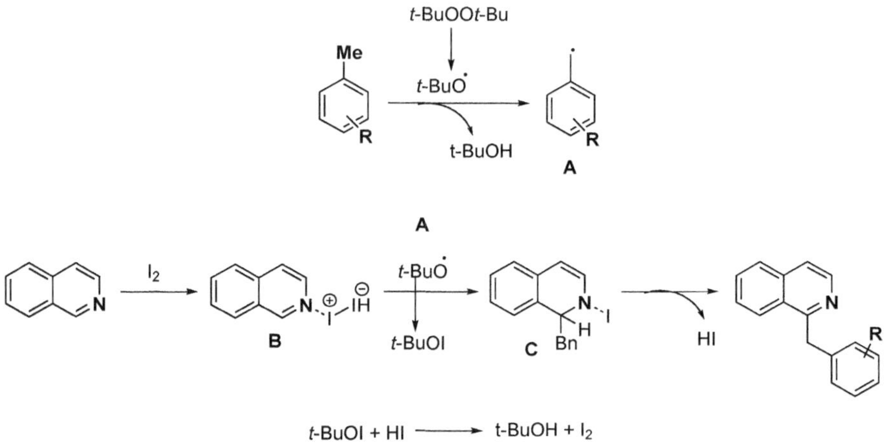

Fig. 16 Mechanism for C1-benzylation of isoquinolines

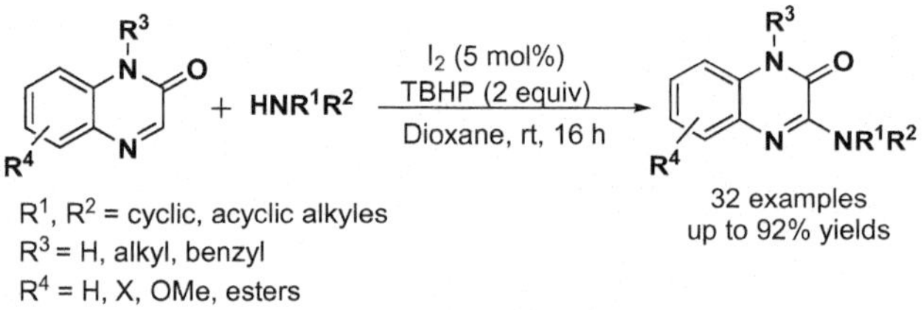

Scheme 66 Iodine-catalyzed synthesis of 3-aminoquinoxalinones via C−N bond formation

Xu, Tan, and co-workers reported a one-pot procedure for synthesis of 2-(piperazine-1-yl)pyrazine derivatives. They treated *N*-substituted piperazines with iodine to give 2-(piperazine-1-yl)pyrazine derivatives; this protocol is attractive since it uses single starting material (Scheme 67) [71]. The authors proposed plausible reaction mechanism for the formation of pyrazine derivatives. *N*-substituted piperazine reacts with iodine form to form *N*-iodo-ammonium intermediate **A**, which forms diimine intermediate **B** by eliminating HI. The diimine intermediate undergoes amination with piperazine to give aminated intermediate **C**. Oxidation and hydrolysis of intermediate **C** give the desired product via intermediates **D** and **E** (Fig. 18).

Following the previous report of Chang and co-workers on the oxidative amination of benzoxazoles [72a], Sun and co-workers developed a method for the synthesis of 2-aminobenzoxazoles by oxidative coupling reaction of secondary amines and benzoxazoles by constructing C–N bond (Scheme 68) [72b].

Barbas III and co-workers developed a C–O bond formation reaction for the synthesis of activated esters. They treated aldehydes with *N*-hydroxyimides in the presence of *n*-Bu_4NBr to give activated esters via CDC process (Scheme 69) [73].

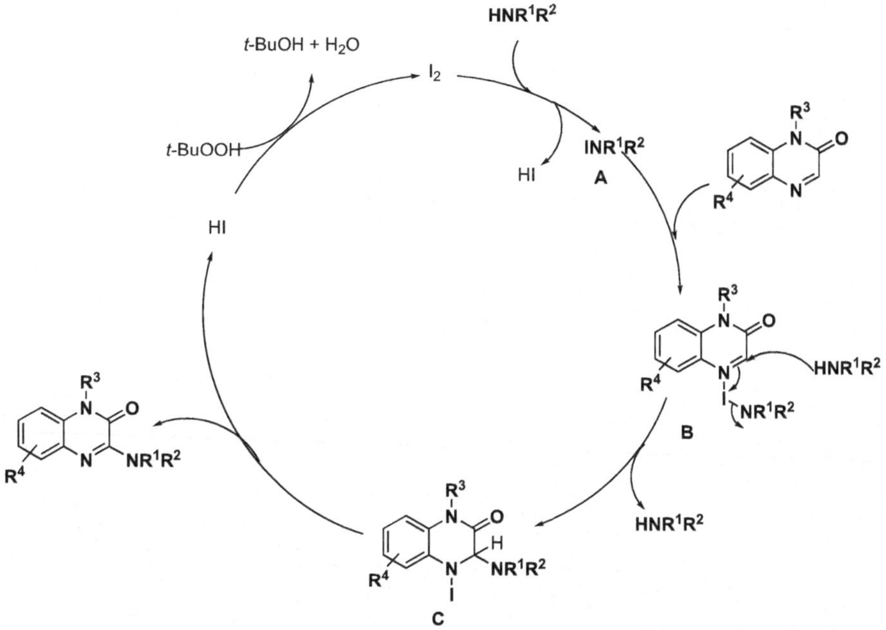

Fig. 17 Mechanism for the synthesis of 3-aminoquinoxalinones

Scheme 67 Iodine-catalyzed one-pot synthesis of 2-(piperazin-1-yl)pyrazine derivatives

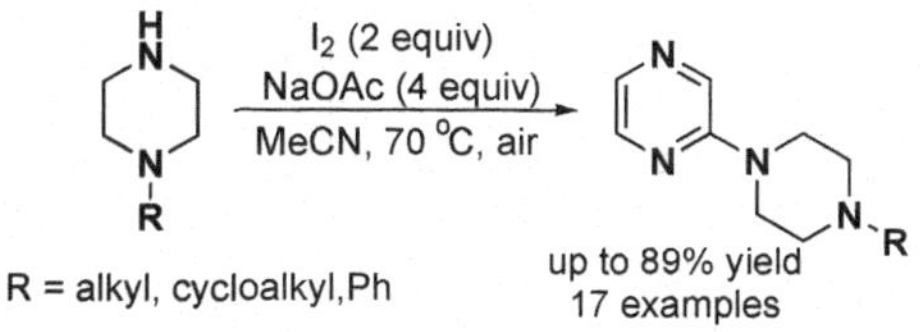

Siddaraju and Prabhu developed a method for regioselective functionalization of C–H bonds on imidazoheterocycles. They treated imidazopyridine derivatives with different kinds of thiols and thianes as coupling partners, in the presence of iodine as a single catalyst, to provide the sulfenylated imidazopyridine derivatives with good regioselectivity. In this iodine-mediated sulfenylation method, new C–S bond was formed selectively and DMSO acted as an oxidant to drive the reaction (Scheme 70) [74]. The authors proposed disulfide intermediate mechanism for this transformation. The thiol or thione forms disulfide intermediate **A** in the presence of iodine/DMSO. The disulfide intermediate reacts with iodine or DMS:I_2 to form iodinated sulfide intermediate **B**, which reacts with imidazopyridine to give the sulfenylated imidazopyridine derivatives (Fig. 19).

Hiebel and Berteina-Raboin developed an iodine-catalyzed regioselective sulfenylation method by constructing C–S bond via CDC process. To deliver sulfenylated indole derivatives, they treated imidazoheterocycles with various

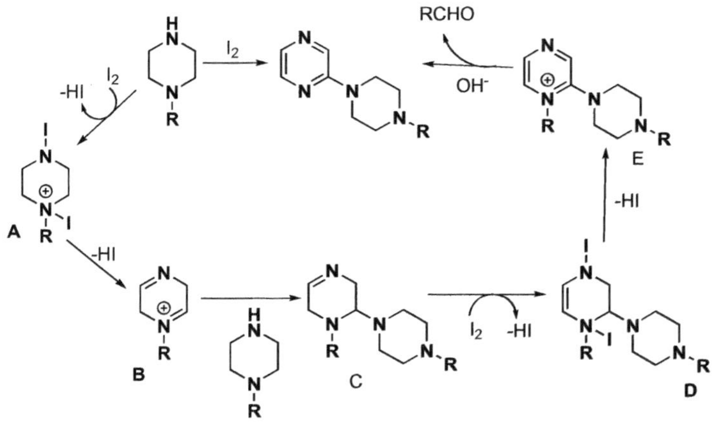

Fig. 18 Mechanism for one-pot synthesis of 2-(piperazin-1-yl)pyrazine derivatives

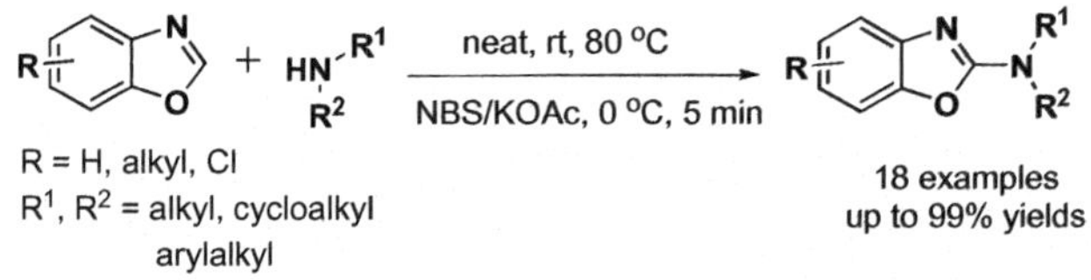

Scheme 68 *N*-Bromosuccinimide-mediated synthesis of 2-aminobenzoxazoles

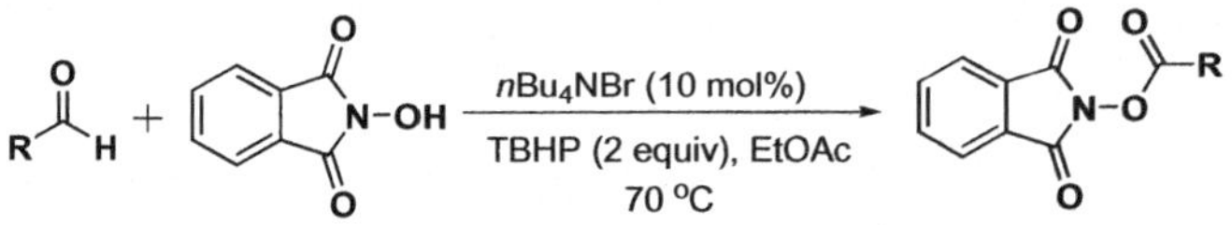

Scheme 69 Organocatalytic amidation and esterification of aldehydes

thiophenols under iodine catalytic system. In this transformation, H_2O_2 was used as an oxidant, and PEG_{400} was used as a solvent (Scheme 71) [75].

Yang, Wang, and co-workers developed an iodine-mediated C–S bond formation reaction via CDC process. They treated pyrazoles with thiols in the presence of iodine, to give C4-sulfenylated pyrazoles. This is one of the highly efficient and practical methods for constructing C–S bond and thio-substituted pyrazole moities as well. In this transformation, DMSO acted as both oxidant and solvent (Scheme 72) [76].

Siddaraju and Prabhu developed an iodine-mediated sulfenylation method for pyrazolones. They treated pyrazolones with a variety of *N*-heterocyclic thiols in the presence of iodine in DMSO to furnish the sulfenylated pyrazolones with wide range of substrate scope. In this protocol, DMSO plays a dual role as an oxidant and

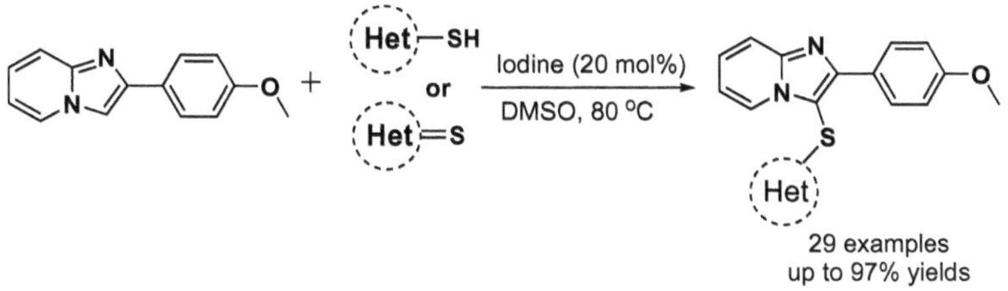

Scheme 70 Iodine-catalyzed regioselective sulfenylation of imidazo heterocycles

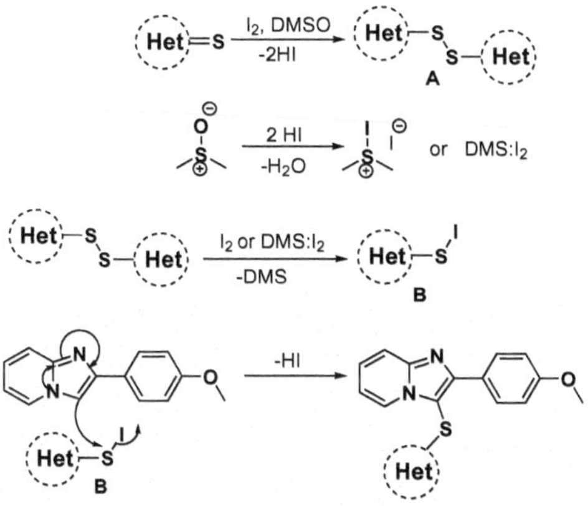

Fig. 19 Mechanism for cross-dehydrogenative sulfenylation of imidazo heterocycles

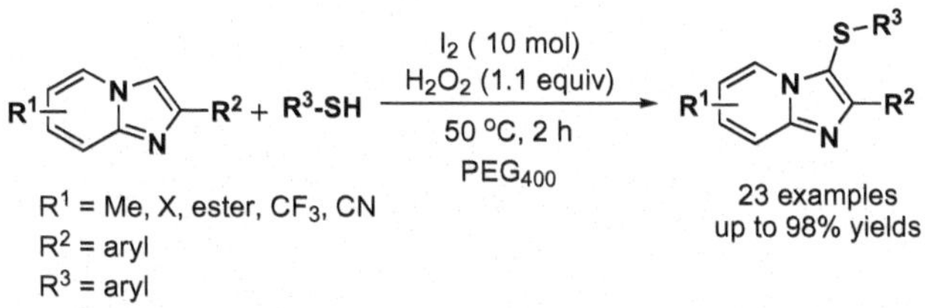

Scheme 71 Iodine-catalyzed regioselective sulfenylation of imidazo heterocycles

Scheme 72 Iodine-catalyzed CDC of pyrazoles and thiols

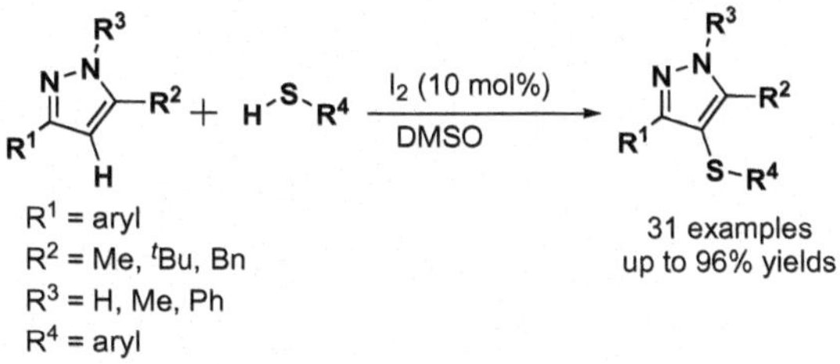

Scheme 73 Iodine-catalyzed sulfenylation of pyrazolones

solvent. The reaction proceeds through thioetherification by forming C–S bond (Scheme 73) [77].

Following the seminal report of Tian and co-workers on the sulfenylation of indoles with hydrazides [78a], Huang and co-workers developed an iodine-catalyzed, regioselective sulfenylation method for indoles. They treated indole derivatives with heteroaromatic thiols in the presence of iodine as a catalyst and H_2O_2 as an oxidant in water to give 3-sulfenyl indoles (Scheme 74) [78b].

Li, Shen and co-workers developed a regioselective C–S bond formation reaction by the demonstration of an iodine-catalyzed sulfenylation of indoles with thiols (Scheme 75) [79].

Wang and co-workers developed a method for the thiolation of indoles. They treated indole derivatives with thiophenols in the presence of iodine along with TBHP to give sulfenylated indoles. In this transformation, iodine was used in

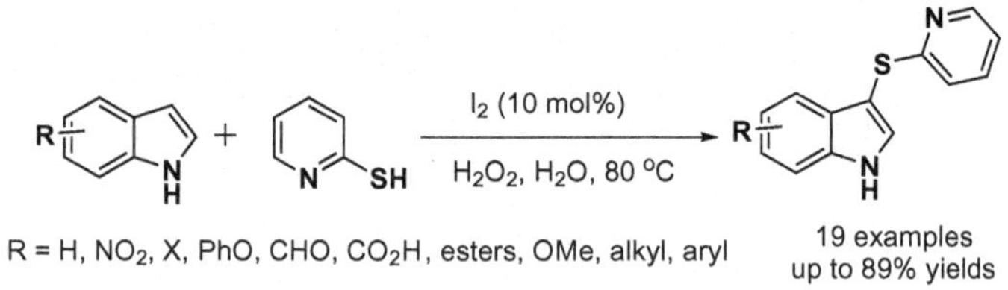

Scheme 74 Iodine-catalyzed regioselective sulfenylation of indoles

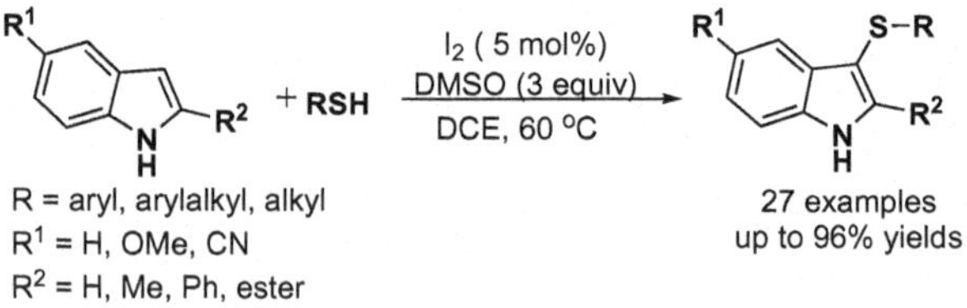

Scheme 75 Iodine-catalyzed regioselective sulfenylation of indoles with thiols

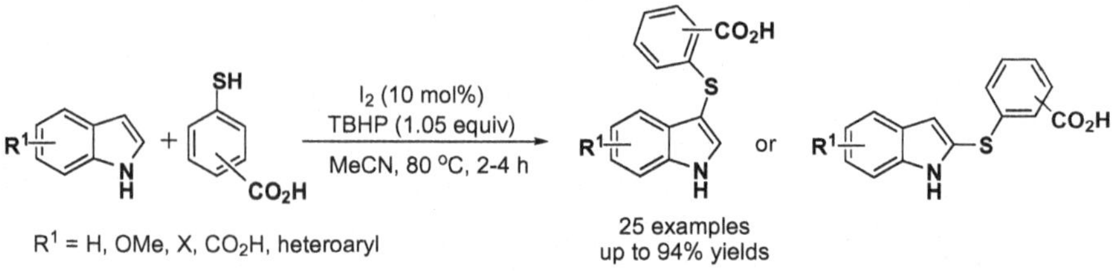

Scheme 76 Iodine-catalyzed controllable sulfenylation of indoles

catalytic amounts to drive the reaction. This method enabled the synthesis of 2- and 3-thio indole derivatives (Scheme 76) [80].

Reddy and co-workers developed a method for the synthesis of C-alkylated 3,4-dihydroisoquinoline derivatives. They demonstrated an oxidative CDC approach for α-C(sp^3)–H functionalization of amines. They treated amine derivatives with nitroalkanes in the presence of KI/TBHP catalytic system to synthesize alkylated amines. They also demonstrated applications of the catalytic system to synthesize 3,4-dihydroquinazoline derivatives, by treating 2-aminobenzamide derivatives with aldehydes to furnish 4-alkyl-3,4-dihydroquinazoline derivatives (Scheme 77) [81].

Li, Meng and co-workers developed a method for constructing new C–N bond by using dimethylamines and dimethylamides as coupling partners. They treated amides with amines in the presence of KI to synthesize coupled products by using TBHP as an oxidant (Scheme 78) [82]. The authors proposed a radical reaction mechanism for this transformation. TBHP reacts with iodine to form *tert*-butyl oxide radical, hydroxyl anion, and iodine species, which further drives the reaction. Amide derivatives react with *tert*-butyl oxide radical to form radical amide **A**, which is further transformed as iminium ion intermediate **B** with iodine, and then, amination of the iminium ion intermediate with amine nucleophile completes the reaction cycle by synthesizing α-amino-substituted amide derivatives (Fig. 20).

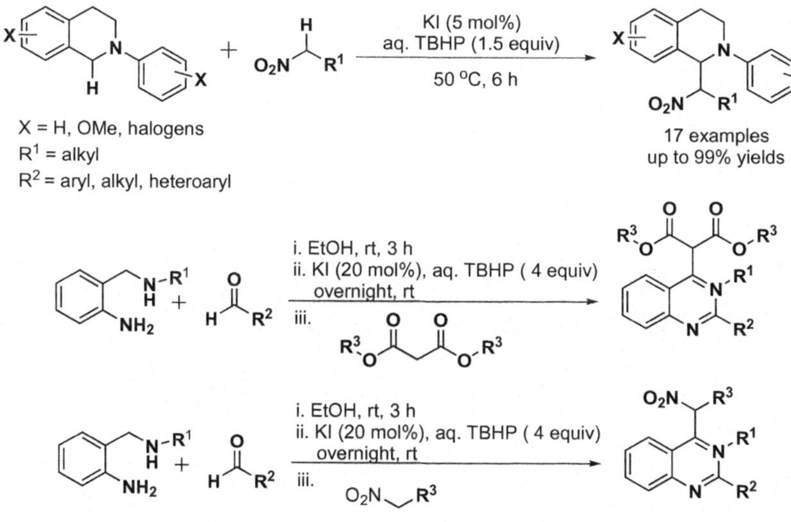

Scheme 77 Oxidative CDC of amines using KI/TBHP catalytic system

Scheme 78 KI-catalyzed imidation of C(sp^3)–H bond

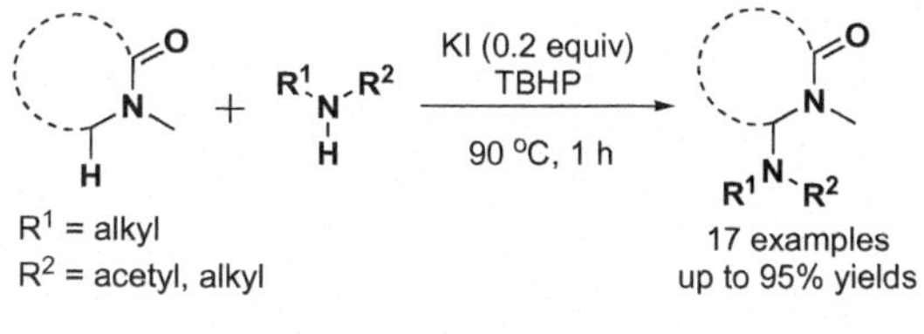

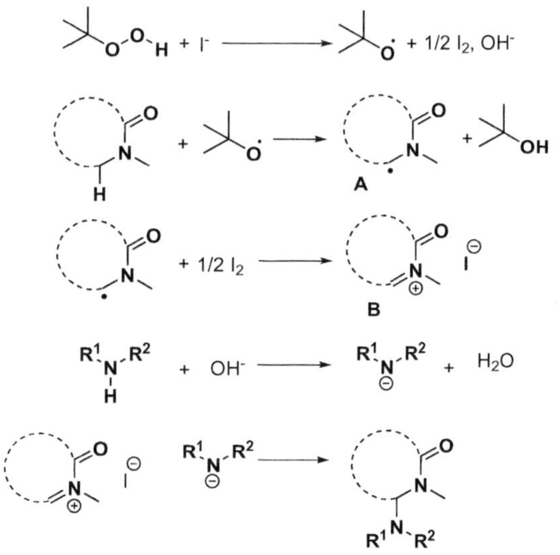

Fig. 20 Proposed mechanism for imidation

3.2 *Functionalization of* N*-Heterocycles via CDC Using Peroxides*

Guo and co-workers developed a simple protocol for the C-benzylation of *N*-heterocyclic oxides with toluene used as a benzyl source via CDC process. This method was applicable to wide range of heterocyclic oxides like isoquinoline oxides, pyridine oxides, and also substituted toluene derivatives. They treated heteroaryl oxides with toluene derivatives in the presence of dicumyl peroxide (DCP), as a single regent to give C1 benzyl heterocycles (Scheme 79) [83].

Qu and co-workers synthesized 3-acylated quinoxalin-2(1*H*)-ones by treating quinoxalin-2(1*H*)-ones with aryl aldehydes in the presence of TBHP. They described an oxidative C–C bond formation reaction for efficient acylation of quinoxalinones (Scheme 80) [84].

Following the metal-free acylation of *N*-heterocycles by Antonchick and co-workers [61], Liu and co-workers developed a CDC process for C-acylation of

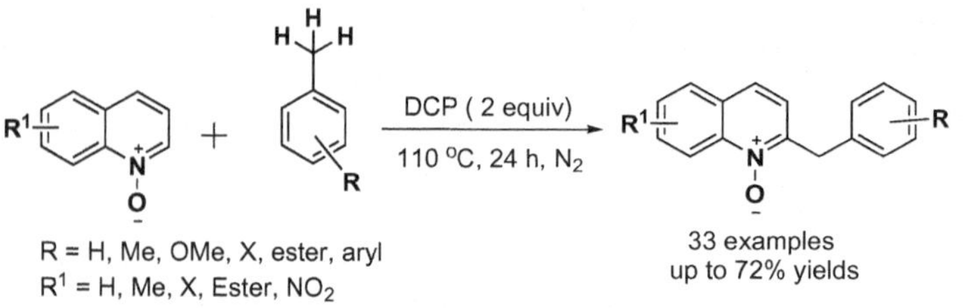

Scheme 79 Cross-dehydrogenative benzylation of heterocyclic *N*-oxides with toluenes

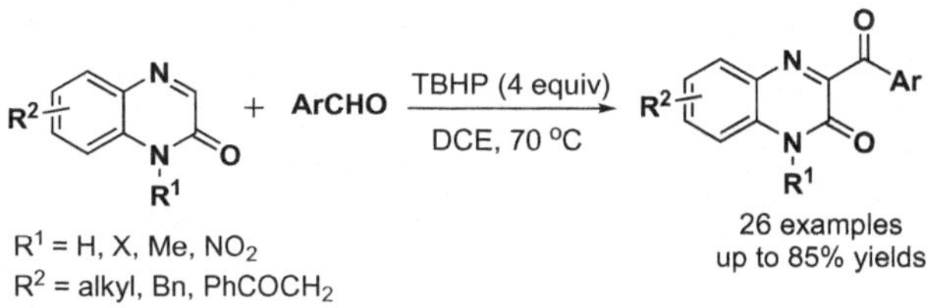

Scheme 80 Synthesis of 3-acylated quinoxalin-2(1*H*)-ones via oxidative coupling of quinoxalin-2(1*H*)-ones and aryl aldehydes

heterocycles with aldehydes in the presence of TBHP along with the presence of Brönsted acids. They treated heterocyclic compounds with aldehydes under the TBHP/TFA reagent system, to give structurally diverse C-acyl-substituted heterocycles. They demonstrated practically enabled protocol for construction of C1 acyl-substituted heterocycles (Scheme 81) [85].

Singh and co-workers developed a method for synthesis of indenoquinolinones, 4-azafluorenones, and fluorenones by forming C–C bond between two sp^2 atoms. They reported a TBHP promoted, eco-friendly, economical, intramolecular CDC method for functionalization of carbonyl C–H bond. TBHP directly functionalized the carbonyl C–H bond by forming radical and initiated the transformation, followed by the formation of C–C bond via CDC method (Scheme 82) [86].

Lei and co-workers developed a visible-light-promoted C–C bond formation reaction. They treated *N*-heterocyclic aromatic compounds with aldehydes under blue LED irradiation, to give acylated *N*-heterocyclic compounds via oxidative acylation (Minisci-type reaction) [61, 87]. Aliphatic, aromatic aldehydes were well tolerated for this transformation. The authors postulate the transient formation of an

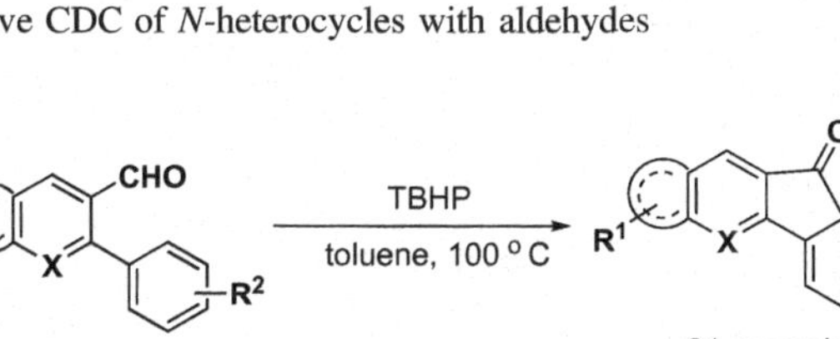

Scheme 81 Oxidative CDC of *N*-heterocycles with aldehydes

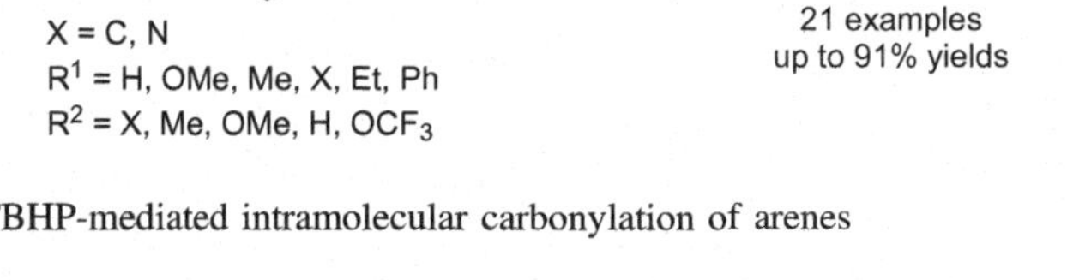

Scheme 82 TBHP-mediated intramolecular carbonylation of arenes

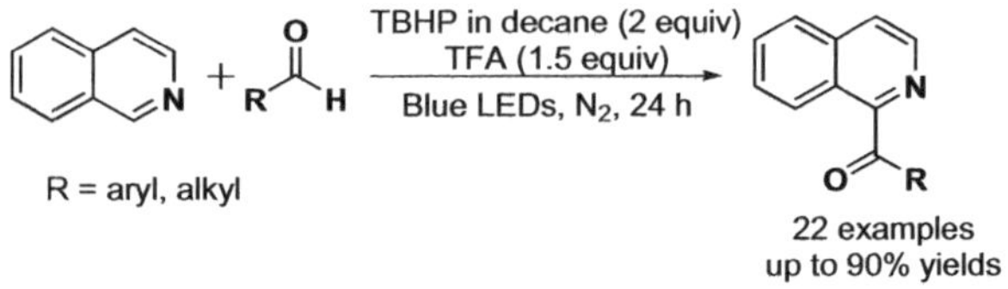

Scheme 83 Visible-light induced oxidative C–H acylation with aldehydes

EDA complex between the protonated heteroarene and ^{t}BuOOH which facilitate the homolytic cleavage of the hydroperoxide (Scheme 83) [87].

Kwong and co-workers developed a method for constructing new C–C bond between C(sp^2)–H and C(sp^3)–H bonds. They treated indoles with cyclic ethers in the presence of DTBP as an oxidant to give C-alkylated indoles by oxidative coupling reaction. This transformation was practically viable for coupling unactivated cyclic ethers with indoles via construction of C–C bonds (Scheme 84) [88].

Cai and co-workers reported a method for the synthesis of 2-substituted azoles by treating azoles with DCP, methyl arenes, and cyclo-alkanes under DTBP, to give 2-substituted azoles. DTBP acted as an oxidant, DCP acted as a methylating agents and oxidant. Wide range of *N*-heterocycles like benzothiazole, benzoxazole, and benzimidazole were methylated with DCP and benzylated with methylarenes in the presence of DTBP with good substrate scope (Scheme 85) [89].

Wang and co-workers developed a CDC method for *N*-acylation with aldehydes. They treated benzimidazoles with aldehydes in the presence of DTBP to give

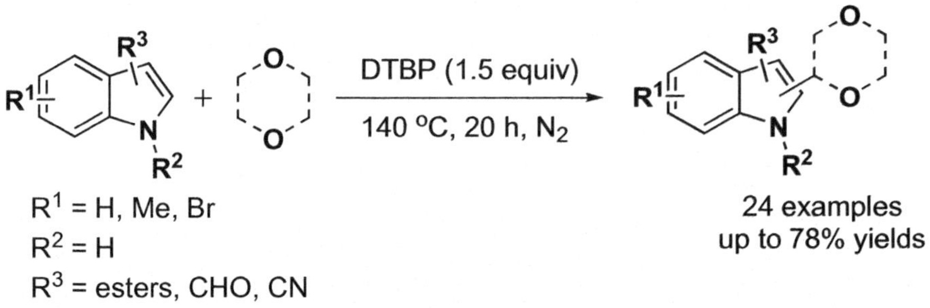

Scheme 84 Oxidative coupling of indoles and cyclic ethers

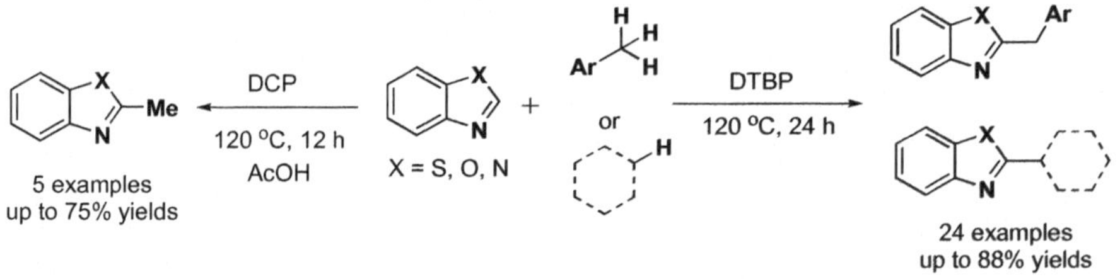

Scheme 85 Radical C–H alkylation of azoles for the synthesis of 2-substituted azoles

N-acylbenzimidazole derivatives. Oxidation of the generated acyl radical intermediate to acyl cation followed by the nucleophilic attack by the benzimidazoles is proposed as key step (Scheme 86) [90].

Singh and co-workers developed a CDC method for C-amination of saturated heterocycles with azaarenes as amine source. They treated *N*-azole derivatives and amides as coupling partners to give C-aminated product by constructing new C–N bond via CDC process. The combination of TBHP/TBAI served as versatile reagent system for the oxidative cross-coupling. This transformation was applicable to various types of azoles like triazoles, tetrazoles, pyrazoles, benzimidazoles, and amides like *N*-methylamides, *N,N'*-dialkylbenzamides, and *N*-alkylpyrrolidines (Scheme 87) [91].

Kumar and co-workers developed a method for synthesizing the aminomethylated imidazoheterocycles via regioselective functionalization of imidazoheterocycles. They treated imidazoheterocycles with 2-amino pyridines in the presence of TBHP, to give aminomethylated imidazoheteroarenes in moderate-to-good yields. In this transformation, TBHP plays a dual role as methyl source and hydrogen acceptor (Scheme 88) [92].

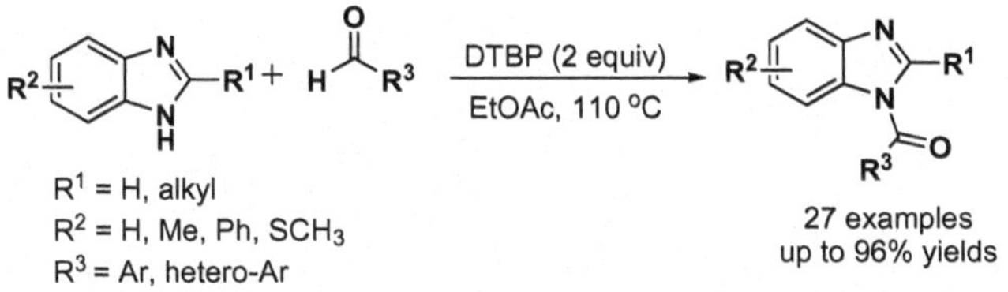

Scheme 86 CDC of benzimidazoles with aldehydes

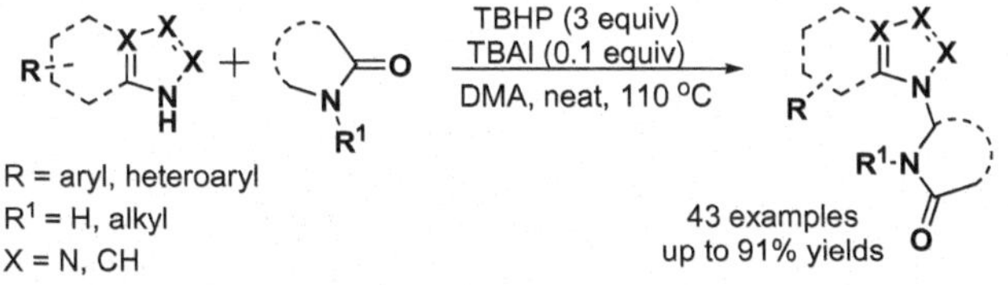

Scheme 87 CDC of HN-azoles with α-C(sp^3)–H amides

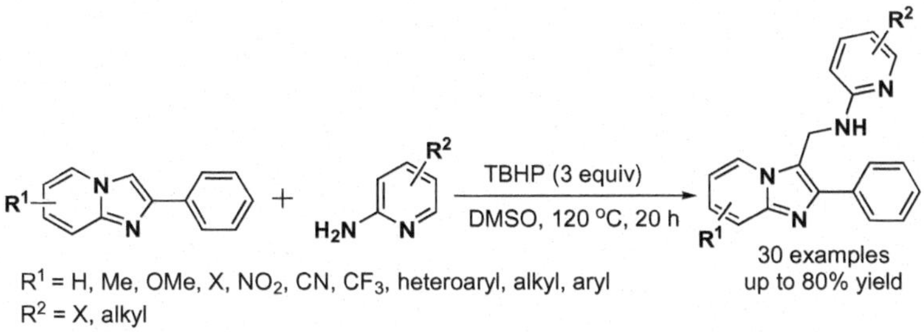

Scheme 88 Synthesis of aminomethylated imidazo heterocycles

3.3 *Functionalization of* N*-Heterocycles via CDC Using Organocatalysts*

Wang and co-workers developed a method for oxidative coupling of α-C(sp^3)–H of tertiary amines and α,β-unsaturated γ-butyrolactam to give Morita–Baylis–Hillman-type coupled products. This method enabled the synthesis of α-substituted, optically active, tetrahydroisoquinoline scaffolds where the organocatalyst facilitates the formation of the nucleophilic 2-hydroxypyrrole intermediates (Scheme 89) [93].

Wang and co-workers developed an enantioselective, oxidative CDC method for constructing C–C bond. They treated tertiary amines with ketones in the presence of chiral organocatalyst to give C1-alkylated tetrahydroisoquinoline derivatives. Taking advantage of the imine/enamine equilibrium in the presence of proline-based organocatalyst, this method enabled the formation of C–C bond between two C(sp^3)–H bonds enantioselectively. This transformation was the first example on metal-free, asymmetric, oxidative coupling reaction of tertiary amines with simple ketones by organocatalysis (Scheme 90) [94].

Liu and co-workers developed a CDC process for the coupling of *N*-carbamoyl cyclic amines with aldehydes, enantioselectively. They treated *N*-protected tetrahydroisoquinolines with aldehydes in the presence of imidazole-derived organocatalyst to construct C–C bond, enantioselectively (Scheme 91) [95].

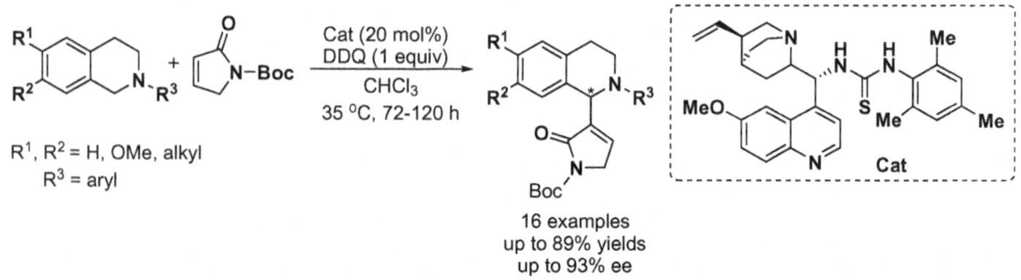

Scheme 89 Asymmetric oxidative coupling of α-C(sp^3)–H of tertiary amines

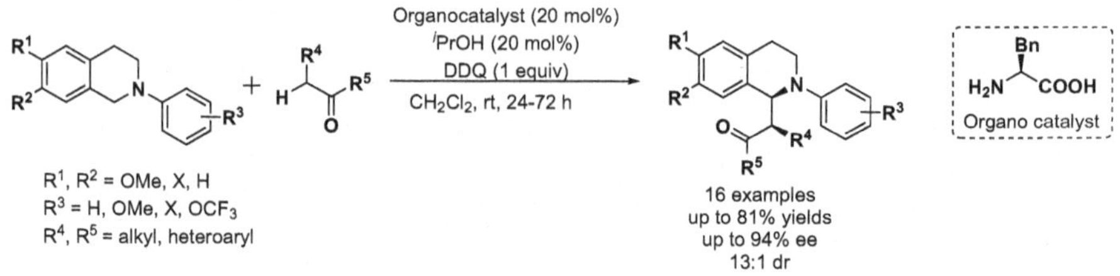

Scheme 90 Enantioselective oxidative CDC of tertiary amines

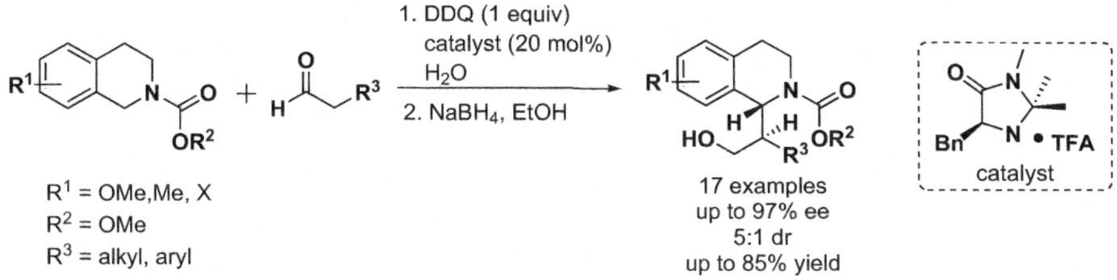

Scheme 91 Enantioselective CDC of *N*-carbamoyl cyclic amines and aldehydes

Jiao and co-workers developed an oxidative CDC method for C–C bond construction between two saturated C–H bonds. They treated acridine derivatives with nitro-olefins in the presence of TEMPO as a catalyst to synthesize C-alkylated acridine derivatives. TEMPO with O_2 is an interesting catalytic system for C–C bond formation reactions. This serves as one of the interesting methods for increasing substitution on ninth position of 9,10-dihydroacridine derivatives under mild, neutral reaction conditions (Scheme 92) [96].

Liu and co-workers synthesized C_{60}-fused heterocycle derivatives by an oxidative dehydrogenative carboannulation reaction by DDQ catalysis. Direct construction of two new C–C bonds, high atom economy, and broad substrate scope with high functional group tolerance are the merits of this transformation (Scheme 93) [97].

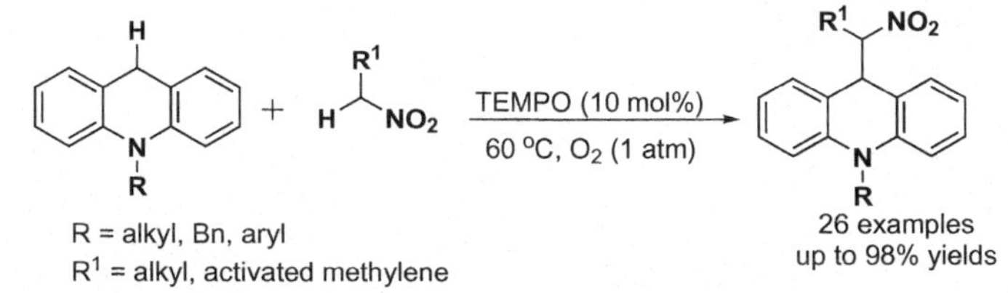

Scheme 92 TEMPO-mediated oxidative C–C bond formation

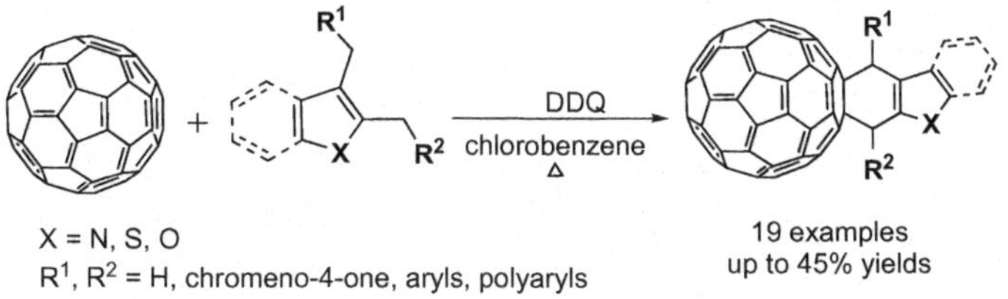

Scheme 93 Dehydrogenative carboannulation for the synthesis of C_{60}-fused heterocycles

Zhang and Wang developed a method for C_α-heteroarylation of ethers with *N*-heterocyclics via CDC process by *N*-hydroxy succinimide (NHS) assistance. Interestingly only 5–10 equivalents of ethers were sufficient to enable the transformations (Scheme 94) [98].

Jana and co-workers reported a method for direct arylation of tetrahydroisoquinolines via CDC process by using organocatalyst. They described a three-component reaction in which tetrahydroisoquinolines, 9-fluorenone imine, and arenes were involved. This method was applicable for wide range of aromatic and heteroaromatic nucleophiles like naphthols, phenols, indoles, and pyrroles. Fluorenyl moiety was crucial for this transformation as it forms iminium ion to drive the reaction (Scheme 95) [99]. Tetrahydroisoquinolines react with organocatalyst to form iminium ion intermediate **A**, which further transforms to new iminium ion intermediate **B** that in turn undergoes aromatic substitution reaction with aryl source to lead to the synthesis of α-arylated tetrahydroisoquinoline derivatives (Fig. 21).

Jiao and co-workers developed a CDC process for constructing C–C bond by using organocatalysis. This protocol enabled the intermolecular, asymmetric CDC process for alkylation of aldehydes where molecular oxygen was used as an oxidant (Scheme 96) [100].

Singh and co-workers developed a method for the functionalization of azoles via C–X (X=C, N) bond formation. They treated azoles with ethers and thioethers under TBHP/TBAI combination for the construction of C–X bond via CDC process. This transformation was applicable for variety of azoles like benzimidazoles, purines, benzotriazoles, triazoles, and pyrazoles and ethers like tetrahydrofuran, tetrahydropyran, 1,4-dioxane, diethyl ether, tetrahydrothiophene, and 1,3-dithiolanes (Scheme 97) [101].

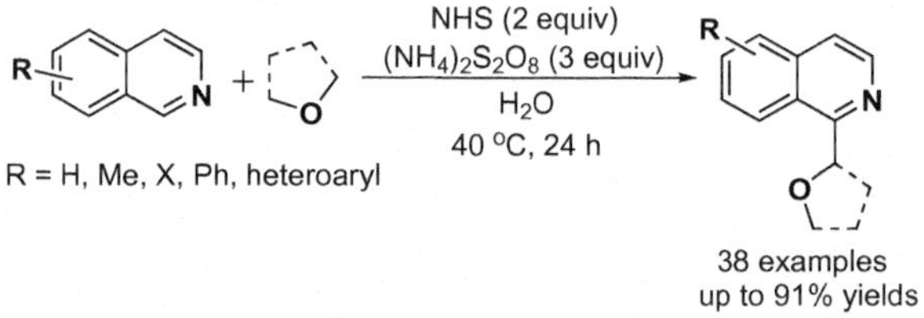

Scheme 94 *N*-Hydroxysuccinimide-mediated CDC reaction

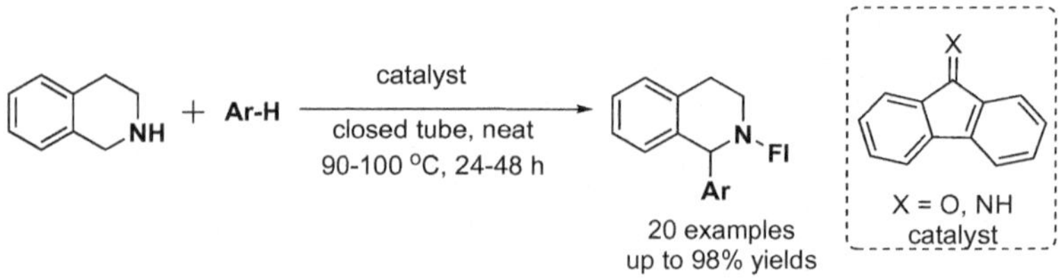

Scheme 95 Metal and oxidant free C(sp^3)–H functionalization for (hetero)arylation of tetrahydroisoquinolines

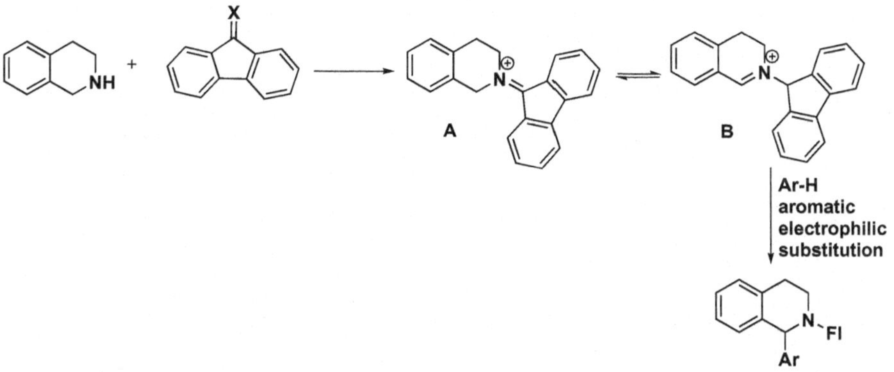

Fig. 21 Mechanism for (hetero)arylation of tetrahydroisoquinolines

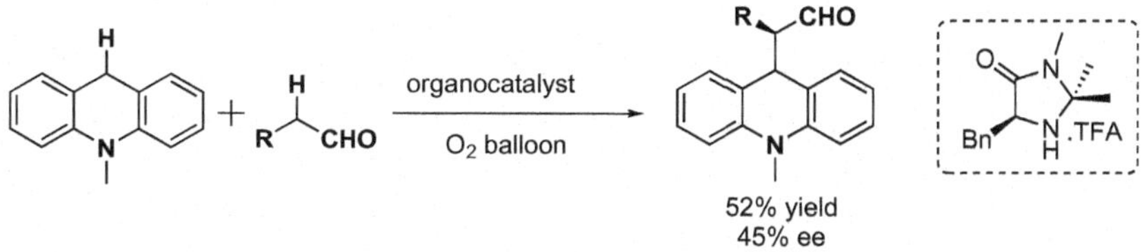

Scheme 96 Asymmetric intermolecular dehydrogenative α-alkylation of aldehydes

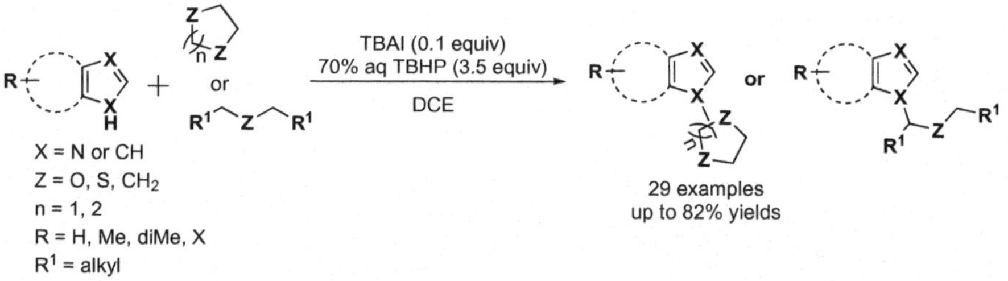

Scheme 97 Cross-dehydrogenative functionalization of azoles

Wang, He and co-workers developed a mediated direct oxidative coupling method for constructing C–N bond by direct amination of ethers. They treated ethers with azoles in the presence of *n*-Bu_4NI to give aminated cyclic ethers. In this transformation, TBAI was used as a catalyst, and TBHP was used as an oxidant (Scheme 98) [102].

Lavilla and co-workers developed a CDC method for synthesizing piperazine-fused bridged compounds. In this transformation, two C–C bonds and one C–N bond were formed via intramolecular oxidative coupling (Scheme 99) [103].

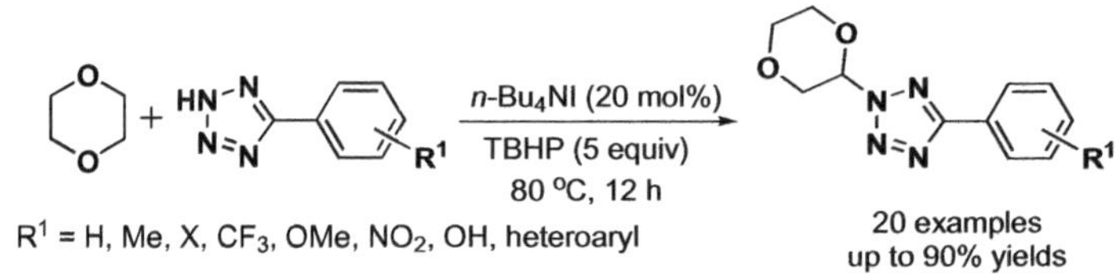

Scheme 98 Cross-dehydrogenative amination of ethers with aryl tetrazoles

Scheme 99 Synthesis of diketopiperazines fused bridged chemotypes

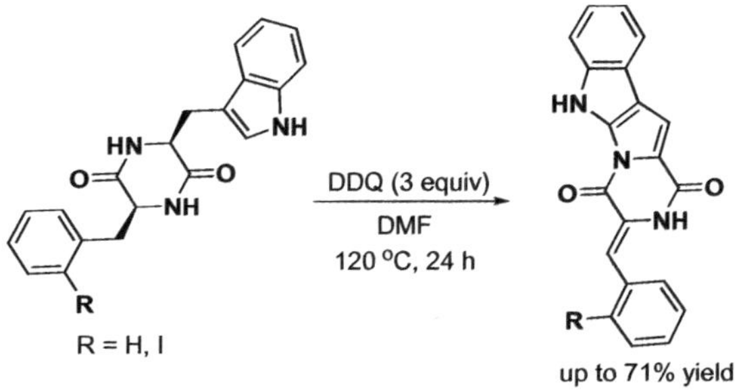

Patel and co-workers developed a CDC protocol for constructing C–N bond. They treated aryl alkyl ethers with tetrazoles in the presence of $Bu_4NI/^tBuOOH$ to give C-aminated product. In this intermolecular C–N bond formation reaction, high level of regioselectivity was achieved (Scheme 100) [104].

Liu and co-workers synthesized 2-aminobenzoxazoles by treating secondary amines with benzoxazoles in the presence of 1-butyl pyridinium iodide as a green catalyst. They showed that the catalyst can be reused by recycling the catalyst without affecting the reactivity (Scheme 101) [105].

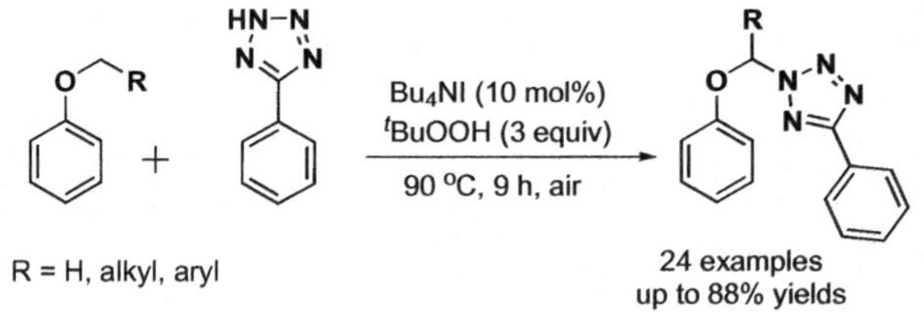

Scheme 100 Cross-dehydrogenative C–N bond formation with aryl ethers and tetrazoles

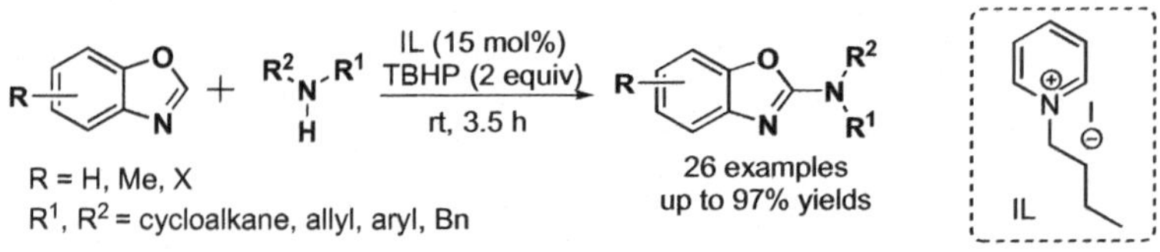

Scheme 101 Ionic liquid-catalyzed synthesis of 2-aminobenzoxazoles

Du and co-workers developed a method for amination of alkyl ethers by using *n*-Bu_4NI as a catalyst. They reported an organocatalyst-mediated intermolecular oxidative CDC process for constructing C–N bond between saturated C–H and N–H bonds. They treated secondary amines with cyclic ethers in the presence of *n*-Bu_4NI/TBHP catalytic system to synthesize hemiaminal ethers. Their studies revealed that hypervalent iodine generated in situ by TBHP mediates the reaction (Scheme 102) [106].

Liu and co-workers developed a method for the formation of C(sp^3)–N bond with the assistance of heterocyclic ionic liquids. They treated azoles with benzylic compounds in the presence of [bpy]I as a catalyst, to form C-aminated product. This is an intermolecular oxidative amination reaction for C–N bond formation between benzylic C(sp^3)–H and *N*-substituted azoles (Scheme 103) [107].

Vidavalur and co-workers developed a CDC method for constructing C–N bond between triazoles and methylarenes. They treated 1*H*-1,2,4-triazoles with methyl arenes in the presence of *n*-Bu_4NI/TBHP, to give *N*-benzyl heterocycle derivatives in moderate-to-good yields (Scheme 104) [108].

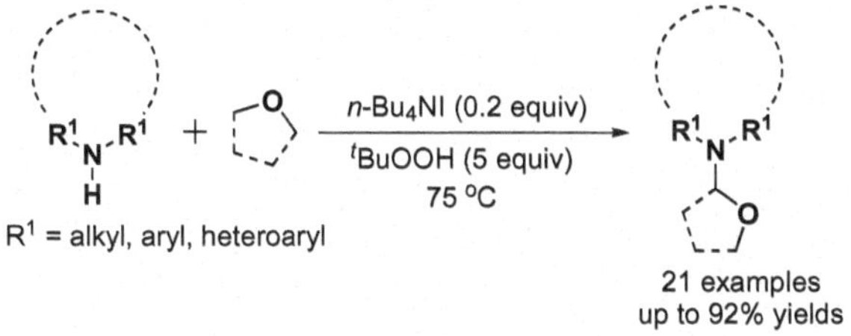

Scheme 102 Intermolecular oxidative C(sp^3)–N bond formation for the synthesis of hemiaminal ethers

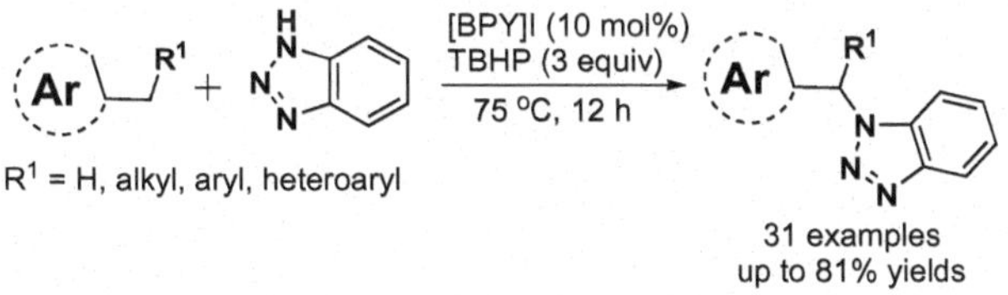

Scheme 103 Ionic liquid-catalyzed oxidative coupling of azoles and benzylic compounds

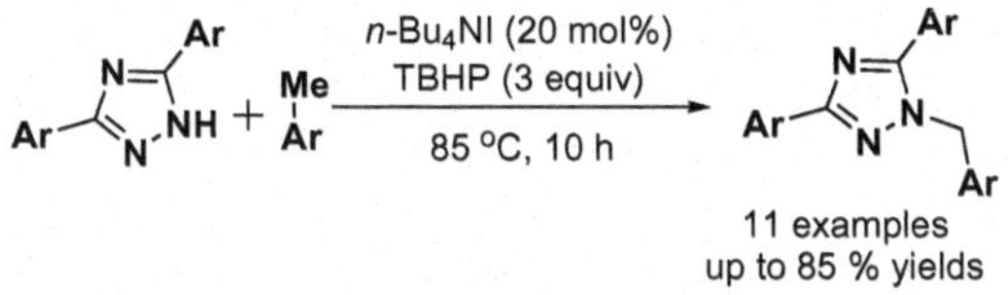

Scheme 104 *n*-Bu_4NI/TBHP-catalyzed CDC for C–N bond formation

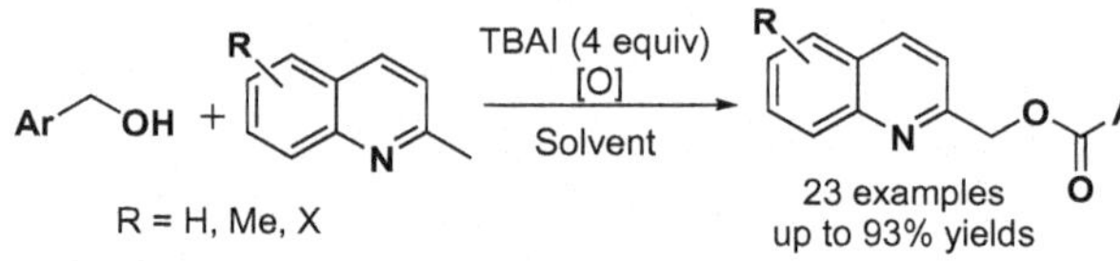

Scheme 105 TBAI-catalyzed Synthesis of benzyl esters

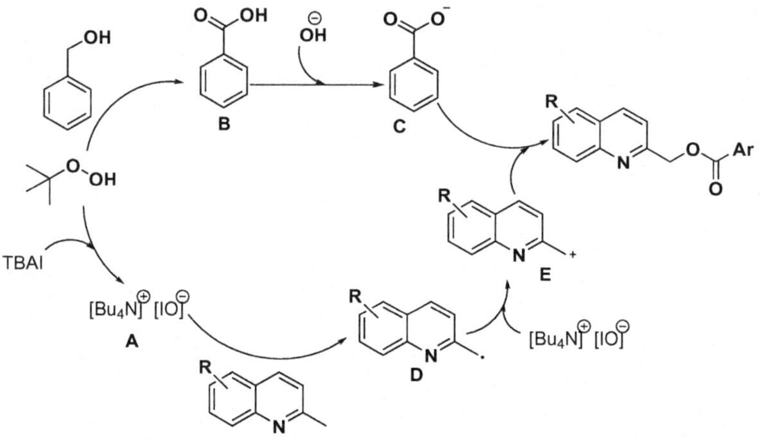

Fig. 22 Mechanism for synthesis of benzyl esters

Wang and co-workers developed a simple method for the synthesis of C-benzylated *N*-heterocycles. They treated alcohols with alkyl-substituted azaarenes in the presence of TBAI to form benzyl esters by C(sp^3)–H bond functionalization via CDC process (Scheme 105) [109]. The authors proposed an interesting radical reaction mechanism. Initially, TBHP reacts with TBAI to form ammonium hypoiodite intermediate **A**, which is the real catalyst in this transformation. In the presence of ammonium hypoiodite intermediate quinoline derivative forms radical intermediate **D**. In the presence of ammonium hypoiodite intermediate benzyl alcohol undergoes oxidation to form benzoic acid, which is further transformed to benzoate anion **C**. The benzoate anion reacts with quinoline radicals to form desired coupled product (Fig. 22).

Huo and co-workers developed a straightforward method for phosphorylation of xanthane derivatives. They treated xanthane derivatives with diarylphosphine oxides in the presence of DDQ to give the corresponding phosphorylated xanthenes at ambient reaction conditions. This transformation provides a direct construction of C–P bond with good functional group tolerance. This was the first report on phosphorylation of C(sp^3)–H bond on xanthane derivatives (Scheme 106) [110]. The authors proposed SET mechanism for this transformation. Xanthane derivatives react with DDQ to form radical ion pair **A**, which further transforms to

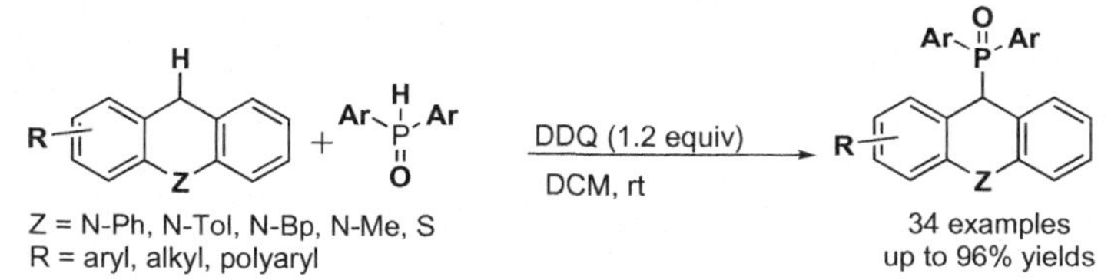

Scheme 106 DDQ-catalyzed C(sp^3)–H phosphorylation of xanthenes derivatives

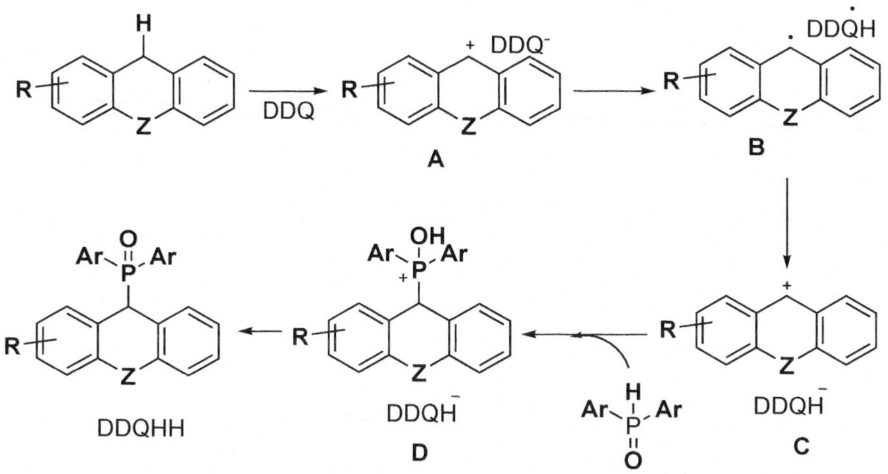

Fig. 23 Mechanism for phosphorylation of xanthenes derivatives

benzylic radical intermediate **B**. The benzylic radical intermediate is coupled with diarylphosphine oxides to form desired product (Fig. 23).

Todd and co-workers developed a method for the synthesis of β-nitroamines by the oxidation of *N*-phenyltertrahydroquinolines catalyzed by DDQ. They treated nitromethane with *N*-phenyltertrahydroquinolines in the presence of DDQ which undergoes rapid oxidation and forms the iminium ions, followed by the addition of nitroalkane, and eventually forms the C-alkylated product (Scheme 107) [111].

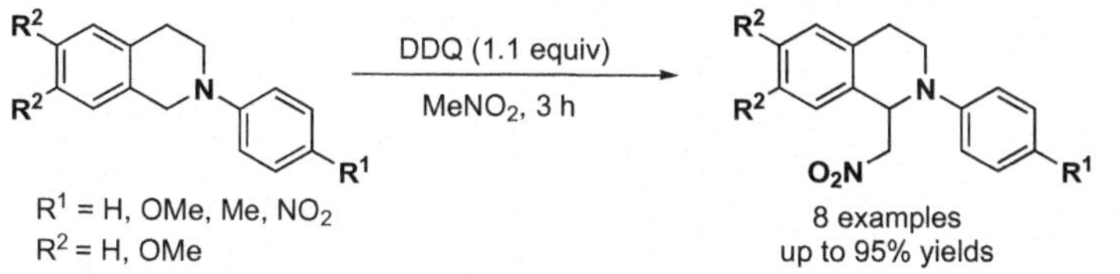

Scheme 107 DDQ-mediated oxidative coupling of *N*-phenyltertrahydroquinolines and nitromethane

3.4 Functionalization of N-Heterocycles via CDC Using Photochemical Reactions

Following the seminal independent reports of König and others on the use of organic dyes as photoredox catalysts to achieve CDC on amines and tetrahydroisoquinoline derivatives [48a–c], Zhou and co-workers developed a method for constructing C–C bonds via CDC protocol promoted by visible light. They treated tertiary amines with diazo compounds in the presence of visible light and rose bengal to give β-amino-α-diazo adducts of tetrahydroisoquinoline (Scheme 108) [112]. The authors proposed SET process for this transformation. *N*-aryl tetrahydroisoquinolines forms radical intermediate **A** in the presence of rose bengal, which is further transformed to iminium ion intermediate **B**. The iminium ion intermediate reacts with diazo compounds to form desired coupled product (Fig. 24).

Hajra and co-workers developed a CDC process for constructing $C(sp^3)$–$C(sp^2)$ bond with the assistance of visible light. They treated tertiary amines with imidazopyridines under the visible light to give the 3-substituted imidazopyridine by using rose bengal in catalytic amounts under aerobic conditions. This method was applicable for wide range of amines like tetrahydroisoquinolines, *N,N′*-dimethyl aniline, and *N*-heterocycles like imidazo[1,2-*a*]pyrimidine, indolizines, indoles, pyrroles, etc. (Scheme 109) [113].

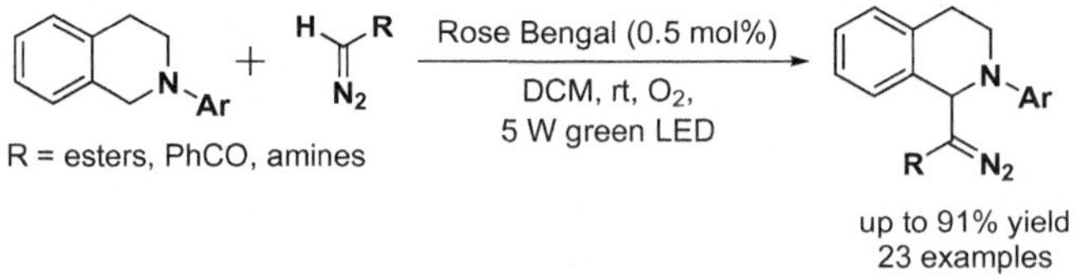

Scheme 108 Visible-light-catalyzed CDC of tertiary amines with diazo compounds

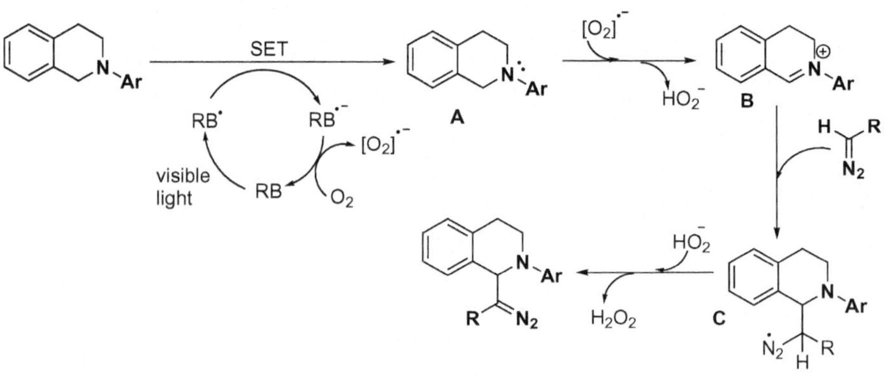

Fig. 24 Mechanism for visible-light-catalyzed CDC of tertiary amines

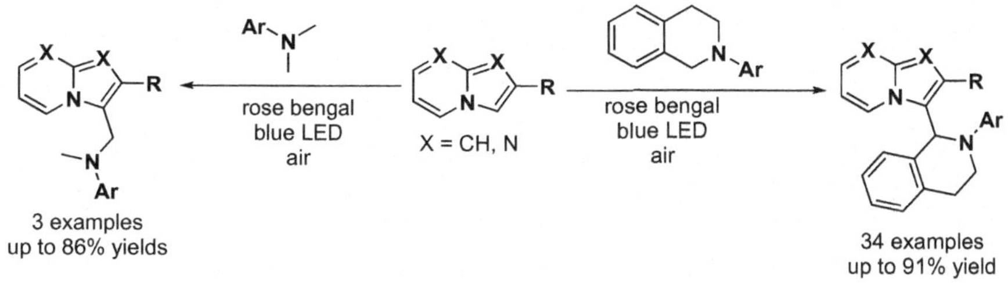

Scheme 109 Visible-light-mediated CDC of tertiary amine with imidazopyridine

Zhao and co-workers developed a photoredox catalysis method for the amination of quinoxalinones with aliphatic amines by constructing C–N bond. They treated primary or secondary amines with quinoxalinones under visible light irradiation to give 3-aminoquinoxalin-2(1*H*)-ones via construction of C–N bond by CDC process (Scheme 110) [114].

Xia and co-workers developed an intermolecular C–N bond formation reaction for amination of phenols under visible light by CDC reaction. They treated phenols with cyclic amines in the presence of visible light and $K_2S_2O_8$ to synthesize aminated products; $K_2S_2O_8$ acted as an external oxidant to oxidize both phenols and amines to generate radical intermediates (Scheme 111) [115].

Lei and co-workers developed visible-light-mediated oxidative amination process for C–N bond formation reaction between privileged tetrahydrofuran moiety and azoles by using air as an oxidant. The reaction proceeds through

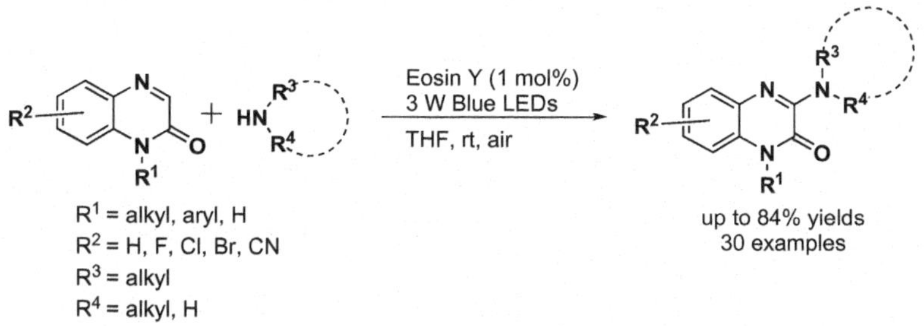

Scheme 110 Visible-light induced CDC of quinoxalinones with aliphatic amines

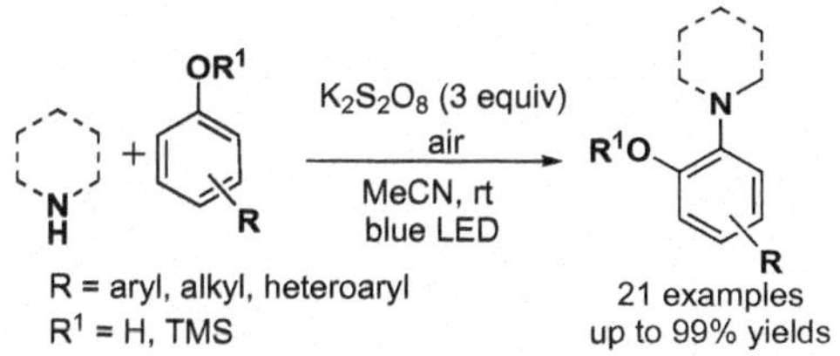

Scheme 111 Visible-light-promoted oxidative CDC of phenols and amines

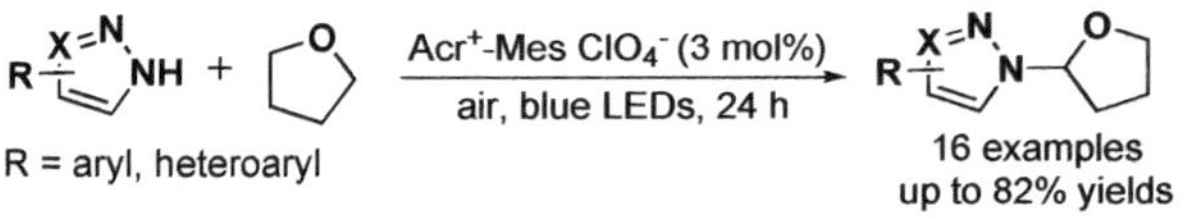

Scheme 112 Visible-light-mediated oxidative cross-coupling of tetrahydrofuran and azoles

functionalization of C(sp^3)–H bond of tetrahydrofuran by visible light catalysis under air followed by the oxidative amination (Scheme 112) [116].

Itoh and co-workers reported a CDC method for C–H amination under aerobic photo-oxidative conditions. 2-*tert*-Butylanthraquinone was used as photocatalyst, and oxygen was used as sole oxidant (Scheme 113) [117]. Phthalimide forms amino radical intermediate **A** in the presence of K_2CO_3, and then, the amino radical reacts with *N*-heterocycles to form 3-amino radical intermediate **B**, which further undergoes aromatization to form 3-amino-substituted *N*-heterocycles (Fig. 25).

Xue and co-workers developed a novel method for synthesizing amides by coupling aromatic aldehydes and amines via oxidative CDC process. BODIPY was used

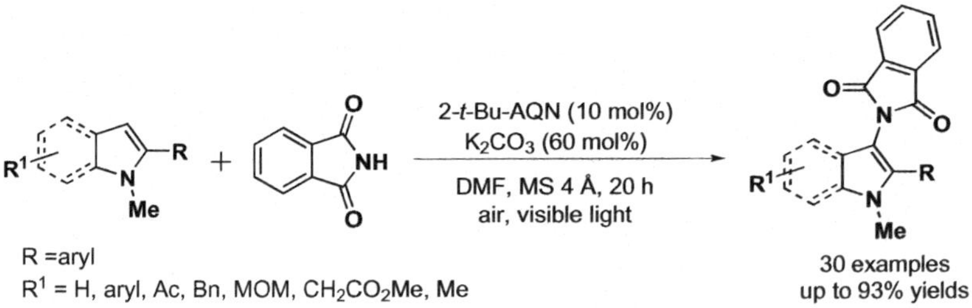

Scheme 113 Photo-oxidative, cross-dehydrogenative C–H amination of indoles

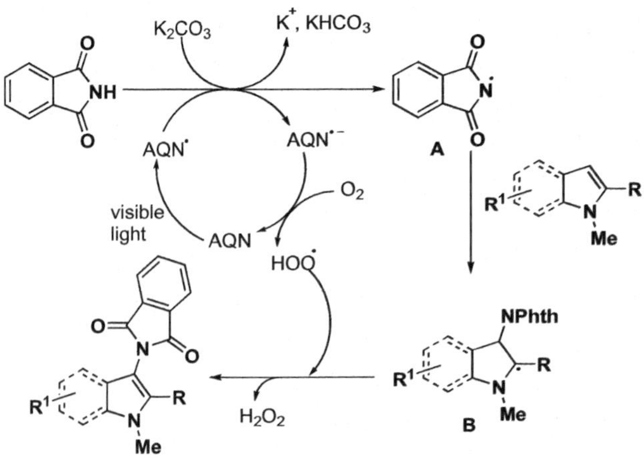

Fig. 25 Mechanism for photo-oxidative, cross-dehydrogenative C–H amination

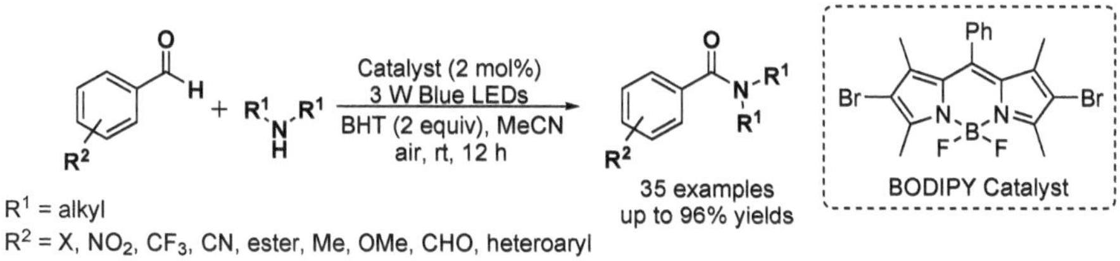

Scheme 114 BODIPY-catalyzed amide synthesis

as a catalyst. This organic dye showed better efficiency for oxidative amidation process than the classical ruthenium- and iridium-based photoredox catalysts. The studies revealed that active peroxides were in situ generated from the succession of electron transfers and single electron transfer on dioxygen. The resulting active peroxide drives the reaction by transforming hemiacetal to amide (Scheme 114) [118].

3.5 Functionalization of N*-Heterocycles via CDC Using Base-Mediated Reactions*

Kumar and co-workers developed a method for the synthesis of 3-aryl oxindoles. They treated oxindoles with arenes in the presence of NaO^{t}Bu to give coupled product by constructing C–C bond. The regio- and chemoselective C3-arylation led to the formation of quaternary center in oxindole substrates (Scheme 115) [119].

Houk, Stoltz, and collaborators reported the silylation of *N*-heteroaromatic compounds by base-mediated CDC process. The studies revealed that the reaction proceeds through radical chain mechanism (Scheme 116) [120].

Wu and co-workers developed a base-catalyzed CDC process for the construction of C–C bond between quinoline *N*-oxide with azoles. They treated quinoline *N*-oxide with azoles in the presence of ^{t}BuOLi to deliver the coupled product (Scheme 117) [121]. In the presence of base benzoxazole anion **A** generates and reacts with quinoline *N*-oxide to form desired coupled product (Fig. 26).

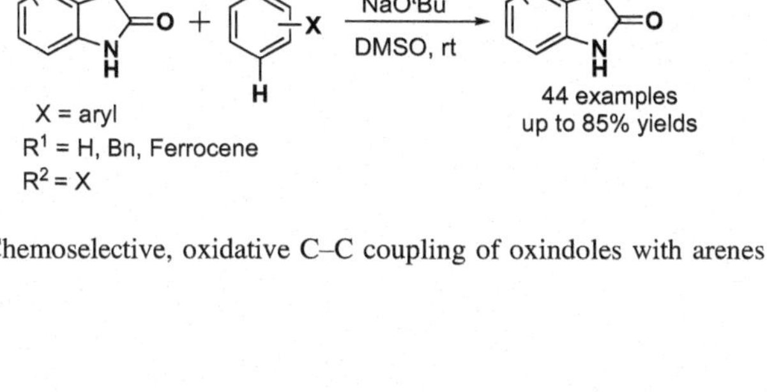

Scheme 115 Chemoselective, oxidative C–C coupling of oxindoles with arenes

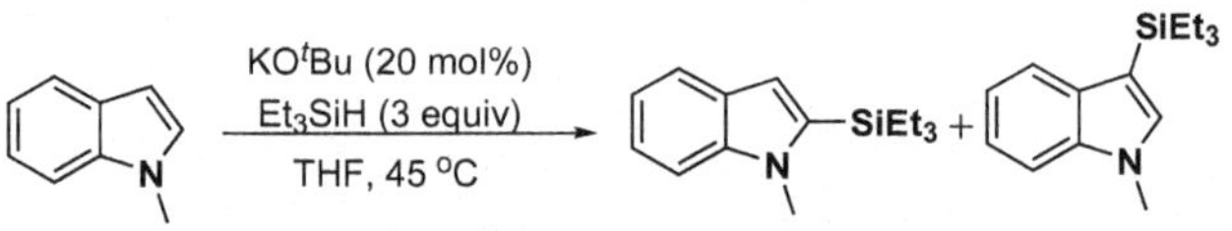

Scheme 116 Base-catalyzed dehydrogenative C−H silylation of heteroaromatics

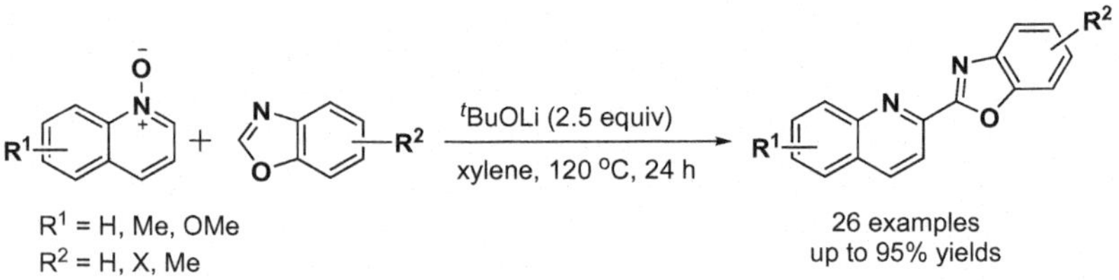

Scheme 117 CDC of quinoline *N*-oxides with 1,3-azoles

Fig. 26 Mechanism for base-mediated coupling of quinoline *N*-oxides with 1,3-azoles

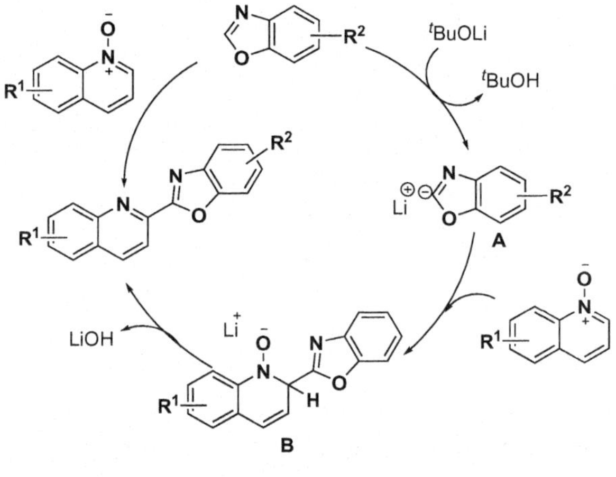

Xiao and co-workers developed a CDC method for constructing C–C bonds between 2-alkyl azaarenes with nitroarenes. They treated nitroarenes with 2-alkylazaarenes in the presence of ^{t}BuOK to form C–C bond by coupling C(sp^3)–H and C(sp^2)–H bonds eventually that leads to the formation of 2-benzyl substituted azaarenes. In this regioselective C(sp^3)–H arylation, DMSO plays a dual role of oxidant and solvent (Scheme 118) [122].

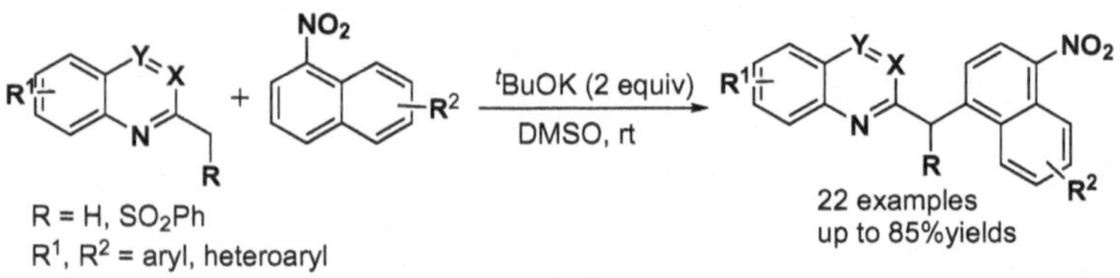

Scheme 118 Oxidative, dehydrogenative arylation of 2-alkylazaarenes with nitroarenes

3.6 Miscellaneous Reactions

Khalymbadzha and co-workers developed a chemo and regioselective reaction for constructing new C–C bond between triazenes and 5,7-dihydroxycoumarins. They treated 5,7-dihydroxy coumarins with electron-deficient trazines in the presence of Lewis acids to give 8-triazinyl 5,7-dihydroxycoumarins (Scheme 119) [123].

Lou and co-workers developed a CDC method for constructing new C–C bond between *N*-carbamoyl tetrahydroisoquinolines with variety of C–H nucleophiles by $Na_2S_2O_8$ catalysis. This transformation was applicable for nucleophiles like simple ketones, aldehydes, and aryl ring (Scheme 120) [124].

Shi and co-workers developed a CDC method for alkylation of pyridines with cyclic ethers in the presence of oxidant. They treated pyridines with cyclic ethers in the presence of $K_2S_2O_8$ as an oxidant to deliver C-alkylated pyridine derivatives (Scheme 121) [125].

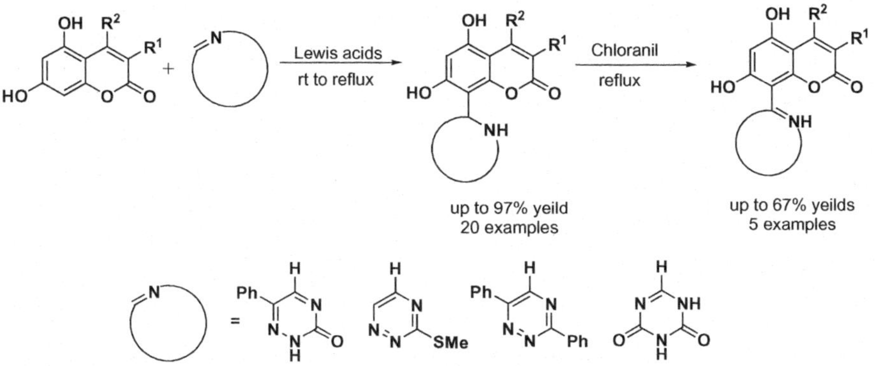

Scheme 119 CDC of triazines with 5,7-dihydroxycoumarins

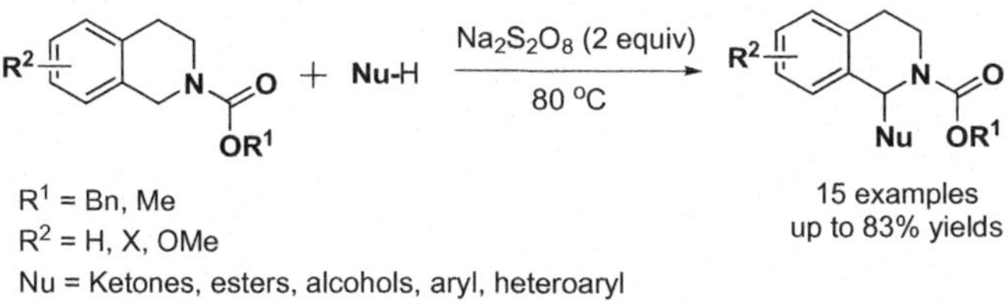

Scheme 120 CDC of *N*-carbamoyl tetrahydroisoquinolines with carbonyl nucleophiles

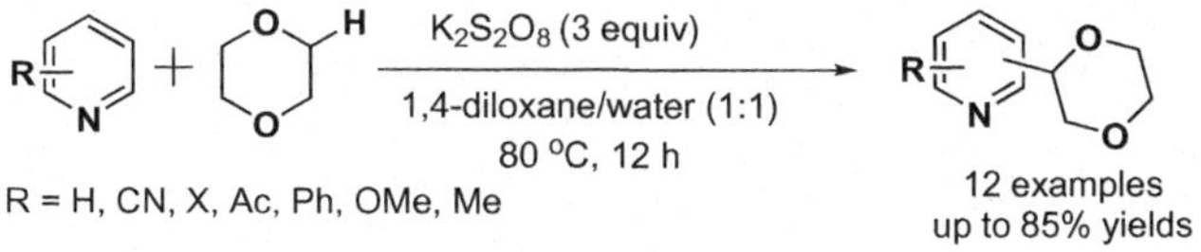

Scheme 121 Cross-dehydrogenative alkylation of pyridines

Yotphan and co-workers developed a chemo and regioselective CDC approach for synthesizing hydroxy-pyrazolylquinoxazolinone derivatives via new C–C bond formation. They treated pyrazolines with quinazolinones in the presence of $K_2S_2O_8$ to give the coupled product (Scheme 122) [126].

Prabhu and co-workers developed a CDC process for acylation of *N*-heterocycles. They treated *N*-heterocycles like isoquinolines, quinolines, and quinoxalines with aldehydes in the presence of substoichiometric amount of TBAB and $K_2S_2O_8$ as oxidants to give acylated *N*-heterocycles. The developed method serves as metal-free versions of Minisci reaction (Scheme 123) [61, 127].

Kobayashi and co-workers developed a CDC process for constructing C–C bonds by using catalytic amount of SO_2Cl_2 under aerobic conditions. Their studies revealed that SO_2Cl_2 acts as a radical initiator and promotes the reaction (Scheme 124) [128].

Adib and co-workers developed a C–C bond formation reaction between quinolines and isoquinolines with aryl methanols. This is an oxidative CDC method for making C–C bonds by using aryl methanol as acylating source. They treated quinolines or isoquinolines with aryl methanols in the presence of $K_2S_2O_8$ and aliquat 336 (methyltrioctylammonium chloride) to get acylated product. $K_2S_2O_8$

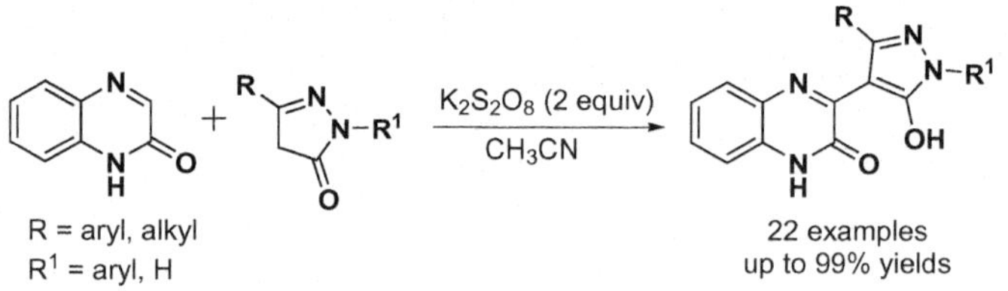

Scheme 122 Oxidative C–C bond formation between pyrazolones and quinoxalinones

Scheme 123 Acylation of heterocycles

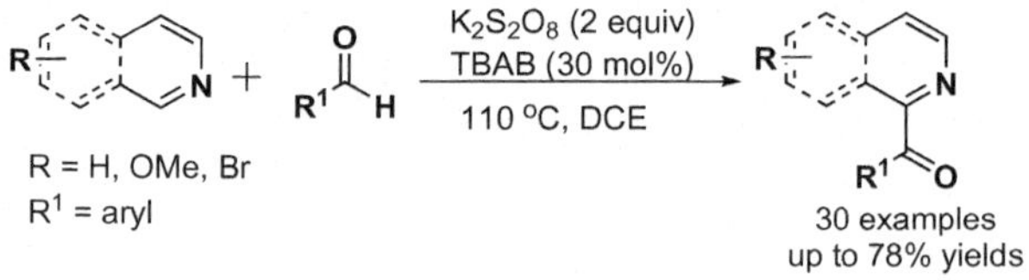

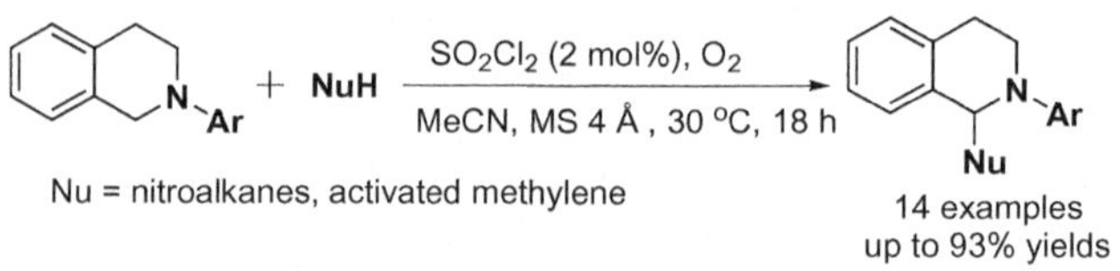

Scheme 124 Aerobic CDC reaction of tertiary amines

was used as an oxidant, and aliquat 336 was used as a transfer agent (Scheme 125) [129].

Prathima and co-workers developed a C–C bond formation reaction between C(sp^2)–H, C(sp^3)–H bonds via CDC process. They treated 2-phenylpyridine with cyclic ethers in the presence of $K_2S_2O_8$ to give C2 alkylated product (Scheme 126) [130].

Yeung and co-workers reported a C–H formylation process for *N*-heterocyclic compounds by C–C bond formation via CDC process. They treated *N*-heterocyclic compounds with trioxane in the presence of $(n\text{-}Bu_4)_4S_2O_8$ under thermal conditions to give the formylated *N*-heterocyclic compounds (Scheme 127) [131].

Prabhu and co-workers developed a catalyst-free method for constructing C–P bond with air as an oxidant by CDC process. They treated *N*-protected tetrahydroisoquinolines with dialkylphosphites in the presence of air to get α-aminophosphonates. Advantages of this method: transformation of C–H bond to C–P bond, open air used as oxidant, water as the only by-product, and catalyst-free

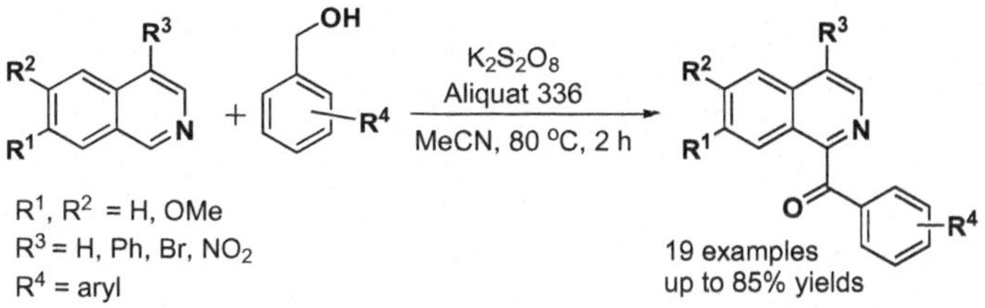

Scheme 125 Oxidative CDC for acylation of heterocycles

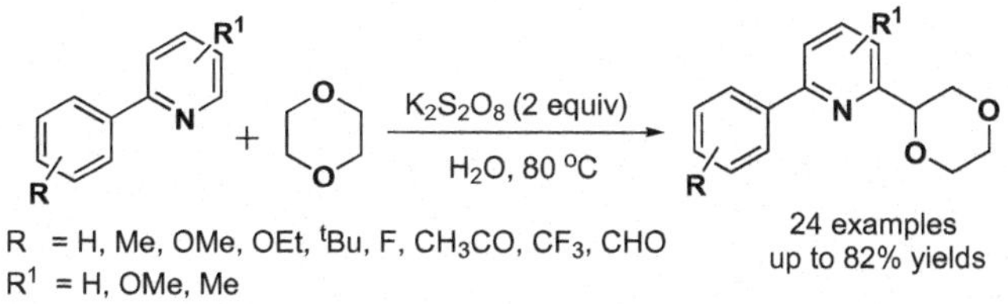

Scheme 126 CDC for C−H bond functionalization of 2-phenyl pyridine

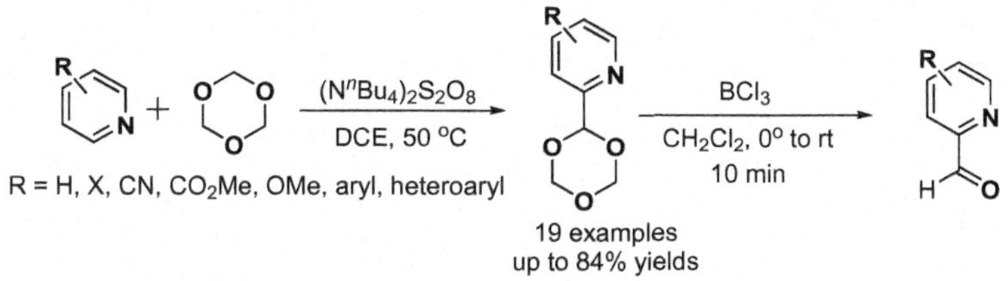

Scheme 127 C−H formylation of nitrogen heterocycles

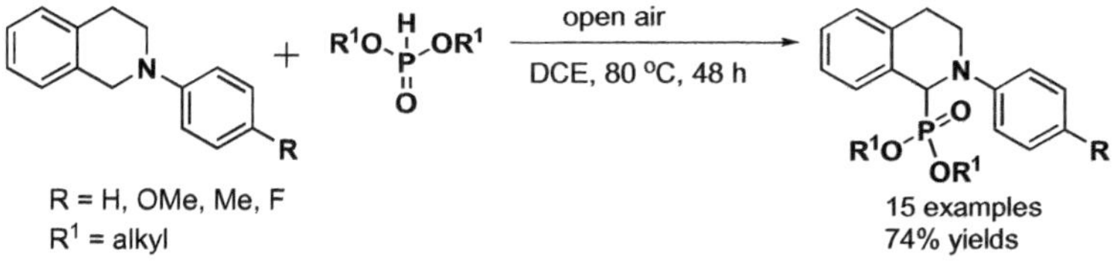

Scheme 128 Cross-dehydrogenative synthesis of α-aminophosphonates

conditions. Mechanistic studies revealed that it is nucleophilic-dependent reaction and specific for dialkylphosphites (Scheme 128) [132].

Finally, the recent reports in the literature have described the metal-free functionalization of saturated cyclic amines in the presence of oxidants. Selective double oxidation of tertiary cyclic amines to 3-alkoxyamine lactams using $NaClO_2$/TEMPO/NaClO system has been reported by Sartillo-Piscil and co-workers [133]. Oxidative β-C(sp^3)–H sulfonylation of cyclic amines in the presence of a large excess of NIS with various sulfinates has been reported by Talbot and collaborators [134].

4 Conclusions

Metal-free approaches toward the synthesis and functionalization of various *N*-heterocycles can be achieved according to various reaction conditions. The cross-dehydrogenative coupling at the α position of THIQs has been mainly used to develop these methodologies. Results demonstrate that the metal-free approaches can compete and, in some cases, outpass the transition metal-catalyzed version of similar transformations. More interestingly, the combination of photocatalysis enables the use of catalytic amount of oxidizing agent. It must be noted that most of the transformations have been limited to the reactivity of the in situ generated iminium intermediates of THIQs.

References

1. (a) Yang L, Huang H (2015) Transition-metal-catalyzed direct addition of unactivated C–H bonds to polar unsaturated bonds. Chem Rev 115:3468–3517. (b) Chen Z, Wang B, Zhang J, Yu W, Liu Z, Zhang Y (2015) Transition metal-catalyzed C–H bond functionalizations by the use of diverse directing groups. Org Chem Front 2:1107–1295. (c) Huang C-Y, Doyle AG (2014) The chemistry of transition metals with three-membered ring heterocycles. Chem Rev 114:8153–8198. (d) Yu J-Q, Shi Z (2009) C–H activation. Springer-Verlag, Berlin, Heidelberg, 2009. (e) Murahashi S-I, Zhang D (2008) Ruthenium catalyzed biomimetic oxidation in organic synthesis inspired by cytochrome P-450. Chem Soc Rev 37:1490–1501. (f) Li C-J (2009) Cross-dehydrogenativecoupling (CDC): exploring C–C bond formations beyond functional group transformations. Acc Chem Res 42(2):335–344

2. (a) Dobereiner GE, Crabtree RH (2010) Dehydrogenation as a substrate-activating strategy in homogeneous transition-metal catalysis. Chem Rev 110:681–703. (b) Guillena G, Ramón DJ, Yus M (2010) Hydrogen autotransfer in the *N*-alkylation of amines and related compounds using alcohols and amines as electrophiles. Chem Rev 110:1611–1641. (c) Batra A, Singh P, Singh KN (2016) Cross dehydrogenative coupling (CDC) reactions of *N*,*N*-disubstituted formamides, benzaldehydes and cycloalkanes. Eur J Org Chem 4927–4947
3. (a) Wang B, Wong HNC (2018) Bromine-mediated cross-dehydrogenative coupling (CDC) reactions. Bull Chem Soc Jpn 91:710–719. (b) Parvatkar PT, Manetsch R, Banik BK (2019) Metal-free cross-dehydrogenative coupling (CDC): molecular iodine as a versatile catalyst/reagent for CDC reactions. Chem Asian J 14:6–30. (c) Li C-J (2015) From C–H to C–C bonds cross-dehydrogenative-coupling. The Royal Society of Chemistry, London
4. Vitaku E, Smith DT, Njardarson JT (2014) Analysis of the structural diversity, substitution patterns, and frequency of nitrogen heterocycles among U.S. FDA approved pharmaceuticals. J Med Chem 57:10257–10274
5. (a) Donald JR, Taylor RJK, Petersen WF (2017) Low-temperature, transition-metal-free cross-dehydrogenative coupling protocol for the synthesis of 3,3-disubstituted oxindoles. J Org Chem 82:11288–11294; (b) Gosh S, De S, Kakde BN, Bhunia S, Adhikary A, Bisai A (2012) Intramolecular dehydrogenative coupling of sp^2 C–H and sp^3 C–H bonds: an expeditious route to 2-oxindoles. Org Lett 14:5864–5867
6. He Z, Liu W, Li Z (2011) I_2-catalyzed indole formation *via* oxidative cyclization of *N*-aryl enamines. Chem Asian J 6:1340–1343
7. Volvoikar PS, Tilve SG (2016) Iodine-mediated intramolecular dehydrogenative coupling: Synthesis of *N*-alkylindolo[3,2-*c*]- and -[2,3-*c*]quinoline iodides. Org Lett 18:892–895
8. He Y, Wang X, Xiao J-A, Pang J, Gan C, Huang Y, Huang C (2018) Metal-free oxidative isocyanides insertion with aromatic aldehydes to aroylated *N*-heterocycles. RSC Adv 8:3036–3040
9. Zheng Z, Dian L, Yuan Y, Zhang-Negrerie D, Du Y, Zhao K (2014) $PhI(OAc)_2$-mediated intramolecular oxidative aryl-aldehyde Csp^2–Csp^2 bond formation: metal-free synthesis of acridone derivatives. J Org Chem 79:7451–7458
10. (a) Deb ML, Baruah B, Bhuyan PJ (2008) A facile synthesis of 6,12-disubstituted 5,7-dihydroindolo[2,3-b]carbazoles from the reaction of 1H-Indole and aldehydes catalyzed by molecular iodine. Synthesis 0286–0292; (b) Xue W-J, Gao Q-H, Wub A-X (2015) Molecular iodine mediated oxidative cross-coupling of sp^3 C–H with sp^2 C–H: direct synthesis of substituted indolo[2,3-*b*]carbazoles *via* formal [2 + 2 + 1 + 1] cyclization. Tetrahedron Lett 56:7115–7119
11. Jiang Y-T, Yu Z-Z, Zhang Y-K, Wang B (2018) *N*-bromosuccinimide-induced C–H bond functionalization: An intramolecular cycloaromatization of electron withdrawing group substituted 1-biphenyl-2-ylethanone for the synthesis of 10-phenanthrenol. Org Lett 20:3728–3731
12. Shang S, Zhang-Negrerie D, Du Y, Zhao K (2014) Intramolecular metal-free oxidative aryl–aryl coupling: an unusual hypervalent-iodine-mediated rearrangement of 2-substituted *N*-phenylbenzamides. Angew Chem Int Ed 53:6216–6219
13. Maiti S, Achar TK, Mal P (2017) An organic intermolecular dehydrogenative annulation reaction. Org Lett 19:2006–2009
14. Maiti S, Mal P (2017) Dehydrogenative aromatic ring fusion for carbazole synthesis *via* C–C/C–N bond formation and alkyl migration. Org Lett 19:2454–2457
15. Long J, Cao X, Zhu L, Qiu R, Au C-T, Yin S-F, Iwasaki T, Kambe N (2017) Intramolecular, site-selective, iodine-mediated, amination of unactivated (sp^3)C–H bonds for the synthesis of indoline derivatives. Org Lett 19:2793–2796
16. Satish G, Polu A, Ramar T, Ilangovan A (2015) Iodine-mediated C–H functionalization of sp, sp^2, and sp^3 carbon: a unified multisubstrate domino approach for isatin synthesis. J Org Chem 80:5167–5175

17. Li Y, Xu H, Xing M, Huang F, Jia J, Gao J (2015) Iodine-promoted construction of polysubstituted 2,3- dihydropyrroles from chalcones and β–enamine ketones (esters). Org Lett 17:3690–3693
18. Ohno M, Spande TF, Witkop B (1970) Cyclization of tryptophan and tryptamine derivatives to 2,3-dihydropyrrolo [2,3-*b*]indole. J Am Chem Soc 92:343–348
19. Yang Z-Y, Tian T, Du Y-F, Li S-Y, Chu C-C, Chen L-Y, Li D, Liub J-Y, Wang B (2017) Direct intramolecular amination of tryptophan esters to prepare pyrrolo[2,3-*b*]indoles. Chem Commun 53:8050–8053
20. Gao Q, Wu X, Li Y, Liu S, Meng X, Wu A (2014) Iodine-promoted sequential C(sp3)-H functionalization reactions: an annulation strategy for the construction of 3-methylthio-4-arylmaleimides. Adv Synth Catal 356:2924–2930
21. Huang H, Ji X, Wu W, Jiang H (2013) Practical synthesis of polysubstitutedimidazoles *via* iodine catalyzed aerobic oxidative cyclization of aryl ketones and benzylamines. Adv Synth Catal 355:170–180
22. Chen Z, Li H, Dong W, Miao M, Ren H (2016) I_2-catalyzed oxidative coupling reactions of hydrazones and amines and the application in the synthesis of 1,3,5-trisubstituted 1,2,4-triazoles. Org Lett 18:1334–1337
23. Yang L, Zhang-Negrerie D, Zhao K, Du Y (2016) Intramolecular functionalization of benzylic methylene adjacent to the ring nitrogen atom in *N*-aryltetrahydroisoquinoline derivatives. J Org Chem 81:3372–3379
24. Yan Y, Zhang Y, Feng C, Zha Z, Wang Z (2012) Selective iodine-catalyzed intermolecular oxidative amination of C(sp3)-H bonds with ortho-carbonyl-substituted anilines to give quinazolines. Angew Chem Int Ed 51:8077–8081
25. (a) Xi L-Y, Zhang R-Y, Shi L, Chen S-Y, Yu X-Q (2016) Iodine-mediated synthesis of 3-acylbenzothiadiazine 1,1-dioxides. Beilstein J Org Chem12:1072–1078. (b) Mohammed S, Vishwakarma R A, Bharate SB (2015) Iodine catalyzed oxidative synthesis of quinazolin-4(3H)-ones and Pyrazolo[4,3-d]pyrimidin-7(6H)-ones *via* Amination of sp^3 C–H bond. J Org Chem 80:6915–6921
26. Yang L, Shi X, Hu B-Q, Wang L-X (2016) Iodine-catalyzed oxidative benzylic C-H bond amination of azaarenes: practical synthesis of quinazolin-4(3H)-ones. Asian J Org Chem 5:494–498
27. Xicun W, Junlinga L, Zhengjuna Q, Linb B (2011) Molecular iodine promoted synthesis of new pyrido[2,3-*d*]pyrimidin-4-ols. Chin J Chem 29:1646–1650
28. Ghorbani-Vaghei R, Malaekehpoor SM (2012) *N*-Bromosuccinimide as an efficient catalyst for the synthesis of indolo[2,3-*b*]quinolines. Tetrahedron Lett 53:4751–4753
29. Nguyen TB, Ermolenko L, Retailleau P, Al-Mourabit A (2016) Molecular iodine-catalyzed aerobic α, β-diamination of cyclohexanones with 2-aminopyrimidine and 2-aminopyridines. Org Lett 18:2177–2179
30. Rong H-J, Yao J-J, Li J-K, Qu J (2017) Molecular iodine-mediated α-C–H oxidation of pyrrolidines to N, O-acetals: synthesis of (±)-preussin by late-stage 2,5-difunctionalizations of pyrrolidine. J Org Chem 82:5557–5565
31. Deb ML, Borpatra PJ, Saikiab PJ, Baruah PK (2017) Iodine/hydrogen peroxide promoted intramolecular oxidative C–O bond formation in ethanol at room temperature: a green approach to 1,3-oxazines. Synlett 28:461–466
32. Liu L, Du L, Zhang-Negrerie D, Du Y (2015) NIS-mediated intramolecular oxidative α-functionalization of tertiary amines: Transition metal-free synthesis of 1,2-dihydro-(4*H*)-3,1-benzoxazin-4-one derivatives. RSC Adv 5:29774–29781
33. Waghmode NA, Kalbandhe AH, Thorat PB, Karade NN (2016) Metal-free new synthesis of 1,3-naphthoxazines *via* intramolecular cross dehydrogenative-coupling reaction of 1-(a-aminoalkyl)-2-naphthols using hypervalent iodine(III) reagent. Tetrahedron Lett 57:680–683
34. Zhao J, Huang H, Wu W, Chen H, Jiang H (2013) Metal-free synthesis of 2-aminobenzothiazoles *via* aerobic oxidative cyclization/dehydrogenation of cyclohexanones and thioureas. Org Lett 15:2604–2607

35. Xu Y, Li B, Zhang X, Fan X (2017) Metal-free synthesis of 2-aminobenzothiazoles *via* iodine-catalyzed and oxygen-promoted cascade reactions of isothiocyanatobenzenes with amines. J Org Chem 82:9637–9646
36. Xue W-J, Li H-Z, Zheng K-l, Wu A-X (2014) Iodine-promoted efficient synthesis of diheteroaryl thioethers *via* the integration of iodination/condensation/cyclization/dehydration sequences. Tetrahedron Lett 55:5544–5547
37. Yu H, Shen J (2015) Dehydrogenative cyclization of *N*-acyl dipeptide esters for the synthesis of imidazolidin-4-ones. RSC Adv 5:9815–9818
38. Verma A, Patel S, Meenakshi Kumar A, Yadav A, Kumar S, Jana S, Sharma S, Prasad CD, Kumar S (2015) Transition metal free intramolecular selective oxidative C(sp^3)–N coupling: synthesis of *N*-aryl-isoindolinones from 2-alkylbenzamides. Chem Commun 51:1371–1374
39. Wang L, Cao J, Chen Q, He M (2015) One-pot synthesis of 2,5-diaryl 1,3,4-oxadiazoles *via* di-*tert*-butyl peroxide promoted *N*-acylation of aryl tetrazoles with aldehydes. J Org Chem 80:4743–4748
40. Zhao D, Wang T, Li J-X (2014) Metal-free oxidative synthesis of quinazolinones via dual amination of sp^3 C-H bonds. Chem Commun 50:6471–6474
41. Wu X, Chen D-F, Chen S-S, Zhu Y-F (2015) Synthesis of polycyclic amines through mild metal-free tandem cross-dehydrogenative coupling/intramolecular hydroarylation of *N*-aryltetrahydroisoquinolines and crotonaldehyde. Eur J Org Chem 468–473
42. Hu W, Lin J-P, Song L-R, Long Y-Q (2015) Direct synthesis of 2-aryl-4-quinolones *via* transition-metal-free intramolecular oxidative C(sp^3)–H/C(sp^3)–H coupling. Org Lett 17:1268–1271
43. Li H, Liu C, Zhang Y, Sun Y, Wang B, Liu W (2015) Green method for the synthesis of chromeno[2,3-*c*]pyrazol-4(1*H*)-ones through ionic liquid promoted directed annulation of 5-(aryloxy)-1*H*-pyrazole-4-carbaldehydes in aqueous media. Org Lett 17:932–935
44. Liang D, Sersen D, Yang C, Deschamps JR, Imler GH, Jiang C, Xue F (2017) One-pot sequential reaction to 2-substitutedphenanthridinones from *N*-methoxybenzamides. Org Biomol Chem 15:4390–4398
45. Neel AJ, Hehn JP, Tripet PF, Toste FD (2013) Asymmetric cross-dehydrogenative coupling enabled by the design and application of chiral triazole-containing phosphoric acids. J Am Chem Soc 135:14044–14047
46. Yi C, Liu S-X, Neels A, Renaud P, Decurtins S (2009) Preparation of zwitterionic hydroquinone-fused [1,4]oxazinium derivatives *via* a photoinduced intramolecular dehydrogenative-coupling reaction. Org Lett 11:5530–5533
47. Zhang H, Muñiz K (2017) Selective piperidine synthesis exploiting iodine-catalyzed Csp^3−H amination under visible light. ACS Catal 7:4122–4125
48. (a) Hari DP, König B (2011) Eosin Y catalyzed visible light oxidative C–C and C–P bond formation. Org Lett 13:3852–3855. (b) Pan Y, Wee CW, Chen L, Tan C–H (2010) Dehydrogenative coupling reactions catalyzed by Rose Bengal using visible light irradiation. Green Chem 13:2682–2685. (c) Rueping M, Vila C, Bootwicha T (2013) Continuous flow organocatalytic C–H functionalization and crossdehydrogenative coupling reactions: visible light organophotocatalysis for multicomponent reactions and C–C, C–P bond formations. ACS Catal 3:1676–1680. (d) Borpatra PJ, Deb ML, Baruah PK. (2017) Visible light-promoted metal-free intramolecular cross dehydrogenative coupling approach to 1,3-oxazines. Tetrahedron Lett 58:4006–4010
49. Xu J, Shao L-D, Li D, Deng X, Liu Y-C, Zhao Q-S, Xia C (2014) Construction of tetracyclic 3-spirooxindole through cross-dehydrogenation of pyridinium: applications in facile synthesis of (±)-corynoxine and (±)-corynoxine B. J Am Chem Soc 136:17962–17965
50. Kumar N, Ghosh S, Bhunia S, Bisai A (2016) Synthesis of 2-oxindoles via 'transition-metal-free' intramolecular dehydrogenative coupling (IDC) of sp^2 C–H and sp^3 C–H bonds. Beilstein J Org Chem 12:1153–1169

51. Reddy BN, Ramana CV (2016) Synthesis of functionalized 6-hydroxy-2-oxindole derivatives by phenoxide cyclization. Org Lett 18:6264–6267
52. Tian H, Yang H, Zhu C, Fu H (2016) Transition metal-free intramolecular regioselective couplings of aliphatic and aromatic C–H bonds. Sci Rep 6:19931
53. Vanjari R, Guntreddi T, Kumar S, Singh KN (2015) Sulphur promoted $C(sp^3)$–$C(sp^2)$ cross dehydrogenative cyclisation of acetophenone hydrazones with aldehydes: efficient synthesis of 3,4,5-trisubstituted 1H-pyrazoles. Chem Commun 51:366–369
54. (a) Andreades S, Zahnow EW (1969) Anodic cyanations of aromatic compounds. J Am Chem Soc 91:4181–4190. (b) Zhao H.-B, Liu Z.-J, Song J, Xu H.-C (2017) Reagent-free C–H/N–H cross-coupling: regioselective synthesis of *N*-heteroaromatics from biaryl aldehydes and NH_3. Angew Chem Int Ed 129:12906–12909
55. Liu X, Zhang J, Ma S, Ma Y, Wang R (2014) Oxidative cross-dehydrogenative coupling between *N*-aryl tetrahydroisoquinolins and 5*H*-oxazol-4-ones through two methodologies: Copper catalysis or a metal-free strategy. Chem Commun 50:15714–15717
56. Nobuta T, Fujiya A, Yamaguchi T, Tada N, Miura T, Itoh A (2013) Molecular-iodine-catalyzed aerobic photooxidative C-C bond formation between tertiary amines and carbon nucleophiles. RSC Adv 3:10189–10192
57. Nobuta T, Tada N, Fujiya A, Kariya A, Miura T, Itoh A (2013) Molecular iodine catalyzed cross-dehydrogenative coupling reaction between two sp3 C-H bonds using hydrogen peroxide. Org Lett 15:574–577
58. Muramatsu W, Nakanoa K, Li C-J (2014) Direct sp^3C-H bond arylation, alkylation, and amidation of tetrahydroisoquinolines mediated by hypervalent iodine(III) under mild conditions Org Biomol Chem 12:2189–2192
59. Xie Z, Liu L, Chen W, Zheng H, Xu Q, Yuan H, Lou H (2014) Practical metal-free C(sp3)-H functionalization: construction of structurally diverse α-Substituted N-benzyl and N-Allyl carbamates. Angew Chem Int Ed 53:3904–3908
60. Dhineshkumar J, Lamani M, Alagiri K, Prabhu KR (2013) A versatile C-H functionalization of tetrahydroisoquinolines catalyzed by iodine at aerobic conditions. Org Lett 15:1092–1095
61. Matcha K, Antonchick AP (2013) Metal-free cross-dehydrogenative coupling of heterocycles with aldehydes. Angew Chem Int Ed 52:2082–2086
62. (a) Beiring L, Antonchick AP (2015) Regioselective metal-free cross-coupling of quinoline *N*-oxides with boronic acids. Org Lett 17:3134–3137. (b) Zhang Z, Pi C, Tong H, Cui X, Wu Y (2017) Iodine-catalyzed direct C–H alkenylation of azaheterocycle *N*-oxides with alkenes. Org Lett 19:440–443
63. Yang K, Bao X, Yao Y, Qu J, Wang B (2018) Iodine-mediated cross-dehydrogenative coupling of pyrazolones and alkenes. Org Biomol Chem 16:6275–6283
64. Huang H-Y, Wu H-R, Wei F, Wang D, Liu L (2015) Iodine-catalyzed direct olefination of 2-oxindoles and alkenes *via* cross-dehydrogenative coupling (CDC) in air. Org Lett 17:3702–3705
65. Yavari I, Hosseinpour R, Skoulika S (2015) Iodine-mediated diastereoselective cyclopropanation of arylidene malononotriles by 2,6-dimethylquinoline. Synlett 26:380–384
66. Sharma R, Abdullaha M, Bharate SB (2017) Oxidant-controlled C–sp^2/sp^3–H cross-dehydrogenative coupling of *N*-heterocycles with benzylamines. J Org Chem 82:9786–9793
67. Jin R, Patureau FW (2016) Mild, periodate-mediated, dehydrogenative C–N bond formation with phenothiazines and phenols. Org Lett 18:4491–4493
68. Morimoto K, Ogawa R, Koseki D, Takahashi Y, Dohi T, Kita Y (2015) Clean synthesis of *N*-pyrrolyl azoles by metal-free oxidative cross-coupling using recyclable hypervalent iodine reagent. Chem Pharm Bull 63:819–824
69. Shi X, Zhang F, Luo W-K, Yang L (2017) Oxidant-triggered C1-benzylation of isoquinoline by iodine-catalyzed cross-dehydrogenative-coupling with methylarenes. Synlett 28:494–498
70. Gupta A, Deshmukh MS, Jain N (2017) Iodine-catalyzed C–N bond formation: synthesis of 3-aminoquinoxalinones under ambient conditions. J Org Chem 82:4784–4792

71. Liu R, Wang Y, Weng Y, Yao C, Zhang Y, Zhu G, He X, Xu K, Tan G (2017) Iodine-mediated one-pot synthesis of 2-(piperazin-1-yl)pyrazine derivatives from *N*-alkyl piperazines. Synlett 28:1083–1086
72. (a) Joseph J, Kim JY, Chang S (2011) A metal-free route to 2-aminooxazoles by taking advantage of the unique ring opening of benzoxazoles and oxadiazoles with secondary amines. Chem Eur J 17:8294–9298. (b) Wang X, Xu D, Miao C, Zhang Q, Sun W (2014) *N*-Bromosuccinimide as an oxidant for the transition-metal-free synthesis of 2-aminobenzoxazoles from benzoxazoles and secondary amines. Org Biomol Chem 12:3108–3113
73. Tan B, Toda N, Barbas CF III (2012) Organocatalytic amidation and esterification of aldehydes with activating reagents by a cross-coupling strategy. Angew Chem Int Ed 51:12538–12541
74. Siddaraju Y, Prabhu KR (2016) Iodine-catalyzed cross dehydrogenative coupling reaction: a regioselective sulfenylation of imidazoheterocycles using dimethyl sulfoxide as an oxidant. J Org Chem 81:7838–7846
75. Hiebel M-A, Berteina-Raboin S (2015) Iodine-catalyzed regioselective sulfenylation of imidazoheterocycles in PEG_{400}. Green Chem 17:937–944
76. Yang D, Sun P, Wei W, Meng L, He L, Fang B, Jiang W, Wang H (2016) Metal-free iodine-catalyzed direct cross-dehydrogenative coupling (CDC) between pyrazoles and thiols. Org Chem Front 3:1457–1461
77. Siddaraju Y, Prabhu KR (2017) Iodine-catalyzed sulfenylation of pyrazolones using dimethyl sulfoxide as an oxidant. Org Biomol Chem 15:5191–5196
78. (a) Yang F-L, Tian S-K (2013) Iodine-catalyzed regioselective sulfenylation of indoles with sulfonylhydrazides. Angew Chem Int Ed 52:4929–4932. (b) He Y, Liu S, Wen P, Tian W, Ren X, Zhou Q, Ma H, Huang G (2016) Iodine-catalyzed regioselective sulfenylation of indoles with thiols in water. Chem Sel 1:1567–1570
79. Yi S, Li M, Mo W, Hu X, Hu B, Sun N, Jin L, Shen Z (2016) Metal-free, iodine-catalyzed regioselective sulfenylation of indoles with thiols. Tetrahedron Lett 57:1912–1916
80. Zhang H, Bao X, Song Y, Qu J, Wang B (2015) Iodine-catalyzed versatile sulfenylation of indoles with thiophenols: controllable synthesis of mono- and bis-arylthioindoles. Tetrahedron 71:8885–8891
81. Kumar RA, Saidulu G, Prasad KR, Kumar GS, Sridhar B, Reddy KR (2012) Transition metal-free α-C(sp^3)-H bond functionalization of amines by oxidative cross dehydrogenative coupling reaction: Simple and direct access to C-4-Alkylated 3,4-dihydroquinazoline derivatives. Adv Synth Catal 354:2985–2991
82. Lao Z-Q, Zhong W-H, Lou Q-H, Li Z-J, Meng X-B (2012) KI-catalyzed imidation of sp^3 C-H bond adjacent to amide nitrogen atom. Org Biomol Chem 10:7869–7871
83. Wan L, Qiao K, Sun XN, Di ZC, Fang Z, Li ZJ, Guo K (2016) Benzylation of heterocyclic *N*-oxides *via* direct oxidative cross-dehydrogenative coupling with toluene derivatives. New J Chem 40:10227–10232
84. Yuan J-W, Fu J-H, Liu S-N, Xiao Y-M, Mao P, Qu L-B (2018) Metal-free oxidative coupling of quinoxalin-2(1*H*)-ones with arylaldehydes leading to 3-acylated quinoxalin-2 (1*H*)-ones. Org Biomol Chem 16:3203–3212
85. Chen J, Wan M, Hua J, Sun Y, Lv Z, Li W, Liu L (2015) TBHP/TFA mediated oxidative cross-dehydrogenative coupling of *N*-heterocycles with aldehydes. Org Biomol Chem 13:11561–11566
86. Mishra K, Pandey AK, Singh JB, Singh RM (2016) Metal free TBHP-promoted intramolecular carbonylation of arenes *via* radical crossdehydrogenative coupling: Synthesis of indenoquinolinones, 4-azafluorenones and fluorenones. Org Biomol Chem 14:6328–6336
87. Zhang L, Zhang G, Li Y, Wang S, Lei A (2018) The synergistic effect of self-assembly and visible-light induced the oxidative C-H acylation of N-heterocyclic aromatic compounds with aldehydes. Chem Commun 54:5744–5747

88. Yang Q, Choy PY, Wu Y, Fan B, Kwong FY (2016) Oxidative coupling between C(sp^2)–H and C(sp^3)–H bonds of indoles and cyclic ethers/cycloalkanes. Org Biomol Chem 14:2608–2612
89. Li Z-l, Jina L-k, Cai C (2017) Efficient synthesis of 2-substituted azoles: radical C-H alkylation of azoles with dicumyl peroxide, methylarenes and cycloalkanes under metal-free condition. Org Chem Front 4:2039–2043
90. Yu L, Wang M, Wang L (2014) Metal-free cross-dehydrogenative coupling of benzimidazoles with aldehydes to *N*-acylbenzimidazoles. Tetrahedron 70:5391–5397
91. Aruri H, Singh U, Kumar M, Sharma S, Aithagani SK, Gupta VK, Mignani S, Vishwakarma RA, Singh PP (2017) Metal-free cross-dehydrogenative coupling of HN-azoles with α-C(sp3)-H amides via C–H activation and its mechanistic and application studies. J Org Chem 82:1000–1012
92. Patel OPS, Nandwana NK, Sah AK, Kumar A (2018) Metal-free synthesis of aminomethylated imidazoheterocycles: dual role of *tert*-butyl hydroperoxide as both an oxidant and a methylene source. Org Biomol Chem 16:8620–8628
93. Ma Y, Zhang G, Zhang J, Yang D, Wang R (2014) Organocatalyzed asymmetric oxidative coupling of α-C(sp3)-H of tertiary amines to α, β-unsaturated γ-butyrolactam: synthesis of MBH-type products. Org Lett 16:5358–5361
94. Zhang G, Ma Y, Wang S, Kong W, Wang R (2013) Chiral organic contact ion pairs in metal-free catalytic enantioselective oxidative cross-dehydrogenative coupling of tertiary amines to ketones. Chem Sci 4:2645–2651
95. Xie Z, Zan X, Sun S, Pan X, Liu L (2016) Organocatalytic enantioselective cross-dehydrogenative coupling of *N*-carbamoyl cyclic amines with aldehydes. Org Lett 18:3944–3947
96. Zhang B, Cui Y, Jiao N (2012) Metal-free TEMPO-catalyzed oxidative C-C bond formation from Csp3–H bonds using molecular oxygen as the oxidant. Chem Commun 48:4498–4500
97. Liu TX, Ma J, Chao D, Zhang P, Ma N, Liu Q, Shi L, Zhang Z, Zhang G (2016) Synthesis of C_{60}-fused tetrahydrocarbazole/dibenzothiophene/ benzothiophene and dibenzofuran derivatives *via* metal-free oxidative dehydrogenative carboannulation. Org Lett 18:4044–4047
98. Liu S, Liu A, Zhang Y, Wang W (2017) Direct Cα-heteroarylation of structurally diverse ethers *via* a mild *N*-hydroxysuccinimide mediated cross-dehydrogenative coupling reaction. Chem Sci 8:4044–4050
99. Haldar S, Jana CK (2019) Direct (het)arylation of tetrahydroisoquinolines *via* a metal and oxidant free C(sp^3)–H functionalization enabled three component reaction. Biomol Chem, Org. https://doi.org/10.1039/c8ob02309c
100. Zhang B, Xiang S-K, Zhang L-H, Cui Y, Jiao N (2011) Organocatalytic asymmetric intermolecular dehydrogenative α-alkylation of aldehydes using molecular oxygen as oxidant. Org Lett 13:5212–5215
101. Aruri H, Singh U, Sharma S, Gudup S, Bhogal M, Kumar S, Singh D, Gupta VK, Kant R, Vishwakarma RA, Singh PP (2015) Cross-dehydrogenative coupling of azoles with α-C (sp^3)–H of ethers and thioethers under metal-free conditions: functionalization of H–N azoles *via* C–H activation. J Org Chem 80:1929–1936
102. Wang L, Zhu K-q, Wu W-t, Chen Q, He M-y (2015) *n*-Bu_4NI-catalyzed direct amination of ethers with aryl tetrazoles and triazoles *via* crossdehydrogenative coupling reaction. Catal Sci Technol 5:2891–2896
103. Mendive-Tapia L, Albornoz-Grados A, Bertran A, Albericio F, Lavilla R (2017) Oxidative couplings on tryptophan-based diketopiperazines leading to fused and bridged chemotypes. Chem Commun 53:2740–2743
104. Rajamanickam S, Majji G, Santra SK, Patel BK (2015) Bu_4NI catalyzed C–N bond formation *via* cross-dehydrogenative coupling of aryl ethers (Csp^3–H) and tetrazoles (N–H). Org Lett 17:5586–5589
105. Zhou Y, Liu Z, Yuan T, Huang J, Liu C (2017) The synthesis of 2-aminobenzoxazoles using reusable ionic liquid as a green catalyst under mild conditions. Molecules 22:576

106. Dian L, Wang S, Zhang-Negrerie D, Du Y, Zhao K (2014) Organocatalytic amination of alkyl ethers *via* n-Bu_4NI/*t*-BuOOH-mediated intermolecular oxidative C(sp^3)–N bond formation: novel synthesis of hemiaminal ethers. Chem Commun 50:11738–11741
107. Liu W, Liu C, Zhang Y, Sun Y, Abdukadera A, Wang B, Li H, Ma X, Zhang Z (2015) Reusable ionic liquid-catalyzed oxidative coupling of azoles and benzylic compounds via sp3 C-N bond formation under metal-free conditions. Org Biomol Chem 13:7154–7158
108. Abebe H, Vidavalur S, Battula VR (2016) n-Bu4NI/TBHP-catalyzed C-N bond formation via cross-dehydrogenative coupling of 1H–1,2,4-triazoles (N–H) and methylarenes (Csp^3–H). RSC Adv 6:82289–82293
109. Dong D-Q, Zhang H, Wang Z-L (2017) Synthesis of benzyl esters from the commercially available alcohols catalyzed by TBAI *via* C(sp^3)–H bond functionalization. RSC Adv 7:3780–3782
110. Chen Q, Wang X, Yu G, Wen C, Huo Y (2018) DDQ-mediated direct C(sp^3)–H phosphorylation of xanthenes derivatives. Org Chem Front 5:2652–2656
111. Tsang AS-K, Todd MH (2009) Facile synthesis of vicinal diamines *via* oxidation of *N*-phenyltetrahydroisoquinolines with DDQ. Tetrahedron Lett 50:1199–1202
112. Xiao T, Li L, Lin G, Mao Z-w, Zhou L (2014) Metal-free visible-light induced cross-dehydrogenative coupling of tertiary amines with diazo compounds. Org Lett 16:4232–4235
113. Kibriya G, Bagdi AK, Hajra A (2018) Visible-light-promoted C(sp^3)–C(sp^2) cross-dehydrogenatived coupling of tertiary amine with imidazopyridine. J Org Chem 83:10619–10626
114. Wei W, Wang L, Bao P, Shao Y, Yue H, Yang D, Yang X, Zhao X, Wang H (2018) Metal-free C(sp^2)–H/N–H cross-dehydrogenative coupling of quinoxalinones with aliphatic amines under visible-light photoredox catalysis. Org Lett 20:7125–7130
115. Zhao Y, Huang B, Yang C, Xia W (2016) Visible-light-promoted direct amination of phenols *via* oxidative cross-dehydrogenative coupling reaction. Org Lett 18:3326–3329
116. Zhang L, Yi H, Wang J, Lei A (2017) Visible-light mediated oxidative C–H/N–H cross-coupling between tetrahydrofuran and azoles using air. J Org Chem 82:10704–10709
117. Yamaguchi T, Yamaguchi E, Itoh A (2017) Cross-dehydrogenative C–H amination of indoles under aerobic photo-oxidative conditions. Org Lett 19:1282–1285
118. Wang X-F, Yu S-S, Wang C, Xue D, Xiao J (2016) BODIPY catalyzed amide synthesis promoted by BHT and air under visible light. Org Biomol Chem 14:7028–7037
119. Sattar M, Rathore V, Prasad CD, Kumar S (2017) Transition-metal-free chemoselective oxidative C-C coupling of the sp3 C-H bond of oxindoles with arenes and addition to alkene: Synthesis of 3-aryl oxindoles, and benzofuro- and indoloindoles. Chem Asian J 12:734–743
120. Liu W-B, Schuman DP, Yang Y-F, Toutov AA, Liang Y, Klare HFT, Nesnas N, Oestreich M, Blackmond DG, Virgil SC, Banerjee S, Zare RN, Grubbs RH, Houk KN, Stoltz BM (2017) Potassium *tert*-butoxide-catalyzed dehydrogenative C–H silylation of heteroaromatics: A combined experimental and computational mechanistic study. J Am Chem Soc 139:6867–6879
121. Chen X, Cui X, Yang F, Wu Y (2015) Base-promoted cross-dehydrogenative coupling of quinoline *N*-oxides with 1,3-azoles. Org Lett 17:1445–1448
122. Li S-S, Fu S, Wang L, Xu L, Xiao J (2017) *t*-BuOK-mediated oxidative dehydrogenative C (sp^3)-H arylation of 2-alkylazaarenes with nitroarenes. J Org Chem 82:8703–8709
123. Khalymbadzha IA, Chupakhin ON, Fatykhov RF, Charushin VN, Schepochkin AV, Kartsev VG (2016) Transition-metal-free cross-dehydrogenative coupling of triazines with 5,7-dihydroxycoumarins. Synlett 27:2606–2610
124. Chen W, Zheng H, Pan X, Xie Z, Zan X, Sun B, Liu L, Lou H (2014) A metal-free cross-dehydrogenative coupling of *N*-carbamoyl tetrahydroisoquinoline by sodium persulfate. Tetrahedron Lett 55:2879–2882
125. Li X, Wang H-Y, Shi Z-J (2013) Transition-metal-free cross-dehydrogenative alkylation of pyridines under neutral conditions. New J Chem 37:1704–1706

126. Toonchue S, Sumunnee L, Phomphrai K, Yotphan S (2018) Metal-free direct oxidative C–C bond coupling of pyrazolones and quinoxalinones. Org Chem Front 5:1928–1932
127. Siddaraju Y, Lamani M, Prabhu KR (2014) A transition metal-free Minisci reaction: Acylation of isoquinolines, quinolines, and quinoxaline. J Org Chem 79:3856–3865
128. Tanoue A, Yoo W-J, Kobayashi S (2014) Sulfuryl chloride as an efficient initiator for the metal-free aerobic cross-dehydrogenative coupling reaction of tertiary amines. Org Lett 16:2346–2349
129. Adib M, Pashazadeh R, Rajai-Daryasarei S, Kabiri R, Gohari SJA (2016) Transition-metal-free acylation of quinolines and isoquinolines with arylmethanols via oxidative cross-dehydrogenative coupling reactions. Synlett 27:2241–2245
130. Babu GVS, Prathima PS, Perumgani PC, Sridhar B, Rao TV, Rao MM (2018) Metal-free cross-dehydrogenative coupling approach for C−H bond functionalization of 2-phenyl pyridine derivatives in water. Heteroatom Chem e21423
131. Ganley JM, Christensen M, Lam Y-H, Peng Z, Angeles AR, Yeung CS (2018) Metal-and acid-free C–H formylation of nitrogen heterocycles: using trioxane as an aldehyde equivalent enabled by an organic-soluble oxidant. Org Lett 20:5752–5756
132. Dhineshkumar J, Samaddar P, Prabhu KR (2017) Catalyst-free cross-dehydrogenative coupling strategy using air as an oxidant: synthesis of α-aminophosphonates. ACS Omega 2:4885–4893
133. Osorio-Nieto U, Chamorro-Arenas D, Quintero L, Höpfl H, Sartillo-Piscil F (2016) Transition metal-free selective double sp3 C-H oxidation of cyclic amines to 3-alkoxyamine lactams. J Org Chem 81:8625–8632
134. Griffiths RJ, Kong WC, Richards SA, Burley GA, Willis MC, Talbot EPA (2018) Oxidative β-C–H sulfonylation of cyclic amines. Chem. Sci. 9:2295–2300

Cross-Dehydrogenative Coupling of Carbonyls for Heterocycle Synthesis

Ryo Yazaki and Takashi Ohshima

Abstract Cross-dehydrogenative coupling of two different carbonyls is regarded as a powerful tool for the synthesis of unsymmetrical 1,4-dicarbonyl, which can be transformed into a variety of synthetically useful heterocycles through Paal–Knorr reaction. In early works, excess amounts of one coupling partner were used for the satisfactory yield of the cross-coupling product over undesired homo-coupling dimer. Recently, several research groups developed an efficient method for the chemoselective cross-coupling of two different carbonyls using an equimolar or slightly excess amount of coupling partner. Furthermore, catalytic cross-dehydrogenative coupling of two different carbonyls was reported. In these reactions, no stoichiometric amount of bases was required. This chapter covers recent progress of intermolecular cross-dehydrogenative coupling of two different carbonyls and overview of the strategy for chemoselective cross-coupling and product transformations into heterocycles.

Keywords Enolate · 1,3-Dicarbonyl · Cross-coupling · Chemoselective reaction

1 Introduction

1,4-Dicarbonyl motifs are widely found in natural products and pharmaceuticals [1–10]. Besides, 1,4-dicarbonyl compounds can be transformed into a variety of 2,3,4,5-substituted five-membered heterocycles through Paal–Knorr synthesis (Fig. 1a). The retrosynthetic analysis of 1,4-dicarbonyl compounds provides two carbonyl synthons [11–13]. In general, carbonyls function as nucleophile via

---

R. Yazaki (✉) · T. Ohshima
Graduate School of Pharmaceutical Sciences, Kyushu University, Maidashi, Higashi-ku, Fukuoka 812-8582, Japan
e-mail: yazaki@phar.kyushu-u.ac.jp

T. Ohshima
e-mail: ohshima@phar.kyushu-u.ac.jp

A. Srivastava and C. K. Jana (eds.), *Heterocycles via Cross Dehydrogenative Coupling*, https://doi.org/10.1007/978-981-13-9144-6_6

deprotonative activation of α-proton. Therefore, umpolung of one carbonyl is required for coupling of two carbonyls [14, 15]. The umpolung approach is typically the coupling of α-brominated carbonyl prepared from bromine and enolate, and α-brominated carbonyl (electrophile) efficiently couples with another enolate (nucleophile) (Fig. 1b left). However, this umpolung approach requires additional steps and more than stoichiometric amount of reagents for pre-activation of both coupling partners, thus reducing the efficiency of the overall synthesis process in terms of atom economy and step economy [16, 17]. In contrast, intermolecular cross-dehydrogenative coupling [18–20] of two carbonyls is regarded as one of the most straightforward methods for the preparation of synthetically useful 1,4-dicarbonyl compounds (Fig. 1b right) [21, 22]. However, due to the concomitant formation of undesired homo-coupling dimers, intermolecular cross-dehydrogenative coupling of two carbonyls suffered from low chemical yields of cross-enolate coupling products. Therefore, extensive efforts over the past several decades have been dedicated to achieve chemoselective formation of cross-enolate coupling products.

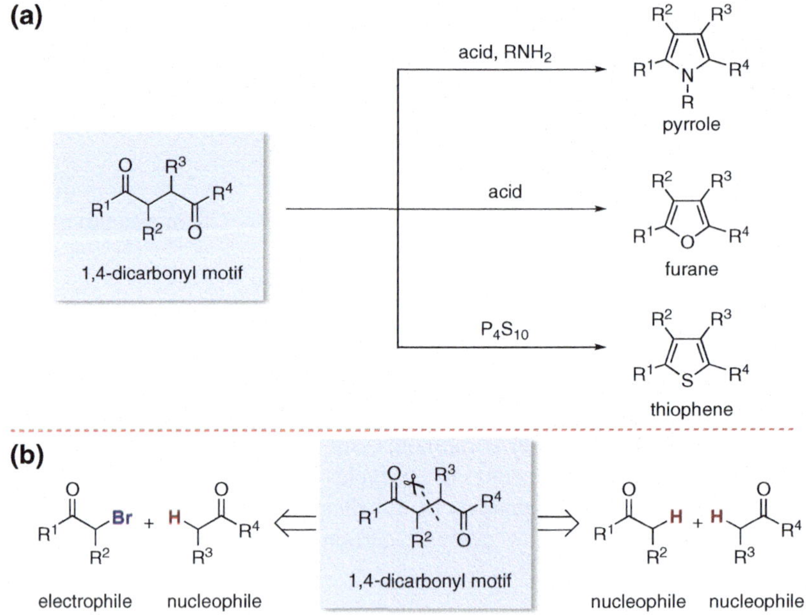

Fig. 1 **a** Heterocycle synthesis from 1,4-dicarbonyls through Paal–Knorr synthesis and **b** 1,4-dicarbonyl synthesis: conventional umpolung approach versus cross-dehydrogenative coupling

2 Cross-Coupling of Metal Enolates

Historically, Ivanoff et al. reported dehydrogenative homo-coupling of sodium salt of carboxylic acid in 1935, by the treatment of Grignard reagent followed by bromine, affording symmetrical 1,4-dicarbonyl compounds (Fig. 2a) [23]. However, these initial works did not gain the attention because the method was only applicable to the homo-coupling of carbonyl and suffered from low chemical yield due to several side product formations, although the method is regarded as the pioneering work as an oxidative enolate coupling in one-pot synthesis. In 1971, Rathke et al. reported homo-dimerization of lithium enolates prepared from lithium bases and esters (Fig. 2b) [24]. The key point of the reaction was the use of copper salts, such as soluble copper(II) bromide and copper(II) valerate, as an oxidant, and the products were isolated in high yield (up to 85% yield). After this advanced report, various oxidative enolate coupling reactions were developed using metal oxidants. In 1975, pioneering work of cross-dehydrogenative coupling of two different carbonyls was achieved by Saegusa (Fig. 2c) [25, 26]. The authors succeeded to obtain the

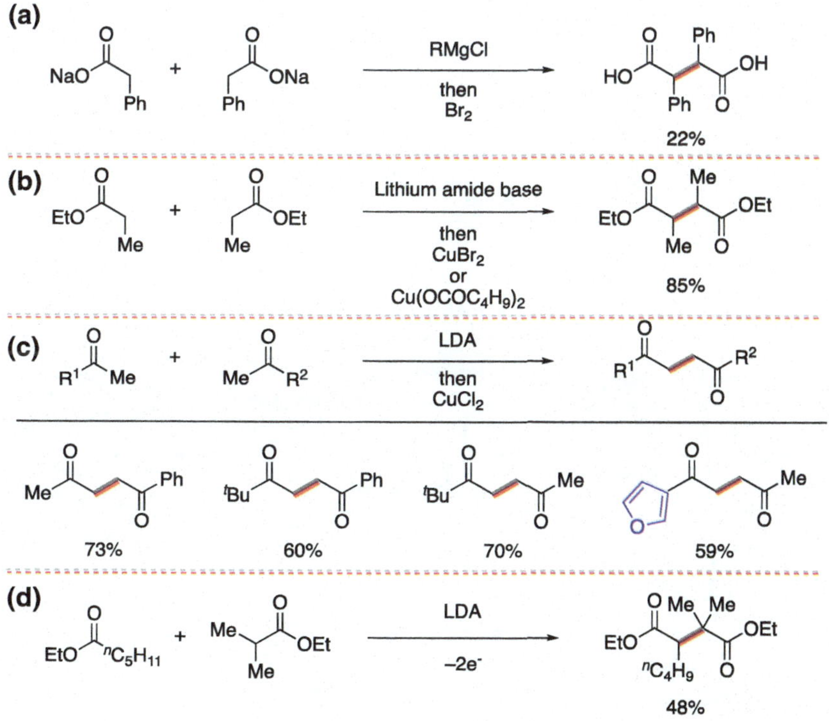

Fig. 2 **a** Oxidative homo-coupling dimer formation, **b** soluble metal oxidant for oxidative homo-coupling, **c** cross-dehydrogenative coupling, **d** cross-dehydrogenative coupling under electrochemical conditions

unsymmetrical 1,4-diketones or 1,4-ketoesters in synthetically useful yields by the treatment of lithium diisopropyl amide followed by $CuCl_2$ as an oxidant. A variety of carbonyls could be used, although excess amount of one carbonyl, 3–3.5 equivalents, was required for good yields due to the competitive formation of homo-coupling dimer. Sensitive furan could be incorporated into the product in high yield under oxidative conditions, highlighting the mildness of the reaction. Itoh et al. reported electrochemical oxidation of lithium enolates (Fig. 2d) [27]. Cross-lithium enolate coupling was achieved under electrochemical conditions, delivering the product in 48% yield using three equivalents of one carbonyl.

Intermolecular approach to suppress undesired homo-coupling dimer was developed by Thomson et al. [28] (Fig. 3). Unsymmetrical silyl bis-enol ethers

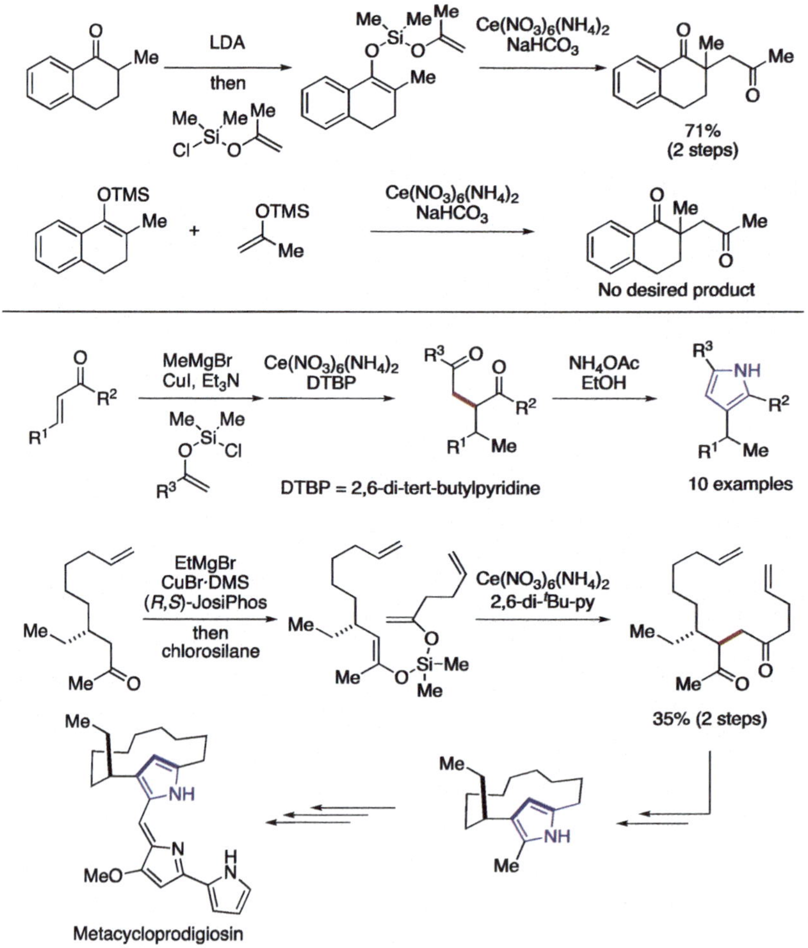

Fig. 3 Intermolecular approach using silyl bis-enol ethers

were selected as enolate substrates. Unsymmetrical silyl bis-enol ethers could be easily prepared by lithium enolate with chlorosilane, and the corresponding cross-coupling product was obtained in high yield using cerium(IV) ammonium nitrate as an oxidant. As a control experiment, equimolar amount of TMS-enol silanes were subjected to the oxidation conditions and the complex reaction mixture was observed, highlighting the usefulness of the unsymmetrical silyl bis-enol ether strategy. The authors applied their strategy to diastereoselective reaction using diisopropylsilyl bis-enol ethers [29]. They also investigated substituted pyrrole synthesis through their developed reaction and achieved total synthesis of meta-cycloprodigiosin and prodigiosin R1 through oxidative cross-enolate coupling followed by pyrrole formation [30]. Recently, they reported synthetic method of stereoselective polycyclic molecules based on oxidative coupling of silyl bis-enol ethers followed by ring-closing metathesis reaction [31].

In 2006, Baran et al. achieved the significant progress of cross-dehydrogenative coupling of two carbonyls. The authors developed practical intermolecular dehydrogenative cross-enolate coupling reaction (Fig. 4a) [32]. The treatment of LDA followed by iron(III) or copper(II) salt as oxidant provided the cross-coupling product in synthetically useful yield using an equimolar amount of starting material. The *N*-acyl oxazolidinones were efficiently coupled with a variety of carbonyl, including ketones and esters. Quaternary carbon center could be constructed in high yield by using oxindole instead of *N*-acyl oxazolidinones. A variety of heterocycles, such as 4-chromanone, 2-chromanone, and indole, could be incorporated. They also demonstrated the synthetic utility of their method by applying total synthesis of (–)-bursehernin via lactone formation. The authors extensively optimize the cross-enolate coupling [33]. Furthermore, they also proposed that the reaction mechanisms were different based on the oxidant used. In case of Fe(III) as an oxidant, ketone was preferentially oxidized over *N*-acyl oxazolidinones, resulting in cross-coupling of iron intermediate derived from ketones with lithium enolate derived from *N*-acyl oxazolidinones. On the other hand, in case of Cu(II) as an oxidant, both enolates are proposed to be transmetalated to a single copper atom, providing pseudo-intramolecular C–C bond formation. Quesnelle et al. applied Baran's method for the synthesis of BMS-9096024 in the enantioselective and diastereoselective manner [34].

Flowers et al. performed a detailed mechanistic analysis of cross-coupling of lithium enolate by ^{7}Li-NMR (Fig. 4b) [35]. The authors proposed that the origin of selective formation of hetero-coupling product was a consequence of heteroaggregation of lithium enolates. The 4-chromanone functionality could be incorporated into the cross-coupling product.

Amaya and Hirao et al. developed different strategies for intermolecular oxidative cross-coupling of enolates (Fig. 5). The key is the use of chemoselective oxidant. Cross-coupling between boron enolates with silyl enolates was achieved by using oxovanadium(V) oxidant ($VO(OEt)Cl_2$) [36]. Oxovanadium(V) salt chemoselectively oxidized boron enolates over silyl enolates [37]. A variety of boron enolates with silyl enolates, including ketones and esters, could be used, and excellent yields were observed using an equimolar amount of enolates.

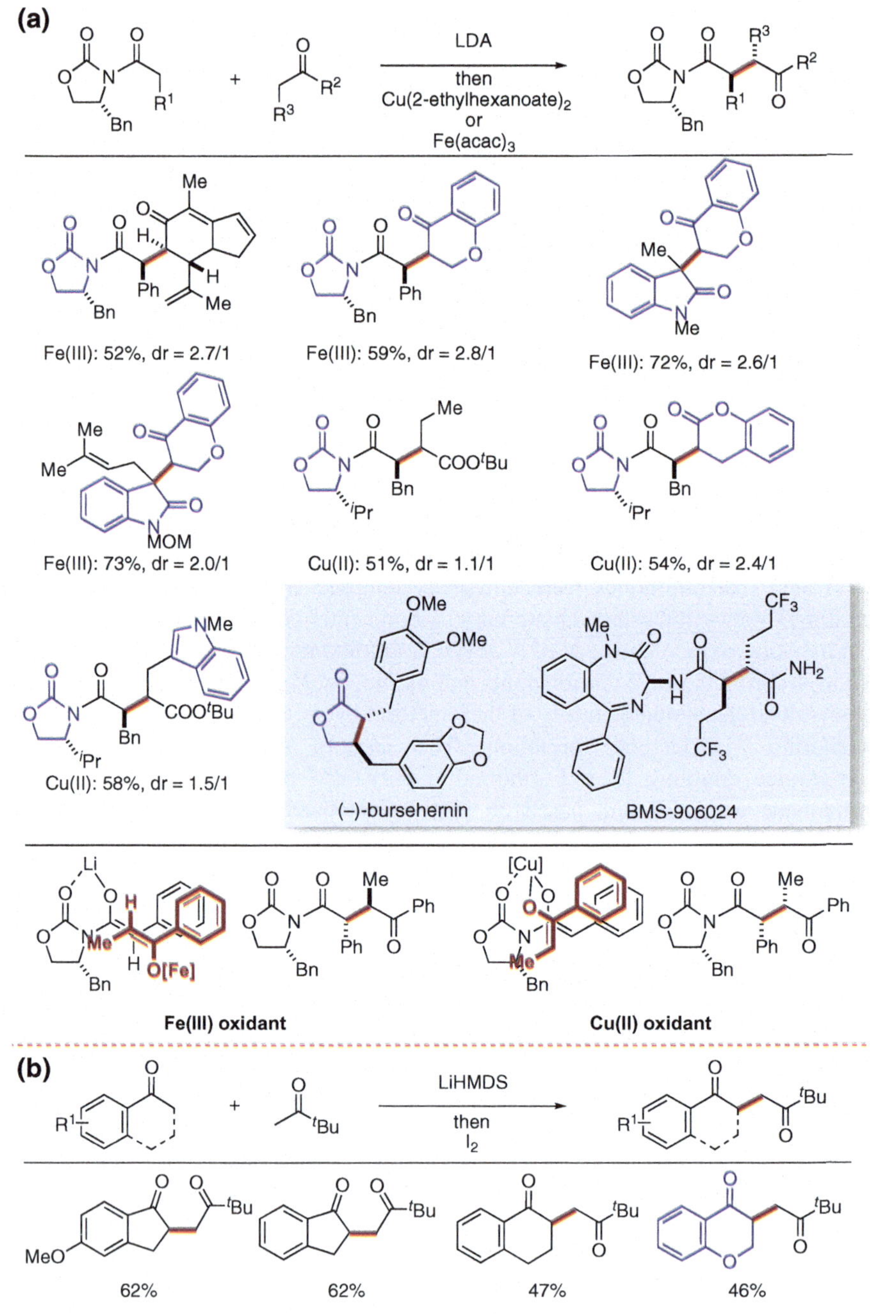

Fig. 4 **a** Oxidative enolate hetero-coupling using *N*-acyl oxazolidinones and **b** cross-coupling of lithium enolates

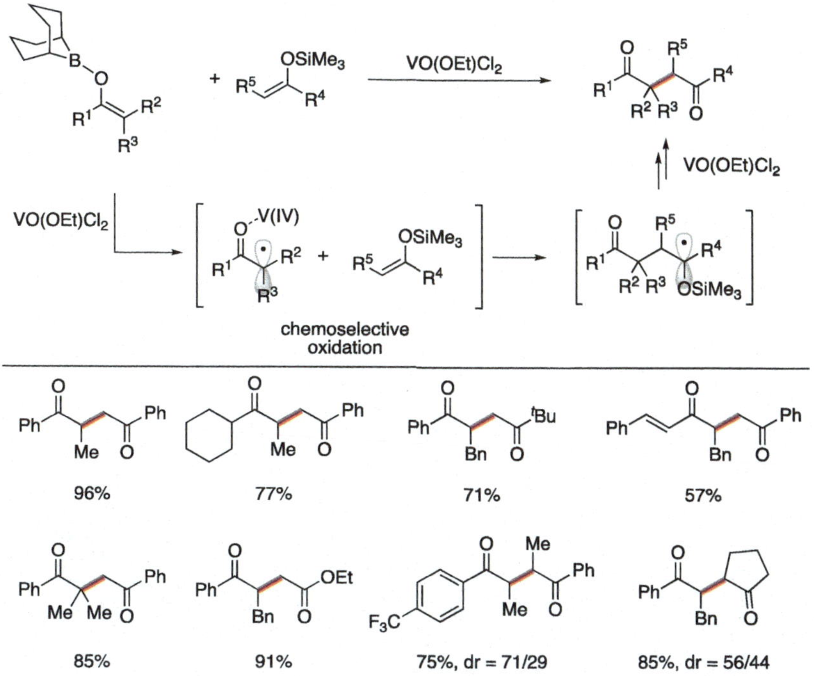

Fig. 5 Vanadium(V)-mediated oxidative cross-enolate coupling

α,α-Dimethyl boron enolate afforded the cross-coupling product bearing quaternary carbon center without any detrimental effect. Other typical metal oxidant, such as $[Ce(NO_2)_6](NH_4)_2$, $FeCl_3$, and $CuCl_2$, showed low efficiency, clearly demonstrating the importance of oxovanadium(V) salt as a chemoselective oxidant. Subsequently, the authors extended substrate scope of the developed oxidative protocol for ketone–ester, ester–ketone, ester–ester, amide–ketone, and amide–ester cross-couplings [38].

3 Cross-Coupling of Enamines

Enamines were also employed in cross-dehydrogenative coupling [39]. As an earliest work, Chiba et al. reported electrochemical oxidative cross-coupling (Fig. 6a) [40]. Preformed enamines were efficiently coupled with β-dicarbonyl compounds in synthetically useful yields. Narasaka et al. developed cross-coupling of enamines with silyl enol ethers under oxidative conditions (Fig. 6b) [41]. Cerium (IV) ammonium nitrate (CAN) was employed as selective oxidant of enamines, generating the radical cation intermediate. MacMillan et al. extended this methodology to catalytic enantioselective reaction (Fig. 6c) [42]. Enamines were

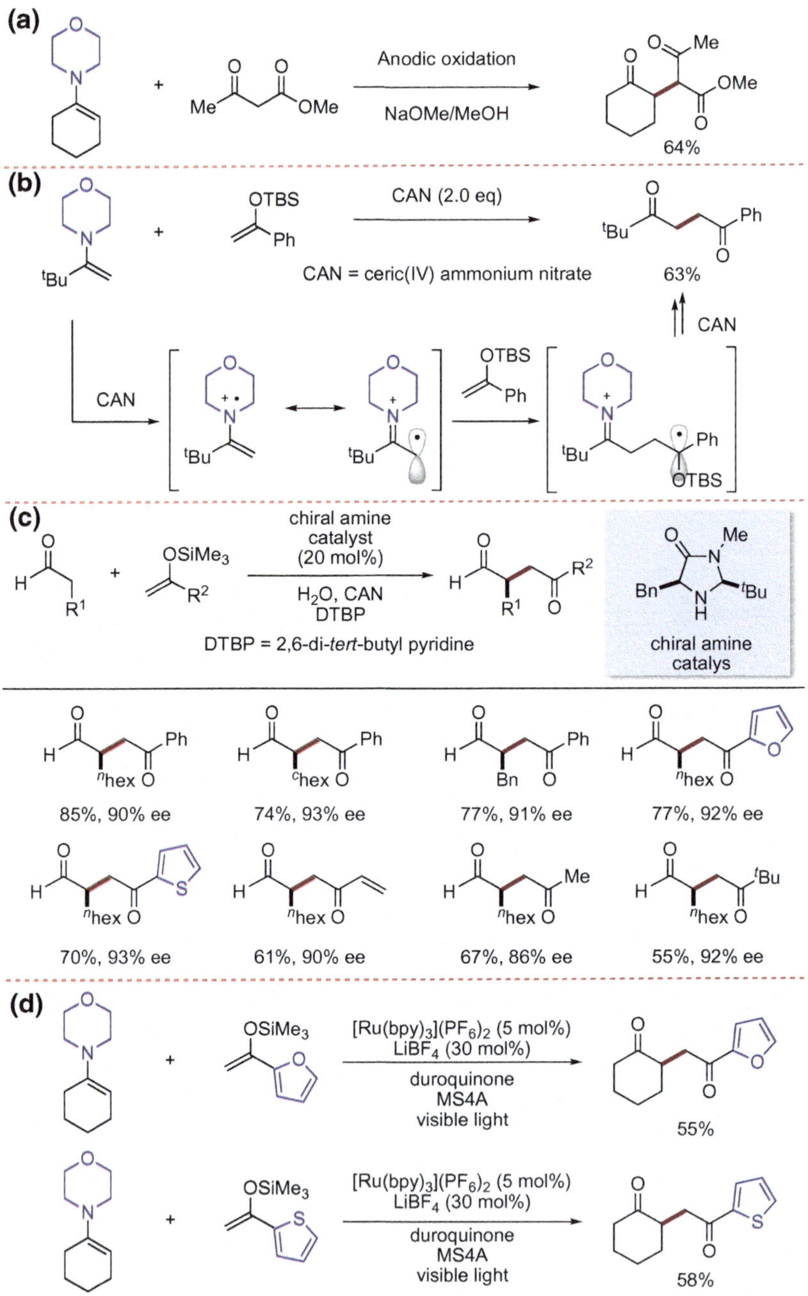

Fig. 6 **a** Electrochemical oxidative coupling of enamines, **b** CAN-mediated oxidative coupling of enamines with silyl enol ethers, **c** chiral amine-catalyzed enantioselective oxidative coupling, **d** photoredox-catalyzed oxidative cross-enolate coupling of enamines

formed in situ from aldehydes with chiral amine catalyst and efficiently coupled with enol silanes in high yield with high enantioselectivity using CAN as an oxidant. Šebesta et al. also reported chiral amine-catalyzed oxidative cross-coupling of aldehydes [43]. A variety of pre-activated silyl compounds, such as cyclic enol silanes, silyl ketene acetals, and silyl ketene thioacetals, could be used. Cyclohexanone also afforded the cross-coupling product, although enantioselectivity was poor. Koike and Akita et al. reported photoredox approach with 2e acceptor [44].[1] Combined use of photoredox $[Ru(bpy)_3](PF_6)_2$ catalyst with duroquinone under visible light irradiation conditions afforded the cross-coupling product. Further improvement of the chemical yield was achieved by the addition of $LiBF_4$, which would lower of the LUMO level of duroquinone. Various enol silanes bearing furan and thiophene could be used as starting materials. Sunlight-promoted reaction was also achieved under similar conditions, and the product was isolated in 61% yield.

4 Organocatalyzed Cross-Coupling of Carbonyls

Organocatalytic approach in cross-dehydrogenative coupling of two carbonyls was recently reported by Odagi and Nagasawa et al. [45] (Fig. 7a). Guanidinium hydroiodide efficiently catalyzed oxidative coupling of β-ketoamides with oxindoles. It is noteworthy that no external base was employed and enolate precursors could be used directly. Hypoiodite (IOH) generated from I^- with cumene hydroperoxide (CHP) was proposed to be an actual catalytic species, which would provide electrophilic intermediate derived from β-ketoamide [46]. Takemoto et al. reported halogen bond donor-catalyzed cross-coupling of iodonium(III) ylides [47] (Fig. 7b). Iodonium(III) ylides were activated by 2-iodobenzimidazolium catalyst though halogen bonding interaction, resulting in increasing electrophilicity of iodonium(III) ylides. The ^{13}C-NMR analysis as well as computational calculation clearly indicated the halogen bonding interaction between iodonium(III) ylides and 2-iodobenzimidazolium catalyst. Various heterocycles, such as indoles, oxindole, and furan, could be incorporated under oxidative conditions.

5 Metal-Catalyzed Dehydrogenative Cross-Coupling

You et al. developed iron-catalyzed directing group-assisted strategies for oxidative cross-coupling (Fig. 8a) [48]. Di-*tert*-butyl peroxide (DTBP) was used as a terminal oxidant. Amino acid derivatives bearing directing group coupled with silyl enol ether or 1,3-dicarbonyls. This method could synthesize various unnatural

---

[1]Yasu Y, Koike T, Akita M (2012) Chem Commun 48:5355.

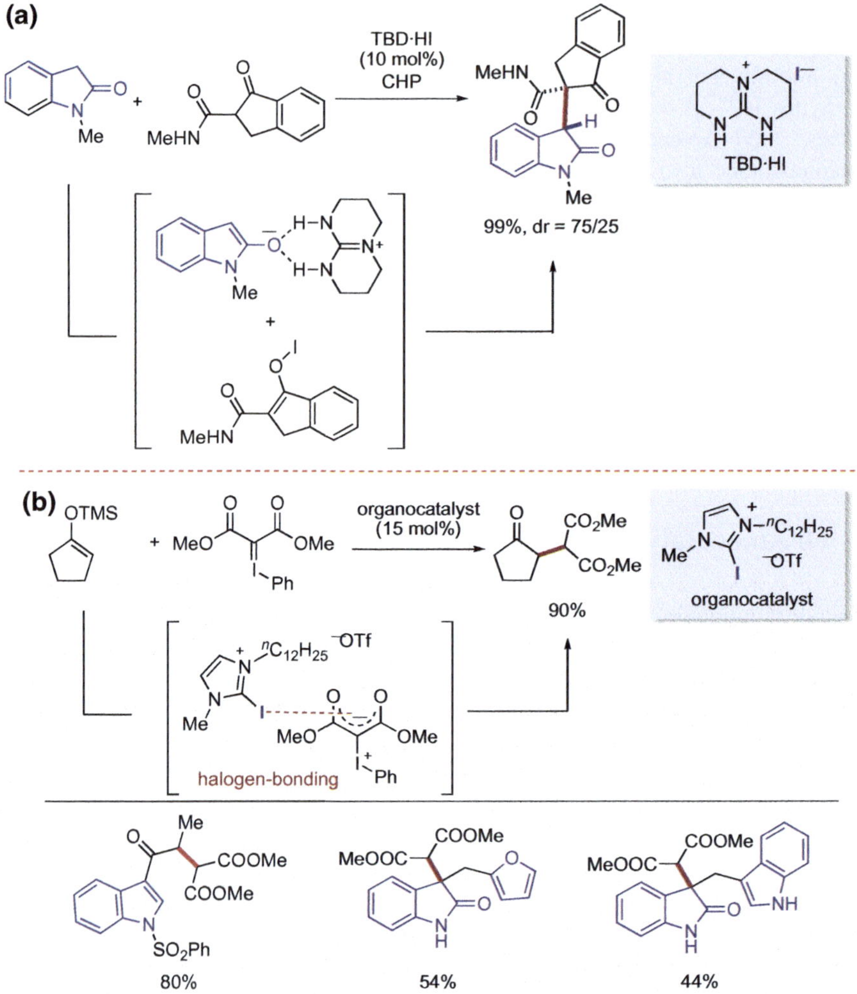

Fig. 7 **a** Guanidinium hydroiodide-catalyzed dehydrogenative cross-coupling and **b** halogen bond donor-catalyzed cross-coupling of iodonium(III) ylides

α,α-dialkylated amino acid derivatives. The proposed catalytic cycle is shown in Fig. 8c. The catalytic cycle would start with the coordination of 2-picolinamido substrate to Fe^{3+} species to afford the intermediate **I**. Then, α-ketoimine intermediate **III** would be generated through radical intermediate **II** and addition of nucleophile to **III** to afford the product. Finally, Fe^{2+} species would be re-oxidized by DTBP or air. Yazaki and Ohshima et al. reported iron-catalyzed oxidative cross-coupling of azlactones with 2-acylimidazoles (Fig. 8c) [49]. A transient homo-coupling dimer strategy, for direct catalytic oxidative cross-enolate coupling, was developed. A dimer of azlactone [50] was firstly generated followed by the

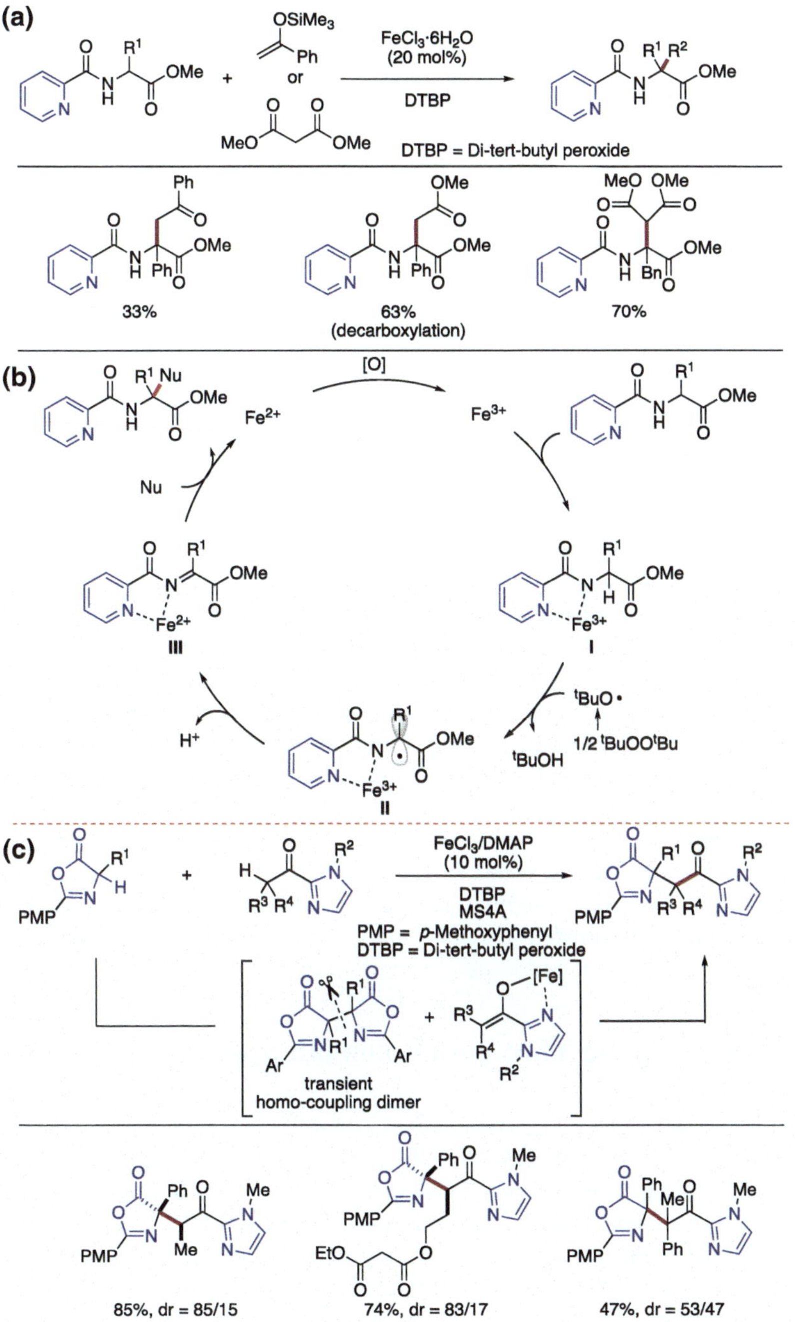

Fig. 8 **a** Directing group-assisted strategy for oxidative cross-coupling, **b** proposed catalytic cycle, **c** transient homo-coupling dimer strategy

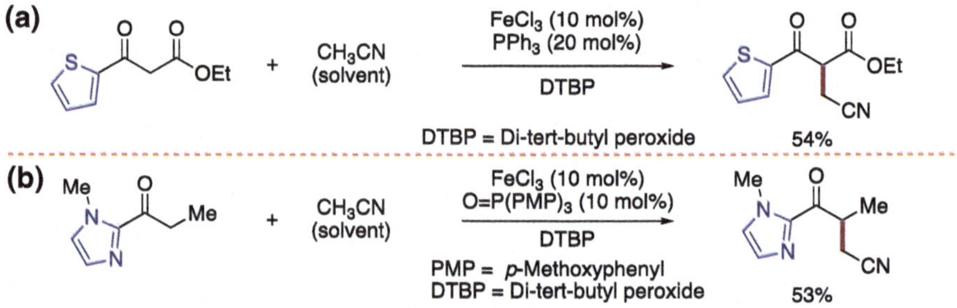

Fig. 9 **a** Catalytic cross-dehydrogenative cyanomethylation of β-dicarbonyls and **b** catalytic cross-dehydrogenative cyanomethylation of 2-acylimidazole

coupling with 2-acylimidazoles. In this reaction, cross-enolate coupling product bearing a contiguous tetrasubstituted carbon center was obtained chemoselectively without the need for stoichiometric amounts of strong bases/metal oxidants. Catalytic chemoselective activation of 2-acylimidazole was achieved even in the presence of more acidic malonate diester. This strategy was also applicable to the oxindole and benzofuranone derivatives [51].

Kim and Wu reported iron-catalyzed cross-dehydrogenative coupling of 1,3-dicarbonyls with acetonitrile using DTBP as an oxidant (Fig. 9a) [52]. A variety of 1,3-dicarbonyls were applicable, affording the α-cyanomethyl-β-dicarbonyls, although the excess amount of cheap acetonitrile was required for synthetically useful yield. Preliminary mechanistic of kinetic isotope effect revealed that both C–H bond cleavage of β-dicarbonyl and acetonitrile have a large contribution to the rate-limiting steps. Yazaki and Ohshima et al. also reported the iron-catalyzed cross-dehydrogenative coupling of 2-acylimidazole, which is regarded as a carboxylic acid equivalent, with acetonitrile (Fig. 9b) [53]. The cross-coupling product derived from two distinct carboxylic acid equivalents was isolated in synthetically useful yield.

6 Cross-Coupling Through Enolonium Species

Maulide et al. developed intermolecular cross-enolate coupling through enolonium intermediate (Fig. 10) [54]. Triflic anhydride and 2-iodopyridine activated the amides, generating keteniminium intermediate. Then, keteniminium intermediate generated was treated with 2,6-lutidine *N*-oxide and enolate nucleophiles under basic conditions, delivering the cross-coupled products in high yields. Amide functionality was chemoselectively activated by Tf_2O/2-iodopyridine conditions over various acidic functionalities, such as esters, nitriles, and ketones. Further transformation of the product by treatment of Tf_2O afforded the trisubstituted furan. The furan underwent cycloaddition/aromatization sequence, providing phthalimide derivative in high yield.

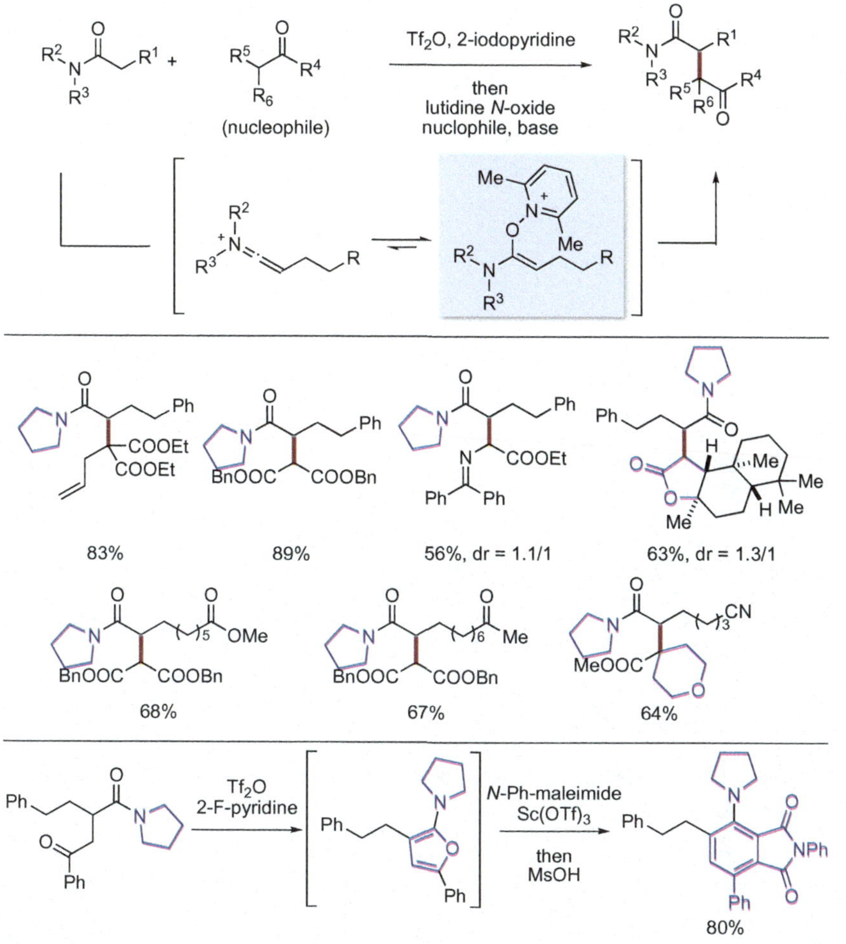

Fig. 10 Chemoselective cross-dehydrogenative coupling of amide via enolonium intermediate

7 Conclusion

Cross-dehydrogenative coupling of two different carbonyls obviously straightforward method to synthesis of synthetically useful and naturally occurring unsymmetrical 1,4-dicarbonyl compounds, which are sources of heterocycles through Paal-Knorr reaction. Throughout this chapter, cross-dehydrogenative coupling of two different carbonyls was described. In the early work, excess amount of one carbonyl and more than stoichiometric amount of strong bases and metal oxidants were required for obtaining synthetically useful chemical yield. In contrast, recent elegant strategies introduced improved the reaction efficiency of cross-dehydrogenative coupling, use of equimolar amount of carbonyls, enantioselective reaction, enolonium intermediate, catalytic reaction without external base. The application to the heterocycle synthesis and total synthesis highlighted the synthetic utilities of the present mythologies. Further efforts for more efficient and practical methods, including oxygen as oxidant or oxidant-free conditions (hydrogen evolution) [55], diastereo- and enantioselective reaction under almost neutral conditions, will be dedicated.

References

1. McDoniel PB, Cole JR (1972) Antitumor activity of *Bursera schlechtendalii* (burseraceae): isolation and structure determination of two new lignans. J Pharm Sci 61:1992–1994. https://doi.org/10.1002/jps.2600611224
2. Seto H, Sato T, Urano S, et al (1976) Utilization of ^{13}C-^{13}C coupling in structural and biosynthetic studies. VII[1)] the structure and biosynthesis of vulgamycin. Tetrahedron Lett 4367–4370. https://doi.org/10.1002/jps.2600611224
3. Omura S, Hirano A, Iwai Y, Mauma R (1979) Herquline, a new alkaloid produced by *Penicillium herquei*: fermentation, isolation and properties. J Antibiot (Tokyo) 32:786–790. https://doi.org/10.7164/antibiotics.32.786
4. Furusaki A, Matsumoto T, Ogura H et al (1980) X-ray crystal structure of herquline, a new biologically active piperazine from *Penicillium herquei* Fg-372. J Chem Soc Chem Commun 698a. https://doi.org/10.1039/c3980000698a
5. Kosuge T, Tsuji K, Hirai K (1981) Isolation and structure determination of a new marine toxin, neosurugatoxin, from the Japanese Ivory Shell, *Babylonia japonica*. Tetrahedron Lett 22:3417–3420. https://doi.org/10.1016/S0040-4039(01)81920-1
6. Whittaker M, Floyd CD, Brown P et al (1999) Design and therapeutic application of matrix metalloproteinase inhibitors. Chem Rev 99:2735–2776. https://doi.org/10.1021/cr9804543
7. Fujisawa T, Igeta K, Odake S et al (2002) Highly water-soluble matrix metalloproteinases inhibitors and their effects in a rat adjuvant-induced arthritis model. Bioorg Med Chem 10:2569–2581. https://doi.org/10.1016/S0968-0896(02)00109-8
8. Li S-H, Wang J, Niu X-M et al (2004) Maoecrystal V, cytotoxic diterpenoid with a novel c_{19} skeleton from *Isodon eriocalyx* (Dunn.) Hara. Org Lett 6:4327–4330. https://doi.org/10.1021/ol0481535
9. Peng F, Danishefsky SJ (2012) Total synthesis of (±)-maoecrystal V. J Am Chem Soc 134:18860–18867. https://doi.org/10.1021/ja309905j
10. Lu P, Mailyan A, Gu Z et al (2014) Enantioselective synthesis of (−)-Maoecrystal V by enantiodetermining C–H functionalization. J Am Chem Soc 136:17738–17749. https://doi.org/10.1021/ja510573v
11. Paal C (1884) Ueber die derivate des acetophenonacetessigesters und des acetonylacetessigesters. Ber Dtsch Chem Ges 17:2756–2767. https://doi.org/10.1002/cber.188401702228
12. Knorr L (1884) Synthese von furfuranderivaten aus dem diacetbernsteinsäureester. Ber Dtsch Chem Ges 17:2863–2870. https://doi.org/10.1002/cber.188401702254
13. Paal C (1885) Synthese von thiophen- und pyrrolderivaten. Ber Dtsch Chem Ges 18:367–371. https://doi.org/10.1002/cber.18850180175
14. Kel'in AV, Kulinkovich OG (1996) A new simple synthesis of aryl-substituted 1,4-diketones. Synthesis 330–332. https://doi.org/10.1055/s-1996-4215
15. Cole DC, Stock JR, Chopra R et al (2008) Acylguanidine inhibitors of β-secretase: optimization of the pyrrole ring substituents extending into the S1 and S3 substrate binding pockets. Bioorg Med Chem Lett 18:1063–1066. https://doi.org/10.1016/j.bmcl.2007.12.010

16. Trost BM (1991) The atom economy—a search for synthetic efficiency. Science 254:1471–1477. https://doi.org/10.1126/science.1962206
17. Wender PA, Verma VA, Paxton TJ, Pillow TH (2008) Function-oriented synthesis, step economy, and drug design. Acc Chem Res 41:40–49. https://doi.org/10.1021/ar700155p
18. Li CJ (2009) Cross-dehydrogenative coupling (CDC): exploring C–C bond formations beyond functional group transformations. Acc Chem Res 42:335–344. https://doi.org/10.1021/ar800164n
19. Liu C, Zhang H, Shi W, Lei A (2011) Bond formations between two nucleophiles: transition metal catalyzed oxidative cross-coupling reactions. Chem Rev 111:1780–1824. https://doi.org/10.1021/cr100379j
20. Girard SA, Knauber T, Li CJ (2014) The cross-dehydrogenative coupling of Csp3–H bonds: a versatile strategy for C–C bond formations. Angew Chemie Int Ed 53:74–100. https://doi.org/10.1002/anie.201304268
21. Guo F, Clift MD, Thomson RJ (2012) Oxidative coupling of enolates, enol silanes, and enamines: methods and natural product synthesis. Eur J Org Chem 4881–4896. https://doi.org/10.1002/ejoc.201200665
22. Murarka S, Antonchick AP (2018) Metal-catalyzed oxidative coupling of ketones and ketone enolates. Synthesis 50:2150–2162. https://doi.org/10.1055/s-0037-1609715
23. Ivanoff D, Spassoff A (1935) Bull Soc Chim Fr 2:76–78
24. Rathke MW, Lindert A (1971) Reaction of ester enolates with copper(II) salts. Synthesis of substituted succinate esters. J Am Chem Soc 93:4605–4606. https://doi.org/10.1021/ja00747a051
25. Ito Y, Konoike T, Saegusa T (1975) Reaction of ketone enolates with copper dichloride. Synthesis of 1,4-diketones. J Am Chem Soc 97:2912–2914. https://doi.org/10.1021/ja00843a057
26. Ito Y, Konoike T, Harada T, Saegusa T (1977) Synthesis of 1,4-diketones by oxidative coupling of ketone enolates with copper(II) chloride. J Am Chem Soc 99:1487–1493. https://doi.org/10.1021/ja00447a035
27. Tokuda M, Shigei T, Itoh M (1975) Electrochemical oxidation of lithium ester enolates. Chem Lett 4:621–664. https://doi.org/10.1246/cl.1975.621
28. Clift MD, Taylor CN, Thomson RJ (2007) Oxidative carbon-carbon bond formation via silyl bis-enol ethers: controlled cross-coupling for the synthesis of quaternary centers. Org Lett 9:4667–4669. https://doi.org/10.1021/ol702306z
29. Avetta CT, Konkol LC, Taylor CN et al (2008) Diastereoselective oxidative carbon-carbon bond formation via silyl bis-enol ethers. Org Lett 10:5621–5624. https://doi.org/10.1021/ol802516z
30. Clift MD, Thomson RJ (2009) Development of a merged conjugate addition/oxidative coupling sequence. Application to the enantioselective total synthesis of metacycloprodigiosin and prodigiosin R1. J Am Chem Soc 131:14579–14583. https://doi.org/10.1021/ja906122g
31. Robinson EE, Thomson RJ (2018) A strategy for the convergent and stereoselective assembly of polycyclic molecules. J Am Chem Soc 140:1956–1965. https://doi.org/10.1021/jacs.7b13234
32. Baran PS, DeMartino MP (2006) Intermolecular oxidative enolate heterocoupling. Angew Chem Int Ed Engl 45:7083–7086. https://doi.org/10.1002/anie.200603024
33. DeMartino MP, Chen K, Baran PS (2008) Intermolecular enolate heterocoupling: scope, mechanism, and application. J Am Chem Soc 130:11546–11560. https://doi.org/10.1021/ja804159y
34. Quesnelle CA, Gill P, Kim S-H et al (2016) A practical approach for enantio- and diastereocontrol in the synthesis of 2,3-disubstituted succinic acid esters: synthesis of the pan-notch inhibitor BMS-906024. Synlett 27:2254–2258. https://doi.org/10.1055/s-0035-1561636
35. Casey BM, Flowers RA (2011) On the nature of the oxidative heterocoupling of lithium enolates. J Am Chem Soc 133:11492–11495. https://doi.org/10.1021/ja205017e

36. Fujii T, Hirao T, Ohshiro Y (1992) Oxovanadium-induced oxidative desilylation for the selective synthesis of 1,4-diketones. Tetrahedron Lett 33:5823–5826. https://doi.org/10.1016/0040-4039(92)89041-A
37. Amaya T, Maegawa Y, Masuda T et al (2015) Selective intermolecular oxidative cross-coupling of enolates. J Am Chem Soc 137:10072–10075. https://doi.org/10.1021/jacs.5b05058
38. Amaya T, Osafune Y, Maegawa Y, Hirao T (2017) Vanadium(V)-induced oxidative cross-coupling of various boron and silyl enolates. Chem - An Asian J 12:1301–1304. https://doi.org/10.1002/asia.201700470
39. Zhu L, Wang D, Jia Z et al (2018) Catalytic asymmetric oxidative enamine transformations. ACS Catal 8:5466–5484. https://doi.org/10.1021/acscatal.8b01263
40. Chiba T, Okimoto H, Hamaguchi H, Imanishi T, Yoshida K (1979) Electrochemical oxidation of enamines in the presence of organic anions. J Org Chem 44:3519–3523. https://doi.org/10.1021/jo01334a016
41. Narasaka K, Okauchi T, Tanaka K, Murakami M (1992) Generation of cation radicals from enamines and their reactions with olefins. Chem Lett 21:2099–2102. https://doi.org/10.1246/cl.1992.2099
42. Jang HY, Hong JB, MacMillan DWC (2007) Enantioselective organocatalytic singly occupied molecular orbital activation: the enantioselective α-enolation of aldehydes. J Am Chem Soc 129:7004–7005. https://doi.org/10.1021/ja0719428
43. Tisovský P, Mečiarová M, Šebesta R (2014) Asymmetric organocatalytic SOMO reactions of enol silanes and silyl ketene (thio)acetals. Org Biomol Chem 12:9446–9452. https://doi.org/10.1039/c4ob01385a
44. Yasu Y, Koike T, Akita M (2012) Sunlight-driven synthesis of γ-diketones via oxidative coupling of enamines with silyl enol ethers catalyzed by $[Ru(bpy)_3]^{2+}$. Chem Commun 48:5355–5357. https://doi.org/10.1039/c2cc31748f
45. Kato T, Yasui K, Odagi M, Nagasawa K (2017) Guanidinium hydroiodide/cumene hydroperoxide-mediated intermolecular oxidative coupling peaction of β–ketoamides with oxindoles. Adv Synth Catal 359:2881–2889. https://doi.org/10.1002/adsc.201700266
46. Yasui K, Kato T, Kojima K, Nagasawa K (2015) Guanidinium iodide-catalyzed oxidative α-nitroalkylation of β-ketoamides. Chem Commun 51:2290–2293. https://doi.org/10.1039/c4cc08581g
47. Saito M, Kobayashi Y, Tsuzuki S, Takemoto Y (2017) Electrophilic activation of iodonium ylides by halogen-bond-donor catalysis for cross-enolate coupling. Angew Chemie Int Ed 56:7653–7657. https://doi.org/10.1002/anie.201703641
48. Li K, Tan G, Huang J et al (2013) Iron-catalyzed oxidative C–H/C–H cross-coupling: an efficient route to α-quaternary α-amino acid derivatives. Angew Chem Int Ed Engl 52:12942–12945. https://doi.org/10.1002/anie.201306181
49. Tanaka T, Tanaka T, Tsuji T et al (2018) Strategy for catalytic chemoselective cross-enolate coupling reaction via a transient homocoupling dimer. Org Lett 20:3541–3544. https://doi.org/10.1021/acs.orglett.8b01313
50. Curto JM, Kozlowski MC (2015) Chemoselective activation of sp^3 vs sp^2 C–H bonds with Pd (II). J Am Chem Soc 137:18–21. https://doi.org/10.1021/ja5093166
51. Sohtome Y, Sugawara M, Hashizume D et al (2017) Reversibility of 3-phenyl-2-oxindole dimer formation: application to construct compounds with two distinct vicinal all-carbon quaternary centers. Heterocycles 95:1030–1040. https://doi.org/10.3987/COM-16-S(S)75
52. Wang C, Li Y, Gong M et al (2016) Method for direct synthesis of α-cyanomethyl-β-dicarbonyl compounds with acetonitrile and 1,3-dicarbonyls. Org Lett 18:4151–4153. https://doi.org/10.1021/acs.orglett.6b01871
53. Tanaka T, Hashiguchi K, Tanaka T et al (2018) Chemoselective catalytic dehydrogenative cross-coupling of 2-acylimidazoles: mechanistic investigations and synthetic scope. ACS Catal 8:8430–8440. https://doi.org/10.1021/acscatal.8b02361

54. Kaiser D, Teskey CJ, Adler P, Maulide N (2017) Chemoselective intermolecular cross-enolate-type coupling of amides. J Am Chem Soc 139:16040–16043. https://doi.org/10.1021/jacs.7b08813
55. Gao XW, Meng QY, Li JX et al (2015) Visible light catalysis assisted site-specific functionalization of amino acid derivatives by C–H bond activation without oxidant: cross-coupling hydrogen evolution reaction. ACS Catal 5:2391–2396. https://doi.org/10.1021/acscatal.5b00093

Inter- and Intramolecular Cross-Dehydrogenative Coupling of Alcohols Through the Hydrogen-Borrowing Approach

Seetharaman Manojveer and Magnus T. Johnson

Abstract Cross-dehydrogenative coupling (CDC) reaction is a versatile method to couple two different alcohols under mild reaction conditions. The chapter summarizes the scientific literature concerning both intra- and intermolecular CDC reactions employing a wide range of catalysts. The various different classes of products that have been obtained are further described in depth. Attention is also given to recent development in the field that employs cooperative ligands.

Keywords Transition metal catalysis · Hydrogen-borrowing · Dehydrogenation · Hydrogenation · Cooperative ligands

1 Introduction

The cross-dehydrogenative coupling (CDC) reaction is a useful tool to make C–C bonds between two different alcohols under mild reaction conditions. Li and other groups have studied extensively and shown the scope of this transformation in achieving various C–C bond formations with different substrates. Recently, it has received much attention in the synthetic community since this bond formation is difficult to achieve or not possible with other strategies. In addition, it is not required to prefunctionalize a substrate to make C–C bonds under CDC reaction conditions but can be utilized directly from abundant and widely available alcohols.

Alkylation of inactivated ketones with alcohols is a challenging process. Traditionally, either ketone or hydroxyl group of alkyl partner is pre-activated to achieve the alkylation of ketones. However, this is an environmentally unfriendly method which generates metal salts, halides and in some cases, hazardous chemical materials as side products. Alkylation of ketones using alcohol as an alkylating agent has been developed in the presence of various transition metals. It would be

S. Manojveer · M. T. Johnson (✉)
Centre for Analysis and Synthesis, Department of Chemistry, Lund University, P.O. Box 124, 22100 Lund, Sweden
e-mail: magnus.johnson@chem.lu.se

A. Srivastava and C. K. Jana (eds.), *Heterocycles via Cross Dehydrogenative Coupling*, https://doi.org/10.1007/978-981-13-9144-6_7

Scheme 1 Schematic overview of hydrogen-borrowing methodology subreactions

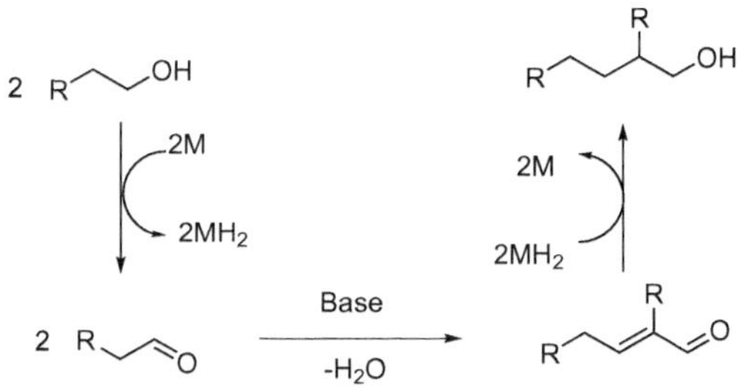

highly desirable to get the same products from alcohol starting materials. As a functional group and starting material, alcohols are easily accessible, highly versatile, widely abundant and often related to renewable sources and green chemistry. The conversion of readily available alcohol derivatives to value-added products under hydrogen-borrowing conditions has become an increasingly important concept in transition metal catalysis [1]. Classically, it is known as the Guerbet reaction which involves the conversion of primary alcohols into the corresponding β-alkylated dimeric alcohol with loss of water as the sole by-product (Scheme 1) [2]. Recently, it is known as hydrogen-borrowing or hydrogen autotransfer process. Using the concept of hydrogen-borrowing, alkylation of ketones and amines with alcohols has been developed using various transition metal catalysts under alkaline reaction conditions. It is an atom-economical and environmentally benign process which eliminates most of the drawbacks associated with traditional alkylation methods.

Initial dehydrogenation is a key step to facilitate the cross-coupling between two alcohols in the presence of metal catalyst and base. Later the formed unsaturated intermediate would undergo hydrogenation to give the final product in the same reaction pot (Scheme 1). This chapter describes the inter- and intramolecular cross-dehydrogenative coupling of alcohols through hydrogen-borrowing approach in the following topics.

1.1 Coupling between primary and secondary alcohols

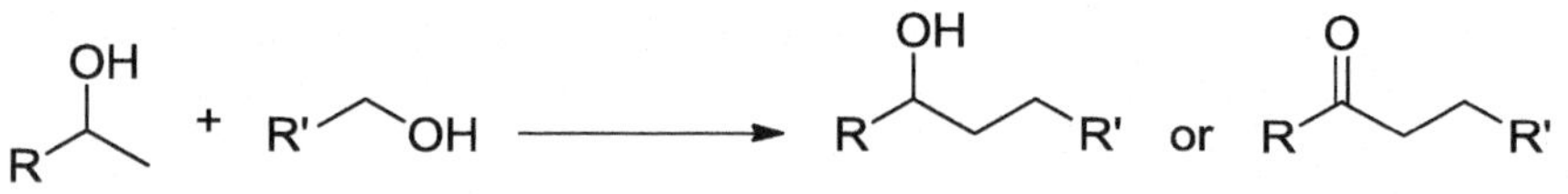

1.2 Homocoupling of secondary alcohol

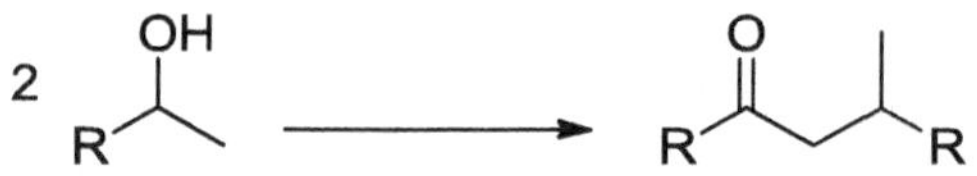

1.3 Homocoupling of primary alcohol

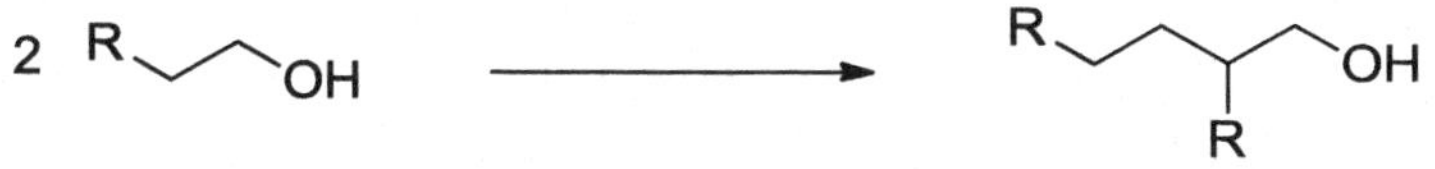

1.4 Coupling between two different primary alcohols

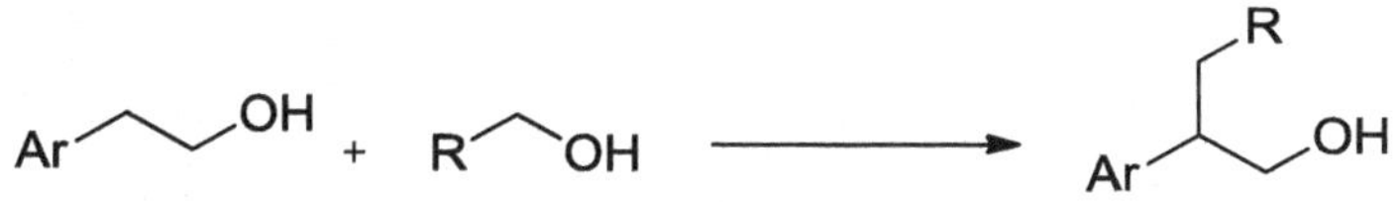

1.5 Complete deoxygenative coupling of primary alcohols

1.6 Coupling between primary alcohols to form esters

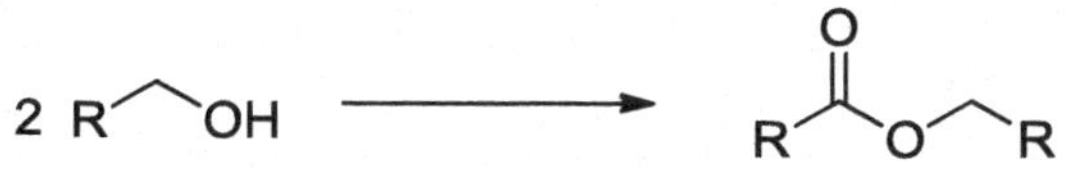

1.1 Coupling Between Primary and Secondary Alcohols

In 2003, Cho and Shim reported an unprecedented one-pot procedure for β-alkylation of secondary alcohols with primary alcohols in the presence of a ruthenium catalyst ($RuCl_2(PPh_3)_3$) along with a base (KOH) and a sacrificial hydrogen acceptor (Scheme 2) [3]. During the optimization study, it was found that the reaction rate was enhanced in the presence of 1-dodecene as a hydrogen acceptor. Here, 1-dodecene is reduced to 1-dodecane by $[Ru]H_2$ and regenerates the active [Ru] for the next catalytic cycle. They mentioned that < 10% of corresponding ketone products were obtained under the reaction conditions. It was also noted that

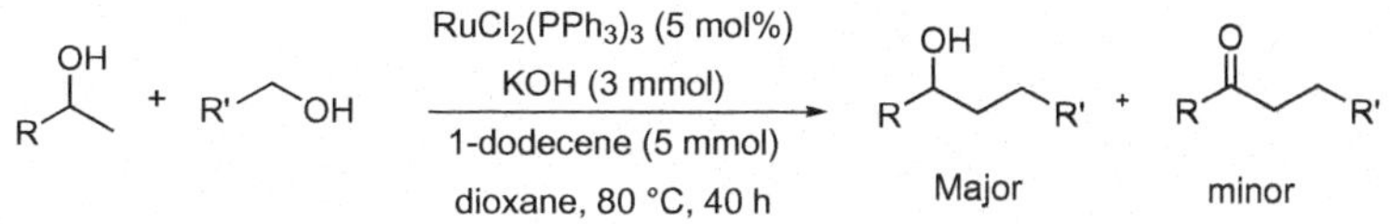

Scheme 2 The first reported one-pot procedure for β-alkylation of secondary alcohols with primary alcohols

Scheme 3 Dehydrogenation using hydrogen acceptors

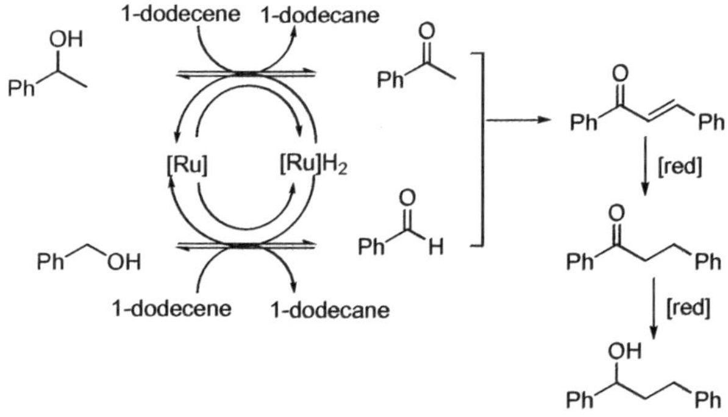

no β,β-dialkylated product was observed, even from the reaction between 1-phenyl-1-propanol and benzyl alcohol under the optimized conditions.

Initially, [Ru]-catalyst dehydrogenates both the secondary and primary alcohols to give ketone and aldehyde, respectively. Then aldol condensation would take place between formed ketone and aldehyde in the presence of base. Subsequent reduction of chalcone intermediate delivers the final ketone or alcohol product (Scheme 3). The same research group reported a similar kind of coupling reaction using catalytic amount of Pd/C and KOH in the presence of 1-decene as a hydrogen acceptor [4].

Two years later, similar kind of alcohol coupling reactions were reported by Fujita and Yamaguchi group using $[Cp^*IrCl_2]_2$ and base through hydrogen auto-transfer process (Scheme 4) [5]. In contrast to previous report by Cho and Shim group, here they did not utilize the sacrificial hydrogen acceptor, and hence, this protocol has more atom economy. Electronic and steric nature of the Cp* ligand plays a crucial role in enhancing the activity of catalyst towards both oxidative and reductive hydrogen transfer reactions.

Since the hydrogen-borrowing methodology as shown in Scheme 5 works flawlessly to reduce the unsaturated aldol product using the hydrogen from the alcohol dehydrogenation, no additional acceptor/donor is required. Using Grubb's catalyst, only one example of β-alkylation of alcohol was shown in the absence of hydrogen acceptor or donor [6].

In the same manner, i.e. without using the hydrogen acceptor, β-alkylation of secondary alcohols with primary alcohols through a hydrogen autotransfer process in the presence of $RuCl_2(DMSO)_4$ catalyst and base was described by Yus and co-workers (Scheme 6) [7]. It was postulated that due to the higher stability of

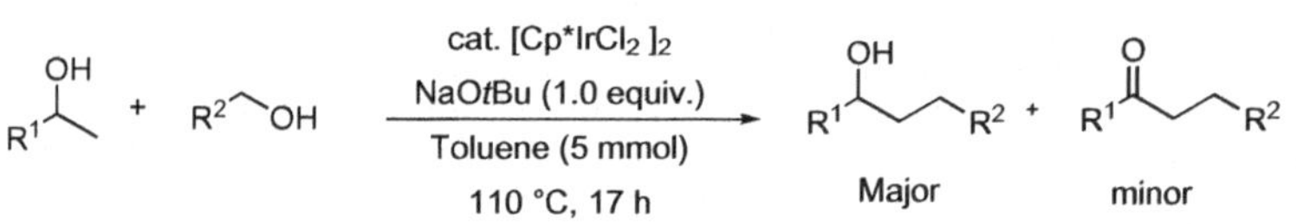

Scheme 4 Fujita and Yamaguchi's Ir-catalyzed hydrogen-borrowing reaction

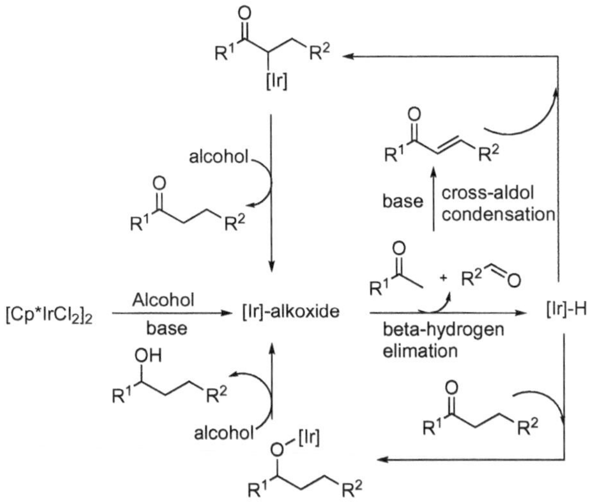

Scheme 5 Schematic mechanism of the Ir-catalyzed hydrogen-borrowing reaction with the hydrogen acceptor

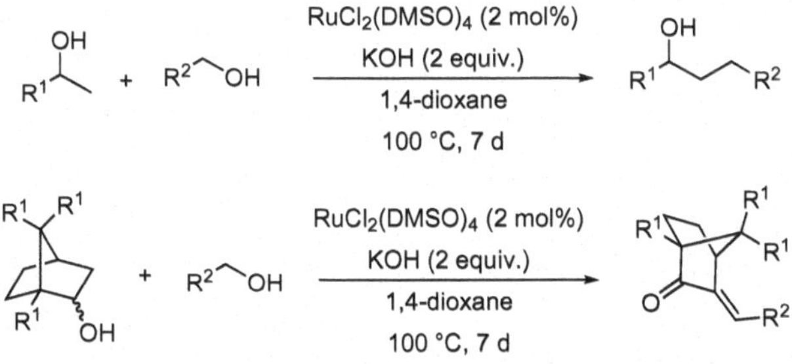

Scheme 6 Yus and coworkers' Ru-catalyzed alkylation reactions

trisubstituted C–C double bond and higher instability of ruthenium bicyclic enolate, α,β-unsaturated ketones will be obtained from methylenic bicyclic alcohols. Otherwise, simple β-alkylated secondary alcohols are formed in good-to-excellent yields from linear alcohol derivatives under this catalytic system. The reaction time could be reduced from 7 to 3 days when it is performed in solvent-free conditions and resulted in slightly lower yield of corresponding coupled product.

In organometallics, ligands play a major role in tuning the activity of catalyst. The chemists started to focus on designing the ligands to increase the activity of metals. A wide variety of metal complexes was prepared with different type of ligands and studied their catalytic behaviour in cross-coupling of secondary and primary alcohols through hydrogen-borrowing approach. Peris and co-workers prepared two [Ru]-complexes containing triazolediylidene (NHC) unit (ditz) and studied their catalytic activity towards the β-alkylation of secondary alcohols with

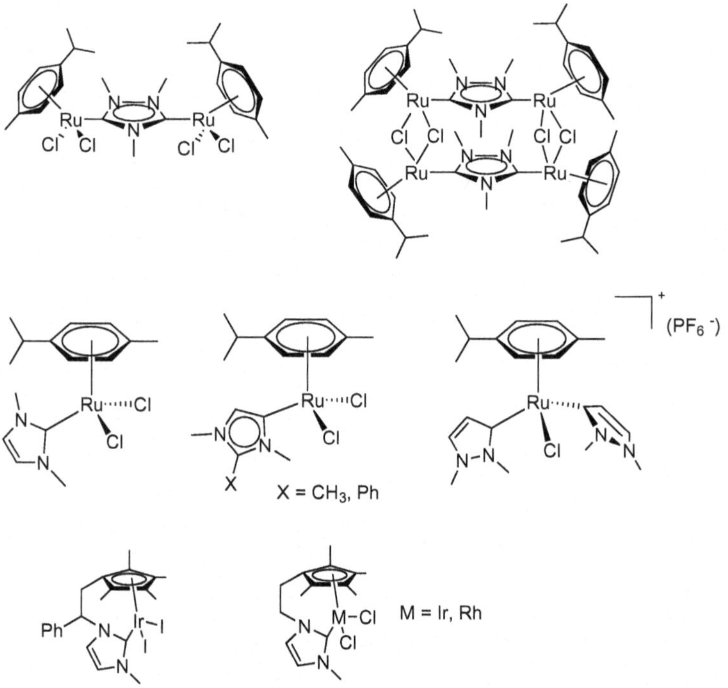

Fig. 1 A variety of functional NHC-based ligands

primary alcohols (Fig. 1) [8]. The reactions were performed using 1:1 molar ratio of two alcohols and 100 mol% KOH in the presence of 1 mol% catalyst in toluene 110 °C. Both the complexes showed similar efficiencies and very selective to the production of the alkylated alcohols. Compared previously reported methods, aliphatic alcohol such as 2-heptanol and isobutyl alcohol afforded good conversation under this catalytic system. Low conversion (11%) to the desired alcohol was obtained when the reaction was carried out using $[RuCl_2(p\text{-cymene})]_2$. This experiment clearly indicates that the introduction of NHC ligand increases the activity of Ru centre towards the hydrogen transfer process. Inspired by this study, later the same group prepared a series of [Ru]-NHC complexes and explored their activities in cross-coupling of secondary and primary alcohols [9]. In addition, the same group showed the preparation of the Ir and Rh-NHC complexes and demonstrated their catalytic activity towards the cross-coupling of secondary and primary alcohols [10, 11]. Later, pyrene-tagged N-heterocyclic carbene complexes of iridium and mono- and dimetallic pyrene–imidazolylidene complexes of iridium (III) were prepared and shown to be effective for the C–C alcohol coupling reaction through hydrogen-borrowing process. It was demonstrated that the presence of π-stacking can significantly affect the outcome of hydrogen-borrowing processes [12]. Furthermore, it has been described that catalyst could effectively be recycled in the alcohol coupling reaction.

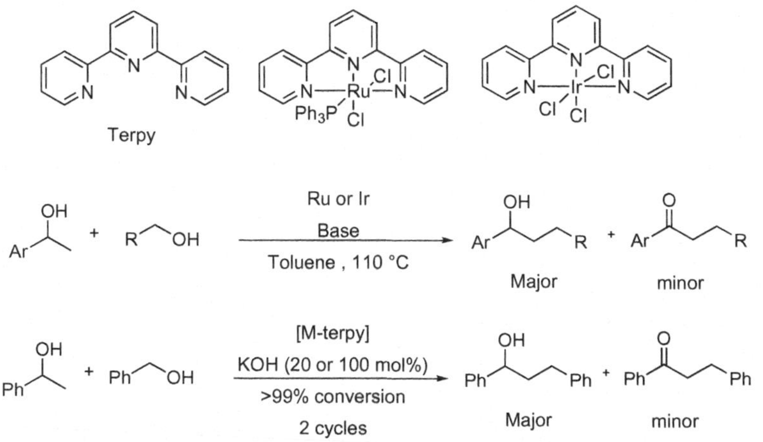

Scheme 7 Activity of terpy-based catalysts in hydrogen-borrowing processes

Since terpy has been considered as a strong π-acceptor ligand, in 2008, Crabtree and co-workers have prepared [(terpy)Ru(PPh_3)Cl_2] (**1**) and [(terpy)IrCl_3] complexes [13]. This metal complex activity was demonstrated in the coupling of primary and secondary benzylic alcohols and found that both the catalysts showed similar and higher activities than previously developed catalytic system. It should be noted that the reaction with [Ru]-complex can be run under air in the presence of 100 mol% KOH, whereas 20 mol% KOH and N_2 atmosphere are required for [Ir] complex (Scheme 7). In addition, the reaction could be conducted under neat conditions with [Ir] catalyst which resulted in better yield than in toluene used as a solvent.

While conducting the experiments in an open vessel, the yields of ketone products could be increased up to 70%. Otherwise, around 10% of ketone products is obtained in a sealed tube. It has also been shown that both [Ru] and [Ir] metal complexes could be employed on two successive catalytic cycles with no loss in activity (Scheme 7).

Cheung et al. synthesized a series of [Ru] complexes with cyclopentadienyl (Cp), hydrotris(pyrazolyl)borato (Tp) and bipyridine ligands and then investigated their efficiency in β-alkylation of secondary alcohols with primary alcohols (Fig. 2) [14]. Among the entire complexes, dicationic bipyridine complex was found to be an active catalyst. Since it does not contain any phosphine ligand, the reaction could not be monitored with ^{31}P NMR spectroscopy. In order to understand the mechanistic pathway of the catalytic reactions, the authors decided to monitor the 1-phenylethanol/benzyl alcohol alkylation reactions catalysed by the Cp and Tp complexes with ^{31}P NMR spectroscopy. During this course of the investigations, they could identify the crucial hydrido complexes ([CpRu(dppm)H] and [TpRu (dppm)H]) and also observed the formation ruthenium carbonyl complexes ([CpRu (PPh_3)(CO)Ph] and [TpRu(PPh_3)(CO)Ph]) via aldehyde decarbonylation process.

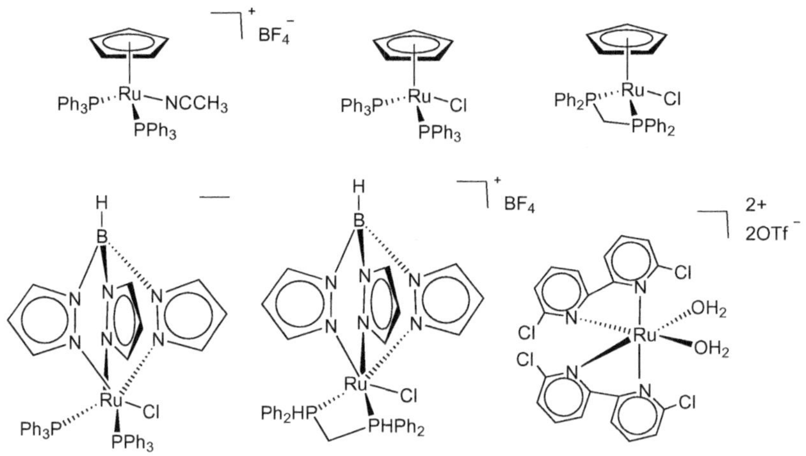

Fig. 2 Cheung's catalysts

Interestingly, ruthenium carbonyl complex, in general it is considered as the cause of catalyst deactivation, was found to be active for the catalytic processes.

Using air-stable iridium and ruthenium complexes chelating with N-heterocyclic carbenes (NHCs), an efficient catalytic system has been developed for alkylation of alcohols by Crabtree and co-workers (Figure 3) [15]. It was observed that iridium complexes are more active than ruthenium complex as it furnished the desired products in good yields with catalyst loading as low as 0.5 mol% in shorter reaction time (3–6 h). The product distribution is highly dependent on the reaction time. The alcohol-coupled products were exclusively obtained with shorter reaction time (3 h), whereas longer reaction time (6 h) favours to the ketone products.

It is known that alkaline-based metal promotes both the oxidation of alcohols to the corresponding carbonyl compounds through an Oppenauer's pathway and reduction of carbonyl derivatives to the respective alcohols via a Meerwein–Ponndorf–Verley (MPV)-type mechanism. Hence, Crabtree group focused on developing a green method for the coupling of alcohols through a hydrogen transfer pathway without using any transition metals. In 2010, they found that stoichiometric amount of base (KOH/NaOH) alone could promote the β-alkylation of secondary alcohols with primary alcohols in good yields at refluxing in toluene

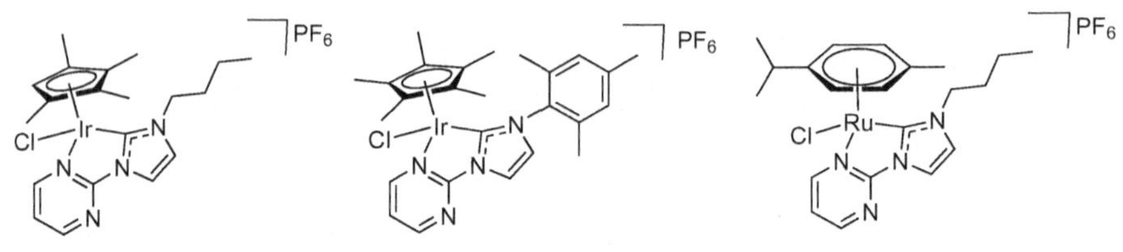

Fig. 3 Crabtree's NHC-based catalysts

under aerobic conditions [16]. It was also observed that reaction showed low conversion and yield under N_2 atmosphere. To check the role of alkali metal, they carried out the reaction in the presence of 18-crown-6 as a potassium chelator and found no reaction. This experiment indicates the involvement of alkali metal in the reaction mechanism.

In 2009, Shimizu and co-workers developed a method for C–C cross-coupling of secondary and primary alcohols using an inexpensive catalytic system composed of Ag/Al_2O_3 (Scheme 8) [17]. This recyclable heterogeneous catalyst requires only the catalytic amount of Cs_2CO_3 base for the cross-coupling of secondary and primary alcohols with high TON (800). This catalytic system is highly selective towards the formation of ketone products rather than the alcohol derivatives. This result contrasts to other previously reported methods where alcohol derivatives are formed as major products. They performed a series of reactions between 1-phenyl ethanol and different benzyl alcohols and found the order of benzyl alcohols was *p*-CH_3O > *p*-CH_3 > *p*-H > *p*-Cl. This result suggests that generation of a positive charge at the α-carbon atom adjacent to the phenyl ring is the rate determining step. Hence, silver cluster assisted the C–H bond cleavage of the alkoxide may be the rate determining step for this *β*-alkylation reaction. It was also found that the reaction is facilitated by cooperation between coordinatively unsaturated silver sites and acid–base sites on the Al_2O_3 support.

Later, a few heterogeneous catalysts based on gold, iridium and copper metals were reported independently by different groups for the cross-dehydrogenative coupling of alcohols [18]. Gold- and copper-based catalysts are more selective to produce ketone derivatives, whereas alcohol products could be obtained in the presence of iridium-based heterogeneous catalyst.

Pd catalysts derived from pincer-type NHCs have been utilized for cross-coupling between secondary and primary alcohols in the presence of catalytic amount of base CsOH (40 mol%) (Fig. 4) [19]. Selectivity towards the alcohol derivatives could be increased either by conducting the experiments in H_2

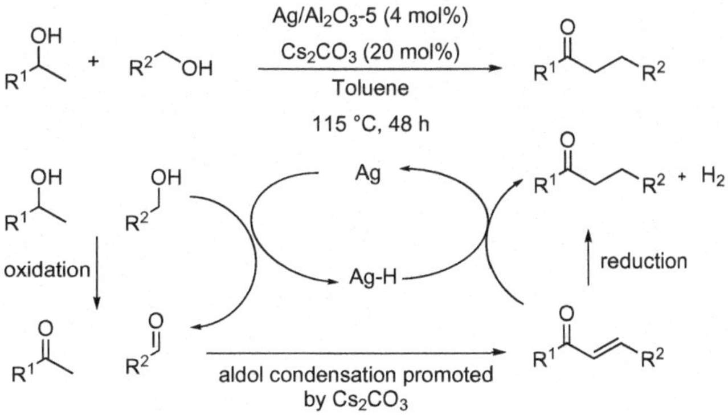

Scheme 8 The inexpensive Ag/alumina catalyst system

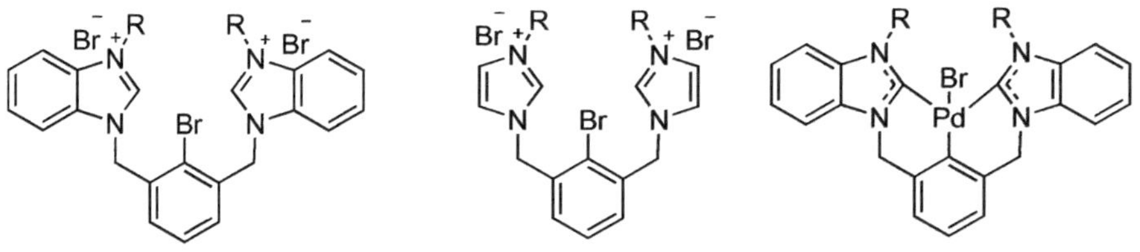

Fig. 4 Pincer-based ligands and catalysts

atmosphere or taking the excess amount (2.0 equiv.) of secondary alcohols relative to primary alcohols.

In continuation of developing a novel Ir-catalytic system, Peris's group reported the preparation of the following Ir complexes and showed their efficiency in *β*-alkylation of secondary alcohols with primary alcohols (Fig. 5) [20]. Catalyst 1 is more efficient than 2 in terms of conversion and selectivity to the final alcohol products. 0.1 mol% of catalyst 1 loading is sufficient to get the maximum conversion of the presence of KOH base. It was noted that in situ generation of catalyst showed lower activity than the actual Ir complex.

β-Alkylation of secondary alcohols with primary alcohols using Ir–thioether–dithiolate complex, [Cp*Ir(η^3-tpdt)] (Cp* = η^5-C_5Me_5, tpdt = $S(CH_2CH_2S-)_2$) was reported by the Pullarkat group (Fig. 6) [21]. This catalytic system gave good yields and selective to the alcohol products in the presence of equimolar amount of two alcohols, 0.1 mol% of catalyst and 50 mol% NaO*t*Bu.

In 2012, the same research group prepared a series of ruthenacycles and demonstrated their potential in the one-pot b-alkylation of secondary alcohols with

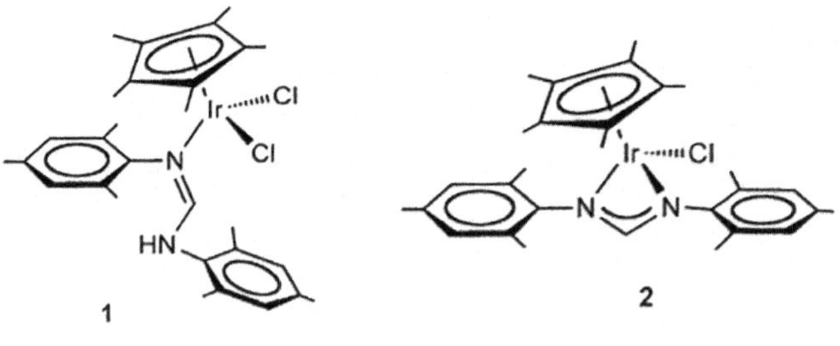

Fig. 5 Peris' catalysts

Fig. 6 Pullarkat's catalysts

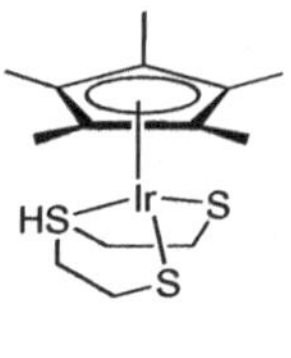

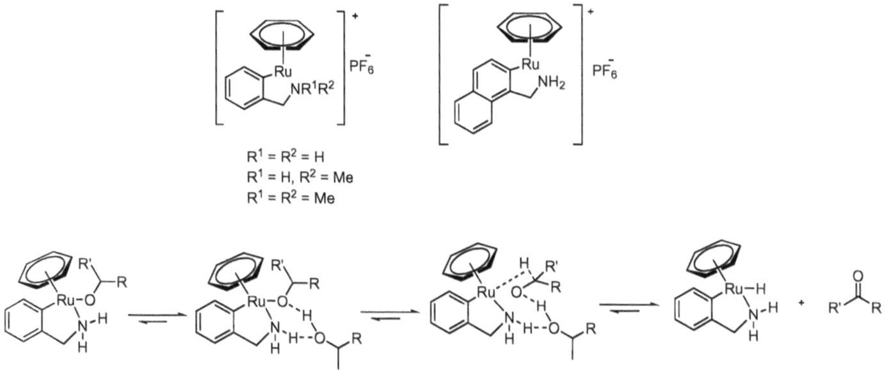

Scheme 9 Cooperative ligand in hydrogen-borrowing catalysis

primary alcohols [22]. Authors paid attention to design a catalytic system in such way that it promotes the hydrogen transfer reactions. Accordingly, a variety of ruthenacycle complexes were easily synthesized by cyclometalation reaction of different benzyl amines with $[(\eta^6\text{-}C_6H_6)RuCl_2]_2$ in the presence of 1 equiv. of NaOH and 1 equiv. of KPF_6 (Scheme 9). Among these, the complexes with acidic NH proton afforded the better results in the cross-coupling of secondary and primary alcohols than the complexes without N–H group. Hence, N–H group plays a crucial role in promoting the hydrogenation transfer reaction through H-bonding interaction as reported by Noyori and co-workers Selectivity of the products under the present catalytic system is highly dependent on the reaction time, as an extended period of time (48 h) favours the formation of ketone derivatives. Interestingly, under these reaction conditions, the substrate scope could be extended to the coupling between unsaturated primary alcohol and aromatic secondary alcohol to get the respective coupled alcohol and ketone products in moderate yield. It was also explained that due to the CH/π attraction and charge–transfer interactions exist between the η^6-benzene ligand of the catalyst and the phenyl moiety of the substrates, aromatic alcohols furnish better yields than aliphatic alcohols (Scheme 9).

It is known in the literature that abnormal NHCs are strong σ-donor and π-acceptor than the normal NHCs. Thus, the complexes chelating with abnormal NHCs have higher activity than the corresponding normal NHCs analogues. Keeping this in mind, Li and co-workers prepared various Ir complexes chelating with abnormal NHCs and investigated their activity in hydrogen transfer reactions (Fig. 7) [23]. For comparison, they also prepared Ir complex with normal NHC. As expected, Ir complexes bearing abnormal NHCs showed higher activity and selectivity than a normal NHC analogue on the β-alkylation of secondary alcohols with primary alcohols and other hydrogen transfer reactions. Like the other previous reports, prolonged reaction favours the formation of ketone products under these reaction conditions.

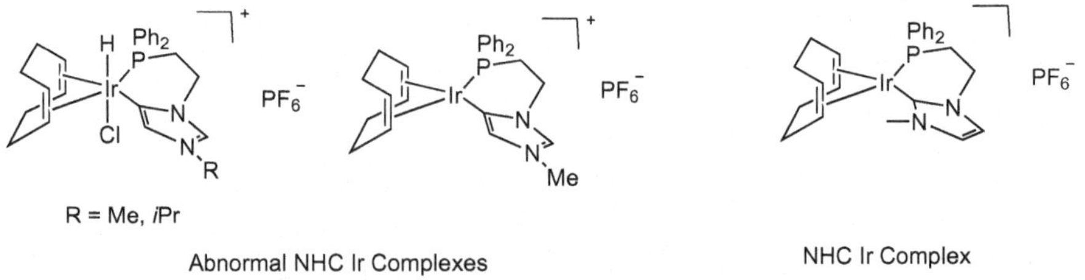

Fig. 7 Normal and abnormal NHC-based catalysts

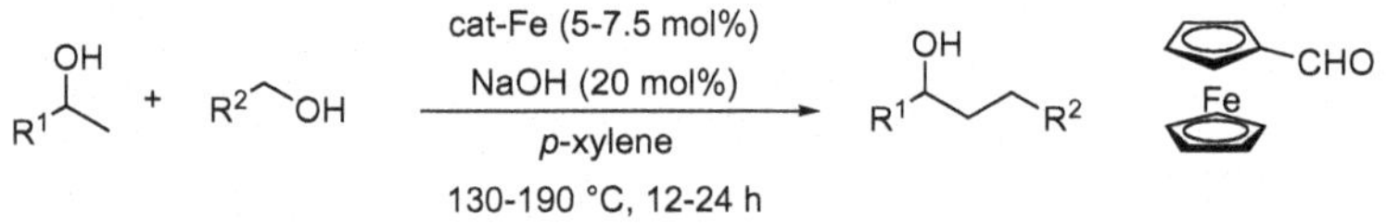

Scheme 10 A rare example of an Fe-catalyzed hydrogen-borrowing reaction

An efficient method for the direct β-alkylation of secondary alcohols with primary alcohols using earth-abundant and commercially available iron catalyst (ferrocenecarboxaldehyde) and sub-stoichiometric amount of base was demonstrated by Sun and co-workers (Scheme 10) [24]. Like other reported noble metals, this iron-catalytic system also showed good conversion and selectivity in the coupling of alcohols but at 130 °C in *p*-xylene. It is necessary to increase the reaction temperature 180 °C for the coupling aliphatic secondary alcohols with aromatic alcohols to get the corresponding alkylated products in good yields. However, aliphatic primary alcohols resulted in moderate yields of products even the temperature was raised to 190 °C. It should be pointed out that the β-alkylation reaction with primary aliphatic alcohols could be performed in the absence of solvent while still obtaining the desired products in satisfying yields.

So far we have seen the coupling between simple secondary and primary alcohols; in 2011, Tang and co-workers developed a new catalytic system for acceptorless β-alkylation of lactic acid with primary alcohols in the presence of nickel(II) acetate tetrahydrate and base to afford the corresponding coupled a-hydroxy carboxylic acids in good-to-excellent yields via a hydrogen transfer process (Scheme 11) [25]. Using this air-stable catalytic system, β-alkylation reaction can be carried out under solvent-free conditions to get the coupled product on 90% yield. Here the reaction temperature has to be maintained more than 155 °C, so that formed potassium lactate during the reaction will be melted and mixed

Scheme 11 A Ni-catalyzed reaction

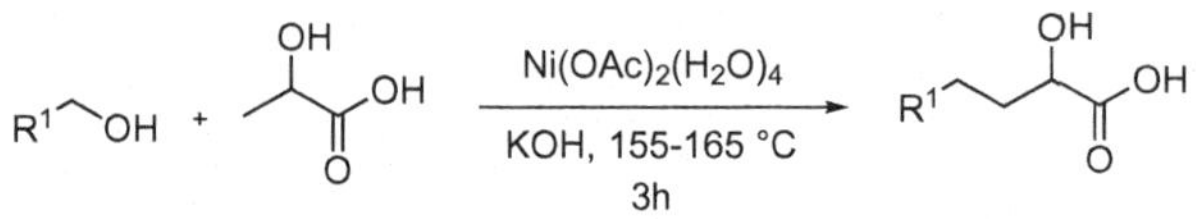

thoroughly with reaction components at this temperature. They observed that when temperature was raised to 180 °C, *β*-alkylation reaction was completed in 30 min in the presence of $Ni(OAc)_2(H_2O)_4$ under aerobic conditions.

In 2011, Saito group demonstrated the C–C bond formation through the coupling of two different alcohols using a simple catalytic system based on $CuBr/H_2/NaOH$ (Scheme 12). Several substrate scopes were studied using CuBr (0.2 mol%), NaOH (4 mol%), equimolar amount of two alcohols refluxing at 135 °C in *p*-xylene under H_2 (1 atm) [26]. The reactions under H_2 significantly enhance the selectivity towards alcohol derivatives. In some cases, it was noted that addition of dppp without purging H_2 was enough to get good conversion. They conducted several studies to elucidate the reaction mechanism and found that the present catalysis follows Meerwein–Pondorf–Verley–Oppenauer (MPV-O) redox processes as reported by Crabtree group rather than following usual transition metal-catalysed hydrogen-borrowing pathway to give the final products. From the mechanistic studies, they observed the following things: (1) iodide-substituted aryl alcohol reacted smoothly to give the final coupled product, and hence, they excluded the role of Cu(0) as active species under the present conditions. (2) To oxidize

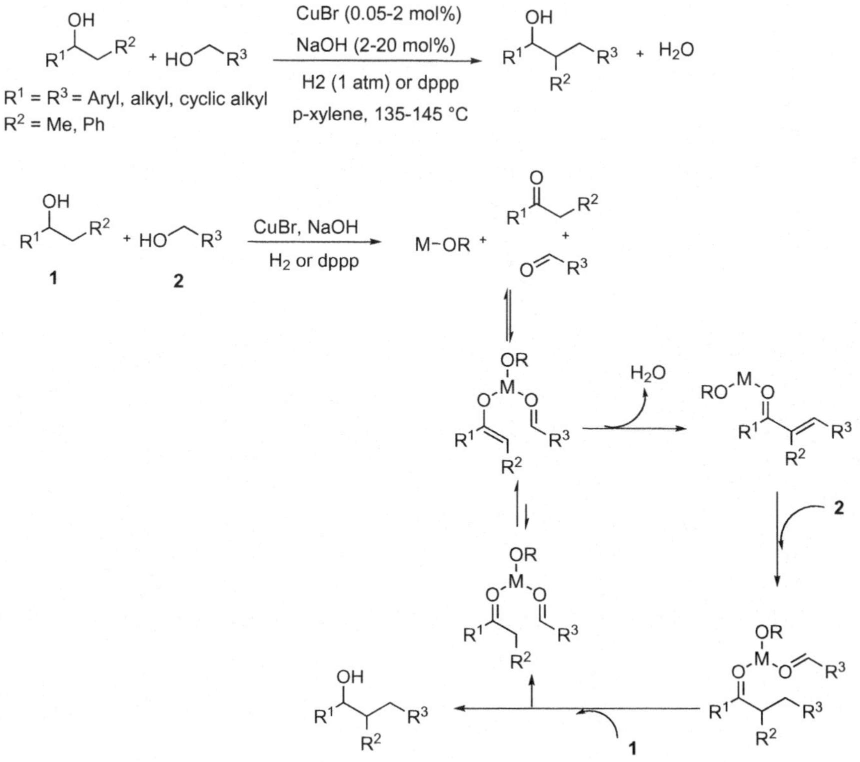

Scheme 12 A mechanistic proposal for the Cu-catalyzed reaction

low-valent Cu species, either O_2 or air was needed. It was reported that NaOH (or KOH)/O_2 promotes the benzylic alcohols in the absence of any transition metals. Thus, the present reaction was checked with other bases such as NaH and Na in place of NaOH under identical reaction conditions with rigorous exclusion of air and found a similar reactivity. (3) The alcohols were allowed to react separately under optimized reaction conditions and resulted in < 5% formation of corresponding carbonyl products. Based on these results, they proposed Meerwein–Pondorf–Verley–Oppenauer (MPV-O) redox pathway for this catalysis. They found that role of CuBr/NaOH/(H_2 or dppp) is to promote initial oxidation of alcohols to the respective carbonyl compounds which later undergo faster cross-aldol reaction to give intermediate enone. Then, reduction of enone would take place through another hydrogen transfer process from alcohols. The formed ketone and aldehyde will follow the same catalytic cycle without transition metals to give the final coupled product (Scheme 12, mechanism). It was also noticed that α,β-unsaturated ketone underwent faster 1,4 reduction than 1,2 reduction under this catalytic system.

An aerobic protocol for the cross-coupling of secondary and primary alcohols using ligand-free copper catalyst was described by Xu group [27]. Under these reaction conditions, various secondary and primary aromatic and aliphatic alcohols underwent selective cross-dehydrogenative coupling to give the corresponding alcohol products in moderate to excellent yields. However, the present catalytic system is not suitable for cyclic secondary alcohols as it resulted on low yield and low selectivity of desired product. Based on the mechanical studies and other reported literature, they proposed a different kind of mechanism based on Meerwein–Pondorf–Verley–Oppenauer (MPV-O) redox pathway rather than hydrogen-borrowing path for the present catalytic system. As observed by Saito and co-workers, here Cu and air play a major role in oxidation of alcohols to the corresponding carbonyl compounds which subsequently undergo base-catalysed cross-aldol condensation to give enone which will be reduced by alcohols via MPV-O process to get the final product.

Another interesting transition metal and solvent-free green protocol for the coupling of secondary and primary alcohols were documented by Xu et al. [28]. They found that an aldehyde could activate the alcohols to form the corresponding carbonyl compounds through Meerwein–Pondorf–Verley–Oppenauer (MPV-O) redox processes which subsequently undergo base-mediated aldol condensation to form α,β-unsaturated carbonyl intermediate followed by another MPV-O redox reaction to give the final coupled products (Scheme 13). Several control experiments were carried out using the different analytical grade of KOH to exclude the involvement of metal trace impurities. Role of alkali metal ion (K) was found to be crucial as the reaction resulted in moderate yields in the presence of K-complexation reagent 18-crown-6. Good yields and higher selectivity were obtained when the reaction was performed using 20 mol% of aldehyde and 30 mol % of KOH refluxing at 110 °C for 32 h under N_2 atmosphere. However, for some other cases such as secondary aliphatic alcohols and heterobenzylic alcohols, higher reagent loading, temperature and longer reaction time are required to get

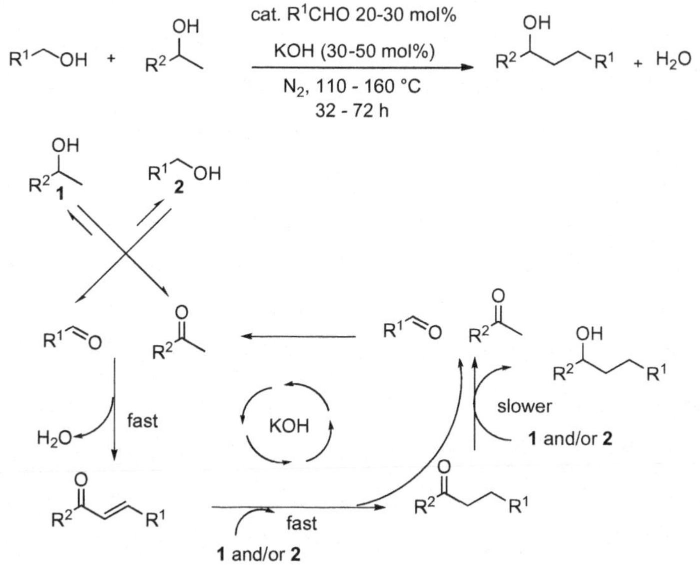

Scheme 13 The MPV-O based process

satisfactory results. It was found that the hydrogen atoms of alcohols were quantitatively consumed in the reaction process to give the final product. Since, α,β-unsaturated ketone was not observed, the reduction of α,β-unsaturated ketone should be faster than the reduction of final ketone under these reaction conditions.

Using the combination of [{Ru(cod)Cl_2}n]/PTA/*t*BuOK, β-alkylation reaction was briefly investigated at low temperature 55 °C [29]. Here PTA (1,3,5-triaza-7-phosphaadamantane) acts as a ligand. Satayanarayan et al. reported a method for selective cross-dehydrogenative coupling of secondary and primary alcohols [30]. They utilized commercially available rhodium catalyst and DABCO ligand to get ketone as a major product in the presence of KOH base.

Iridium(I)–cyclooctadiene complexes having a NHC ligand with a O- or N-functionalised wingtip have been synthesized and studied their catalytic activity on oxidation of alcohols and β-alkylation of secondary and primary alcohols (Fig. 8) [31]. It was found that Ir complexes with 2-methoxy-benzyl wingtip efficiently catalysed the β-alkylation reactions than other prepared complexes. Substrate scope was explored using both neutral and cationic Ir complexes and

Fig. 8 Ir-NHC complexes

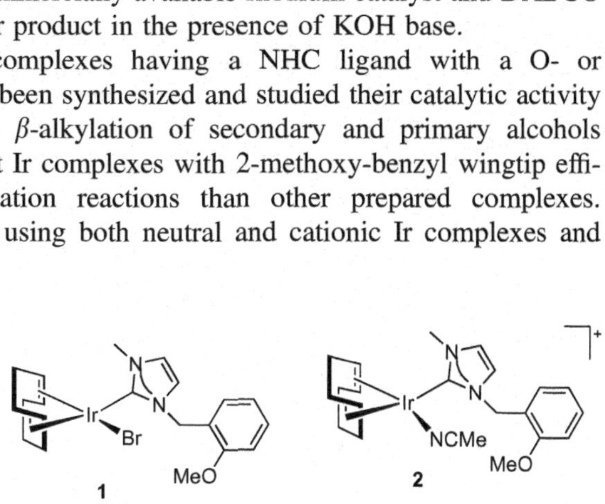

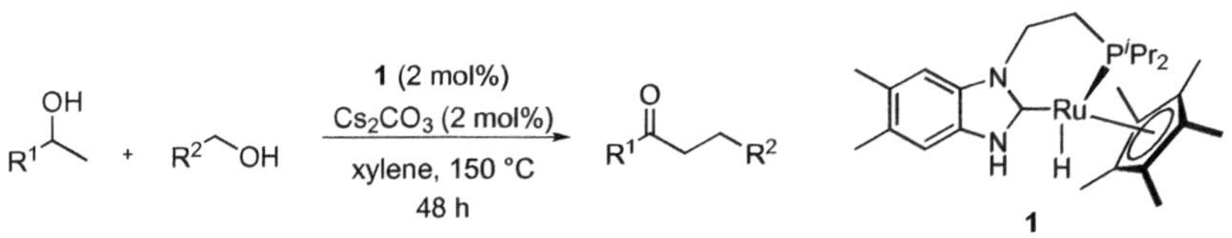

Scheme 14 Song's catalytic reaction that produces a ketone from two alcohols

observed that cationic catalyst precursor is more active than neutral precursor. Under this catalysis, a range of secondary and primary alcohols underwent cross-dehydrogenative coupling reaction to produce the corresponding *β*-alkylated alcohols in moderate selectivity. They achieved the complete selectivity to alcohols by using excess amount of base Cs_2CO_3 (150 mol%).

Song and co-workers reported the preparation a variety of protic NHC Ru complexes ligated with a phosphine-tethered imidazole moiety which could effectively catalyse the acceptorless dehydrogenative coupling reaction of secondary and primary alcohols to get ketone as major products in the presence only 2 mol% of Cs_2CO_3 (Scheme 14) [32].

The metal–ligand cooperation (MLC) strategy, where both metal and ligand are directly involved in the bond activation process, has become increasingly important in transition metal catalysis. Based on this approach, a different catalytic system has been developed for the cross-coupling of two different alcohols. Gelman and co-workers described an efficient and selective method for the cross-dehydrogenative coupling of alcohols using a new metal–ligand-cooperating system [33]. They prepared [Ir] and [Ru] complexes and examined their activity in both home and cross-coupling of alcohols (Scheme 15). It was observed that [Ir] complex is more efficient than Ru-based catalyst, even using excess amount. The complex is believed to activate the alcohols through metal–ligand cooperation, i.e. interaction between OH group of ligand and metal hydride to eliminate H_2, as shown in the mechanism (Scheme 15, mechanism) (Scheme 16).

2-Hydroxypyridine (2-HP) is an interesting ligand which could undergo tautomerization between lactim and lactam form, which in turn depends on solvent polarity and the reaction medium. It has been shown that 2-HP could promote both the hydrogenation and dehydrogenation process through metal–ligand cooperation. The Li group developed a tandem process for the acceptorless *β*-alkylation of secondary alcohols using metal–ligand bifunctional iridium complex [Cp*Ir(2,2′-bpyO)(H_2O)] (Scheme 17) [34]. Compared to Cp*Ir complexes bearing a hydroxylpyridine, the Cp*Ir complex bearing a bipyridonate ligand showed excellent reactivity and selectivity to get more exclusively the ketone products in the presence of catalytic amount of base Cs_2CO_3 (10 mol%) in two steps. First step involves the conversion of secondary alcohol to ketone in the presence of Ir-catalyst (Schemes 16 and 17). After forming the ketone product, primary alcohol and base will be added to the reaction mixture in the second step which is allowed to reflux in for another 6 h to get more selectively ketone derivatives in good-to-excellent

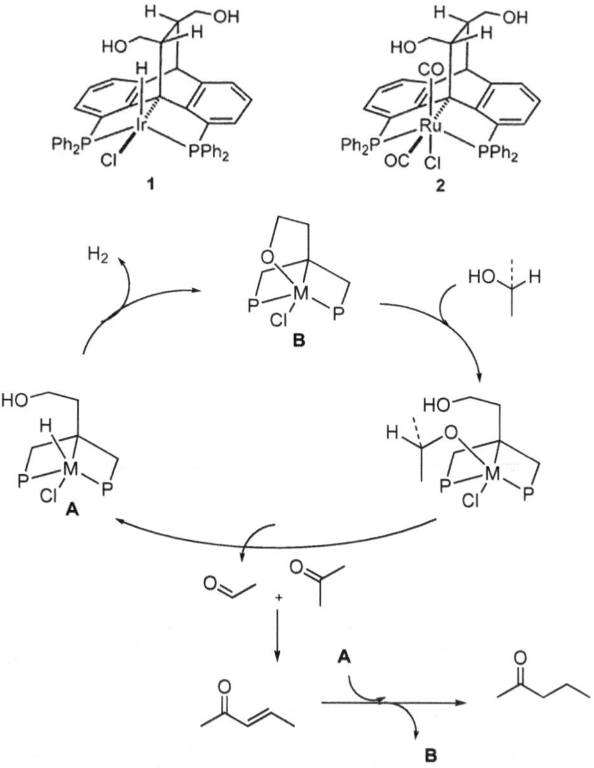

Scheme 15 An MLC-based process with pincer complexes as catalysts

yields (Scheme 18). However, the present catalytic system showed low selectivity for the direct coupling reaction of secondary and primary alcohols. The catalytic cycle proceeds through ligand-promoted hydrogen transfer process as shown in the mechanism.

Kundu and co-workers showed that a bifunctional Ru(II) complex containing 2-HP skeleton could efficiently catalyse the cross-dehydrogenative coupling of secondary and primary alcohols through MLC approach (Scheme 19) [35]. In addition to β-alkylation of secondary alcohols with primary alcohols, they explored the efficiency of catalyst on double alkylation of cyclopentanol using significantly low catalyst loading (0.1 mol%) and got the desired products in good yields (Scheme 20). Other Ru complexes were synthesized and checked their activity in similar alkylation reaction. It was found that complexes without 2-HP displayed less activity than a complex 2-HP moiety. This result clearly suggests that involvement OH group of ligand in the bond activation process through metal–ligand cooperation process as shown in the mechanism. While performing the experiments in the presence of excess PPh_3, it was noted that the reaction of complex without 2HP gave low yield of final product, whereas no significant observed with a complex bearing 2-HP system. Hence, catalytic cycle of complex with 2-HP system follows the outer sphere mechanism (Scheme 21), whereas other

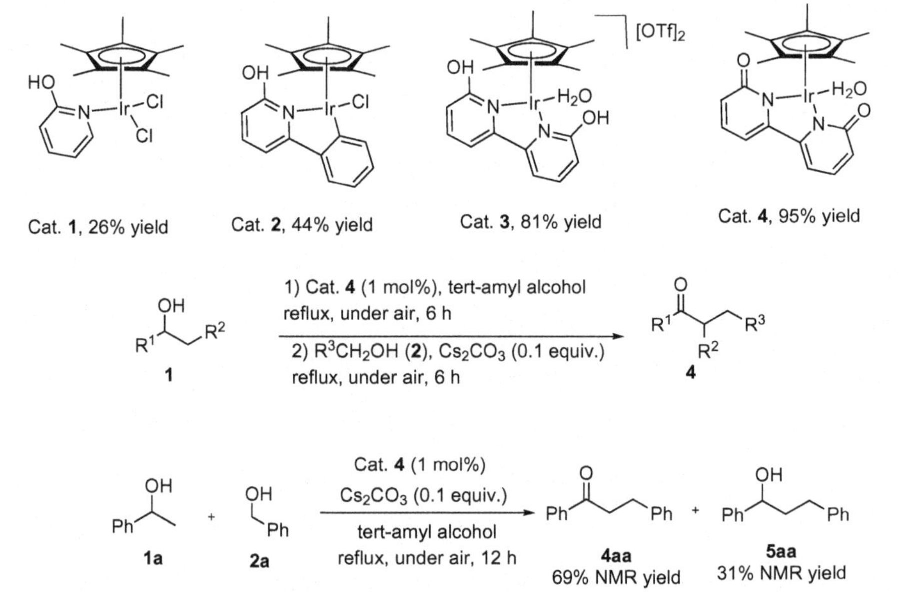

Scheme 16 2-hydroxypyridine-based catalysts and their reactions

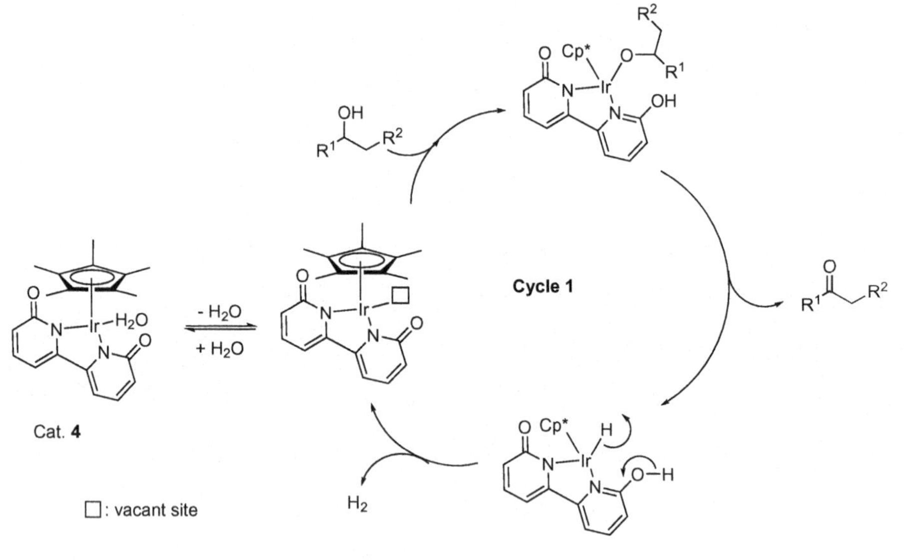

Scheme 17 Mechanism for the cooperative catalysis

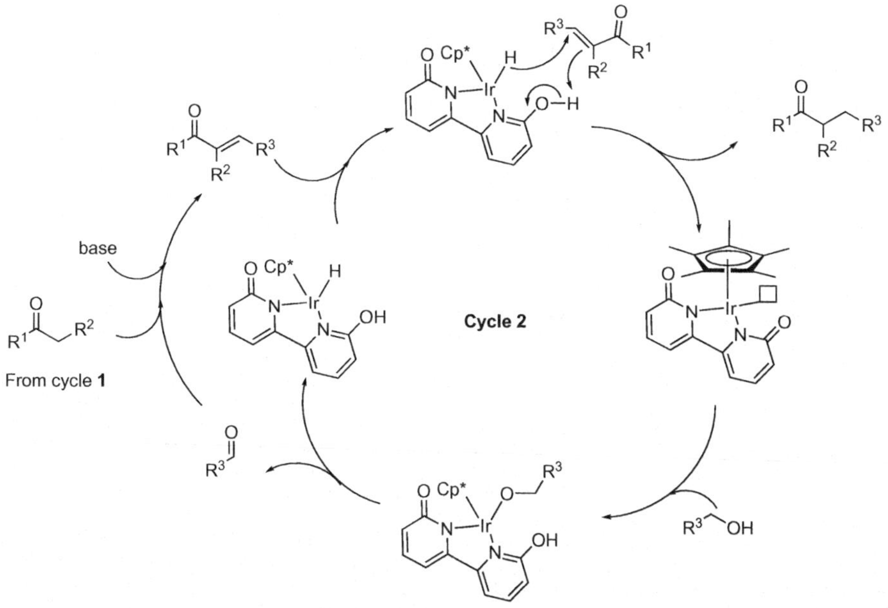

Scheme 18 The continued mechanism from Scheme 17

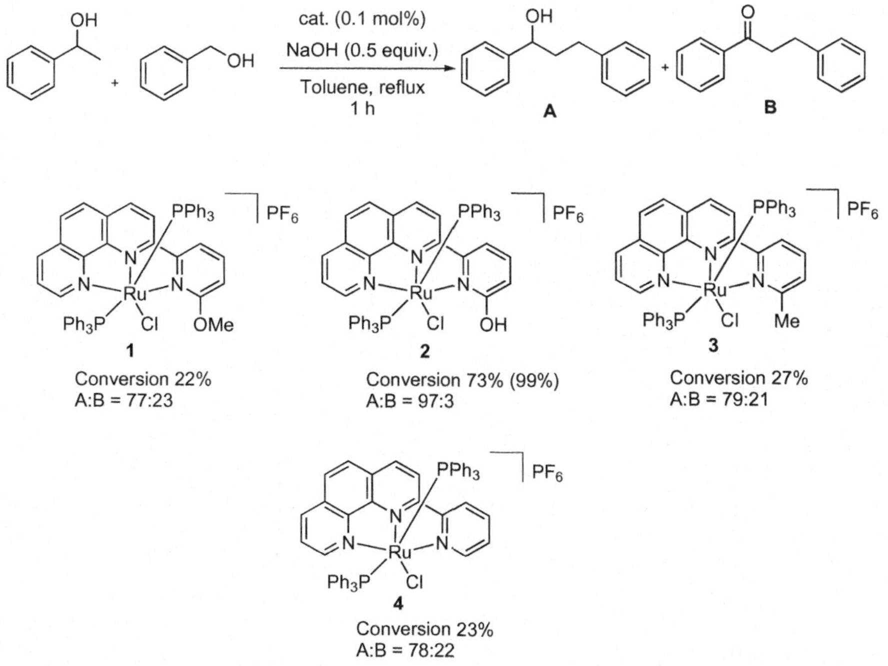

Scheme 19 Kundu's Ru-catalyzed alkylation reactions

Scheme 20 The dialkylation reaction by Kundu

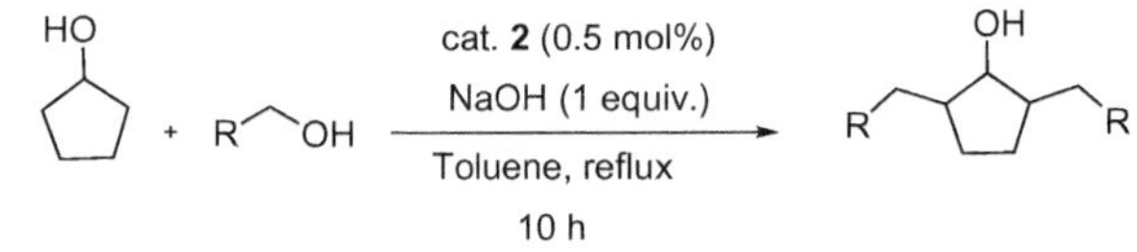

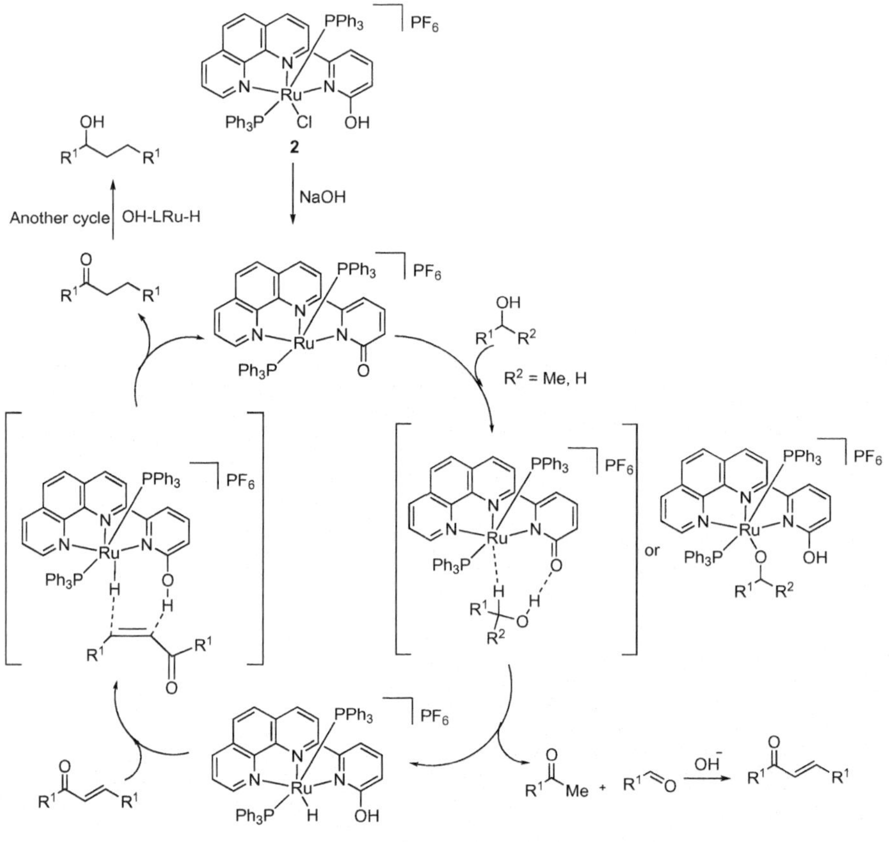

Scheme 21 The cooperative mechanism for the alkylation reaction

complex follows the inner sphere mechanism to activate the alcohols under the present reaction conditions. As previous report, it was also noted that hydrogenation of α,β-unsaturated ketone is faster than the reduction of carbonyl functionality. Inspired by this work, they prepared another Ru complex and explored their activity in the same heterocoupling reaction (Fig. 9).

In 2017, the same group described this similar transformation using newly prepared highly reactive Ru(II)NHC complex with 0.001 mol% catalyst loading (Scheme 22) [36]. This catalytic system worked under solvent-free conditions to

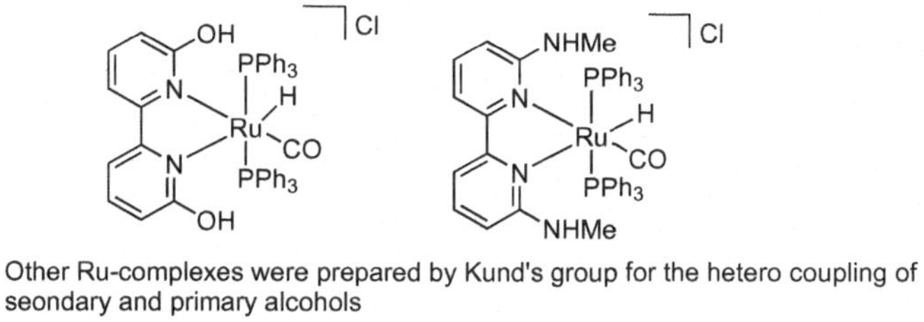

Fig. 9 Cooperative Ru-catalysts based on either O- och N-atoms

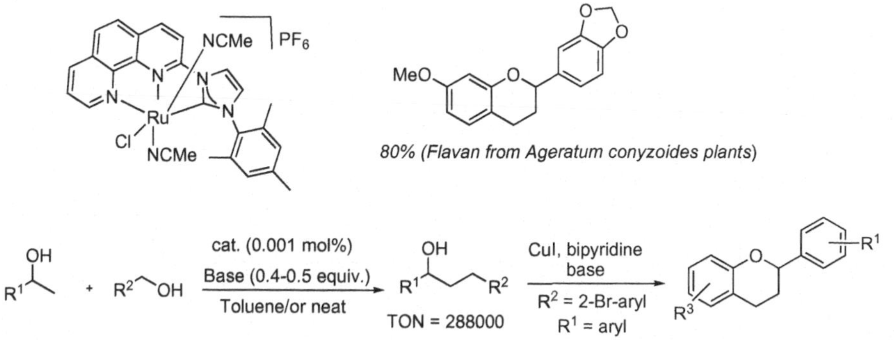

Scheme 22 The highly active system based on a tridentate NHC-ligand

afford the desired products in good yields. TON of present catalysis is 288,000 after 24 h by using 0.00016 mol% complex A. The application of the present protocol has been shown by synthesizing a range of bioactive flavan derivatives. In this one-pot process, initially two alcohols undergo Ru-catalysed cross-dehydrogenative coupling to give the coupled alcohol derivatives followed by intramolecular Cu-catalysed C–C coupling reaction to produce the final flavan products. A naturally occurring flavan derivative, which is found in Ageratum conyzoides plants, was prepared using this one-pot procedure. DFT calculation studies in the same report found that the lower activation energy barriers for the dehydrogenation of alcohols as well as hydrogenation of the α,β-unsaturated ketones signified the higher efficiency and selectivity of this catalytic system.

Another Ru-based-complex-bearing 2-HP system was shown to be efficient for the cross-dehydrogenative coupling of secondary and primary alcohols through metal–ligand cooperation strategy (Scheme 23). This catalytic system is more selective to the ketone products [37]. Interestingly, using this catalytic system, the selective tandem alkylation of isopropanol by sequential addition of different alcohols and α,α-dialkylation of ketone product which is formed from first cross-coupling reaction was achieved. In addition, a reaction using ethanol as an alkylation agent to get the corresponding alkylated product in 75% yield under the present reaction conditions was also performed.

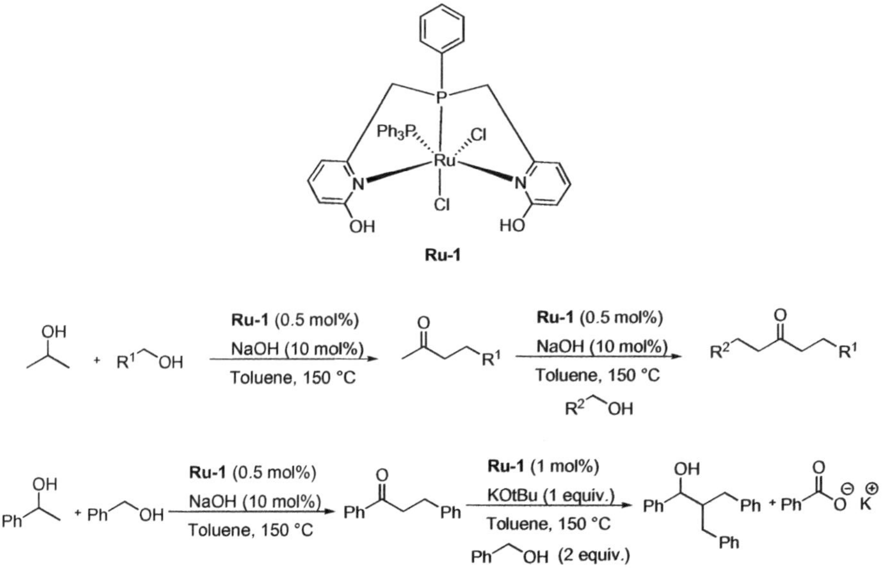

Scheme 23 The selective hydroxypyridine-based system for alcohol alkylation

Kempe group developed a method for the heterocoupling of secondary and primary alcohols using a base or non-precious more abundant cobalt metal which is chelated with triazine-based PN_5P pincer ligand (Scheme 24) [38]. Maximum yield of alcohol products will be obtained in the presence of 5 mol% Co-complex 1.1 equiv. of KHMDS base and refluxing for 20 h in toluene. Substrate scope was examined with a wide variety of secondary and primary alcohols having different substituted aromatic rings and aliphatic chains.

By introducing a proton-responsive ligand such as 2-HP systems to the metal centre, a series of ruthenium complexes were prepared to study the hydrogen transfer alcohol coupling reaction through metal–ligand cooperation approach [39]. Among these, metal complexes 1 and 2 were found to be more active and selective for the β-alkylation reactions to get alcohols as major products in the presence of KOtBu (0.5 equiv.) under air (Fig. 10). Since catalysts 1 and 2 exhibit similar activities, it was suggested that complex 2 is formed as a catalytic intermediate of 1

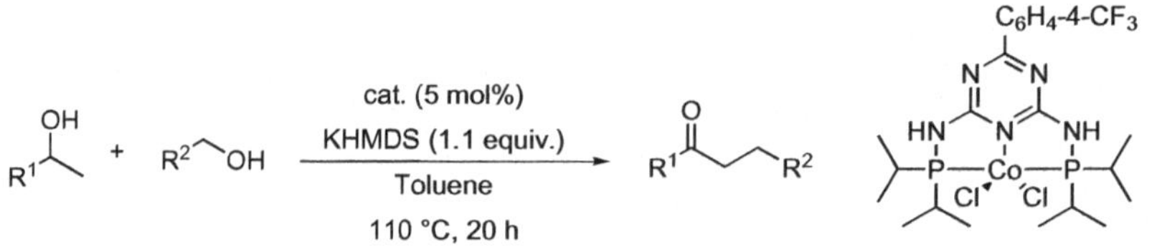

Scheme 24 Kempe's alkylation method using a Co-catalyst

Fig. 10 Base deprotonation of the hydroxypyridine-moiety

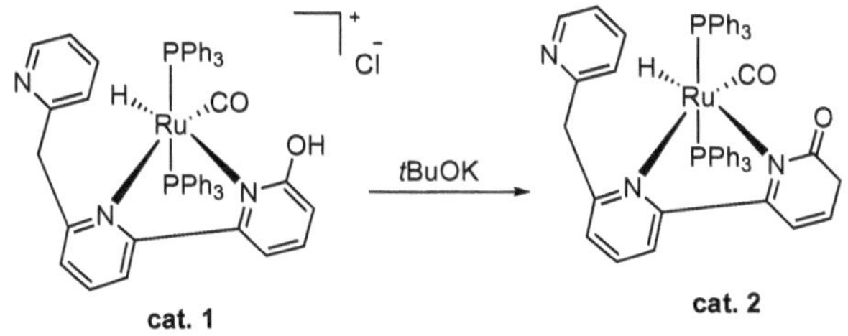

in the presence of base. Thus, the substrate scope was evaluated using catalyst 1, and it was found that introducing electron-withdrawing group at the para-position of 1-phenyl ethanol did not influence the product outcome, while electron-donating group shows negative effect on both the activity and selectivity of present transformation. To investigate the involvement of PPh_3 dissociation from the metal centre in the catalytic cycle, the reaction was carried out in the presence of excess PPh_3 and observed decreased in the conversion of 1-phenylethanol. Hence, it was concluded that elimination of PPh_3 plays a crucial role in this catalytic transformation.

Using copper(I) 4,6-dimethylpyrimidine-2-thiolate cluster as a catalyst, Li and Lang group developed an efficient and convenient method for acceptorless dehydrogenative coupling of alcohols to produce exclusively the ketone derivatives in good-to-excellent yields (Scheme 25) [40]. The substrate scope was shown on a wide variety of secondary and primary alcohols, notably ferrocenylmethanol nicely underwent cross-coupling with benzyl alcohol to give the corresponding alkylated product in 96% yield under the optimized reaction conditions. It should be pointed out that they carried out a reaction in gram scale (10 mmol) and obtained the desired alkylated ketone product in 72% yield using standard reaction conditions. Substrate scope has been extended to γ-amino and 2-aminobenzyl alcohols to get pyridines and quinolines, respectively, through cross-coupling annulation process in one pot. It was also observed that SH group of ligand 4,6-dimethylpyrimidine-2-thiolate plays a key role in controlling activity and selectivity of the present transformation through metal–ligand cooperation.

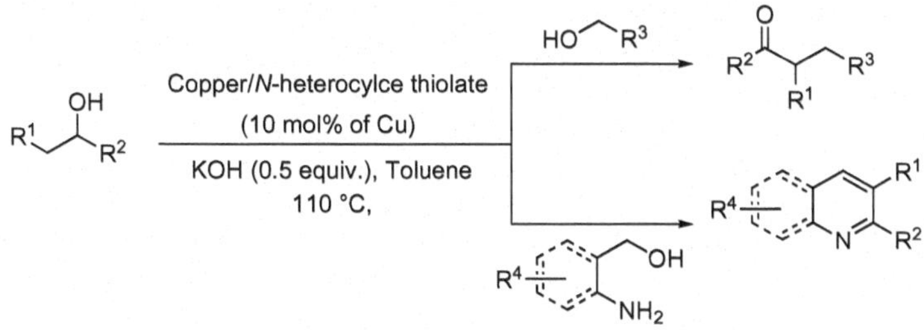

Scheme 25 The Cu-catalyzed system by Li and Lang

Scheme 26 The Ir-NHC system for alkohol alkylation to produce ketones

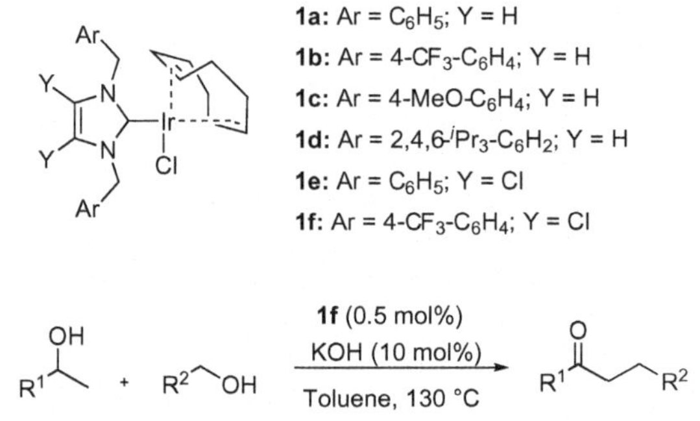

A series of air- and moisture-stable [IrCl(COD)(NHC)] complexes have been prepared and investigated their catalytic activity in alcohol cross-coupling reaction (Scheme 26) [41]. It was found that the complex bearing 1,3-bis(4-trifluorobenzyl)-4,5-dichloroimidazol-2-ylidene ligand (**1f**) provided the highest selectivity to the ketone products, whereas more electron-deficient complex **1d** was shown to be selective to alcohol products. Hence, ligand plays a major role in controlling the product selectivity. This catalytic system showed high TOF (98 h^{-1}) for the direct cross-dehydrogenative coupling of secondary and primary alcohols under air in the presence of catalytic amount of base.

1.2 Homocoupling of Secondary Alcohol

A ruthenium-catalysed dehydrogenative homocoupling of secondary alcohol to furnish ketone dimer as major products was reported by Madsen and co-workers [42]. The acceptorless dehydrogenative coupling of primary alcohols to give esters using the same catalyst which is discussed in this chapter under the separate topic was also reported. In the presence of a Ru complex, ligand $PCy_3{\cdot}HBF_4$ and excess amount of KOH, a range of secondary alcohols underwent a self-coupling reaction in refluxing toluene to give its dimeric ketone in good yields (Scheme 27). Halogenated substrates resulted in several mixtures of ketones due to partial dehalogenation process. A mixture of mono and α,α′-dialkylated products obtained from cyclopentanol and cyclohexanol substrates, whereas cycloheptanol produced exclusively monoalkylated product in good yield. However, the present catalytic system is not suitable for the homocoupling of more substituted secondary alcohols

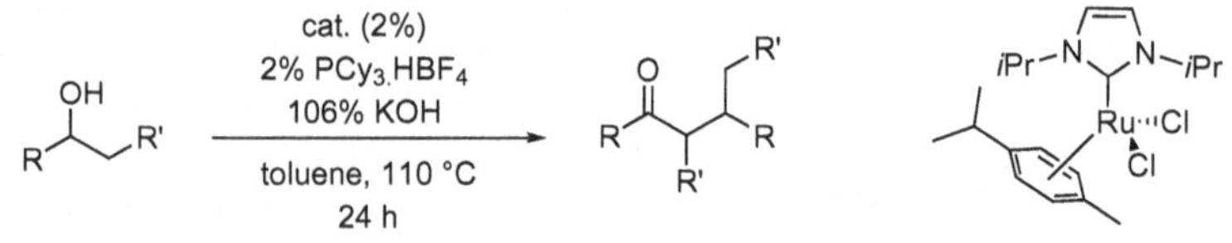

Scheme 27 Ru-catalyzed homocoupling of secondary alcohols

to get corresponding dimers. Since mechanical study revealed that dehydrogenation of alcohol is a reversible process as it was observed rapid scrambling of deuterium and hydrogen in the α-position, ruthenium species formed during the reaction is a dihydride.

A similar kind of coupling reaction was reported by different groups using homogeneous and heterogeneous catalyst while studying the hetero coupling of secondary and primary alcohols [18c, 33].

1.3 Homocoupling of Primary Alcohols

First homocoupling of primary alcohols was reported by Guerbet and co-workers It is classically known as Guerbet reaction which involves the conversion of primary alcohols into the corresponding *β*-alkylated dimeric alcohol with loss of water as the sole by-product. Primary alcohols such as *n*-butanol and other long-chain aliphatic alcohols underwent homocoupling in the presence of $[IrCl(cod)]_2$ and $[Cp^*IrCl_2]_2$ complexes, to give the corresponding α-alkylated dimer alcohols (Guerbet alcohols) in good yields (Scheme 28) [43]. For instance, 93% yield of 2-ethyl1-hexanol was obtained from the reaction of 1-butanol in the presence of $[Cp^*IrCl_2]_2$ (1 mol%), *t*BuOK (40 mol%) and 1,7-octadiene (10 mol%) after 4 h heating at 120 °C in *p*-xylene.

n-Butanol is being considered as an alternative biofuel to ethanol. Since butanol has more advantages such as non-corrosive, immiscible with water and its energy density closer to gasoline, the drawbacks associated with ethanol can be eliminated. So, n-butanol has been considered as an advanced biofuel. Hence, upgrading of ethanol to 1-butanol has received attention in this area. Initially, the Guerbet reaction was carried under very harsh reaction conditions at high temperature and pressure. Later, a different type of homogeneous and heterogeneous [44] catalyst has been developed for the ethanol-to-butanol conversion under mild reaction conditions of 110–140 °C at atmospheric pressure using various transition metal complexes. Initially, Ishii and Wass groups have studied the upgrading process using Ir and Ru metal complexes, respectively (Scheme 29). Under Ir-catalytic conditions, they could achieve the selectivity for n-butanol up to 67 at 12% conversion [45]. Using the (Ru complex) protocol developed by Wass and co-workers, the selectivity for n-butanol can be reached > 90% at 31 conversion of ethanol [46].

A bifunctional Ir complex has been shown to be highly selective (> 99%) for the conversation of ethanol (up to 37%) to butanol in the presence of bulky nickel or copper hydroxides by Jones and co-workers (Scheme 30) [47]. In this Guerbet

Scheme 28 Ir-catalyzed homocoupling of primary alcohols

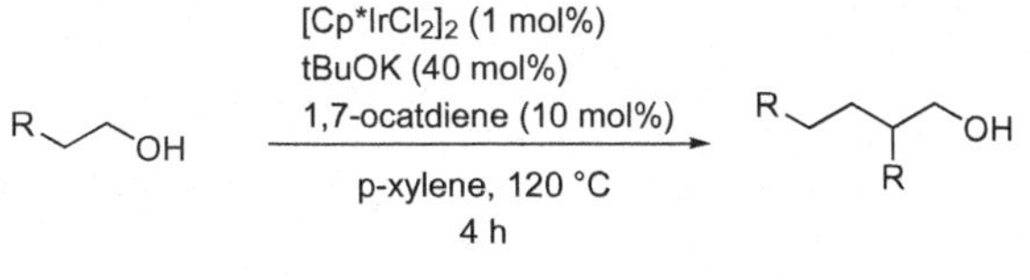

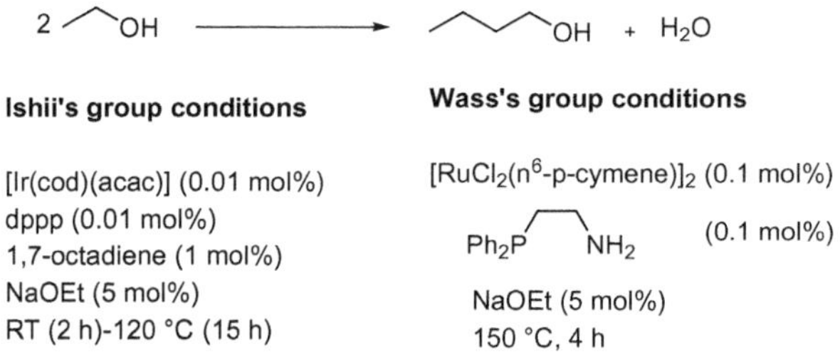

Scheme 29 The first steps towards homogeneously catalyzed homocoupling of ethanol to produce butanol with relevance for biofuel production

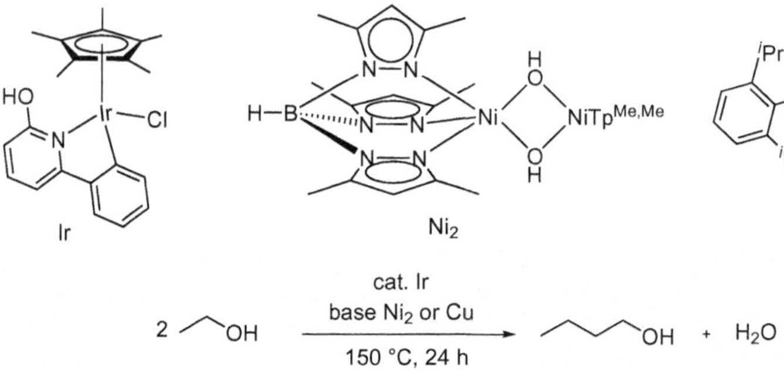

Scheme 30 Further developed systems based on iridium for ethanol homocoupling to produce butanol

process, sterically crowded nickel and copper hydroxides play a key role in selective conversion of acetaldehyde to the C4 coupling product, crotonaldehyde. Thus, high selectivity is observed in the conversion of ethanol to butanol under these catalytic reaction conditions. It has also been demonstrated upgrading ethanol to more selectively (>99%) butanol even at 100 °C.

Milstein groups showed a Ru–pincer complex which could effectively catalyse the ethanol to *n*-butanol through Guerbet process with TON 18 209 and 86.1% selectivity (Scheme 31) [48]. Using this protocol, 73.4% of ethanol conversion

Scheme 31 Milstein catalysts for ethanol coupling towards butanol

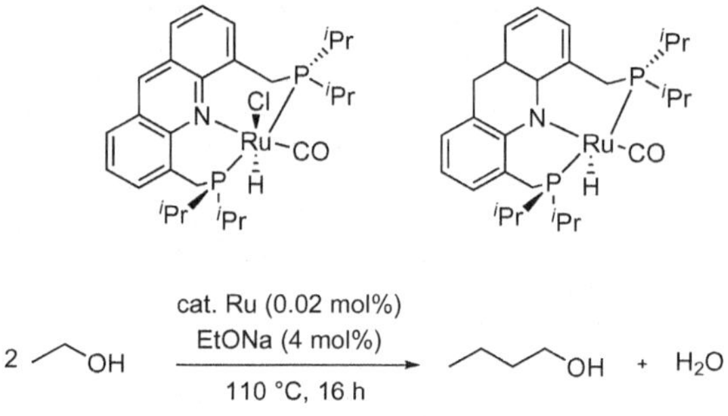

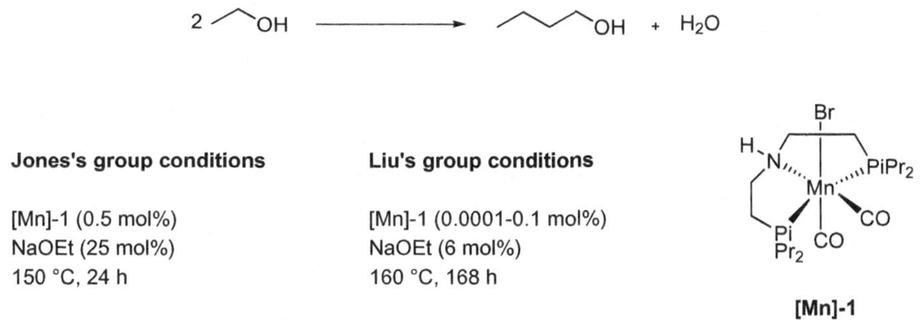

Scheme 32 Comparison of Jones and Liu's conditions for the Mn-catalyzed coupling

could be achieved which includes the significant amounts of C_4 (35.8% yield), C_6 (28.2% yield) and C_8 (9.4% yield) alcohols.

Liu [49] and Jones [50] groups have reported a method for ethanol to n-butanol conversion using homogeneous non-noble metal catalyst based on Mn complexes (Scheme 32). Both the groups identified that N–H moiety present in the manganese catalyst plays a key role in the catalytic activity.

1.4 Coupling Between Two Different Primary Alcohols

The selective cross-coupling between two different primary alcohols such as aryl ethanol and benzylic alcohol derivatives is relatively underexplored. The cross-coupling between 2-arylethanols and methanol was demonstrated by Beller and co-workers [51]. They have used Ru-MACHO and Shvo's diruthenium complex bimetallic system and catalytic amount of base (0.1 equiv.) to get the maximum yield of desired alkylated products (Scheme 33). While doing the mechanistic

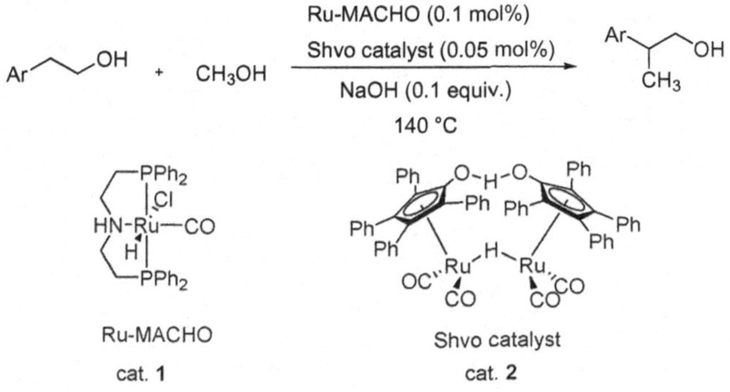

Scheme 33 Ru-MACHO catalyzed methylation with methanol

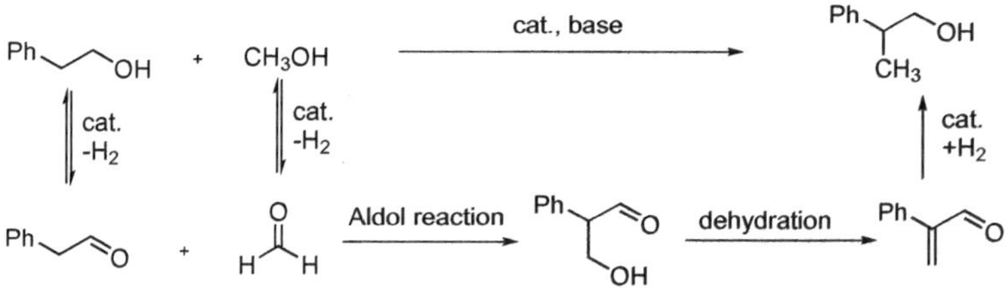

Scheme 34 Reaction substeps in the methylation reaction

studies, it was observed that Ru-MACHO catalyst does not favour the dehydrogenation of 2-phenylethanol, but favours the dehydrogenation methanol to formaldehyde. Thus, they have added second catalyst which promotes the selective dehydrogenation of 2-phenylethanol (Scheme 34). Compared to hydroformylation and reduction process to get the same products, the current protocol has advantages such as avoiding toxic CO and extra hydrogen source or high-pressure infrastructure.

The cross-coupling between 1-arylethanol/2-arylethanol and methanol was developed by Obora groups using DMF-stabilized iridium nanoclusters under heterogeneous conditions [52]. Liu and co-workers have reported a similar kind of methylation using methanol as C1 source in the presence of heterogeneous iridium clusters encapsulated in carbon nanospheres in water medium [53].

In 2012, Ramón and co-workers reported a method for cross-alkylation of primary alcohols using recyclable impregnated iridium oxide on magnetite as a catalyst under heterogeneous conditions (Scheme 35) [54]. Good-to-excellent yields of cross-coupled products from a variety of 1-arylethanols and benzyl alcohol derivatives were obtained in the presence of 0.14 mol% IrO_2–Fe_3O_4 heterogeneous catalyst and 1.0 equiv. of KOH in toluene for 4 days heating at 110 °C. Even the reaction worked with aliphatic alcohols and benzyl alcohol derivatives. However, the reaction failed when two different aliphatic alcohols were used, resulting in the recovery of the starting materials.

The cross-dehydrogenative coupling between two different primary alcohols to α,β-unsaturated aldehydes through in situ formation of enolates using a ruthenium catalyst, silica-grafted amine and crotononitrile as a hydrogen acceptor was demonstrated by Porcheddu et al. [55] under microwave conditions (Scheme 36). Here, enolates were prepared in situ from alcohols rather from unstable aldehydes in the presence of Ru catalyst through hydrogen transfer process. Moreover, a silica-grafted

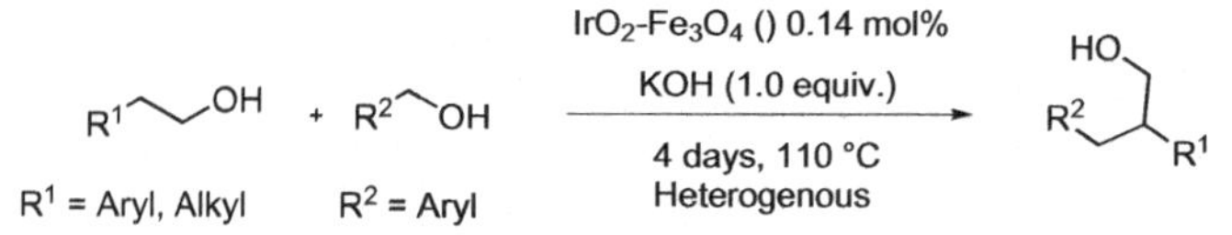

Scheme 35 The cross dehydrogenative coupling of primary alcohols using a heterogeneous iridium magnetite catalyst

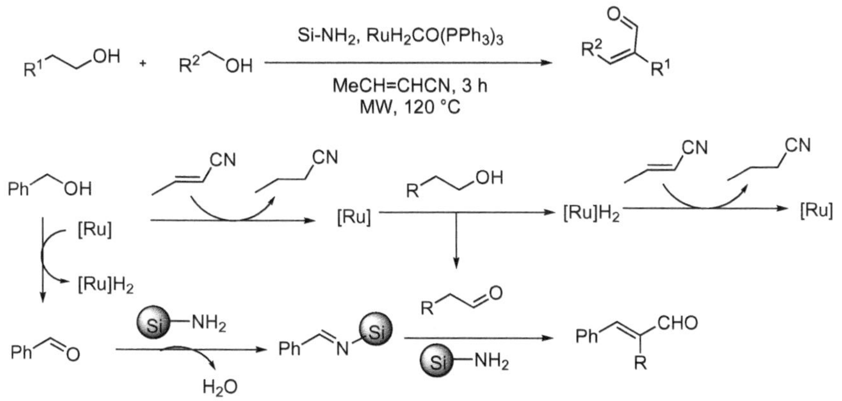

Scheme 36 Pocheddu's coupling using the silica grafted amine

amine acts a as phase-switch tag leading to a selective one-pot process in favour of cross-dehydrogenative coupling products. Interestingly, it was observed that the MW heating cuts down the reaction time from 24 h to a mere 1 h.

A method has been developed for the cross-dehydrogenative coupling between two different primary alcohols using readily available $RuCl_2(PPh_3)_3$ as a precatalyst through the hydrogen-borrowing approach by Johnson and co-workers (Scheme 37) [56]. Linear, cyclic and heterocyclic alcohols could be used in the present method to prepare a wide variety of Guerbet alcohol derivatives. This eventually led them to develop an operationally simple method for the synthesis of commercially available fragrances such as Rosaphen and Cyclamenaldehyde from highly abundant alcohol derivatives in good yields. In general, the preparation method for these fragrances involves the condensation of the corresponding aldehydes followed by hydrogenation.

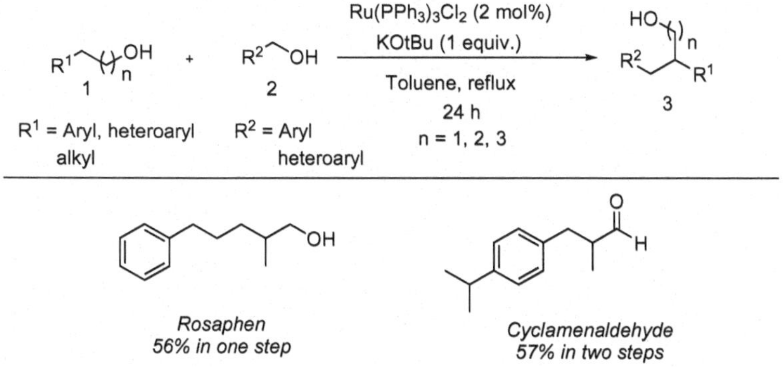

Scheme 37 Ru-catalyzed coupling which demonstrates potential application towards fragrance synthesis

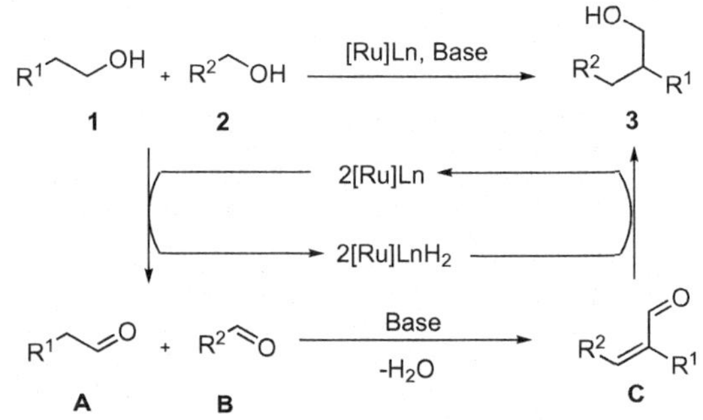

Scheme 38 Subreactions including dehydrogenation, aldol reaction and condensation and rehydrogenation

A plausible mechanism for the cross-coupling between two primary alcohols has been shown in the Scheme 38. Initially, alcohols **1** and **2** would undergo dehydrogenation in the presence of $Ru(PPh_3)_3Cl_2$ and base to give the respective aldehydes **A** and **B**. Base-promoted aldol condensation would take place between aldehydes **A** and **B** to afford an intermediate **C** which subsequently undergoes hydrogenation reaction to give the final Guerbet alcohol derivative **3** in the presence of Ru catalyst.

1.5 Complete Deoxygenative Coupling of Primary Alcohols

In comparison to traditional hydrogen-borrowing, alkylation chemistry, in which the heteroatoms are not removed, the challenge in such a reaction lies in the double deoxygenation. Given the abundance and diversity of alcohols, a general and fully deoxygenative coupling to produce a carbon chain directly would be desirable. First, Ishii, Obora and co-workers reported the iridium-catalysed coupling of 2-arylethanols to produce hydrocarbons in good yields (Scheme 39) [57]. Under the same reaction conditions, 3-arylpropanol gave the corresponding hydrocarbon product in less yield. Then, they found a two-step method giving similar products with

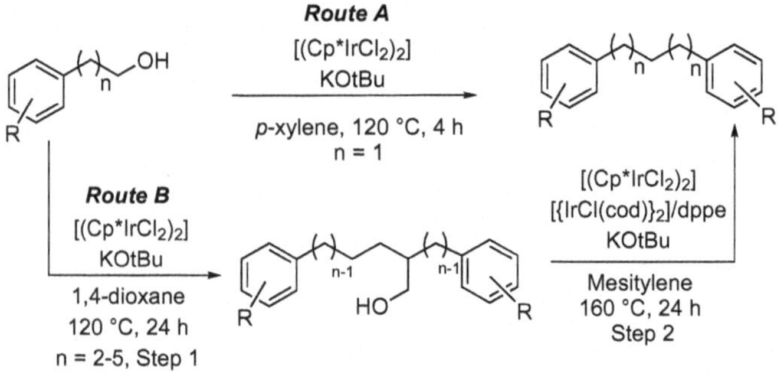

Scheme 39 Ishii and Obora's deoxygenative coupling reaction

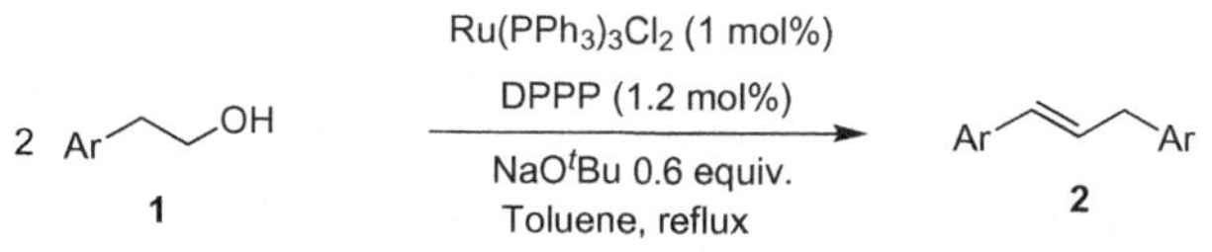

Scheme 40 The Ru-catalyzed net decarbonylative deoxygenative coupling of 2-arylethanols

improved yields from longer-chain alcohols such as 3-arylpropanol and 4-arylbutanol.

Later, Johnson and co-workers developed the similar method using versatile and commercially available [$RuCl_2(PPh_3)_3$]. Under the Ru-catalysed reaction conditions, 2-arylethanol underwent fully deoxygenative coupling to afford the respective alkene as a sole/major product (Scheme 40) [58]. They carried out a few control experiments to understand the mechanistic pathway and found that the base has an important role in the second deoxygenation or net decarbonylation to produce the hydrocarbon product. Further, in the presence of dppp, the olefin was always obtained as the exclusive product, whereas otherwise the corresponding 1,3-diarylpropane often formed as a side product.

Liu and co-workers have reported the first example of the catalytic dual deoxygenation of alcohols using a non-noble metal catalyst such as manganese [59]. Here they have shown that both deoxygenative homocoupling of 2-arylethanols and their deoxygenative cross-coupling with other primary alcohols proceeded smoothly to form the corresponding alkenes by a dehydrogenation and deformylation reaction sequence in the presence of well-defined Mn/PNP pincer complex (Scheme 41). During the cross-coupling reaction, a mixture of alkene products was obtained due to non-selective aldol condensation of the two aldehyde intermediates. The regioisomeric alkenes were subjected to Ni-catalysed hydrogenation conditions to get the single alkane product.

Few control experiments and kinetic studies were carried out to understand the mechanistic pathway. It was found that sodium formate was formed as a by-product in this transformation, and this cascade reaction involves a sequence of catalytic dehydrogenation, aldol condensation and base-promoted deformylation processes (Scheme 42).

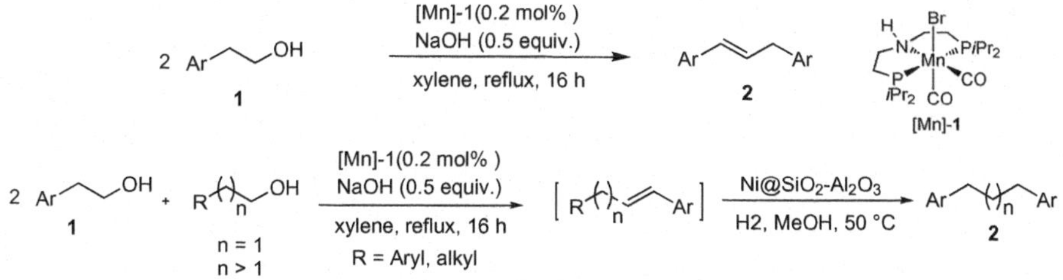

Scheme 41 Liu's Mn-pincer complex catalyzed version of the coupling of arylethanols

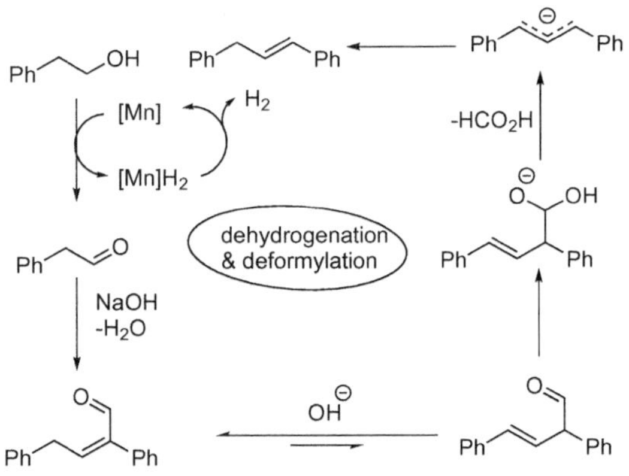

Scheme 42 Mechanistic pathway for Mn-catalysed fully deoxygenative coupling of alcohols

1.6 Coupling Between Primary Alcohols to Form Ester

Another interesting type of acceptorless dehydrogenative coupling of alcohols involves the formation of an ester through dehydrogenation process in the presence of transition metal catalyst. Initially, transition metal-catalysed dehydrogenation of an alcohol takes place to give an aldehyde which reacts with another molecule of alcohol to form a hemiacetal. Then second dehydrogenation would happen on hemiacetal to afford an ester (Scheme 43).

Intramolecular oxidative cyclization of diols to lactones using various stoichiometric oxidants and heterogeneous metal catalysts is known in the literature. Homogeneous catalysts such as $Ru_3(CO)_{12}$ and $Pd(OAc)_2$ have also been reported to catalyse the cyclization reaction of diols to lactones in the presence of stoichiometric amount of oxidants such as diphenylacetylene [60] and bromobenzene [61], respectively. Along these lines, Shvo [62] and Murahasi [63] group reported the Ru-catalysed dehydrogenative coupling of primarily alcohols to esters in the absence of any hydrogen acceptor (Scheme 44). Good conversion in esters formation was obtained when the reactions were performed in open system, as hydrogen gas evolution during this transformation shifts the equilibrium towards ester formation (forward reaction).

In Murahasi report, it is mentioned that the yields of esters from sterically hindered alcohols can be increased in the presence of diphenylacetylene hydrogen

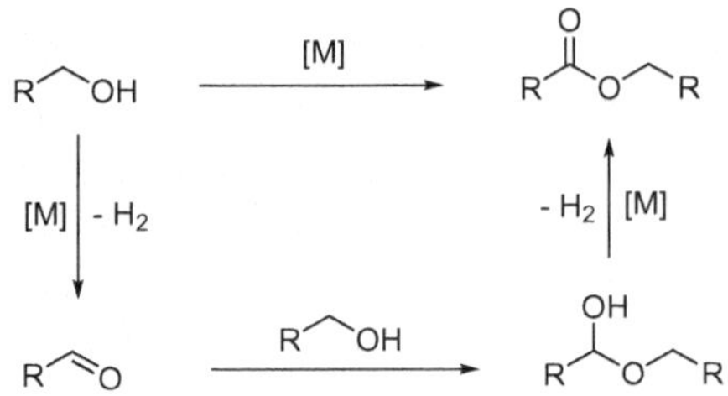

Scheme 43 Acceptorless dehydrogenative coupling of alcohols to esters

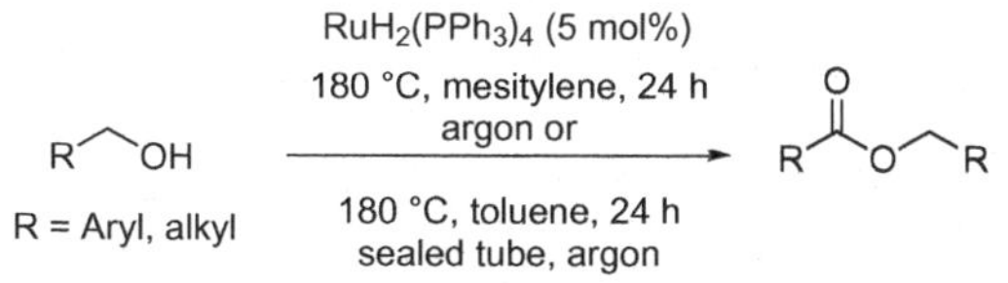

Scheme 44 Ru-catalysed dehydrogenative coupling of primarily alcohols to esters

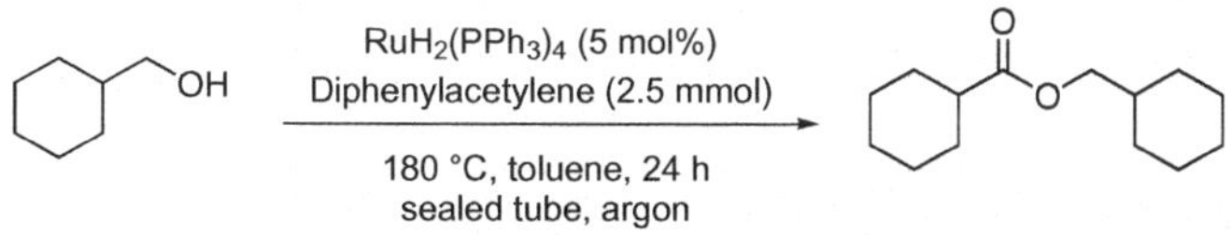

Scheme 45 Coupling of cyclohexanemethanol to cyclohexylmethyl cyclohexanecarboxylate

acceptor. For instance, cyclohexylmethyl cyclohexanecarboxylate was obtained in 93% yield from cyclohexanemethanol in the presence of diphenylacetylene (Scheme 45). Without using the hydrogen acceptor, the same reaction resulted in only 30% yield.

The application of this methodology has been shown in the synthesizing of perfume esters such as 2-methylpentyl, 2-methylpentanoate, 3-methylbutyl-3-methylbutanoate (apple-like fragrance) and 2-phenylethyl, 2-phenylacetate (hyacinth-like fragrance) from readily available alcohol materials (Fig. 11).

The cross-dehydrogenative coupling of two different alcohols has also been examined under the same reaction conditions. It was found that the reaction was specific towards formation of the homocoupled product benzyl benzoate in 65% yield from the reaction of benzyl alcohol and 2-butanol. It did not give any cross-coupled 1-methylpropyl benzoate product (Scheme 46). However, the reaction of 1-octanol with benzyl alcohol gave a mixture all possible four esters.

In addition, the same catalytic system has been utilized to study the intramolecular cyclization of diols to form lactones through dehydrogenative coupling process. It is noteworthy to mention that this intramolecular reaction works well in the presence of

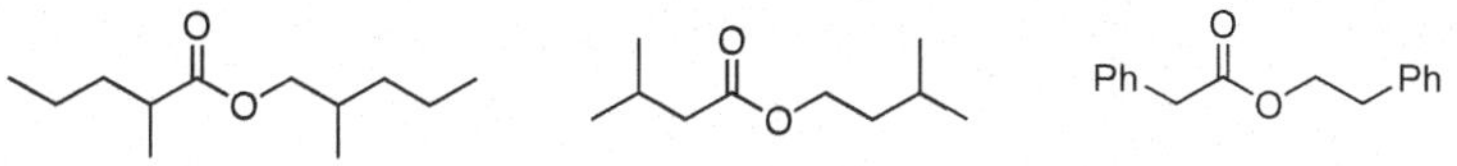

Fig. 11 Examples of perfume esters

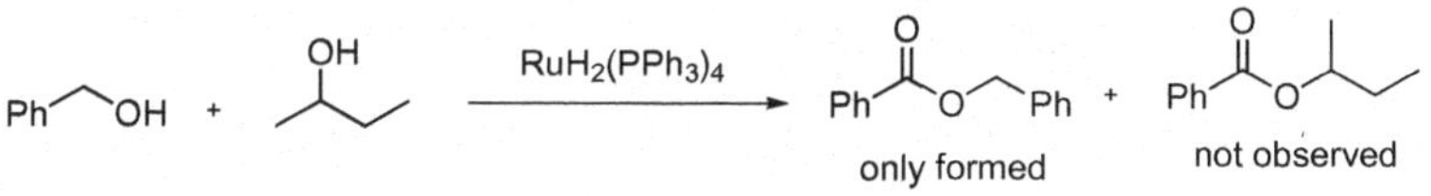

Scheme 46 Cross dehydrogenative coupling of two different alcohols, with indication of the specificity

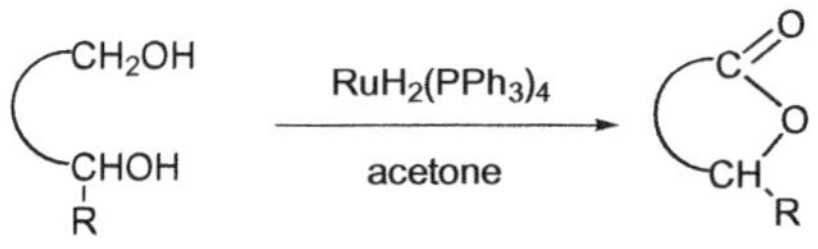

Scheme 47 Intramolecular cyclization of diols to form lactones

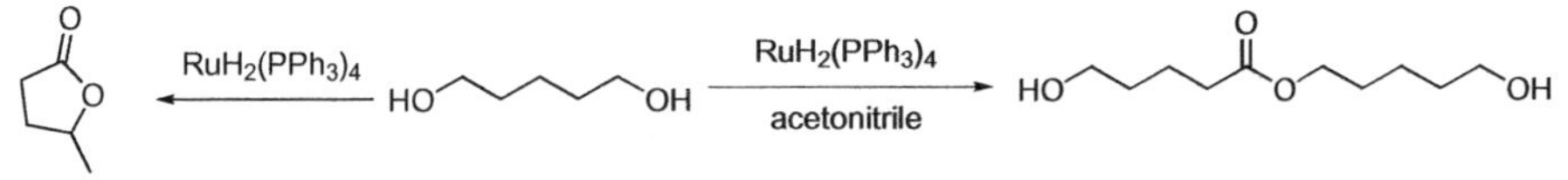

Scheme 48 Intra- and intermolecular ADC of 1,5-pentanediol

acetone as a hydrogen acceptor (Scheme 47). Another interesting thing was observed while studying the cyclization of 1,5-pentanediol that it gave 82% yield of δ-valerolactone. On the other hand, the same reaction underwent an intermolecular esterification to afford 5-hydroxypentyl-5-hydroxypentanoate in 41% yield in the presence of 1.0 equiv. of acetonitrile (Scheme 48). The reason may be due to the strong coordination of acetonitrile avoids the bidentate chelation of diols and subsequent ring closing reaction.

Furthermore, the catalytic system is more chemoselective in the oxidation of primary alcohols in the presence of secondary alcohols. Hence, γ-valerolactone and γ-octanolactone are exclusively formed from respective 1,4-pentanediol and 1,4-octanediol through chemoselective oxidation of primary alcohol groups followed by lactone formation with secondary alcohol groups (Scheme 49). Not only the chemoselectivity, a regioselective oxidation of sterically less hindered alcohols can be achieved using $RuH_2(PPh_3)_4$ catalyst. The following lactone can be prepared in 95% yield by regio- and chemoselective oxidation of less hindered primary alcohol group followed by cyclization with secondary alcohol group using $RuH_2(PPh_3)_4$ catalyst (Scheme 50). Similar kind of regio- and chemoselective dehydrogenation followed by lactonization was observed using iridium and ruthenium polyhydrides as catalyst under mild and neutral reaction conditions.

An efficient method has been described for oxidative lactonization of 1,4- or 1,5-diols using a new amino alcohol-based Ir bifunctional complex catalyst in the

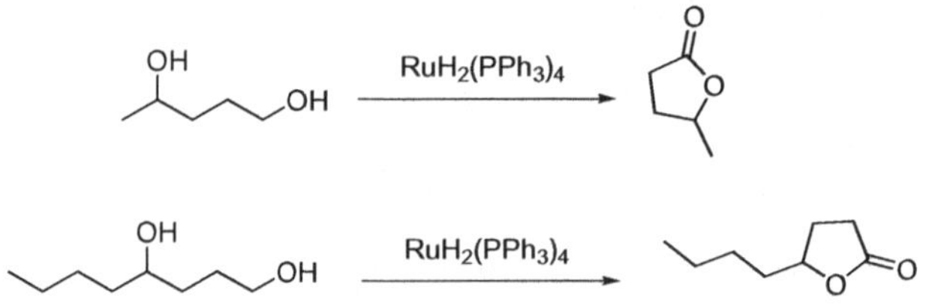

Scheme 49 Example for the chemoselective oxidation and lactonization of diols

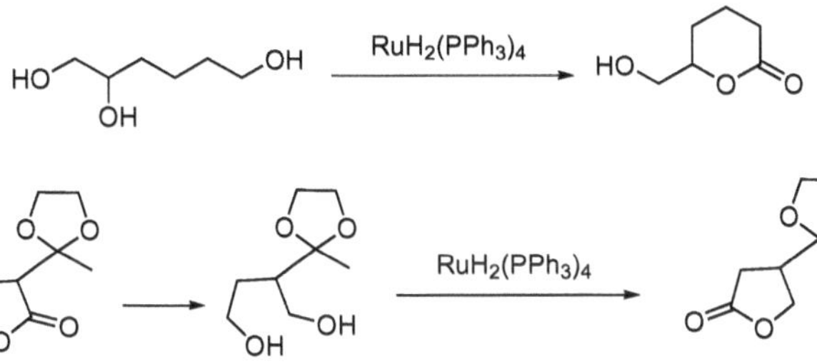

Scheme 50 Regio- and chemoselective dehydrogenation followed by lactonization

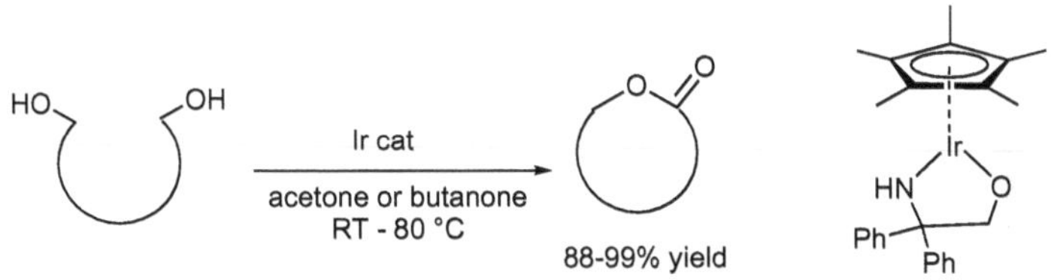

Scheme 51 Ir-catalysed lactonization of 1,4- or 1,5-diols

presence of hydrogen acceptor such as acetone or butanone (Scheme 51) [64]. Generally, these reports required either high temperature or hydrogen acceptor to get the lactones product from diols.

Because alkylphophines are more stable than aryphosphine at higher temperatures, Hartwig and co-workers designed a new Ru complex bearing alkylphosphine ligand [65]. Eventually, they showed an acceptor and solvent-free method for an intramolecular dehydrogenative coupling 1,4-butanediol to γ-butyrolactone with high selectivity and high turnover numbers (17,000) using an alkylphosphine version of ruthenium bis-phosphine diamine catalyst (Scheme 52).

Milstein's group prepared a several Ru complexes bearing electron-rich PNP and PNN ligands and explored their activity towards the acceptorless dehydrogenative coupling of primary alcohols to the esters (Fig. 12) [66]. It was found that a catalytic amount of base KOH is necessary to catalyse the present transformation. Upon treatment of complex **2** with a base, later they synthesized another complex **3**. Interestingly, this complex **3** itself could catalyse the reaction of esters formation from the corresponding primary alcohols in the absence of base under mild and neutral conditions. Excellent yields of benzyl benzoate and butyl butyrate are formed from heating the reaction of benzyl alcohol and 1-hexanol respectively with

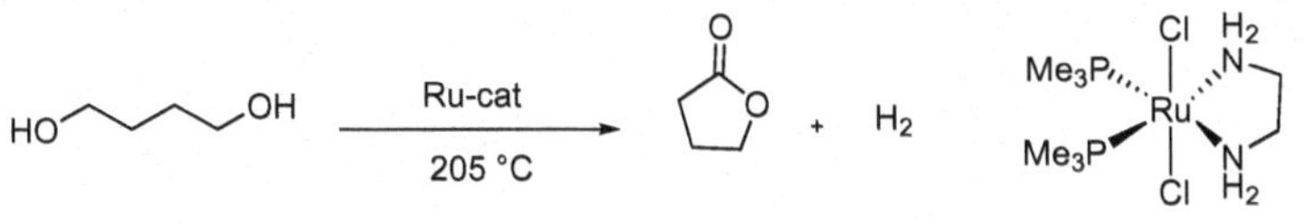

Scheme 52 1,4-Butanediol to γ-butyrolactone using Ru-complex bearing alkylphosphine ligand

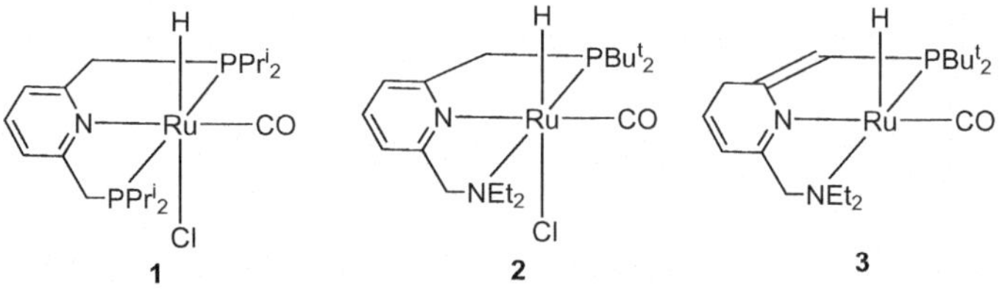

Fig. 12 Ru-complexes from Milstein's group

complex **3** at 115 °C in toluene. To elucidate reaction mechanism, benzaldehyde alone was subjected to the standard reaction conditions and found that no benzyl benzoate formation. On the other hand, benzaldehyde and benzyl alcohol were allowed to react in the presence complex **3** in toluene at 115 °C for 12 h which resulted in formation of benzyl benzoate in 100% yield (Scheme 53). This study revealed that the present reaction may proceed through the dehydrogenation of alcohol to aldehyde followed by hemiacetal formation and subsequent dehydrogenation gives the final ester rather. Here Ru complex activates the alcohols through metal–ligand cooperation process. Hence, Tischenko-type disproportionation involving the aldehyde pathway may be ruled out for the present catalytic transformation of alcohols to esters.

Later, they prepared the N_2-bridged Ru(II) dinuclear complex [(PNN)Ru$(Cl)_2]_2(\mu\text{-}N_2)$ **1** and (PNN)Ru(CO)Cl_2 **2** (Fig. 13) [67]. Upon addition of base, these

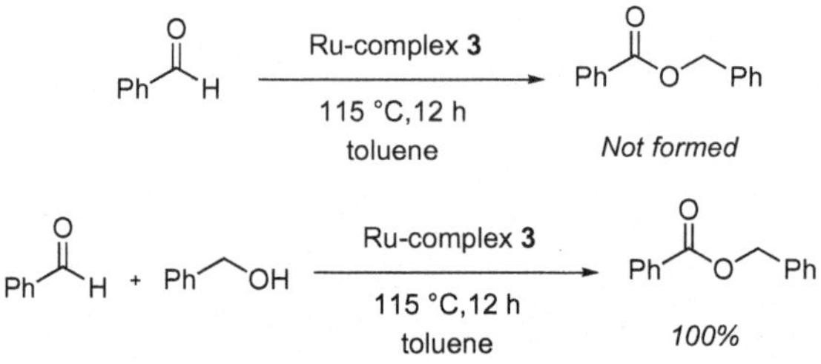

Scheme 53 Control experiments using Ru-complex 3

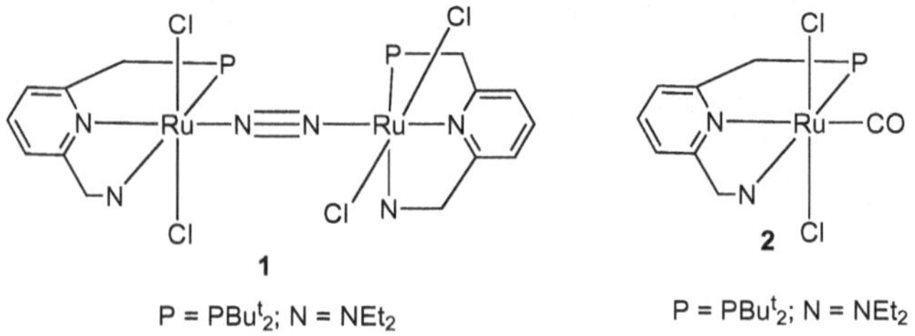

Fig. 13 PNN-Ru-pincer complexes

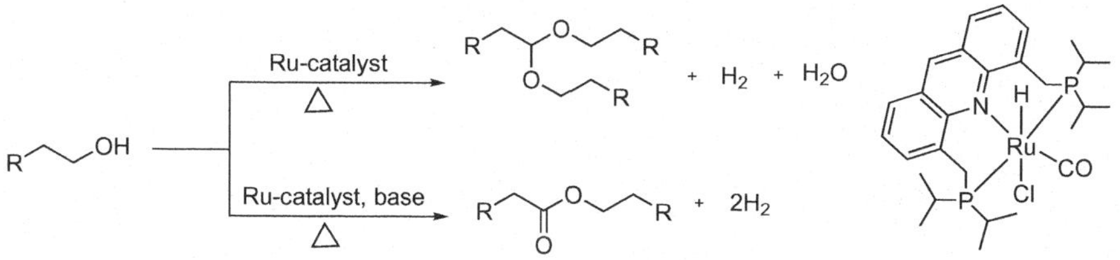

Scheme 54 Acridine-based ruthenium pincer complex catalysed ADC

complexes could catalyse the dehydrogenation of primary and secondary alcohols to the corresponding esters and ketones, respectively, in good yields and high selectivity.

A novel acridine-based ruthenium pincer complex is shown to efficiently catalyse the dehydrogenation of alcohols to acetals under neutral conditions and to the esters in the presence of base (Scheme 54) [68]. Here formed hemiacetal intermediate undergoes water elimination to form enol ether. Upon addition of another alcohol molecule to the C = C of enol ether provides the final acetal product (Scheme 55). Since enol ether formation is not possible from the reaction of benzyl alcohol, it gives only the benzyl benzoate ester rather than the corresponding acetal (benzaldehyde dibenzylacetal) under neutral reaction conditions.

The same group described again the similar kind of selective formation of acetals and esters from alcohols through the dehydrogenative coupling using bis(acetonitrile)bis(triphenylphosphine) ruthenium(II) sulphate $[Ru(PPh_3)_2(NCCH_3)_2(SO_4)]$ (**1**) and bis(triphenylphosphine)ruthenium(II) acetate $[RuPPh_3)_2(OAc)_2]$ (**2**), respectively (Scheme 56) [69]. In contrast to their earlier report, here alcohols lacking β-hydrogens such as benzyl alcohol form benzaldehyde dibenzylacetal in the presence of the Ru (**1**) catalyst. Hence, this reaction does not proceed via enol ether intermediate like acridine-based ruthenium catalysis (Route A). Instead, it follows the direct substitution of hydroxyl group of hemiacetal intermediate by an alcohol molecule to give acetal products (Route B) (Scheme 57). The reason might be the increasing the Lewis acidity of Ru-1 complex by the sulphate counter-anion and formation of vacant coordination site from easy dissociation of acetonitrile

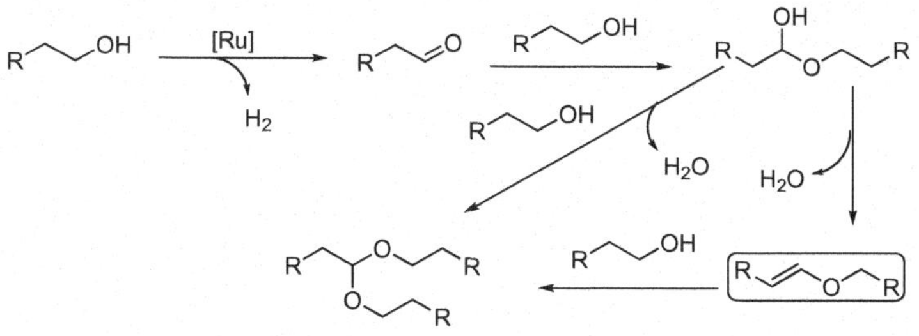

Scheme 55 Mechanistic pathway for acetal and ester formation

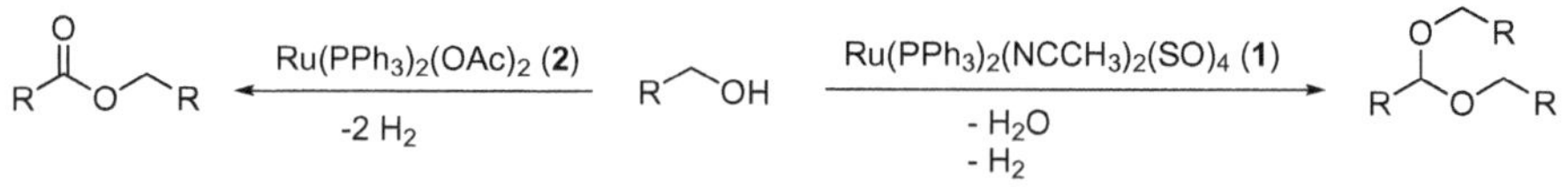

Scheme 56 Esters and acetals formation from alcohols under Ru-catalysis

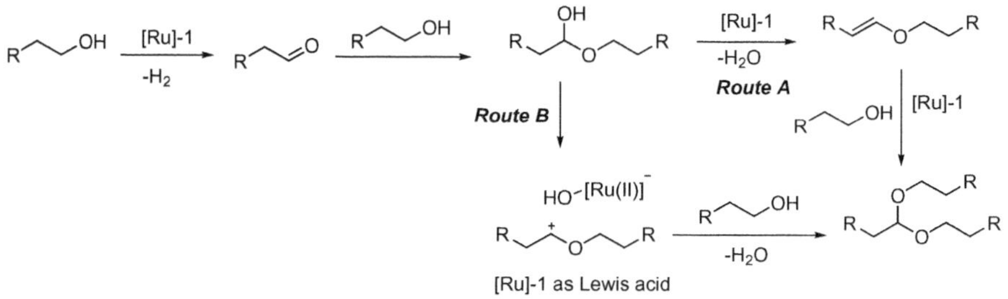

Scheme 57 Mechanistic pathway for acetal and ester formation

ligand. So the OH group of hemiacetal could easily form Lewis-acidic complex with Ru-1 which is eventually substituted by alcohol to give an acetal. Due to strong coordination of acetate, the Ru-2 complex could not make any vacant site to form Lewis-acidic complex with hemiacetal OH group. Thus, ester will be formed from hemiacetal intermediate in the presence of Ru-2 catalyst.

Due to high oxidation potential, methanol would less readily undergo oxidation than most other primary alcohols. Hence, a selective cross-dehydrogenative coupling could be happened between primary alcohols and methanol to form the corresponding methyl esters. Accordingly, Williams and co-workers have shown the method to prepare a wide variety of methyl esters through selective cross-coupling between alcohols and methanol using Ru catalyst in the presence of crotononitrile as a hydrogen acceptor (Scheme 58) [70]. The same chemoselective cross-coupling reaction between primary alcohols and methanol has been shown to produce methyl ester by utilizing Rh catalyst [71].

Aerobic oxidation of primary alcohols to the corresponding esters has been achieved using homogeneous iridium [72] and palladium[12–14] catalysts or with heterogeneous gold catalysts [73]. Beller and co-workers have developed a palladium catalytic system containing air-stable and bulky phosphine ligands for the oxidative dimerization of alcohols to esters under oxygen atmosphere (Scheme 59)

Scheme 58 Chemoselective cross coupling reaction between primary alcohols and methanol

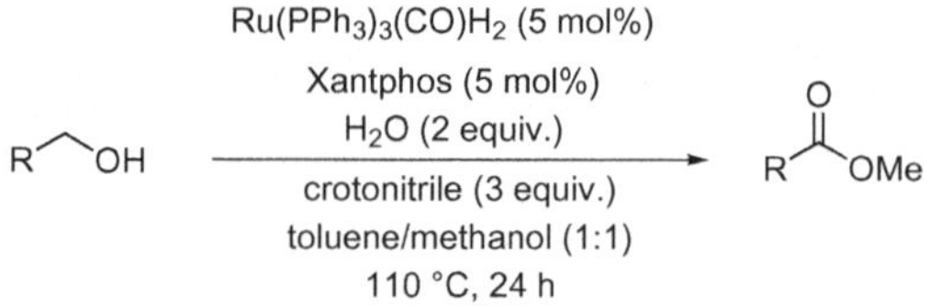

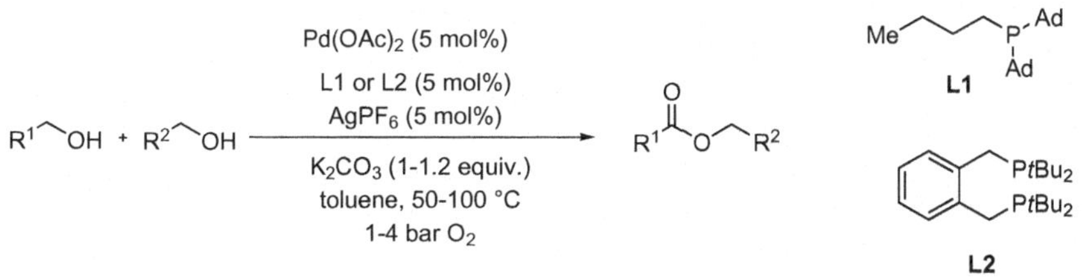

Scheme 59 Palladium catalysed dimerization of alcohols to esters

[74]. Chemoselective oxidation to esters can be achieved by using bulky ligand and $AgPF_6$. Interestingly, only the formation of benzaldehyde was observed from the reaction of benzyl alcohol with same conditions in the absence of $AgPF_6$. This catalytic system works for the both homocoupling and selective heterocoupling of alcohols to the corresponding esters in good yields under mild reaction conditions.

Lei and co-workers developed a palladium-catalysed dehydrogenative coupling of alcohols to esters. This transformation was carried out in the presence of $[PdCl_2(CH_3CN)_2]$, $AgBF_4$ and NaOtBu using O_2 balloon at 45 °C (Scheme 60) [75]. Here silver salt helps to make the palladium centre more electron-deficient, so that *β*-hydride elimination would be faster. Especially good results were obtained when using highly electron-donating and electron-withdrawing substituents on the benzyl alcohol derivatives. It should be noteworthy that selective esterification with long-chain alcohols could be accomplished by using a P-olefin ligand.

A cross-oxidative coupling of benzylic alcohols and phenols has been described to get the aryl benzoate derivatives using Pd-NHC catalytic system under aerobic conditions (Scheme 61) [76].

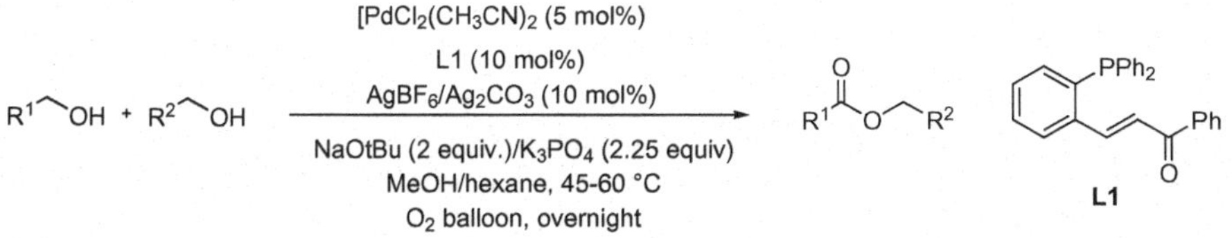

Scheme 60 Dehydrogenative coupling of alcohols to esters under Pd-catalysis

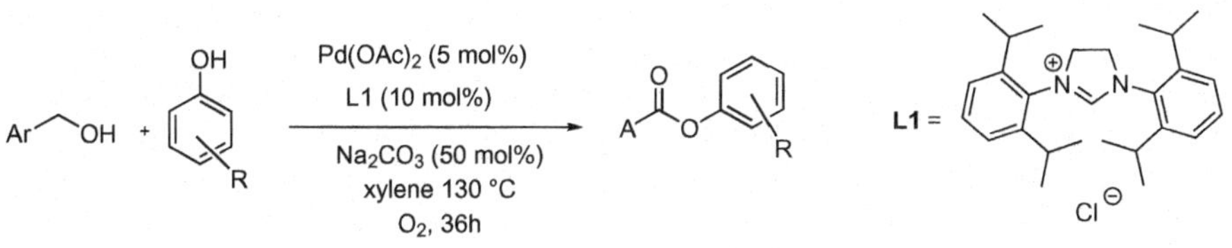

Scheme 61 Coupling of benzylic alcohols and phenols

Scheme 62 Ru-NHC complex catalysed homocoupling of alcohols to esters

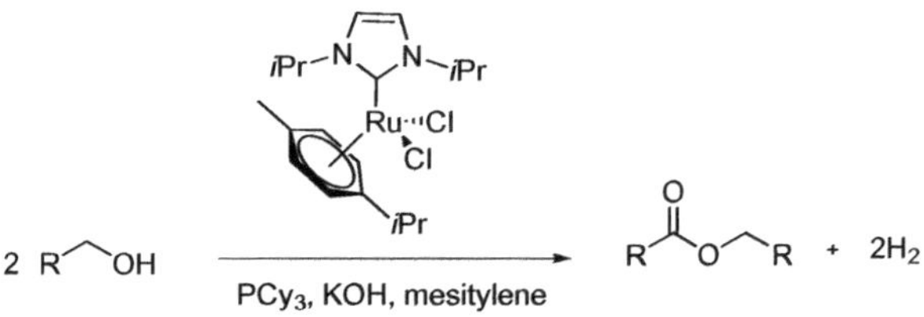

Madsen group synthesized a new Ru-NHC complex and explored its activity on coupling of alcohols to produce esters [77]. This Ru complex works for homo coupling of long-chain aliphatic alcohols to form the corresponding esters in good yields in the presence of catalytic amount of ligand PCy_3 (2.5–9 mol%) and KOH (10 mol%) (Scheme 62). However, similar reaction conditions were not suitable for a heterocoupling of two different alcohols, as it resulted in all possible four ester derivatives. An experiment was conducted with deuterated benzyl alcohol (benzyl alcohol-α,α-$d2$), and it was found that the formation of benzyl benzoate with 64% hydrogen and 36% deuterium in the benzylic position after 18 h. The same experiment was stopped after 2.5 h and observed the scrambling of hydrogen and deuterium at the α-position of the starting material. This experiment suggests that benzaldehyde formation from alcohols by β-hydride elimination is a reversible reaction and catalytic active species is a ruthenium dihydride.

Milstein's research group developed a protocol for the selective cross-dehydrogenative coupling between primary and secondary alcohols to get mixed esters in high yields using a bipyridyl-based PNN ruthenium(II) pincer catalyst under neutral conditions (Scheme 63) [78]. Good yields of heterocoupled esters were obtained from the reaction between primary alcohols and cyclic secondary alcohols. On the other hand, the reaction of primary alcohols with 1-phenyl ethanol gave lower yield of corresponding cross-ester, due to the fast dehydrogenation of 1-phenyl ethanol to acetophenone.

Electron-rich PNN- and PNP-Ru borohydride complexes have been shown to catalyse the acceptorless dehydrogenative coupling of primary alcohols to esters under mild base-free neutral conditions (Scheme 64) [79]. Among these, catalyst **2** is highly active and gives the high yields of ester products from primary alcohols with 1000 TON. The same catalyst provides good yields of lactone derivatives from a variety of diols through intramolecular dehydrogenation and cyclization processes.

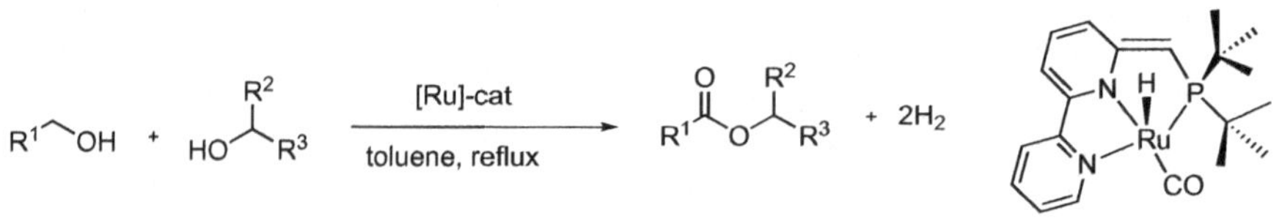

Scheme 63 Cross dehydrogenative coupling between primary and secondary alcohols

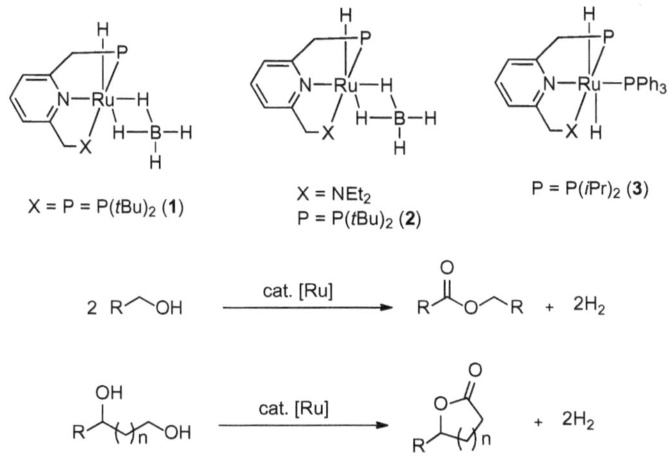

Scheme 64 PNN- and PNP-Ru borohydride complexes catalysed dehydrogenative coupling

Ethyl acetate is being widely used as a green solvent in both industry and academic researches. Apart from these, it is an important building block in the food industry and application in making inks, glues, perfumes, etc. Either acetic acid or acetaldehyde is a precursor to make ethyl acetate in bulk scale. These precursors are derived primarily from fossil fuels, and hence, it is limited. The conversion of ethanol to ethyl acetate can be achieved using heterogeneous catalysts. However, it requires higher temperatures, and yields will be up to only 56% [80]. Thus, Beller's group has demonstrated a protocol to prepare ethyl acetate in bulk scale from ethanol which is cheap and readily accessible from biomass [81]. Under neat and mild reaction conditions, ethanol undergoes acceptorless dehydrogenative coupling to produce ethyl acetate in the presence of ruthenium-based PNP pincer catalyst (Scheme 65). High yield of ethyl acetate was obtained with TON of 15,400 by performing the reaction using only 50 ppm of Ru catalyst and 0.6 mol% of NaOEt at 90 °C. Using the same catalytic system, 1-octanol could also be employed to prepare the corresponding esters in good yields.

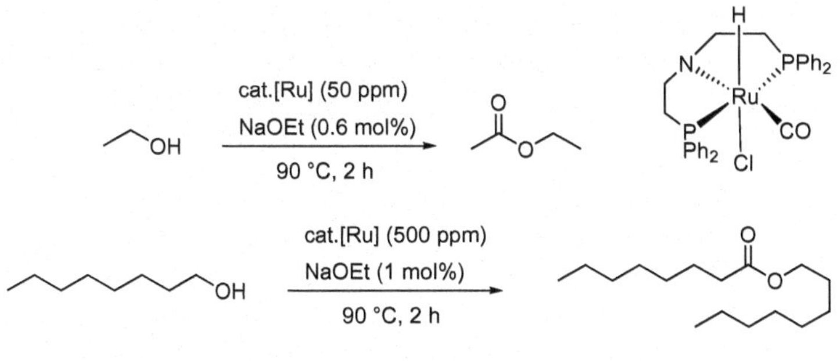

Scheme 65 Ruthenium-based PNP pincer complex catalyzed ADC

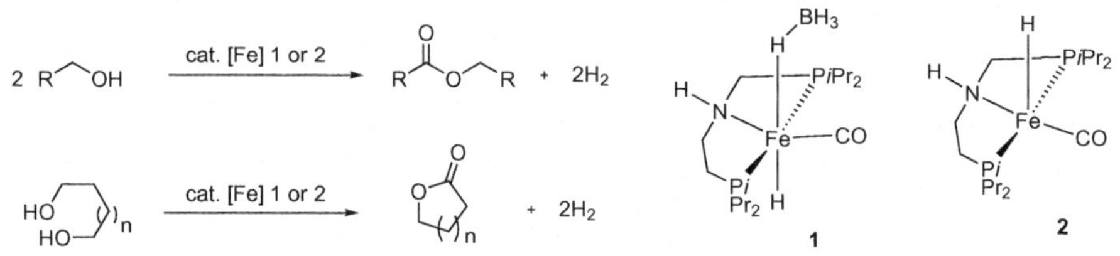

Scheme 66 Iron-catalysed inter- and intramolecular dehydrogenative coupling of alcohols

A well-defined and inexpensive iron complex supported by a cooperating PNP pincer ligand was synthesized and explored its activity in dehydrogenation of alcohols by Jones and Schneider groups [82]. Since, it was realized that complex **2** might be an active catalyst Fe-complex was prepared for studying similar transformations. Primary alcohols and diols furnished the respective esters and lactones in good yields in the presence of Fe complex (Scheme 66).

The selective formation of esters and acetals from primary alcohols has been shown by using Ru complex bearing a proton-responsive ligand such as 2-hydroxy pyridine (Scheme 67) [83]. Ru complex could catalyses the conversion of benzyl alcohol to acetal more selectively in the absence of base, albeit in low yield by following outer sphere mechanism (Scheme 68). On the other hand, the same reaction resulted in formation of benzyl benzoate under basic conditions via inner sphere mechanism (Scheme 69). This catalysis takes place through following metal–ligand-cooperating mechanistic pathway.

Aliphatic PNP pincer-supported earth-abundant manganese(I) complex has been demonstrated to catalyse the acceptorless dehydrogenative coupling of primary alcohols to esters under base-free conditions (Fig. 14) [84]. Good-to-excellent yields of esters from aliphatic linear and benzylic alcohols were obtained while

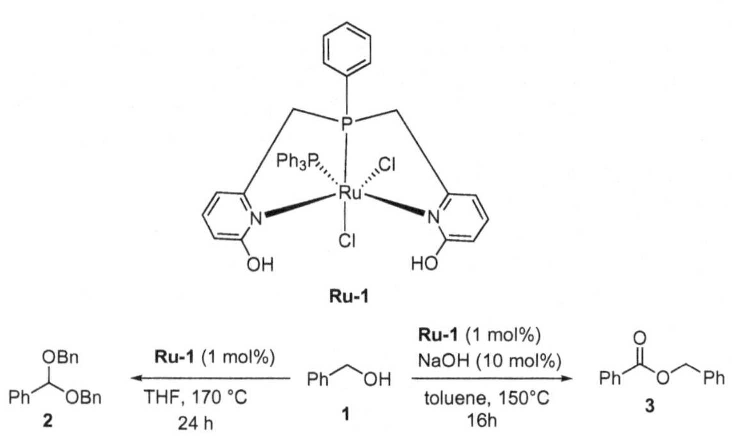

Scheme 67 Selective formation of esters and acetals from primary alcohols under Ru-catalysis

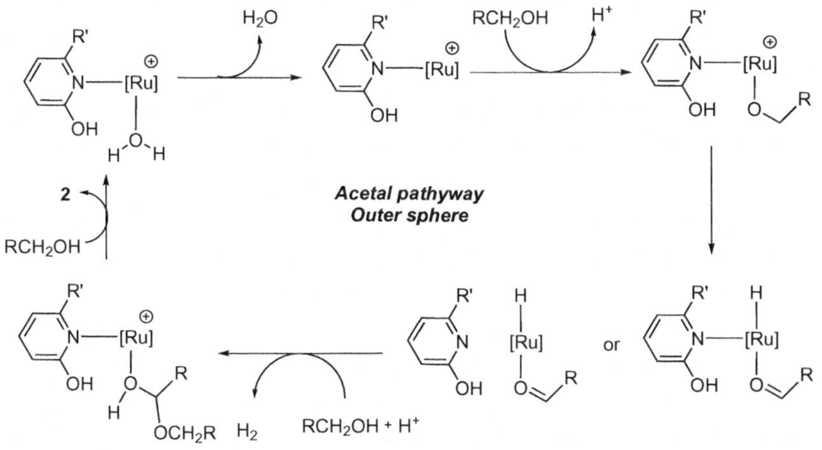

Scheme 68 Mechanism for acetal formation

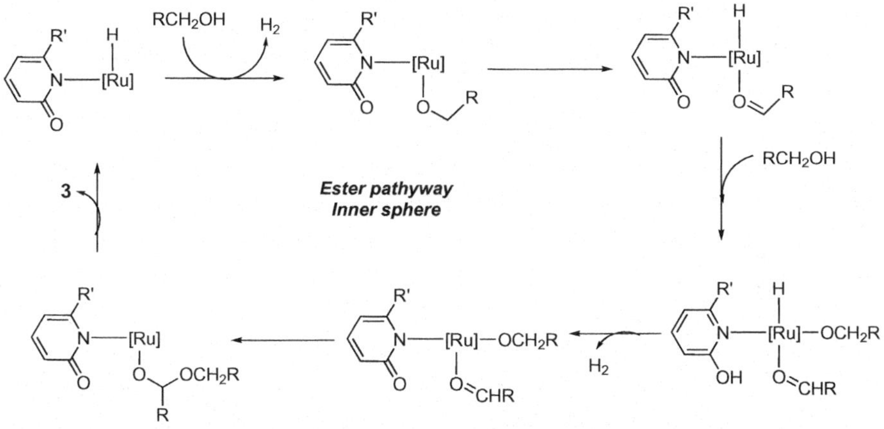

Scheme 69 Mechanism for ester formation

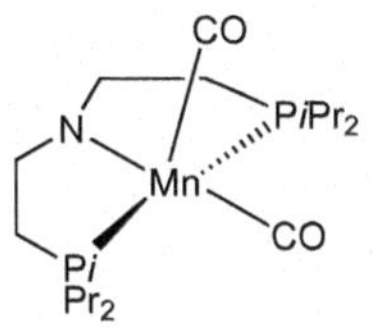

Fig. 14 PNP pincer-manganese(I) complex

conducting the experiments in the presence of 0.6 mol% Mn complex at 150 °C under neat conditions. For lighter alcohols such as 1-propanol and 1-butanol, the reaction temperature has to be maintained 110 and 130 °C to get the satisfactory yields of the corresponding esters after 72 h. Under this catalytic system, branched aliphatic alcohols give lower yields of the esters than their corresponding linear aliphatic alcohols. For instance, 2-methylpropan-1-ol afforded isobutyl isobutyrate in 70% yield which is 19% lower than that obtained with the linear isomer (n-butanol). This catalytic cycle proceeds through the outer sphere mechanism for dehydrogenative of alcohols via metal–ligand cooperative concept.

2 Conclusions

To conclude, a wide variety of both noble and non-noble transition metal complexes have been shown to act as catalysts for effective coupling of alcohols under mild reaction conditions. In addition, under certain conditions, a transition metal-free protocol is also functional with good-to-excellent yields. The ligand plays a major role in controlling the selectivity of the reactions. Metal complexes bearing proton-responsive ligand could effectively catalyse the alcohol coupling reactions through the metal–ligand cooperation (MLC) process. Based on the control experiments, it is demonstrated that CDC reaction under noble metal catalysis follows the hydrogen-borrowing pathway, whereas non-noble metal complexes proceed through Meerwein–Pondorf–Verley–Oppenauer (MPV-O) redox pathway to form the final alcohol-coupled products. It has been observed that the reduction of α,β-unsaturated ketone is faster than the reduction of final ketone under the reaction conditions. Tischenko-type disproportionation reaction of aldehyde can be ruled out for the acceptorless dehydrogenative coupling of alcohols to esters. Applications of this CDC protocol have shown by synthesizing a different type of heterocyclic compounds, longer-chain alcohols, hydrocarbons, fragrances, etc. Though a large variety of metal complexes have been shown for the effective coupling of alcohols, the corresponding asymmetric version is not known in the literature. Hence, there is more scope in designing and development of metal complexes for the highly challenging asymmetric version of alcohols cross-dehydrogenative coupling reactions.

References

1. (a) Corma A, Navas J, Sabater MJ (2018) Chem Rev 118:1410–1459 and references cited therein. (b) Chelucci G (2017) Coord Chem Rev 331:1–36. (c) Huang F, Liu ZQ, Yu ZK (2016) Angew Chem Int Ed 55:862–875. (d) Gunanathan C, Milstein D (2013) Science 341:1229712. (e) Bähn S, Imm S, Neubert L, Zhang M, Neumann H, Beller M (2011) ChemCatChem 3:1853–1864

2. (a) Matsu-ura T, Sakaguchi S, Obora Y, Ishi Y (2006) J Org Chem 71:8306–8308. (b) Gregorio G, Pregaglia GF, Ugo R (1972) J Organometal Chem 37:385–387. (c) Veibel S, Nielsen JI (1967) Tetrahedron 23:1723–1733. (d) Guerbet, M (1909) C R Hebd Seances Acad Sci 149:129–132. (e) Guerbet M (1899) C R Hebd Seances Acad Sci 128:1002–1004
3. Cho CS, Kim BT, Kim H-S, Kim T-J, Shim SC (2003) Organometallics 22:3608–3610
4. Cho CS, Ren WX, Shim SC (2005) Bull Korean Chem Soc 26:1611–1613
5. Fujita K-I, Asai C, Yamaguchi T, Hanasaka F, Yamaguchi R (2005) Org Lett 7:4017–4019
6. Adair GRA, Williams JMJ (2005) Tetrahedron Lett 46:8233–8235
7. (a) Martinez R, Ramon DJ, Yus M (2006) Tetrahedron 62:8988–9001. (b) Martinez R, Ramon DJ, Yus M (2006) Tetrahedron 62:8982–8987
8. Viciano M, Sanaú M, Peris E (2007) Organometallics 26:6050–6054
9. Prades A Viciano M Sanau M, Peris E (2008) Organometallics 27:4254–4259
10. Viciano M, Sanaú M, Merino MS, Tejeda J, Peris E, Royo B (2008) Organometallics 27:1305–1309
11. Costa AP, Sanaú M, Peris E, Roy B (2009) Dalton Trans 6960–6966
12. (a) Ruiz-Botella S, Peris E (2015) Chem Eur J 21:15263–15271. (b) Ibañez S, Poyatos M, Peris E (2016) Dalton Trans 45:14154–14159
13. Gnanamgari D, Leung CH, Schley ND, Hilton ST, Crabtree RH (2008) Org Biomol Chem 6:4442–4445
14. Cheung HW, Lee TY, Lui HY, Yeung CH, Lau CP (2008) Adv Synth Catal 350:2975–2983
15. Gnanamgari D, Sauer ELO, Schley ND, Butler C, Incarvito CD, Crabtree RH (2009) Organometallics 28:321–325
16. Allen LJ, Crabtree RH (2010) Green Chem 12:1362–1364
17. Shimizu K-I, Sato R, Satsuma A (2009) Angew Chem Int Ed 48:3982–3986
18. (a) Liu X, Ding R-S, He L, Liu Y-M, Cao Y, He H-Y, Fan K-N (2013) ChemSusChem, 6:604–608. (b) Wang D, Guo X-Q, Wang C-X, Wang Y-N, Zhong R, Zhu X-H, Cai L-H, Gao Z-W, Hou X-F (2013) Adv Synth Catal 355:1117–1125. (c) Xu J, Yue H, Liu S, Wang H, Du Y, Xu C, Dong W, Liu C (2016) RSC Adv 6:24164–24174
19. Kose O, Saito S (2010) Org Biomol Chem 8:896–900
20. Segarra C, Mas-Marzá E, Mata JA, Peris E (2011) Adv Synth Catal 353:2078–2084
21. Xu C, Goh LY, Pullarkat SA (2011) Organometallics 30:6499–6502
22. Xu C, Chuan LW, Yongxin L, Pullarkat SA (2012) Tetrahedron Lett 53:1450–1455
23. Gong X, Zhang H, Li X (2011) Tetrahedron Lett 52:5596–5600
24. Yang J, Liu X, Meng D-L, Chen H-Y, Zong Z-H, Feng T-T, Sun K (2012) Adv Synth Catal 354:328–334
25. Tanga G, Cheng C-H (2011) Adv Synth Catal 353:1918–1922
26. Miura T, Kose O, Li F, Kai S, Saito S (2011) Chem Eur J 17:11146–11151
27. Liao S, Yu K, Li Q, Tian H, Zhang Z, Yu X, Xu Q (2012) Org Biomol Chem 10:2973–2978
28. Xu Q, Chen J, Liu Q (2013) Adv Synth Catal 355:697–704
29. Jumde VR, Gonsalvi L, Guerriero A, Peruzzini M, Taddei M (2015) Eur J Org Chem 1829–1833
30. Satyanarayana P, Reddy GM, Maheswaran H, Kantam ML (2013) Adv Synth Catal 355:1859–1867
31. Jiménez MV, Fernández-Tornos J, Modrego FJ, Pérez-Torrente JJ, Oro LA (2015) Chem Eur J 21:17877–17889
32. Chang W, Gong X, Wang S, Xiao L-P, Song G (2017) Org Biomol Chem 15:3466–3471
33. Musa S, Ackermann L, Gelman D (2013) Adv Synth Catal 355:3077–3080
34. Wang R, Ma J, Li F (2015) J Org Chem 80:10769–10776
35. (a) Chakrabarti K, Paul B, Maji M, Roy BC, Shee S, Kundu S (2016) Org Biomol Chem 14:10988–10997. (b) Roy BC, Chakrabarti K, Shee S, Paul B, Kundu S (2016) Chem Eur J 22:18147–18155. (c) Roy BC, Debnath S, Chakrabarti K, Paul B, Maji M, Kundu S (2018) Org Chem Front 5:1008–1018
36. Shee S, Paul B, Panja D, Roy BC, Chakrabarti K, Ganguli K, Das A, Das GK, Kundu S (2017) Adv Synth Catal 359:3888–3893

37. Sahoo AR, Lalitha G, Murugesh V, Bruneau C, Sharma GVM, Suresh S, Achard M (2017) J Org Chem 82:10727–10731
38. Freitag F, Irrgang T, Kempe R (2017) Chem Eur J 23:12110–12113
39. Shi J, Hu B, Ren P, Shang S, Yang X, Chen D (2018) Organometallics 37:2795–2806
40. Tan D-W, Li H-X, Zhu D-L, Li H-Y, Young DJ, Yao J-L, Lang J-P (2018) Org Lett 20:608–611
41. Genç S, Günnaz S, Çetinkaya B, Gülcemal S, Gülcemal D (2018) J Org Chem 83:2875–2881
42. Makarov IS, Madsen R (2013) J Org Chem 78:6593–6598
43. Matsu-ura T, Sakaguchi S, Obora Y, Ishii Y (2006) J Org Chem 71:8306–8308
44. Riittonen T, Toukoniitty EK, Madnani D, Leino A-R, Kordas K, Szabo M, Sapi A, Arve K, Warna J, Mikkola J-P (2012) Catalyst 2:68–84
45. Koda K, Matsu-ura T, Obora Y, Ishii Y (2009) Chem Lett 38:838–839
46. (a) Dowson GRM, Haddow MF, Lee J, Wingad RL, Wass DF (2013) Angew Chem Int Ed 52:9005–9008. (b) Wingad RL, Gates PJ, Street STG, Wass DF (2015) ACS Catal 5:5822–5826
47. Chakraborty S, Piszel PE, Hayes CE, Baker RT, Jones WD (2015) J Am Chem Soc 137:14264–14267
48. Xie Y, Ben-David Y, Shimon LJW, Milstein D (2016) J Am Chem Soc 138:9077–9080
49. Fu S, Shao Z, Wang Y, Liu Q (2017) J Am Chem Soc 139:11941–11948
50. Kulkarni NV, Brennessel WW, Jones WD (2018) ACS Catal 8:997–1002
51. Li Y, Li HQ, Junge H, Beller M (2014) Chem Commun 50:14991–14994
52. Oikawa K, Itoh S, Yano H, Kawasaki H, Obora Y (2017) Chem Commun 53:1080–1083
53. Liu Q, Xu G, Wang Z, Liu X, Wang X, Dong L, Mu X, Liu H (2017) Chemsuschem 10:4748–4755
54. Cano R, Yus M, Ramon DJ (2012) Chem Commun 48:7628–7630
55. Mura MG, Luca LD, Taddei M, Williams JMJ, Porcheddu A (2014) Org Lett 16:2586–2589
56. Manojveer S, Salahi S, Wendt OF, Johnson MT (2018) J Org Chem 83:10864–10870
57. Obora Y, Anno Y, Okamoto R, Matsu-ura T, Ishii Y (2011) Angew Chem Int Ed 50:8618–8622
58. Manojveer S, Forrest SJK, Johnson MT (2018) Chem Eur J 24:803–807
59. Wang Y, Shao Z, Zhang K, Liu Q (2018) Angew Chem Int Ed 57:15143–15147
60. (a) Blum Y, Reshef D, Shvo Y (1981) Tetrahedron Lett 22:1541–1544. (b) Shvo Y, Blum Y, Reshef D, Menzin M (1982) J Organomet Chem 226:C21–C24. (c) Cottier L, Descotes G, Sabadie J (1980) J Mol Catal 7:337–348
61. Tamaru Y, Yamada Y, Inoue K, Yamamoto Y, Yoshida Z (1983) J Org Chem 48:1286–1292
62. Blum Y, Shvo Y (1985) J Organomet Chem 282:C7–C10
63. Murahashi S-I, Naota T, Ito K, Maeda Y, Taki H (1987) J Org Chem 52:4319–4327
64. Suzuki T, Morita K, Tsuchida M, Hiroi K (2002) Org Lett 4:2361–2363
65. Zhao J, Hartwig JF (2005) Organometallics 24:2441–2446
66. Zhang J, Leitus G, Ben-David Y, Milstein D (2005) J Am Chem Soc 127:10840–10841
67. Zhang J, Gandelman M, Shimon LJW, Milstein D (2007) J Chem Soc Dalton Trans 107–113
68. Gunanathan C, Shimon LJW, Milstein D (2009) J Am Chem Soc 131:3146–3147
69. Kossoy E, Diskin-Posner Y, Leitus G, Milstein D (2012) Adv Synth Catal 354:497–504
70. Owston NA, Parker AJ, Williams JMJ (2008) Chem Commun 624–625
71. Zweifel T, Naubron J-V, Grützmacher H (2009) Angew Chem Int Ed 48:559–563
72. (a) Arita S, Koike T, Kayaki Y, Ikariya T (2008) Chem Asian J 3:1479–1485. (b) Izumi A, Obora Y, Sakaguchi S, Ishii Y (2006) Tetrahedron Lett 47:9199–9201
73. (a) Kaizuka K, Miyamura H, Kobayashi S (2010) J Am Chem Soc 132:15096–15098. (b) Xu B, Haubrich J, Freyschlag CG, Madix RJ, Friend CM (2010) Chem Sci 1:310–314. (c) Su F-Z, Ni J, Sun H, Cao Y, He H-Y, Fan K-N (2008) Chem Eur J 14:7131–7135. (d) Nielsen IS, Taarning E, Egeblad K, Madsen R, Christensen CH (2007) Catal Lett 116:35–40
74. Gowrisankar S, Neumann H, Beller M (2011) Angew Chem Int Ed 50:5139–5143
75. Liu C, Wang J, Meng L, Deng Y, Li Y, Lei A (2011) Angew Chem Int Ed 50:5144–5148

76. Luo F, Pan C, Cheng J, Chen F (2011) Tetrahedron 67:5878–5882
77. Sølvhøj A, Madsen R (2011) Organometallics 30:6044–6048
78. Srimani D, Balaraman E, Gnanaprakasam B, Ben-David Y, Milstein D (2012) Adv Synth Catal 354:2403–2406
79. Zhang J, Balaraman E, Leitus G, Milstein D (2011) Organometallics 30:5716–5724
80. (a) Wang L, Zhu W, Zheng D, Yu X, Cui J, Jia M, Zhang W, Wang Z (2010) React Kinet Mech Catal 101:365. (b) Inui K, Kurabayashi T, Sato S, Ichikawa N (2014) J Mol Catal A 216:147. (c) Zonetti PC, Celnik J, Letichevsky S, Gaspar AB, Appel LG (2011) J Mol Catal A 334:29. (d) Medeiros PRS, Eon JG, Appel LG (2000) Catal Lett 69:79
81. Nielsen M, Junge H, Kammer A, Beller M (2012) Angew Chem Int Ed 51:5711–5713
82. Chakraborty S, Lagaditis PO, Förster M, Bielinski EA, Hazari N, Holthausen MC, Jones WD, Schneider S (2014) ACS Catal 4:3994–4003
83. Sahoo AR, Jiang F, Bruneau C, Sharma GVM, Suresh S, Roisnel T, Dorcet V, Achard M (2017) Catal Sci Technol 7:3492–3498
84. Nguyen DH, Trivelli X, Capet F, Paul J-F, Dumeignil F, Gauvin R (2017) M ACS Catal 7:2022–2032

Synthesis and Functionalization of S-heterocycles via CDC Reactions

Nadale K. Downer-Riley

Abstract Cross-dehydrogenative coupling has become a powerful tool in organic synthesis. The current developments in the assembly and functionalization of S-heterocycles using this strategy are highlighted in this chapter. The preparation of benzothiophenes, benzothiazoles and cyclic sultams and sulfoxides by intramolecular C(sp^2)–H/S–H and C(sp^2)–H/C(sp^2)–H coupling is discussed. In addition, the coupling of S-heterocycles to arenes and aliphatic compounds is also presented.

Keywords Benzothiazole · Benzothiophene · Cross-dehydrogenative coupling · Cyclic sulfoxide · Cyclic sultam

1 Introduction

Sulfur-containing heterocyclic compounds have maintained widespread interest among researchers due to their presence in many natural products as well as their applications in the fields of drug development, agriculture and material science. For example, famotidine (heartburn relief), raloxifene (osteoporosis treatment), meticrane (diuretic) and tazarotene (acne treatment) are currently utilized as pharmaceutical drugs while thiabendazole (antifungal and antihelminthic), isofetamid (antifungal) and endosulfan (insecticidal) are used in crop protection (Fig. 1).

Recently, there has been renewed interest in the development of synthetic methodologies for S-containing compounds which, as a group, encompass a wide diversity of structures. Typically, the formation of the carbon-sulfur bond from cross-coupling reactions between halides and thiols is seen. Although many of these synthetic methods represent noteworthy approaches toward these structures, in this chapter, we will focus on the preparation of S-heterocycles using cross-dehydrogenative coupling reactions. Cross-dehydrogenative coupling

N. K. Downer-Riley (✉)
Department of Chemistry, The University of the West Indies, Mona, Kingston 7, Jamaica
e-mail: nadale.downer02@uwimona.edu.jm

A. Srivastava and C. K. Jana (eds.), *Heterocycles via Cross Dehydrogenative Coupling*, https://doi.org/10.1007/978-981-13-9144-6_8

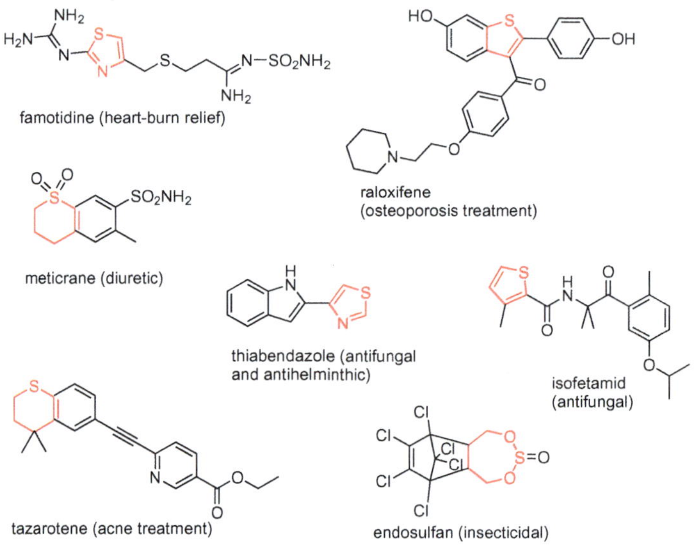

Fig. 1 Examples of useful S-heterocyclic compounds

(CDC) reactions have become a powerful synthetic tool for the construction of carbon–carbon and carbon–heteroatom bonds with the advantages of higher atom economy and reduced synthetic steps since prefunctionalization for the sole purpose of activation of reaction sites is avoided. However, successful reports of these reactions with respect to S-heterocycles are not as abundant as other heterocyclic motifs. This may be attributed to the fact that many synthetic approaches for CDC reactions utilize transition metals as catalysts. Divalent sulfur has strong coordination properties and so there is the potential for some metal catalysts to be deactivated by the formation of strong and sometimes irreversible S-metal bonds [29]. Even with these challenges, many successful approaches utilizing reagents such as palladium salts or mild oxidants have been employed to assemble S-heterocycles, such as thiophenes and benzothiazoles, via cross-dehydrogenative coupling reactions.

2 Synthesis of S-heterocycles

2.1 Synthesis of S-heterocycles via an Intramolecular C(sp^2)–H/S–H Coupling

There have been quite a few reports of the assembly of S-heterocycles via C–H/S–H activation especially with regards to the preparation of benzothiophenes using palladium salts. For example, in a straightforward approach, aryl ethenethiols were

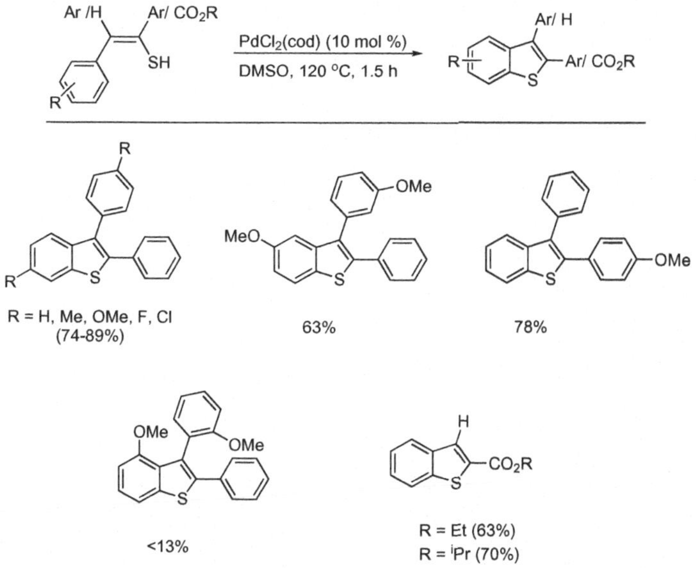

Scheme 1 Conversion of aryl ethenethiols to benzothiophenes

converted to the corresponding benzothiophenes in good yield using $PdCl_2$(cod) in DMSO as shown in Scheme 1 [23]. It is worth noting that this reaction did not show appreciable conversion in solvents other than DMSO nor when an additive oxidant, such as $Cu(OAc)_2$, was present. In addition, while the reaction conditions are tolerated by various functional groups, the presence of *ortho*-methoxy groups on geminal aryl substituents appears to sterically hinder the reaction.

Based on mechanistic investigations, it was suggested that the thiol is first converted to the disulfide based on the fact that the disulfide is observed as a minor by-product. Additionally, heating the thiol in DMSO only resulted in some conversion to the disulfide which, on isolation, could be converted to the benzothiophene on exposure to a palladium catalyst. Insertion of palladium between the weak S–S bond of the disulfide and subsequent electrophilic attack of the aryl ring then gives the 6-membered palladacycle while regenerating a molecule of the thiol. Reductive elimination then provides the benzothiophene as shown in Scheme 2.

Similarly, it was found that palladium chloride in DMSO was ideal for cyclization of 2-biphenyl sulfides to dibenzothiophenes [44, 79]. The robustness of this method was demonstrated by the good yields (85–95%) obtained using substrates with both electron-withdrawing and electron-donating groups (Scheme 3). Zhang and coworkers proposed a similar mechanism to that in Scheme 2 for the transformation of the phenyl sulfide to the dibenzothiophene after conducting experiments which also confirmed the disulfide to be a likely intermediate. Furthermore, Noshino, Ogiwara and Sakai reported several examples of aryl

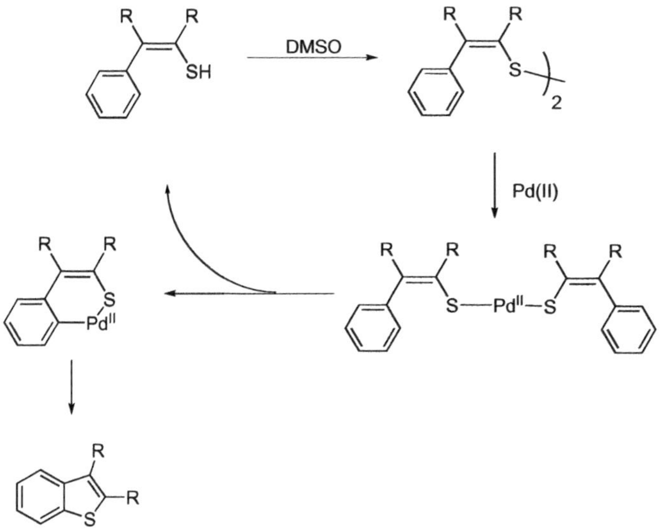

Scheme 2 Mechanism for the cyclization of aryl ethenethiols

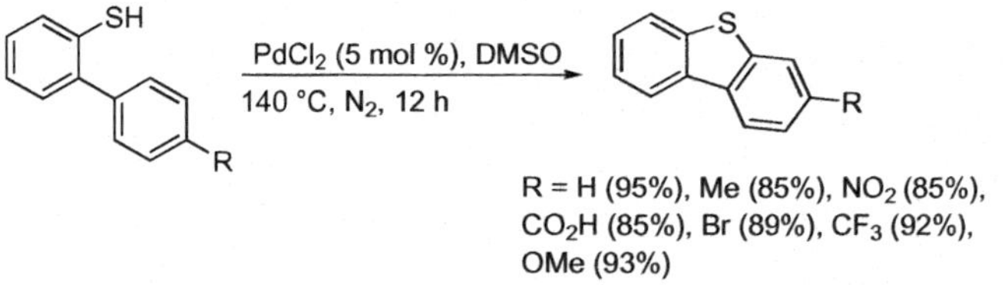

Scheme 3 Selected examples for the conversion of 2-biphenyl sulfides to dibenzothiophenes

disulfides being converted to the corresponding dibenzothiophene under similar reaction conditions.

While the atom economy of CDC approaches is attractive, the execution of such conversions using a metal-free approach is always an added advantage. Nishino et al. [43] demonstrated a facile approach to dibenzothiophenes from 2-biphenyl sulfides using molecular iodine which is known to be environmentally safe, economical and easily handled. Although their investigation focused on the cyclization of disulfides using iodine in the presence of sodium carbonate, they reported excellent transformation of 2-phenylbenzenethiol to dibenzothiophene using this protocol in which the sulfenyl iodide is a likely reaction intermediate (Scheme 4).

Interestingly, in an attempt to prepare dihydrobenzothiophene and thiochromans, thioacetates were employed as masked thiols to allow for cyclization in the presence of a Pd(II) catalyst [9]. For this conversion, it was revealed that the free thiol was an unlikely intermediate as treating 2-phenylethanethiol to the standard conditions provided the expected product in only small quantities (Scheme 5). These results

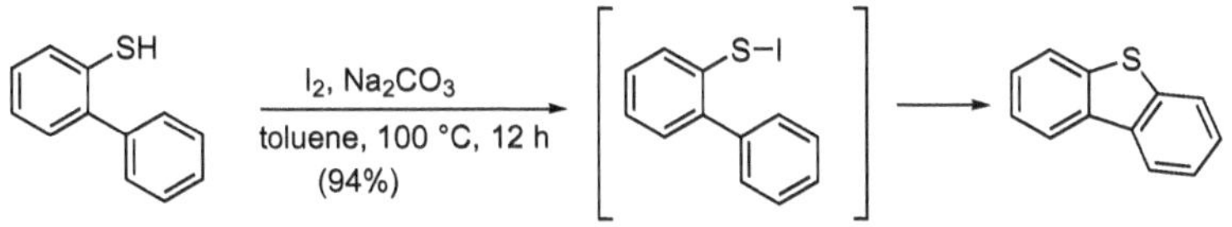

Scheme 4 Conversion of 2-biphenyl sulfides to dibenzothiophenes using iodine

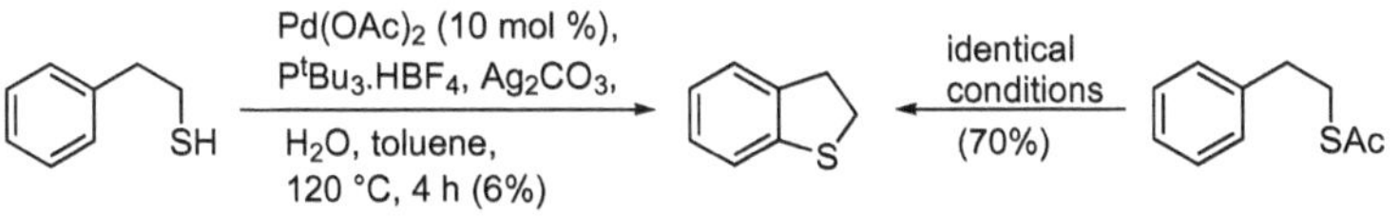

Scheme 5 Conversion of 2-phenylethanethiol to dihydrobenzothiophene

highlight the challenges presented by the presence of the free S–H moiety which seems to have deactivated the palladium catalyst.

Unlike most S-heterocycles, synthetic methodologies for the preparation of benzothiazoles have received a lot of attention [47, 75]. This may be due to their prevalence in biologically and medicinally important compounds such as vitamin B_1 (thiamine), the anticonvulsant drug riluzole and the anticancer agent Phortress. In several instances, thiobenzanilides or phenyl thioureas have served as substrates for these reactions which have utilized various oxidants including molecular iodine, bromine, PIFA or transition metal salts of Pd(II), Fe(III), Mn(III) or Ce(IV) to name a few. These reactions proceed via various intermediates inclusive of the thiol radical, thioimidic cation or aryl radical cation with the end result being a cross dehydrogenation to produce the benzothiazole (Scheme 6).

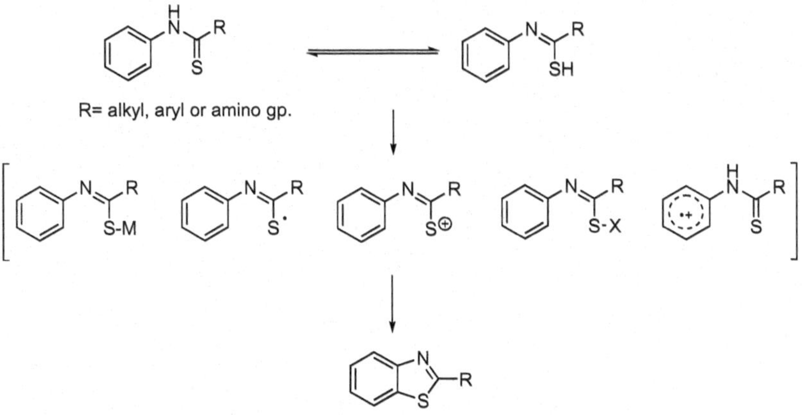

Scheme 6 Cyclization of thioanilides to benzothiazoles

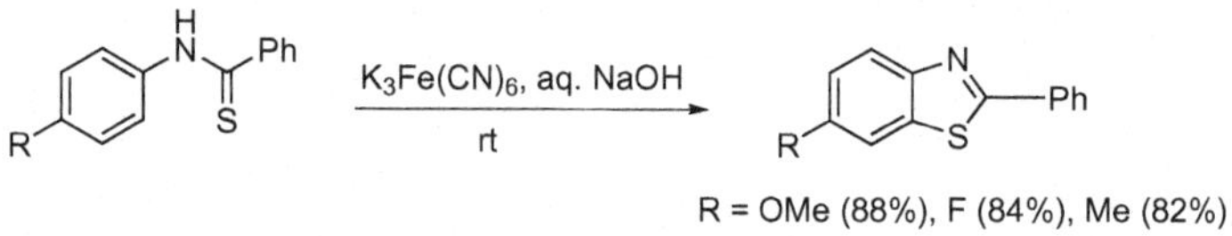

Scheme 7 Selected examples of Jacobson's cyclization [64]

The most common approach for the cyclization of thioanilides to benzothiazoles involves the use of potassium ferricyanide under basic conditions, commonly referred to as Jacobson's cyclization (Scheme 7). There have been two mechanisms proposed for this reaction, one which proceeds via the thiol radical [59] and the other via the thioimidic cation [37]. This reaction, however, is not useful neither when strongly deactivating groups are present on the primary ring nor when there are base-sensitive groups present on the molecule [24].

The thiol radical is also thought to be an intermediate for similar reactions mediated by DDQ [4], Dess–Martin periodinane [2] and Manganese triacetate in the presence of microwave [42]. Like Jacobson's cyclization, these methods are low yielding when strongly electron-withdrawing nitro groups are present. One noteworthy addition to the cadre of reagents which can be used for the cyclization of thioanilides is the combination of sodium persulfate and catalytic ferric chloride in dimethylsulfoxide and pyridine [66]. Unlike those reagents previously mentioned, this system effects cyclization even in the presence of electron-withdrawing groups on the primary ring.

The common reagent for the conversion of phenylthioureas to 2-aminobenzo thiazoles is molecular bromine which is sometimes replaced by tribromide salts [26], NBS [19] or hydrobromic acid [51]. These reactions are known to proceed via a sulfenylbromide which is readily attacked by the aromatic ring to yield the product. When dithiobiurets are used as substrates for these reactions, however, they are preferentially converted to the dithiazoles as opposed to the corresponding benzothiaozle even when the aryl ring is electron rich [61] (Scheme 8).

The cyclization of thioanilides using organic ammonium tribromide salts (OATB), a useful alternative to liquid bromine, has been reported [40]. The use of

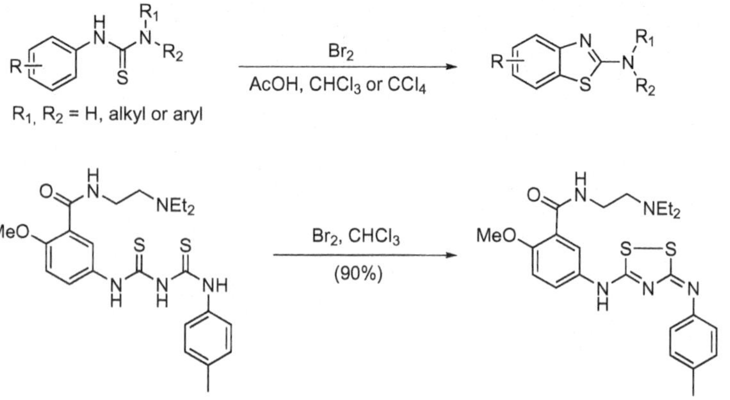

Scheme 8 Bromine-mediated cyclizations

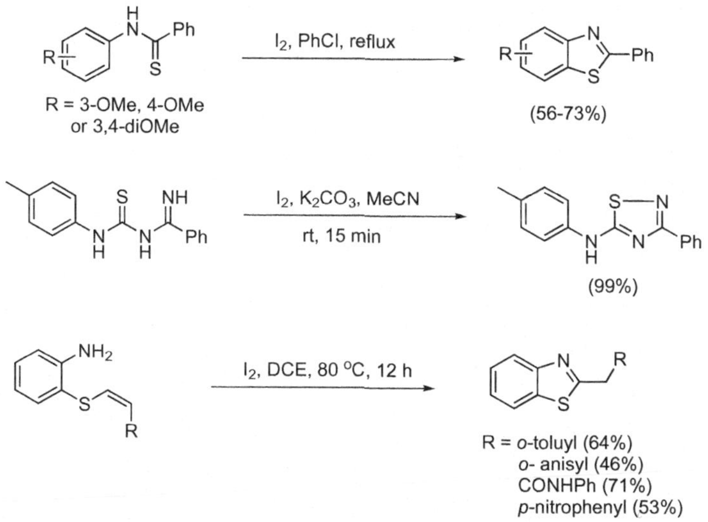

Scheme 9 Iodine-mediated cyclizations

iodine in the cyclization of thiobenzanilides to give 2-arylbenzothiazoles is analogous to bromine-mediated cyclization of phenylureas and is a useful alternative as bromination of the primary aromatic ring is a competing reaction when bromine or NBS is used [13]. Iodine has also been used in an analogous reaction to prepare benzothiazoles from 2-(styrylthio)anilines via C(sp^2)–H/N–H coupling [80] (Scheme 9). It is interesting to note that under the conditions of iodine in the presence of bicarbonate imidoylthioureas are converted to the corresponding thiadiazole indicating that its formation is more favorable than that of the thiazole [67, 68]. A similar transformation can also be achieved using phenyliodine(III) (bis) trifluoroacetate [36].

Transition metal salts of palladium and copper have been used with better success for the cyclization of thioanilides and thioureas [25, 27, 54]. In general, these reactions are efficient even in the presence of electron-withdrawing groups while being mild enough to be used when groups prone to hydrolysis, such as a methyl carboxylate group, are present. One such example is the combination of $PdCl_2$ and CuI in the presence of tetrabutylammonium bromide (Bu_4NBr) which is useful for the cyclization of both thiobenzanilides and thioureas (Scheme 10) [24]. Both $PdCl_2$ (10 mol%) and CuI (50 mol%) are used in catalytic amounts, and there is no conversion in the absence of Bu_4NBr. It was therefore proposed that both the CuI and Bu_4NBr are involved in the reoxidation of Pd(0) in the catalytic cycle.

The use of cerium(IV) ammonium nitrate (CAN) or phenyliodine(III) bis (trifluoroacetate) is thought to provide a unique pathway toward benzothiazoles via an aryl radical cation [14]. Like Pd(II)/Cu(I) catalyzed reactions, these reaction conditions are mild and tolerate strongly electron-withdrawing groups such as the methyl carboxylate substituent (Scheme 11).

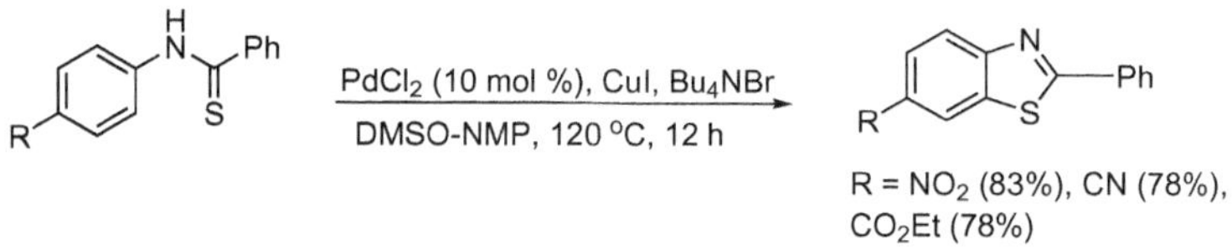

Scheme 10 Cyclization of thiobenzanilides by Pd(II)/Cu(I)

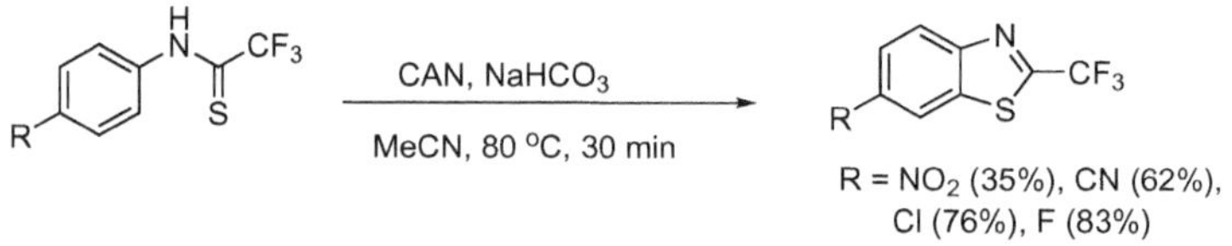

Scheme 11 Examples of CAN-mediated cyclization of thioanilides [82, 83]

The use of oxidants and proton acceptors in many of the reactions reported is sacrificial and produces unwanted and wasteful by-products from the reactions. Furthermore, these reactions tend to be carried out at high temperatures. A warranted development in dehydrogenative cross-coupling reactions is achieving conversions under oxidant-free conditions. In that regard, photocyclizations have been investigated. The ability of $Ru(bpy)_3^{2+}$ to act as a photosensitizer was established decades ago [5–7]. Since then, researchers have utilized this catalyst in the visible-light catalyzed preparation of benzothiazoles via an intramolecular cross-coupling [11, 78]. These reactions employ stoichiometric quantities of molecular oxygen as the effective oxidant and generate hydrogen gas as a by-product. A typical protocol involves irradiation at room temperature in the presence of 1–3 mol% $Ru(bpy)_3(PF_6)_2$, 1 molar equivalent of base, acetonitrile or DMF as solvent, and in some instances, a catalytic amount of cobaloxime is also used which serves as the proton-reducing catalysts. While this protocol tolerates both electron-withdrawing and electron-donating functionalities on the primary ring, dehalogenation occurs when halide groups are present. The mechanism for this transformation involves deprotonation of the thioamide to give the thiolate anion which is oxidized to the thiol radical by $Ru(bpy)_3^{3+}$-generated from oxidation of $Ru(bpy)_3^{2+}$ by molecular oxygen in the presence of light (Scheme 12). Cyclization of the thiol radical and rearomatization, aided by the oxygen radical anion generated in the previous step, then yields the benzothiazole [11]. Photo-induced cyclization of thioanilides has also been reported using chloranil as photosensitizer and hydrogen acceptor [53].

A simpler ruthenium catalytic system, $RuCl_3$ and oxone, has also been used to successfully effect the cyclization of *N*-arylthioureas to 2-aminobenzothiazoles [58]. The photoredox mechanism in which the thiol radical is generated as an intermediate was ruled out as free radical scavengers had no effect on the reaction outcome. The mechanism shown in Scheme 13 was therefore proposed in which S-ruthenation of the thiourea generates a coordination complex subsequent to which

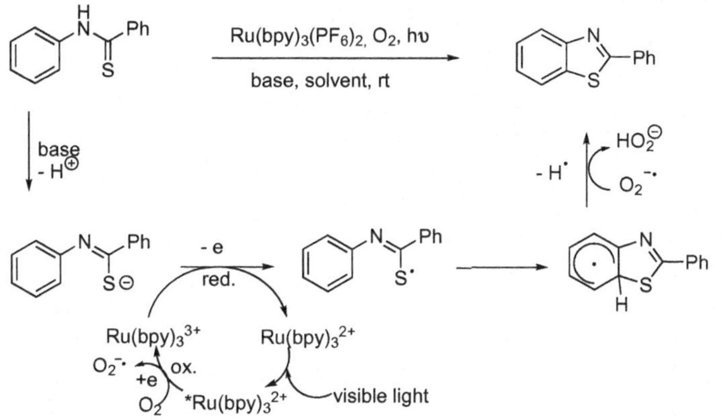

Conditions
a: $Ru(bpy)_3(PF_6)_2$ (1 mol %), hυ, DBU, 5% O_2, DMF, rt (84%) - Cheng et al. 2012
b: $Ru(bpy)_3(PF_6)_2$ (3 mol %), Co(III) cat. (8 mol %), sodium-Gly, CH_3CN, blue LED, Ar, 12 h (99%) - Zhang et al. 2015

Scheme 12 Photo-catalyzed cyclization of thioanilides

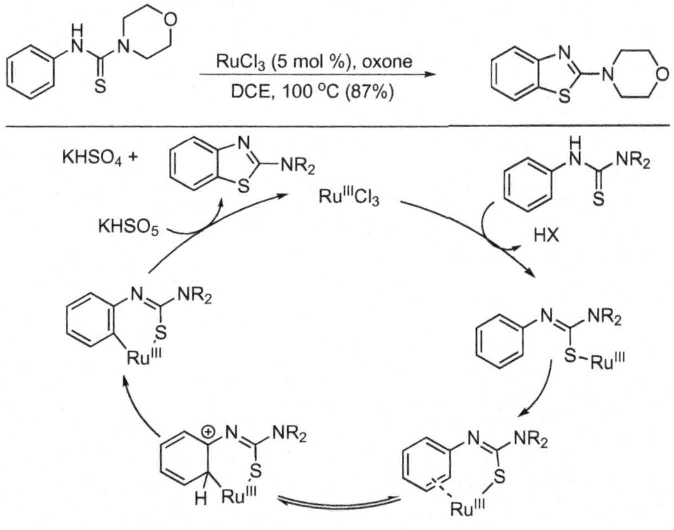

Scheme 13 Ru(III)-catalyzed cyclization of *N*-arylthioureas

the π-complex is formed. Electrophilic insertion to produce the cycloruthenated intermediate and subsequent reductive elimination then gives the benzothiazole product.

Qian et al. [48] have developed a very environmentally friendly electrochemical cross-dehydrogenative coupling reaction toward benzothiazoles and thiazolopyridines

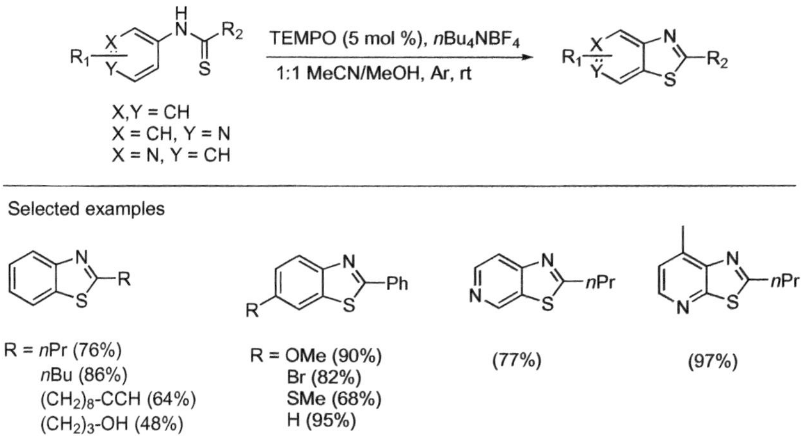

Scheme 14 TEMPO-catalyzed cyclization of thioamides

using 2,2,6,6-tetramethylpiperidine-*N*-oxyl radical (TEMPO) as catalyst. While the usefulness of this reaction was demonstrated with the efficient conversion of several thioamide substrates bearing a variety of functional groups such as alkynes, esters and alcohols, those bearing nitro substituents decomposed under the reaction conditions (Scheme 14). Recently, there have even been reports from this research team of these reactions being carried out with just electricity [16]. There have also been reports of the thermal cyclization of thioanilides to benzothiazoles by heating in nitrobenzene [1].

The DMP-mediated intramolecular cyclization of *o*-aminothiophenol Schiff bases to generate benzothiazoles has also been reported [3]. It is likely that under the reaction conditions, DMP dehydrogenates the thiazoline intermediate as shown in Scheme 15. $Pd(OAc)_2$ in imidazolium ionic liquids has also been used for this transformation [28].

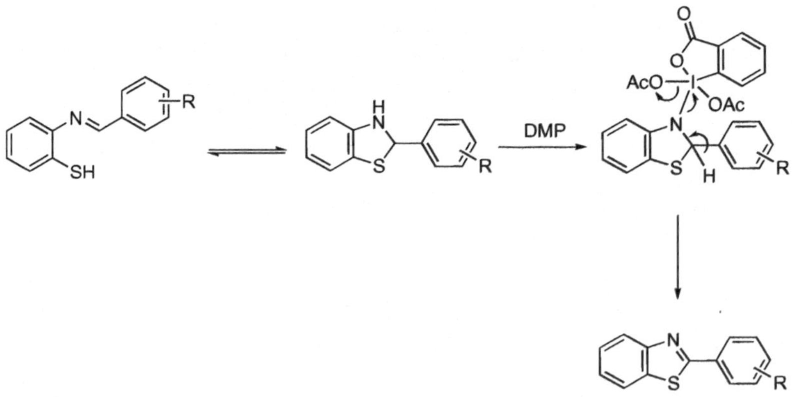

Scheme 15 Cyclization of *o*-aminothiophenol Schiff bases using Dess–Martin periodinane (DMP)

Scheme 16 Cyclization of diphenyl sulfide

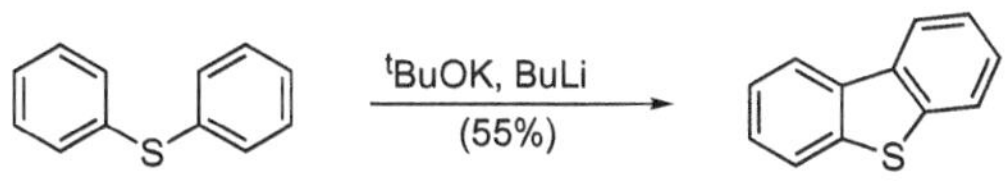

2.2 *Synthesis of S-heterocycles via an Intramolecular C(sp^2)–H/C(sp^2)–H Coupling*

It is only recently that there has been significant synthetic interest in the assembly of S-heterocycles via the intramolecular dehydrogenative coupling of two C–H bonds. One early reaction in which dibenzothiophene was prepared from the cyclization of diphenyl sulfide under harshly basic conditions of tBuLi/tBuOK has not been proven to be synthetically useful as the methodology was only suitable for the construction of unsubstituted polycyclic thiophenes (Scheme 16) [45]. Under these reaction conditions, methoxy- and methyl-substituted diphenyl sulfides decomposed, and undoubtedly, many functional groups would not be able to tolerate such super basic conditions.

Another early report for the preparation of dibenzothiophenes from the corresponding diaryl sulfide utilized molecular iodine and light under an atmosphere of nitrogen [77]. This method has been proven to be useful in the synthesis of a benzothiophene required as a precursor to a polycyclic photoswitch, albeit in low yield (Scheme 17) [70]. The exploration of efficient methods for such transformations was therefore necessary.

Recently, transition metals, especially palladium salts, have risen to the forefront for these intramolecular dehydrogenative C–H/C–H cross-coupling reactions. Since a challenge in the construction of S-heterocycles resides in the poisoning effect of sulfides on transition metals, sulfones and sulfoxides have been employed in this respect [22, 30, 65]. The work carried out by Samanta and Antonchick was a breakthrough in the assembly of dibenzothiophenes from benzyl phenyl sulfoxides via double C–H activation using a Pd(II)/Ag(I) catalytic system [55]. This reaction was also noteworthy as it tolerated a wide range of functional groups including halides, trifluoromethyl, methoxy and methyl in the *meta* and *para* positions of the thiophenol ring (Scheme 18). While *ortho* substitution at the sulfoxide end of the molecule did not impede the reaction, *ortho* substitution at the thiophenol end led to only trace amounts of the desired product. Mechanistically, it is assumed that the sulfoxide moiety directed the first palladation to give the 5-membered palladacycle. Subsequent oxidative addition of aryliodide facilitates a reductive elimination and then another C–H activation to produce the 7-membered intermediate, which yields the cyclic sulfoxide after reductive elimination. The cyclic sulfoxide is not isolated

Scheme 17 Preparation of benzothiophene via iodine-promoted cyclization

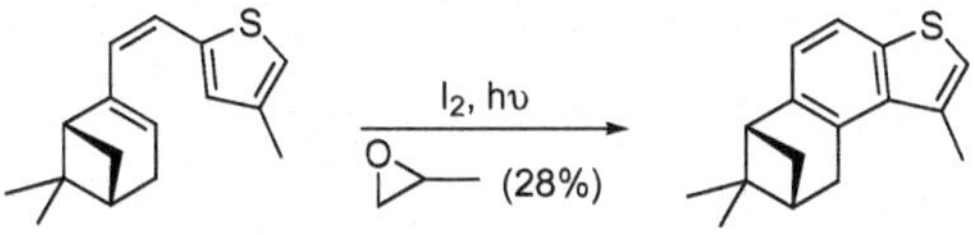

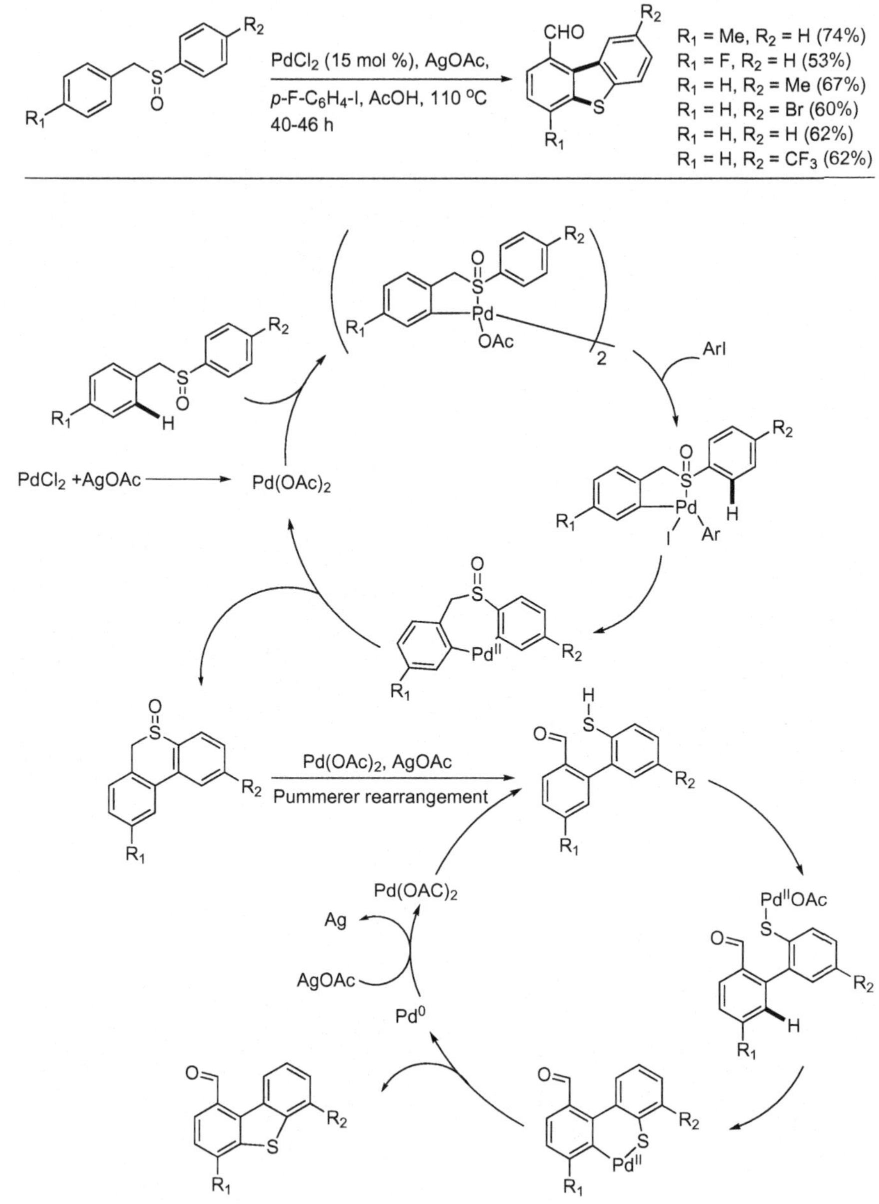

Scheme 18 Synthesis of dibenzothiophenes from benzyl phenyl sulfoxides

however as under the reaction conditions, in the presence of silver acetate and acetic acid, a Pummerer-type rearrangement then ensues to give the corresponding benzothiophene carbaldehyde via the mercapto-aldehyde.

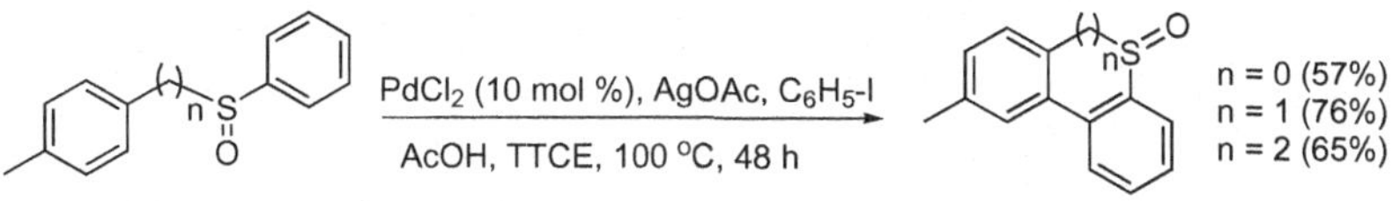

Scheme 19 Synthesis of polycyclic sulfoxides

Wang et al. [65] also explored the palladium-catalyzed cyclization of sulfoxides using a modified protocol to that above which included 1,1,2,2-tetrachloroethane as solvent, allowing for the isolation of 5-, 6- and 7-membered sulfoxide-bridged polycycles in moderate yields (Scheme 19). However, *ortho*-substituted derivatives were unconverted presumably due to steric interactions with the sulfoxide group. Steric hindrance of the sulfoxide moiety impedes the reaction because it is through chelation with the sulfoxide moiety that the *ortho* position is activated to generate a palladacycle complex. It is worth noting that if the 6-membered sulfoxide product is treated to the conditions reported by Samanta and Antonchick ($PdCl_2$ and AgOAc in AcOH at 110 °C), it undergoes a Pummerer rearrangement to yield the dibenzothiophene.

Huang et al. [22] investigated similar transformations using a Rh(III)/Ag(I) catalytic system which had the added advantage of deoxygenating the product to yield the dibenzothiophene directly. The reaction tolerates alkyl and methoxy functional groups but is low yielding with halogens and does not proceed with the strongly electron-withdrawing nitro group (Scheme 20). Unlike the palladium-mediated cyclization of sulfoxides, this method also tolerated methyl groups at both *ortho* positions. The mechanism is thought to involve the usual oxidative addition of the Pd(II) species to the *ortho* position of the sulfoxide followed by a second intramolecular palladation and reductive elimination to give the dibenzothiophene oxide. Deoxygenation with rhodium trifluoroacetate, a species

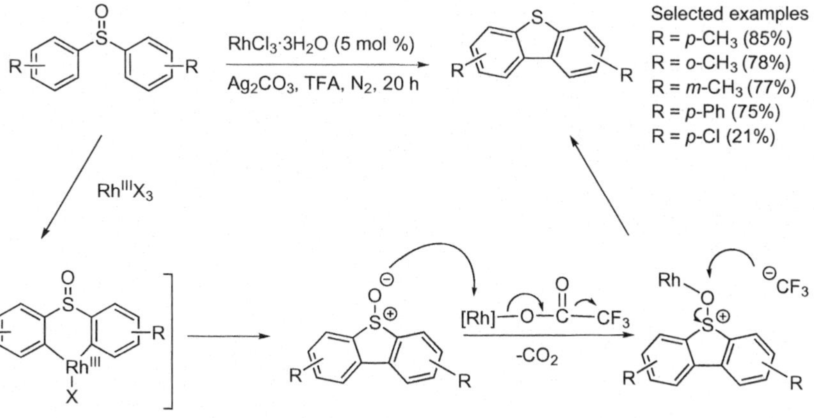

Scheme 20 Rh(III)-catalyzed cyclization of diphenyl sulfoxides

likely to be present in the reaction mixture, then yields the dibenzothiophene with concomitant loss of carbon dioxide.

The work reported by Che et al. [8] revealed that like sulfoxides, diaryl sulfides could be cleanly converted to the corresponding dibenzothiophene using a Pd(II)/Ag(I) catalytic system. After optimization studies, the ideal condition for this conversion was found to be 10 mol% $Pd(tfa)_2$ in the presence of 4 molar equivalents of AgOAc and 1 molar equivalent K_2CO_3 in PivOH at 130 °C for 24 h. The generality of this method was demonstrated using substrates bearing functional groups including Me, OMe, F, Cl, *t*Bu which were converted in yields of up to 88%. Substrates bearing strong electron-withdrawing nitro and acyl groups however remained unconverted. The mechanism likely proceeds as expected by way of an initial electrophilic palladation of the diarylsulfide *ortho* to the sulfur followed by a second palladation to generate the 6-membered palladacycle in what is thought to be the rate-determining step. Subsequent reductive elimination then yields the targeted dibenzothiophene [8]. Interestingly, Tobisu et al. [62] reported that for substrates such as biphenylphenylthiol, the outcome of the cyclization could be regulated by the use of the silver oxidant (Scheme 21). In the presence of both the Pd(II) and Ag(I) ions, i.e., Che's conditions, the dibenzothiophene via C–H/C–H coupling is predominantly obtained. However, using as palladium diacetate in the presence of a carboxylic acid ligand in toluene at 130 °C yielded the product from cyclization with loss of benzene (Scheme 21). This clearly highlights that the C–H bond activated in these reactions is dependent on whether or not the oxidant is present.

Laha et al. [32] described the Pd-catalyzed intramolecular coupling of aryl sulfanilides to give biaryl sultams. While appreciable yields were obtained for compounds which possessed electron-withdrawing groups, the presence of a methoxy group at the *ortho* or *para* position on the aniline ring gave no biaryl sultam (Scheme 22). Additionally, when there is a phenyl substituent on the sulfonamide nitrogen, cyclization occurs to give the indole as the sole product. This method can also be extended to the preparation of sultones as shown in Scheme 22. It is uncertain whether or not the mechanism for this reaction involves palladation

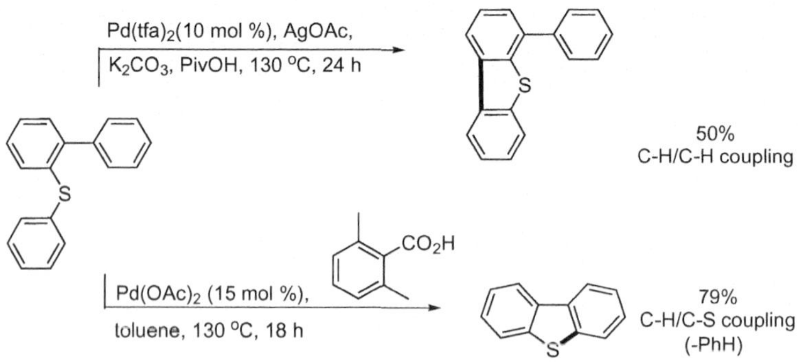

Scheme 21 Regulating the outcome of the Pd-mediated cyclization of a diaryl thiol

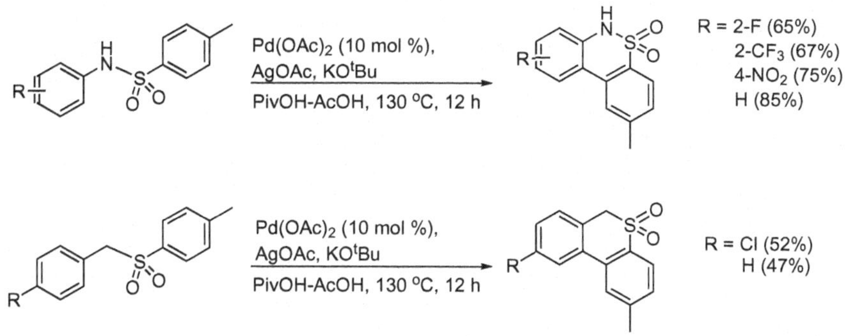

Scheme 22 Synthesis of biaryl sultams

of the aniline or benzenesulfonyl ring in the initial step. However, regardless of which ring is first palladated, cyclopalladation and subsequent transmetalation, then reductive elimination would give the product.

The copper-mediated cyclization of sulfones to give the corresponding dibenzothiophene-S,S-dioxide has also been a useful addition to methods for C–H/C–H coupling [30]. Although there are only a few examples of these reactions, it is worth highlighting that this method tolerates halogen functional groups (Scheme 23).

2.3 *Synthesis of S-heterocycles via N–H/C–H or N–H/S–H Coupling*

The preparation of benzothiazoles by iodine-mediated dehydrogenative coupling to N–H was previously discussed in the chapter (Scheme 9). Another useful addition to this group of reactions is the intramolecular amidation of sulfonamides using a chiral dirhodium(II) catalyst with phenyliodine diacetate in the presence of magnesium oxide [17]. The reaction hinges on the phenyliodinane generated in situ and was found to proceed best in dichloromethane. Intramolecular amidation reactions can also be accomplished using sulfamate esters (Scheme 24). Another noteworthy approach to the preparation of S-heterocycles is the intramolecular copper-catalyzed cyclization of 2-mercaptobenzamides to yield benzo[d]isothiazol-3(2H)-ones [63].

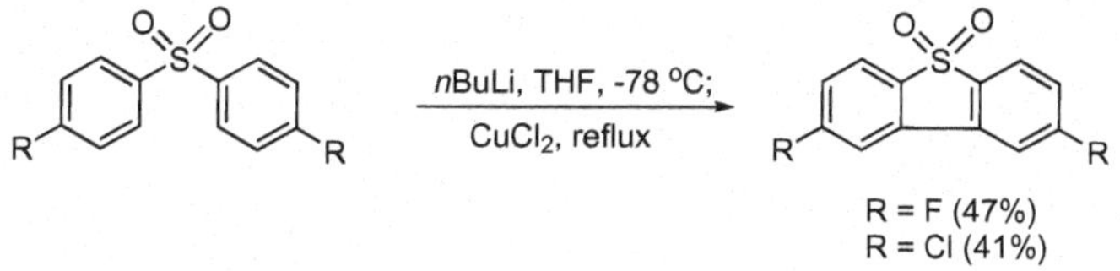

Scheme 23 Copper-mediated cyclization of sulfones

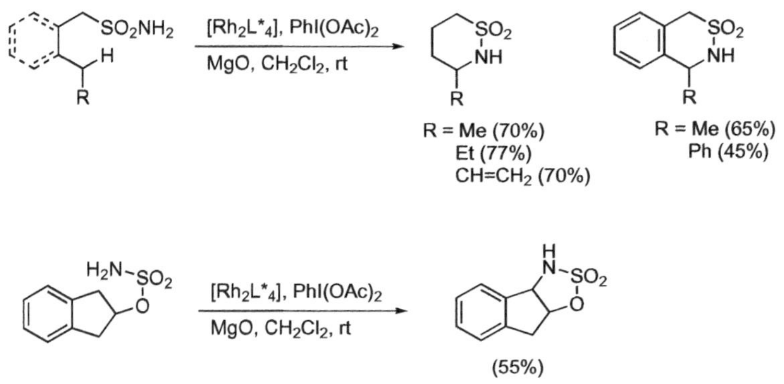

Scheme 24 Intramolecular amidation reactions

3 Functionalization of S-heterocycles

After assembly of the S-heterocyclic motif, it is often necessary to further functionalize in an effort to create synthetically and medicinally useful compounds. Functionalization by C–H activation has received significant attention and is dominated by the use of Pd(II) catalytic systems to allow for dehydrogenative coupling with arenes and aliphatic compounds.

3.1 Coupling of S-heterocycles to Heteroarenes via C(sp²)–H/C(sp²)–H Activation

The preparation of biheteroaryls has attracted significant attention by synthetic, medicinal and material chemists. A major challenge in the dual C–H activation of two heteroarenes to produce biheterocyclic systems is the competing homocoupling reaction. The success of these coupling reactions therefore relies on differences in the electron density or the C–H bond strength of the two arenes to be coupled. One of the first reports of the coupling of S-heterocycles with an unactivated arene was by He and coworkers in which thiophenes were coupled to polyfluoroarenes using palladium diacetate as catalyst and silver carbonate as oxidant [21]. A more practical palladium-catalyzed coupling of thiophenes was later revealed in which copper (II) actetate was employed as oxidant for a successful coupling to xanthines and even *N*-oxides (Scheme 25) [73]. The dehydrogenative coupling of thiophenes with benzamides as the unactivated arene has also been reported using CuOAc, Li_2CO_3, $Zn(OAc)_2$ and $AgNO_3$ in DMF [81].

Similarly, the cross-coupling of thiazoles and pyridine *N*-oxides can be accomplished by Pd(II) catalysis [18]. It was recognized that the addition of Cu $(OPiv)_2$ as an oxidant and cesium carbonate as base along with copper bromide was

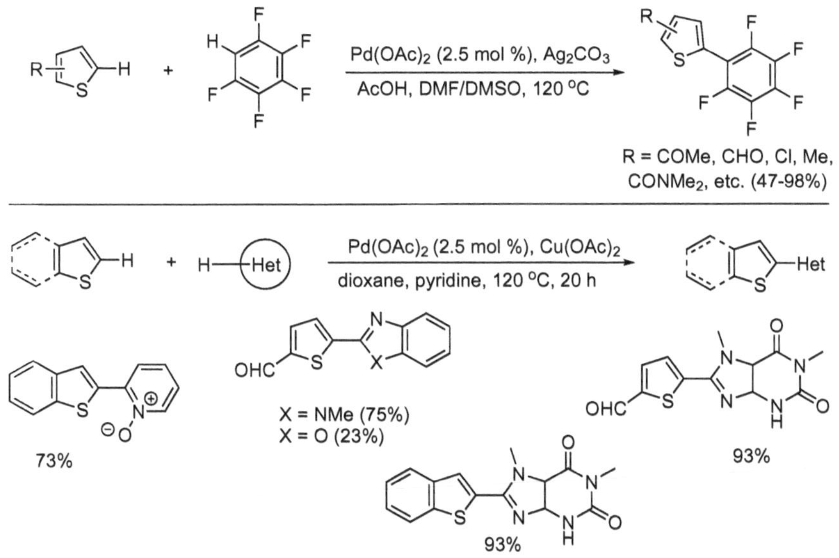

Scheme 25 Coupling of thiophenes to electron-deficient arenes

ideal to suppress the undesired homocoupling side reaction. As such, the reaction of thiazoles and benzothiazoles with *N*-oxides bearing electron-withdrawing or electron-donating groups resulted in the biheteroaryl systems in good yield (Scheme 26). Although the precise mechanism for this reaction remains unknown, it is thought that $Cu(OPiv)_2$, in addition to acting as an oxidant, also promotes the activation of the C–H bond since it was found to be superior to $Cu(OAc)_2$. Subsequent to this finding, Willis and Smith [71] demonstrated the regioselective C5–H coupling of thiazoles with pyridine *N*-oxides using a $Pd(OAc)_2/Ag_2CO_3$ catalytic system.

Although Pd(II) catalysis is a useful strategy for the coupling of S-heterocycles with *N*-oxides, the contamination of products with the metal is an unwanted challenge when preparing pharmaceutical compounds. Chen et al. [10] have demonstrated the base-promoted cross-coupling of quinoline *N*-oxides with thiazoles under ambient and metal-free conditions. The success of these conversions hinges on the fact that thiazoles are π-electron-rich systems bearing acidic C–Hs while *N*-oxides are electron-deficient due to the electron-withdrawing effect of the NO moiety. For these reactions, deprotonation at the C2-position of the benzazole by ^{t}BuOLi and subsequent attack of the *ortho* position of quinoline *N*-oxide by the resulting anion yields the coupled product with concomitant loss of lithium hydroxide (Scheme 27). Similar reactions have subsequently been accomplished using pyrazine *N*-oxides and thiazoles [57]. It was also recognized that benzothiazole was a superior coupling partner when compared to methyl or dimethylbenzothiazole.

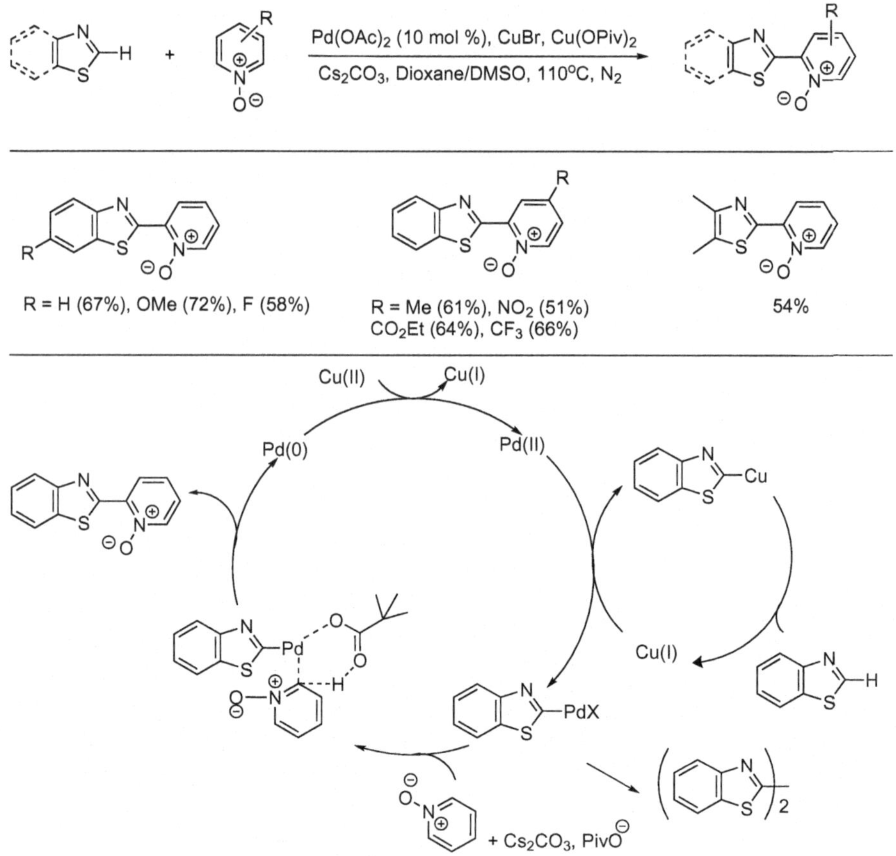

Scheme 26 Pd-catalyzed coupling of thiazoles and *N*-oxides

Unlike the coupling of π-electron-rich thiazoles with pyridine *N*-oxides, the coupling of two azoles which share similar π-electronic characteristics is especially challenged by homocoupled by-products. Homocoupling is also more often observed with benzothiazoles than thiazoles as they are more π-electron rich and more acidic at C-2 [12]. The transition metal-catalyzed homocoupling of thiophenes and thiazoles by C2–H activation has been reported using Pd(II) salts, Cu(I) salts and Cu(II) salts under various conditions [21, 33, 41, 82, 83]. The C3–H/C3–H homocoupling of thiophenes using DDQ in the presence of methane sulfonic acid is a noteworthy addition to the reagents which are capable of effecting homocoupling of benzothiophenes [72].

In 2011, the palladium-catalyzed cross-coupling of benzothiazole with various 2-unsubstituted azole systems under oxidative conditions was reported [20]. Using

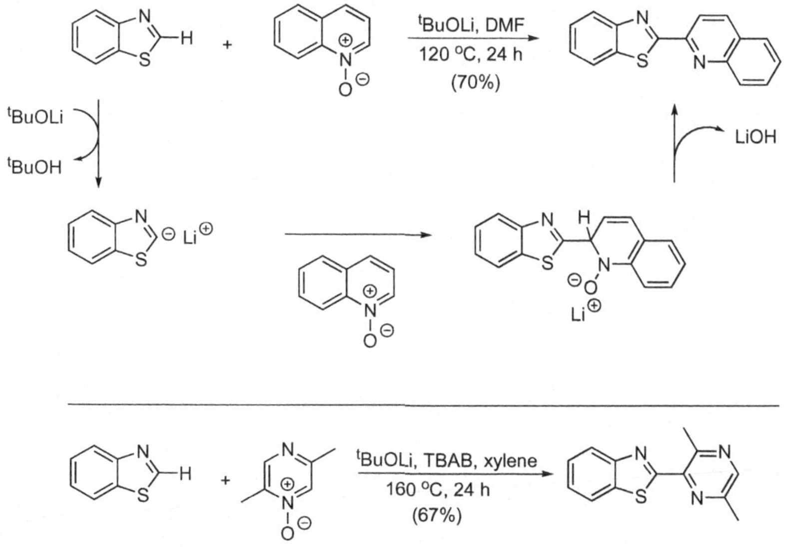

Scheme 27 Base-mediated coupling of thiazoles with *N*-oxides

$Pd(OAc)_2$ in combination with AgF and $Cu(OAc)_2$ in a polar aprotic solvent, the corresponding heterocoupled products were obtained even when a halide functional group was present in the substrate (Scheme 28). Although the mechanism for this conversion is not fully understood, the presence of Ag^+ ions is essential in suppressing the formation of the homocoupled product and may execute this function by aiding cleavage of the second C–H bond.

A modified protocol which used $PdCl_2$ and $Cu(OAc)_2$ in the presence of PPh_3 and CuCl was found to allow for the regioselective C2–C2′ coupling of two non-benzofused azoles [12]. This reaction occurred very quickly and tolerated a variety of functional groups (Scheme 29). The regioselectivity of the reaction is demonstrated by the good conversion to the C2-heterocoupled products even when the thiazole was fully unsubstituted. Mechanistic investigations of this reaction confirmed the necessity of $Cu(OAc)_2$ as opposed to other copper(II) salts and oxidants, confirming the findings of Han and coworkers that the acetate ion is essential. The triphenylphosphine is thought to stabilize the arylmetal intermediate.

It was later revealed that the heterocoupling of two azoles, including nonbenzofused thiazoles with oxazoles and imidazoles, could be successfully achieved using similar systems without the palladium(II) salt [15, 49]. Treating equimolar amounts of thiazoles and oxazoles or imidazoles with $Cu(OAc)_2$, Ag_2CO_3 and pyridine and heating at 140 °C for a day under oxygen atmosphere gave similar yields of biheteroaryl compounds to those shown in Scheme 29 [49]. The coupling of benzothiazoles with thiazoles was also found to be feasible under a protocol which utilizes excess thiazole, ᵗBuLi and Ag_2CO_3 with catalytic quantities of $CuCl_2$ [15]. Alternately, Mao et al. [35] reacted equimolar quantities of azoles with molar

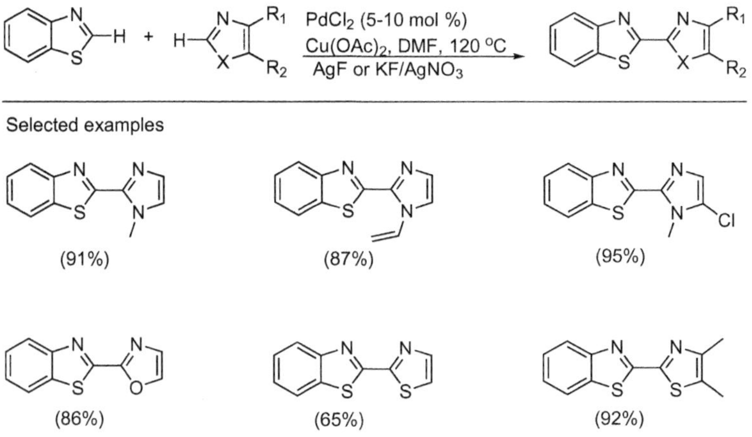

Scheme 28 Pd-catalyzed coupling of benzothiazole with azoles

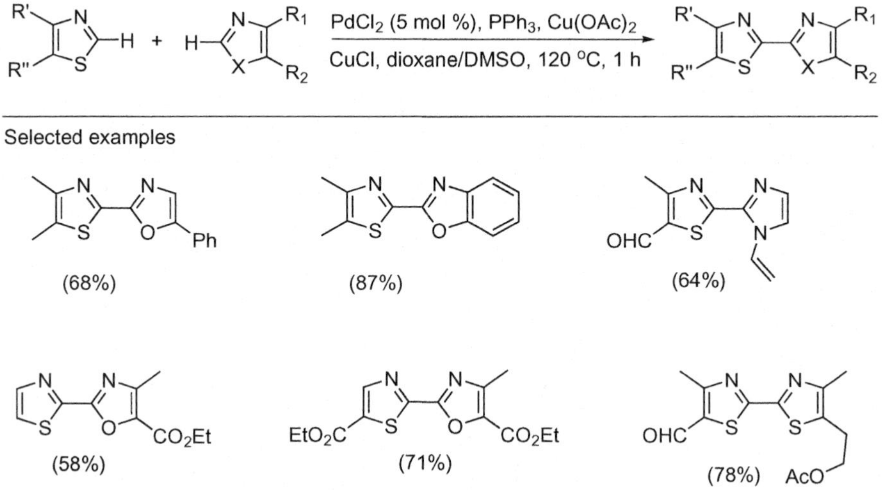

Scheme 29 Pd-catalyzed heterocoupling of two nonbenzofused azoles

equivalents of $Cu(OAc)_2$ in DMSO under inert conditions to afford azolyl-substituted thiazoles (Scheme 30).

As observed with thiazoles, Pd(II) catalytic systems have been successfully applied to the heterocoupling of thiophenes with an azole [76]. The simple catalytic system included the use of catalytic quantities of $Pd(OAc)_2$ (10 mol%) in the presence of Ag_2CO_3 and pyridine. The reaction of thiophenes or benzothiophene with 1-benzyl-1,2,3-triazole afforded the coupled product in good yield. Regioselectivity of the reaction was however somewhat compromised when attempts were made to couple 1-benzyl-1,2,3-triazole with unsubstituted thiophene

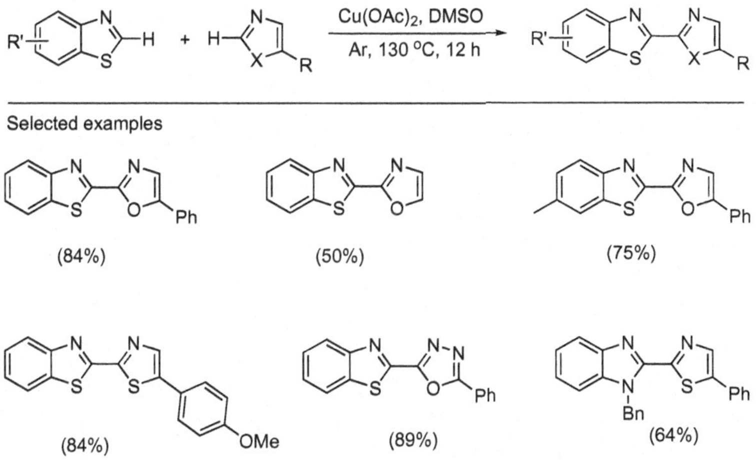

Scheme 30 Cu-catalyzed coupling of azoles

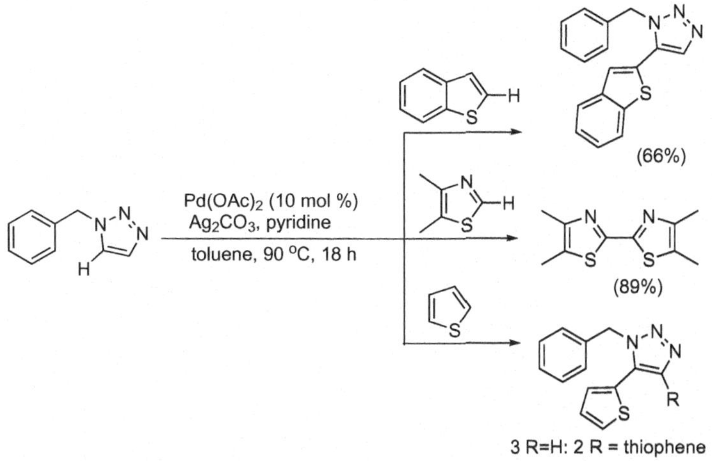

Scheme 31 Pd-catalyzed dehydrogenative coupling of thiophene

or 4,5-dimethylthiazole as the mono- and disubstituted triazole or bithiazole were obtained as products, respectively (Scheme 31).

In 2012, the first report of the successful application of $[RhCp^*Cl_2]_2$ for the dehydrogenative cross-coupling of two heterocycles, thiophene and furan, was reported [31]. The Rh(III) catalyst was used with $AgSbF_6$ as cocatalyst and copper bis (2-ethylhexanoate) as oxidant to allow for 2,2′-cross-coupled products in good yield (Scheme 32). Although the reaction could be successfully carried out in the absence of $AgSbF_6$, its addition resulted in a slight improvement in yield. The protocol could also efficiently couple the C3-position of indole and pyrrole to C2 of thiophenes and

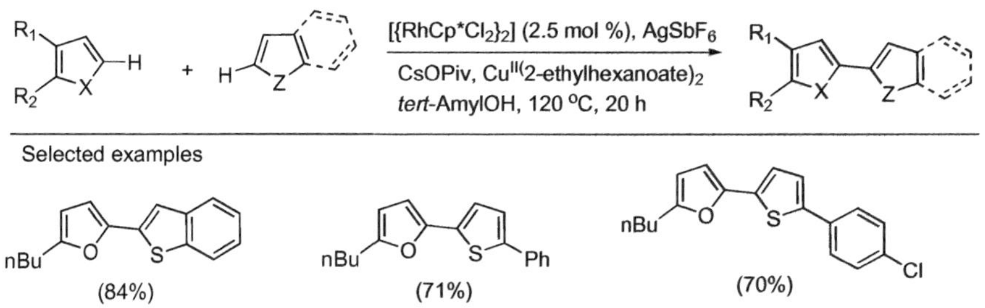

Scheme 32 Rh(III)-catalyzed heterocoupling reactions

furans. A similar protocol ($[RhCp^*Cl_2]_2$, $AgSbF_6$, Ag_2CO_3, PivOH, DMF at 150 °C for 24 h under nitrogen) would later be used for C2-dehydrogenative coupling of thiophene and thiazoles with *N*-pyrimidinyl indoles [50].

3.2 Coupling of S-heterocycles to Aliphatic Compounds

The C5-alkenylation of thiazoles using a palladium catalyst has also been accomplished [39]. A similar transformation has also been reported for the reaction of thiophenes with cyclic enones in the presence of Pd(II)/Ag(I) (Scheme 33) [69]. Wen and coworkers proposed that for these reactions, the palladium initially adds to the heterocycle. A subsequent migratory insertion of the resulting Pd(II) species into the alkene would then eventually lead to the coupling product.

A unique radical-mediated coupling of a heterocycles with a nitrile was recently reported by Liu and Li [34]. Remarkably, the use of catalytic quantities of CuCl (10 mol%) with dicumyl peroxide was able to effect regioselective coupling of alkyl nitriles with furans and indoles to give products in good yields. Although their

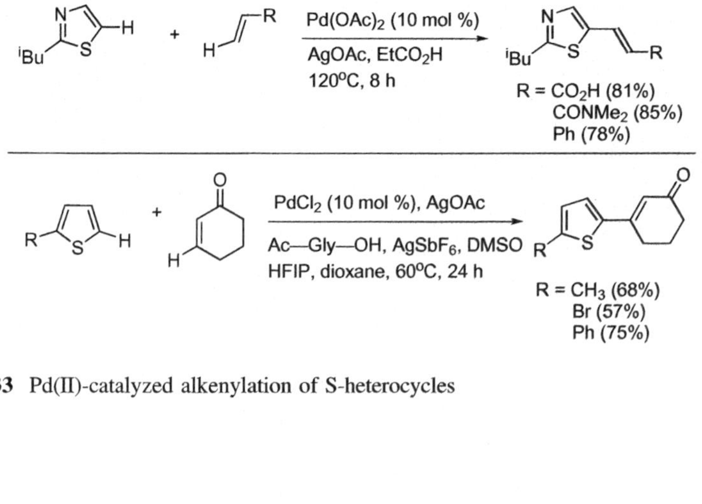

Scheme 33 Pd(II)-catalyzed alkenylation of S-heterocycles

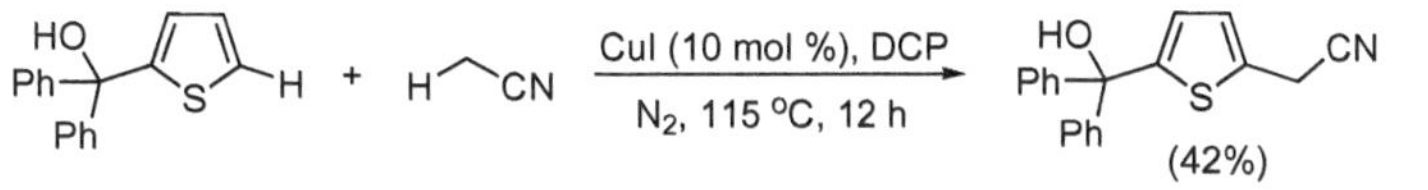

Scheme 34 Radical-mediated coupling of thiophene and alkyl nitrile

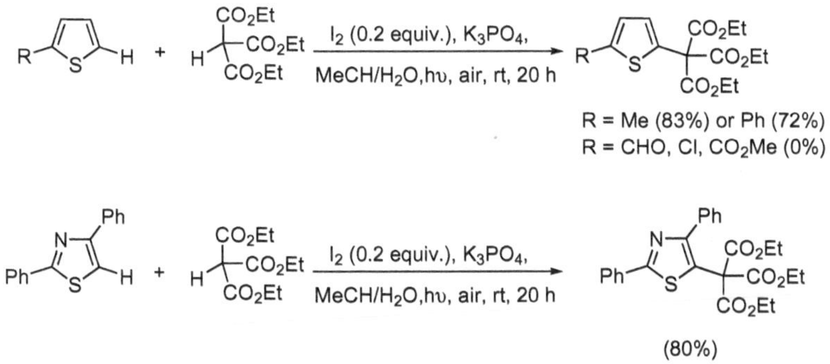

Scheme 35 Iodine-catalyzed coupling of thiophenes to carbonyl compounds

investigation focused on indoles and furans, the reagents were also successfully utilized for the coupling of a thiophene with a nitrile (Scheme 34).

Thiophenes have also been coupled successfully to carbonyl compounds via a photo-oxidative reaction using a catalytic amount of molecular iodine and molecular oxygen as a terminal oxidant [60]. Although this reaction was typically high yielding, the method was not found to be useful for electron-poor thiophenes (Scheme 35). The method could however be extended to thiazoles as well.

The Cu(II)-catalyzed coupling of cyclic thioethers to alkynes and carbonyl compounds is a significant addition to CDC reactions [46, 56]. Pan and coworkers reported the coupling of a cyclic benzylic thioether with acetophenone using $CuBr_2$ and $Na_2S_2O_8$, albeit in only fair yield (Scheme 36). However, the C-alkynylation of 4-thiazolidinone at the position adjacent to sulfur with a terminal alkyne was achieved using CuBr and *tert*-butyl hydroperoxide in good yield [56]. Although the full mechanisms are unknown, both reactions are thought to proceed by a radical pathway.

The direct C5-alkylation of thiazoles with ethers and cyclic alkanes by way of a cobalt catalyst was achieved recently (Scheme 37) [67, 68]. Coupling occurs adjacent to the oxygen atom of the ether, and the reaction works equally well with oxazoles and tolerates halide functional groups. Although the mechanism for this reaction is not yet fully understood, it is also thought to proceed via a radical pathway.

The Cu(II)-catalyzed C-2 alkylation of thiazoles with cyclic ethers in the presence of potassium persulfate represented a practical approach synthetically useful intermediates such as the acetals of 2-thiazolecarboxaldehydes when dioxolane was used as the ether [74]. The method was found to be high yielding and regioselective

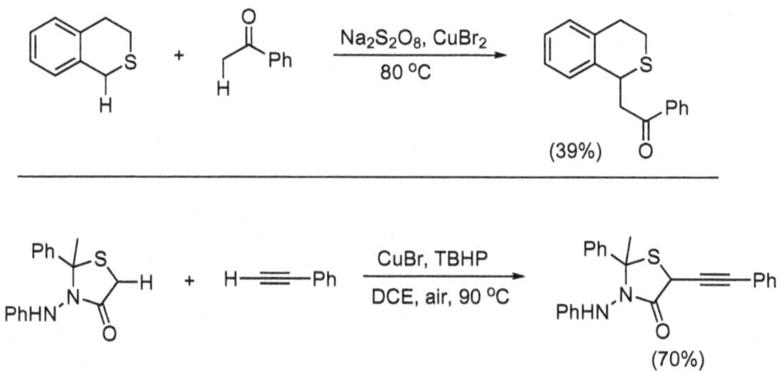

Scheme 36 Copper-mediated coupling to cyclic thioethers

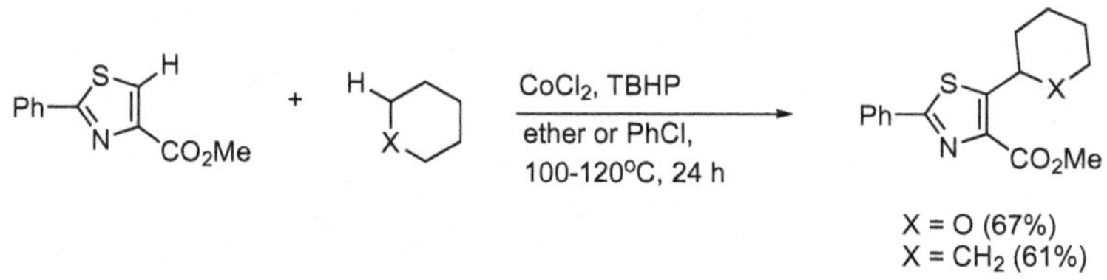

Scheme 37 Cobalt-mediated alkylation of thiazoles

when either thiazoles and benzothiazoles are coupled to tetrahydrofuran, 1,4-dioxane or 1,3-dioxolane (Scheme 38). Mechanistic investigations confirmed the presence of a radical intermediate for this transformation which is likely to be generated after dehydrogenation by a sulfate radical anion. Addition of the ether radical to the organocopper species and subsequent intramolecular radical elimination then generated the Cu(II)-aryl species which would eliminate Cu(I)Otf with the formation of the product.

Copper salts have also found utility in the direct C-2 thiolation [52] and amination of thiazoles [38]. Although the thiolation of thiazoles is done using equimolar quantities of CuI/Bipy, the protocol is applicable to both benzothiazoles and unsubstituted thiazoles and the thiol coupling partners may be aromatic or aliphatic. Mechanistically, it is postulated that a Cu–thiol complex (as opposed to a Cu–thiazole complex) is initially formed which subsequently adds to the thiazole. Molecular oxygen then abstracts hydrogen from the C-2 position of the thiazole while forming a Cu-hydroperoxo compound which breaks down to yield the product (Scheme 39).

The coupling of thiazole and thiophene derivatives with amides and amines was found to occur in the presence of Cu(II) acetate and an oxidant such as molecular oxygen or Ag_2CO_3 [38]. The reaction proceeded well in various solvents and tolerated the presence of ester functional groups and halogens (Scheme 40). Although this reaction was not successful with some amines, the reaction scope was

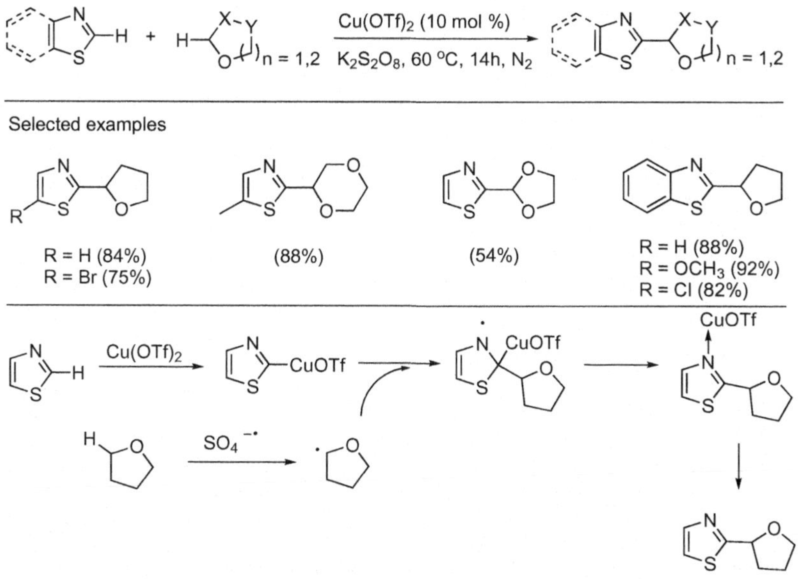

Scheme 38 Cu-catalyzed C-2 alkylation of thiazoles

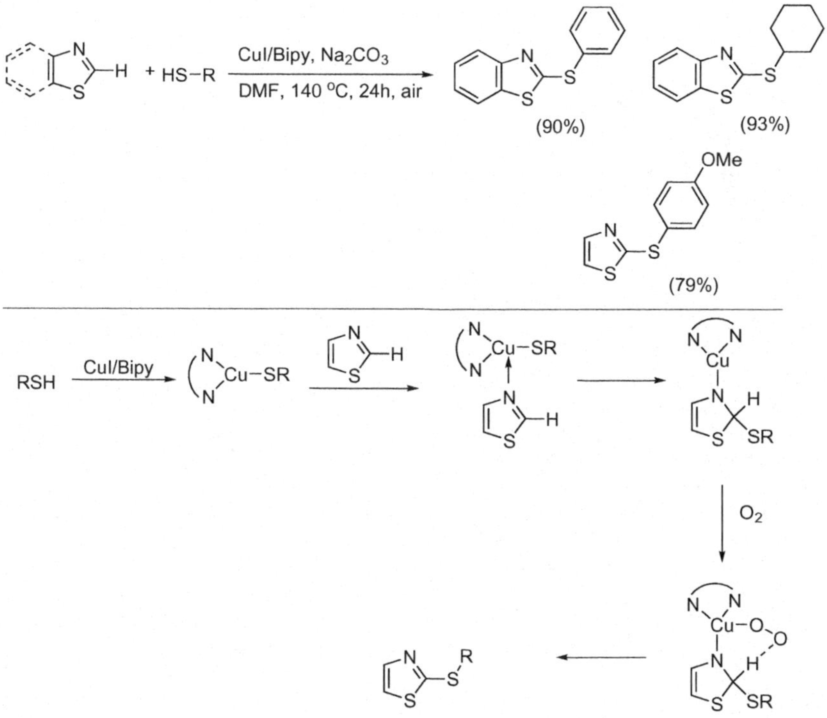

Scheme 39 Cu-mediated C-2 thiolation of thiazoles

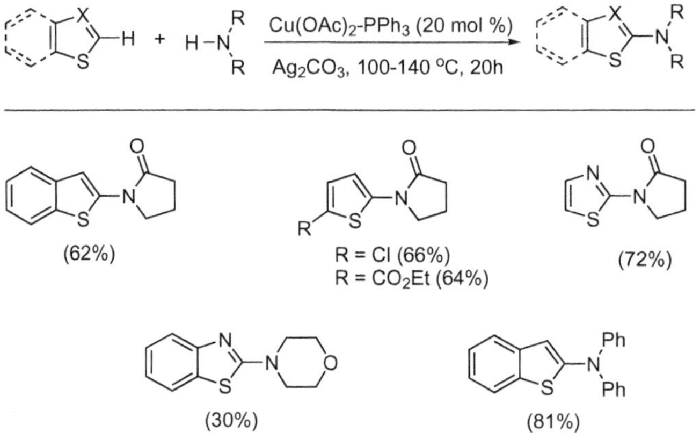

Scheme 40 Cu-catalyzed C-2 amination of thiophenes and thiazoles

extended to the amination of oxazoles and imidazoles using a similar protocol. This reaction represented a noteworthy addition to the methods available for accessing *N*-substituted S-heterocycles.

4 Conclusion

Recently, cross-dehydrogenative coupling reactions for the assembly and functionalization of S-heterocycles have received focused attention by synthetic chemists. The overview given of current successful approaches, in which their strengths and weaknesses are highlighted, should generate even more interest by the synthetic community and ultimately lead to more cost-effective and efficient approaches for these conversions.

References

1. Barrett OV, Downer-Riley NK, Jackson YA (2012) thermally induced cyclization of electron-rich *N*-Arylthiobenzamides to benzothiazoles. Synthesis 44(16):2579–2586
2. Bose DS, Idrees M (2006) Hypervalent iodine mediated intramolecular cyclization of thioformanilides: expeditious approach to 2-substituted benzothiazoles. J Org Chem 71 (21):8261–8263
3. Bose DS, Idrees M (2010) Dess-Martin periodinane mediated intramolecular cyclization of phenolic azomethines: a solution-phase strategy toward benzoxazoles and benzothiazoles. Synthesis 2010(03):398–402
4. Bose DS, Idrees M, Srikanth B (2007) Synthesis of 2-arylbenzothiazoles by DDQ-promoted cyclization of thioformanilides; a solution-phase strategy for library synthesis. Synthesis 2007 (06):819–823

5. Cano-Yelo H, Deronzier A (1984) Photocatalysis of the Pschorr reaction by tris-(2,2′-bipyridyl) ruthenium (II) in the phenanthrene series. J Chem Soc Perkin Trans 2(6):1093–1098
6. Cano-Yelo H, Deronzier A (1984) Photo-oxidation of some carbinols by the Ru (II) polypyridyl complex-aryl diazonium salt system. Tetrahedron Lett 25(48):5517–5520
7. Cano-Yelo H, Deronzier A (1984) Photo-oxidation of tris (2,2′-bipyridine) ruthenium (II) by para-substituted benzene diazonium salts in acetonitrile. Two-compartment photoelectrochemical cell applications. J Chem Soc Faraday Trans 1: Phys Chem Condens Phases 80 (11):3011–3019
8. Che R, Wu Z, Li Z, Xiang H, Zhou X (2014) Synthesis of dibenzothiophenes by Pd-catalyzed dual C–H activation from Diaryl sulfides. Chem A Eur J 20(24):7258–7261
9. Chen S, Wang M, Jiang X (2018) Pd-catalyzed C–S cyclization via C–H functionalization strategy: access to sulfur-containing Benzoheterocyclics. Chin J Chem 36(10):921–924
10. Chen X, Cui X, Yang F, Wu Y (2015) Base-promoted cross-dehydrogenative coupling of Quinoline *N*-Oxides with 1,3-azoles. Org Lett 17(6):1445–1448
11. Cheng Y, Yang J, Qu Y, Li P (2012) Aerobic visible-light photoredox radical C–H functionalization: catalytic synthesis of 2-substituted benzothiazoles. Org Lett 14(1):98–101
12. Dong J, Huang Y, Qin X, Cheng Y, Hao J, Wan D, Li W, Liu X, You J (2012) Palladium (II)-catalyzed oxidative C–H/C–H cross-coupling between two structurally similar azoles. Chem A Eur J 18(20):6158–6162
13. Downer-Riley NK, Jackson YA (2007) Iodine-mediated cyclisation of thiobenzamides to produce benzothiazoles and benzoxazoles. Tetrahedron 63(41):10276–10281
14. Downer-Riley NK, Jackson YA (2008) Conversion of thiobenzamides to benzothiazoles via intramolecular cyclization of the aryl radical cation. Tetrahedron 64(33):7741–7744
15. Fan S, Chen Z, Zhang X (2012) Copper-catalyzed dehydrogenative cross-coupling of benzothiazoles with thiazoles and polyfluoroarene. Org Lett 14(18):4950–4953
16. Folgueiras-Amador AA, Qian XY, Xu HC, Wirth T (2018) Catalyst-and supporting-electrolyte-free electrosynthesis of benzothiazoles and thiazolopyridines in continuous flow. Chem A Eur J 24(2):487–491
17. Fruit C, Müller P (2004) Intramolecular asymmetric amidations of sulfonamides and sulfamates catalyzed by chiral dirhodium (II) complexes. Helv Chim Acta 87(7):1607–1615
18. Fu XP, Xuan QQ, Liu L, Wang D, Chen YJ, Li CJ (2013) Dual C-H activations of electron-deficient heteroarenes: palladium-catalyzed oxidative cross coupling of thiazoles with azine N-oxides. Tetrahedron 69(22):4436–4444
19. Ganín J, Melendez E, Merchan FL, Merino P, Orduna J, Tejero T (1990) Chemoselectivity in the oxidation of unsymmetrical Thioureas with NBS/sulfuric acid: Benzothiazoles vs. 1,2,4-Thiadiazoles. Synth Commun 20(15):2327–2334
20. Han W, Mayer P, Ofial AR (2011) Palladium-catalyzed Dehydrogenative cross-couplings of benzazoles with azoles. Angew Chem Int Ed 50(9):2178–2182
21. He CY, Fan S, Zhang X (2010) Pd-catalyzed oxidative cross-coupling of perfluoroarenes with aromatic heterocycles. J Am Chem Soc 132(37):12850–12852
22. Huang Q, Fu S, Ke S, Xiao H, Zhang X, Lin S (2015) Rhodium-catalyzed sequential dehydrogenation/deoxygenation in one-pot: efficient synthesis of dibenzothiophene derivatives from diaryl sulfoxides. Eur J Org Chem 2015(30):6602–6605
23. Inamoto K, Arai Y, Hiroya K, Doi T (2008) Palladium-catalysed direct synthesis of benzo [b] thiophenes from thioenols. Chem Commun 43:5529–5531
24. Inamoto K, Hasegawa C, Hiroya K, Doi T (2008) Palladium-catalyzed synthesis of 2-substituted benzothiazoles via a C–H functionalization/intramolecular C–S bond formation process. Org Lett 10(22):5147–5150
25. Inamoto K, Nozawa K, Kondo Y (2012) Palladium-catalyzed C–H cyclization in water: a milder route to 2-arylbenzothiazoles. Synlett 23(11):1678–1682
26. Jordan AD, Luo C, Reitz AB (2003) Efficient conversion of substituted aryl thioureas to 2-aminobenzothiazoles using benzyltrimethylammonium tribromide. J Org Chem 68 (22):8693–8696

27. Joyce LL, Batey RA (2009) Heterocycle formation via palladium-catalyzed intramolecular oxidative C–H bond functionalization: an efficient strategy for the synthesis of 2-aminobenzothiazoles. Org Lett 11(13):2792–2795
28. Kalkhambkar RG, Laali KK (2012) Pd (OAc) 2 catalyzed synthesis of 2-aryl-and 2-heteroaryl-benzoxazoles and benzothiazoles in imidazolium ionic liquids (ILs) without additives and with recycling/reuse of the IL. Tetrahedron Lett 53(32):4212–4215
29. Kaur N, Jangid NK, Sharma V (2018) Metal-and nonmetal-catalyzed synthesis of five-membered S. N-heterocycles. J Sulfur Chem 39(2):193–236
30. Kortan AM, Cannizzaro RJ, Robb MJ, Knauss DM (2016) Poly (ether sulfone) s using a rigid dibenzothiophene dioxide heterocycle. J Polym Sci, Part A: Polym Chem 54(19):3127–3131
31. Kuhl N, Hopkinson MN, Glorius F (2012) Selective rhodium (III)-catalyzed cross-dehydrogenative coupling of furan and thiophene derivatives. Angew Chem Int Ed 51(33):8230–8234
32. Laha JK, Jethava KP, Dayal N (2014) Palladium-catalyzed intramolecular oxidative coupling involving double C (sp2)–H bonds for the synthesis of annulated Biaryl Sultams. J Org Chem 79(17):8010–8019
33. Li NN, Zhang YL, Mao S, Gao YR, Guo DD, Wang YQ (2014) Palladium-catalyzed C–H homocoupling of furans and thiophenes using oxygen as the oxidant. Org Lett 16(10):2732–2735
34. Liu ZQ, Li Z (2016) Radical-promoted site-specific cross dehydrogenative coupling of heterocycles with nitriles. Chem Commun 52(99):14278–14281
35. Mao Z, Wang Z, Xu Z, Huang F, Yu Z, Wang R (2012) Copper (II)-mediated dehydrogenative cross-coupling of heteroarenes. Org Lett 14(15):3854–3857
36. Mariappan A, Rajaguru K, Merukan Chola N, Muthusubramanian S, Bhuvanesh N (2016) Hypervalent Iodine (III) mediated synthesis of 3-substituted 5-amino-1, 2, 4-thiadiazoles through intramolecular oxidative S-N bond formation. J Org Chem 81(15):6573–6579
37. Metzger J, Planck H (1956) Methods of synthesis of benzothiazoles. Chim Ind 75:929–939
38. Mitsuda S, Fujiwara T, Kimigafukuro K, Monguchi D, Mori A (2012) Copper-catalyzed oxidative C–H, N–H coupling of azoles and thiophenes. Tetrahedron 68(18):3585–3590
39. Miyasaka M, Hirano K, Satoh T, Miura M (2010) Palladium-catalyzed direct oxidative alkenylation of azoles. J Org Chem 75(15):5421–5424
40. Moghaddam FM, Boeini HZ (2005) Oxidative cyclization of thiobenzanilides to benzothiazoles using *N*-benzyl-DABCO tribromide under mild conditions. Synlett 2005(10):1612–1614
41. Monguchi D, Yamamura A, Fujiwara T, Somete T, Mori A (2010) Oxidative dimerization of azoles via copper (II)/silver (I)-catalyzed CH homocoupling. Tetrahedron Lett 51(5):850–852
42. Mu XJ, Zou JP, Zeng RS, Wu JC (2005) Mn (III)-Promoted cyclization of substituted thioformanilides under microwave irradiation: a new reagent for 2-substituted benzothiazoles. Tetrahedron Lett 46(25):4345–4347
43. Nishino K, Ogiwara Y, Sakai N (2017) Green preparation of dibenzothiophene derivatives using 2-Biphenylyl disulfides in the presence of molecular iodine and its application to Dibenzoselenophene synthesis. Eur J Org Chem 2017(39):5892–5895
44. Nishino K, Ogiwara Y, Sakai N (2018) Palladium (II)-catalyzed synthesis of dibenzothiophenes from 2-Biphenylyl Disulfides by C-H functionalization. Chem Eur J 4(43):10971–10974
45. Ohgaki H, Mitsuhashi H, Suzuki H (2003) Tert-BuOK/BuLi-Induced facile cyclodehydrogenation of diphenyl sulfide to dibenzothiophene. J Chem Res 2003(5):264–265
46. Pan X, Hu Q, Chen W, Liu X, Sun B, Huang Z, Zeng Z, Wang L, Zhao D, Ji M, Liu L (2014) Copper (II) catalyzed cross-dehydrogenative coupling of cyclic benzylic ethers with simple carbonyl compounds by $Na_2S_2O_8$. Tetrahedron 70(21):3447–3451
47. Prajapati NP, Vekariya RH, Borad MA, Patel HD (2014) Recent advances in the synthesis of 2-substituted benzothiazoles: a review. Rsc Adv 4(104):60176–60208

48. Qian XY, Li SQ, Song J, Xu HC (2017) TEMPO-catalyzed electrochemical C–H thiolation: synthesis of benzothiazoles and thiazolopyridines from thioamides. ACS Catal 7(4): 2730–2734
49. Qin X, Feng B, Dong J, Li X, Xue Y, Lan J, You J (2012) Copper (II)-catalyzed dehydrogenative cross-coupling between two azoles. J Org Chem 77(17):7677–7683
50. Qin X, Liu H, Qin D, Wu Q, You J, Zhao D, Qiang G, Huang X, Lan J (2013) Chelation-assisted Rh (III)-catalyzed C2-selective oxidative C–H/C–H cross-coupling of indoles/pyrroles with heteroarenes. Chem Sci 4(5):1964–1969
51. Ra CS, Jung BY, Park G (2004) The fungicidal benzothiazole methoxyacrylates: Synthesis, conformational analysis and fungicidal activity. Heterocycles 62:793–802
52. Ranjit S, Lee R, Heryadi D, Shen C, Wu JE, Zhang P, Huang KW, Liu X (2011) Copper-mediated C–H activation/C–S cross-coupling of heterocycles with thiols. J Org Chem 76(21):8999–9007
53. Rey V, Soria-Castro SM, Argüello JE, Peñéñory AB (2009) Photochemical cyclization of thioformanilides by chloranil: an approach to 2-substituted benzothiazoles. Tetrahedron Lett 50(33):4720–4723
54. Sahoo SK, Jena HS, Majji G, Patel BK (2014) Formation of Imidazolidine-benzothiazole–Copper (II) complexes via a copper-mediated room-temperature C–H activation of imidazolidinecarbothioamides. Synthesis 46(14):1886–1900
55. Samanta R, Antonchick AP (2011) Palladium-catalyzed double C–H activation directed by sulfoxides in the synthesis of dibenzothiophenes. Angew Chem Int Ed 50(22):5217–5220
56. Shaikh MM, Patel AP, Chikhalia KH (2019) Copper catalysed cross-dehydrogenative coupling (CDC) reaction of 4-thiazolidinone with terminal alkyne. Tetrahedron 75(4): 475–485
57. Shan Y, Lai M, Li R, Wu Z, Zhao M (2017) Synthesis of pyrazine-thiazole biheteroaryl compounds through base-promoted oxidative cross-dehydrogenative Coupling reactions. Asian J Org Chem 6(12):1715–1718
58. Sharma S, Pathare RS, Maurya AK, Gopal K, Roy TK, Sawant DM, Pardasani RT (2016) Ruthenium catalyzed intramolecular C–S coupling reactions: synthetic scope and mechanistic insight. Org Lett 18(3):356–359
59. Stevens MF, McCall CJ, Lelievald P, Alexander P, Richter A, Davies DE (1994) Structural studies on bioactive compound. 23. Synthesis of polyhydroxylated 2-phenylbenzothiazoles and a comparison of their cytotoxicities and pharmacological properties with genistein and quercetin. J Med Chem 37(11):1689–1695
60. Sudo Y, Yamaguchi E, Itoh A (2017) Photo-oxidative cross-dehydrogenative coupling-type reaction of thiophenes with α-position of carbonyls using a catalytic amount of molecular iodine. Org Lett 19(7):1610–1613
61. Tayade DT, Thombare RD, Waghmare SA (2017) Synthesis of 3-[4-Amino-*N*-Substituted-o-AnisamidO-5-yl] Amino-5-Substituted-1,2,4-Dithiazole. Indo Am J Pharm Sci 4(10):3890–3894
62. Tobisu M, Masuya Y, Baba K, Chatani N (2016) Palladium (II)-catalyzed synthesis of dibenzothiophene derivatives via the cleavage of carbon–sulfur and carbon–hydrogen bonds. Chem Sci 7(4):2587–2591
63. Wang Z, Kuninobu Y, Kanai M (2013) Copper-catalyzed intramolecular N–S bond formation by oxidative dehydrogenative cyclization. J Org Chem 78(14):7337–7342
64. Wang R, Liu D, Ren H, Zhang T, Wang X, Li J (2011) Homoleptic tris-cyclometalated iridium complexes with 2-phenylbenzothiazole ligands for highly efficient orange OLEDs. J Mater Chem 21(39):15494–15500
65. Wang B, Liu Y, Lin C, Xu Y, Liu Z, Zhang Y (2014) Synthesis of sulfur-bridged polycycles via Pd-catalyzed dehydrogenative cyclization. Org Lett 16(17):4574–4577
66. Wang J, Zong Y, Zhang X, Gao Y, Li Z, Yue G, Quan Z, Wang X (2014) Synthesis of *N*-benzothiazol-2-yl-amides by an iron-catalyzed oxidative C (sp2)–H functionalization. Synlett 25(15):2143–2148

67. Wang B, Meng Y, Zhou Y, Ren L, Wu J, Yu W, Chang J (2017) Synthesis of 5-amino and 3, 5-diamino substituted 1,2,4-thiadiazoles by I2-mediated oxidative N–S bond formation. J Org Chem 82(11):5898–5903
68. Wang X, Lei B, Ma L, Zhu L, Zhang X, Zuo H, Zhuang D, Li Z (2017) Cobalt-catalyzed cross-dehydrogenative c (sp2)–C (sp3) coupling of oxazole/thiazole with ether or cycloalkane. Chem Asian J 12(21):2799–2803
69. Wen ZK, Song TT, Liu YF, Chao JB (2018) Palladium-catalyzed dehydrogenative coupling of cyclic enones with thiophenes: a rapid access to β-heteroarylated cyclic enones. Chem Commun 54(29):3668–3671
70. Wigglesworth TJ, Sud D, Norsten TB, Lekhi VS, Branda NR (2005) Chiral discrimination in photochromic helicenes. J Am Chem Soc 127(20):7272–7273
71. Willis NJ, Smith JM (2014) An operationally simple, palladium catalysed dehydrogenative cross-coupling reaction of pyridine *N*-oxides and thiazoles "on water". RSC Adv 4 (22):11059–11063
72. Wirtanen T, Muuronen M, Hurmalainen J, Tuononen HM, Nieger M, Helaja J (2016) Intermolecular oxidative dehydrogenative 3,3′-coupling of benzo [b] furans and benzo [b] thiophenes promoted by DDQ/H + : total synthesis of shandougenine B. Org Chem Front 3 (12):1738–1745
73. Xi P, Yang F, Qin S, Zhao D, Lan J, Gao G, Hu C, You J (2010) Palladium (II)-catalyzed oxidative C–H/C–H cross-coupling of heteroarenes. J Am Chem Soc 132(6):1822–1824
74. Xie Z, Cai Y, Hu H, Lin C, Jiang J, Chen Z, Wang L, Pan Y (2013) Cu-catalyzed cross-dehydrogenative coupling reactions of (benzo) thiazoles with cyclic ethers. Org Lett 15 (17):4600–4603
75. Yadav PS, Devprakash SG, Senthilkumar GP (2011) Benzothiazole: different methods of synthesis and diverse biological activities. Int J Pharm Sci Drug Res 3(1):01–07
76. Yu X, Huang Z, Liu W, Shi S, Kuang C (2015) Palladium-catalyzed oxidative C–H/C–H cross-coupling of 1-substituted 1,2,3-triazoles with furans and thiophenes. Org Biomol Chem 13(15):4459–4465
77. Zeller KP, Petersen H (1975) Photochemische Herstellung von Dibenzofuranen und dibenzothiophenen. Synthesis 1975(08):532–533
78. Zhang G, Liu C, Yi H, Meng Q, Bian C, Chen H, Jian JX, Wu LZ, Lei A (2015) External oxidant-free oxidative cross-coupling: a photoredox cobalt-catalyzed aromatic C–H thiolation for constructing C–S bonds. J Am Chem Soc 137(29):9273–9280
79. Zhang T, Deng G, Li H, Liu B, Tan Q, Xu B (2018) Cyclization of 2-Biphenylthiols to dibenzothiophenes under PdCl2/DMSO catalysis. Org Lett 20(17):5439–5443
80. Zhao DY, Guo XK, Li JH, Tang RY (2012) Iodine-mediated intramolecular oxidative cyclization of 2-(Styrylthio) anilines: synthesis of 2-substituted benzothiazoles. Synthesis 44 (06):927–933
81. Zhao S, Yuan J, Li YC, Shi BF (2015) Copper-catalyzed oxidative C–H/C–H cross-coupling of benzamides and thiophenes. Chem Commun 51(64):12823–12826
82. Zhu J, Xie H, Li S, Chen Z, Wu Y (2011) Switching reaction pathways of trifluoromethylated thiobenzanilides by choice of different oxidative systems. J Fluor Chem 132(5):306–309
83. Zhu M, Fujita KI, Yamaguchi R (2011) Efficient synthesis of biazoles by aerobic oxidative homocoupling of azoles catalyzed by a copper (I)/2-pyridonate catalytic system. Chem Commun 47(48):12876–12878

Oxidative P-C Couplings and Related Reactions

Alajos Grün and György Keglevich

Abstract The oxidative P-C coupling is a special method for the preparation of phosphinoylated substrates. This reaction was utilized to make available a wide range of compounds, including *N*-, *P*- and *O*-heterocycles and is expected to spread in the future. In this chapter, we give an overview of this relatively new approach to reveal the potential of the oxidative P-C coupling serving as a "green" method. The oxidative approach was regarded in a broader sense.

Keywords Oxidative P-C couplings · Secondary phosphine oxides · Tertiary phosphine oxides · Cascades dehydrogenative cyclizations · Heterocycles cascades · Heterocycles

1 Introduction

These days, the C–C and C-heteroatomic coupling reactions are of utmost importance, and the variations in dehydrogenations including the coupling of sp^3 CH and sp CH moieties, sp^3 CH and sp^2 CH units, or two sp^3 carbon atoms form special cases [1]. It is also possible to couple an aromatic CH and a vinylic CH moiety, or two aromatic rings. The coupling under discussion may require the presence of special catalysts and oxygen. Light-assisted cross-dehydrogenative couplings are also known.

The Hirao reaction, i.e., the P-C coupling between vinyl halides or aryl halides and >P(O)H reagents in the presence of $Pd(PPh_3)_4$ or other Pd/Ni catalysts was discovered more than thirty years ago [2, 3]. This protocol affording phosphonates, phosphinates and tertiary phosphine oxides is in the focus these days [4–7].

In this chapter, we wish to summarize the somewhat analogous reactions comprising the oxidative couplings of secondary phosphine oxides and a wide

A. Grün · G. Keglevich (✉)
Department of Organic Chemistry and Technology, Budapest University of Technology and Economics, Budapest 1521, Hungary
e-mail: gkeglevich@mail.bme.hu

A. Srivastava and C. K. Jana (eds.), *Heterocycles via Cross Dehydrogenative Coupling*, https://doi.org/10.1007/978-981-13-9144-6_9

variety of substrates in a broader sense, showing also examples incorporating addition reactions of the >P(O)H species followed by an oxidative cyclization. Even more complicated cascades may also occur. The synthetic procedures taken from the last eight years give rise to interesting *P*-, *N*- and *O*-heterocyclic or carbocyclic derivatives.

2 Simple Oxidative P-C Coupling Reactions

A typical example for the oxidative phosphinoylation is the reaction of substituted pyridines with secondary phosphine chalcogenides to furnish the corresponding $-P(X)Y_2$ pyridines. The acetylene derivative served as a hydrogene scavenger (Scheme 1) [8].

Additional oxidative P-C couplings starting from substituted benzenes, 2-naphthol derivatives and indole are shown in Scheme 2/(1), (2) and (3) [9].

2-Phenylpyridine took part in a P-C coupling reaction with diarylphosphine oxides using 1 equivalent of 1,4-benzoquinone as the oxidant, and $Pd(OAc)_2$ as the catalyst at 120 °C. A heterocycle-directed *ortho*-palladation is responsible for the selectivity (Scheme 3) [10].

N-aryl-tetrahydroisoquinolines were the subject of a gold-catalyzed oxidative coupling with diarylphosphine oxides to provide the corresponding phosphinoyl derivatives (Scheme 4) [11].

Another variation of the P-C coupling reactions is the phosphinoylation of quinoxalin-2(1*H*)-ones (Scheme 5) [12], and that of other quinoxaline derivatives (Scheme 6) [13]. The two reactions mentioned were carried out in the presence of $K_2S_2O_8$ and O_2 as the oxidants, respectively.

This series of oxidative P-C coupling reactions is continued with the silver-mediated phosphinoylation of benzothiazoles (Scheme 7) [14]. The course of the reaction is the following. The thiazole ring is activated via the coordination of its N atom with the Ag^+. Then, $Ph_2P(O)H$ is added on the C=N moiety. The so formed dihydro derivative is stabilized by oxidative aromatization (Scheme 8) [14].

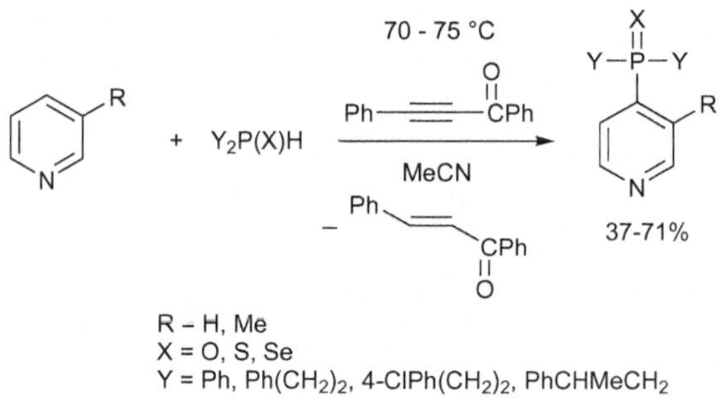

Scheme 1 Oxidative P-C coupling of pyridines and secondary phosphine chalcogenides

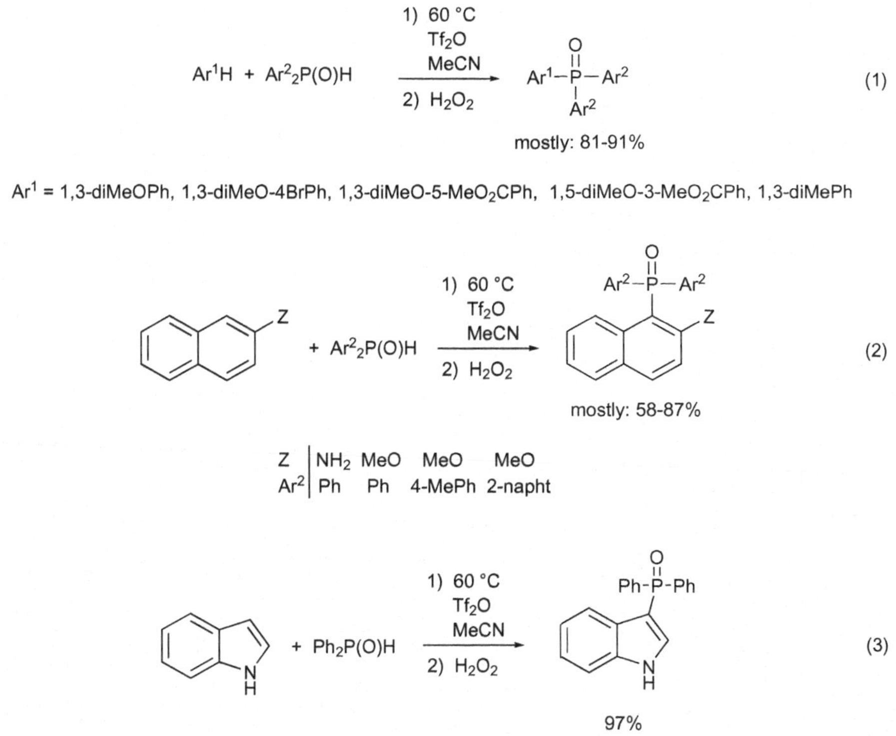

Scheme 2 Oxidative P-C coupling of aromatic substrates and diarylphosphine oxides

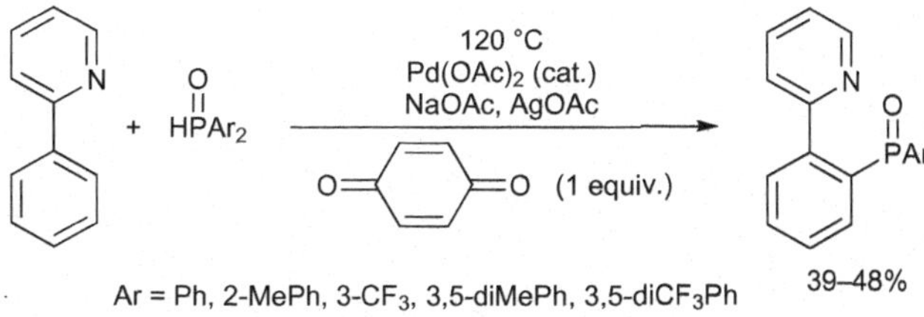

Scheme 3 A heterocycle-directed ortho oxidative coupling

2*H*-indazol underwent oxidative phosphinoylation with diarylphosphine oxides under photochemical conditions (Scheme 9) [15].

The phosphinoylation of benzothiazoles was also performed under air in the presence of $Mn(OAc)_3$ in a ball mill (Scheme 10) [16].

According to a plausible mechanism, a phosphinoyl radical is formed that is added on the C=N unsaturation. Finally, the adduct is stabilized by the loss of a hydrogen radical (Scheme 11) [16].

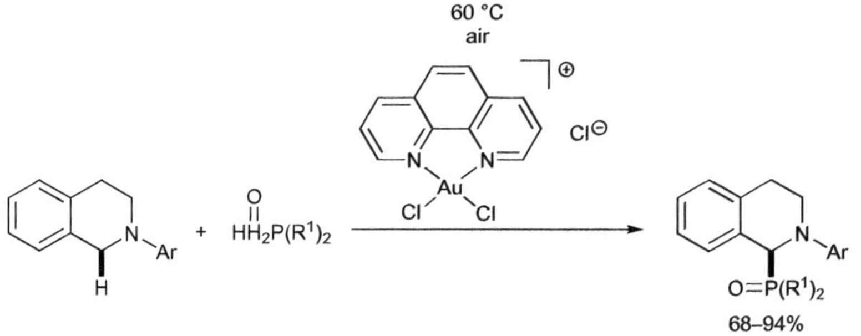

Ar = C_6H_5, 4-MeOC_6H_4, 2-MeOC_6H_4, 4-MeC_6H_4, 4-FC_6H_4, 4-ClC_6H_4, 4-BrC_6H_4
R^1 = C_6H_5, 4-FC_6H_4, 3-FC_6H_4, 4-ClC_6H_4, 3-ClC_6H_4, 4-MeC_6H_4

Scheme 4 A gold-catalyzed oxidative coupling

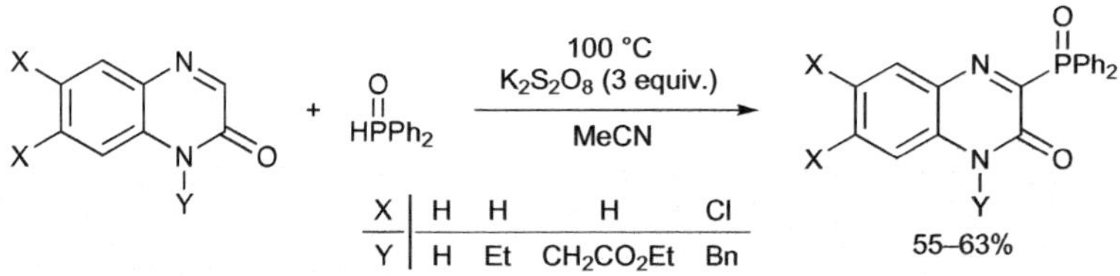

Scheme 5 Oxidative P-C coupling by the effect of $K_2S_2O_8$

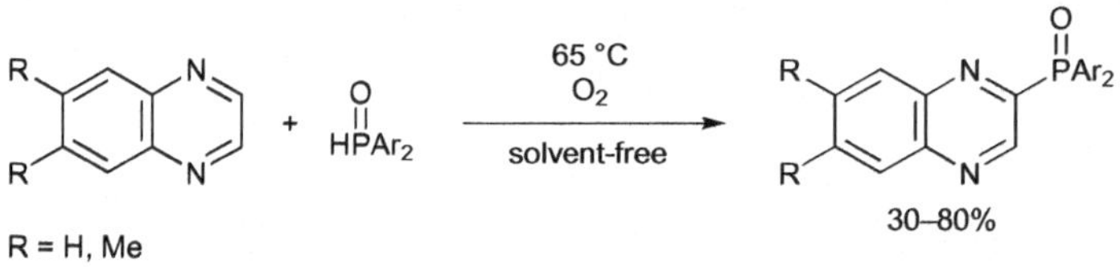

R = H, Me
Ar = Ph, 4-MeOPh, 3,5-diMePh, 4-MePh, 3-MePh, 2-MePh, 4-FPh, 3-FPh, 4-CF_3Ph

Scheme 6 Oxidative P-C coupling by the effect of O_2

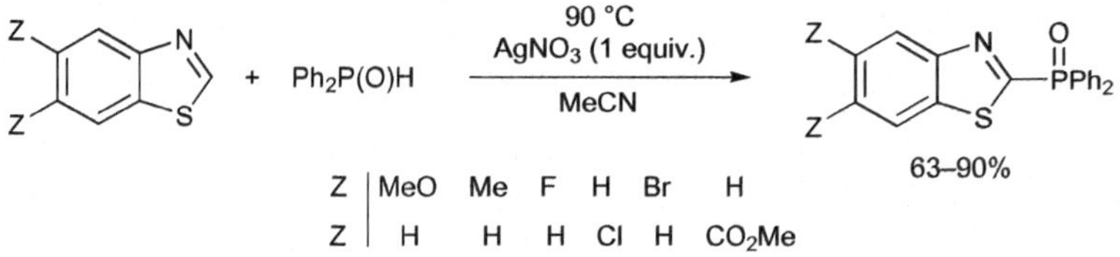

Scheme 7 Silver-mediated oxidative coupling

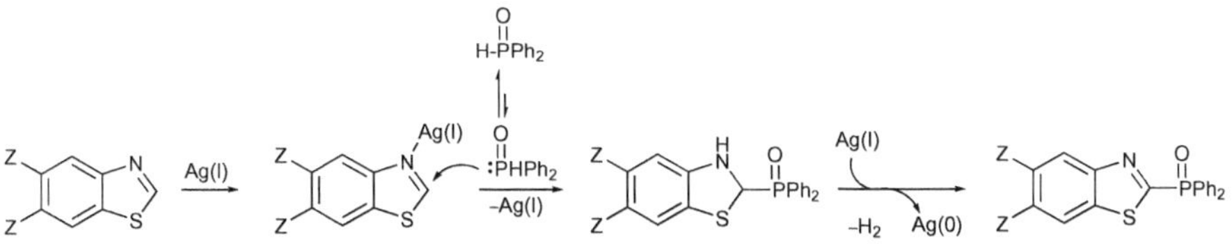

Scheme 8 Mechanism of the silver-mediated coupling of thiazoles and diphenylphosphine oxide

77–90%

Ar^1	4-MeOPh	4-MePh	3-MePh	4-FPh	4-ClPh	3-ClPh	4-BrPh
Ar^2	Ph	Ph	Ph	Ph	Ph	Ph	Ph

Ar^1	4-MePh	4-ClPh	4-MePh	4-FPh	4-BrPh	4-MePh	4-MePh	4-MePh
Ar^2	4-MeOPh	4-MeOPh	3-MeOPh	4-MePh	4-MePh	3-MePh	2-MePh	4-FPh

Scheme 9 P-C coupling under photochemical conditions

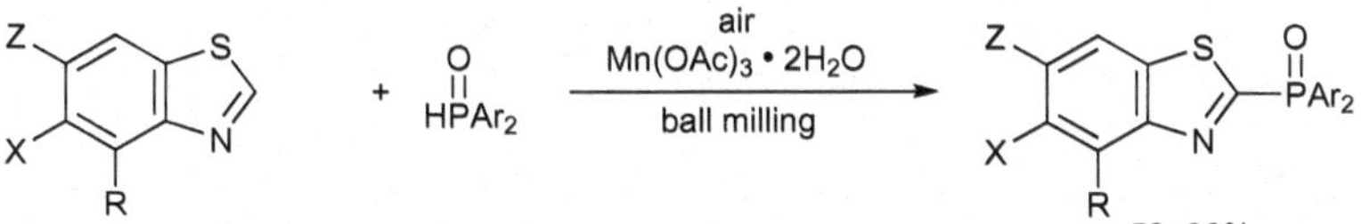

58–96%

Z	MeO	Me	F	Cl	Br	CO_2Et	CN	H	H
X	H	H	H	H	H	H	H	Cl	H
R	H	H	H	H	H	H	H	H	Me

Ar = Ph, 3,5-diMePh, 4-MePh, 4-FPh, 4-BrPh, 2-naphthyl, ^{c}Hex, ^{n}Bu

Scheme 10 P-C coupling in the presence of air/$Mn(OAc)_3$

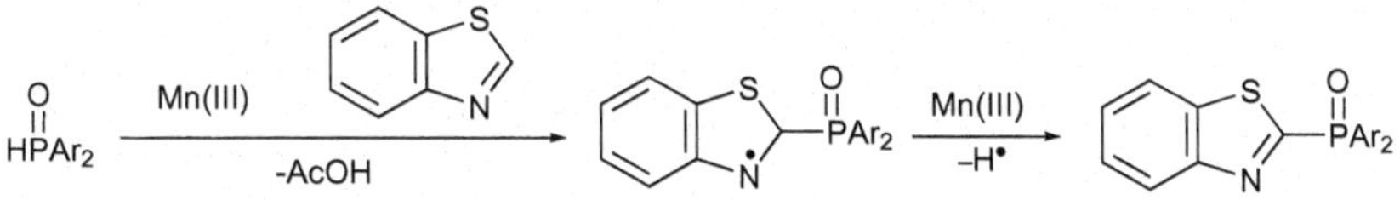

Scheme 11 Mechanism of the $Mn(OAc)_3$-promoted coupling

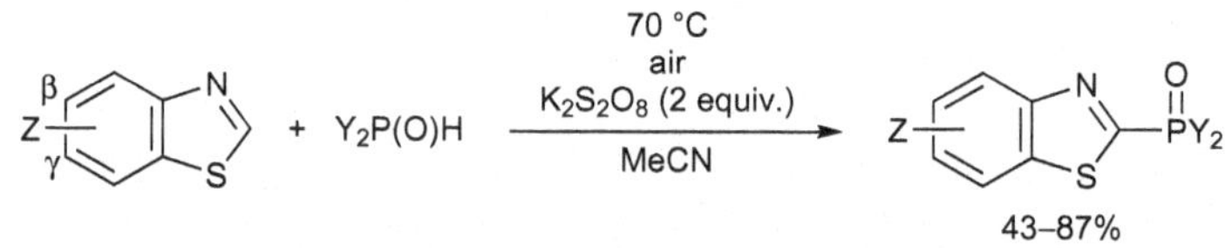

Z = H, EtO (γ), MeO (γ), Me (γ), F (γ), Br (γ), Cl (β), MeO_2C (β), O_2N (β)
Y = Ph, 3,5-diMePh, 4-MePh, 4-FPh, 4-BrPh, 4-PhPh, 2-naphthyl, ^{n}Bu, ^{c}Hex

Scheme 12 Oxidative P-C coupling of benzothiazoles and secondary phosphine oxides

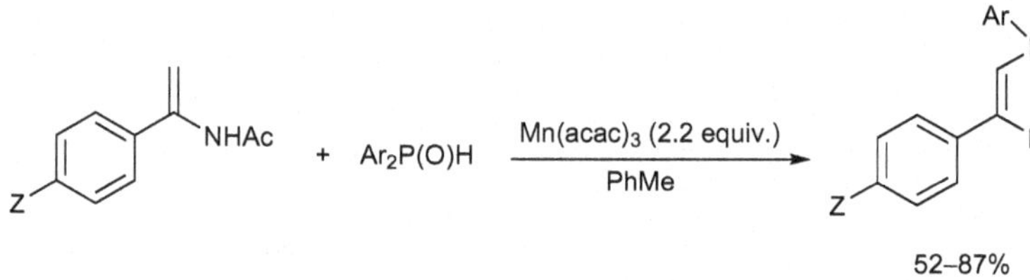

Z	H	MeO	Me	F	Cl	Br	I	CO_2Me	NO_2
Ar	Ph	Ph	Ph	Ph	Ph	Ph	Ph	Ph	Ph

Z	H	H	H	H	H	H
Ar	3,5-diMePh	4-MePh	3-MePh	4-FPh	4-ClPh	2-ClPh

Scheme 13 Phosphinoylation of enamides

In another series of reactions, benzothiazoles were phosphinoylated using $K_2S_2O_8$ under air (Scheme 12) [17].

Enamides could also be involved in oxidative phosphinoylation using $Mn(acac)_3$ (Scheme 13) [18].

Di-, tri- and tetraphosphorylated quinone derivatives were synthesized by subsequent oxidative substitutions starting from 2-diphenylphosphinoylhydroquinone, and using >P(O)H species as shown in Scheme 14 [19].

The radical oxidative reaction of [60]fullerene with diphenylphosphine oxide and acetic acid catalyzed by manganese(III) acetate afforded three different types of products. One of them is shown in Scheme 15 [20].

2,5-Bis(diarylphosphinoyl)-3,6-bis(arylamino)cyclohexa-2,5-diene-1,4-diones were synthesized through the regioselective dual phosphination and amination of benzoquinone in a one-pot manner. A typical example is shown in Scheme 16 [21].

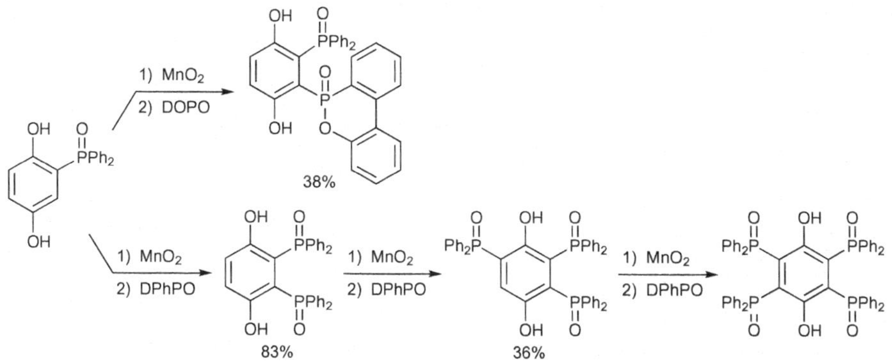

Scheme 14 Di-, tri-, and tetraphosphinoylation of 2-diphenylphosphinoylhydroquinone

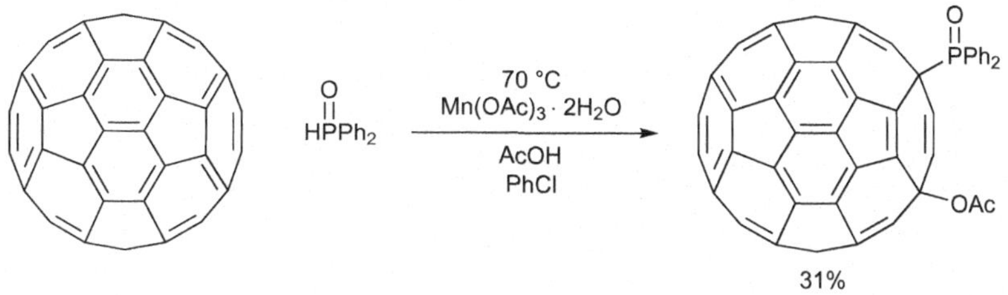

Scheme 15 Oxidative phosphinoylation of [60]fullerene

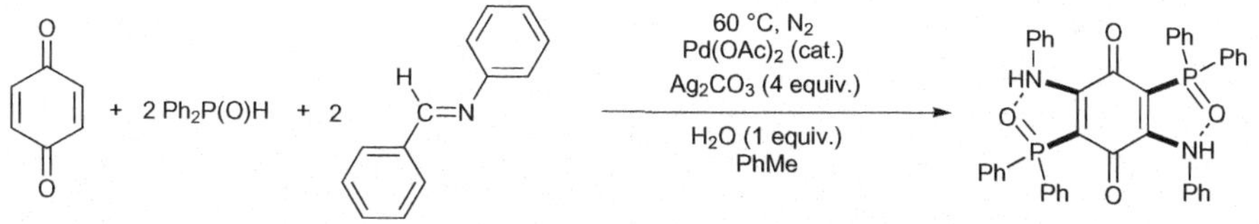

Scheme 16 The synthesis of 2,5-bis(diphenylphosphinoyl)-3,6-bis(phenylamino)cyclohexa-2,5-diene-1,4-diones

3 The Synthesis of Dibenzophosphole Oxides by Intramolecular Dehydrogenative Cyclization

Dibenzophosphole oxides were obtained by the intramolecular cyclization of secondary phosphine oxides having a biphenyl moiety. The dehydrogenation was accomplished in the presence of Pd(II) acetate (Scheme 17/(1)) [22]. Modification of the biphenyl unit of the starting material led to other derivatives (Scheme 17/(2) and (3)) [22].

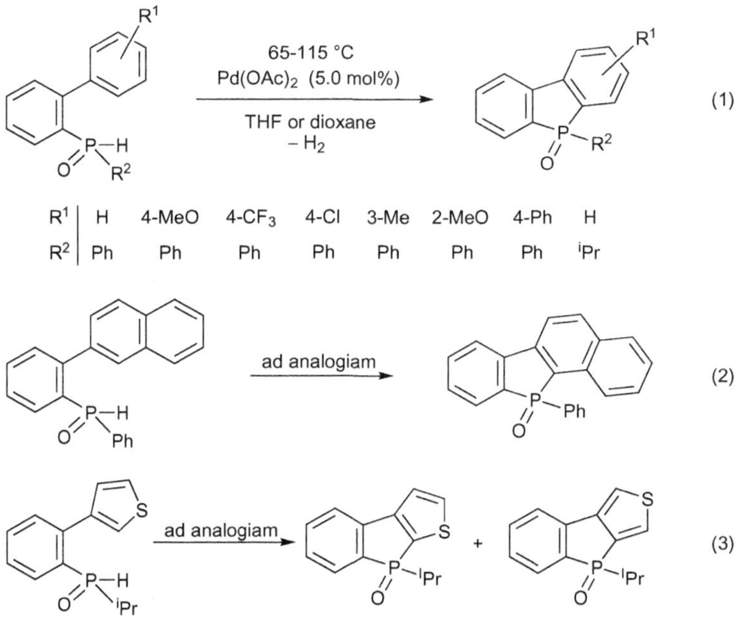

Scheme 17 Synthesis of fused phosphole derivatives by dehydrogenation cyclization

A double oxidative intramolecular cyclization of bis(phenylphosphinyl)-*p*-terphenyl afforded the diastereomers of "twin" bis(dibenzophosphole oxides) that were also converted to the bis(sulfide derivatives) (Scheme 18) [23].

4 The Formation of Benzo[b]phosphole Oxides by an Addition—Oxidative Cyclization Sequence

A series of benzo[b]phosphole oxides was synthesized from diphenylphosphine oxide and disubstituted acetylene derivatives via the silver salt-mediated dehydrogenative annulation involving a C–C and a P-C bond formation. The first step is obviously an addition that is followed by an oxidative cyclization (Scheme 19) [24].

Then, the reaction was extended to other substituted model compounds using AgOAc or Ag_2O (Schemes 20 and 21) [25, 26].

The reaction of disubstituted alkynes and $Ph_2P(O)H$ was also carried out in the presence of *tert*-butyl hydroperoxide and a Cu catalyst to afford the corresponding benzophosphole oxides (Scheme 22) [27].

Then, Zhao used $K_2S_2O_8$ as another oxidant, and extended further the scope of the reaction (Scheme 23) [28].

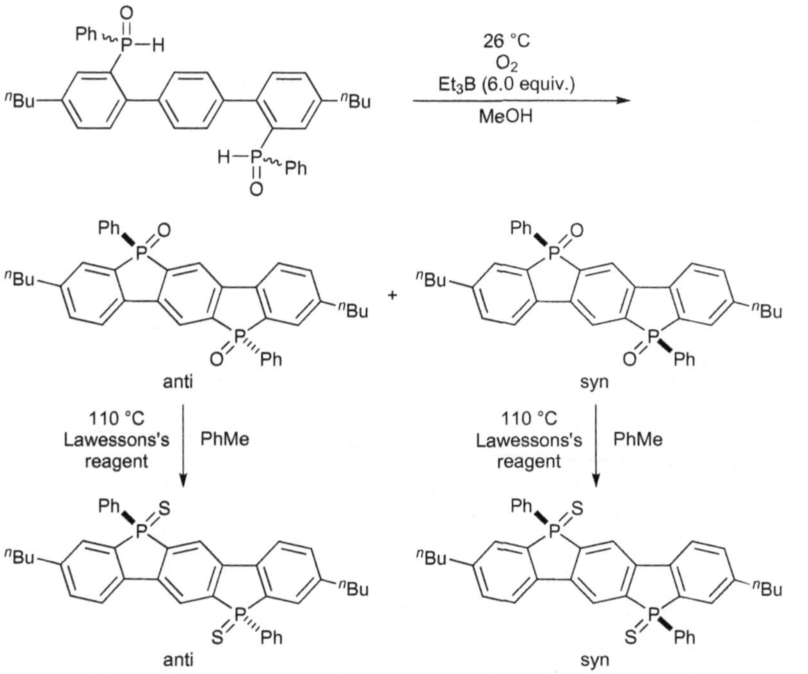

Scheme 18 Dehydrogenative cyclization of a triphenyl derivative followed by re-functionalization

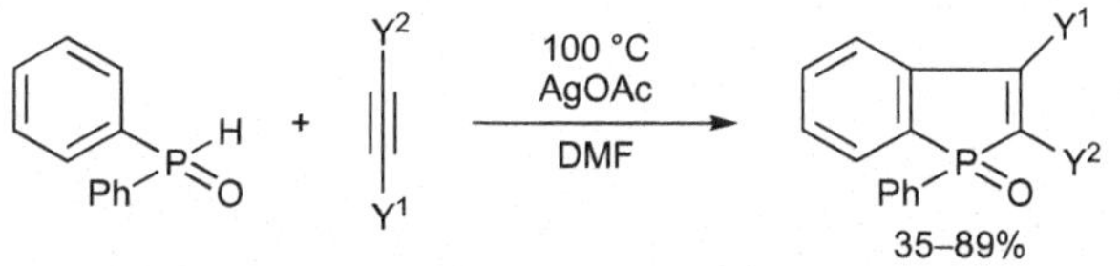

Y^1	4-MeOPh	4-MePh	4-NMe$_2$Ph	4-Cl	4-Br	^{n}Pr	Ph	Ph	Ph	Ph	Ph	Ph
Y^2	4-MeOPh	4-MePh	4-NMe$_2$Ph	4-Cl	4-Br	^{n}Pr	Me	Bu	TMS	CO$_2$Et	Ac	P(O)(OEt)$_2$

Scheme 19 Synthesis of benzophosphole oxides from diphenylphosphine oxide and acetylenic derivatives

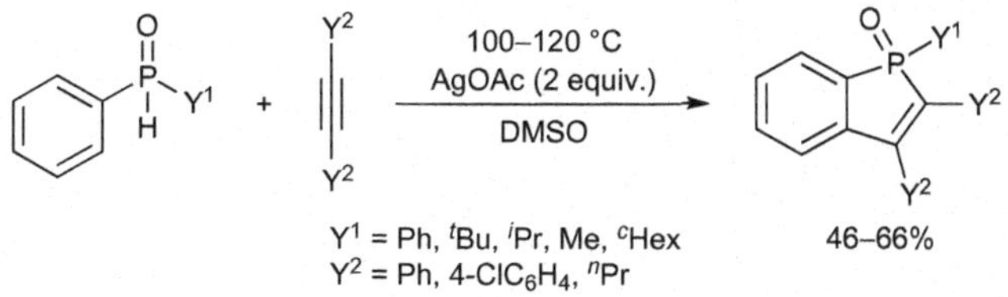

Scheme 20 Further benzophosphole oxides by the addition-ring closure approach

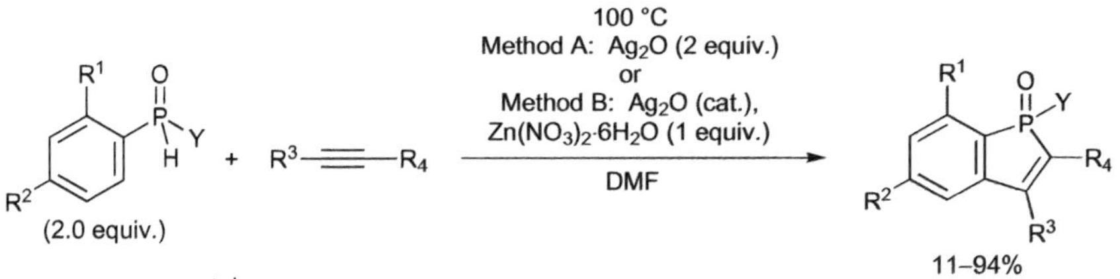

R^1	H	H	H	H	H	H	Me
R^2	H	H	H	F	Cl	MeO	H
Y	Ph	^{t}Bu	Cyclohex	4-FPh	4-ClPh	4-MeOPh	2-MePh

R^3	Ph	4-FPh	Ph	Ph	Ph	Ph	4-FPh	^{n}Pr
R_4	Ph	4-FPh	^{n}Pr	Me	CO_2Et	CN	Me	^{n}Pr

Scheme 21 The synthesis of substituted benzophosphole oxides

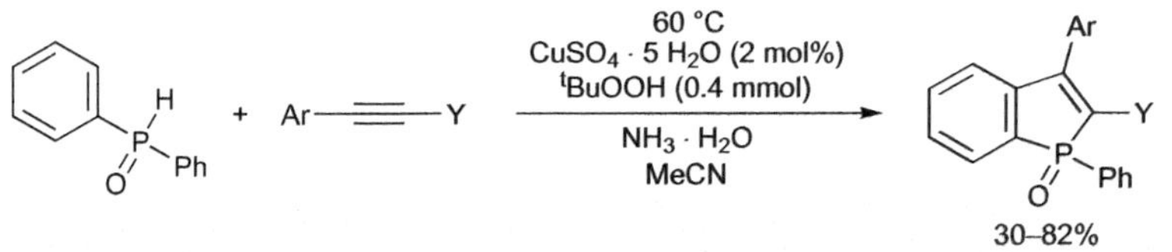

Ar	Ph	4-MeOPh	4-MePh	3-MePh	4-FPh	4-ClPh	4-BrPh	Ph	Ph	Ph	Ph
Y	Ph	4-MeOPh	4-MePh	3-MePh	4-FPh	4-ClPh	4-BrPh	TMS	^{n}Bu	CO_2Et	CONHPh

Ar	Ph	4-MeOPh	4-MePh	4-FPh	4-ClPh	4-BrPh	Ph	Ph
Y	$P(O)Ph_2$	$P(O)Ph_2$	$P(O)Ph_2$	$P(O)Ph_2$	$P(O)Ph_2$	$P(O)Ph_2$	$P(O)(OEt)_2$	$P(O)(O^iPr)_2$

Scheme 22 Benzophosphole oxides from diphenylphosphine oxide and acetylenic derivatives

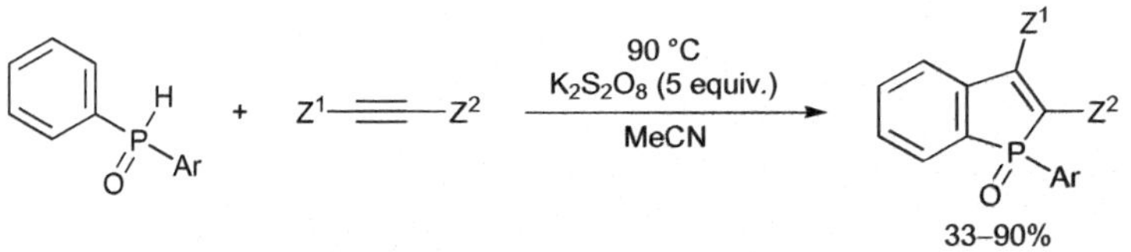

Z^1	Ph	4-MeOPh	4-MePh	4-EtPh	4-^{t}BuPh	4-FPh	4-ClPh	4-BrPh	4-CF_3Ph
Z^2	Ph	4-MeOPh	4-MePh	4-EtPh	4-^{t}BuPh	4-FPh	4-ClPh	4-BrPh	4-CF_3Ph

Z^1	Ph	Ph	4-MeOPh	4-MePh	3-MePh	4-FPh	4-BrPh	4-CF_3Ph	Ph
Z^2	CO_2Me	CO_2Et	CO_2Et	CO_2Et	CO_2Et	CO_2Et	CO_2Et	CO_2Et	CO_2Et

Scheme 23 Further extension of the synthesis by the addition-ring closure protocol

Gaumont studied the tandem reaction under discussion on UV irradiation, and prepared a series of benzophosphole oxides (Scheme 24) [29].

The next example comprises an addition–decarboxylation oxidation cascade followed by another oxidative phosphinoylation. The interaction of arylpropiolic

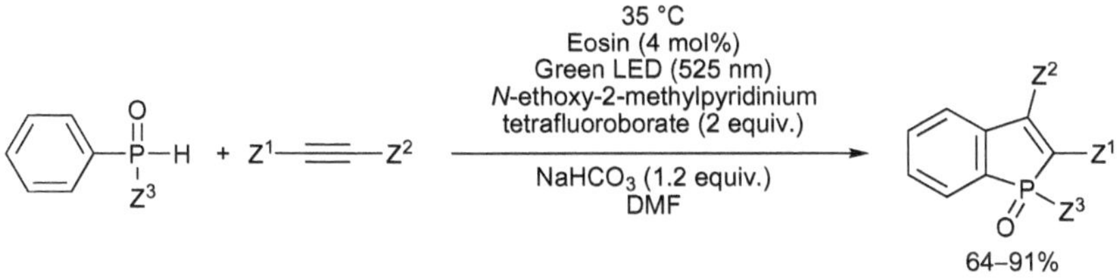

Z^1	Ph	4-MeOPh	4-FPh	4-ClPh	4-CF_3Ph	Me	Et	CO_2Et	^{n}Pr	SPh	CH_2OH	P(O)Ph_2
Z^2	Ph	4-MeOPh	4-FPh	4-ClPh	4-CF_3Ph	Ph	Ph	Ph	^{n}Pr	Ph	CH_2OH	Ph

Z^3 = Ph, Et, ^{t}Bu

Scheme 24 The synthesis of benzophosphole oxides under photochemical conditions

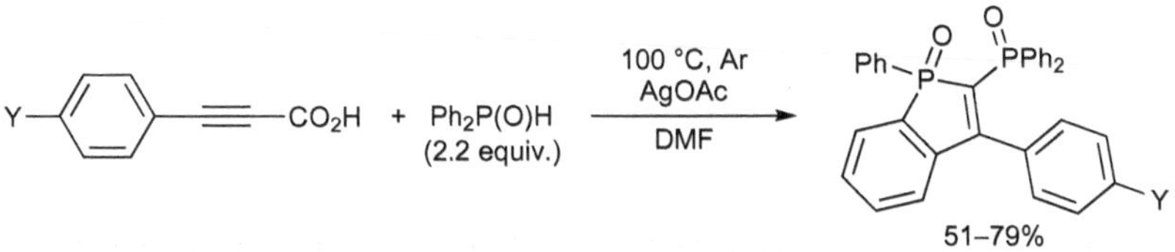

Y = H, MeO, ^{t}Bu, Me, F, Cl, Br, F_3CO, CF_3, Ac, OHC, EtO_2C, HO_2C, NC

Scheme 25 The synthesis of phosphinoyl-substituted benzophosphole oxides

acids and two equivalents of diphenyl phosphine oxide in the presence of AgOAc at 100 °C afforded eventually 2-phosphinoyl-benzo[*b*]phosphole oxides (Scheme 25) [30].

Starting from *ortho*-arylalkynylanilines and diarylphosphine oxides, a series of cascade reactions formed tribenzo[*b*,*e*,*g*]phosphindole oxides as valuable products. The synthesis is exemplified starting from *ortho*-phenylalkynylaniline (Scheme 26) [31].

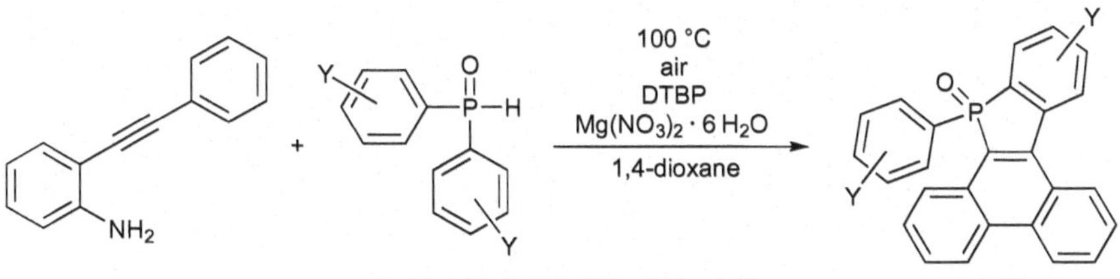

Y = H, 4-MeO, 3,5-diMe, 4-Me, 4-Cl

Scheme 26 The synthesis of tribenzophosphindole oxides via cascade reactions

5 The Synthesis of Different *N*-Heterocycles with Exocyclic P-Function

Two independent research groups elaborated the synthesis of aza-phenanthridin-6-yl-diarylphosphine oxides by the oxidative cyclization of 2-isocyanobiphenyls with diarylphosphine oxides (Schemes 27 and 28) [32, 33].

Another cascade reaction of an α,β-unsaturated amide and $Ph_2P(O)H$ in the presence of $AgNO_3$ as the oxidant gave eventually dihydroquinolin-2-(1*H*)-ones (Scheme 29). The adduct formed in the first step undergoes an oxidative cyclization [34].

The tandem addition—cyclization protocol was utilized in the synthesis of 3-phosphinoylquinolines (Scheme 30) [35] and 2-phosphinoyl-9*H*-pyrrolo[1,2-*a*] indoles (Scheme 31) [36]. In the latter case, aromatization was the final step. In the first step, a phosphinoyl radical is formed that is added on the triple bond of the propargyl moiety. This is followed by cyclization and by the loss of a hydrogen radical. The final pyrroloindole is formed after an oxidative deprotonation and isomerization (Scheme 32) [36].

An analogous tandem addition and oxidative cyclization was described in the reaction of *N*-cyanamide olefinic starting materials with $Ph_2P(O)H$. Depending on

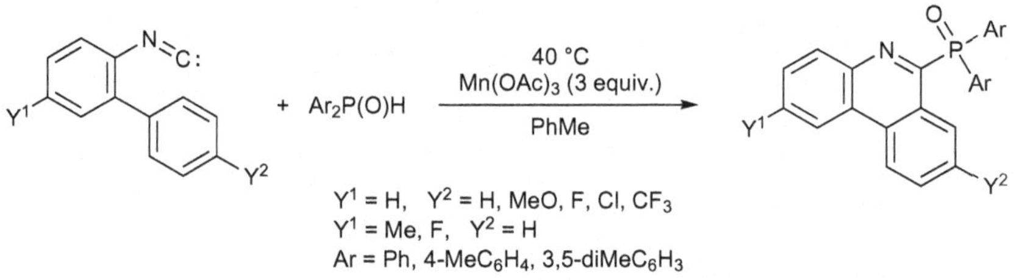

Scheme 27 The synthesis of phosphinoyl-azaphenanthridines

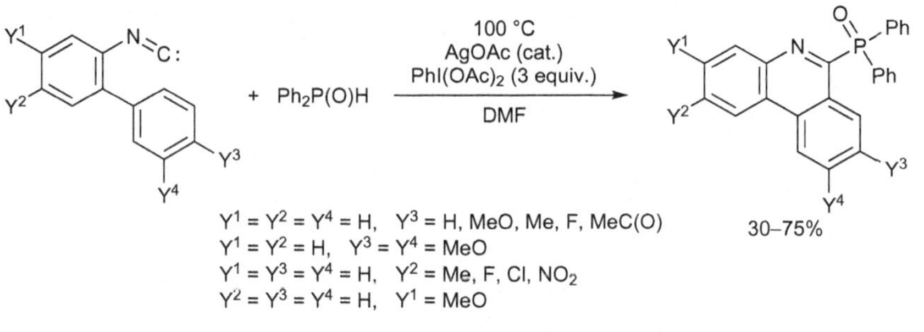

Scheme 28 The synthesis of polysubstituted phosphinoyl-azaphenanthridines

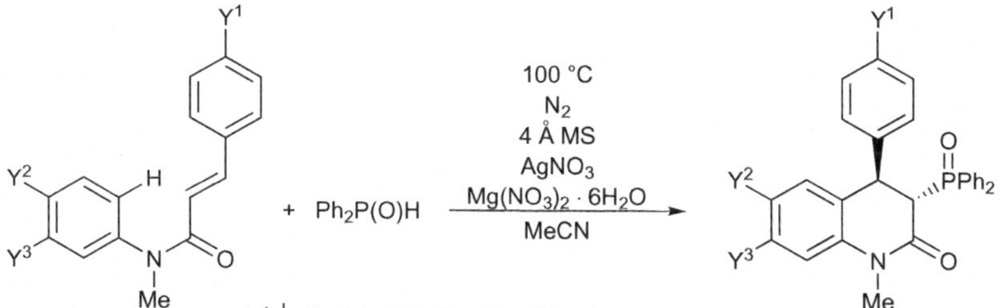

Y^1	H	Me	Cl	Br	Ph	CF_3	H	H	H	H
Y^2	H	H	H	H	H	H	F	CF_3	H	H
Y^3	H	H	H	H	H	H	H	H	Me	F

Scheme 29 The synthesis of 3-phosphinoyl-dihydroquinolinones

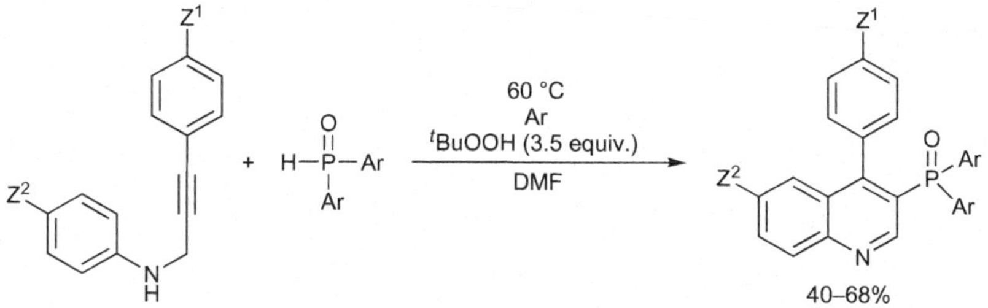

Z^1 = H, MeO, ^{t}Bu, Me, F, Cl, Br, CF_3, MeC(O), Z^2 = H, Ar = Ph;
Z^1 = H, Z^2 = MeO, Me, F, Cl, Br, EtO_2C, Ar = Ph;
Z^1 = Z^2 = H, Ar = 4-MePh, 4-ClPh

Scheme 30 The synthesis of 3-phosphinoylquinolines

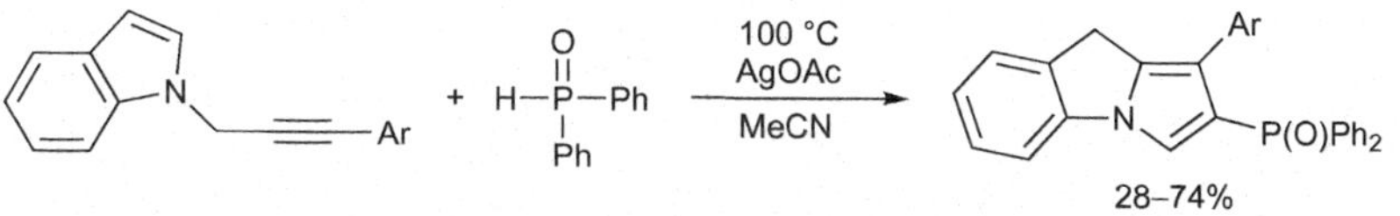

Ar = Ph, 4-MeOPh, 4-^{t}BuPh, 4-MePh, 3-MePh, 2-MePh, 4-PhPh, 4-FPh, 4-ClPh, 4-BrPh, 4-IPh, 4-CF_3Ph, 4-C(O)Me

Scheme 31 The synthesis of phosphinoyl-pyrroloindoles

the starting materials, dihydroisoquinolinones or quinazolinones were obtained (Scheme 33/1 and 2) [37].

The copper-catalyzed hydrophosphinoylation of 1-(3-phenylprop-2-yn-1-yl)-1*H*-indole initiated tandem reactions (double-bond rearrangement, oxidative cyclization and another double-bond rearrangement) that ended up with 2-phosphinoylpyrrolo[1,2-*a*]indole (Scheme 34) [38].

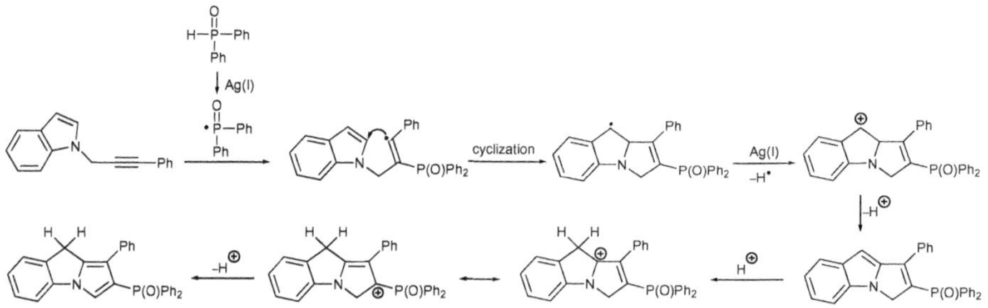

Scheme 32 Mechanism for the formation phosphinoyl-pyrroloindoles

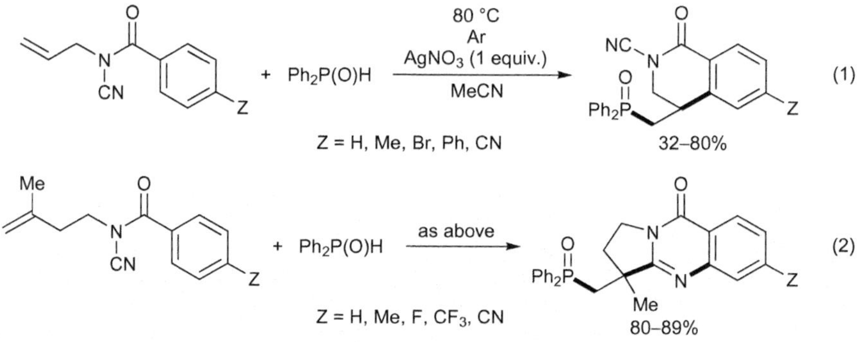

Scheme 33 Synthesis of dihydroisoquinolines and quinazolinones by tandem addition-oxidative cyclization

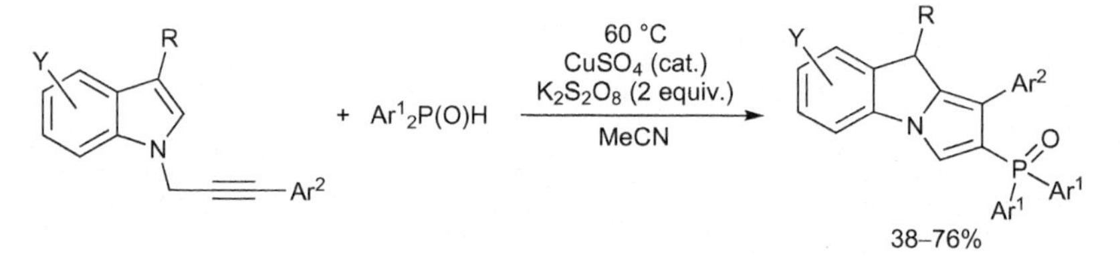

Ar^1	Ph	Ph	Ph	Ph	Ph	Ph	Ph	Ph	Ph	Ph	Ph	Ph
Ar^2	Ph	Ph	Ph	Ph	4-MeOPh	4-CF_3OPh	4-^{t}BuPh	4-^{t}BuPh	4-MePh	4-MePh	3-MePh	2-MePh
R	Me	H	H	H	H	H	H	H	H	H	H	H
Y	H	H	Me(α)	Cl(γ)	H	H	H	MeO(β)	H	MeO(β)	H	H

Ar^1	Ph	Ph	4-MeOPh	4-MeOPh	4-MePh	4-MePh	4-MePh	4-MePh	4-BrPh
Ar^2	4-FPh	4-MeC(O)Ph	Ph	2-thienyl	4-^{t}BuPh	4-^{t}BuPh	4-MePh	4-MePh	Ph
R	H	H	H	H	H	H	H	H	H
Y	H	H	H	H	H	MeO(β)	MeO(β)	H	H

Scheme 34 2-Phosphinoyl-pyrroloindoles via tandem reactions

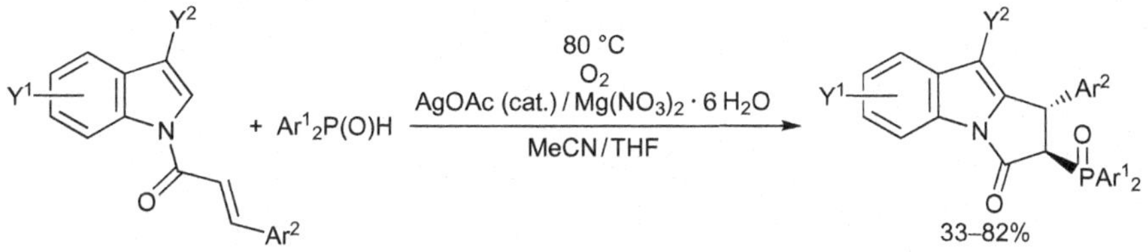

Y^1	H	H	H	H	H	H	MeO(α)	Me(α)	F(β)	Cl(β)
Y^2	Ph	Ph	Ph	4-MePh	Me	H	Ph	Ph	Ph	Ph
Ar^2	Ph	4-FPh	2-ClPh	Ph	Ph	Ph	Ph	Ph	Ph	Ph
Ar^1	Ph	Ph	Ph	Ph	Ph	Ph	Ph	Ph	Ph	Ph

Y^1	H	H	H	H	H	H	H	H
Y^2	Ph	Ph	Ph	Ph	Ph	Ph	Ph	Ph
Ar^2	Ph	Ph	Ph	Ph	Ph	Ph	Ph	Ph
Ar^1	4-MeOPh	2-MeOPh	3,4-diMePh	3,5-diMePh	4-MePh	2-MePh	4-FPh	4-ClF

Scheme 35 Fused phosphinoylindoles via cascade reactions

A cascade involving the addition of secondary phosphine oxides to 1-(1*H*-indol-1-yl)-3-phenylprop-2-en-1-ones followed by oxidative cyclization furnished eventually a fused indole derivative with a phosphinoyl substituent (Scheme 35) [39]. A phosphinoyl radical is added on the α,β-unsaturation of the starting material that is followed by cyclization and the loss of a hydrogen radical. Oxidative deprotonation leads to the final pyrroloindole (Scheme 36) [39].

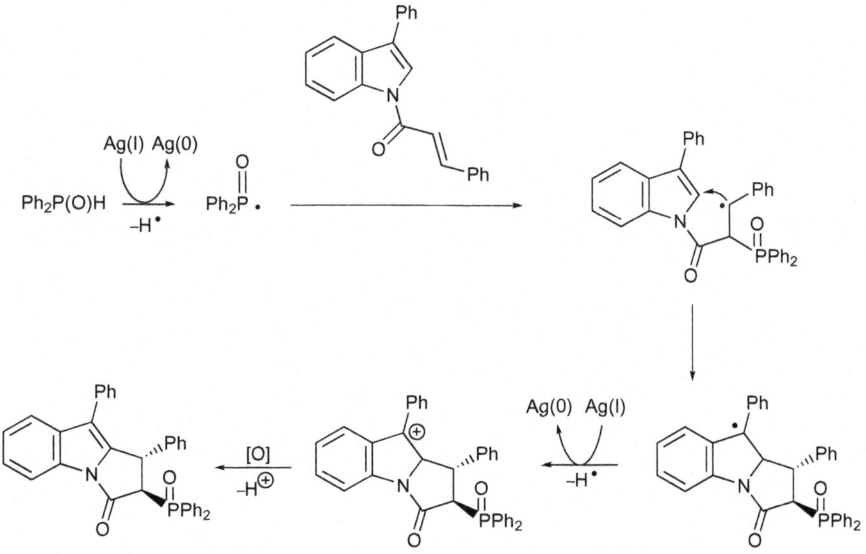

Scheme 36 Mechanism for the formation of fused phosphinoylindoles

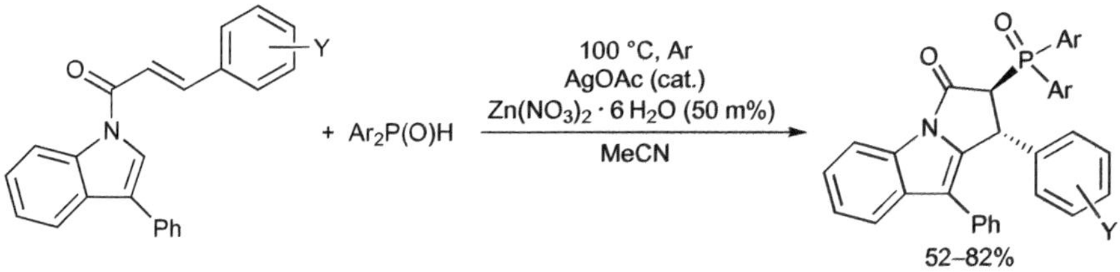

Ar	Ph	Ph	Ph	Ph	Ph	Ph	Ph	Ph	3,5-diMePh	4-MePh	4-ClPh
Y	4-MeO	4-Me	4-F	4-F-3-Br	4-Cl	3-Cl	4-Br	3-Br	Ph	Ph	Ph

Scheme 37 A tandem reaction affording 2-phosphinoyl-pyrroloindoles

A similar tandem sequence starting with the interaction of somewhat different cinnamyl-substituted indole derivatives and diarylphosphine oxides led eventually to 2-phosphinoyl-3*H*-pyrrolo[1,2-*a*]indoles (Scheme 37) [40]. It is unclear, why the authors worked under argon atmosphere. The mechanism is similar to that shown in the previous case.

A nice example is the analogous tandem reaction of propargylamide-substituted indoles and $Ph_2P(O)H$ to afford at the end indoloazepinones. Depending on the conditions, two isomers may be formed (Scheme 38/(1) and (2)) [41]. The application of inert atmosphere is again a guess.

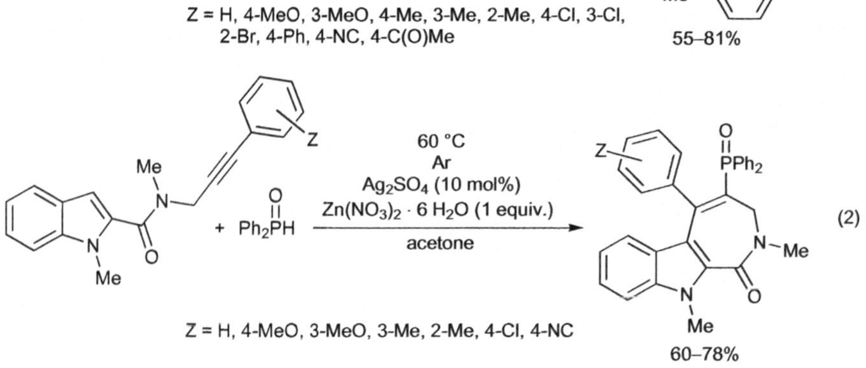

Scheme 38 The synthesis of indoloazepinones

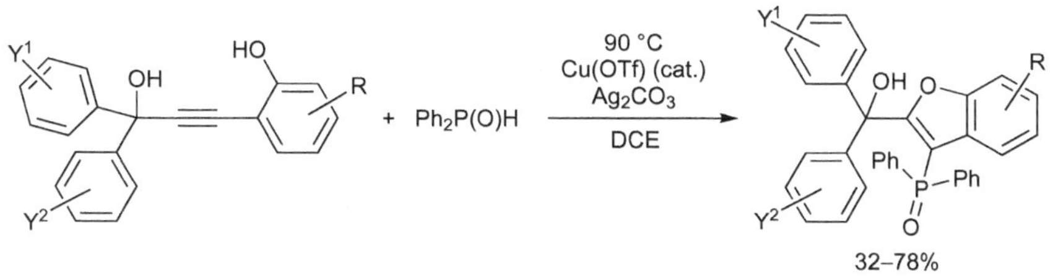

Y^1	H	4-MeO	4-Me	3-Me	2-Me	4-F	4-Cl	3-Cl	4-Ph	4-Me	4-F	4-Cl
Y^2	H	H	H	H	H	H	H	H	H	4-Me	4-MeO	4-Cl

Scheme 39 The synthesis of phosphinoyl-benzofuranes

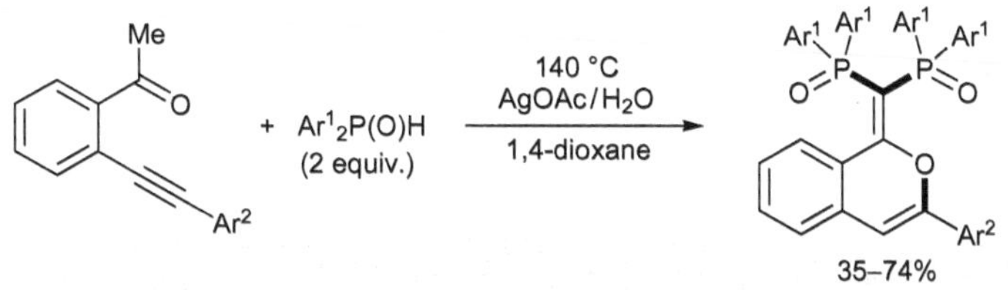

Ar^1	Ph	Ph	Ph	Ph	Ph	Ph	Ph	Ph	Ph
Ar^2	Ph	4-MeOPh	3-MeOPh	4-^{t}BuPh	4-EtPh	4-MePh	3-MePh	4-FPh	3-FPh

Ar^1	Ph	Ph	Ph	3,5-diMePh	4-MePh	4-ClPh
Ar^2	4-ClPh	3-ClPh	4-BrPh	Ph	Ph	Ph

Scheme 40 Bis(phosphinoyl)methylene-isochromenes via a bisphosphinoylation-cyclization approach

6 The Synthesis of Different *O*-Heterocycles with Exocyclic P-Function

Cascade annulation of triarylpropargylic alcohols with diphenylphosphine oxide comprising two additions and a dehydrogenation gave rise to phosphinoylbenzofurans (Scheme 39) [42].

Bisphosphinoylation of α-alkynylaryl methyl ketones connected with cyclization afforded bis(phosphinoyl)methylene-isochromenes (Scheme 40) [43].

7 The Synthesis of Carbocycles with Exocyclic P-Function

The reaction of 1,6-enynes with secondary phosphine oxides under appropriate conditions resulted in the formation of phosphinoylfluorene derivatives as the end product. The addition of the >P(O)H species to the acetylenic moiety was followed

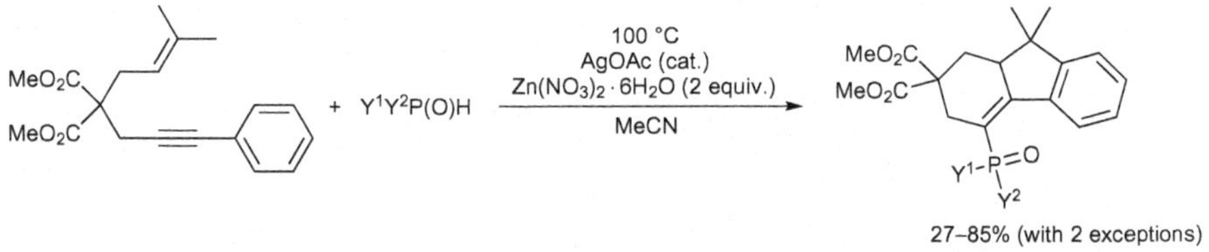

Y^1	4-MeOC_6H_4	4-MeC_6H_4	4-FC_6H_4	3-MeOC_6H_4	3-MeC_6H_4	2-MeC_6h_4	$3,5\text{-diMeC}_6H_3$	$3\text{-F-}4\text{-MeC}_6H_3$	Ph	^{n}Bu
Y^2	4-MeOC_6H_4	4-MeC_6H_4	4-FC_6H_4	3-MeOC_6H_4	3-MeC_6H_4	2-MeC_6h_4	$3,5\text{-diMeC}_6H_3$	$3\text{-F-}4\text{-MeC}_6H_3$	^{n}Bu	^{n}Bu

Scheme 41 The synthesis of phosphinoylfluorenes

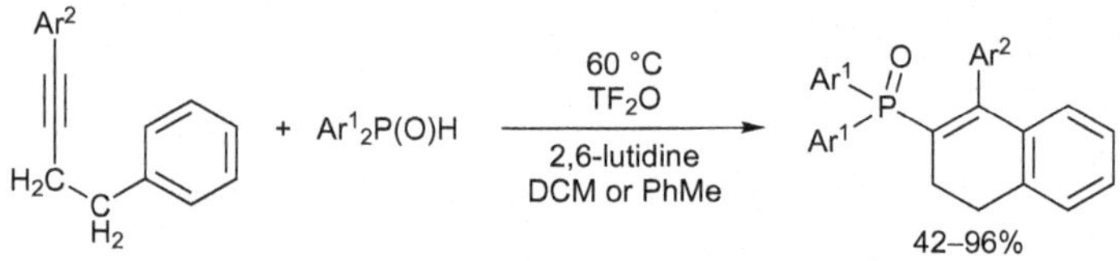

Ar^1	Ph	Ph	Ph	Ph	Ph	Ph	Ph	Ph
Ar^2	Ph	4-MeOPh	4-CF_3Ph	4-NCPh	2-MePh	2-BrPh	1-naphthyl	2-thienyl

Ar^1	4-MeO-3,5-di^{t}BuPh	2-MePh	4-BrPh	1-naphthyl	2-naphthyl
Ar^2	Ph	Ph	Ph	Ph	Ph

Scheme 42 The synthesis of phosphinoyl-dihydronaphthalenes

by two intramolecular cyclizations that are an addition and a dehydrogenation (Scheme 41) [44].

The reaction cascade involving an addition and a dehydrogenative cyclization of methylene-tethered arylalkynes and diarylphosphine oxides ended up with phosphinoyl-dihydronaphthalenes (Scheme 42) [45].

8 Conclusions

In summary, oxidative P-C coupling reactions between >P(O)H species and a series of substrates, along with different variations including addition and subsequent oxidation often combined with cyclization are surveyed to afford valuable P-functionalized products embracing heterocyclic and cyclic compounds as well. Mechanism of the cross-dehydrogenative couplings is mostly radical, and the multifaceted transformations generally require an oxidizing agent and a catalyst. The reactions surveyed represent green transformations, as H_2O or H_2 is the by-product instead of typically hydrogen halides or other species in classical cross-couplings. Many challenges remain for the future in respect of the dehydrogenative C–C and C-P couplings.

References

1. "From C–H to C–C bonds: Cross dehydrogenative coupling" (Part 26. of the "RSC Green Chemistry" series) ed. C.-J. Li, Royal Society of Chemistry, Cambridge, UK, 2015
2. Hirao T, Masunaga T, Ohshiro Y, Agawa T (1980) Tetrahedron Lett 21:3595
3. Hirao T, Masunaga T, Yamada N, Ohshiro Y, Agawa T (1982) Bull Chem Soc Jpn 55:909
4. Jablonkai E, Keglevich G (2014) Curr Org Synth 11:429
5. Jablonkai E, Keglevich G (2015) Curr Green Chem 2:379
6. Henyecz R, Keglevich G (2019) Curr Org Synth 16:523
7. Henyecz R, Keglevich G (2018) P-C couplings by the Hirao reaction pp. 158–178. In: Keglevich G (szerk.) Organophosphorus chemistry: novel developments. Németország/Walter de Gruyter, Berlin/Boston, p 315
8. Trofimov BA, Volkov PA, Khrapova KO, Telezhkin AA, Ivanova NI, Albanov AI, Gusarovaa NK, Chupakhin ON (2018) Chem Commun 54:3371
9. Yuan T, Huang S, Cai C, Lu G (2018) Org Biomol Chem 16:30
10. Feng C-G, Ye M, Xiao K-J, Li S, Yu J-Q (2013) J Am Chem Soc 135:9322
11. Xie J, Li H, Xue Q, Cheng Y, Zhu C (2012) Adv Synth Catal 354:1646
12. Gao M, Li Y, Xie L, Chauvin R, Cui X (2016) Chem Commun 52:2846
13. Luo K, Chen Y-Z, Chen L-X, Wu L (2016) J Org Chem 81:4682
14. Zhang H-J, Lin W, Wu Z, Ruan W, Wen T-B (2015) Chem Commun 51:3450
15. Singsardar M, Dey A, Sarkar R, Hajra A (2018) J Org Chem 83:12694
16. Li L, Wang J-J, Wang G-W (2016) J Org Chem 81:5433
17. Lin W, Su F, Zhang H-J, Wen T-B (2017) Eur J Org Chem 2017:1757
18. Liu Y, Liu Z, Zhang Y, Xiong C (2018) Adv Synth Catal 360:3492
19. Müller P, Fuhr O, Döring M (2013) Heteroatom Chem 22:252
20. Wang G-W, Wang C-Z, Zou J-P (2011) J Org Chem 76:6088
21. Chang Y-C, Yuan P-T, Hong F-E (2017) Eur J Org Chem 2017:2441
22. Kuninobu Y, Yoshida T, Takai K (2011) J Org Chem 76:7370
23. Furukawa S, Haga S, Kobayashi J, Kawashima T (2014) Org Lett 16:3228
24. Unoh Y, Hirano K, Satoh T, Miura M (2013) Angew Chem Int Ed 52:12975
25. Ma W, Ackermann L (2014) Synthesis 46:2297
26. Chen Y-R, Duan W-L (2013) J Am Chem Soc 135:16754
27. Zhang P, Gao Y, Zhang L, Li Z, Liu Y, Tang G, Zhao Y (2016) Adv Synth Catal 358:138
28. Ma D, Chen W, Hu G, Zhang Y, Gao Y, Yin Y, Zhao Y (2016) Green Chem 18:3522
29. Quint V, Morlet-Savary F, Lohier J-F, Lalevée J, Gaumont A-C, Lakhdar S (2016) J Am Chem Soc 138:7436
30. Hu G, Zhang Y, Su J, Li Z, Gao Y, Zhao Y (2015) Org Biomol Chem 13:8221
31. Li J, Zhang W-W, Wei X-J, Hao W-J, Li G, Tu S-J, Jiang B (2017) Org Lett 19:4512
32. Li Y, Qiu G, Ding Q, Wu J (2014) Tetrahedron 70:4652
33. Cao J-J, Zhu T-H, Gu Z-Y, Hao W-J, Wang S-Y, Ji S-J (2014) Tetrahedron 70:6985
34. Zhang H, Gu Z, Li Z, Pan C, Li W, Hu H, Zhu C (2016) J Org Chem 81:2122
35. Zhang P, Zhang L, Gao Y, Tang G, Zhao Y (2016) RSC Adv. 6:60922
36. Chen S, Zhang P, Shu W, Gao Y, Tang G, Zhao Y (2016) Org Lett 18:5712
37. Zheng J, Zhang Y, Wang D, Cui S (2016) Org Lett 18:1768
38. Zhang H, Li W, Zhu C (2017) J Org Chem 82:2199
39. Liu JM, Zhao SS, Song WW, Li R, Guo XY, Zhuo KL, Yue YY (2017) Adv Synth Catal 359:609
40. Xu J, Yu X, Song Q (2017) Org Lett 19:980
41. Hua H-L, Zhang B-S, He Y-T, Qiu Y-F, Wu X-X, Xu P-F, Liang Y (2016) Org Lett 18:216

42. Li X-S, Han Y-P, Zhu X-Y, Li M, Wei W-X, Liang Y-M (2017) J Org Chem 82:11636
43. Sun J, Qiu J-K, Wu Y-N, Hao W-J, Guo C, Li G, Tu S-J, Jiang B (2017) Org Lett 19:754
44. Zhou Z-Z, Jin D-P, Li L-H, He Y-T, Zhou P-X, Yan X-B, Liu X-Y, Liang Y-M (2014) Org Lett 16:5616
45. Unoh Y, Hirano K, Miura M (2017) J Am Chem Soc 139:6106

Mechanistic Pathways Toward the Synthesis of Heterocycles Under Cross-Dehydrogenative Conditions

Ahmad A. Almasalma and Esteban Mejía

Abstract Due to the richness of the field of cross-dehydrogenative coupling reactions and the plethora of applications toward the synthesis and derivatization of heterocyclic molecules, there is an increasing amount of synthetic methodologies which spread across different catalytic fields and hence, yield to a huge variety of operating mechanisms. Although all the pathways follow the same underlying principles, they can be easily differentiated according to the type of activation (thermal of photocatalyzed) and the type of catalyst (metallic or metal-free). In this account, we have summarized the most recent and relevant examples under the aforementioned categories, focusing especially on those presenting the most compelling and better understood mechanisms.

Keywords Reaction mechanisms · Heterocycles · Organometallic catalysis · Oxidative coupling · Photocatalysis

1 Introduction and Scope

Throughout this chapter, we will present representative examples of the synthesis and functionalization of heterocyclic compounds by means of cross- dehydrogenative coupling (CDC) reactions, focusing on the employed conditions and the underlying reaction mechanisms. This account, apart from presenting up-to-date advances in the field, is complementary to the excellent review compiled by Todd and co-workers in 2015 [146] since in here the focus is mainly laid on those CDC reactions which result on the creation of an heterocycle, and to a lesser extent on those merely dealing with the functionalization of pre-formed heterocyclic rings.

In general, a CDC reaction can be described as the coupling of a (pre)electrophile with a (pre)nucleophile, both devoid of any prefunctionalization of leaving

---

A. A. Almasalma · E. Mejía (✉)
Leibniz Institute for Catalysis (LIKAT), Albert-Einstein-Str. 29a, 18059 Rostock, Germany
e-mail: Esteban.mejia@catalysis.de

A. Srivastava and C. K. Jana (eds.), *Heterocycles via Cross Dehydrogenative Coupling*, https://doi.org/10.1007/978-981-13-9144-6_10

groups with the intermediacy of a chemical oxidant. Hence, the synthesis of heterocycles by electrochemical methods will be not covered in this chapter [49]. In strong contrast with *classical cross-coupling,* where the regioselectivity of the reaction is determined by a leaving group (where the reaction takes place) or by a directing group (making a specific atom more reactive) [20, 87], *CDC reactions* rely on the stability of the intermediates (neutral radical or cation) and bond dissociation energies [115, 170]. For this reason, most of the C(sp^3–H) carbons functionalized via CDC reactions presented in this review are allylic [5, 69], benzylic [22], or adjacent to heteroatom (oxygen or nitrogen), since these can stabilize the reactive intermediates and foster the expected regioselectivity [14, 21, 28, 35, 168]. Taking into account that one of the most important features of the CDC reaction is the ability to use unfunctionalized reactants, any transformation for heterocyclic or synthesis that includes pre-functionalized reagents will be out of our focus [25]. Also, oxidative cross-coupling between two (hetero)arenes will not be considered [27, 172]. Moreover, since many CDC reactions can be effectively catalyzed by complexes of cheap and abundant metals, reports using precious metals (like palladium) [177] for this transformation are also excluded from this chapter. Importantly, reactions involving two-electron oxidative addition of a C–H bond at a metal center, formally described as C–H activation rather than C–H functionalization, are out of the scope of this book, and hence not covered in this chapter. Some reports discussing oxidative annulation via single electron transfer processes have been included in this account due to its relevance for the mechanistic discussion [80, 146, 156, 178].

The transformations presented in this review have been classified in two main categories according to the electronic energy estate of the reactants (and catalysts): thermal reactions (Sect. 2), where all molecules throughout the catalytic cycle remain in their electronic ground states; photochemical reactions (Sect. 3), where the productive cycle is triggered by electronic excitation of reactants or catalysts. Both sections are further subdivided by the nature of the catalyst.

2 Thermal CDC Reactions

The most common type mechanism of CDC reactions involves the thermal generation of radical species, typically using a redox-active metal catalyst and a chemical oxidant. Due to the high temperatures generally required to achieve homolytic bond scission, these transformations commonly possess challenges in terms of regio- and stereoselectivity control. In spite of these limitations, starting from the initial investigations using copper catalysts, CDC reactions have been successfully achieved using other non-noble metal catalysts including ruthenium, iron cobalt, and vanadium, among others. Interesting developments have been also made on metal-free systems using iodine and organic oxidants like DDQ and TEMPO.

2.1 Ruthenium Catalysts

There are many reports of the use of ruthenium catalysts on CDC reactions, although the metal complexes have been attributed different roles in the mechanism [1–3, 89–91, 100, 127, 132, 140, 154] including the generation of the reactive intermediates (oxidation), its stabilization by coordination (generally an iminium ion), generation of the nucleophile or fostering of its addition to the electrophile. In all cases, the unavoidable use of an oxidant (oxygen of hydrogen peroxide) is necessary to oxidize the substrate, to regenerate the reduced catalyst, or both.

In 2015, Rajaka and co-workers reported the synthesis of N-fused polycyclic heteroarynes by one-pot two-steps reaction [72], using oxygen as terminal oxidant (Fig. 1) instead of the previously reported $Cu(OAc)_2$ [78]. Intermolecular competition experiments showed that electron-rich derivatives are more reactive than electron-poor ones. Furthermore, they found that a directing group is essential for the transition metal catalyzed annulation process and that the oxidative-dehydrogenative step occurred first, followed by cross-coupling/annulation. The first step, dehydrogenation of **I**, begins with the more basic nitrogen binding to $LRuCl_2$(L = (p-cymene)) to form complex B, which then undergoes β-H elimination to afford quinazolinone **II** and [RuCl]-H. Subsequently, intermediate complex C decomposes to a Ru(0)species, which is oxidized by O_2 in presence of HCl to regenerate the active Ru(II) catalyst. In a parallel cycle, the cross-coupling/annulation of quinazolinone with alkyne takes place, beginning with coordination of quinazolinone to the Ru(II) catalyst to afford ruthenacycle**III**, followed by

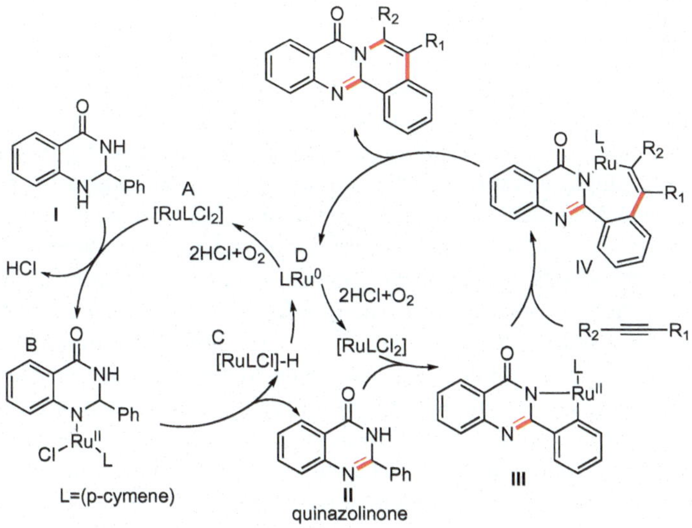

Fig. 1 Synthesis of N-fused polycyclic heteroarynes with using Ru^{II} catalyst as reported by Rajaka [72]

insertion of the alkyne to form complex **IV**. This metallacycle undergoes reductive elimination to produce the desired product and a Ru(0)complex, which is reactivated by oxygen and HCl.

In 2017, Mohit and co-workers reported the synthesis of symmetrical and unsymmetrical bis(indoyl)methane, where tetramethylurea (TUM) act as methylene source assisted by microwave using $RuCl_3 \cdot 3H_2O$/TBHP (tert-butyl hydroperoxide) [26]. They proposed a radical mechanism due the decreased yields observed when a radical scavenger was added to the mixture under the optimal reaction conditions. Their proposed mechanism starts by a single electron transfer (SET) from TUM to the Ru(III) complex to generate a radical cation, which is oxidized to an iminium ion by TBHP. A nucleophilic addition reaction on the previous intermediate by *N*-substituted indole followed by trimethyl urea elimination produced an alkylideneindoleninium ion. This ion further attacked by NH-indole or N-substituted indole to furnish the desired product, and reactivate the catalyst by oxidation with TBHP. The main difference between these two examples is the different roles of the oxidants and the Ru catalyst in the generation of the iminium intermediate.

2.2 *Iron Catalysts*

Iron catalysts are among the most attractive in organic synthesis due to the abundance of iron, low cost, non-toxicity, and environmentally harmless character. It has resulted in important applications in the synthesis of bioactive molecules and pharmaceuticals. In CDC reactions, iron catalysts have been mainly used for the formation of C–C, C–N, and C–O bonds [9, 57, 65, 77, 79, 94, 111, 161, 175, 183]. Here, we will present some representative examples of iron-catalyzed CDC reactions for which there is a consensus on the operating mechanism. In general, there are two mechanistic pathways depending on the nature of the pre-electrophile (Fig. 2): (A) In the case of benzylic or allylic carbons, generation of a carbon radical by reaction with the oxidant is considered to be the first step, which then is reoxidized via single electron transfer (SET) by the iron catalyst to produce a carbocation, which is finally trapped by the nucleophile (pre-nucleophile) to provide the desired product; (B) functionalization of a $C(sp^3)$–H carbon adjacent to heteroatom starting with formation of a radical cation by single electron transfer (SET) from the heteroatom to the Iron catalyst, followed by an oxidation step where the oxidant forms stable carbenium intermediate (oxonium or nitronium), which is then trapped by the nucleophile to furnish the desired product and regenerating the active catalyst by reaction with the oxidant. In both pathways, deprotonation of the NuH to generate Nu^- can happen before or after the trapping step depending on its acidity. Functionalization of $C(sp^3)$–H bonds adjacent to heteroatoms has been reported by Li and co-workers for different types of ethers, sulfides and amines with different types of iron catalyst. Kinetic isotope effect studies, (KIE, by means of a competition reaction between THF and $[D_8]$THF), showed that C–H bond cleavage is the rate-determining step [70]. Synthesis of quinoline-fused lactones was reported

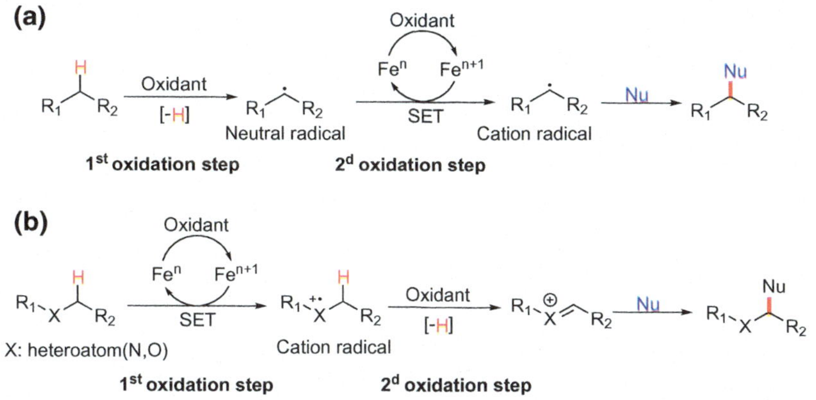

Fig. 2 Proposed mechanistic pathways for the iron-catalyzed in CDC reactions

by Huo and co-workers in 2015 by dual oxidative-dehydrogenative tandem annulation of glycine derivatives and tetrahydrofurans. Their mechanism study indicates that imine and THF could be involved in the intermediate; initially, in the presence of $FeCl_2$/TBHP, glycine ester and THF are converted to aryl imine and DHF, then an oxidative imino Diels–Alder [4 + 2] reaction occurs to have tetrahydro-quinazoline as intermediate, which is transformed to 2,3-disubstituted quinoline intermediate under acidic conditions. Finally, aromatization of the intermediate and intramolecular ester exchange ring closure followed to afford the product [46].

Li and co-workers also reported the iron-catalyzed synthesis polysubstituted benzofurans from simple phenol and β-ketoester under CDC conditions [36], the proposal mechanism rely on the basis of KIE and cross-over experiments to proceed via a radical-based oxidative step initially and then a cyclization step, where iron acts as transition-metal catalyst in the oxidative coupling step and as Lewis acid in the condensation step. The KIE indicated that aromatic C–H bond cleavage is not involved in the rate-determining steps of this reaction.

Very recently, Zhong reported an iron-catalyzed radical [3 + 2] cyclization, which allows coupling of different *N*-sulfonylanilines with various α-substitued styrenes (Fig. 3) [34]. Mechanistically, EPR studies implied that reaction is likely initiated by N–H hydrogen abstraction to generate an N-centered neutral radical **I** (Fig. 3). The electrophilicity of aniline radical **I** is not sufficient to attack styrene **II** directly; however, activation by iron makes the addition possible. As a result, a carbon radical **III** is generated and the cleavage of *ortho* C–H bond is proposed as the rate-determining step. The desired product is produced by re-aromatization of the phenyl ring with simultaneous release of $FeCl_3$.

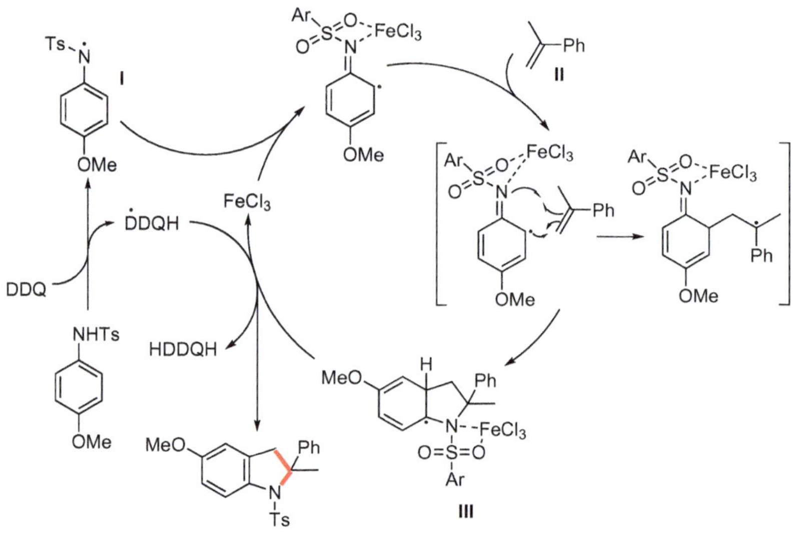

Fig. 3 Iron-catalyzed [3 + 2] cyclization under oxidative conditions according to Zhong. Ghosh et al. [34]

2.3 *Vanadium and Cobalt Catalysts*

Even though vanadium is non-toxic, readily available, cheap, and also present in biological molecules [120], its application in CDC reactions has not been explored as widely as other transition metals like Ru, Rh, Fe, and Cu [42, 43].

Vanadium-catalyzed oxidative C–C bond formation reactions can be classified in two categories: synthesis of biaryls by homo-coupling phenolic compounds [15, 51, 55, 74, 118, 119, 138] and aminoalkylation [52, 53, 60, 86]. The Mechanism of the vanadium-catalyzed CDC reaction follows a pathway analogous to those of iron systems used for C–Sp^3–H functionalization adjacent an heteroatom (nitrogen), characterized by the formation of an iminium intermediate preceded by amine coordination to the metal center accompanied by SET/HAT (hydrogen atom transfer) steps [129]. The oxidative functionalization of α-carbons to nitrogen has witnessed a faster growth than for other heteroatom such as oxygen and sulfur due to the higher oxidation potential of these α-carbons, although, in general, they follow almost the same reaction mechanism [30].

Punniyamurthy and co-workers reported the condensation of α,β-unsaturated ketones with aryl hydrazines to synthesize different types of functionalized pyrazoles using air as an oxidant [120]. In their proposed mechanism, the condensation of a ketone with an aryl hydrazine is followed by reaction with $VOSO_4$ to furnish the intermediate. Concerted metalation–deprotonation of this intermediate gives a metallacyclic intermediate, which undergoes reductive elimination to produce the desired product. Regeneration of the active catalyst is done by the peroxo-vanadium species to close the catalytic cycle.

In general, the use of cobalt catalysts for the synthesis of heterocycles via CDC transformations [159] is not as wide as it is for C–H bond activation reactions [88]. Cobalt(II)-catalyzed CDC of *N*-aryl tetrahydroisoquinolines with various pro-nucleophiles was reported recently by Patil and co-workers, using both oxygen and *N*-hydroxyphthalimide (NHPI) as oxidants [105]. In the proposed mechanism, an iminium intermediate is generated via SET from the nitrogen atom in the substrate to a Co(III) species, which is formed by oxidation of Co(II) by oxygen. Then, HAT from the α-carbon of nitrogen in the substrate to the reduced species of NHPI (PINO) takes place. Finally, the nucleophile attacks the iminium ion to form the α-substituted CDC product.

Cobalt-catalyzed CDC of 2-arylimidazo[1,2-a]pyridines with isochroman using molecular oxygen as oxidant to produce 2-arylimidazo[1,2-a]pyridine derivatives was reported by Wang and co-workers [171]. In their investigations, a regioselective functionalization of α-sp^3 carbon atom to oxygen (benzylic) was achieved. They report two possible reaction mechanisms: in the first one, the active intermediate (oxonium) is formed by one-step oxidation of isochroman by O_2 and Co(II) to yield an oxonium–cobalt intermediate; in the second proposed pathway, the oxonium intermediate is obtained by SET from isochroman to Co(II) followed by HAT. On a final step, the CDC product results from the attack of the nucleophile to the oxonium intermediate.

2.4 Copper Catalysts

Copper catalysts are arguably the most popular choice for CDC reactions, having a growing number of examples in the literature for the synthesis and functionalization of different types of heterocyclic compounds which, in lieu of space, will not be discussed in here [37, 47, 63, 76, 101, 109, 125, 128, 139, 160, 162, 166, 179].

The mechanism for the copper-catalyzed CDC derives from the pioneering works of Li and co-workers, mostly dealing with the derivatization of C(sp^3–H) adjacent to nitrogen. In their proposed mechanism, the starting material is oxidized by an oxidant (generally oxygen or TBHP) to produce the cationic iminium intermediate, which is stabilized by reaction with the copper catalyst forming a π-imine-copper complex. The nucleophile (pre-nucleophile) is generated by coordination with the copper, which reacts with the iminium intermediate to produce the CDC product [8, 66–68, 124].

Alternatively, a radical-ionic mechanism was discussed as a possible pathway to generate the iminium intermediate. Two steps are proposed in the radical mechanism: HAT and SET. In the ionic mechanism, water is eliminated from the copper–nitrogen (amine) complex. The evidence of one mechanism over the other was obtained by using radical scavenger reagents (BHT, TEMPO, etc.). Unfortunately, the oxidative nature of these molecules could bias the results by accelerating of inhibiting the reaction [32, 128, 176].

Klussmann and co-workers reported an experimental study of the mechanism proposed by Li [10, 50]. They were able to isolate the iminium intermediate as the dichlorocuprate complex and to characterize it by X-Ray crystallography. From their experiment, the ionic character of the bond between the copper and the iminium ion was established, instead of the π–iminium–copper complex interaction proposed by Li. They also studied the effect of methanol to effectively stabilize the iminium ion and it was shown to be important in an oxidative allylation reaction. Moreover, they observed by NMR experiments that the nucleophilic addition to the iminium–copper intermediate depends on the counter ions: weekly coordinated counter ions lead to faster formation of the CDC product and vice versa.

Borpatra and co-workers have developed a 3-component route for the synthesis of 1,3-oxazines using methanol as C1 feedstock. The methylene carbon of the product originates from methanol. The reaction is carried out at room temperature using $Cu(AcO)_2{\cdot}H_2O$ as catalyst and tert-butyl hydroperoxyde (TBHP) as oxidant. On the basis of their control experiments, a free-radical mechanism is proposed (Fig. 4) [11]. The catalytic cycle starts with reduction of the copper (II) catalyst by TBHP to tert-butyl peroxy radical (**I**). A reaction between **I** and methanol generates a hydroxymethyl radical (**II**). Afterward, TBHP reacts with Cu(I) to produce tert-butoxy radical (**III**), which in the presence of a hydroxymethyl radical gives

Fig. 4 Copper-catalyzed synthesis of 1,3-oxazines using methanol as C1 synthon

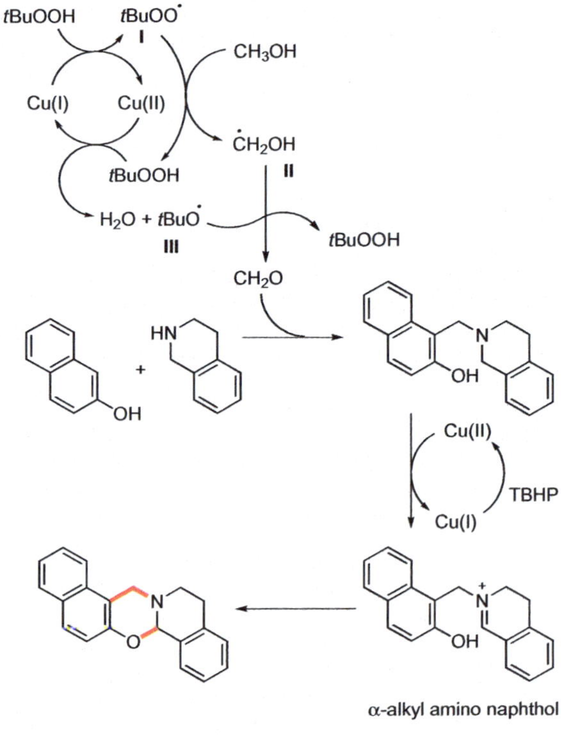

formaldehyde. The 3-component Mannich reaction of naphthol, amine, and formaldehyde gives α-alkylaminonaphthol, which is then converted into the corresponding iminium ion in the presence of Cu(II) and finally cyclized to the desired product. Interestingly, only Cu(I or II) catalysts were able to catalyze the reaction, while no reaction was observed using Fe or Ru catalysts.

Peng and co-workers reported the synthesis of dicoumarin-fused [1, 5] -diazocines from aryl methyl ketones and 4-aminocoumarins via sequential oxidative cross-coupling, C–C bond cleavage, and cyclization processes using a $CuCl_2/O_2$ system [106]. The proposed reaction mechanism (Fig. 5) involves radical intermediates as demonstrated by control experiments in the presence of radical scavengers (TEMPO and BHT). The two-steps oxidation of benzophenone to 2-hydroxy-1-phenylethan-1-one (**I**) and 2,2-dihydroxy-1-phenylethan-1-one (**II**) by $CuCl_2/O_2$ is followed by the addition of water to **II** to produce the more stable dihyrox-1-phenylethanone. Nucleophilic addition of aminocoumarin to the alcoholic carbon produces the first C–N bond with elimination water. The third oxidation reaction to convert the last intermediate to diketone is produced, where the next C–N bond is formed by a second nucleophilic addition. SET from the two nitrogen atoms to copper(II) formed the nitronium intermediate, which is converted to the neutral radical by losing two H^+. Finally, the radical undergoes intramolecular C–C bond cleavage and cyclization to furnish the desired product and regenerating the active copper(II) catalyst.

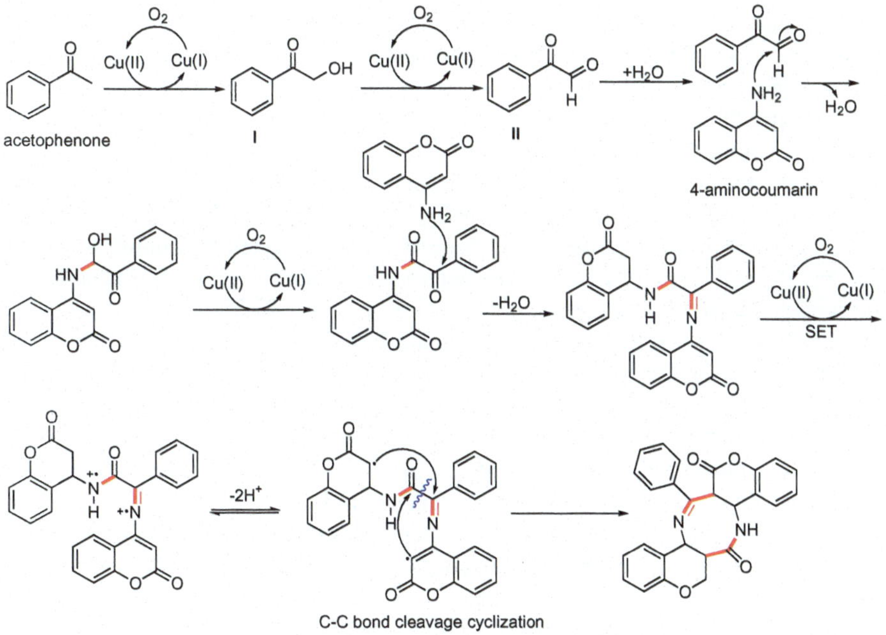

Fig. 5 Synthesis of dicoumarin-Fused [1,5]-diazocines by Cu(II)/O_2 system

2.5 Metal-Free Systems

In general, metal-free catalytic reaction is an interesting and promising research area taking into account the urges toward the development of environmentally friendly and sustainable "green" processes. Typically, the drawbacks of using transition metal (homogeneous) catalysts are their elevated price, the often difficult separation from the reaction products and toxicity concerns, especially in the synthesis of products for food and pharmaceutical applications. Among these, CDC reactions have had recently a lively growth [18, 44, 45, 48, 59, 82, 96, 122, 143, 149, 167, 182].

Among the first reports of metal-free CDC reactions are those employing potassium iodide and TBHP mixtures in which the ability to mediate the reaction and their synergetic roles were proposed to produce the reactive intermediate [58, 61]. The role of KI was proposed to be either as the source of iodine, which oxidizes the substrate to the iminium intermediate or generates *tert*-butyloxide, which abstracts hydrogen from a carbon next to the heteroatom which is further oxidized by iodine to form the iminium intermediate.

Hypervalent iodine compounds have been also used in metal-free CDC reactions [48, 81, 92, 151, 163], although large amount of iodobenzene as by-product are generated. Consequently, with the aim of developing "greener" CDC reactions, I_2 catalyzed transformations in the presence of DMSO, O_2, or peroxide as terminal oxidant have gain growing popularity. Iodine has low toxicity, high availability, easily handling and low cost compared to transition-metal-based and hypervalent iodine catalysts. Additionally, iodine has various oxidation states (−1,0, +1, +3, +5, and +7), and thus, its structural properties and reactivity are in many ways reminiscent of those of transition metals. Furthermore, compared to meta-based catalysts, iodine operates under milder reaction conditions and has better substrate compatibility, higher tolerance to moisture and air, easier removal from the reaction mixture (washing with reducing agent), and higher reactivity both in concentrated and diluted reaction media. For these reasons, iodine has been widely used as catalyst and reagent in chemical synthesis [4, 73, 103, 104, 136, 164].

The roles of I_2 in these reactions depend on the targeted functionalized bond (Fig. 6): (a) activate the α-carbon (to heteroatom or carbonyl) to form the active intermediate or (b) decompose the peroxide initiating the reaction, or (c) heteroatom (X) activation to form X–I bond intermediate [103].

Different types of heterocyclic compounds such as indolo[2,3-b]carbazoles [169], different furans [142], indolizines derivatives [110, 142], dihydrofurans [141], 3,3-disubstituted-2-oxindoles [45], indole derivatives [38], and oxazoles [31] have been synthesized and functionalized [19, 29, 96, 123, 126, 150, 173, 181] under metal-free CDC conditions with using I_2 [103].

Iodine-catalyzed CDC reactions have many advantages but also some limitations. They work smoothly with electron-rich substrates and less efficiently with the electron-poor ones. This is a consequence of the required coordination of the substrate to the iodine as prerequisite of regioselectivity. Hence, while this process

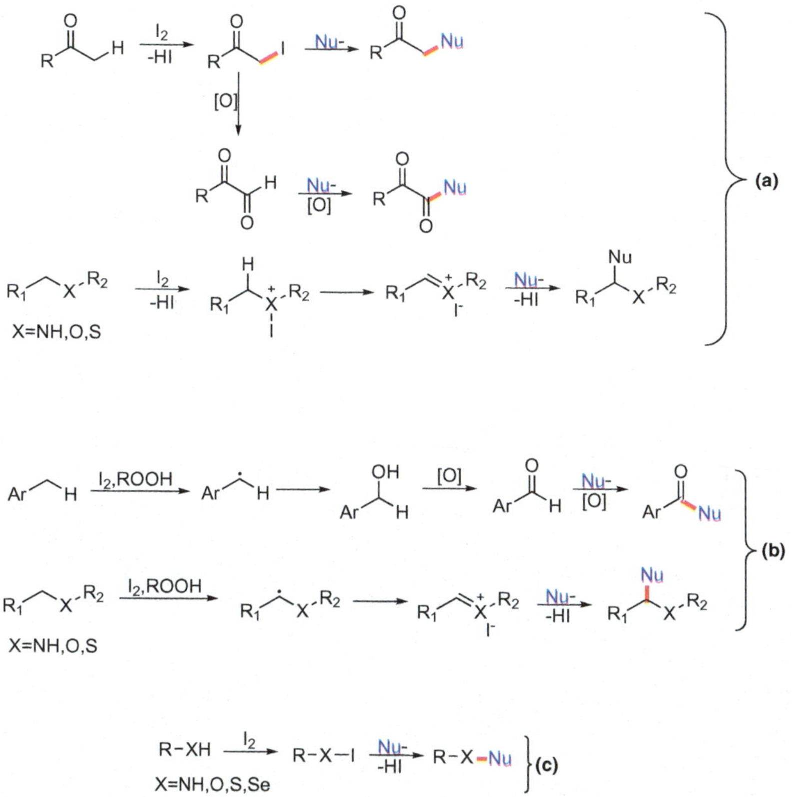

Fig. 6 General mechanisms of I_2 as catalyst in metal-free CDC reactions

is rather straightforward with electron-rich substrates, it is hampered with electron-poor substrates, resulting in low yield and selectivity.

Another oxidant that has gained a lot of attention in the recent year in the field of metal-free CDC is 2,3-Dichloro-5,6-Dicyanobenzoquinone (DDQ), for which many successful applications have been reported [71, 131]. It follows two mechanism pathways: radical [155] or ionic [7, 64, 144, 145, 147]. In the ionic mechanism, two steps are proposed, hydride abstraction by the DDQ (slow step) to form $DDQH^-$ followed by proton transfer (fast step) to form $DDQH_2$. In the radical mechanism, there are three steps, hydrogen abstraction to form DDQH (radical) and then SET to form $DDQH^-$, and finally, proton transfer to produce $DDQH_2$.

Recently, the use 2,2,6,6-tetramethylpiperidine-1-oxy (TEMPO) as an oxidant in the free metal CDC reaction was successfully reported to synthesis heterocyclic compounds such as 2-Aryl-4-quinolones [44] and Pyrrolin-4-ones [184].

3 Photocatalyzed CDC Reactions

The dawn of the twenty-first century has witnessed a dramatic increase on research toward the utilization of light, ultimately from the sun, as the driving force for a myriad of chemical processes, of course motivated by the foretold scarcity of energy sources looming the chemical industry in the future. Combination between the advantages of photocatalysis (generally at ambient temperature) and CDC reactions (direct use of underivatized reactants) have resulted in highly attractive research field in organic synthesis. Nevertheless, when compared to the bulk of reports on CDC reactions, the reports on the synthesis of heterocycles using photocatalytic systems remains largely underdeveloped.

Photocatalysis is by definition a process in which the energy derived from photons is absorbed by a metal-based photocatalyst, an organic dye [113] or photosensitizer [97, 98, 113] and utilized to either generate a reactive species (initiator) to start the reaction or to facilitate the catalytic turnover by giving access to activated intermediates. In its excite state, the photocatalyst is a better electron donor and electron acceptor compared to its ground state, so the photoexcited state can initiate a single electron transfer (SET) from the substrate (reductive quenching) or to it (oxidative quenching) to form radical cation or radical anion, respectively, triggering the subsequent chemical transformation [83].

Different types of C(sp^3)–H bonds [114] adjacent to amino group [158, 186], ether [165], as well as C(sp^2)–H bonds in aromatic [112], and alkenyl systems [75] have been successfully functionalized photocatalytically. Most of the reports in this field deal with the formation of C–C or C-heteroatom bonds using pre-functionalized substrates (leaving group) and radical precursors (arylhalide, aryl diazonium salt, iodonium salt, etc.) as coupling partners [153].

Many complexes with d^6-electron configuration based on Re^{I}, Ir^{III}, and Os^{II} have long-lived excited states with redox properties completely different from their ground state. These complexes have been exploited widely as light absorbers and catalyst for photochemical transformations [93, 121, 174]. Nevertheless, in this section, we will focus on photocatalysts based on more abundant metals [62] as this has become a dynamic research field, offering alternatives to the well-studied precious metal-based photocatalysts [23, 33, 108, 130, 148]. The basic limitation of using complexes of base and abundant metals is the lifetime of their excited states, which is very short compared with those of the precious metals complexes. Nevertheless, via wiser modification of the accompanying ligands, it is possible to extend their lifetimes on the exited state [180].

There are four types of electron-transfer processes upon light absorption in the metal complexes: (1) from a*d* orbital at the metal to an unoccupied one, also at the metal (metal–metal charge transfer, MMCT); (2) metal to ligand charge transfer (MLCT), when an electron moves from a *d* orbital at the metal to an unoccupied π^* orbital at the ligand; (3) ligand to metal charge transfer (LMCT), when an electron from an orbital at the ligand is promoted to an unoccupied orbital in the metal; (4) π–π^* electron transfers from a π orbital in the ligand to unoccupied one, also at the ligand (LLCT). Commonly, *d–d* MMCT and π–π* LLCT are followed by rapid excited state relaxation since they have large wave function overlap with the ground state, and consequently, have low efficiency in the SET reactions with the substrate [157]. Metal complexes with a d^0 configuration do not have any electron in the *d* orbitals; therefore, they do not have any possibility for MMCT or MLCT.

In general, photocatalyzed CDC processes can be classified in two types (Fig. 7): (a) using a sacrificial oxidant (peroxide, oxygen, etc.) and (b) using proton reduction reagent (dual catalyst). Both approaches share the same general mechanism; (1) photocatalytic activation of the substrate to form the key active intermediate and (2) reaction between the intermediate and pre-nucleophile to produce the desired product. The difference lays in the nature of the by-product, which is in ideal conditions hydrogen peroxide for the first type (when using oxygen as oxidant), and hydrogen gas in the second one [17].

Just as in the thermal CDC reactions, most of the reported functionalized carbons are (sp^3–H) adjacent to nitrogen, since the SET step form the heteroatom to the excited photocatalyst (oxidative quenching) easily yields to a cationic nitrogen, which facilitates the hydrogen abstraction from the α-carbon (by lowering the Pk_a value and bond dissociation energy of α C–H bond) forming carbon radical. This carbon radical has a low oxidation potential and can readily lose an electron producing the iminium intermediate.

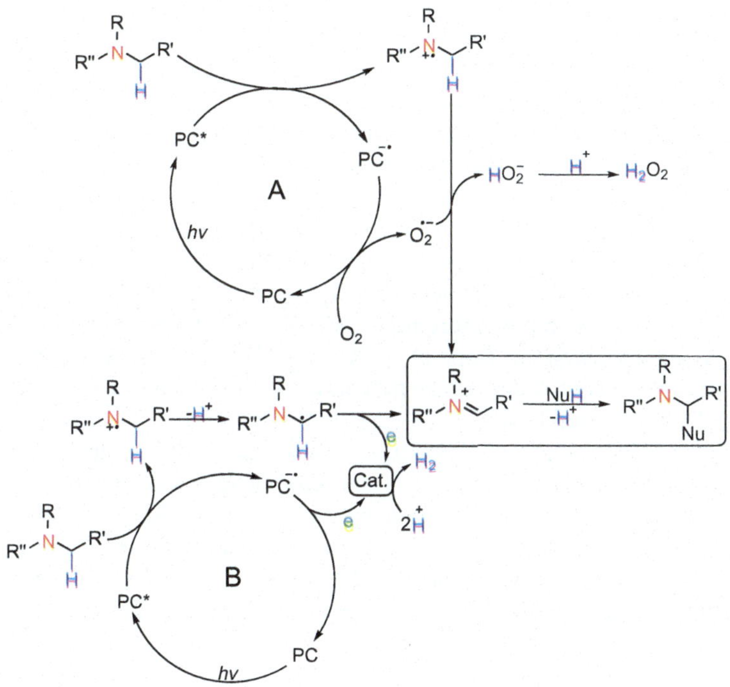

Fig. 7 General mechanisms for photocatalyzed CDC reactions. **a** With using sacrificial oxidant. **b** With using proton reducing reagent

3.1 Chromium and Iron Catalyst

Recently, chromium-based photocatalyst have been used successfully under CDC conditions to catalyze [4 + 2] Diels–Alder reactions using oxygen as oxidant [13, 41, 133–135], and for the cyanation of *tert*-amines with using trimethylsilyl cyanide (TMSCN) as precursor [99]. Although Cr-based systems are still underdeveloped, these reports will surely elicit future investigations towards the synthesis different types of compounds.

Iron-based photocatalysts have been used successfully on the CDC synthesis of carbazoles using oxygen as oxidant under flow conditions [102]. $Fe(phen)_3(NTf_2)_2$ was the best photocatalyst for this reaction being, under optimal conditions, better than Ru(II) and Ir(III) photosensitizers [16, 84]. The mechanism of this reaction is still not well understood and further investigations are required to foster the application of iron-based photocatalyst for organic synthesis under CDC conditions.

3.2 Copper Catalysts

In spite of the growing interest on copper as cheaper and abundant metal source for photocatalysis, the research on this field remains underexplored compared to their Ru and Ir counterparts [137]. In the late 1980s, Sauvage and co-workers reported pioneering investigations on the long-lived ^{3}MLCT excited state of Bis(phenanthroline)Copper(I) and its successful application as photoredox catalyst to synthesize bis(*p*-nitro)bibenzyl from *p*-nitrobenzylbromide [54]. The field had a revival when Reiser and co-workers used copper photocatalyst for the C–C bond formation via atom transfer radical addition (ATRA) between different types of alkenes and organ-halogen compounds [107] utilizing the photoreductant ability of homoleptic $[Cu(dap)_2]^{2+}$(dap=2,9-di(p-anisyl)-1,10-phenanthroline) (−1.43 V vs. SCE) with a ^{3}MLCT lifetime of 270 ns in de-aerated CH_2Cl_2 at 25 °C [6]. These ATRA reactions are redox neutral and therefore no sacrificial oxidant is required. Furthermore, the same group extended the scope of ATRA and allylation reactions [56] by using heteroleptic copper complexes, which they have stronger excited state reductant and longer ^{3}MLCT life time [24].

The syntesis of carbazoles from triarylamines using heteroleptic Cu(I) complexes in situ with a combination of I_2 and propylene oxide as oxiadtion system utilizing continuous flow was reported by collins and co-workers [39, 40]. The proposed mechanism starts with oxidative excited state quenching of $[Cu(Xantphos)(dmp)]^+$ by iodine to form Cu(II) species, I^- and I. Cu(II) oxidized the trialkylamine to form trialkylamine cation and subsequent secondary oxidation occurred with the iodine radical. Finally, a re-aromatization step was proposed by oxygen or iodine to produce the carbazoles.

Che and co-workers reported the ability of heteroleptic complexes of copper with phenanthroline and monoanionic nido-carborane-diphosphine ligands to mediate theaza-Henrry-CDC reaction between tetrahydroisoquinoline and nitroalkanes [152]. The steric bulkiness of the carborane ligand was predicted to prevent significant excited state distortions, the excited life time of the complex in degassed CH_2Cl_2 was thus 1.4 μs, which is shorter than for the related $[Cu(P^\wedge P)(N^\wedge N)]^+$ complexes [24], and it has excited state reduction potential of 1.01 V versus Ag\Agcl. Moreover, utilizing this strong reduction potental, different aza-Henry reactions of tertiary amines with, different nitroalkylsunder aerobic conditions were faciliated [152].

$[Cu(dap)_2]^+$ was used successfully to synthesize tetrahydroquinlines and tetrahydroisoquinolines by reaction between *N,N*-dialkylanilines and *N*-aryltetrahydroisoquinolines with electron-deficient alkenes at room temperature under aerobic condition (Fig. 8) [95]. From NMR studies, the crucial role of the TFA was observed to mediate the oxidation of Cu(I) photocatalyst excitation state to Cu(II) by oxygen, which does not proceed without it. In the mechanism, after photoexcitation, the TFA-mediated aerobic oxidation of $[Cu^*(dap)_2]^+$ provides $[Cu(dap)_2]^{2+}$(**I**). The formed Cu(II) species could oxidize *N,N*-dimethylamine (**II**) with simultaneous generation of the Cu(I) species. HAT from the *N,N*-dimethylamine radical cation leads to the formation of an α-amino radical intermediate (**III**). Reaction between **III** and *N*-phenylmaleimide via radical addition produces the first C–C bond. Concomitant intramolecular cyclization (**V**) and aerobic oxidation (**VI**) deliver the desired annulation product.

Hwang and co-workers reported the one-pot synthesis of quinoxaline derivatives by reaction between *o*-phenylenediamine and a terminal alkyne induced by visible light under aerobic conditions. They conclude that the key factor in the reaction is the direct photoexcitation of the in situ generated copper arylacetylide (λ_{abs} = 420–480 nm), which reacts with *o*-phenylenediamine to generate the quinoxaline products upon blue LED light irradiation. The addition of a base was critical to produce the active photocatalyst [116]. Further extension of this methodology was used to synthesize different types of indoles by reaction between arylamines with terminal alkynes and benzoquinone [117]. Just as in the previous application, the copper arylacetylide is the key intermediate responsible for the light absorption and drives the reaction. No reaction was observed upon addition of TEMPO suggesting the involvement of radicals in the reaction pathway. Furthermore, luminescence quenching experiment indicates SET from the photocatalyst excited state of the copper-aryl acetylated to the benzoquinone to form benzoquinone radical and copper(II), which was confirmed by EPR measurements. The reaction mechanism is assumed to start with photoinduced SET from the excited state of Cu(I)-phenylacetylide (MLCT) to benzoquinone, oxidizing the copper catalyst and fostering the formation of a C–C bond between benzoquinone and the alkyne which results in the regeneration of the Cu(I) catalyst. Nucleophilic attack at the triple bond by aniline at the β-carbon atom is facilitated by a copper(I)-alkyne complex, in which the

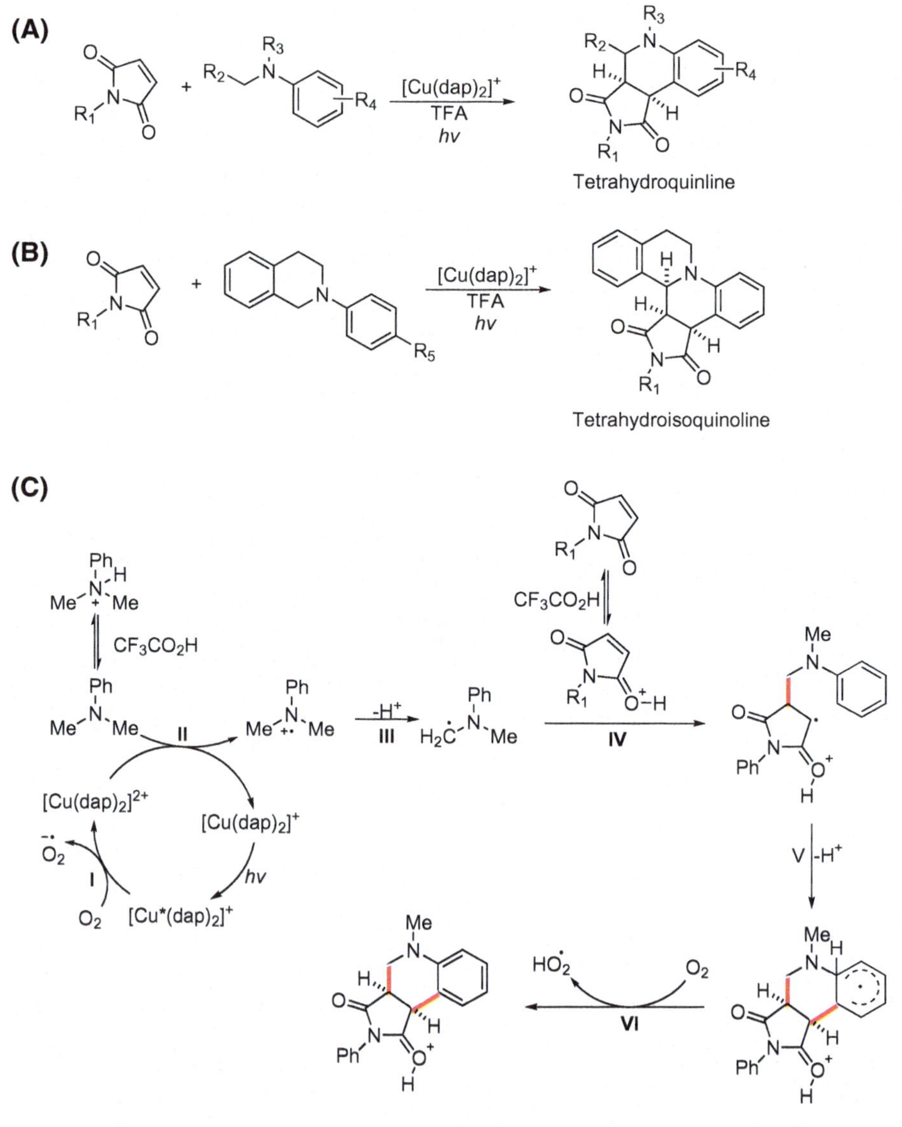

Fig. 8 Synthesis of tetrahydroisoquinolines (**a**) and tetrahydroisoquinolines (**b**) and the proposed reaction mechanism (**c**)

triple bond more electron-deficient. The final product was produced after a Friedel–Crafts cyclization liberating the catalyst with re-aromatization.

Wu and co-workers identified the key intermediates generated in situ for the Cu (II)-catalyzed synthesis of a range of quinolone, indole[3,2-c]quinolone, β-amino acid, and 1,4-dihydropyridine derivatives under air. The mechanistic study revealed that the Cu(II) ion associates with secondary amine by SET to form a visible light absorbing intermediate [Cu(I)-$NH^{+\cdot}$]. Directly, the N–H and C–H bonds of amine are activated and further transformed into the corresponding imine by $^1O_2$, generated by reaction of the intermediate [Cu(I)-$NH^{+\cdot}$] with O_2 under visible light. The second complex intermediate formed also by SET from the imine to Cu(II), which increases the electrophilicity of the imine for the next nucleophile to form the desired product. The nature of the intermediates was confirmed by XANSE, DFT, and ESR [85]. Analogous to the previous mechanism, the synthesis of α-heteroaryl substituted α-amino acid derivatives by a reaction between *N*-arylglycine ethers and imidazo[1,2-a] pyridines was reported by Le and co-workers [185]. In their proposed mechanism, also an imine intermediate is formed after SET from the amine to Cu(II). Nucleophilic addition on the imine intermediate furnished the desired product formation and regenerating the Cu(II) from Cu(I)by oxidation with air.

3.3 Molybdenum Catalysts

Molybdenum(0) complexes have been used successfully in the CDC reaction to synthesize heterocyclic compounds via intramolecular rearrangement. The Wengar group recently reported homoleptic Mo(0) complexes with chelating bidentate isocyanide ligands that show long-life ^{3}MLCT excited states, which could be utilized in photoredox chemistry [12]. The complex [$Mo(CNAr_3NC)_3$] was able to afford 2,3-dihydrofuran from an acyl cyclopropane without the use of any sacrificial oxidant (redox neutral reaction). The mechanism starts with SET from the Mo-complex excited state to the substrate to form an enolate radical. The enolate could be oxidized by the Mo(I) species generating the catalyst and forming the second radical. The internal radical–radical coupling furnished the desired product. The key feature in this reaction is the strong oxidation potential of the catalyst excited state (−2.6 V vs. Fc^+/Fc), displaying quantum yields and lifetimes similar to those of the benchmark photosensitizer $Ru(bpy)_3^{2+}$.

4 Conclusions and Outlook

Since the beginning of the twenty-first century, with the pioneering reports by Li and colleagues, the field of cross-dehydrogenative coupling has witnessed an exponential growth, being applied in a myriad of transformations with various catalytic systems. Although the underlying principles of reactivity are the same, the way these transformations lead to the CDC product varies with the nature of the catalyst and the activation process in every case, resulting in a very rich and diverse research field.

A great deal of work has been devoted to the understanding of the underlying reaction mechanisms for both thermally driven and photocatalyzed processes.

Nevertheless, due to the elusive nature and high reactivity of the open-shell reaction intermediates, there are still many open questions that wait for a definite answer. Hence, all the proposed mechanisms (included in this review or not) have to be treated exactly as *proposals* which, being as they are, a reflection of the state-of-the-art at the time of their conception, are subject to constant evolution and improvement, along with our understanding of Nature's works.

Interestingly, in spite of the maturity of the field, the number of reports of CDC reactions for the synthesis of heterocycles remains somewhat modest. Hence, in the light of the inspiring work at hand, one can foresee the upcoming development of many new heterocycle syntheses employing the versatile cross-dehydrogenative coupling reaction.

References

1. Ackermann L (2014) Carboxylate-assisted ruthenium-catalyzed alkyne annulations by C–H/Het-H bond functionalizations. Acc Chem Res 47:281–295. https://doi.org/10.1021/ar3002798
2. Ackermann L, Lygin AV, Hofmann N (2011) Ruthenium-catalyzed oxidative synthesis of 2-pyridones through C–H/N–H bond functionalizations. Org Lett 13:3278–3281. https://doi.org/10.1021/ol201244s
3. Ackermann L, Pospech J (2011) Ruthenium-catalyzed oxidative C–H bond alkenylations in water: expedient synthesis of annulated lactones. Org Lett 13:4153–4155. https://doi.org/10.1021/ol201563r
4. Aggarwal T, Kumar S, Verma AK (2016) Iodine-mediated synthesis of heterocycles via electrophilic cyclization of alkynes. Org Biomol Chem 14:7639–7653. https://doi.org/10.1039/c6ob01054g
5. Almasalma AA, Mejia E (2018) Copper-catalyzed allylic C–H alkynylation by cross-dehydrogenative coupling. Chem Eur J 24:12269–12273. https://doi.org/10.1002/chem.201801772
6. Armaroli N (2001) Photoactive mono- and polynuclear Cu(i)–phenanthrolines. A viable alternative to Ru(ii)–polypyridines? Chem Soc Rev 30:113–124. https://doi.org/10.1039/b000703j
7. Barnard JR, Jackman LM (1960) 622. Hydrogen transfer. Part X. The dehydrogenation of hydroaromatic hydrocarbons by quinones: theoretical calculations for possible intermediates. J Chem Soc 3110. https://doi.org/10.1039/jr9600003110
8. Baslé O, Li C-J (2007) Copper catalyzed oxidative alkylation of sp^3 C–H bond adjacent to a nitrogen atom using molecular oxygen in water. Green Chem 9:1047. https://doi.org/10.1039/b707745a
9. Bi HP, Chen WW, Liang YM, Li CJ (2009) A novel iron-catalyzed decarboxylative Csp^3–Csp^2 coupling of proline derivatives and naphthol. Org Lett 11:3246–3249. https://doi.org/10.1021/ol901129v
10. Boess E, Sureshkumar D, Sud A, Wirtz C, Fares C, Klussmann M (2011) Mechanistic studies on a Cu-catalyzed aerobic oxidative coupling reaction with *N*-phenyl tetrahydroisoquinoline: structure of intermediates and the role of methanol as a solvent. J Am Chem Soc 133:8106–8109. https://doi.org/10.1021/ja201610c
11. Borpatra PJ, Deb ML, Baruah PK (2018) Copper-catalyzed tandem multi-component approach to 1,3-oxazines at room temperature by cross-dehydrogenative coupling using methanol as C1 feedstock. Synlett 29:1171–1175. https://doi.org/10.1055/s-0036-1591775

12. Buldt LA, Guo X, Prescimone A, Wenger OS (2016) A molybdenum(0) isocyanide analogue of Ru(2,2′-Bipyridine)3 (2+): a strong reductant for photoredox catalysis. Angew Chem Int Ed Engl 55:11247–11250. https://doi.org/10.1002/anie.201605571
13. Buldt LA, Wenger OS (2017) Chromium complexes for luminescence, solar cells, photoredox catalysis, upconversion, and phototriggered NO release. Chem Sci 8:7359–7367. https://doi.org/10.1039/c7sc03372a
14. Campos KR (2007) Direct sp^3 C–H bond activation adjacent to nitrogen in heterocycles. Chem Soc Rev 36:1069–1084. https://doi.org/10.1039/b607547a
15. Carrick WL, Karapinka GL, Kwiatkowski GT (1969) Oxidative coupling of phenols using vanadium tetrachloride and vanadium oxytrichloride. J Org Chem 34:2388–2392. https://doi.org/10.1021/jo01260a029
16. Chang HR, McCusker JK, Toftlund H, Wilson SR, Trautwein AX, Winkler H, Hendrickson DN (1990) [Tetrakis(2-pyridylmethyl)ethylenediamine]iron(II) perchlorate, the first rapidly interconverting ferrous spin-crossover complex. J Am Chem Soc 112:6814–6827. https://doi.org/10.1021/ja00175a012
17. Chen B, Wu L-Z, Tung C-H (2018) Photocatalytic activation of less reactive bonds and their functionalization via hydrogen-evolution cross-couplings. Acc Chem Res. https://doi.org/10.1021/acs.accounts.8b00267
18. Chen W, Xie Z, Zheng H, Lou H, Liu L (2014) Structurally diverse alpha-substituted benzopyran synthesis through a practical metal-free $C(sp^3)$–H functionalization. Org Lett 16:5988–5991. https://doi.org/10.1021/ol503004a
19. Chen WL, Yan RL, Tang D, Guo SB, Meng X, Chen BH (2012) Iodine-induced regioselective direct alkylation of azoles via in situ formed alkyliodide. Tetrahedron 68:7956–7959. https://doi.org/10.1016/j.tet.2012.07.008
20. Chen ZK, Wang BJ, Zhang JT, Yu WL, Liu ZX, Zhang YH (2015) Transition metal-catalyzed C–H bond functionalizations by the use of diverse directing groups. Org Chem Front 2:1107–1295. https://doi.org/10.1039/c5qo00004a
21. Cheng G-J, Song L-J, Yang Y-F, Zhang X, Wiest O, Wu Y-D (2013) Computational studies on the mechanism of the copper-catalyzed sp^3–C–H cross-dehydrogenative coupling reaction. ChemPlusChem 78:943–951. https://doi.org/10.1002/cplu.201300117
22. Correia CA, Li CJ (2010) Copper-catalyzed cross-dehydrogenative coupling (CDC) of alkynes and benzylic C–H bonds. Adv Synth Catal 352:1446–1450. https://doi.org/10.1002/adsc.201000066
23. Cuthbertson JD, MacMillan DW (2015) The direct arylation of allylic sp(3) C–H bonds via organic and photoredox catalysis. Nature 519:74–77. https://doi.org/10.1038/nature14255
24. Cuttell DG, Kuang SM, Fanwick PE, McMillin DR, Walton RA (2002) Simple Cu(I) complexes with unprecedented excited-state lifetimes. J Am Chem Soc 124:6–7. https://doi.org/10.1021/ja012247h
25. Daugulis O, Do HQ, Shabashov D (2009) Palladium- and copper-catalyzed arylation of carbon-hydrogen bonds. Acc Chem Res 42:1074–1086. https://doi.org/10.1021/ar9000058
26. Deb ML, Borpatra PJ, Saikia PJ, Baruah PK (2017) Introducing tetramethylurea as a new methylene precursor: a microwave-assisted $RuCl_3$-catalyzed cross dehydrogenative coupling approach to bis(indolyl)methanes. Org Biomol Chem 15:1435–1443. https://doi.org/10.1039/c6ob02671k
27. DeBoef B, Porter AL (2015) Aryl–aryl coupling via cross-dehydrogenative-coupling reactions. In: Li C-J (ed) From C–H to C–C bonds: cross-dehydrogenative-coupling. Green Chemistry Series, vol 26. The Royal Society of Chemistry, pp 114–132. https://doi.org/10.1039/9781782620082-00114
28. Deng G-J, Xiao F, Yang L (2015) Cross-dehydrogenative-coupling reactions involving allyl, benzyl and alkyl C–H bonds. In: Li C-J (ed) From C–H to C–C bonds: cross-dehydrogenative-coupling. Green Chemistry Series, vol 26. The Royal Society of Chemistry, pp 93–113. https://doi.org/10.1039/9781782620082-00093

29. Dhineshkumar J, Lamani M, Alagiri K, Prabhu KR (2013) A versatile C–H functionalization of tetrahydroisoquinolines catalyzed by iodine at aerobic conditions. Org Lett 15:1092–1095. https://doi.org/10.1021/ol4001153
30. Faisca Phillips AM, Pombeiro AJL (2018) Recent developments in transition metal-catalyzed cross-dehydrogenative coupling reactions of ethers and thioethers. ChemCatChem 10:3354–3383. https://doi.org/10.1002/cctc.201800582
31. Gao Q-H et al (2013) Metal-free dual sp3 C-H functionalization: I2-promoted domino oxidative cyclization to construct 2,5-disubstituted oxazoles. Tetrahedron 69:22–28. https://doi.org/10.1016/j.tet.2012.10.072
32. García Mancheño O, Stopka T (2013) TEMPO derivatives as alternative mild oxidants in carbon-carbon coupling reactions. Synthesis 45:1602–1611. https://doi.org/10.1055/s-0033-1338480
33. Ghosh I, Marzo L, Das A, Shaikh R, Konig B (2016) visible light mediated photoredox catalytic arylation reactions. Acc Chem Res 49:1566–1577. https://doi.org/10.1021/acs.accounts.6b00229
34. Ghosh S, Pahovnik D, Kragl U, Mejia E (2018) Isospecific copolymerization of cyclohexene oxide and carbon dioxide catalyzed by dialkylmagnesium compounds. Macromolecules 51:846–852. https://doi.org/10.1021/acs.macromol.7b02463
35. Girard SA, Knauber T, Li CJ (2014) The cross-dehydrogenative coupling of C(sp^3)–H bonds: a versatile strategy for C–C bond formations. Angew Chem Int Ed Engl 53:74–100. https://doi.org/10.1002/anie.201304268
36. Guo X, Yu R, Li H, Li Z (2009) Iron-catalyzed tandem oxidative coupling and annulation: an efficient approach to construct polysubstituted benzofurans. J Am Chem Soc 131:17387–17393. https://doi.org/10.1021/ja907568j
37. Guo XX, Gu DW, Wu Z, Zhang W (2015) Copper-catalyzed C–H functionalization reactions: efficient synthesis of heterocycles. Chem Rev 115:1622–1651. https://doi.org/10.1021/cr500410y
38. He Z, Liu W, Li Z (2011) I2-catalyzed indole formation via oxidative cyclization of *N*-aryl enamines. Chem Asian J 6:1340–1343. https://doi.org/10.1002/asia.201100045
39. Hernandez-Perez AC, Caron A, Collins SK (2015) Photochemical synthesis of complex carbazoles: evaluation of electronic effects in both UV- and visible-light methods in continuous flow. Chemistry 21:16673–16678. https://doi.org/10.1002/chem.201502661
40. Hernandez-Perez AC, Collins SK (2013) A visible-light-mediated synthesis of carbazoles. Angew Chem Int Ed Engl 52:12696–12700. https://doi.org/10.1002/anie.201306920
41. Higgins RF et al (2016) Uncovering the roles of oxygen in Cr(III) photoredox catalysis. J Am Chem Soc 138:5451–5464. https://doi.org/10.1021/jacs.6b02723
42. Hirao T (1997) Vanadium in modern organic synthesis. Chem Rev 97:2707–2724. https://doi.org/10.1021/cr960014g
43. Hirao T (2007) Synthetic transformations via vanadium-induced redox reactions. In: Vanadium: the versatile metal, vol 974. ACS Symposium Series, vol 974. American Chemical Society, pp 2–27. https://doi.org/10.1021/bk-2007-0974.ch001
44. Hu W, Lin J-P, Song L-R, Long Y-Q (2015) Direct synthesis of 2-aryl-4-quinolones via transition-metal-free intramolecular oxidative C(sp^3)–H/C(sp^3)–H coupling. Org Lett 17:1268–1271. https://doi.org/10.1021/acs.orglett.5b00248
45. Huang HY, Wu HR, Wei F, Wang D, Liu L (2015) Iodine-catalyzed direct olefination of 2-oxindoles and alkenes via cross-dehydrogenative coupling (CDC) in air. Org Lett 17:3702–3705. https://doi.org/10.1021/acs.orglett.5b01662
46. Huo C, Chen F, Yuan Y, Xie H, Wang Y (2015) Iron catalyzed dual-oxidative dehydrogenative (DOD) tandem annulation of glycine derivatives with tetrahydrofurans. Org Lett 17:5028–5031. https://doi.org/10.1021/acs.orglett.5b02504
47. Hurst TE, Taylor RJK (2017) A Cu-catalysed radical cross-dehydrogenative coupling approach to acridanes and related heterocycles. Eur J Org Chem 2017:203–207. https://doi.org/10.1002/ejoc.201601336

48. Ito H, Ueda K, Itami K (2015) Cross-dehydrogenative-coupling reactions without metals. In: Li C-J (ed) From C–H to C–C bonds: cross-dehydrogenative-coupling. Green Chemistry Series, vol 26. The Royal Society of Chemistry, pp 153–196. https://doi.org/10.1039/9781782620082-00153
49. Jiang Y, Xu K, Zeng C (2018) Use of electrochemistry in the synthesis of heterocyclic structures. Chem Rev 118:4485–4540. https://doi.org/10.1021/acs.chemrev.7b00271
50. Jones KM, Klussmann M (2012) Oxidative coupling of tertiary amines: scope, mechanism and challenges. Synlett 2012:159–162. https://doi.org/10.1055/s-0031-1290117
51. Kang H et al (2017) Asymmetric oxidative coupling of phenols and hydroxycarbazoles. Org Lett 19:5505–5508. https://doi.org/10.1021/acs.orglett.7b02552
52. Kaswan P, Nandwana NK, DeBoef B, Kumar A (2016) Vanadyl acetylacetonate catalyzed methylenation of imidazo[1,2-a]pyridines by using dimethylacetamide as a methylene source: direct access to Bis(imidazo[1,2-a]pyridin-3-yl)methanes. Adv Synth Catal 358:2108–2115. https://doi.org/10.1002/adsc.201600225
53. Kaswan P, Porter A, Pericherla K, Simone M, Peters S, Kumar A, DeBoef B (2015) Oxidative cross-coupling of sp(3)- and sp(2)-hybridized C–H bonds: vanadium-catalyzed aminomethylation of imidazo[1,2-a]pyridines. Org Lett 17:5208–5211. https://doi.org/10.1021/acs.orglett.5b02539
54. Kern J-M, Sauvage J-P (1987) Photoassisted C–C coupling via electron transfer to benzylic halides by a bis(di-imine) copper(I) complex. J Chem Soc, Chem Commun: 546–548 https://doi.org/10.1039/c39870000546
55. Kim HY, Takizawa S, Sasai H, Oh K (2017) Reversal of enantioselectivity approach to BINOLs via single and dual 2-naphthol activation modes. Org Lett 19:3867–3870. https://doi.org/10.1021/acs.orglett.7b01734
56. Knorn M, Rawner T, Czerwieniec R, Reiser O (2015) [Copper(phenanthroline)(bisisonitrile)(+)-complexes for the visible-light-mediated atom transfer radical addition and allylation reactions. ACS Catal 5:5186–5193. https://doi.org/10.1021/acscatal.5b01071
57. Kshirsagar UA, Regev C, Parnes R, Pappo D (2013) Iron-catalyzed oxidative cross-coupling of phenols and alkenes. Org Lett 15:3174–3177. https://doi.org/10.1021/ol401532a
58. Kumar RA, Saidulu G, Prasad KR, Kumar GS, Sridhar B, Reddy KR (2012) Transition metal-free α-C(sp^3)–H bond functionalization of amines by oxidative cross dehydrogenative coupling reaction: simple and direct access to C-4-alkylated 3,4-dihydroquinazoline derivatives. Adv Synth Catal 354:2985–2991. https://doi.org/10.1002/adsc.201200679
59. Laha JK, Jethava KP, Patel S (2015) Scope of successive C–H functionalizations of the methyl group in 3-picolines: intramolecular carbonylation of arenes to the metal-free synthesis of 4-azafluorenones. Org Lett 17:5890–5893. https://doi.org/10.1021/acs.orglett.5b03071
60. Langeslay RR, Kaphan DM, Marshall CL, Stair PC, Sattelberger AP, Delferro M (2018) Catalytic applications of vanadium: a mechanistic perspective. Chem Rev https://doi.org/10.1021/acs.chemrev.8b00245
61. Lao ZQ, Zhong WH, Lou QH, Li ZJ, Meng XB (2012) KI-catalyzed imidation of sp^3 C–H bond adjacent to amide nitrogen atom. Org Biomol Chem 10:7869–7871. https://doi.org/10.1039/c2ob26430g
62. Larsen CB, Wenger OS (2018) Photoredox catalysis with metal complexes made from earth-abundant elements. Chem Eur J 24:2039–2058. https://doi.org/10.1002/chem.201703602
63. Lee A, Betori RC, Crane EA, Scheidt KA (2018) An enantioselective cross-dehydrogenative coupling catalysis approach to substituted tetrahydropyrans. J Am Chem Soc 140:6212–6216. https://doi.org/10.1021/jacs.8b03063
64. Lewis ES, Perry JM, Grinstein RH (1970) Mechanism of hydride transfer. III. Rates and isotope effects in the quinone oxidation of leuco triphenylmethane dyes. J Am Chem Soc 92:899–905. https://doi.org/10.1021/ja00707a027

65. Li K, You J (2016) Cascade oxidative coupling/cyclization: a gateway to 3-amino polysubstituted five-membered heterocycles. J Org Chem 81:2327–2339. https://doi.org/10.1021/acs.joc.5b02838
66. Li Z, Bohle DS, Li CJ (2006) Cu-catalyzed cross-dehydrogenative coupling: a versatile strategy for C–C bond formations via the oxidative activation of sp(3) C–H bonds. Proc Natl Acad Sci USA 103:8928–8933. https://doi.org/10.1073/pnas.0601687103
67. Li Z, Li CJ (2004) CuBr-catalyzed efficient alkynylation of sp^3 C–H bonds adjacent to a nitrogen atom. J Am Chem Soc 126:11810–11811. https://doi.org/10.1021/ja0460763
68. Li Z, Li CJ (2005) Highly efficient copper-catalyzed nitro-Mannich type reaction: cross-dehydrogenative-coupling between sp^3 C–H bond and sp^3 C–H bond. J Am Chem Soc 127:3672–3673. https://doi.org/10.1021/ja050058j
69. Li Z, Li CJ (2006) Catalytic allylic alkylation via the cross-dehydrogenative-coupling reaction between allylic sp^3 C–H and methylenic sp^3 C–H bonds. J Am Chem Soc 128:56–57. https://doi.org/10.1021/ja056541b
70. Li Z, Yu R, Li H (2008) Iron-catalyzed C–C bond formation by direct functionalization of C–H bonds adjacent to heteroatoms. Angew Chem Int Ed Engl 47:7497–7500. https://doi.org/10.1002/anie.200802215
71. Lingamurthy M, Jagadeesh Y, Ramakrishna K, Rao BV (2016) DDQ-promoted benzylic/allylic sp(3) C–H activation for the stereoselective intramolecular C–N bond formation: applications to the total synthesis of (−)-codonopsinine, (+)-5-epi-codonopsinine, (+)-radicamine B, and (−)-codonopsinol. J Org Chem 81:1367–1377. https://doi.org/10.1021/acs.joc.5b02275
72. Lingayya R, Vellakkaran M, Nagaiah K, Nanubolu JB (2015) Ruthenium as a single catalyst for two steps: one-pot ruthenium(II)-catalyzed aerobic oxidative dehydrogenation of dihydroquinazolinones and cross-coupling/annulation to give N-fused polycyclic heteroarenes. Asian J Org Chem 4:462–469. https://doi.org/10.1002/ajoc.201500025
73. Liu D, Lei A (2015) Iodine-catalyzed oxidative coupling reactions utilizing C–H and X–H as nucleophiles. Chem Asian J 10:806–823. https://doi.org/10.1002/asia.201403248
74. Liu L, Carroll PJ, Kozlowski MC (2015) Vanadium-catalyzed regioselective oxidative coupling of 2-hydroxycarbazoles. Org Lett 17:508–511. https://doi.org/10.1021/ol503521b
75. Liu WQ et al (2017) Visible light promoted synthesis of indoles by single photosensitizer under aerobic conditions. Org Lett 19:3251–3254. https://doi.org/10.1021/acs.orglett.7b01367
76. Long H et al (2017) Regio- and diastereoselective cross-dehydrogenative coupling of tetrahydropyridines with 1,3-dicarbonyl compounds. Org Lett 19:2146–2149. https://doi.org/10.1021/acs.orglett.7b00787
77. Lou J, Wang Q, Wu K, Wu P, Yu Z (2017) Iron-catalyzed oxidative C–H functionalization of internal olefins for the synthesis of tetrasubstituted furans. Org Lett 19:3287–3290. https://doi.org/10.1021/acs.orglett.7b01431
78. Lu H, Yang Q, Zhou Y, Guo Y, Deng Z, Ding Q, Peng Y (2014) Cross-coupling/annulations of quinazolones with alkynes for access to fused polycyclic heteroarenes under mild conditions. Org Biomol Chem 12:758–764. https://doi.org/10.1039/c3ob41955j
79. Lv L, Li Z (2016) Fe-catalyzed cross-dehydrogenative coupling reactions. Top Curr Chem 374:38. https://doi.org/10.1007/s41061-016-0038-y
80. Maes J, Maes BUW (2016) Chapter five—a journey through metal-catalyzed CH functionalization of heterocycles: insights and trends. In: Scriven EFV, Ramsden CA (eds) Advances in heterocyclic chemistry, vol 120. Academic Press, pp 137–194. https://doi.org/10.1016/bs.aihch.2016.04.005
81. Maiti S, Achar TK, Mal P (2017) An organic intermolecular dehydrogenative annulation reaction. Org Lett 19:2006–2009. https://doi.org/10.1021/acs.orglett.7b00562
82. Maiti S, Mal P (2017) Dehydrogenative aromatic ring fusion for carbazole synthesis via C–C/C–N bond formation and alkyl migration. Org Lett 19:2454–2457. https://doi.org/10.1021/acs.orglett.7b01117

83. Marzo L, Pagire SK, Reiser O, Konig B (2018) Visible-light photocatalysis: does it make a difference in organic synthesis? Angew Chem Int Ed Engl 57:10034–10072. https://doi.org/10.1002/anie.201709766
84. McCusker JK, Walda KN, Dunn RC, Simon JD, Magde D, Hendrickson DN (1993) Subpicosecond ^{1}MLCT -$^{5}T_2$ intersystem crossing of low-spin polypyridyl ferrous complexes. J Am Chem Soc 115:298–307. https://doi.org/10.1021/ja00054a043
85. Meng QY et al (2017) Identifying key intermediates generated in situ from Cu(II) salt-catalyzed C–H functionalization of aromatic amines under illumination. Sci Adv 3: e1700666. https://doi.org/10.1126/sciadv.1700666
86. Mitchell D, Cole KP, Pollock PM, Coppert DM, Burkholder TP, Clayton JR (2012) Development and a practical synthesis of the JAK2 inhibitor LY2784544. Org Process Res Dev 16:70–81. https://doi.org/10.1021/op200229j
87. Miyaura N, Suzuki A (1995) Palladium-catalyzed cross-coupling reactions of organoboron compounds. Chem Rev 95:2457–2483. https://doi.org/10.1021/Cr00039a007
88. Moselage M, Li J, Ackermann L (2016) Cobalt-catalyzed C–H activation. ACS catal 6:498–525. https://doi.org/10.1021/acscatal.5b02344
89. Murahashi S, Komiya N, Terai H, Nakae T (2003) Aerobic ruthenium-catalyzed oxidative cyanation of tertiary amines with sodium cyanide. J Am Chem Soc 125:15312–15313. https://doi.org/10.1021/ja0390303
90. Murahashi S, Naota T, Yonemura K (1988) Ruthenium-catalyzed cytochrome-P-450 type oxidation of tertiary-amines with alkyl hydroperoxides. J Am Chem Soc 110:8256–8258. https://doi.org/10.1021/Ja00232a060
91. Murahashi SI, Naota T, Miyaguchi N, Nakato T (1992) Ruthenium-catalyzed oxidation of tertiary-amines with hydrogen-peroxide in the presence of methanol. Tetrahedron Lett 33:6991–6994. https://doi.org/10.1016/S0040-4039(00)60914-0
92. Narayan R, Manna S, Antonchick AP (2015) Hypervalent iodine(III) in direct carbon-hydrogen bond functionalization. Synlett 26:1785–1803. https://doi.org/10.1055/s-0034-1379912
93. Narayanam JM, Stephenson CR (2011) Visible light photoredox catalysis: applications in organic synthesis. Chem Soc Rev 40:102–113. https://doi.org/10.1039/b913880n
94. Narute S, Pappo D (2017) Iron phosphate catalyzed asymmetric cross-dehydrogenative coupling of 2-naphthols with beta-ketoesters. Org Lett 19:2917–2920. https://doi.org/10.1021/acs.orglett.7b01152
95. Nicholls TP, Constable GE, Robertson JC, Gardiner MG, Bissember AC (2015) Brønsted acid cocatalysis in copper(I)-photocatalyzed α-amino C–H bond functionalization. ACS Catal 6:451–457. https://doi.org/10.1021/acscatal.5b02014
96. Nobuta T, Tada N, Fujiya A, Kariya A, Miura T, Itoh A (2013) Molecular iodine catalyzed cross-dehydrogenative coupling reaction between two sp^3 C–H bonds using hydrogen peroxide. Org Lett 15:574–577. https://doi.org/10.1021/ol303389t
97. Ohkubo K, Fujimoto A, Fukuzumi S (2013) Photocatalytic monofluorination of benzene by fluoride via photoinduced electron transfer with 3-cyano-1-methylquinolinium. The journal of physical chemistry A 117:10719–10725. https://doi.org/10.1021/jp408315a
98. Ohkubo K, Kobayashi T, Fukuzumi S (2011) Direct oxygenation of benzene to phenol using quinolinium ions as homogeneous photocatalysts. Angew Chem Int Ed Engl 50:8652–8655. https://doi.org/10.1002/anie.201102931
99. Otto S et al (2017) Photo-chromium: sensitizer for visible-light-induced oxidative C–H bond functionalization-electron or energy transfer? ChemPhotoChem 1:344–349. https://doi.org/10.1002/cptc.201700077
100. Pan B, Liu B, Yue E, Liu Q, Yang X, Wang Z, Sun W-H (2016) A ruthenium catalyst with unprecedented effectiveness for the coupling cyclization of γ-amino alcohols and secondary alcohols. ACS Catal 6:1247–1253. https://doi.org/10.1021/acscatal.5b02638
101. Pan J, Li X, Qiu X, Luo X, Jiao N (2018) Copper-catalyzed oxygenation approach to oxazoles from amines, alkynes, and molecular oxygen. Org Lett 20:2762–2765. https://doi.org/10.1021/acs.orglett.8b00992

102. Parisien-Collette S, Hernandez-Perez AC, Collins SK (2016) Photochemical synthesis of carbazoles using an $[Fe(phen)_3](NTf_2)_2/O_2$ catalyst system: catalysis toward sustainability. Org Lett 18:4994–4997. https://doi.org/10.1021/acs.orglett.6b02456
103. Parvatkar PT, Manetsch R, Banik BK (2018) Metal-free cross-dehydrogenative coupling (CDC): molecular iodine as a versatile catalyst/reagent for CDC reactions. Chem Asian J 0. https://doi.org/10.1002/asia.201801237
104. Parvatkar PT, Parameswaran PS, Tilve SG (2012) Recent developments in the synthesis of five- and six-membered heterocycles using molecular iodine. Chem Eur J 18:5460–5489. https://doi.org/10.1002/chem.201100324
105. Patil MR, Dedhia NP, Kapdi AR, Kumar AV (2018) Cobalt(II)/ *N*-hydroxyphthalimide-catalyzed cross-dehydrogenative coupling reaction at room temperature under aerobic condition. J Org Chem 83:4477–4490. https://doi.org/10.1021/acs.joc.8b00203
106. Peng F, Li LL, Liu J, Chen ZW (2018) Copper-catalyzed oxidative cross-coupling/C–C bond cleavage/cyclization of aryl methyl ketones with 4-aminocoumarins: domino synthesis of dicoumarin-fused [1,5]-diazocines. Asian J Org Chem 7:1667–1673. https://doi.org/10.1002/ajoc.201800306
107. Pirtsch M, Paria S, Matsuno T, Isobe H, Reiser O (2012) [Cu(dap)2Cl] as an efficient visible-light-driven photoredox catalyst in carbon-carbon bond-forming reactions. Chem Eur J 18:7336–7340. https://doi.org/10.1002/chem.201200967
108. Prier CK, Rankic DA, MacMillan DW (2013) Visible light photoredox catalysis with transition metal complexes: applications in organic synthesis. Chem Rev 113:5322–5363. https://doi.org/10.1021/cr300503r
109. Ray D, Manikandan T, Roy A, Tripathi KN, Singh RP (2015) Ligand-promoted intramolecular dehydrogenative cross-coupling using a Cu catalyst: direct access to polycyclic heteroarenes. Chem Commun 51:7065–7068. https://doi.org/10.1039/c5cc01817j
110. Reddy NNK, Donthiri RR, Ravi C, Adimurthy S (2016) Iodine-catalyzed [3 + 2] cyclization of 2-pyridylesters and chalcones: metal-free approach for the synthesis of substituted indolizines. Tetrahedron Lett 57:3243–3246. https://doi.org/10.1016/j.tetlet.2016.05.083
111. Richter H, Garcia Mancheno O (2011) TEMPO oxoammonium salt-mediated dehydrogenative Povarov/oxidation tandem reaction of *N*-alkyl anilines. Org Lett 13:6066–6069. https://doi.org/10.1021/ol202552y
112. Romero NA, Margrey KA, Tay NE, Nicewicz DA (2015) Site-selective arene C–H amination via photoredox catalysis. Science 349:1326–1330. https://doi.org/10.1126/science.aac9895
113. Romero NA, Nicewicz DA (2016) Organic photoredox catalysis. Chem Rev 116:10075–10166. https://doi.org/10.1021/acs.chemrev.6b00057
114. Roslin S, Odell LR (2017) Visible-light photocatalysis as an enabling tool for the functionalization of unactivated $C(sp^3)$-substrates. Eur J Org Chem 2017:1993–2007. https://doi.org/10.1002/ejoc.201601479
115. Roudesly F, Oble J, Poli G (2017) Metal-catalyzed CH activation/functionalization: the fundamentals. J Mol Catal A: Chem 426:275–296. https://doi.org/10.1016/j.molcata.2016.06.020
116. Sagadevan A, Ragupathi A, Hwang KC (2013) Visible-light-induced, copper(I)-catalysed C–N coupling between o-phenylenediamine and terminal alkynes: one-pot synthesis of 3-phenyl-2-hydroxy-quinoxalines. Photochemical & photobiological sciences: Official journal of the European Photochemistry Association and the European Society for Photobiology 12:2110–2118. https://doi.org/10.1039/c3pp50186h
117. Sagadevan A, Ragupathi A, Hwang KC (2015) Photoinduced copper-catalyzed regioselective synthesis of indoles: three-component coupling of arylamines, terminal alkynes, and quinones. Angew Chem Int Ed Engl 54:13896–13901. https://doi.org/10.1002/anie.201506579

118. Sako M, Takeuchi Y, Tsujihara T, Kodera J, Kawano T, Takizawa S, Sasai H (2016) Efficient enantioselective synthesis of oxahelicenes using redox/acid cooperative catalysts. J Am Chem Soc 138:11481–11484. https://doi.org/10.1021/jacs.6b07424
119. Sako M, Takizawa S, Yoshida Y, Sasai H (2015) Enantioselective and aerobic oxidative coupling of 2-naphthol derivatives using chiral dinuclear vanadium(V) complex in water. Tetrahedron: Asymmetry 26:613–616 https://doi.org/10.1016/j.tetasy.2015.05.002
120. Sar D, Bag R, Yashmeen A, Bag SS, Punniyamurthy T (2015) Synthesis of functionalized pyrazoles via vanadium-catalyzed C–N dehydrogenative cross-coupling and fluorescence switch-on sensing of BSA protein. Org Lett 17:5308–5311. https://doi.org/10.1021/acs.orglett.5b02669
121. Schultz DM, Yoon TP (2014) Solar synthesis: prospects in visible light photocatalysis. Science 343:1239176. https://doi.org/10.1126/science.1239176
122. Sha Q, Arman H, Doyle MP (2015) Three-component cascade reactions with 2,3-Diketoesters: a novel metal-free synthesis of 5-vinyl-pyrrole and 4-hydroxy-indole derivatives. Org Lett 17:3876–3879. https://doi.org/10.1021/acs.orglett.5b01855
123. Sharma R, Abdullaha M, Bharate SB (2017) Oxidant-controlled C-sp(2)/sp(3)-H cross-dehydrogenative coupling of *N*-heterocycles with benzylamines. J Org Chem 82:9786–9793. https://doi.org/10.1021/acs.joc.7b00856
124. Shen Y, Li M, Wang S, Zhan T, Tan Z, Guo CC (2009) An efficient copper-catalyzed oxidative Mannich reaction between tertiary amines and methyl ketones. Chem Commun 953–955. https://doi.org/10.1039/b819657e
125. Shi X, Chen X, Wang M, Zhang X, Fan X (2018) Regioselective synthesis of acylated *N*-heterocycles via the cascade reactions of saturated cyclic amines with 2-Oxo-2-arylacetic acids. J Org Chem 83:6524–6533. https://doi.org/10.1021/acs.joc.8b00805
126. Shi X, Zhang F, Luo W-K, Yang L (2017) Oxidant-triggered C1-benzylation of isoquinoline by iodine-catalyzed cross-dehydrogenative-coupling with methylarenes. Synlett 13:494–498. https://doi.org/10.1055/s-0036-1588331
127. Shuai Q, Yang L, Guo X, Basle O, Li CJ (2010) Rhodium-catalyzed oxidative C–H arylation of 2-arylpyridine derivatives via decarbonylation of aromatic aldehydes. J Am Chem Soc 132:12212–12213. https://doi.org/10.1021/ja105396b
128. Shukla G, Srivastava A, Yadav D, Singh MS (2018) Copper-catalyzed one-pot cross-dehydrogenative thienannulation: chemoselective access to naphtho[2,1-b]thiophene-4,5-diones and subsequent transformation to benzo[a]thieno[3,2-c]phenazines. J Org Chem 83:2173–2181. https://doi.org/10.1021/acs.joc.7b03092
129. Singhal S, Jain SL, Sain B (2009) An efficient aerobic oxidative cyanation of tertiary amines with sodium cyanide using vanadium based systems as catalysts. Chem Commun 2371–2372. https://doi.org/10.1039/b820402k
130. Skubi KL, Blum TR, Yoon TP (2016) Dual catalysis strategies in photochemical synthesis. Chem Rev 116:10035–10074. https://doi.org/10.1021/acs.chemrev.6b00018
131. Song C, Dong X, Yi H, Chiang C-W, Lei A (2018) DDQ-catalyzed direct $C(sp^3)$–H amination of alkylheteroarenes: synthesis of biheteroarenes under aerobic and metal-free conditions. ACS Catal 8:2195–2199. https://doi.org/10.1021/acscatal.7b04434
132. Srimani D, Ben-David Y, Milstein D (2013) Direct synthesis of pyridines and quinolines by coupling of gamma-amino-alcohols with secondary alcohols liberating H2 catalyzed by ruthenium pincer complexes. Chem Commun 49:6632–6634. https://doi.org/10.1039/c3cc43227k
133. Stevenson SM, Higgins RF, Shores MP, Ferreira EM (2017) Chromium photocatalysis: accessing structural complements to Diels-Alder adducts with electron-deficient dienophiles. Chem Sci 8:654–660. https://doi.org/10.1039/c6sc03303b
134. Stevenson SM, Shores MP, Ferreira EM (2015) Photooxidizing chromium catalysts for promoting radical cation cycloadditions. Angew Chem 127:6606–6610. https://doi.org/10.1002/ange.201501220

135. Stevenson SM, Shores MP, Ferreira EM (2015) Photooxidizing chromium catalysts for promoting radical cation cycloadditions. Angew Chem Int Ed Engl 54:6506–6510. https://doi.org/10.1002/anie.201501220
136. Sunil UT, Sushma SK, Satish AD, Swapnil RS, Rajendra PP (2012) Molecular iodine: an efficient and versatile reagent for organic synthesis. Curr Org Chem 16:1485–1501. https://doi.org/10.2174/138527212800672574
137. Suva Paria OR (2018) Visible light and copper complexes: a promising match in photoredox catalysis. In: Stephenson C, Yoon T, MacMillan DWC (eds) Visible light photocatalysis in organic chemistry. Wiely. https://doi.org/10.1002/9783527674145.ch7
138. Takizawa S, Kodera J, Yoshida Y, Sako M, Breukers S, Enders D, Sasai H (2014) Enantioselective oxidative-coupling of polycyclic phenols. Tetrahedron 70:1786–1793. https://doi.org/10.1016/j.tet.2014.01.017
139. Tan DW, Li HX, Zhu DL, Li HY, Young DJ, Yao JL, Lang JP (2018) Ligand-controlled copper(I)-catalyzed cross-coupling of secondary and primary alcohols to alpha-alkylated ketones, pyridines, and quinolines. Org Lett 20:608–611. https://doi.org/10.1021/acs.orglett.7b03726
140. Tan Z, Jiang H, Zhang M (2016) Ruthenium-catalyzed dehydrogenative beta-benzylation of 1,2,3,4-tetrahydroquinolines with aryl aldehydes: access to functionalized quinolines. Org Lett 18:3174–3177. https://doi.org/10.1021/acs.orglett.6b01390
141. Tang S, Liu K, Long Y, Gao X, Gao M, Lei A (2015) Iodine-catalyzed radical oxidative annulation for the construction of dihydrofurans and indolizines. Org Lett 17:2404–2407. https://doi.org/10.1021/acs.orglett.5b00912
142. Tang S, Liu K, Long Y, Qi X, Lan Y, Lei A (2015) Tuning radical reactivity using iodine in oxidative C(sp(3))-H/C(sp)-H cross-coupling: an easy way toward the synthesis of furans and indolizines. Chem Commun 51:8769–8772. https://doi.org/10.1039/c5cc01825k
143. Tanoue A, Yoo WJ, Kobayashi S (2014) Sulfuryl chloride as an efficient initiator for the metal-free aerobic cross-dehydrogenative coupling reaction of tertiary amines. Org Lett 16:2346–2349. https://doi.org/10.1021/ol500661t
144. Trost BM (1967) Dehydrogenation mechanisms. On the mechanism of dehydrogenation of acenaphthene by quinones. J Am Chem Soc 89:1847–1851. https://doi.org/10.1021/ja00984a017
145. Tsang ASK, Jensen P, Hook JM, Hashmi ASK, Todd MH (2011) An oxidative carbon–carbon bond-forming reaction proceeds via an isolable iminium ion. Pure Appl Chem 83:655–665. https://doi.org/10.1351/pac-con-11-01-01
146. Tsang ASK, Park SJ, Todd MH (2015) Mechanisms of cross-dehydrogenative-coupling reactions. In: Li C-J (ed) From C–H to C–C bonds: cross-dehydrogenative-coupling. RSC Green Chemistry Series, vol 26. The Royal Society of Chemistry, pp 254–294. https://doi.org/10.1039/9781782620082-00254
147. Tsang ASK, Todd MH (2009) Facile synthesis of vicinal diamines via oxidation of *N*-phenyltetrahydroisoquinolines with DDQ. Tetrahedron Lett 50:1199–1202. https://doi.org/10.1016/j.tetlet.2008.12.101
148. Twilton J, Le C, Zhang P, Shaw MH, Evans RW, MacMillan DWC (2017) The merger of transition metal and photocatalysis. Nat Rev Chem 1:0052. https://doi.org/10.1038/S41570-017-0052
149. Ueda H, Yoshida K, Tokuyama H (2014) Acetic acid promoted metal-free aerobic carbon-carbon bond forming reactions at alpha-position of tertiary amines. Org Lett 16:4194–4197. https://doi.org/10.1021/ol5018883
150. Vuppalapati SVN, Lee YR (2012) Iodine-catalyzed efficient synthesis of azaarene substituted 3-hydroxy-2-oxindole derivatives through sp^3 C–H functionalization. Tetrahedron 68:8286–8292. https://doi.org/10.1016/j.tet.2012.07.051
151. Waghmode NA, Kalbandhe AH, Thorat PB, Karade NN (2016) Metal-free new synthesis of 1,3-naphthoxazines via intramolecular cross dehydrogenative-coupling reaction of 1-(α-aminoalkyl)-2-naphthols using hypervalent iodine(III) reagent. Tetrahedron Lett 57:680–683. https://doi.org/10.1016/j.tetlet.2015.12.117

152. Wang B et al (2015) Long-lived excited states of zwitterionic copper(I) complexes for photoinduced cross-dehydrogenative coupling reactions. Chem Eur J 21:1184–1190. https://doi.org/10.1002/chem.201405356
153. Wang CS, Dixneuf PH, Soule JF (2018) Photoredox catalysis for building C–C bonds from C(sp(2))-H bonds. Chem Rev 118:7532–7585. https://doi.org/10.1021/acs.chemrev.8b00077
154. Wang L, Ackermann L (2013) Versatile pyrrole synthesis through ruthenium(II)-catalyzed alkene C–H bond functionalization on enamines. Org Lett 15:176–179. https://doi.org/10.1021/ol303224e
155. Waters WA (1946) Evidence for the dehydrogenation theory of oxidation. Trans Faraday Soc 42:184–190. https://doi.org/10.1039/TF9464200184
156. Wendlandt AE, Suess AM, Stahl SS (2011) Copper-catalyzed aerobic oxidative C–H functionalizations: trends and mechanistic insights. Angew Chem Int Ed Engl 50:11062–11087. https://doi.org/10.1002/anie.201103945
157. Wenger OS (2018) Photoactive complexes with earth-abundant metals. J Am Chem Soc https://doi.org/10.1021/jacs.8b08822
158. Wu CJ, Zhong JJ, Meng QY, Lei T, Gao XW, Tung CH, Wu LZ (2015) Cobalt-catalyzed cross-dehydrogenative coupling reaction in water by visible light. Org Lett 17:884–887. https://doi.org/10.1021/ol503744a
159. Wu X-F, Beller M (2014a) Cobalt-catalyzed heterocycle synthesis. In: Economic synthesis of heterocycles: zinc, iron, copper, cobalt, manganese and nickel catalysts. RSC Catalysis Series. The Royal Society of Chemistry, pp 349–385. https://doi.org/10.1039/9781782620839-00349
160. Wu X-F, Beller M (2014b) Copper-catalyzed heterocycle synthesis. In: Economic synthesis of heterocycles: zinc, iron, copper, cobalt, manganese and nickel catalysts. RSC Catalysis Series. The Royal Society of Chemistry, pp 159–348. https://doi.org/10.1039/9781782620839-00159
161. Wu X-F, Beller M (2014c) Iron-catalyzed heterocycle synthesis. In: Economic synthesis of heterocycles: zinc, iron, copper, cobalt, manganese and nickel catalysts. RSC Catalysis Series. The Royal Society of Chemistry, pp 59–158. https://doi.org/10.1039/9781782620839-00059
162. Wu X, Zhao P, Geng X, Wang C, Wu YD, Wu AX (2018) Synthesis of pyrrole-2-carbaldehyde derivatives by oxidative annulation and direct Csp^3–H to C=O oxidation. Org Lett 20:688–691. https://doi.org/10.1021/acs.orglett.7b03821
163. Wu Y, Arenas I, Broomfield LM, Martin E, Shafir A (2015) Hypervalent activation as a key step for dehydrogenative ortho C–C coupling of iodoarenes. Chem Eur J 21:18779–18784. https://doi.org/10.1002/chem.201503987
164. Xiang L et al (2014) I2-mediated oxidative cyclization for synthesis of substituted indolizines. J Org Chem 79:10641–10647. https://doi.org/10.1021/jo5019574
165. Xiang M, Meng Q-Y, Gao X-W, Lei T, Chen B, Tung C-H, Wu L-Z (2016) Reactivity and mechanistic insight into the cross coupling reaction between isochromans and β-keto esters through C–H bond activation under visible light irradiation. Org Chem Front 3:486–490. https://doi.org/10.1039/C5QO00412H
166. Xie J, Huang Y, Song H, Liu Y, Wang Q (2017) Copper-catalyzed aerobic oxidative [2 + 3] cyclization/aromatization cascade reaction: atom-economical access to tetrasubstituted 4,5-Biscarbonyl Imidazoles. Org Lett 19:6056–6059. https://doi.org/10.1021/acs.orglett.7b02767
167. Xie Z, Zan X, Sun S, Pan X, Liu L (2016) Organocatalytic enantioselective cross-dehydrogenative coupling of *N*-carbamoyl cyclic amines with aldehydes. Org Lett 18:3944–3947. https://doi.org/10.1021/acs.orglett.6b01625
168. Xing Y, Wang N-X, Zhang W (2015) Advances in transition-metal-catalyzed direct sp^3-carbon–hydrogen bond functionalization. Synlett 26:2088–2098. https://doi.org/10.1055/s-0034-1381031
169. Xue W-j, Gao Q-H, Wu A-x (2015) Molecular iodine mediated oxidative cross-coupling of sp^3 C–H with sp^2 C–H: direct synthesis of substituted indolo[2,3-b]carbazoles via formal

[2 + 2+1 + 1] cyclization. Tetrahedron Lett 56:7115–7119. https://doi.org/10.1016/j.tetlet.2015.11.026

170. Xue XS, Ji P, Zhou B, Cheng JP (2017) The essential role of bond energetics in C–H activation/functionalization. Chem Rev 117:8622–8648. https://doi.org/10.1021/acs.chemrev.6b00664
171. Yang Q, Li S, Wang J (2018) Cobalt-catalyzed cross-dehydrogenative coupling of imidazo[1,2-a]pyridines with isochroman using molecular oxygen as the oxidant. Org Chem Front 5:577–581. https://doi.org/10.1039/C7QO00875A
172. Yang Y, Lan J, You J (2017) Oxidative C–H/C–H coupling reactions between two (hetero) arenes. Chem Rev 117:8787–8863. https://doi.org/10.1021/acs.chemrev.6b00567
173. Yavari I, Hosseinpour R, Skoulika S (2015) Iodine-mediated diastereoselective cyclopropanation of arylidene malononotriles by 2,6-dimethylquinoline. Synlett 26:380–384. https://doi.org/10.1055/s-0034-1379496
174. Yoon TP, Ischay MA, Du J (2010) Visible light photocatalysis as a greener approach to photochemical synthesis. Nat Chem 2:527–532. https://doi.org/10.1038/nchem.687
175. Yu JB, Zhang Y, Jiang ZJ, Su WK (2016) Mechanically induced Fe(III) catalysis at room temperature: solvent-free cross-dehydrogenative coupling of 3-benzylic indoles with methylenes/indoles. J Org Chem 81:11514–11520. https://doi.org/10.1021/acs.joc.6b02197
176. Zhang B, Cui Y, Jiao N (2012) Metal-free TEMPO-catalyzed oxidative C–C bond formation from Csp^3–H bonds using molecular oxygen as the oxidant. Chem Commun 48:4498–4500. https://doi.org/10.1039/c2cc30684k
177. Zhang C, Li TL, Wang LG, Rao Y (2017) Synthesis of diverse heterocycles via one-pot cascade cross-dehydrogenative-coupling (CDC)/cyclization reaction. Org Chem Front 4:386–391. https://doi.org/10.1039/c6qo00522e
178. Zhang C, Tang C, Jiao N (2012) Recent advances in copper-catalyzed dehydrogenative functionalization via a single electron transfer (SET) process. Chem Soc Rev 41:3464–3484. https://doi.org/10.1039/c2cs15323h
179. Zhang R, Qin Y, Zhang L, Luo S (2017) Oxidative synthesis of benzimidazoles, quinoxalines, and benzoxazoles from primary amines by ortho-quinone catalysis. Org Lett 19:5629–5632. https://doi.org/10.1021/acs.orglett.7b02786
180. Zhang Y, Schulz M, Wachtler M, Karnahl M, Dietzek B (2018) Heteroleptic diimine-diphosphine Cu(I) complexes as an alternative towards noble-metal based photosensitizers: design strategies, photophysical properties and perspective applications. Coord Chem Rev 356:127–146. https://doi.org/10.1016/j.ccr.2017.10.016
181. Zhang Z, Pi C, Tong H, Cui X, Wu Y (2017) Iodine-catalyzed direct C–H alkenylation of azaheterocycle *N*-oxides with alkenes. Org Lett 19:440–443. https://doi.org/10.1021/acs.orglett.6b03399
182. Zhao J, Fang H, Qian P, Han J, Pan Y (2014) Metal-free oxidative $C(sp^3)$–H bond functionalization of alkanes and conjugate addition to chromones. Org Lett 16:5342–5345. https://doi.org/10.1021/ol502524d
183. Zhao M-N, Yu L, Hui R-R, Ren Z-H, Wang Y-Y, Guan Z-H (2016) Iron-catalyzed dehydrogenative [4 + 2] cycloaddition of tertiary anilines and enamides for the synthesis of tetrahydroquinolines with amido-substituted quaternary carbon centers. ACS Catal 6:3473–3477. https://doi.org/10.1021/acscatal.6b00849
184. Zhao X, Liu T-X, Ma N, Zhang G (2017) In situ generated TEMPO oxoammonium salt mediated tandem cyclization of β-oxoamides with amine hydrochlorides for the synthesis of pyrrolin-4-ones. The J Org Chem 82:6125–6132. https://doi.org/10.1021/acs.joc.7b00686
185. Zhu Z-Q, Xiao L-J, Zhou C-C, Song H-L, Xie Z-B, Le Z-G (2018) A visible-light-promoted cross-dehydrogenative-coupling reaction of *N*-arylglycine esters with imidazo[1,2-a]pyridines. Tetrahedron Lett 59:3326–3331. https://doi.org/10.1016/j.tetlet.2018.07.047
186. Zou YQ, Lu LQ, Fu L, Chang NJ, Rong J, Chen JR, Xiao WJ (2011) Visible-light-induced oxidation/[3 + 2] cycloaddition/oxidative aromatization sequence: a photocatalytic strategy to construct pyrrolo[2,1-a]isoquinolines. Angew Chem Int Ed Engl 50:7171–7175. https://doi.org/10.1002/anie.201102306

Metal-Free Synthesis of Heterocycles via Cross-Dehydrogenative Coupling

Anju Modi and Bhisma K. Patel

Abstract The synthesis of heterocyclic molecules has always been a central theme in organic chemistry since centuries. Numerous methodologies have been developed towards the construction of complex heterocyclic molecules. The traditional approaches generally involve the use of preactivated substrates as the essential requirement, thus making the overall method tedious and expensive. Working towards the development of greener and sustainable approach, metal-free cross-dehydrogenative coupling strategies have emerged as a boon to the synthetic chemists. The method is straightforward eliminating the pre-functionalization steps and avoiding the toxicity imposed by the metal impurities. This chapter illustrates various metal-free cross-dehydrogenative tactics for the synthesis of different *N*, *O*, and *S* based heterocyclic molecules with few representative examples in each case along with the mechanism demonstrating the synthesis of respective molecule.

Keywords Heterocycle synthesis · Metal-free CDC strategies · *N*-heterocycles · *O*-heterocycles · *N,O*-heterocycles · *S*-heterocycles · *N,O,S*-heterocycles

1 Introduction

Heterocycles represent a predominant class of organic molecules having important physiological and biological properties. Due to their omnipresent nature, synthesis of heterocycles has always attracted synthetic chemists. They form the core skeleton of the majority of natural products, agrochemicals and drug molecules exhibiting antibacterial, antifungal, antimalarial, antiviral, antimicrobial, antitumor, anti-inflammatory, antidepressant, herbicidal, fungicidal and insecticidal activities. Heterocyclic compounds have found enormous applications in the field of material science as fluorescent sensors, analytical reagents, organic conductors, semiconductors, photovoltaic cells and

A. Modi · B. K. Patel (✉)
Department of Chemistry, Indian Institute of Technology Guwahati, Guwahati 781039, Assam, India
e-mail: patel@iitg.ac.in

A. Srivastava and C. K. Jana (eds.), *Heterocycles via Cross Dehydrogenative Coupling*, https://doi.org/10.1007/978-981-13-9144-6_11

organic light-emitting diodes (OLEDs). Heterocycles are the vital components of bio-macromolecules such as DNA, RNA, protein and enzymes. Besides, these heterocyclic compounds are also utilized as the synthetic intermediates, organocatalysts, protecting groups, chiral auxiliaries and as ligands in various synthetic methodologies. Owing to the incredible diverse assets of the heterocyclic scaffold, designing novel, efficient and economical strategies for their preparation is highly desirable.

In this regard, the evolution of cross-dehydrogenative couplings over the past decades for the construction of C–C and C–X bond (X=N, O, S, P, Si) has revolutionized the synthetic chemists' arsenal. The strategy obviates the need for the pre-functionalization of coupling partners, thereby rendering high atom economy and reducing reaction times. Cross-dehydrogenative coupling chemistry is a vast area of research that encompasses transition metals, non-transition metals and metal-free reaction conditions. However, transition-metal-catalyzed reactions are invariably associated with several drawbacks such as high cost of the metal catalysts and the requirement of supporting ligands, toxicity of heavy metals and their removal from the desired product which limits their application in the pharmaceutical industry. Also, metal catalysts are moisture and air sensitive, and in many cases, special additives and co-catalysts are essential to promote the efficiency and selectivity of the reaction. As an alternative pathway, transition-metal-free coupling reactions are highly appealing and provide much greener and sustainable approach towards the construction of complex molecules. These transition-metal-free CDC reaction are generally catalyzed or promoted by reagents such as (diacetoxyiodo)benzene (PIDA) or (bis(trifluoroacetoxy)iodo)benzene (PIFA), I_2, 2,3-dichloro-5,6-dicyano-1,4-benzoquinone (DDQ), benzoyl peroxide (BPO), *tert*-butyhydroperoxide (TBHP), S_8, elemental sulphur and photocatalysts such as Eosin Y and rose bengal. This chapter gives a brief account of various metal-free approaches reported for the synthesis of a diverse biologically important heterocyclic scaffolds. The chapter is divided into five sections based on the nature of the heterocycle synthesized.

2 Synthesis of *N*-Heterocycles

The wealth of nitrogen-containing heterocycles in biologically active molecules has provided great impetus for chemists to develop increasingly efficient methods towards their synthesis. The *N*-heterocycles have the ability to form salts due to their basic nature and enhance their solubility which is important for oral absorption and bioavailability. Hence, most of the drugs possess at least one nitrogen heterocycle within their structures and are thus considered as privileged targets by medicinal chemists. Below are the few examples of various *N*-heterocycles synthesized under metal-free conditions via cross-dehydrogenative coupling (CDC) strategy.

2.1 Synthesis of Aziridine

Aziridines, the smallest three-membered nitrogen-containing heterocycles, are useful building blocks in chemistry for decades. The ring strain in the molecule and its

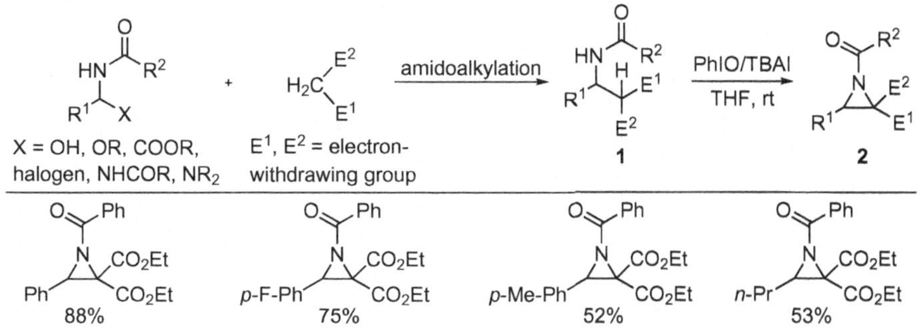

Scheme 1 Synthesis of aziridine

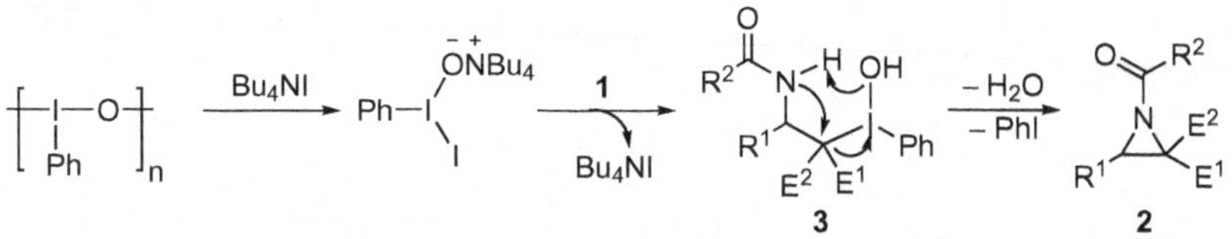

Scheme 2 Plausible reaction pathway for the synthesis of aziridine

unique reactivity have made them attractive synthetic targets and gained significant attention for developing efficient methods for its synthesis. Following an oxidative cyclization, Fan group reported the synthesis of *N*-benzoyl aziridines **2** from amidoalkylation adducts **1** of activated methylene compounds using iodosobenzene and catalytic amount of TBAI under a neutral condition (Scheme 1) [1].

A proposed reaction mechanism is shown in Scheme 2. The polymeric iodosobenzene is depolymerized with TBAI to generate reactive iodine(III) species, which reacts with **1** forming the intermediate **3**. A [1,3] intramolecular nucleophilic displacement by nitrogen atom affords the corresponding product **2**.

2.2 Synthesis of Azetidine

Azetidines are an important class of compound unveiling remarkable medicinal and biological activities. Natural products such as mugineic acid, nicotianamine, thrombin inhibitor melagatran and polyoxins have azetidine carboxylic acid units as their core structure. They undergo various transformations to afford several nitrogen-containing compounds and are utilized as chiral ligands for the asymmetric synthesis. Numerous transition-metal-catalyzed and electrochemical methods are available towards the synthesis of azetidine derivatives. Fan et al. reported a facile stereoselective synthesis of highly functionalized azetidines **5** from 2-aminomalonates and chalcones (Scheme 3) [2]. The reaction proceeds via grind-promoted solvent-free Michael addition and PhIO/TBAI-mediated oxidative cyclization.

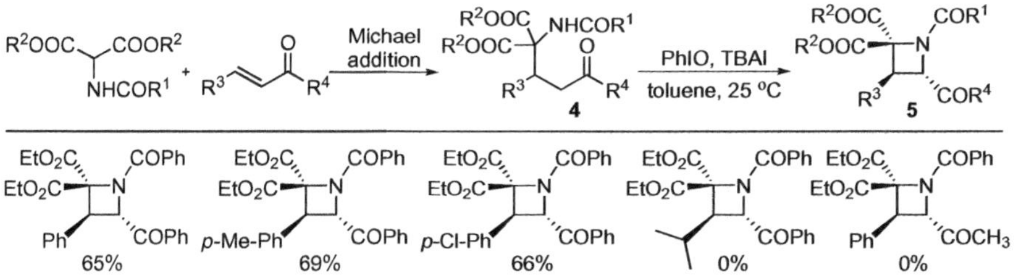

Scheme 3 Synthesis of azetidine

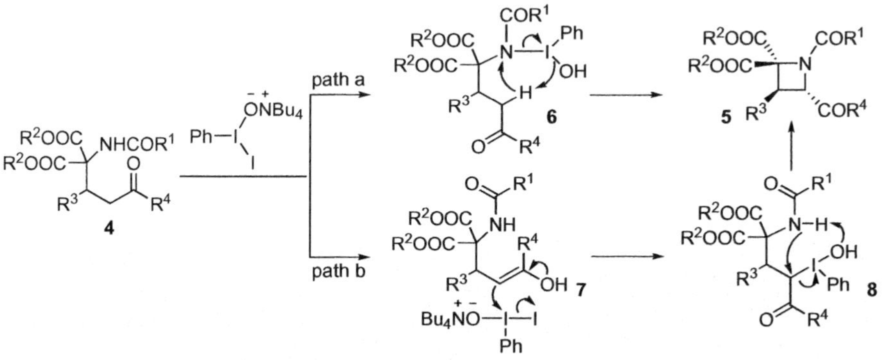

Scheme 4 Plausible reaction pathway for the synthesis of azetidine

The active iodine(III) species generated by the depolymerization of iodosobenzene by TBAI reacts with the Michael adduct **4** to form **6**, which undergoes an intramolecular reductive elimination to afford azetidine **5** (Scheme 4, path a). In an alternative pathway (Scheme 4, path b), α-hyperiodination of the methylene of the phenyl ketone in **4** generates species **8** via intermediate **7**, which by an intramolecular attack of the nitrogen gives the desired product accompanied by the regeneration of PhI.

2.3 *Synthesis of Pyrrole*

Pyrroles are among the simplest five-membered *N*-heterocycles that exist in several natural products and bioactive compounds. Many synthetic methods are reported towards the synthesis of pyrroles of which the classical Paal–Knorr and Hantzsch reactions are most frequently adopted. Transition-metal-catalyzed coupling and multi-component reactions have also been established in the past decades for the synthesis of pyrroles. Among the metal-free conditions, oxidative cyclization of enamines using various oxidants such as $Pb(OAc)_4$, $PhI(OAc)_2$ and CAN provides an alternative approach towards the synthesis of polysubstituted pyrroles. In 2016, Gao, Guan and co-workers developed a novel and efficient $K_2S_2O_8$-promoted

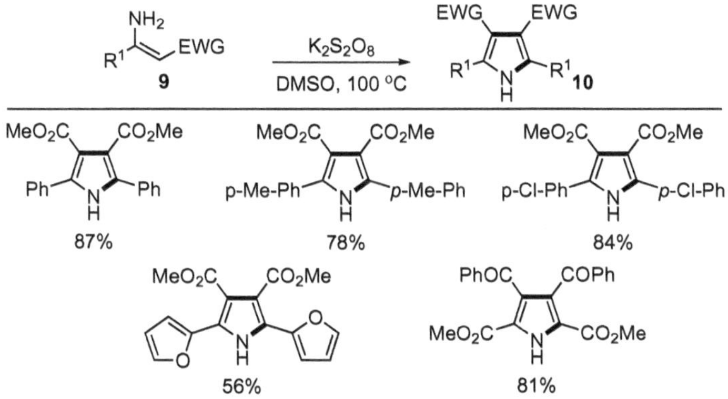

Scheme 5 Synthesis of polysubstituted pyrroles

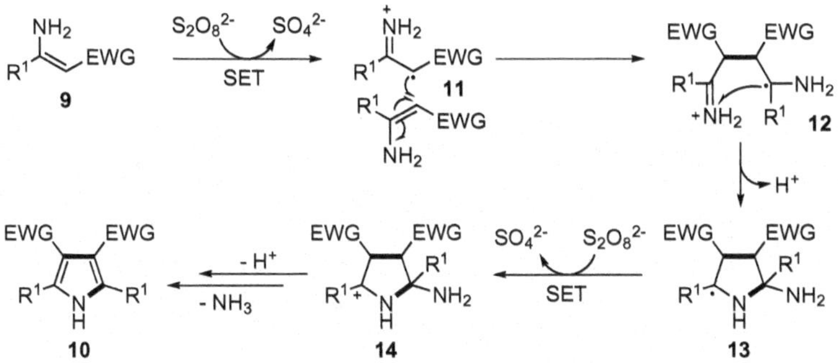

Scheme 6 Plausible reaction mechanism

oxidative cyclization of enamines for the synthesis of polysubstituted pyrroles in DMSO at 100 °C (Scheme 5) [3].

The reaction takes place by the initial oxidation of enamine by $K_2S_2O_8$ generating an amino radical **11** (Scheme 6). This radical then reacts with a second enamine molecule forming the intermediate **12**. An intramolecular C–N bond formation leads to intermediate **13** which is oxidized to **14** and finally, a sequential elimination of H^+ and NH_3 from **14** results into the desired polysubstituted pyrrole **10**.

2.4 Synthesis of Pyrazoles

Pyrazoles constitute another class of *N*-heterocycles with two adjacent nitrogen atoms. They are vital building blocks of many natural products and pharmaceuticals. The prevailing methods for the synthesis of pyrazoles involve the reaction of hydrazines with 1,3-dicarbonyl compounds/unsaturated hydrocarbons or 1,3-dipolar cycloaddition of diazoalkanes with alkenes or alkynes. Singh group

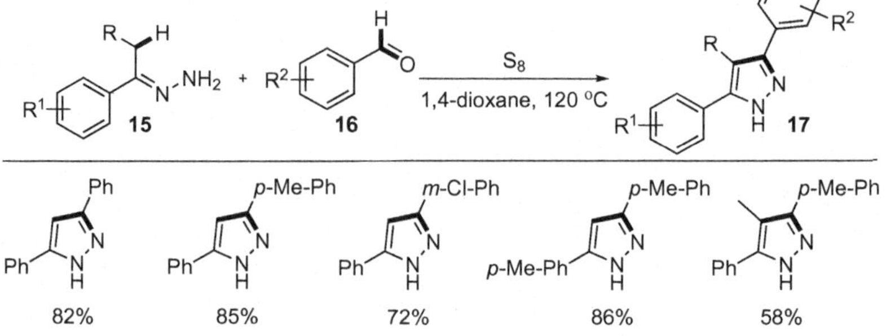

Scheme 7 Synthesis of substituted 1*H*-pyrazoles

demonstrated a unique cross-dehydrogenative coupling of acetophenone hydrazones **15** with aldehydes **16** resulting in the synthesis of substituted pyrazoles **17** (Scheme 7) [4]. The reaction was performed using elemental sulphur as the promoter cum hydrogen acceptor leading to the formation of C(sp^3)–C(sp^2) bond.

The acetophenone hydrazone **15** reacts with aldehyde **16** at room temperature without the addition of S_8 forming intermediate **18**. This intermediate isomerizes to its enamine form **19** which activates sulphur affording **20**. Finally, with the loss of H_2S from **20** and rearrangement gives substituted pyrazole **17** (Scheme 8). Radical-trapping experiments conducted using BHT (3,5-di-*tert*-butyl-4-hydroxytoluene) and TEMPO (2,2,6,6-tetramethylpiperidine-*N*-oxyl) as radical scavengers did not inhibit the reaction although the product yield was slightly lower.

Zhu group reported another metal- and additive-free scalable method for the synthesis of pyrazoles **22** via oxidative C–H cycloamination of vinyl hydrazones **21** in the presence of phenyliodine(III) diacetate (PIDA) as the oxidant in hexafluoroisopropanol (HFIP) solvent at 25 °C (Scheme 9) [5]. The reaction proceeds instantly with high yield showing wide a range of functional group tolerance. Initially, nucleophilic

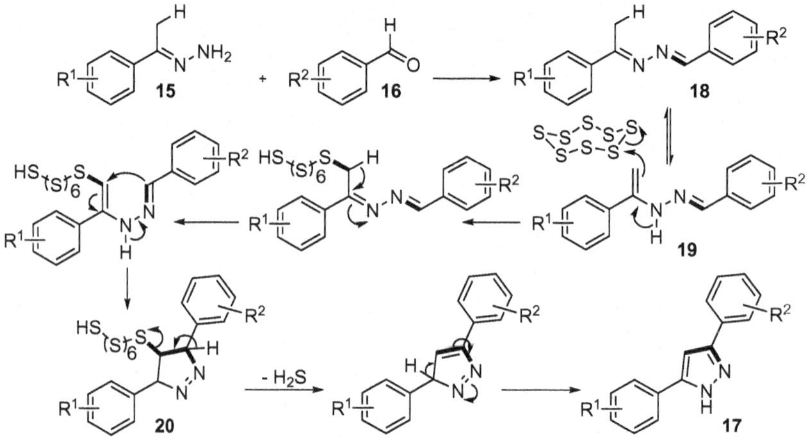

Scheme 8 Plausible mechanism for the synthesis of pyrazole

Scheme 9 Iodine(III)-promoted synthesis of pyrazole

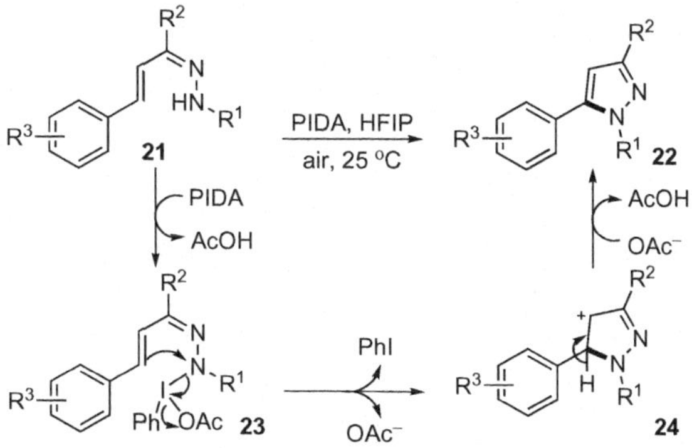

substitution of the hydrazone nitrogen on the iodine(III) centre in PIDA forms an electrophilic *N*-iodo intermediate **23** and acetic acid. Electrophilic addition of the alkene to the hydrazone nitrogen gives intermediate **24** with the concurrent release of PhI and AcO$^-$. Finally, deprotonation affords the desired product **22**.

2.5 Synthesis of Indoles

Indoles are the basic structural elements in a wide variety of naturally occurring biologically important compounds and have diverse application in chemistry, biology and material sciences. Consequently, the development of new and efficient methods to synthesize indoles has always attracted the synthetic community. Although various well-designed transition-metal-catalyzed protocols show high efficiency, the metal-free process can circumvent the toxicity issues associated with the metal catalysis. Therefore, the development of a practical and effective metal-free process for indole synthesis is highly desirable considering the requisite purity in biological and medicinal research. In 2014, Muñiz group developed iodine (III)-mediated intramolecular amination of alkenes to the corresponding indoles **26** from 2-vinyl anilines **25** (Scheme 10) [6]. The reaction involves the use of modified

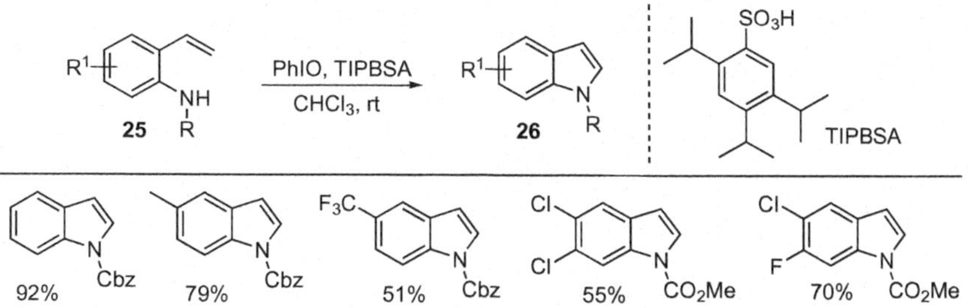

Scheme 10 Iodine(III)-promoted synthesis of indoles

Köser reagent generated from sterically crowded 2,4,5-tris-isopropylbenzene sulphonic acid (TIPBSA) and an iodosobenzene, either in stoichiometric amount or as a catalyst in the presence of *m*CPBA as the terminal oxidant.

The proposed mechanism is depicted in Scheme 11, where the reaction is initiated by the interaction of the modified Köser reagent with an alkene in **25**, leading to an alkene-iodine(III) adduct **27** and then to intermediate **28**. The 1,2-iodooxygenated product **28** forms a cyclopropyl phenonium ion **29** which gains stabilization from the amino group. The spiro-cyclopropyl ring in **29** can undergo two possible ring opening paths: (i) an attack at the methylene position resulting in the formation of **30**, which is aromatized to an indole product; (ii) ring opening at the oxygenated carbon provided 2-oxygenated indoline **31** and/or the corresponding iminium derivative **31′**, which give the desired product with the elimination of arylsulphonic acid.

Youn group described an effective metal-free synthesis of substituted indoles via C–H amination of *N*-Ts-2-alkenylanilines **32** using DDQ as the oxidant (Scheme 12) [7]. The reaction mechanism involves a radical cation generated by SET followed by a migratory process via a phenonium ion intermediate.

In the case of *β*,*β*-disubstituted 2-alkenylanilines **32**, under a standard reaction condition, the electron-rich aryl group undergoes preferential migration, supporting a cationic rearrangement path through a phenonium ion intermediate **38** (Scheme 13, path b) affording the substituted indoles **34**. The use of BHT or TEMPO as additives had a deleterious effect on the efficiency of indole formation although the corresponding trapping products were not observed in both the cases. This suggests that the oxidation of carbon radical **36** resulting from the intramolecular nucleophilic attack by the *o*-sulphonamide group towards an olefinic radical cation gives the benzylic carbocation intermediate **37**, which upon further deprotonation yielded the desired indole **33** (Scheme 13, path a).

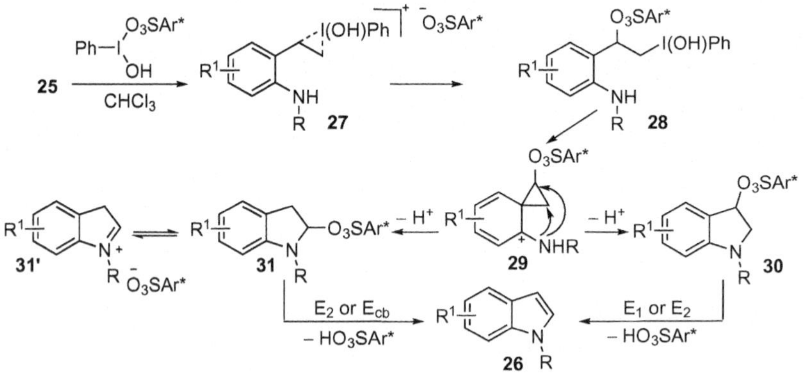

Scheme 11 Mechanism for iodine(III)-promoted synthesis of indoles

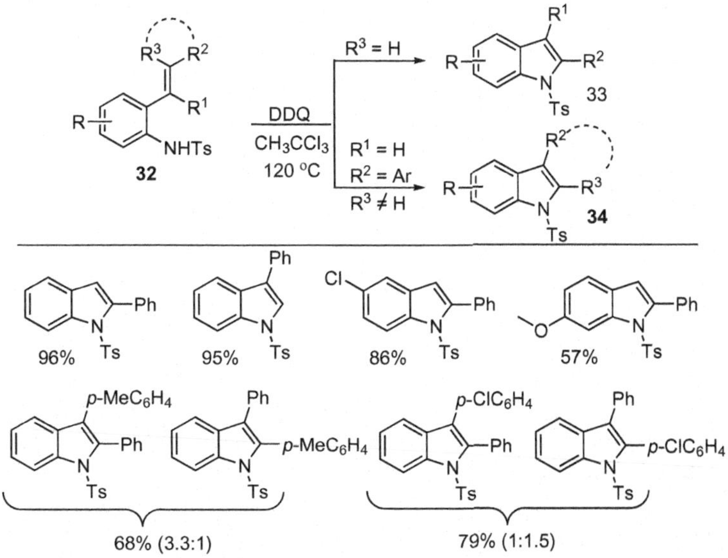

Scheme 12 DDQ-promoted synthesis of substituted indoles

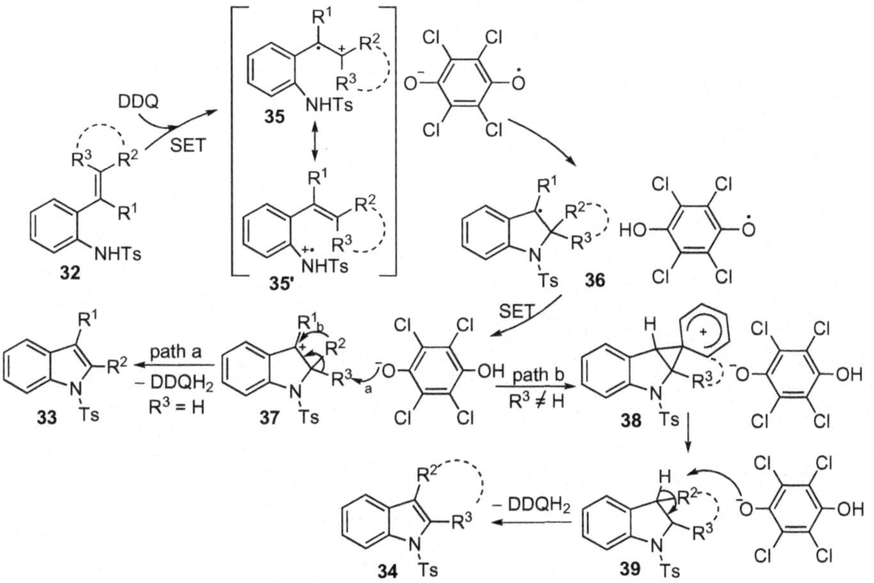

Scheme 13 Mechanistic pathway for the synthesis of indoles

2.6 Synthesis of Oxindoles

Oxindoles are important heterocyclic scaffolds found in a wide variety of bioactive compounds, natural alkaloids and pharmaceutically active molecules. Its derivatives are revealed to have significant potential biological applications such as NMDA antagonist, calcium channel blockers, antiangiogenic, anticancer agent and vasopressin V2 receptor antagonist. In particular, 3,3-disubstituted 2-oxindoles are attractive targets for synthesis and research programmes due to their significant biological properties. A few approaches to their synthesis involve functionalization of monosubstituted 3-oxindoles and isatin derivatives, transition-metal-catalyzed arylation of 3-oxindoles and derivatization of preformed indoles. Towards the metal-free approach, Bisai group achieved an intramolecular-dehydrogenative-coupling (IDC) strategy to synthesize a variety of 2-oxindoles bearing an all-carbon quaternary stereocentre at the pseudobenzylic position (Scheme 14) [8]. The reaction involves a facile one-pot C-alkylation of β-*N*-arylamido esters **40** with alkyl halides using potassium *tert*-butoxide, which undergoes concomitant oxidative coupling in the presence of stoichiometric I_2.

It was observed that the intramolecular dehydrogenative coupling is feasible when the reaction is carried out either after alkylation or when the carbon atom α to the amide having two substituents. This led to the possibility of a radical-mediated process involving a single-electron transfer (SET) as shown in Scheme 15. After

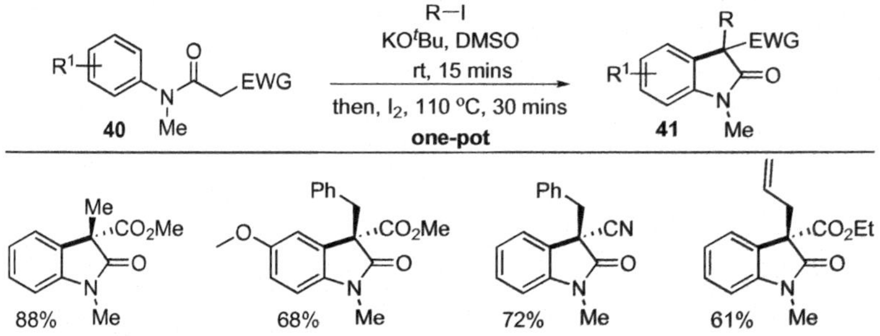

Scheme 14 Synthesis of 3,3-disubstituted-2-oxindole

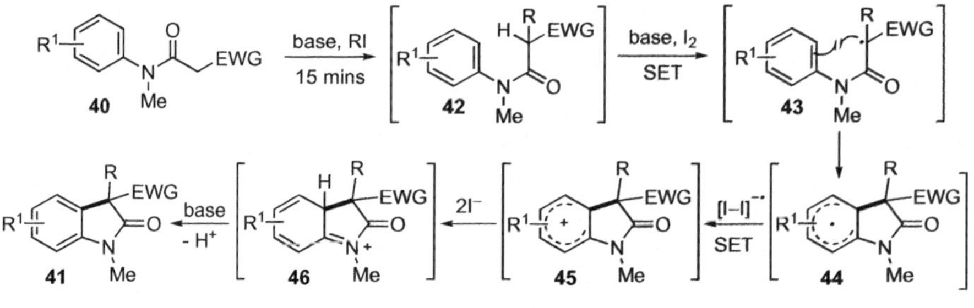

Scheme 15 Plausible mechanism for base promoted synthesis of 3,3-disubstituted-2-oxindole

alkylation, an initial SET leads to **43**, which, in turn, forms an intermediate aryl radical **44**. This radical transfers one electron to the oxidant to form an aryl carbocation **45**, which is stabilized by the amide nitrogen forming **46**. Eventually, rearomatization of **46** in the presence of base affords the oxidative coupling product **41**.

Later, Roy et al. reported a similar reaction from β-*N*-arylamido nitrile promoted by di-*tert*-butyl peroxide (DTBP) via intramolecular C(sp^2)–C(sp^3) coupling [9]. It provided a convenient approach to access oxindoles in moderate to good yields in DCE at 110 °C avoiding the use of base or iodine reagents.

Despite many methods available for the synthesis of 2-oxindoles, the overwhelming majority of these techniques require elevated temperature. This prevents the use of thermally sensitive substrates and making the development of an asymmetric variant of these reactions quite challenging. In view of this, Taylor, Peterson and co-workers recently developed a metal-free low-temperature protocol for the synthesis of 3,3-disubstituted oxindoles via cross-dehydrogenative coupling (Scheme 16) [10]. The use of KHMDS as the strong and non-reversible base has a dramatic effect on the reaction condition relative to the existing procedures. The mechanism proposed above by Bisai's group featured single-electron oxidation of the enolate by I_2 to form a α-carbonyl radical **43**, followed by cyclization (Scheme 15). However, reaction performed at −78 °C using KHMDS (Scheme 16) leads to the formation of C-iodinated anilide **47**, which was not observed in Bisai's work probably due to high reaction temperature (110 °C). By bringing the reaction to room temperature, the iodide undergoes slow thermal homolysis back to the α-carbonyl radical **48**, presumably by halogen bond cleavage with the solvent and/or HMDS followed by cyclization affording the desired oxindoles. A significantly lower reaction temperature offers conditions more appropriate for the stereoselective synthesis and asymmetric CDC routes to 2-oxindoles.

Further, there are several reports towards the synthesis of 2-oxindole frameworks with a tetrasubstituted carbon stereocentre at 3-position via cross-dehydrogenative coupling of *N*-arylacrylamides with different coupling partners as illustrated in Scheme 17.

Song, Li and co-workers described a metal-free oxidative cascade coupling of activated alkenes, i.e. *N*-arylacrylamides with aldehydes using TBHP as the oxidant (Scheme 17, Eq. i) [11]. The reaction proceeds via a radical pathway affording 3-

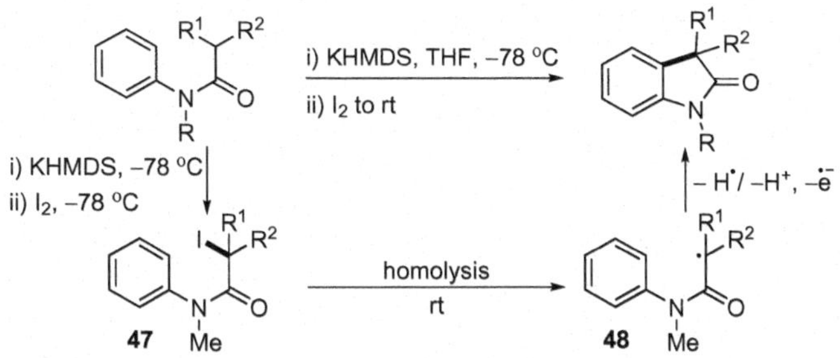

Scheme 16 Metal-free synthesis of 2-oxindoles

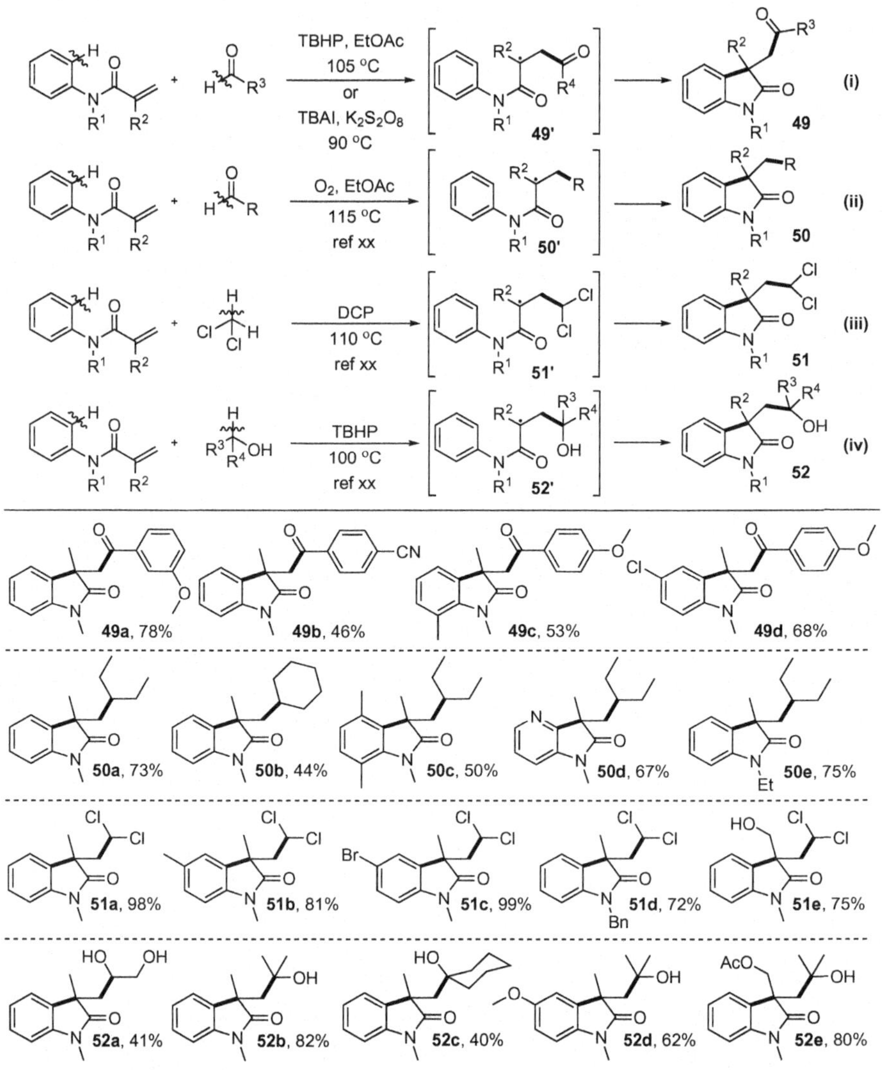

Scheme 17 Synthesis of diverse 3,3-disubstituted 2-oxindoles from *N*-arylacrylamides

(2-oxoethyl)indolin-2-ones **49**. In the presence of TBHP, aldehyde forms an aldehydic radical that adds to the starting substrate forming intermediate **49′**. Intramolecular cyclization with the aryl ring and finally a proton abstraction furnishes the product **49**. Similar reaction was reported by Guo group using TBAI/$K_2S_2O_8$ combination for the synthesis of carbonyl oxindoles [12].

Guin and co-workers also developed a radical-cascade alkylation/cyclization of *N*-arylacrylamide with aldehydes using molecular oxygen as the sole oxidant (Scheme 17, Eq. ii) [13]. In this strategy, the acyl radical generated from aldehyde undergoes auto-oxidation delivering the corresponding alkyl radical through decarbonylation. The alkyl radical thus formed reacts with electron-deficient *N*-alkyl-*N*-arylacrylamide providing an amidyl radical intermediate **50′** which undergoes intramolecular cyclization with the aromatic ring and subsequent deprotonation furnishing the desired oxindoles **50**.

Liu group demonstrated efficient access to dichloromethylated oxindoles via coupling of *N*-arylacrylamide with dichloromethane in the presence of dicumyl peroxide (DCP) (Scheme 17, Eq. iii) [14]. The reaction involves a free-radical addition/cyclization cascade mechanism, in which homolysis of DCP forms cumyloxy radical that abstracts the H-atom from DCM generating a dichloromethyl radical. This radical then adds to *N*-acrylamide generating the radical intermediate **51′** which cyclizes to the aromatic core and eventually undergoes hydrogen abstraction to give the final product **51**.

In yet another report, Guo, Duan and co-workers demonstrated a metal-free tandem oxidative hydroxyalkylation of *N*-arylacrylamide by direct sp^3 C–H functionalization of alcohols in the presence of TBHP (Scheme 17, Eq. iv) [15]. The reaction takes place in similar fashion as reported above, where TBHP promotes the oxidation of alcohol forming a α-hydroxyethylradical, which undergoes addition to the activated alkene generating the radical intermediate **52′**. Finally, cyclization and dehydrogenation give the hydroxylated oxindoles **52**.

2.7 Synthesis of Isoindolinones

Isoindolinone scaffold having a $C(sp^3)$–N bond is found in many natural and synthetic drug molecules possessing various biological activities like inhibitors for the production of tumour necrosis factor (TNF-α), MGR-1 antagonist, antitumour and anti-inflammatory activities. In view of the potential application of isoindolinone, Kumar group established a metal-free synthesis of *N*-aryl-isoindolinones

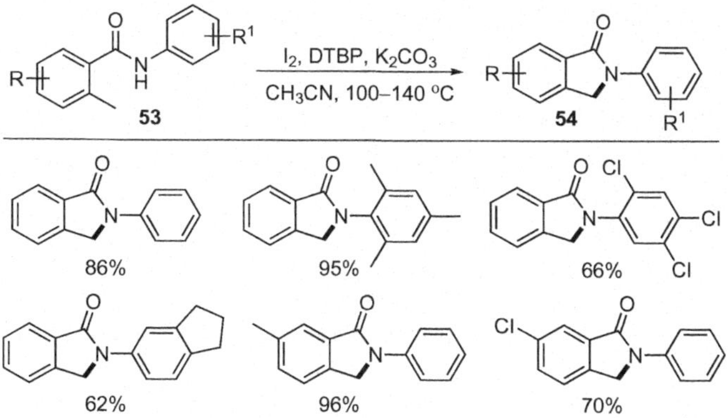

Scheme 18 Synthesis of *N*-aryl-isoindolinones

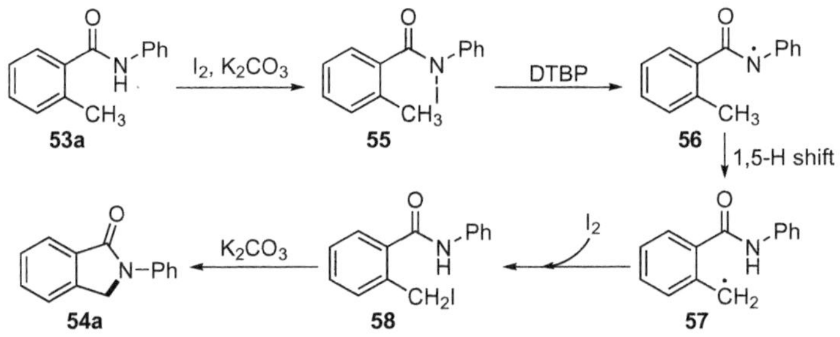

Scheme 19 Proposed reaction mechanism

54 from 2-alkylbenzamide **53** utilizing K_2CO_3, I_2 and di-*tert*-butyl peroxide (DTBP) in acetonitrile (Scheme 18) [16].

According to the proposed mechanism, iodine reacts with N–H bond in the substrate forming an N–I intermediate **55**. This intermediate reacts with DTBP forming an amide radical **56** which undergoes 1,5-H shift generating a benzyl radical **57**. Reaction of this radical with I_2 leads to 2-(iodomethyl)-*N*-phenylbenzamide **58**, which finally undergoes nucleophilic substitution to furnish isoindolinone **54a** (Scheme 19).

2.8 *Synthesis of Quinoline*

Quinoline scaffold occurs in several natural compounds such as cinchona alkaloids and pharmacologically active substances displaying a broad range of biological activities. A variety of well-established methods are available, among which CDC reaction under a metal-free condition for the synthesis of quinoline derivatives is also reported. In 2015, Huo et al. reported a novel synthesis of quinoline derivatives **60** via dehydrogenative Povarov/aromatization of glycine derivatives **59** with alkenes mediated by CBr_4 (Scheme 20) [17].

Based on the results of the control experiments performed, a plausible reaction mechanism has been proposed as depicted in Scheme 21. The starting material **59** reacts with CBr_4 to form intermediate **61** which undergoes a homolytic cleavage of weak N–Br bond forming a radical cation intermediate **62** and bromine radical. The Br radical abstracts a proton from the secondary amine forming intermediate **63**, which reacts with O_2 to give the peroxide radical **64**. Further hydrogen abstraction from the amine by the peroxide radical gives a hydrogen peroxide **65**, and the propagation continues until the starting material is consumed. Subsequently, a Povarov reaction followed by aromatization resulted in the desired quinoline derivative **60**.

The same group later demonstrated a novel CBr_4-mediated double-oxidative dehydrogenative (DOD) cyclization/acidic ring opening/aromatization reaction of

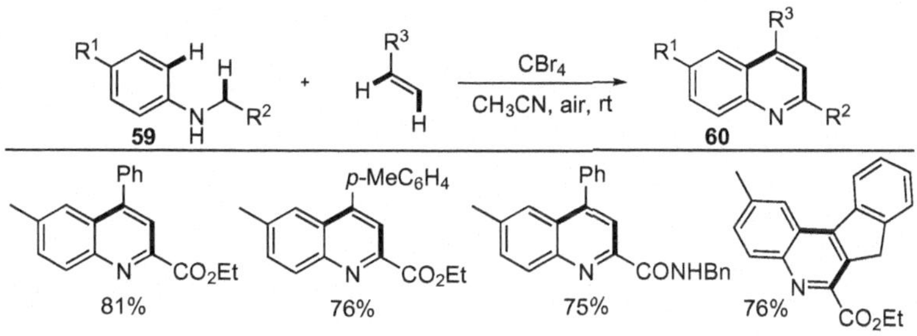

Scheme 20 CBr_4-mediated CDC reaction of glycine derivatives with alkenes

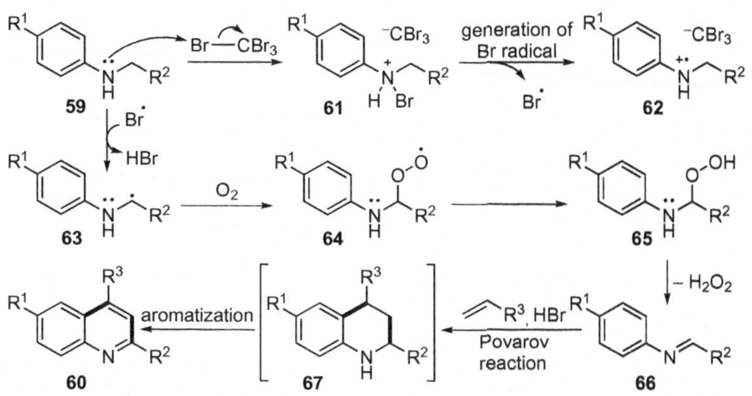

Scheme 21 Mechanism for CBr_4 mediated synthesis of quinolines

glycine derivatives **68** with dioxane **69**, leading to quinoline derivatives **70**. This work again proved CBr_4 as an efficient reagent to initiate the radical reaction under mild reaction conditions (Scheme 22) [18].

The reaction between PPh_3 and CBr_4 generates a PPh_3 radical cation and bromine radical. The bromine radical abstracts protons from **68** and **69** to generate radicals **71** and **75**, respectively, which react further with O_2 to provide peroxide radicals **72** and **76**. These radicals abstract a proton from **68** or **69** to form hydroperoxide **73** and **77**. Elimination of H_2O_2 followed by a [4 + 2] Povarov reaction furnishes the adduct **79** which is unstable under acidic conditions and undergoes an acid-mediated ring opening to give **80**. Subsequent aromatization affords the desired quinazoline **70** (Scheme 23).

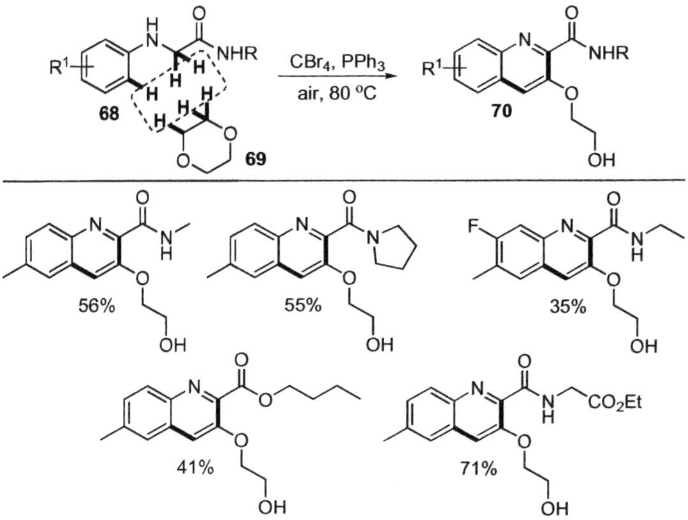

Scheme 22 CBr_4-promoted aerobic double-oxidative dehydrogenative (DOD) cyclization reaction of glycine esters with dioxane

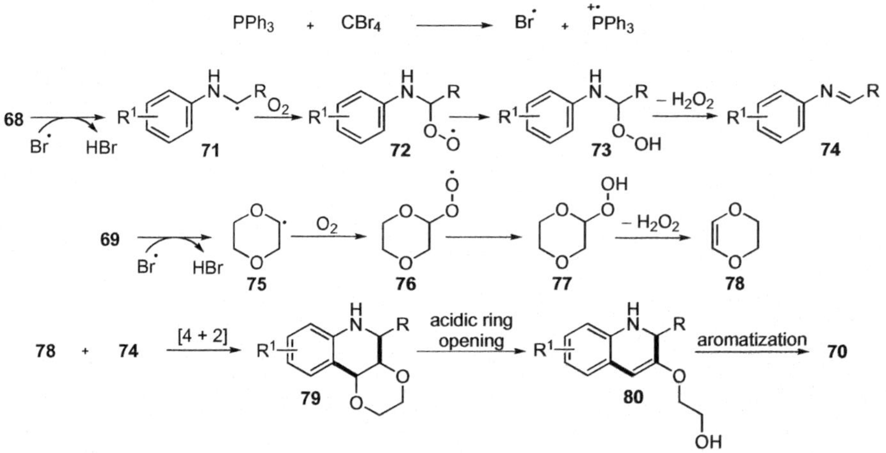

Scheme 23 Mechanistic pathway for CBr_4–PPh_3 promoted synthesis of quinolines

2.9 *Synthesis of Dihydroquinolin-2(1*H*)-Ones and Dihydroisoquinolin-1(2*H*)-One*

Dihydroquinolin-2(1*H*)-ones are another class of important heterocycles found in a myriad of bioactive molecules and natural products. Several substituted 3,4-dihydroquinolin-2(1*H*)-ones have attracted considerable attention due to their

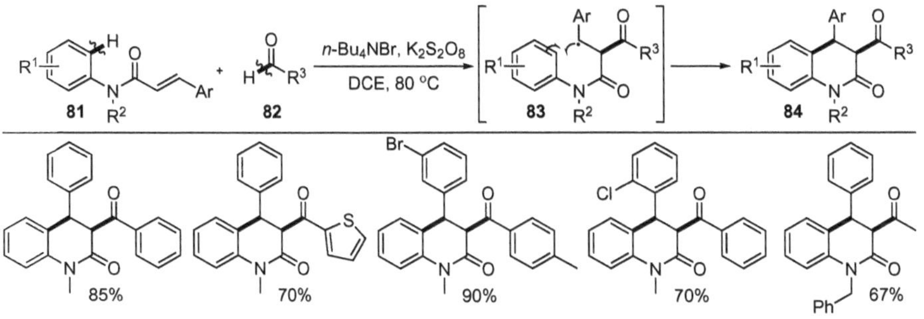

Scheme 24 Synthesis of 3-acyl-4-arylquinolin-2(1*H*)-ones from *N*-arylcinnamide and aldehydes

anti-inflammatory, cardiovascular and phosphodiesterase inhibitory activities. Consequently, the development of practical and highly efficient methods for the construction of 3,4-dihydroquinolin-2(1*H*)-one is desirable. Mai, Lu and co-workers established a TBAB/$K_2S_2O_8$ promoted synthesis of disubstituted dihydroquinolin-2(1*H*)-ones **84** via an intermolecular radical tandem addition/cyclization of *N*-arylcinnamides **81** with aldehydes **82** under a mild reaction condition (Scheme 24) [19]. In this metal-free difunctionalization process, the acyl radical generated from the aldehyde **82** attacks the α-position of the double bond in *N*-arylcinnamide **81** resulting in the radical intermediate **83**. This intermediate undergoes a tandem 6-*endo*-trig cyclization and forms two C–C bonds in one step by SET process and successive rearomatization afforded 3-acyl-4-arylquinolin-2(1*H*)-ones **84**.

Owing to the importance of isoquinolinone as a vital scaffold in many natural products, Han group described a di-*tert*-butyl peroxide-promoted radical cyclization of *N*-allylbenzamides **85** with alcohols **86** resulting in the synthesis 4-substituted 3,4-dihydroisoquinolin-1(2*H*)-one derivatives **88** (Scheme 25) [20]. The reaction proceeds via cascade C(sp^3)–H bond cleavage, followed by an oxyalkylation of the double bond and an intramolecular cyclization.

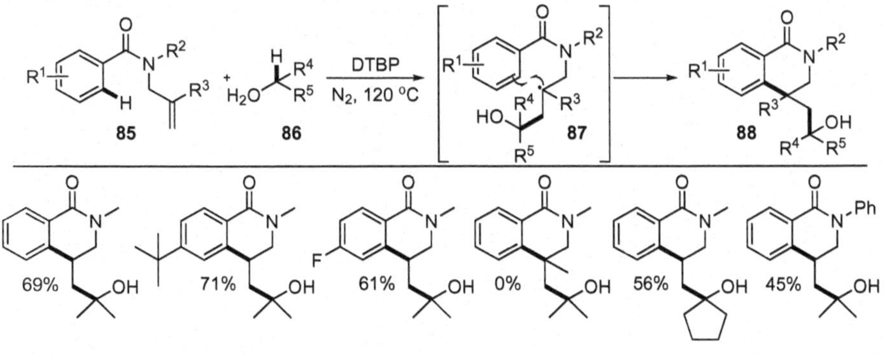

Scheme 25 Synthesis of 3,4-dihydroisoquinolin-1(2*H*)-one

2.10 Synthesis of Phenanthridine

Phenanthridines are ubiquitous units found in several bioactive natural alkaloids and medicinally relevant compounds exhibiting antibacterial, antitumour, antileukemic and cytotoxic activities. So far, many powerful strategies involving radical isonitrile insertion towards the synthesis of a series of 6-substituted phenanthridines are reported. A *tert*-butyl hydroperoxide (TBHP)-promoted carboxamidation and cyclization of isonitriles **89** with formamides have been reported by Yu group resulting in the formation of phenanthridine 6-carboxamides **90** (Scheme 26, Eq. i) [21]. Han and co-workers demonstrated the synthesis of 6-amidophenanthridines **91** by reacting 2-isocyanobiphenyls with acetamides in the presence of *tert*-butyl peroxybenzoate (TBPB) as the oxidant (Scheme 26, Eq. ii) [22]. The reaction gave excellent yields with a variety of 2-isocyanobiphenyl derivatives. Reaction of monomethyl-substituted acetamides gave the desired products in excellent

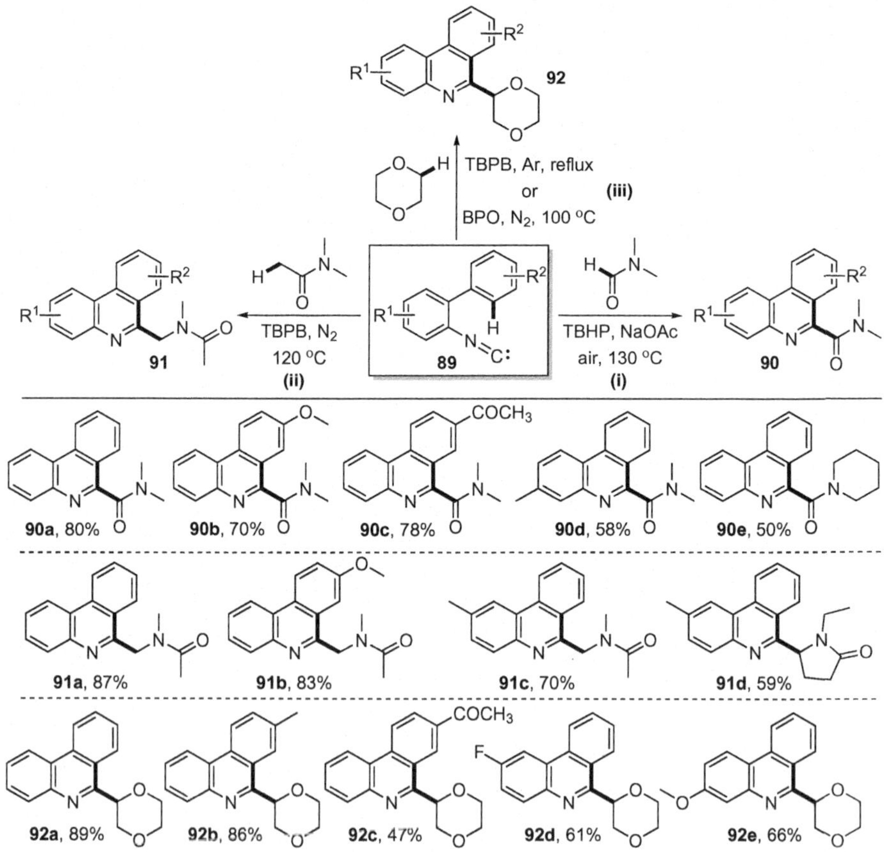

Scheme 26 Metal-free synthesis of C-6 functionalized phenanthridines from 2-isocyanobiaryl compounds

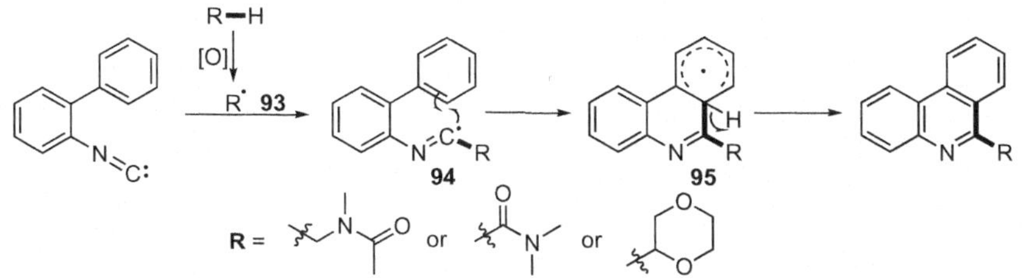

Scheme 27 Mechanistic pathway for the synthesis of functionalized phenanthridines

chemoselectivity. Interestingly reaction with DMF gave two isomeric products via C(sp^3)–H as well as C(sp^2)–H bond cleavages in decent yields.

In a report by Ji group, construction of 6-alkyl phenanthridines **92** was achieved via 2-isocyanobiaryl insertion with 1,4-dioxane under a reflux condition mediated by *tert*-butyl peroxybenzoate (TBPB) in an argon atmosphere (Scheme 26, Eq. iii) [23]. A similar result, i.e. α-phenanthridinylation of ethers by isocyanides, was also demonstrated by Cheng et al. promoted by benzoyl peroxide in N_2 atmosphere (Scheme 26, Eq. iii) [24]. In both cases, good to moderate yields of 6-alkyl phenanthridines were obtained. However, in the former case reaction with other cyclic ethers such tetrahydrofuran did not give the desired alkynylated product, rather yielded simple phenanthridine. Interestingly, in the latter case, the reaction underwent smoothly with THF as well as other acyclic ethers giving a decent yield of the desired products.

All the reactions mentioned in Scheme 26, proceeds via a common reaction pathway (Scheme 27), where the coupling partner such as amides or ethers undergo a homolytic cleavage in the presence of a suitable oxidant generating a radical species **93**. This radical adds to isonitrile producing another radical intermediate **94**, which undergoes intramolecular radical cyclization with the aromatic ring giving **95**. Finally, hydrogen radical abstraction gives the phenanthridine derivatives.

2.11 *Synthesis of Quinazolines*

Quinazoline derivatives have gained significant attention for their various biological and pharmacological activities. Among various metal-free strategies adopted towards the synthesis of quinazoline derivatives, a few are discussed here. Wang et al. described a simple and efficient approach for the synthesis of 2-arylquinazolines via the coupling of 2-aminobenzoketones **96** with arylmethanamines **97** catalyzed by I_2 and TBHP as the oxidant at 90 °C (Scheme 28) [25].

A plausible mechanism has been proposed as shown in Scheme 29. At first, intermediate **99** is generated from 2-aminobenzoketone **96** and aryl methanamine **97** which undergo subsequent oxidation to give **101** via sp^3 C–H functionalization

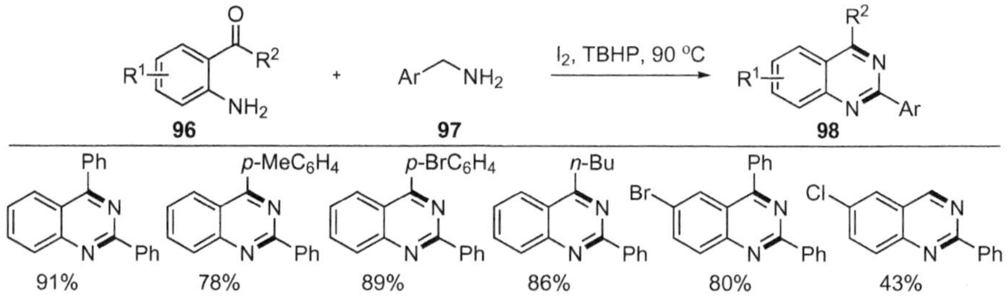

Scheme 28 I_2-catalyzed synthesis of 2-arylquinazolines

under the given reaction condition. Then, the intermediate **101** upon tandem intramolecular cyclization and oxidation gives the desired quinazoline derivative **98**.

A similar synthesis of 2-arylquinolines was reported by Han, Yu and co-workers using 4-hydroxy-tempo radical as the catalyst in *o*-xylene under O_2 atmosphere at 140 °C [26]. The reaction proceeds in a similar fashion as depicted in Scheme 29.

Li group developed a novel three-component approach towards the synthesis of 2-arylquinolines in the presence of KI and TBHP from *ortho*-carbonyl-substituted anilines, NH_4OAc, as the nitrogen source and sp^3 carbon of methylarenes as C_1 source (Scheme 30) [27].

Based on control experiments and literature report, a plausible reaction mechanism has been proposed as illustrated in Scheme 31. Initially, toluene **104a** undergoes radical oxidation in the presence of TBHP forming benzaldehyde **104a′**.

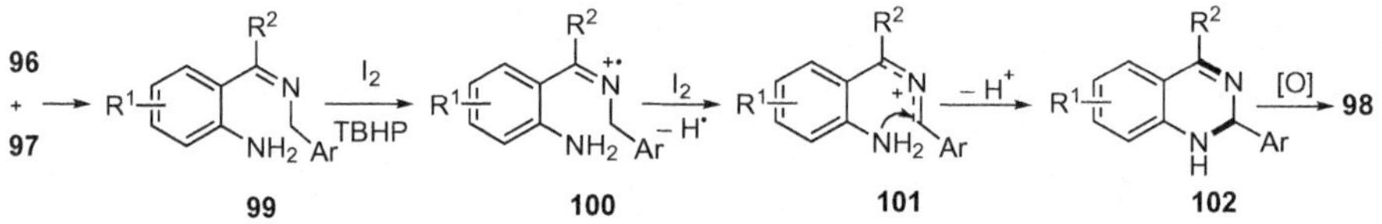

Scheme 29 Plausible mechanism for I_2-catalyzed synthesis of 2-arylquinazolines

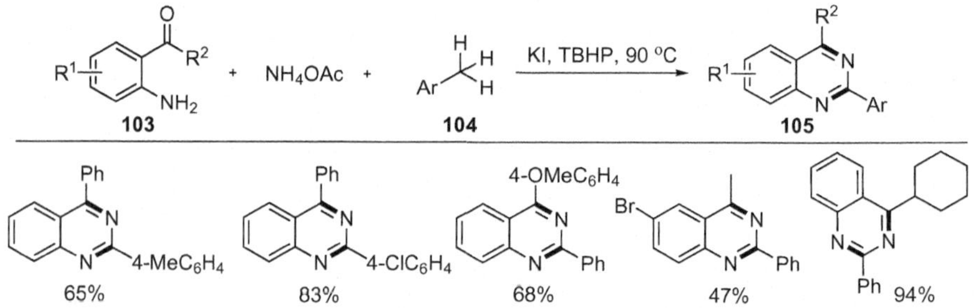

Scheme 30 Three-component synthesis of quinazolines from 2-aminophenyl ketones, nitrogen and C_1 source

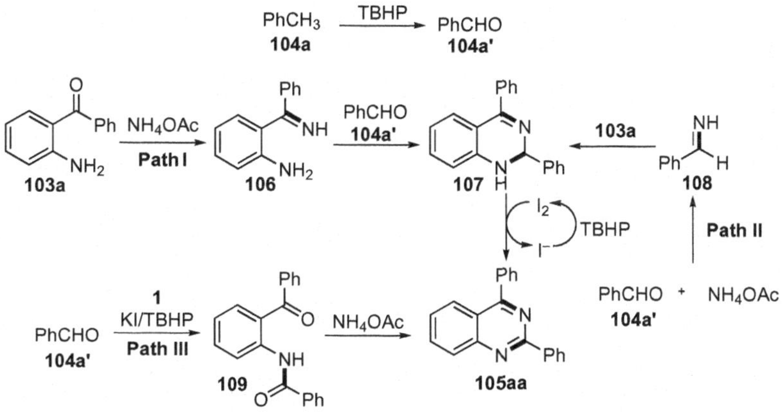

Scheme 31 Plausible mechanism of quinazolines synthesis via three-component reaction

The carbonyl group in 2-aminobenzophenone **103a** or benzaldehyde **104a′** reacts with NH_4OAc to give ketimine **106** or aldimine **108**, respectively, which after cyclization generates **107** (paths I and II). Subsequently, **107** get converted to quinazoline **105aa** via iodine-promoted oxidative aromatization. Alternatively, a third pathway might occur (path III), where an intermediate **109** could be formed via amidation under KI/TBHP catalytic system and subsequent condensation with NH_4OAc would afford the desired product **105aa**.

An unusual solvent- or oxidant-switchable direct synthesis of multi-substituted quinazolines was achieved by Long et al. from *N*-alkyl-*N′*-arylamidines **110**. Using iodobenzene diacetate, a broad range of multi-substituted quinazolines was achieved via an intramolecular oxidative C(sp^3)–C(sp^2) bond formation in non-polar solvents (Scheme 32) [28]. Also, the use of $K_2S_2O_8$ as the oxidant and TEMPO as the additive in CH_3CN at 120 °C resulted in the synthesis of quinazolines **111**. The TEMPO-catalyzed synthesis in polar solvent displayed superior reaction efficiency compared to $PhI(OAc)_2$-mediated synthesis in a non-polar solvent.

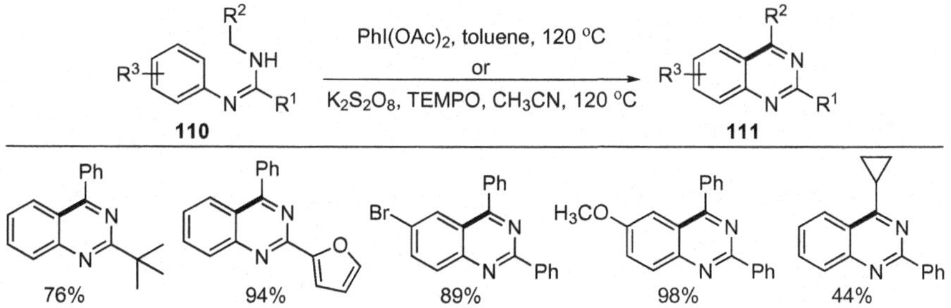

Scheme 32 Iodine(III)-promoted synthesis of quinazolines

A plausible mechanism has been proposed where the initial treatment of **101** with $PhI(OAc)_2$ furnishes an *N*-(phenylacetoxyiodo)imidamido species **112** (Scheme 33). In non-polar solvent, the imidamido species **112** is prone to conversion to the neutral benzylidene-*N*′-phenylcarboxamidine **113** via an intramolecular attack of the acetate ion on the benzylic $C(sp^3)$–H of **112**. The intermediate **113** is transformed into a benzylic cationic species **114** by the proton acid formed in situ, which further undergoes an intramolecular Friedel–Craft reaction to afford **116**. Finally, the oxidation of **116** afforded the desired product quinazoline **111**.

A visible light-mediated synthesis of multi-substituted quinazolines **118** was achieved by Tang et al. from *N*-alkyl-*N*′-arylamidines **117** via oxidative $C(sp^3)$–C(sp^3) bond formation, catalyzed by rose bengal, a photoredox organocatalyst (Scheme 34) [29]. The reaction takes place in the presence of CBr_4 as the oxidant, K_2CO_3 as the base in DMSO solvent using 18 W fluorescent lamp as the visible light source.

On the basis of the control experiments and literature reports, a possible mechanism is proposed (Scheme 35). In the absence of rose bengal, CBr_4 is disproportionate to a CBr_3 radical and bromide radical upon irradiation of visible light.

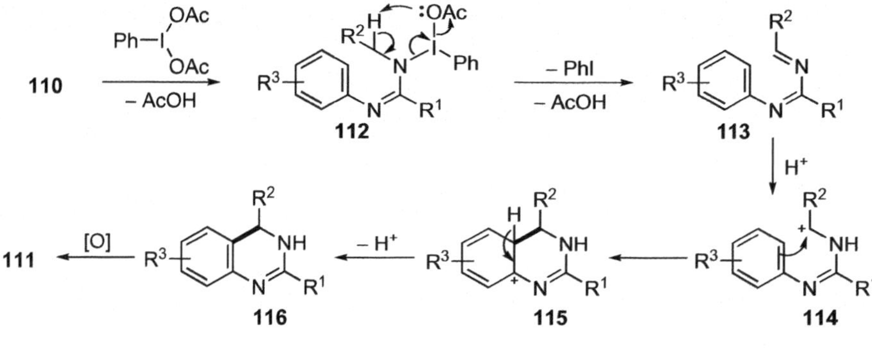

Scheme 33 Plausible mechanism for iodine(III)-mediated synthesis of quinazolines

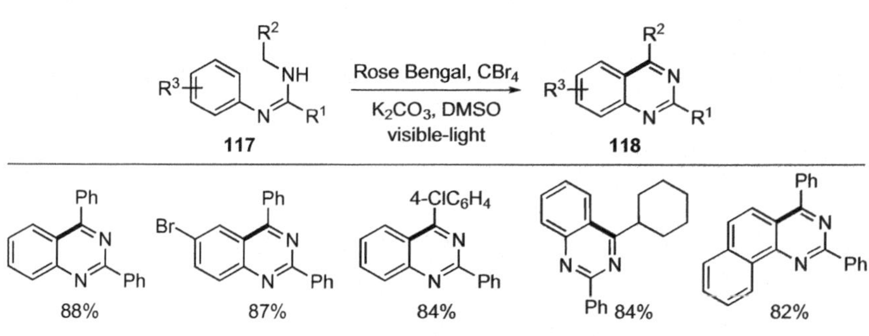

Scheme 34 Visible light-mediated synthesis of quinazolines

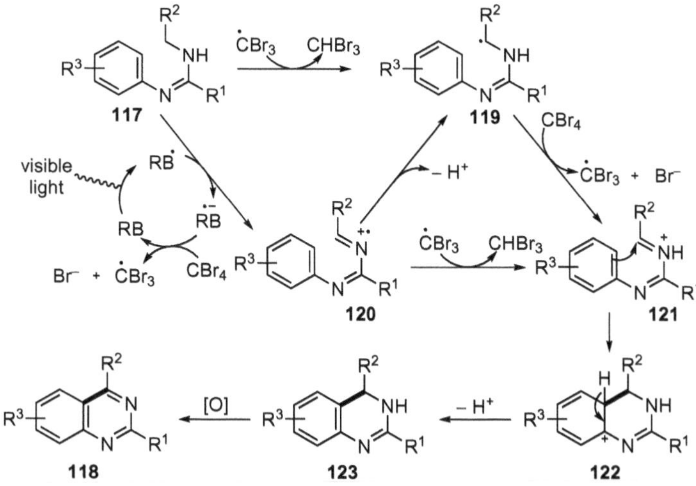

Scheme 35 Mechanism for visible light-mediated synthesis of quinazolines

The CBr_3 radical abstracts a hydrogen atom from the benzimidamide **117** to produce a α-amino radical intermediate **119**, which once again enters the radical chain process with CBr_4 to generate an iminium ion **121**. The iminium ion undergoes an intramolecular Friedel–Craft reaction to give intermediate **122**. Further, dehydrogenation affords hydroquinone **123**, which finally aromatizes to give the stable product quinazoline **118**. In another path, the oxidative quenching of the visible light-excited rose bengal (RB*) by benzimidamide **117** results in the formation of a rose bengal radical anion ($RB^{\cdot -}$) and a radical cation **120**. Upon transfer of an electron to CBr_4 from the superoxide radical anion, regenerates the rose bengal, thus completing the photoredox cycle. On the other hand, the radical cation **120** gives a hydrogen atom, presumably to the rose bengal radical anion to afford a CBr_3 radical, a bromide anion and the radical intermediate **119**. Yet another pathway has been assumed in which CBr_3 radical abstracts a hydrogen atom from **120** to generate an iminium ion **121** directly following the same procedure to give the stable product quinazoline **118**.

2.12 *Synthesis of Quinazolinones*

Quinazolinone core occurs widely in natural products and synthetic pharmaceuticals. They have a significant role in medicinal chemistry and have been extensively studied as anti-inflammatory, antiulcer, anticancer and as hypolipidemic agent. Many catalytic strategies are available for their synthesis which mainly involves condensation of 2-aminobenzamides with carboxylic acids or aldehydes followed by oxidation. To overcome the drawbacks associated with metal-catalyzed reactions and to develop greener and sustainable approach, Wei et al. reported iodine-catalyzed one-pot synthesis of quinazolinones **126** from primary alcohols **125** and *o*-aminobenzamides **124** using DMSO as the oxidant in dimethyl carbonate (DMC) solvent (Scheme 36) [30].

A plausible mechanism has been proposed as illustrated in Scheme 37. At first, the alcohol is oxidized to the corresponding aldehyde with iodine as the oxidant. The in situ generated aldehyde reacts with **124** forming an imine intermediate **127**. In the presence of iodine, the imine intermediate cyclizes to form an *N*-iodinated species **128**. Finally, the elimination of HI produces the final product **126**. Iodine can be regenerated via the oxidation of HI with DMSO as the oxidant.

In yet another report, Li group described a metal-free approach for the synthesis of quinazolinones **131** via dual benzylic C(sp^3)–H bond amination, using *o*-aminobenzamides **129** and methylarenes **130** as the accessible starting materials. The reaction takes place in the presence of DTBP as the oxidant, *p*-toluenesulphonic acid as the additive in DMSO at 110 °C (Scheme 38) [31].

Several control experiments were performed which ruled out the possibility of benzaldehyde as the reaction intermediate. Reactions performed in the presence of radical scavengers BHT or TEMPO had a negative influence on the product yield suggesting the possibility of a radical mechanism. Thus, based on literature reports, it was assumed that the *tert*-butoxy intermediate **132** reacts with **129** forming **133**. Subsequent intramolecular amination–oxidation results in the formation of the desired quinazolinone **131** (Scheme 39).

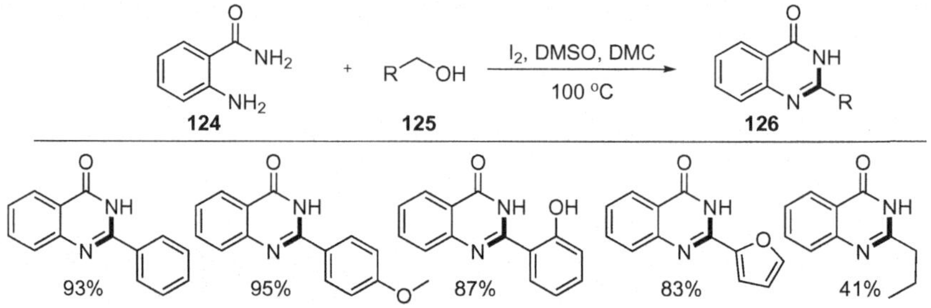

Scheme 36 I_2-catalyzed synthesis of quinazolinones

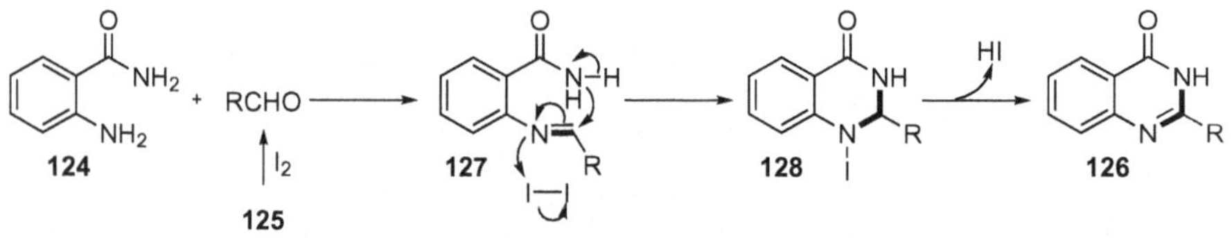

Scheme 37 Plausible mechanism for I_2-catalyzed synthesis of quinazolinones

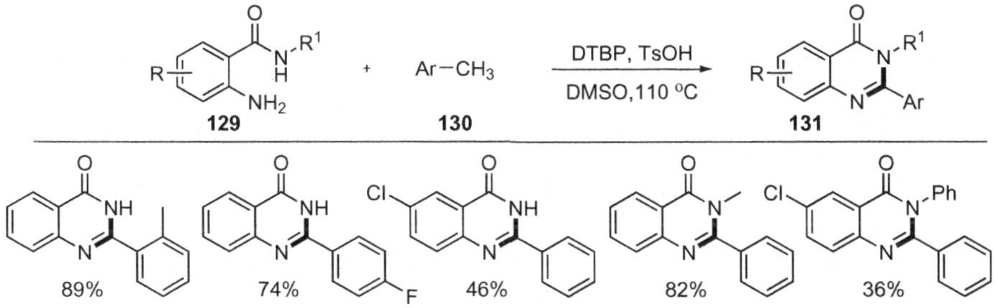

Scheme 38 DTBP-promoted synthesis of quinazolinones

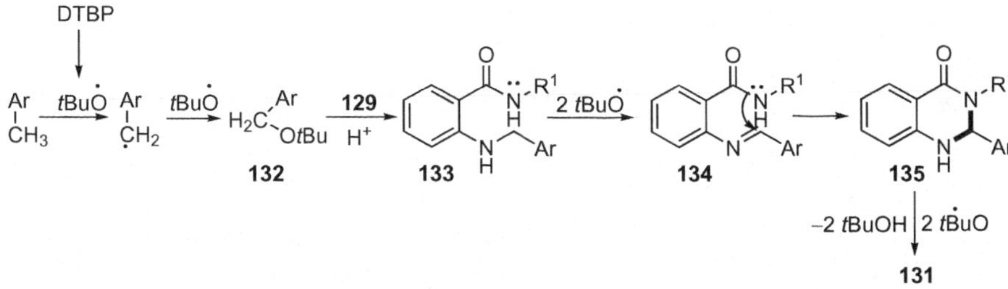

Scheme 39 Mechanistic pathway for DTBP-promoted synthesis of quinazolinones

2.13 Synthesis of Carbazoles

Carbazoles are tricyclic *N*-heterocycles having an indole skeleton and serve as key intermediates in the synthesis of many alkaloids and drug molecules. They are extensively used as pharmaceuticals, photoconducting materials and as organic dyes. Many methods have been reported for the construction of carbazole skeleton promoted by transition metals. Among the metal-free approaches, Antonchick and co-workers demonstrated a highly efficient, atom-economical, environmentally friendly organocatalytic method for the synthesis of carbazoles **137** from 2-aminobiphenyls **136** via intramolecular C–H amination (Scheme 40) [32]. The use of unprotected or alkyl-protected 2-aminobiphenyls under the given reaction condition was unproductive.

According to the control experiment, the organocatalyst **A** is oxidized by peracetic acid to generate the active catalyst **A′** which facilitates the amination of 2-aminobiphenyl **136a**. The (diacetoxy)iodobenzene reacts with amide to give intermediate **138**, which is then transformed into a nitrenium ion **139** via an oxidative process. Finally, the attack of the nucleophilic arene to the electron-deficient nitrenium ion gives the desired product **137a** (Scheme 41).

Recently, Mal group described a hypervalent iodine(III)-mediated intermolecular dehydrogenative annulation reaction for the synthesis of *N*-substituted carbazoles

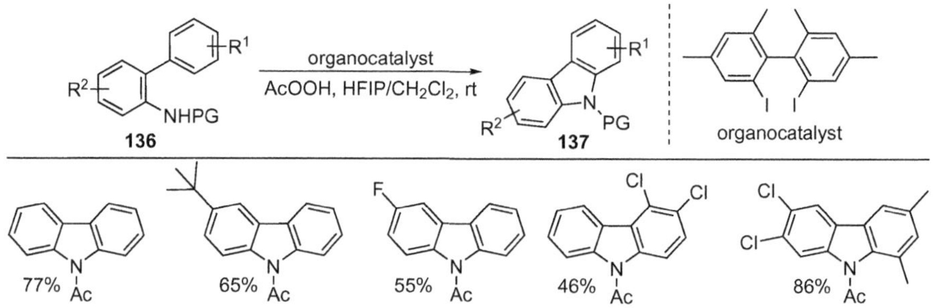

Scheme 40 Organocatalyzed synthesis of carbazoles

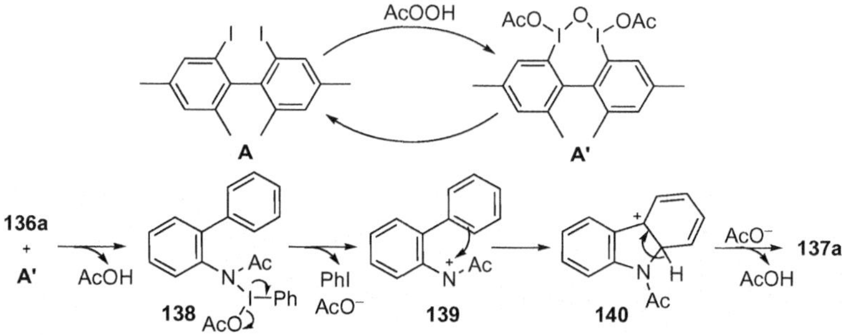

Scheme 41 Proposed mechanism for carbazole synthesis

143 via tandem C–C/C–N bond formation (Scheme 42) [33]. The reaction involves the coupling of non-pre-functionalized arenes **142** and anilides **141** in the presence of phenyliodine(III) diacetate (PIDA) as the sole oxidant in a fluorinated solvent at room temperature. The organocatalytic method utilizes iodobenzene and *m*CPBA under the same condition.

Based on a control experiment, a plausible mechanism has been proposed (Scheme 43). The reaction of anilides **141** with PIDA generates a nitrenium ion intermediate **144**. The nitrenium ion gets stabilized by charge delocalization to form carbenium ion intermediate **145**. Nucleophilic attack by arene takes place either at the carbonic ion (C–C bond formation) or at nitrenium ion (C–N bond formation). However, studies reveal the formation of C–C bond over C–N bond. From intermediate **146**, the oxidative C–N bond formation gives the desired carbazole **143**. In an organocatalytic approach, the iodobenzene is oxidized by *m*CPBA to generate trivalent organoiodine species, which enables intermolecular annulation of arenes **142** and anilides **141** to afford the final product **143**.

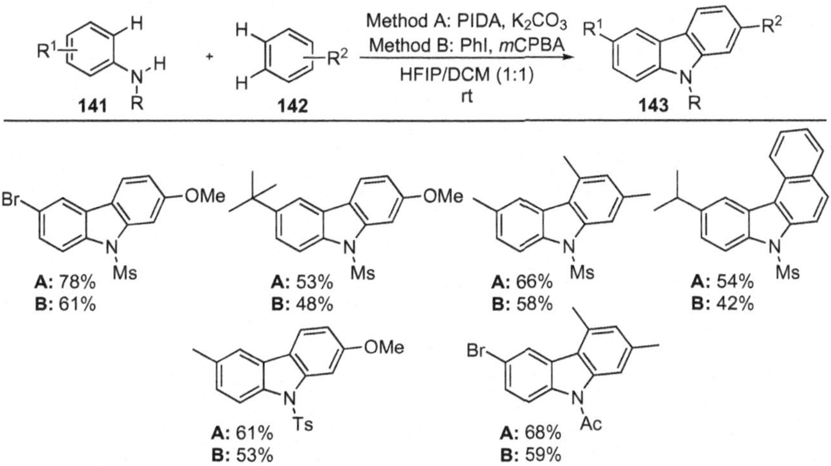

Scheme 42 Iodine(III)-mediated intermolecular dehydrogenative synthesis of carbazoles

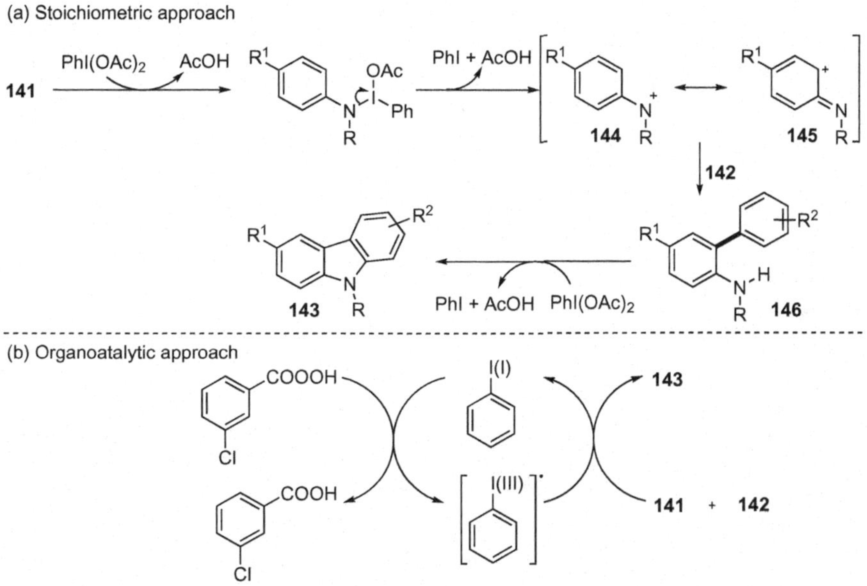

Scheme 43 Plausible mechanism for annulation

2.14 *Synthesis of Benzimidazole*

Benzimidazoles are ubiquitous units found in natural products and synthetic molecules with diverse bioactivities. Zhu et al. reported a general and efficient method for the construction of 2-substituted benzimidazole via PIDA-mediated

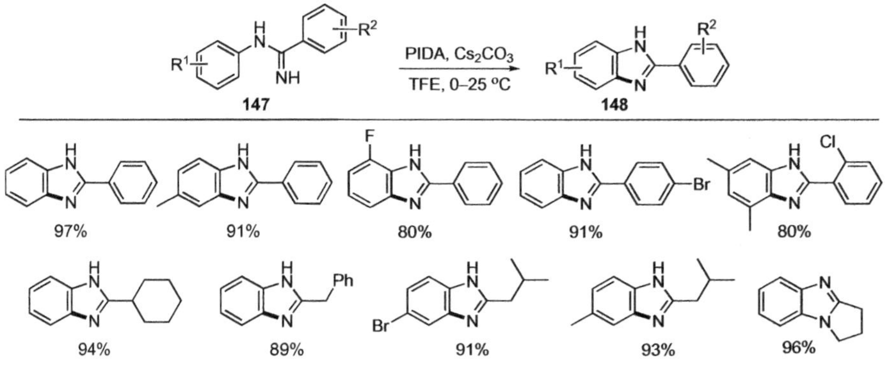

Scheme 44 PIDA-mediated synthesis of benzimidazole

intramolecular oxidative imidation of aromatic C–H bonds in *N*-arylamidines **147** (Scheme 44) [34]. The reaction takes place at 0 °C or ambient temperature in a short time.

Control experiments performed with a radical scavenger 1,1-diphenylethylene inhibited the reaction suggesting a radical pathway. Initially, PIDA reacts with amidines **147** to form an *N*-iodo species **149** with the release of AcOH. Further, homolysis generates the *N*-centred free radical intermediate **150** and hypervalent iodine(III)-centred radical **151**. Subsequent addition of the imidyl radical to the aromatic ring generates **152**, which undergoes oxidation to cyclohexadienyl cation **153** via the intermediacy of **151** involving SET. Finally, a proton abstraction by the acetate radical gives the desired product **148** (Scheme 45).

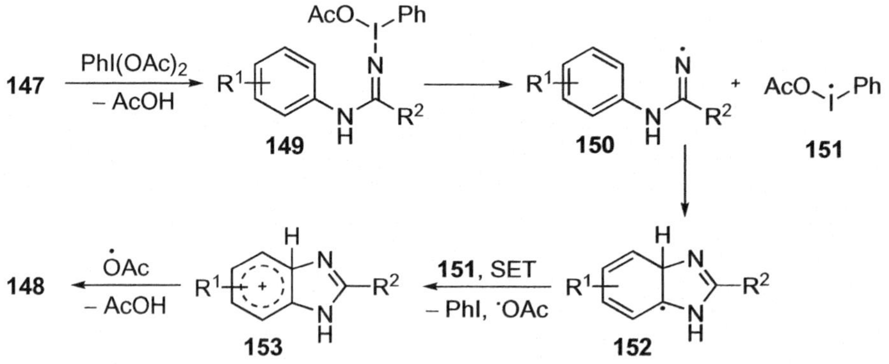

Scheme 45 Proposed mechanism for PIDA-mediated synthesis of benzimidazole

2.15 Synthesis of Acridone

Du, Zhao and co-workers reported a metal-free approach towards the synthesis of acridone derivatives. The protocol involves a direct aryl-aldehyde C(sp^2)–C(sp^2) bond formation via PhI(OAc)$_2$-mediated intramolecular cross-dehydrogenative coupling of various 2-(*N*-arylamino)aldehydes **154** (Scheme 46) [35]. The method does not require any pre-activation of the aldehyde and is applicable to a large variety of functionalized substrates.

A mechanism entailing a radical pathway has been proposed as illustrated in Scheme 47. The reaction is initiated by heating BPO forming a benzoyloxyl radical **B**, which reacts with iodosobenzene diacetate **C** to generate radical **D**. The radical **D** abstracts the aldehydic hydrogen generating an acyl radical **156** which

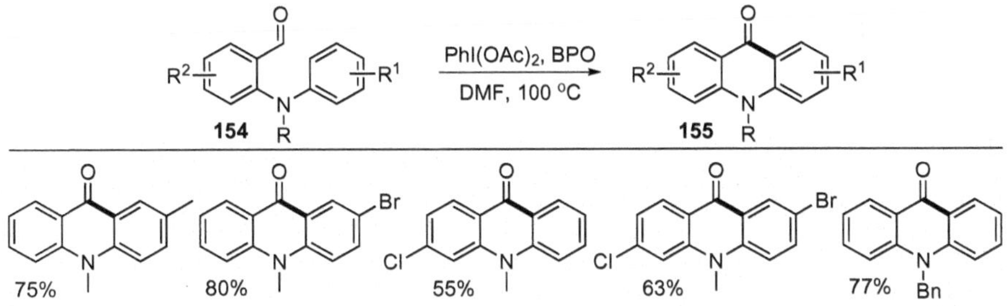

Scheme 46 PIDA-mediated synthesis of acridone derivative

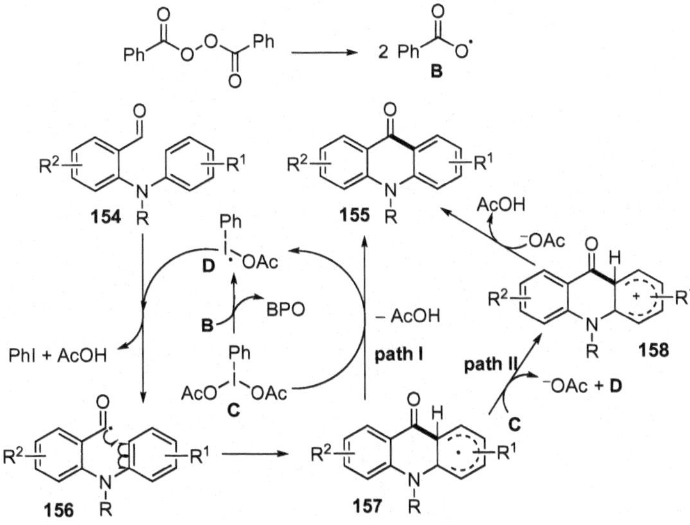

Scheme 47 Proposed mechanism for acridone synthesis

subsequently breaks a pi-bond generating cyclohexadienyl radical **157**. Finally, a second H-abstraction connected to the sp^3-carbon gives the desired product by regaining the aromaticity. Alternatively, single-electron oxidation of **157** gives cation **158**, which upon deprotonation by the acetate ion affords the product.

2.16 Synthesis of Azafluorenones

A convenient metal-free approach to 4-azafluorenone was demonstrated by Laha group by TBHP-mediated intramolecular carbonylation of arenes in 2-aryl-3-picolines **159** via oxidative C–H functionalization of the methyl group (Scheme 48) [36].

According to the mechanism, TBHP upon heating forms radical **i** and **ii** which are assumed to initiate the reaction (Scheme 49). A proton abstraction from the methyl group forms the 3-picoline radical, which on subsequent proton abstraction forms a primary alcohol **161**. Oxidation of alcohol gives an aldehyde **162** which reacts with TBHP forming an acyl carbonyl radical **163**. Finally, acylation with the arene delivers the product.

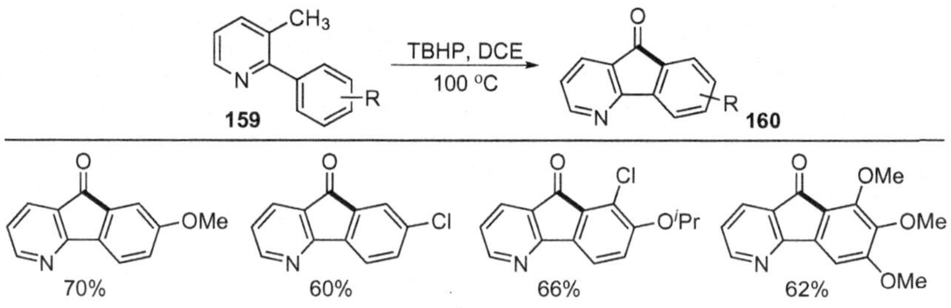

Scheme 48 TBHP-mediated synthesis of 4-azafluorenones

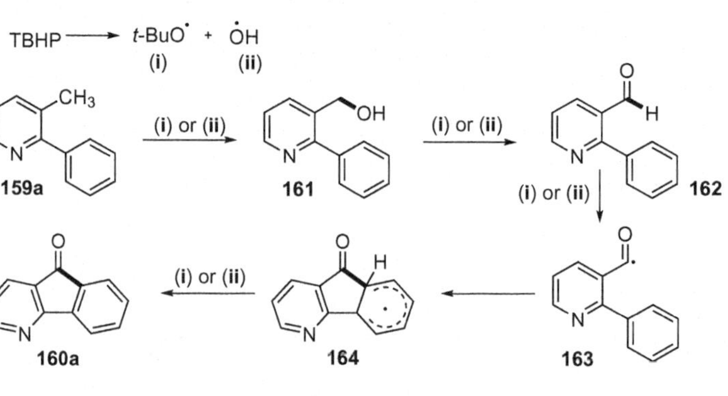

Scheme 49 Plausible mechanism for the synthesis of 4-azafluorenones

2.17 *Synthesis of Indolizines*

Indolizines are eminent fused-ring heterocycles found in many natural products and synthetic drugs. They form the core skeleton of many pharmacological compounds and have also found application in the field of material sciences. Yan group established an I_2-mediated oxidative tandem cyclization between alkylpyridines **165** and aldehydes **166** towards the synthesis of substituted indolizines **167** via C–N/C–C bond formation (Scheme 50) [37].

As per the proposed mechanism, substrate **165a** reacts with aldehyde **166a** to give an aldol condensation product **168**, which upon dehydration gives **169** and **169′** (Scheme 51). During this transformation, **168** is converted to **169′** via inversion of configuration under the optimized reaction conditions. An electrophilic substitution of **169′** with iodine gives intermediate **170**, which finally undergoes

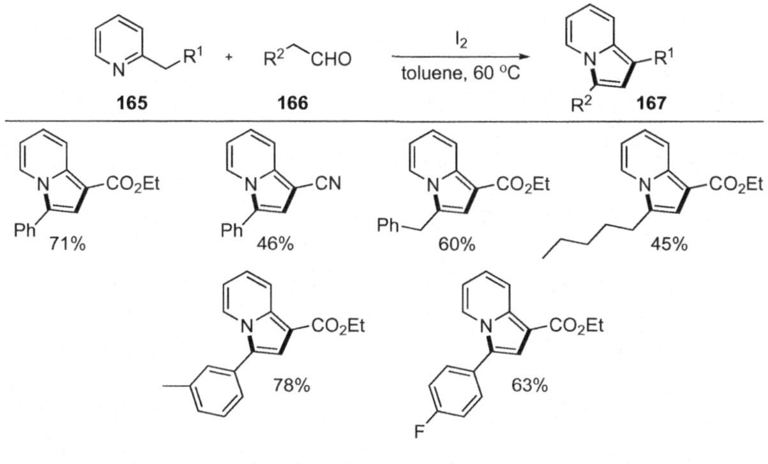

Scheme 50 I_2-mediated synthesis of indolizines

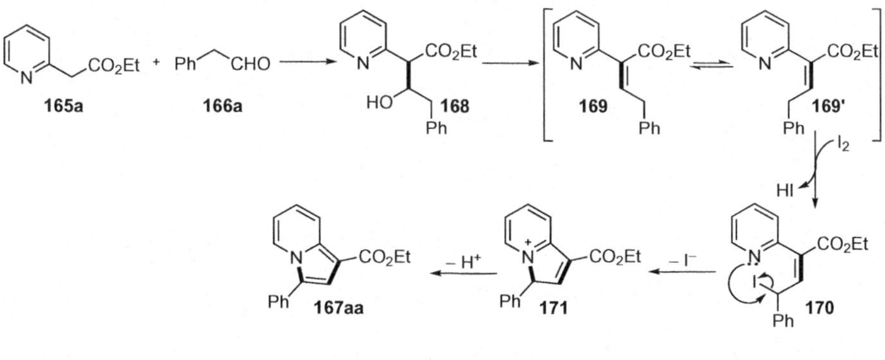

Scheme 51 Mechanistic pathway for I_2-mediated synthesis of indolizines

intramolecular nucleophilic substitution to give intermediate **171**. Subsequent proton elimination affords the desired product **167aa**.

Adimurthy and co-workers developed a similar method for the synthesis of substituted indolizines **173** from 2-pyridylesters **172** and chalcones **173** catalyzed by iodine and DTBP as the oxidant at 120 °C in 1,2-dichlorobenzene as the solvent (Scheme 52) [38].

Based on the results obtained from control experiments, the substrate **172a** forms a radical intermediate **175** in the presence of DTBP which reacts with **173a** to form radical **176**. This undergoes intramolecular radical addition followed by oxidation to give **178** from which an H-abstraction and subsequent aromatization afford the final product **174aa** (Scheme 53).

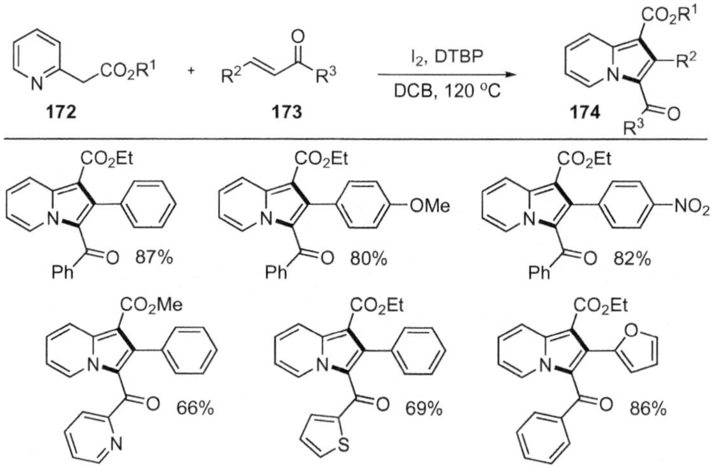

Scheme 52 Iodine-mediated synthesis of indolizines from 2-pyridylesters and chalcones

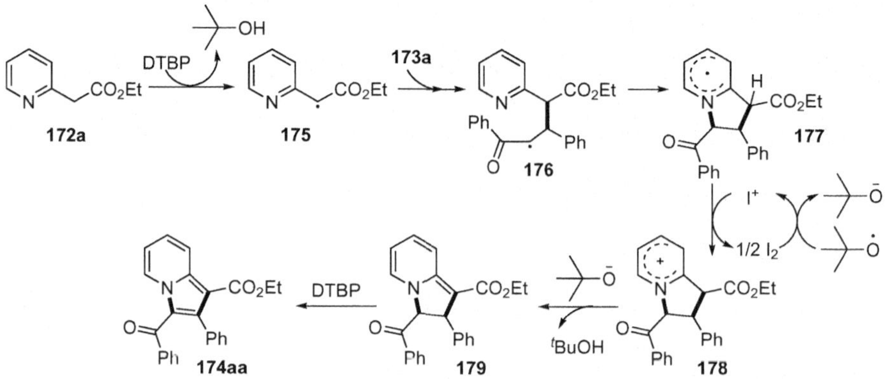

Scheme 53 Radical pathway for I_2-mediated synthesis of indolizines

2.18 Synthesis of Imidazo[1,5-a]Pyridines

Imidazo[1,5-*a*]pyridines constitute a class of aza-fused heterocycle existing in many pharmacologically important compounds. Molecules with this skeleton have potential applications in organic light-emitting diodes (OLEDs) and organic thin-layer field effect transistors (FETs). In 2013, Wang et al. developed an iodine-promoted multi-substituted synthesis of imidazo[1,5-*a*]pyridines **182** from 2-pyridylesters **180** and benzylamines **181** via C(sp^3)–H bond amination with water and *t*-BuOH as the only waste product (Scheme 54) [39]. The reaction involves two oxidative C–N coupling and one oxidative dehydrogenation process with the loss of six hydrogen atoms.

To gain insights into the mechanism, the reaction was performed in the presence of radical scavenger TEMPO and BHT which inhibited the reaction implying a radical pathway. Initially, H-abstraction from the benzylic position of 2-pyridylester **180a** generates the carbon radical **183** which undergoes SET to give a benzylic intermediate **184**. Nucleophilic substitution by **181a** to **184** gives **185** which upon oxidation gives imine intermediate **186**. Finally, **186** provides intermediate **187** via similar radical process to the initial amination, which upon removal of hydrogen ion affords the product **182aa** (Scheme 55).

Liu, Wei and co-workers also reported a similar oxidative cyclization of 2-pyridylesters and benzylamine leading to various imidazo[1,5-*a*]pyridines using elemental sulphur as the oxidant in DMSO at 110 °C in moderate to excellent yields [40]. The method avoids the use of peroxides and other additives which make the reaction attractive for potential synthetic application.

Chen group described the synthesis of imidazo[1,5-*a*]pyridines via the coupling of ethyl (*E*)-2-(benzylideneamino)acetates **188** with pyridines **189** using I_2 as the catalyst, TBPB as the oxidant and Cs_2CO_3 as the base in CH_3CN at 80 °C (Scheme 56) [41]. Based on literature reports, a plausible reaction mechanism has

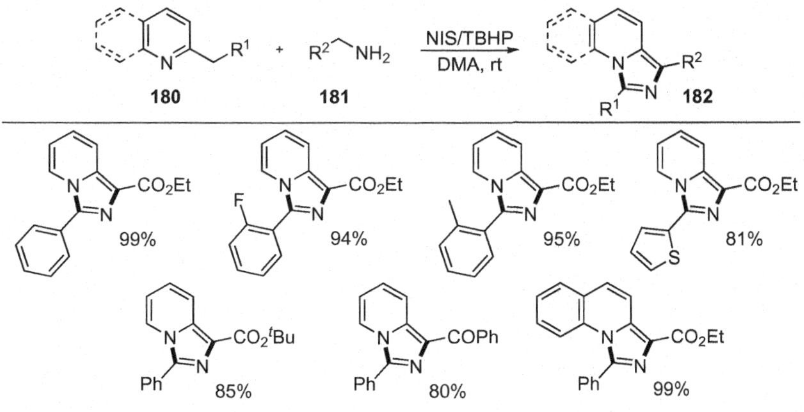

Scheme 54 I_2-mediated synthesis of imidazo[1,5-*a*]pyridines

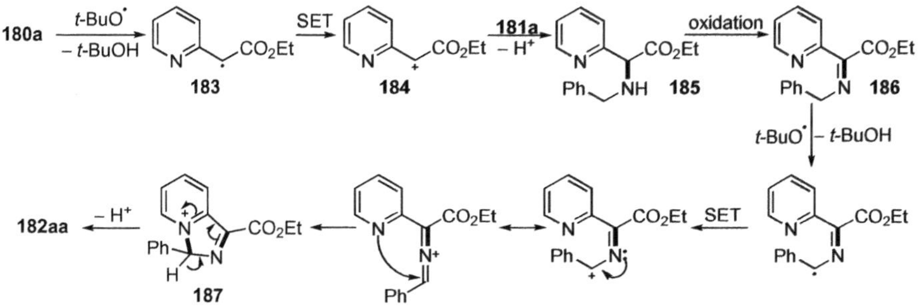

Scheme 55 Mechanistic pathway for I_2-mediated synthesis of imidazo[1,5-*a*]pyridines

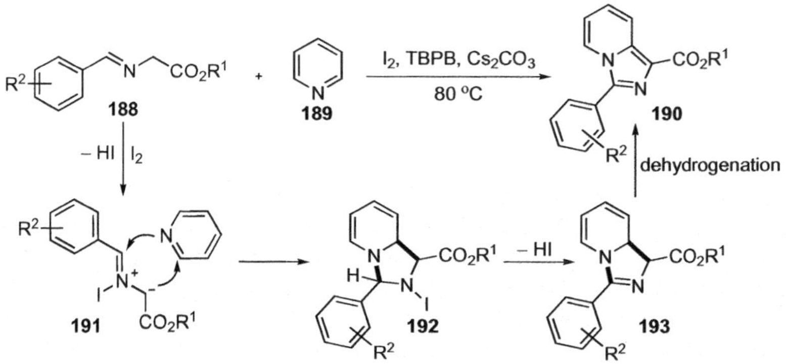

Scheme 56 Synthesis of imidazo[1,5-*a*]pyridines from ethyl (*E*)-2-(benzylideneamino)acetate and pyridine

been proposed (Scheme 56). Iodine acts as the Lewis acid and coordinates with *N*-atom of the ethyl (*E*)-2-(benzylideneamino)acetates **188**, leading to the formation of **191** with the elimination of HI. The base Cs_2CO_3 might help to promote the removal of HI. Next, the intermediate **191** and pyridine undergo a [3 + 2] cycloaddition leading to cycloadduct **192** followed by transformation to **193** with the release of HI. In the presence of TBPB, the HI is oxidized to I_2, and the catalyst is regenerated and finally with the elimination of hydrogen from **193** affords the product **190**.

2.19 Synthesis of Imidazo[1,2-a]Pyridines

Other important *N*-fused heterocycles are imidazo[1,2-*a*]pyridines which are widely distributed in natural products and biologically important molecules. Their derivatives exhibit important activities such as antibacterial, antiviral, antiulcer, antipyretic, anti-inflammatory and anticancer. Many drugs featuring imidazo[1,2-*a*] pyridines skeleton are commercially available in the market. Yu, Han and co-workers developed a metal-free approach towards the synthesis of imidazo[1,2-*a*]pyridines **196** involving TBAI-catalyzed oxidative coupling of 2-aminopyridines **194** with β-keto esters **195** or 1,3-diones via C–N bond coupling (Scheme 57 and Scheme 58) [42]. The reaction was performed using TBHP as the oxidant and $BF_3.OEt_2$ as the promoter in CH_3CN at 80 °C.

Initially, n-Bu_4NI gets oxidized by TBHP to n-$Bu_4N^+[IO_2]^-$ (**E**), and latter reacts with **195** to give intermediate **197**. The 2-aminopyridine **194** attacks **197** to afford **198**, which gives the product **196**. The released n-$Bu_4N^+[IO]^-$ (**F**) is reoxidized to **E** by TBHP $BF_3.OEt_2$ helps in increasing the electrophilicity of **A** as well as facilitates the removal of **F** from **197**. The other possibility is the active oxidant **F** reacts with **195** generating **199**. However, based on control experiments, the second pathway is less likely to occur.

Recently, Das et al. reported the synthesis of 2-functionalized imidazopyridine derivatives **201** from 2-aminopyridines **194** and chalcones **200** via regioselective C–N bond formation (Scheme 59) [43]. The reaction is promoted by I_2 and NH_4OAc as the additive in $CHCl_3$ solvent under a reflux condition.

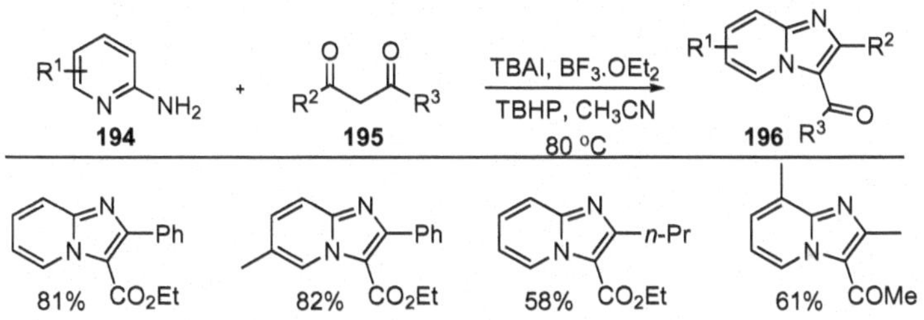

Scheme 57 TBAI-catalyzed synthesis of imidazo[1,2-*a*]pyridine

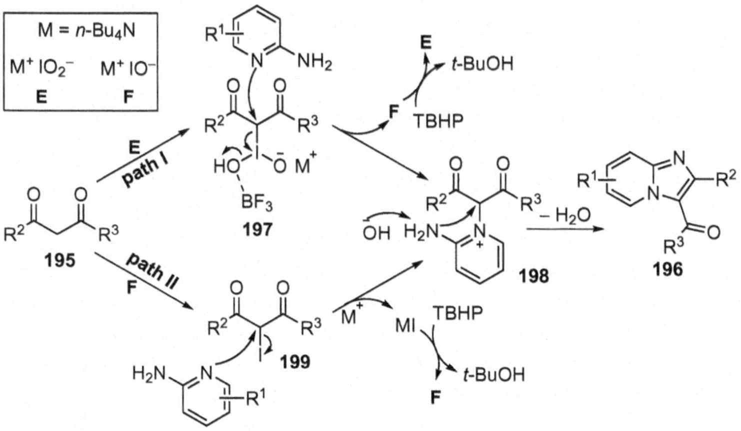

Scheme 58 Plausible mechanism for the synthesis of imidazo[1,2-*a*]pyridine

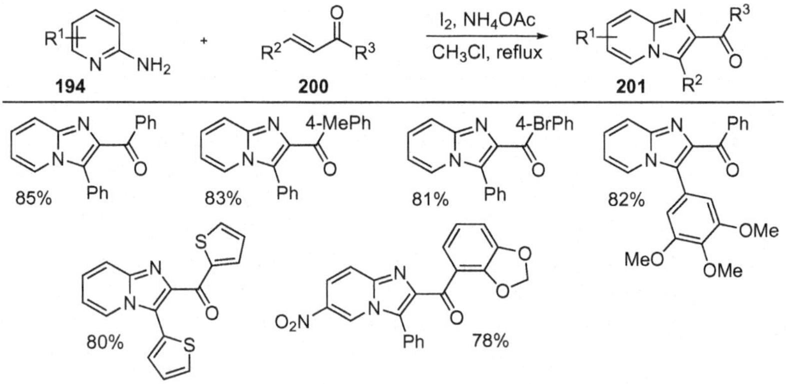

Scheme 59 I_2-mediated synthesis of imidazo[1,2-*a*]pyridine

The proposed mechanism is depicted in Scheme 60, and accordingly, iodine reacts with **200a** to form an iodonium intermediate **202**. 2-Aminopyridine**194a** attacks the β-position of the iodonium intermediate forming intermediate **204** via Ortoleva–King type intermediate **203**. Next, an intramolecular cyclization of **204** gives **205**, which undergoes aerial oxidation to afford the desired product **201aa**.

2.20 Synthesis of 1,2,4-Triazoles

1,2,4-Triazoles heterocycles are privileged scaffolds having broad application in biological, pharmaceutical and material sciences. With the potent biological activities such as antifungal, antibacterial and antihypertensive, its synthesis has gained considerable attention among medicinal chemists. Towards the metal-free approach, Chen, Ren and co-workers established an I_2 and TBHP-promoted synthesis of 1,3,5-trisubstituted 1,2,4-triazoles **208** from hydrazones **206** and aliphatic amines **207** under aerobic oxidative conditions (Scheme 61) [44]. The reaction

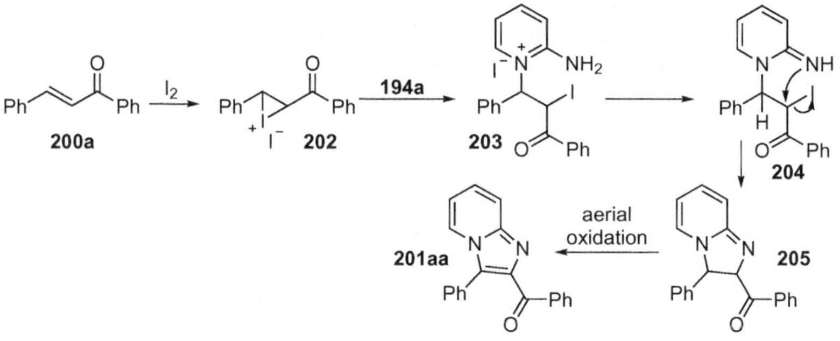

Scheme 60 Proposed mechanism for I_2-mediated synthesis imidazo[1,2-*a*]pyridine

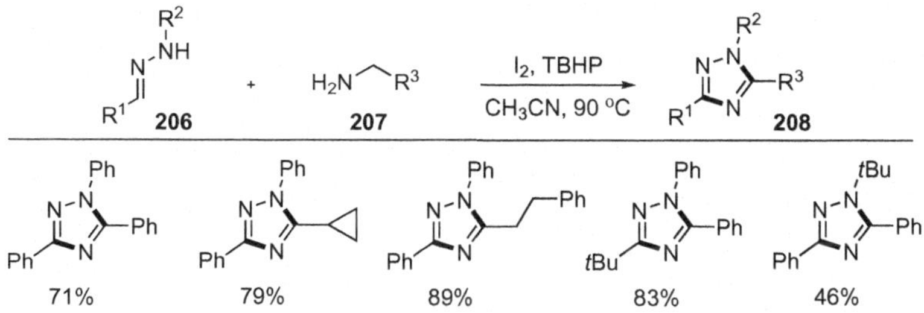

Scheme 61 I_2-mediated synthesis of 1,2,4-triazoles

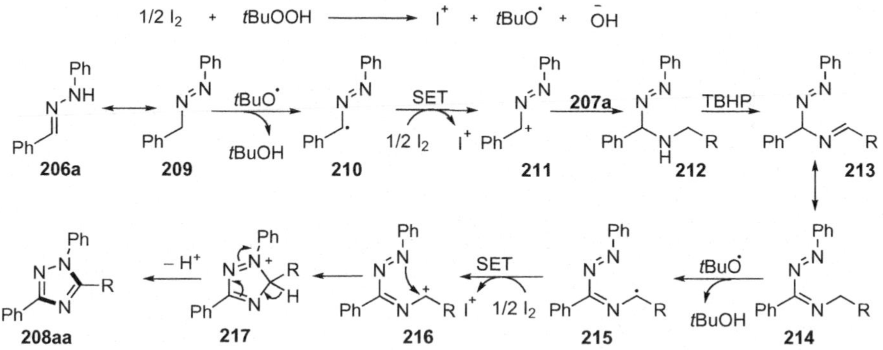

Scheme 62 Mechanism for I_2-mediated synthesis of 1,2,4-triazoles

proceeds via cascade C–H functionalization followed by double C–N bond formation and oxidative aromatization.

A plausible mechanism has been proposed as illustrated in Scheme 62. Initially, hydrazone **206a** tautomerizes via 1,3-H shift to the diazo compound **209** from which an H-abstraction takes place from the benzylic C–H bond generating radical **210**. Then, a SET from **210** gives a benzylic intermediate **211** which undergoes nucleophilic addition with an amine to generate **212**. An oxidative dehydrogenative process results in intermediate **213** which tautomerizes to **214**. Finally, **214** undergoes a similar radical process giving rise to intermediate **217**, which by subsequent dehydrogenative aromatization process affords 1,2,4-triazole **208aa**.

3 Synthesis of *O*-Heterocycles

Oxygen-containing heterocycles are of great significance in pharmaceutical industries and in organic synthesis as these compounds possess important biological activities and form the core structure of many natural products. The synthesis of oxygen-containing heterocycles has always attracted synthetic chemists. Below are the few examples of oxygen-containing heterocyclic compounds.

3.1 Synthesis of Oxetanes

Oxetanes are four-membered *O*-heterocyclic compounds. They are useful compounds as their rigid scaffolds appear as substructures in many medicinally and biologically active natural and unnatural molecules. They undergo several transformations to afford useful synthetic building blocks. An efficient solvent-controlled synthesis of oxetanes **219** was demonstrated by Fan et al. via oxidative cyclization of Michael adducts **218** obtained by the reaction of malonates with chalcones using iodosobenzene and tetrabutylammonium iodide (TBAI) combination in H_2O at 30 °C (Scheme 63) [45].

A plausible mechanistic pathway involving ligand exchange and elimination reaction of organic iodine(III) compounds has been depicted in Scheme 64.

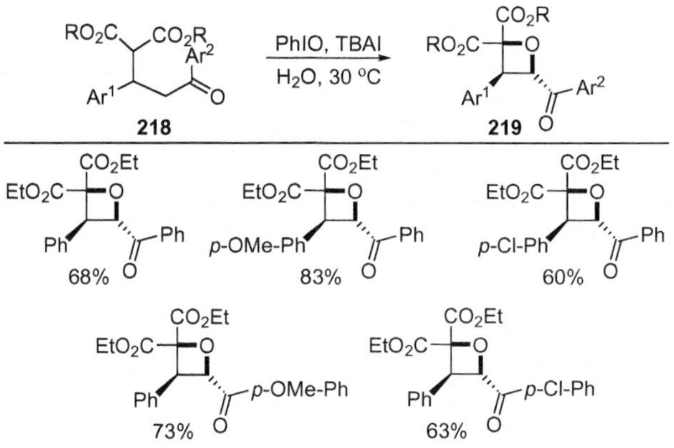

Scheme 63 Iodosobenzene-mediated synthesis of oxetanes

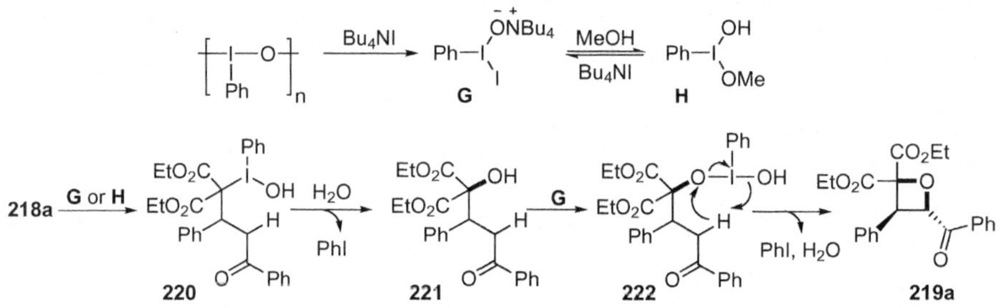

Scheme 64 Mechanistic pathway for the synthesis of oxetanes

The polymeric iodosobenzene is depolymerized in the presence of TBAI generating an oxidized monomeric species **G** which is transformed to **H** in MeOH. The Michael adduct **218a** reacts with the active species **G** or **H** via ligand exchange reaction to form intermediate **220** which in the presence of H_2O yields 2-hydroxy compound **221**. A second ligand exchange with **G** generates intermediate **222** which upon subsequent elimination affords the desired oxetane **219a**.

3.2 Synthesis of Dihydrofuran Derivatives

Fused dihydrofurans often occur as core substructure in diverse classes of natural products. Due to the important biological activities exhibited by functionalized dihydrofuran derivatives, substantial efforts have been devoted to their synthesis. Fan and co-workers reported an aqueous PhIO/Bu_4NI-mediated stereoselective oxidative construction of functionalized fused dihydrofurans via cyclization of Michael adducts **223** of 1,3-dicarbonyl compounds with chalcones (Scheme 65) [46].

A hypothesized reaction pathway is illustrated in Scheme 66. The reactive iodine(III) species **J** generated by the depolymerization of the polymeric iodosobenzene by TBAI reacts with the enolate of Michaels adduct **223** via ligand exchange. With the assistance of Bu_4NOH, **223** gives intermediate **225**, which upon intramolecular reductive elimination afford the fused dihydrofuran **224**.

Recently, Anand et al. demonstrated the synthesis of 2,3-diarylbenzo[*b*]furan derivatives **228** via *N*-heterocyclic carbene-catalyzed reaction between aromatic

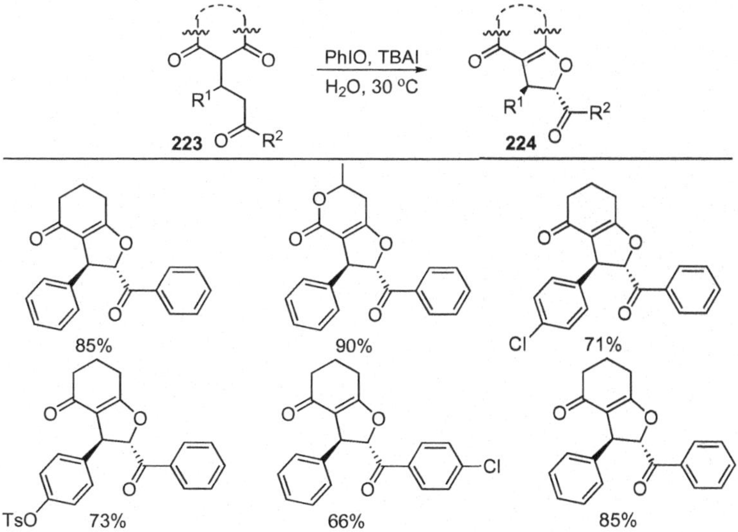

Scheme 65 Synthesis of fused dihydrofuran derivatives

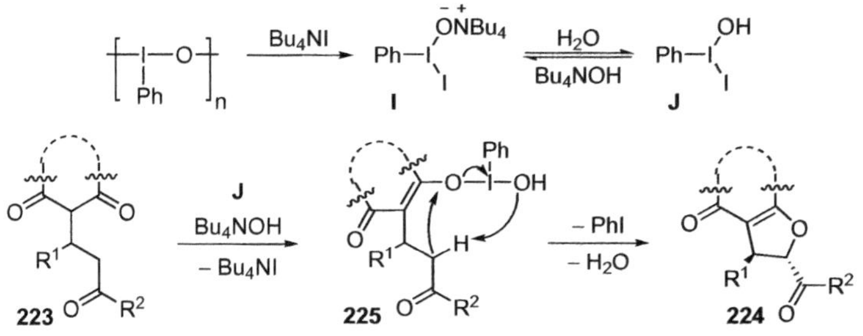

Scheme 66 Mechanistic pathway for the synthesis of fused dihydrofuran derivatives

aldehydes **227** and 2-hydroxyphenyl-substituted *para*-quinone methides **226** (Scheme 67) [47].

The group proposed a hypothetical mechanistic pathway according to which the reaction proceeds via 1,6-conjugate addition of the aromatic aldehyde **227** to 2-hydroxyphenyl-substituted *p*-quinone methides **226** catalyzed by *N*-heterocyclic carbene. Further, acid-mediated dehydrative aromatic annulation affords 2,3-diarylbenzo[*b*]furan **228** as shown in Scheme 68.

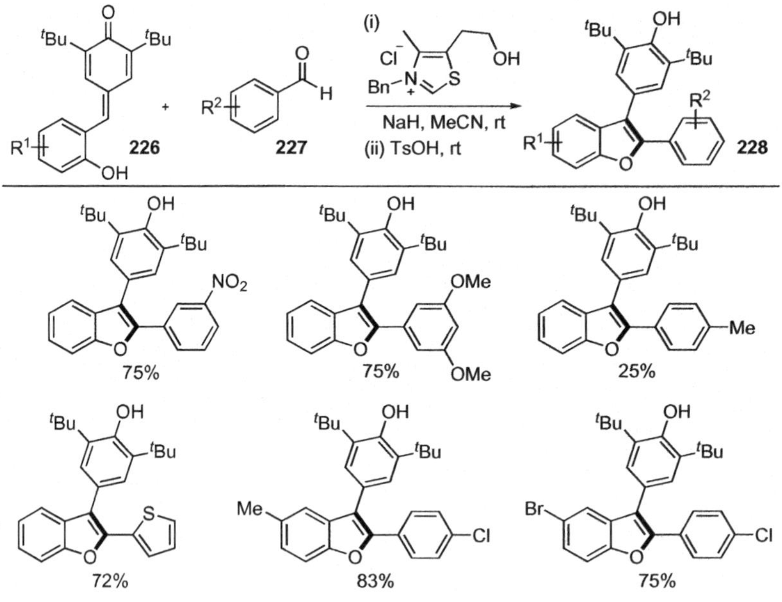

Scheme 67 *N*-heterocyclic carbene-catalyzed synthesis of dihydrofuran derivatives

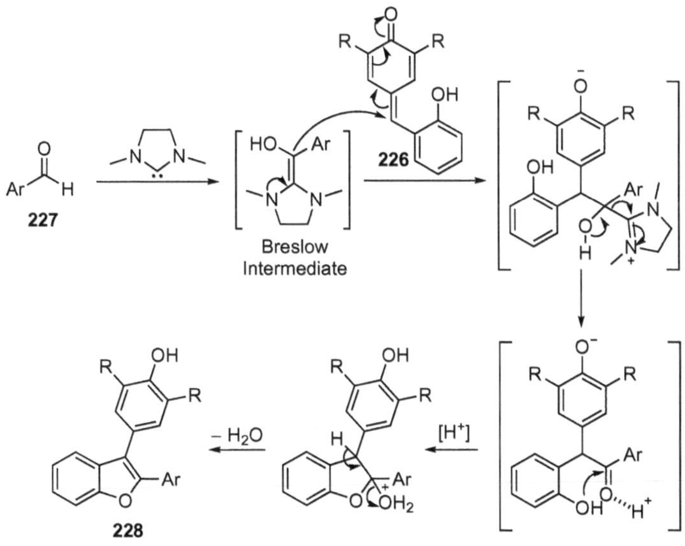

Scheme 68 Mechanistic pathway for the synthesis of 2,3-diarylbenzo[*b*]furan

3.3 Synthesis of Pyran Derivatives

Cheng group reported a metal-free cross-dehydrogenative coupling of 2-hydroxy-1,4-naphthoquinone **229** with 1,3-diarylpropene **230** in the presence of DDQ resulting into the synthesis of pyranonaphthaquinones **231** and pyranocoumarins **231′** having potent pharmaceutical applications (Scheme 69) [48].

Control experiments performed with radical scavenger TEMPO decreased the product yield indicating a single-electron transfer (SET) mechanism for the reaction. In the presence of DDQ, 1,3-diarylpropene **230** generates an allylic cation which is attacked by 2-hydroxy-1,4-naphthoquinone **229** to give the coupled product **232**. The allylic cation was rearranged between α- and γ-positions, giving a mixture of α- and γ-isomers **232a** and **232a′** ith unsymmetrical 1,3-diarylpropene **230**. Further, the reaction of the coupled product with DDQ gives allylic cation followed by an intramolecular nucleophilic attack of the hydroxyl group to produce a mixture of cyclic products **231a** and **231a′** (Scheme 70).

3.4 Synthesis of Coumarin Derivatives

Coumarins are the important class of oxygen heterocycles commonly found in many natural products, pharmaceuticals and dyes. They are also used as versatile building blocks in organic synthesis. Among coumarins, 3-acyl derivatives exhibit important activities like antioxidant, antimalarial, antitumour, anti-inflammatory,

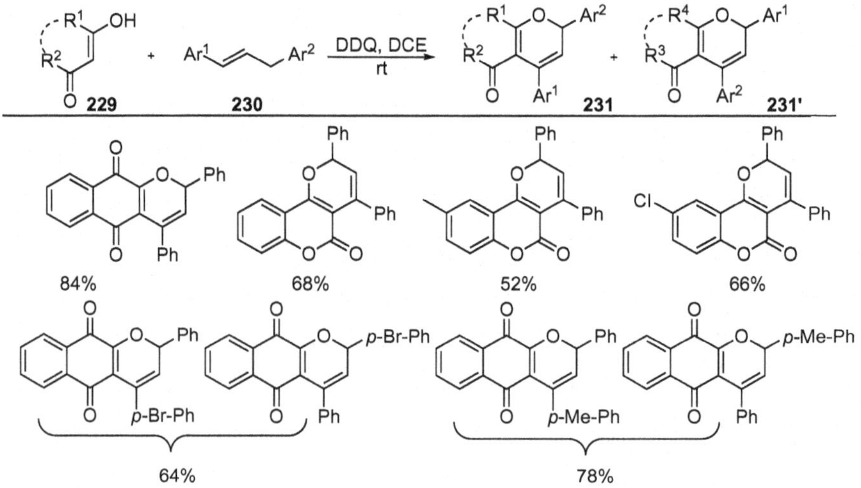

Scheme 69 DDQ-mediated synthesis of pyranonaphthaquinones and pyranocoumarins

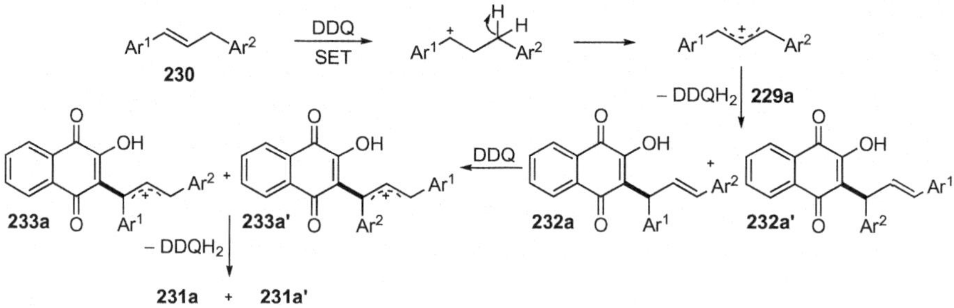

Scheme 70 Mechanistic pathway for the synthesis of pyranonaphthaquinones and pyranocoumarins

monoamine oxidase inhibitors. Due to its significant applications, several methods have been reported towards its synthesis. A metal-free synthesis of 3-acyl-4-arylcoumarins has been reported by Huang, Wu and co-workers via tandem acylation/cyclization of alkynoates **234** with aldehydes **235** using $K_2S_2O_8$ as the oxidant and TBAB as the additive in DCE at 90 °C in N_2 atmosphere (Scheme 71) [49].

The reaction is assumed to proceed via a radical pathway as depicted in Scheme 72. Initially, peroxydisulphate reacts with TBAB to generate the bis(tetrabutylammonium) peroxydisulphate which could be readily converted into tetrabutylammonium sulphate radical anion at a higher temperature. The tetrabutylammonium sulphate radical reacts with aldehyde **235a** to form an acyl radical. The acyl radical adds selectively to the α-position of the C=O in alkynoate **234a**

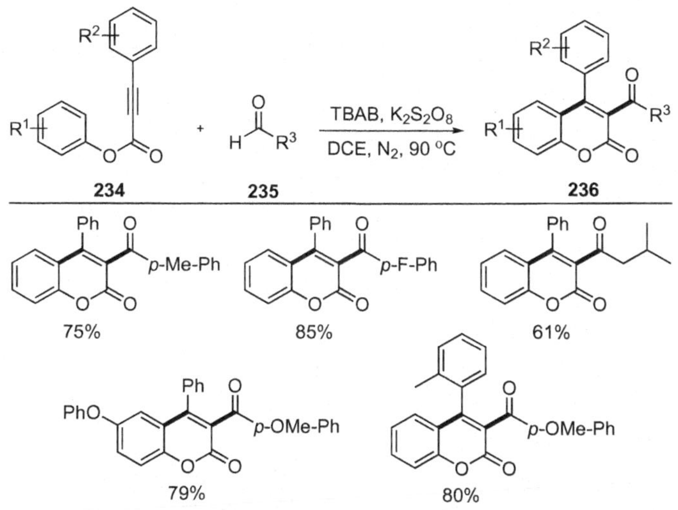

Scheme 71 Synthesis of 3-acylcoumarin derivatives

generating the vinyl radical **237**. The intermediate **237** then cyclizes to arene to form the radical intermediate **238** which upon single-electron transfer forms the cationic intermediate **239**. Finally, deprotonation by sulphate dianion affords the product **236aa**.

Xu group established a metal-free, visible-light-promoted difunctionalization of alkynoates **234** resulting in the synthesis of 3-phosphorylated coumarins **238**. The reaction was performed using Eosin Y as the photocatalyst and TBHP as the oxidant in DMSO at 25 °C in green LED under N_2 atmosphere (Scheme 73 and Scheme 74) [50].

According to the proposed mechanism, the irradiation of Eosin Y by visible light at room temperature produces the photo-excited state of Eosin Y (EY*).

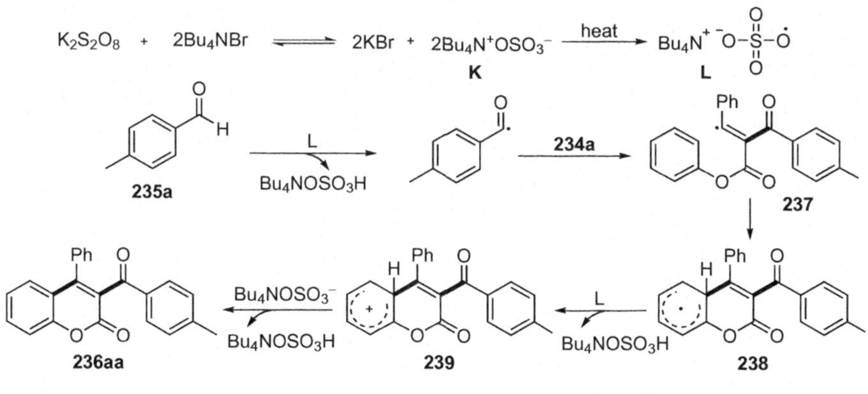

Scheme 72 Mechanistic pathway for the synthesis of 3-acylcoumarins

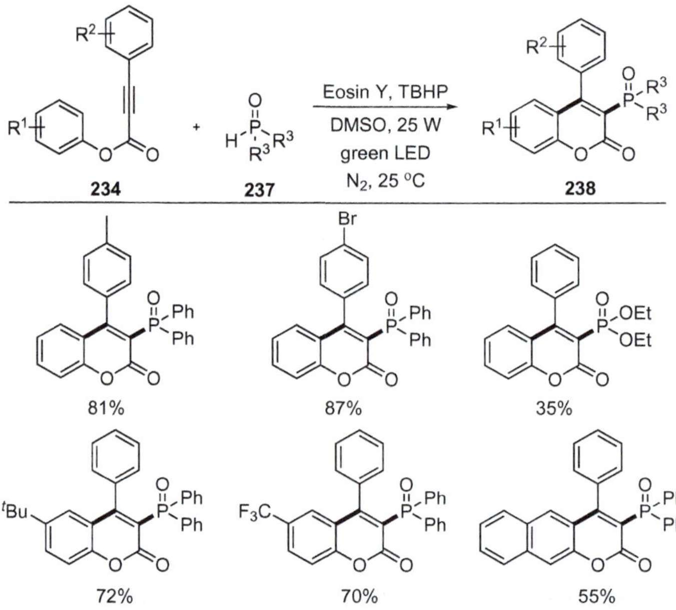

Scheme 73 Synthesis of 3-phosphorylated coumarin

The phosphinous acid **237a′** is oxidized by excited EY* to give phosphoryl radical, which adds at the α-position of the C=O in alkynoate **234a** to produce the vinyl radical **239**. The vinyl radical is highly reactive and trapped by the aryl ring forming intermediate **240** rapidly. The rearomatization of **240** by oxidation and deprotonation affords the phosphorylated coumarin **238aa**.

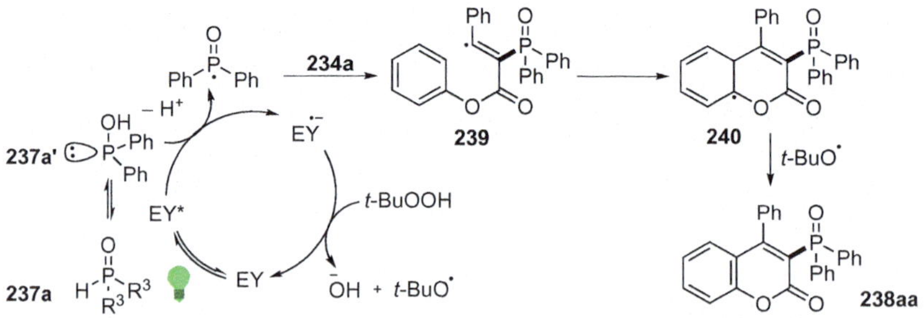

Scheme 74 Proposed mechanism for the synthesis of 3-phosphorylated coumarin

3.5 Synthesis of Xanthones

Xanthones form the core skeleton of many naturally occurring compounds exhibiting antibacterial, antioxidant and anticancer activities. A TBAB-promoted direct annulation of 2-aryloxybenzaldehydes **241** to xanthones **242** was demonstrated by Rao, Li and co-workers (Scheme 75) [51]. The reaction occurs via the oxidative coupling of aldehyde C–H bond and aromatic C–H bond under the metal-free condition in the aqueous medium.

A tentative mechanism has been proposed for this TBAB-promoted oxidative coupling as shown in Scheme 76. Initially, *tert*-butoxy radical is generated in the presence of TBAB that abstracts a proton from aldehyde **241a** forming an acyl radical **243**. The radical adds to the aryloxy unit giving radical **244**, which is further transformed to **245** by the SET process. Finally, OH^- abstracts a proton providing the desired annulated xanthone **242a**.

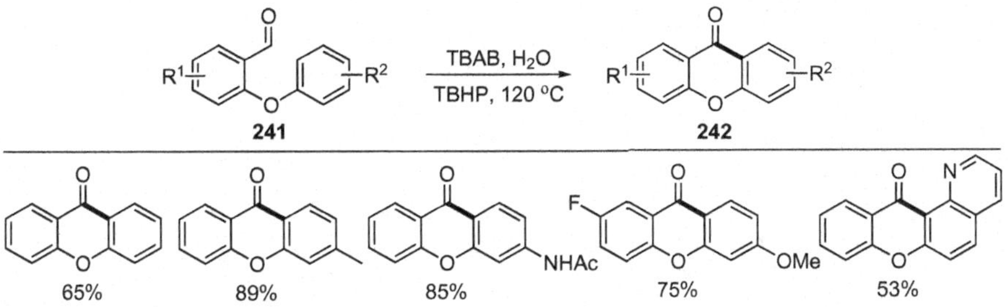

Scheme 75 TBAB-promoted synthesis of xanthones

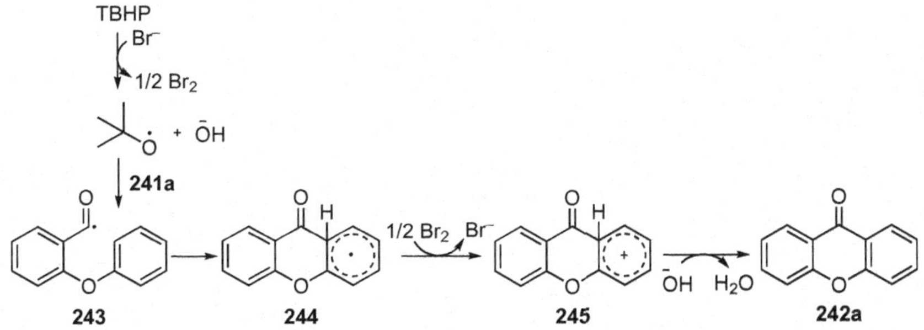

Scheme 76 Plausible mechanism for the synthesis of xanthones

4 Synthesis of *N,O*-Heterocycles

4.1 Synthesis of Oxazoles

Oxazoles are five-membered heterocycles having oxygen and nitrogen heteroatoms. They exist in biologically active drug molecules and are significant units of many natural, pharmaceuticals and agrochemical products. Molecules having oxazole scaffold are used as fluorescent dyes, in polymer industries and are indispensable chemical entities in organic synthesis. The importance of oxazole moieties has led to the development of many methods among which, Jiang group reported a metal-free cascade synthesis of polysubstituted oxazoles **248** from aryl alkenes **246** and benzylic amines **247** via TBHP/I_2-mediated oxidative cyclization (Scheme 77) [52].

According to the proposed mechanism, the alkene **246a** is oxidized to **249** in the presence of TBHP and I_2, which undergoes further oxidation to **250** via Kornblum oxidation. The benzylamine adds to **250** forming **251** which isomerize to its enol form **252**. An intramolecular nucleophilic addition of **252** gives **253**, which leads the final product **248aa** by deprotonation and oxidation (Scheme 78).

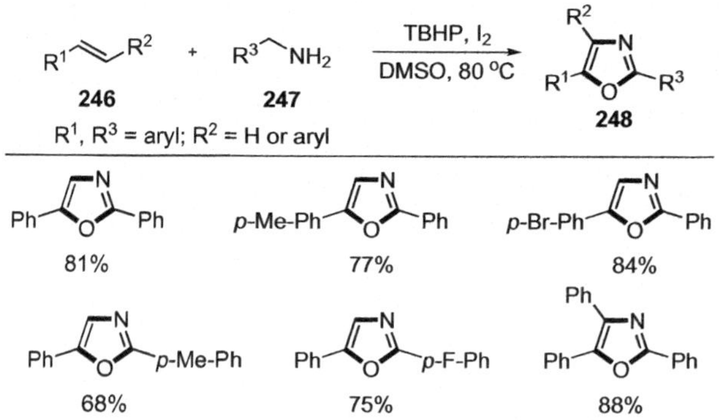

Scheme 77 TBHP/I_2-mediated synthesis of oxazoles

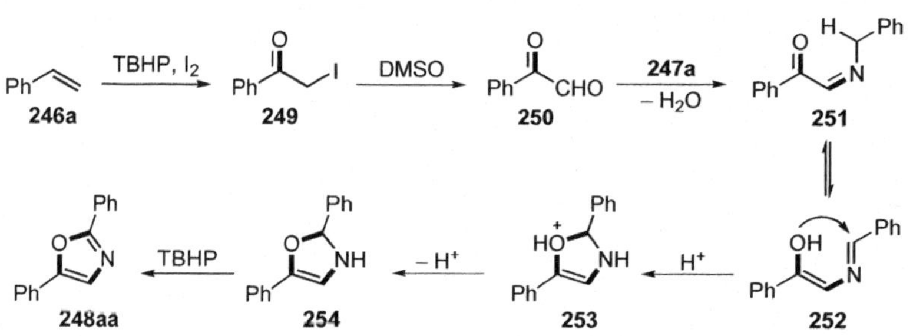

Scheme 78 Plausible mechanism for TBHP/I_2-promoted synthesis of oxazoles

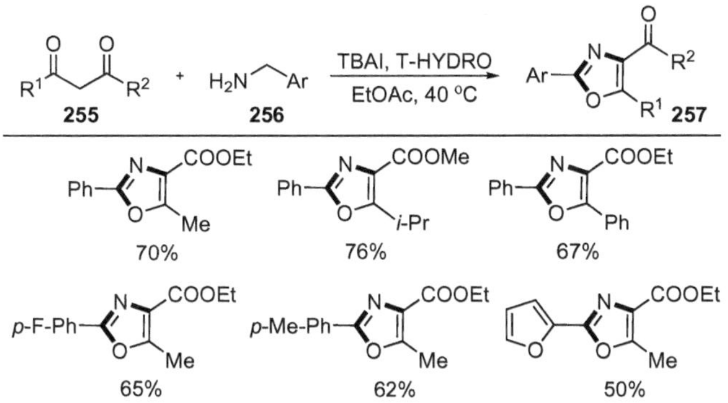

Scheme 79 Organocatalytic cascade synthesis of oxazoles

Zhu and co-workers demonstrated an organocatalytic cascade synthesis of oxazoles **257** from 1,3-dicarbonyl compounds **255** and arylmethanamines **256** (Scheme 79) [53]. The reaction takes place in the presence of TBAI as the catalyst, T-HYDRO (*tert*-butyl hydroperoxide) as the oxidant in EtOAc at 40 °C, involving the formation of C–N, C–O and C=N bonds in a single process via dual sp^3 C–H bond activation.

Reactions of 1,3-dicarbonyl compounds and arylmethanamines performed in the presence of radical inhibitors such as BHT (2,6-di-*tert*-butyl-4-methylphenol) and TEMPO (2,2,6,6-tetra-methylpiperidine-*N*-oxyl) ruled out the possibility of any radical mechanism. A plausible mechanism has been proposed according to which, the active iodine species ammonium hypoiodite **M** ($[n\text{-}Bu_4N]^+[IO]^-$) or iodite **N** ($[n\text{-}Bu_4N]^+[IO_2]^-$) plays an important role in this cascade reaction (Scheme 80). The intermediate **258**, formed in situ from ethyl acetoacetate (**255a**) and benzylamine (**256a**), reacts with **N** ($[n\text{-}Bu_4N]^+[IO_2]^-$) to form **259**. Subsequent nucleophilic attack of **259** by benzylamine gives **260** which is oxidized to **261**. Next, the

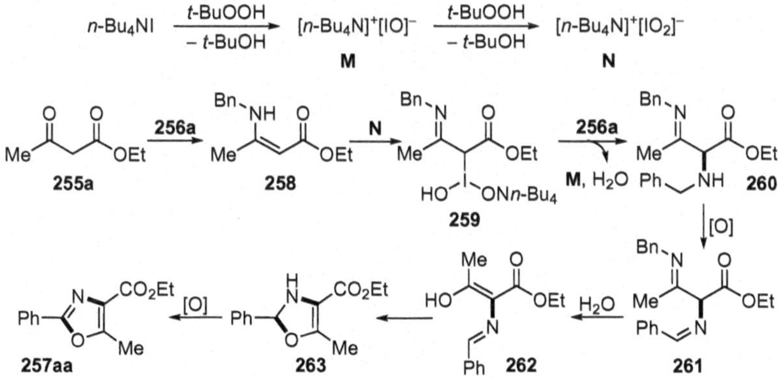

Scheme 80 Plausible mechanism for organoctalyzed synthesis of oxazoles

hydrolysis of **261** generates **262** that undergo an intramolecular nucleophilic addition to afford intermediate **263**. Finally, oxidation of **263** by either **M** or **N** produces the polysubstituted oxazole **257aa**.

4.2 Synthesis of Benzoxazoles

Yu group demonstrated the synthesis of benzoxazoles via intramolecular oxidative C–O coupling of electron-rich benzamides **264** in the presence of PIFA as the oxidant and TMSOTf as catalyst (Scheme 81) [54]. The reaction is sensitive to the structural feature of the substrate as *N*-(4-alkoxy-phenyl)benzamides are converted into the corresponding benzoxazole in high yields while *N*-(4-alkoxy-phenyl)acetanilides are unreactive.

Two different mechanisms were accounted for the formation of benzoxazole. First, the attack of PIFA on the amide moiety generates intermediate **266** which is converted to the nitrenium ion intermediate **267** by the cleavage of N–I bond (Scheme 82, path I). The formation of nitrenium ion is favoured in the presence of

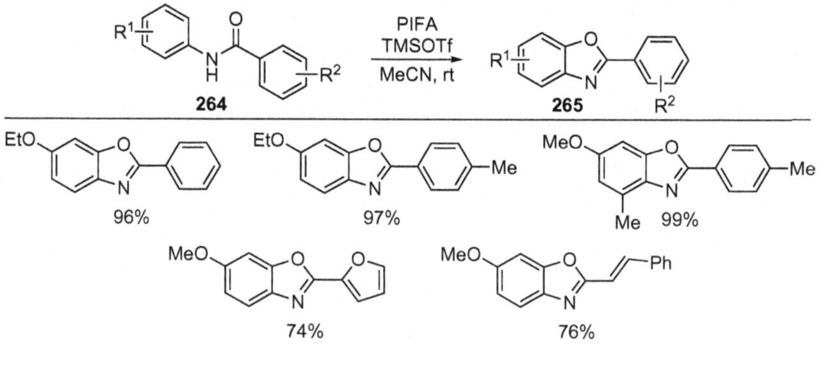

Scheme 81 Synthesis of benzoxazoles

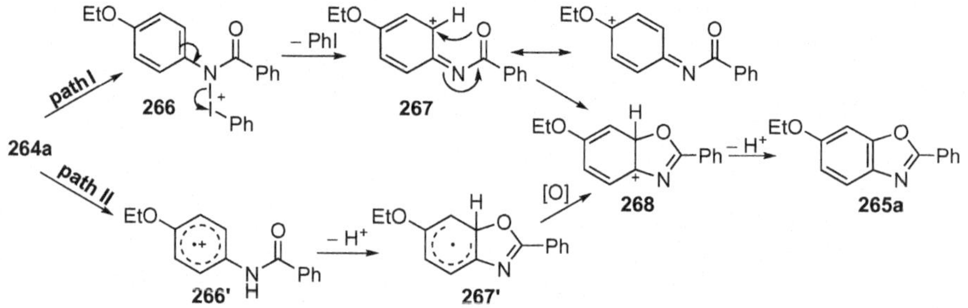

Scheme 82 Mechanistic pathway for the Synthesis of benzoxazoles

para alkoxy group. Next, the nucleophilic attack by the carbonyl oxygen generates the cyclic intermediate **268**, which after deprotonation gives the desired benzoxazole **265a**. The other possible pathway involves single-electron transfer (SET) from **264a** by PIFA (Scheme 82, path II). This mechanism has been proposed to explain inter- and intramolecular oxidative coupling of electron-rich aromatic compounds.

4.3 Synthesis of 1,3,4-Oxadiazole

Molecules with 1,3,4-oxadiazole skeleton show a wide variety of biological activities like analgesic, antiviral, antibacterial, antifungal, anticancer, anticonvulsant, antihypertensive and antidiabetic. They have also found applications as scintillators, fluorescent agents and photographic materials. Yadav and co-workers developed a photocatalytic oxidative heterocyclization of semicarbazones **269** to 5-substituted 2-amino-1,3,4-oxadiazoles **270** (Scheme 83) [55]. The reaction utilizes CBr_4 in combination with Eosin Y in CH_3CN under air in green LED.

On the basis of literature reports and control experiments, the reaction is assumed to proceed via a radical pathway as depicted in Scheme 84. The irradiation of Eosin Y with visible light generates an active triplet state of Eosin Y which

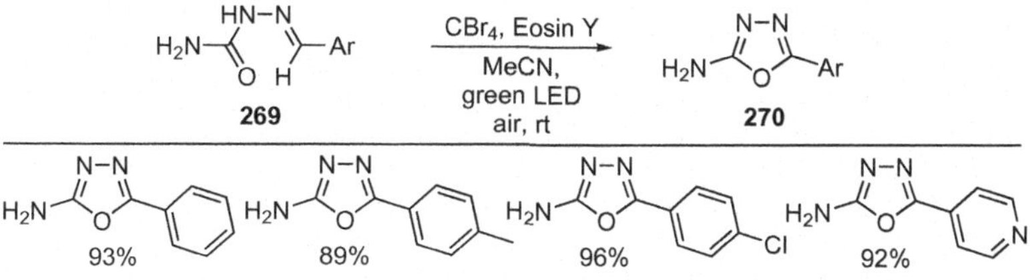

Scheme 83 Synthesis of 1,3,4-oxadiazoles

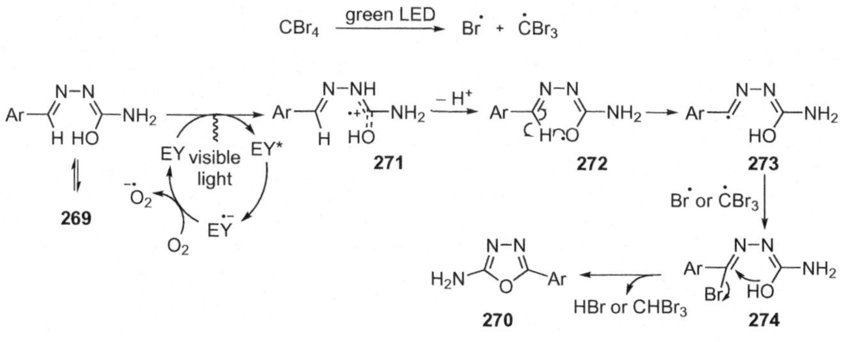

Scheme 84 Mechanistic pathway for the synthesis of 1,3,4-oxadiazoles

participates in the photoredox cycle via reductive quenching of semicarbazone **269** to radical cation **271**. This radical cation upon deprotonation gives the resonance stabilized radical **272**. Further, abstraction of a proton from **272** generates radical **273**, which undergo radical bromination to form **274**. Finally, a 5-*endo*-trig cyclization affords the desired 1,3,4-oxadiazole **270**.

*4.4 Synthesis of Chromeno[2,3-c]Pyrazol-4(1*H*)-Ones*

Chromones exhibit a variety of biological and pharmaceutical activities. The chromone structure containing a pyrazole ring is important pharmacophore having numerous applications such as A_2-subtype selective adenosine receptor antagonist A^3. Thus, significant attention has been paid towards the synthesis of chromone with a pyrazole skeleton. In this regard, Liu group reported the synthesis of chromeno[2,3-*c*]pyrazol-4(1*H*)-ones **276** via ionic liquid directed annulation of 5-aryloxy-1*H*-pyrazole-4-carbaldehydes **275** (Scheme 85) [56]. The reaction involves an intramolecular cross-coupling promoted by 1,3-dibutyl-1*H*-benzo[*d*][1,2,3]-triazol-3-ium bromide as the ionic liquid (IL) that can be easily recycled and reused with the same efficiency for at least five cycles in an aqueous medium.

A radical mechanism has been proposed as shown in Scheme 84. The ionic liquid promotes the generation of *tert*-butoxyl radical that abstracts hydrogen radical from the aldehyde forming an acyl radical **277**. The acyl radical adds to the aryloxy unit generating radical **278**, which via a single-electron-transfer process gives intermediate **279**. Finally, the previously formed hydroxyl anion acts as the H-abstractor from **279** affording the desired annulated product **276a**.

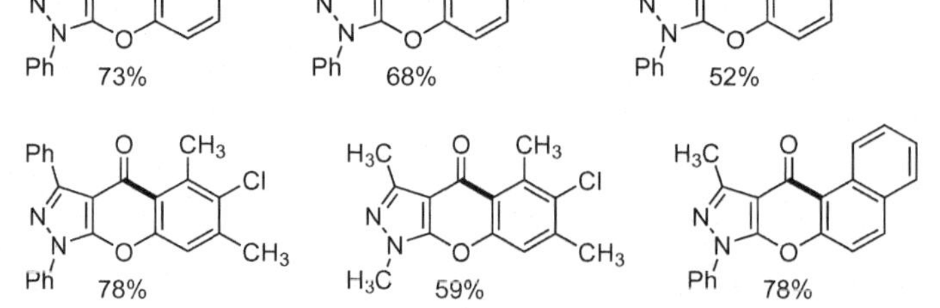

Scheme 85 Ionic liquid-promoted synthesis of chromeno[2,3-*c*]pyrazol-4(1*H*)-ones

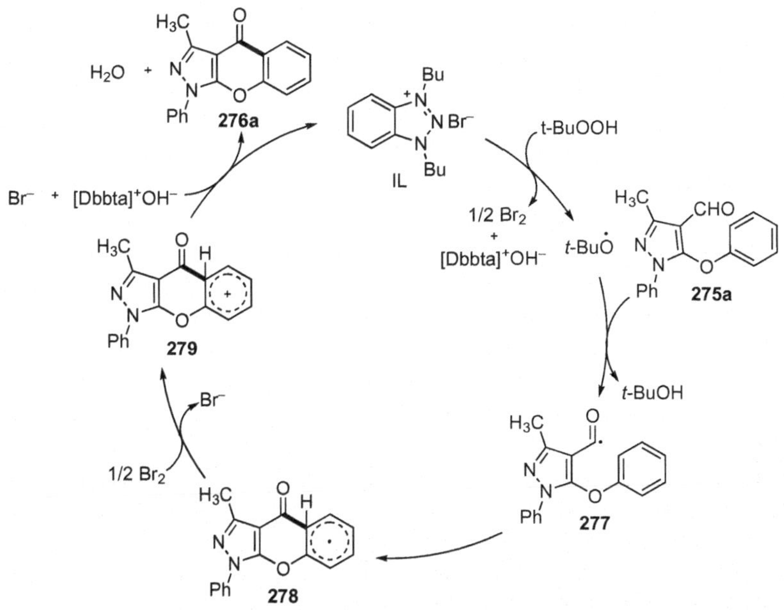

Scheme 86 Plausible mechanism for the synthesis of chromeno[2,3-*c*]pyrazol-4(1*H*)-ones

5 Synthesis of *S*-Heterocycles

5.1 Synthesis of Benzothiazole

Benzothiazoles are an important class of heterocycles existing in a variety of natural products. They display potent biological and medicinal activities. Wu group described an I_2-promoted synthesis of 2-acylbenzothiazoles **282** from simple and readily available aromatic ketones/unsaturated methyl ketones **280** and *o*-aminobenzenethiols **281** (Scheme 87) [57].

A plausible mechanism has been proposed using acetophenone **280a** and 2-aminobenzenethiol **281a** (Scheme 88). Initially, acetophenone is converted to **283** in the presence of I_2, which subsequently forms phenylglyoxal **284** in the presence of DMSO. Finally, **284** react with **281a** via condensation, Michael addition and oxidative dehydrogenation sequence to afford the desired product **282aa**.

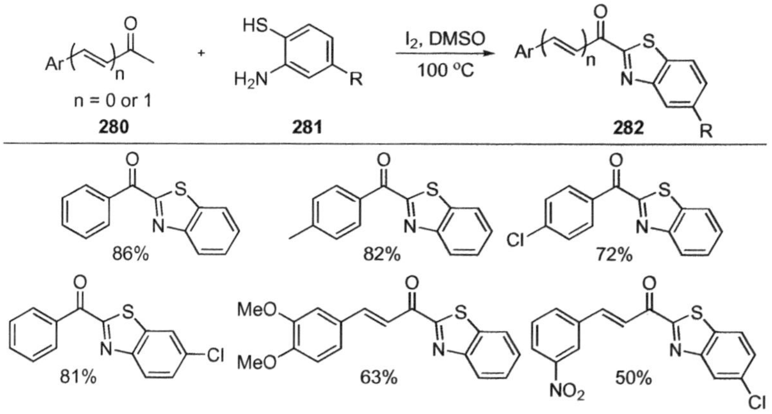

Scheme 87 Iodine-promoted synthesis of 2-acylbenzothiazoles

6 Synthesis of *N,O* and *S*-Heterocycles

6.1 Synthesis of 1,4,2-Oxathiazoles

1,4,2-Oxathiazoles are five-membered ring heterocycles containing three different heteroatoms that have been very barely reported in the literature. These heterocycles can be of great interest for the medicinal chemists as a new scaffold for structural activity relationship. Upon thermal decomposition, they serve as useful precursors for isothiocyanate. Pierce et al. reported a metal-free oxidative synthesis of 1,4,2-oxathiazoles **289** from thiohydroximic acids **288** in the presence of DDQ as the oxidant, *p*-TsOH.H_2O as the additive in DMF at 150 °C (Scheme 89) [58].

Based on control experiments, the reaction proceeds via the oxidation of the oxime **288a** to an iminoxyl radical **291**, followed by 1,5-H atom abstraction of the activated C–H bond (Scheme 90). Iminoxyl radical are relatively stable that undergoes hydrogen shift producing a more reactive carbon radical **292** in a

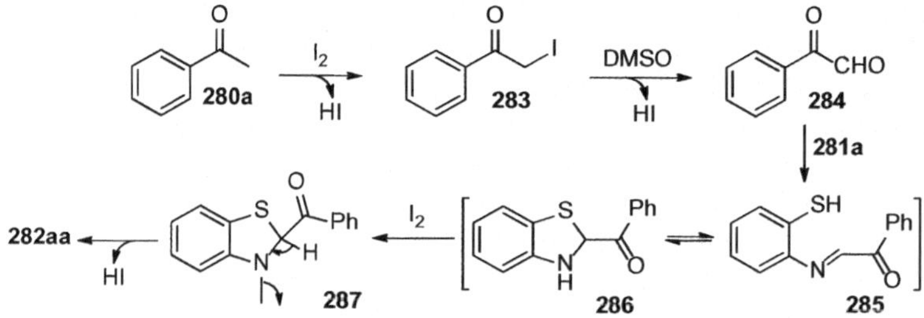

Scheme 88 Plausible mechanism for the synthesis of 2-acylbenzothiazoles

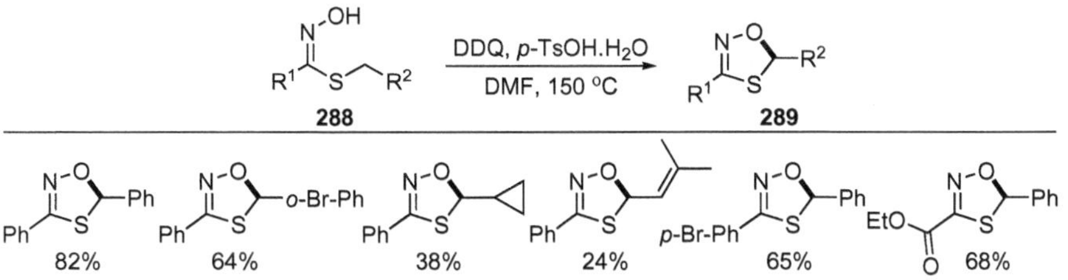

Scheme 89 Oxidative synthesis of 1,4,2-oxathiazoles

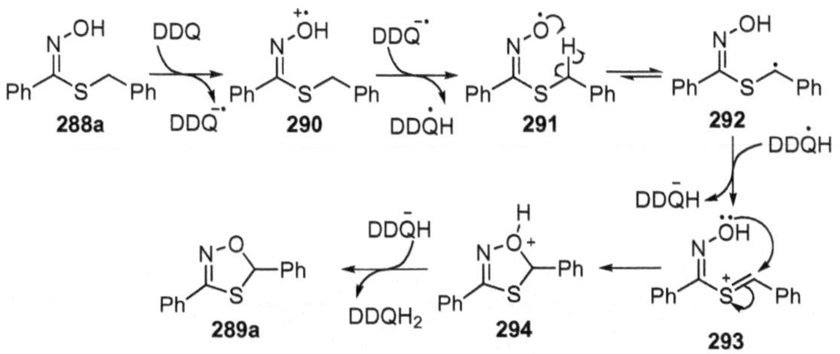

Scheme 90 Plausible mechanism for the synthesis of 1,4,2-oxathiazoles

reversible manner. Further intramolecular cyclization results into the desired 1,4,2-oxathiazole **289a**.

7 Conclusion

In conclusion, the ubiquitous nature of heterocyclic scaffold predominates the synthetic chemist community. In this regard, cross-dehydrogenative coupling reactions have emerged as a great strategy towards the synthesis of heterocyclic molecules as it eliminates the requirement for the pre-functionalization of the starting materials, giving high atom economy in less time. Transition-metal-free CDC reactions are even more promising compared to metal-catalyzed reactions as they overcome the drawbacks associated with transition metals that limits their use in pharmaceutical industries. Metal-free CDC reactions provide a greener and sustainable approach towards the construction of complex molecules.

References

1. Fan R, Wang H, Ye Y, Gan J (2010) PhIO/Bu_4NI mediated oxidative cyclization of amidoalkylation adducts for the synthesis of *N*-benzoyl aziridines and oxazolines. Tetrahedron Lett 51:453
2. Ye Y, Wang H, Fan R (2010) Stereoselective construction of highly functionalized azetidines via a [2 + 2]-cycloaddition. Org Lett 12:2802
3. Gao P, Wang J, Bai Z-J, Shen L, Yan Y-Y, Yang D-S, Fan M-J, Guan Z-H (2016) Synthesis of polycarbonyl pyrroles via $K_2S_2O_8$-mediated oxidative cyclization of enamines. Org Lett 18:6074
4. Vanjari R, Guntreddi T, Kumar S, Singh KN (2015) Sulphur promoted C(sp^3)–C(sp^2) cross dehydrogenative cyclization of acetophenone hydrazones with aldehydes: efficient synthesis of 3,4,5-trisubstituted 1*H*-pyrazoles. Chem Commun 51:366
5. Liang D, Zhu Q (2015) A facile synthesis of pyrazoles through metal-free oxidative C(sp^2)–H cycloamination of vinyl hydrazones. Asian J Org Chem 4:42
6. Fra L, Millán A, Souto JA, Muñiz K (2014) Indole synthesis based on a modified Koser reagent. Angew Chem Int Ed 53:7349
7. Jang YH, Youn SW (2014) Metal-free C–H amination for indole synthesis. Org Lett 16:3720
8. Ghosh S, De S, Kakde BN, Bhunia S, Adhikary A, Bisai A (2012) Intramolecular dehydrogenative coupling of sp^2 C–H and sp^3 C–H bonds: an expeditious route to 2-oxindoles. Org Lett 14:5864
9. Mondal B, Roy B (2015) Di-tert-butyl peroxide (DTBP) promoted dehydrogenative coupling: an expedient and metal-free synthesis of oxindoles via intramolecular C(sp^2)–H and C(sp^3)–H bond activation. RSC Adv 5:69119
10. Donald JR, Taylor RJK, Petersen WF (2017) Low-temperature, transition-metal-free cross-dehydrogenative coupling protocol for the synthesis of 3,3-disubstituted oxindoles. J Org Chem 82:11288
11. Zhou M-B, Song R-J, Ouyang X-H, Liu Y, Wei W-T, Deng G-B, Li J-H (2013) Metal-free oxidative tandem coupling of activated alkenes with carbonyl C(sp^2)–H bonds and aryl C(sp^2)–H bonds using TBHP. Chem Sci 4:2690
12. Ji P-Y, Zhang M-Z, Xu J-W, Liu Y-F, Guo C-C (2016) Transition-metal-free TBAI-facilitated addition–cyclization of *N*-methyl-*N*-arylacrylamides with arylaldehydes or benzenesulfonohydrazides: access to carbonyl- and sulfone-containing *N*-methyloxindoles. J Org Chem 81:5181
13. Biswas P, Paul S, Guin J (2016) Aerobic radical-cascade alkylation/cyclization of α,β-unsaturated amides: an efficient approach to quaternary oxindoles Angew. Chem Int Ed 55:7756
14. Tian Y, Liu Z-Q (2014) Metal-free radical cascade dichloromethylation of activated alkenes using CH_2Cl_2: highly selective activation of the C–H bond. RSC Adv 4:64855
15. Meng Y, Guo L-N, Wang H, Duan X-H (2013) Metal-free oxidative hydroxyalkylarylation of activated alkenes by direct sp^3 C–H functionalization of alcohols. Chem Commun 49:7540
16. Verma A, Patel S, Meenakshi, Kumar A, Yadav A, Kumar S, Jana S, Sharma S, Prasad CD, Kumar S (2015) Transition-metal-free intramolecular selective oxidative C(sp^3)–N coupling: synthesis of *N*-aryl-isoindolinones from 2-alkylbenzamides. Chem Commun 51:1371
17. Huo C, Xie H, Wu M, Jia X, Wang X, Chen F, Tang J (2015) CBr_4-mediated cross-dehydrogenative coupling reaction of amines. Chem Eur J 21:5723
18. Huo C, Xie H, Chen F, Tang J, Wang Y (2016) Double-oxidative dehydrogenative (DOD) cyclization of glycine derivatives with dioxane under metal-free aerobic conditions. Adv Synth Catal 358:724
19. Mai W-P, Wang J-T, Xiao Y-M, Mao P, Lu K (2015) A novel direct synthesis of 3-acyl-4-aryldihydroquinolin-2(1*H*)-ones via metal-free radical tandem cyclization between *N*-arylcinnamamides and aldehydes. Tetrahedron 71:8041

20. Zhou W, Ni S, Mei H, Han J, Pan Y (2015) Hydroxyalkylation-initiated radical cyclization of *N*-allylbenzamide for direct construction of isoquinolinone. Org Lett 17:2724
21. Feng X, Zhu H, Wang L, Jiang Y, Cheng J, Yu J-T (2014) TBHP-promoted sequential carboxamidation and aromatisation of aryl isonitriles with formamides. Org Biomol Chem 12:9257
22. Fang H, Zhao J, Ni S, Mei H, Han J, Pan Y (2015) Metal-free oxidative functionalization of a C(sp^3)–H bond adjacent to nitrogen and intramolecular aromatic cyclization for the preparation of 6-amidophenanthridines. J Org Chem 80:3151
23. Cao J-J, Zhu T-H, Wang S-Y, Gu Z-Y, Wang X, Ji S-J (2014) tert-Butyl peroxybenzoate (TBPB)-mediated 2-isocyanobiaryl insertion with 1,4-dioxane: efficient synthesis of 6-alkyl phenanthridines via C(sp^3)–H/C(sp^2)–H bond functionalization. Chem Commun 50:6439
24. Wang L, Sha W, Dai Q, Feng X, Wu W, Peng H, Chen B, Cheng J (2014) The benzoyl peroxide promoted dual C–C bond formation via dual C–H bond cleavage: α-phenanthridinylation of ether by isocyanide. Org Lett 16:2088
25. Zhang J, Zhu D, Yu C, Wan C, Wang Z (2010) A simple and efficient approach to the synthesis of 2-phenylquinazolines via sp^3 C−H functionalization. Org Lett 12:2841
26. Han B, Wang C, Han R-F, Yu W, Duan X-Y, Fang R, Yang X-L (2011) Efficient aerobic oxidative synthesis of 2-aryl quinazolinesviabenzyl C–H bond amination catalyzed by 4-hydroxy-TEMPO. Chem Commun 47:7818
27. Zhao D, Shen Q, Li J-X (2015) Potassium iodide-catalyzed three-component synthesis of 2-arylquinazolines via amination of benzylic C−H bonds of methylarenes. Adv Synth Catal 357:339
28. Lin J-P, Zhang F-H, Long Y-Q (2014) Solvent/oxidant-switchable synthesis of multisubstituted quinazolines and benzimidazoles via metal-free selective oxidative annulation of arylamidines. Org Lett 16:2822
29. Shen Z-C, Yang P, Tang Y (2016) Transition metal-free visible light-driven photoredox oxidative annulation of arylamidines. J Org Chem 81:309
30. Ge W, Zhuab X, Wei Y (2013) Iodine-catalyzed oxidative system for cyclization of primary alcohols with *o*-aminobenzamides to quinazolinones using DMSO as the oxidant in dimethyl carbonate. RSC Adv 3:10817
31. Zhao D, Wang T, Li J-X (2014) Metal-free oxidative synthesis of quinazolinones via dual amination of sp^3 C–H bonds. Chem Commun 50:6471
32. Antonchick AP, Samanta R, Kulikov K, Lategahn J (2011) Organocatalytic, oxidative, intramolecular C–H bond amination and metal-free cross-amination of unactivated arenes at ambient temperature. Angew Chem Int Ed 50:8605
33. Maiti S, Achar TK, Mal P (2017) An organic intermolecular dehydrogenative annulation reaction. Org Lett 19:2006
34. Huang J, He Y, Wang Y, Zhu Q (2012) Synthesis of benzimidazoles by PIDA-promoted direct C(sp^2)–H imidation of *N*-arylamidines. Chem Eur J 18:13964
35. Zheng Z, Dian L, Yuan Y, Zhang-Negrerie D, Du Y, Zhao K (2014) PhI$(OAc)_2$-mediated intramolecular oxidative aryl-aldehyde Csp^2–Csp^2 bond formation: metal-free synthesis of acridone derivatives. J Org Chem 79:7451
36. Laha JK, Jethava KP, Patel S (2015) Scope of successive C–H functionalizations of the methyl group in 3-picolines: intramolecular carbonylation of arenes to the metal-free synthesis of 4-azafluorenones. Org Lett 17:5890
37. Xiang L, Yang Y, Zhou X, Liu X, Li X, Kang X, Yan R, Huang G (2014) I_2-mediated oxidative cyclization for synthesis of substituted indolizines. J Org Chem 79:10641
38. Reddy NNK, Donthiri RR, Ravi C, Adimurthy S (2016) Iodine-catalyzed [3 + 2] cyclization of 2-pyridylesters and chalcones: metal-free approach for the synthesis of substituted indolizines. Tetrahedron Lett 57:3243
39. Yan Y, Zhang Y, Zha Z, Wang Z (2013) Mild metal-free sequential dual oxidative amination of C(sp^3)–H bonds: efficient synthesis of imidazo[1,5-*a*]pyridines. Org Lett 15:2274

40. Sheng J, Liu J, Zhao H, Zheng L, Wei X (2018) Metal-free synthesis of imidazo[1,5-*a*] pyridines via elemental sulphur mediated sequential dual oxidative Csp^3–H amination. Org Biomol Chem 16:5570
41. Qin M, Tian Y, Guo X, Yuan X, Yang X, Chen B (2018) I_2/TBPB mediated oxidative reaction to construct of imidazo[1,5-*a*]pyridines under metal-free conditions. Asian J Org Chem 7:1591
42. Ma L, Wang X, Yu W, Han B (2011) TBAI-catalyzed oxidative coupling of aminopyridines with β-keto esters and 1,3-diones-synthesis of imidazo[1,2-*a*]pyridines. Chem Commun 47:11333
43. Kour D, Gupta A, Kapoor KK, Gupta VK, Rajnikant, Singh D, Das P (2018) Iodine–NH_4OAc mediated regioselective synthesis of 2-aroyl-3-arylimidazo[1,2-*a*]pyridines from 1,3-diaryl-prop-2-en-1-ones. Org Biomol Chem 16:1330
44. Chen Z, Li H, Dong W, Miao M, Ren H (2016) I_2-Catalyzed oxidative coupling reactions of hydrazones and amines and the application in the synthesis of 1,3,5-trisubstituted 1,2,4-triazoles. Org Lett 18:1334
45. Ye Y, Zheng C, Fan R (2009) Solvent-controlled oxidative cyclization for divergent synthesis of highly functionalized oxetanes and cyclopropanes. Org Lett 11:3156
46. Ye Y, Wang L, Fan R (2010) Aqueous iodine(III)-mediated stereoselective oxidative cyclization for the synthesis of functionalized fused dihydrofuran derivatives. J Org Chem 75:1760
47. Singh G, Goswami P, Sharma S, Anand RV (2018) A one-pot approach to 2,3-diarylbenzo [*b*]furans through *N*-heterocyclic carbene-catalyzed 1,6-conjugate addition followed by acid mediated dehydrative annulation. J Org Chem 83:10546
48. Cheng D, Wu L, Lv H, Xu X, Yan J (2017) CDC reaction and subsequent cyclization for the synthesis of 2-hydroxy-3-alkyl-1,4-naphthoquinones and pyranonaphthoquinones. J Org Chem 82:1610
49. Mi X, Wang C, Huang M, Wu Y, Wu Y (2015) Preparation of 3-acyl-4-arylcoumarins via metal-free tandem oxidative acylation/cyclization between alkynoates with aldehydes. J Org Chem 80:148
50. Liu D, Chen J-Q, Wang X-Z, Xu P-F (2017) Metal-free, visible-light-promoted synthesis of 3-phosphorylated coumarins via radical C−P/C−C bond formation. Adv Synth Catal 359:2773
51. Rao H, Ma X, Liu Q, Li Z, Cao S, Li C-J (2013) Metal-free oxidative coupling: xanthone formation via direct annulation of 2-aryloxybenzaldehyde using tetrabutylammonium bromide as a promoter in aqueous medium. Adv Synth Catal 355:2191
52. Jiang H, Huang H, Cao H, Qi C (2010) TBHP/I_2-mediated domino oxidative cyclization for one-pot synthesis of polysubstituted oxazoles. Org Lett 12:5561
53. Xie J, Jiang H, Cheng Y, Zhu C (2012) Metal-free, organocatalytic cascade formation of C–N and C–O bonds through dual sp^3 C–H activation: oxidative synthesis of oxazole derivatives. Chem Commun 48:979
54. Yu Z, Ma L, Yu W (2012) Phenyliodine bis(trifluoroacetate) mediated intramolecular oxidative coupling of electron-rich *N*-phenyl benzamides. Synlett 23:1534
55. Kapoorr R, Singh SN, Tripathi S, Yadav LDS (2015) Photocatalytic oxidative heterocyclization of semicarbazones: an efficient approach for the synthesis of 1,3,4-oxadiazoles. Synlett 26:1201
56. Li H, Liu C, Zhang Y, Sun Y, Wang B, Liu W (2015) Green method for the synthesis of chromeno[2,3-*c*]pyrazol-4(1*H*)-ones through ionic liquid promoted directed annulation of 5-(aryloxy)-1*H*-pyrazole-4-carbaldehydes in aqueous media. Org Lett 17:932
57. Zhu Y-P, Lian M, Jia F-C, Liu M-C, Yuan J-J, Gao Q-H, Wu A-X (2012) I_2 Promoted domino oxidative cyclization for one-pot synthesis of 2-acylbenzothiazoles via metal-free sp^3 C–H functionalization. Chem Commun 48:9086
58. Lemercier BC, Pierce JG (2015) Synthesis of 1,4,2-oxathiazoles via oxidative cyclization of thiohydroximic acids. Org Lett 17:4542

Photo-Driven Catalytic Cross-Dehydrogenative Coupling (CDC)-Type Reactions

Eiji Yamaguchi, Saki Maejima and Akichika Itoh

Abstract The main focus of this chapter is the recent innovation of a new synthetic method that can be used to form carbon–carbon bonds from two C–H bonds driven by visible light. This photo-driven cross-dehydrogenative coupling (CDC) reaction of heteroarenes is a good alternative that can be used in place of traditional cross-coupling reactions.

Keywords CDC reactions · Photoreaction · C–C bond-forming reaction

1 Introduction

The transition-metal-catalyzed cross-coupling reaction of organometallic reagents with organic (pseudo)halides is one of the most powerful methods available to form carbon–carbon bonds and is widely used in the field of organic chemistry (Eq. 1) [1–7]. However, this type of cross-coupling reaction requires at least a stoichiometric amount of organometallic reagent, which, in turn, produces a stoichiometric amount of metal waste. Over the past couple of decades, enormous progress has been made in the transition-metal-catalyzed direct arylation reaction of unactivated C–H bonds. In particular, the coupling reaction of arene C–H bonds using organometallic reagents or organic (pseudo)halides is now a substitute for traditional cross-coupling reactions (Eq. 2) [8–10]. However, since this type of C–C bond-forming reaction requires a prefunctionalized coupling partner, transition-metal-mediated double C–H bond activation and subsequent C–C bond-forming catalysis have attracted much attention [11–13]. This type of reaction is known as a cross-dehydrogenative coupling (CDC) reaction and, as aforementioned, can be used as an alternative method to traditional cross-coupling reactions,

E. Yamaguchi (✉) · S. Maejima · A. Itoh (✉)
Gifu Pharmaceutical University, 1-25-4, Daigaku-Nishi, Gifu 501-1196, Japan
e-mail: yamaguchi@gifu-pu.ac.jp

A. Itoh
e-mail: itoha@gifu-pu.ac.jp

A. Srivastava and C. K. Jana (eds.), *Heterocycles via Cross Dehydrogenative Coupling*, https://doi.org/10.1007/978-981-13-9144-6_12

as it is a more atom economical and straightforward method (Eq. 3). On the other hand, the CDC reaction generally requires harsh reaction conditions, such as a high reaction temperature and/or a stoichiometric amount of metal oxidant.

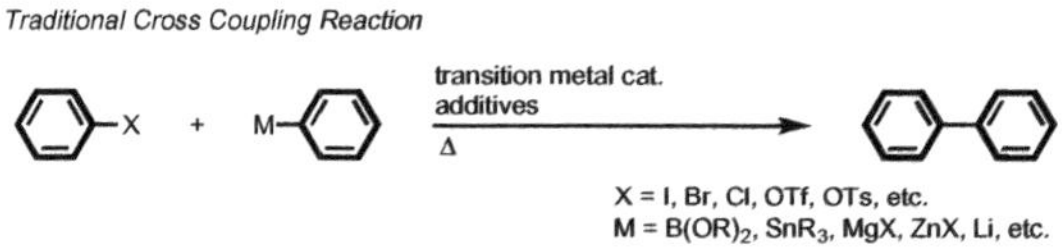

(1)

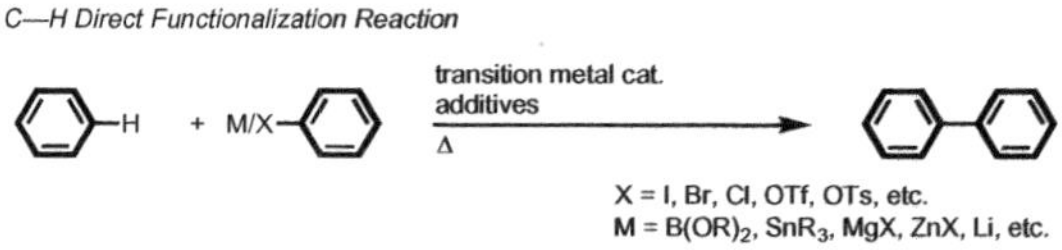

(2)

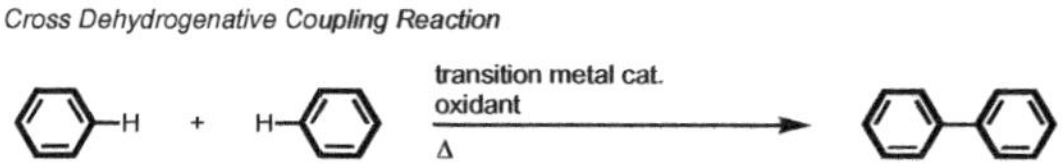

(3)

In contrast to coupling reactions that proceed through the thermal activation of a catalyst or substrate, photoactivation of catalysts or substrates is now attracting attention because of the high reactivity and selectivity of these reactions [14–18]. Therefore, photo-driven CDC-type reactions make synthetic schemes more efficient and environmentally friendly and are highly desirable to use in new C–H/C–H coupling reactions. This chapter will briefly discuss the recent evolution of photo-driven CDC reactions.

2 Traditional Photo-Driven CDC Reaction Method

The oxidative photocyclization of stilbenes was discovered during an investigation on the photoisomerization of stilbenes [19]. After Mallory discovered that iodine could be used to catalyze the reaction under air as a co-oxidant, it became feasible to use the reaction as a synthetic tool to prepare phenanthrenes and other polyaromatic hydrocarbons (PAHs; Eq. 4) [20, 21].

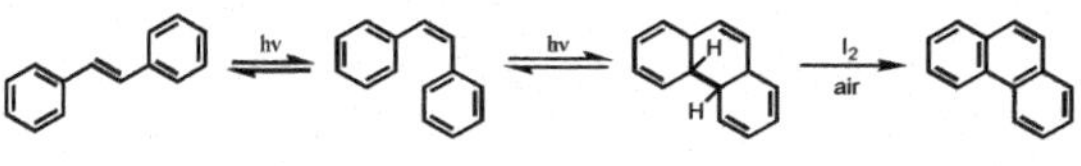

(4)

Further to this, the reaction mechanism was investigated in detail by Mallory and co-workers, and the Mallory reaction is now one of the most versatile solutions that can be used to construct PAHs through an intramolecular CDC-type process to furnish chrysenes, benzopyrenes, helicenes, and further π-expanded PAHs [22–26].

For example, the irradiation of cyclophane derivatives with a high-pressure Hg lamp in the presence of a catalytic amount of molecular iodine gives the Mallory reaction product in 42% yield, which is then converted to [7] circulene (Scheme 1) [27].

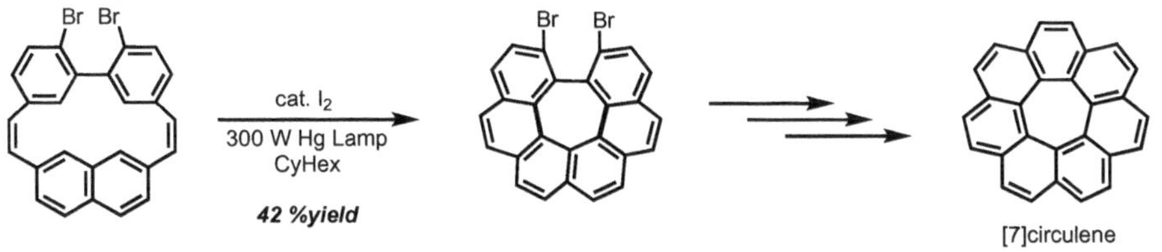

Scheme 1 Synthesis of [7] circulene by means of the Mallory reaction

3 Photo-Driven CDC Reaction via Photoredox Catalysis

3.1 Ruthenium-Based Photoredox Catalysis

In the past, ruthenium-based photoredox catalysts have been extensively investigated and used in laboratories [28, 29]. Their versatile reactivity, as a result of their chemical stability, redox potential, and excited state lifetime, still attracts attention, and many chemists have studied these systems by applying them in photoredox catalysis [30].

The CDC reaction employing a ruthenium polypyridyl complex was first reported by Stephenson and co-workers in 2010 in the Aza–Henry reaction of tetrahydroisoquinoline **I** with nitromethane (Scheme 2) [31]. The authors proposed a plausible reaction mechanism for the visible-light-driven ruthenium catalysis as follows: Ru(II)* is formed by the visible-light irradiation of Ru(II) in the ground state. Radical cation **II** is formed by the reductive quenching of the generated Ru(II)* species by **I** to furnish Ru(I), a strong reductant. The formed Ru(I) reacts with nitromethane or oxygen to give its radical anion, and this radical anion can then react with **II** to furnish iminium ion **III**. Finally, the addition of nitromethane to the iminium ion forms the desired product.

Although this type of oxidative CDC reaction transformation has already been explored under thermal reaction conditions, these reactions generally require stoichiometric amounts of potentially explosive reagents as oxidants (Scheme 3) [12, 32, 33].

By contrast, photocatalysis can produce an oxidized product under mild conditions without the use of the abovementioned oxidizing reagent. The excited Ru (II)* species undergoes reductive quenching to amine **I** and can furnish Ru (I) and oxidized radical cation species **II** (Scheme 4). It is easier to deprotonate or abstract a hydrogen from the formed product **II** than neutral amine **I**. It is well known that the pK_a value of a molecule is affected by its oxidation state. For example, amine radical cation **II** has a pK_a value of approximately 15, which makes it significantly more acidic than neutral amine **I** [34], whereas Wayner and co-workers reported that the bond-dissociation energy of the α-amino C–H bond in neutral amines is

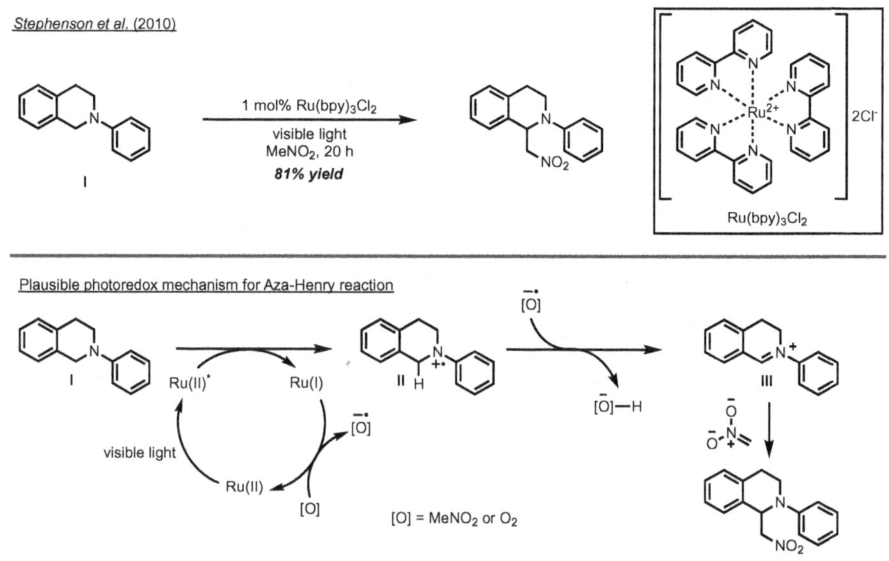

Scheme 2 Visible-light-driven CDC reaction: Aza–Henry reaction via ruthenium catalysis

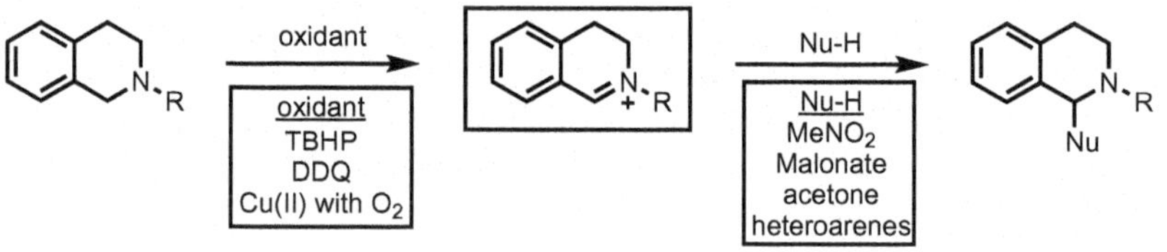

Scheme 3 Traditional Aza–Henry reaction of tetrahydroisoquinolines: using stoichiometric amounts of oxidant

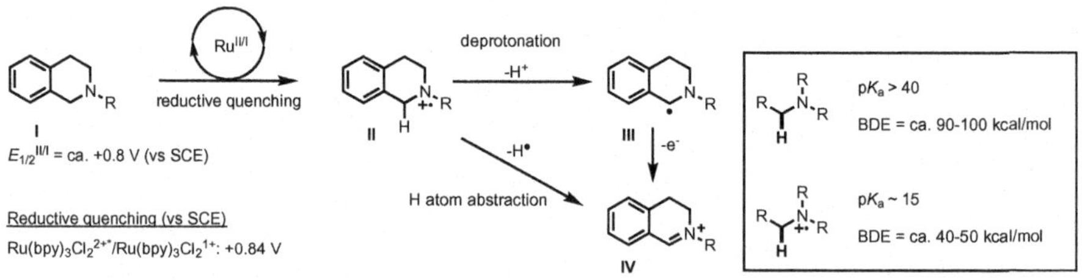

Scheme 4 Rational explanation of amine oxidation by photoredox catalysis

40–50 kcal/mol higher than that in oxidized amines [35]. Therefore, photoredox catalysis can easily generate iminium ions from neutral amines without requiring any stoichiometric amounts of oxidant, such as peroxide, and harsh reaction conditions.

Subsequent investigations related to CDC reactions via ruthenium photocatalysis have shown that the iminium ion generated from tetrahydroisoquinolines can be reacted with various nucleophiles.

Rueping and co-workers reported the direct Mannich reaction of tetrahydroisoquinolines with ketones via visible-light photoredox catalysis (Scheme 5) [36]. They developed a dual catalytic process involving ruthenium-based photoredox catalysis with proline catalysis to achieve an oxidative Mannich reaction under visible-light irradiation. In the reaction, the rapid formation of the iminium ion resulted in a decrease in the yield of the corresponding product, accompanied by the formation of an amide by-product, generated from the overoxidation of the iminium ion intermediate. The findings show that the nucleophilic addition to the iminium ion is not fast enough under these conditions.

In related experiments, the indirect CDC reaction of a silyl enol ether with tetrahydroisoquinoline was reported by Xia et al. (Eq. 5) [37]. Interestingly, they revealed that the use of MeOH as a solvent is effective in reactions that involve the rapid formation of a reactive iminium ion intermediate under the conditions. Their developed reaction took place rapidly to furnish the desired Mannich-type products in good to excellent yield.

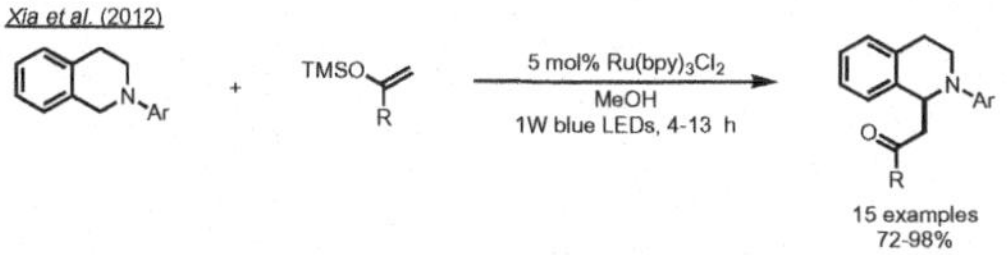

(5)

Moreover, direct alkynylation of tetrahydroisoquinoline was also developed by Rueping (Eq. 6) [38]. The CDC reaction of a terminal alkyne with amines was

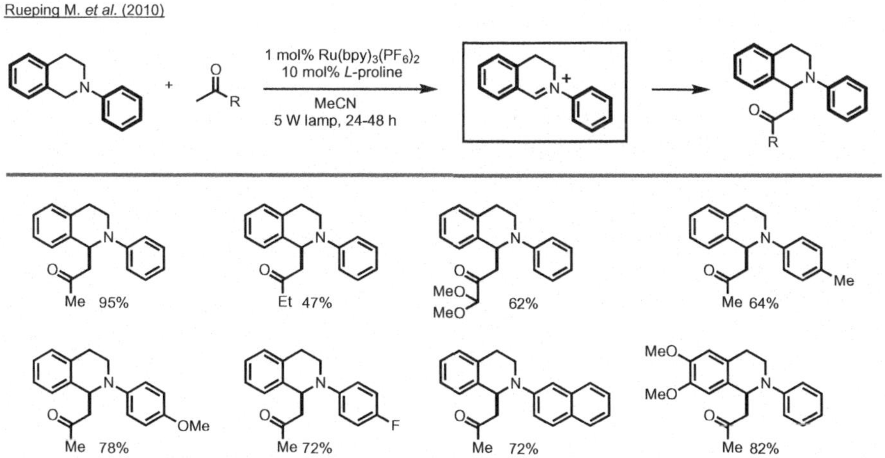

Scheme 5 CDC reaction of ketones with tetrahydroisoquinolines: the direct Mannich reaction via photoredox catalysis

achieved using the more electron-rich photoredox catalyst $Ru(bpy)_2(dtbpy)(PF_6)_2$ in dichloromethane (DCM). Using $(MeCN)_4Cu(PF_6)$ as a co-catalyst, a variety of aromatic alkynes could be efficiently added to the photocatalytically generated iminium ion intermediate in 43–95% yield. When 4-^{t}Bu-ethynylbenzene was employed as a coupling agent, only the decomposition of the starting material was observed. In this case, the use of a silver salt was effective in promoting the reaction progress and gave the corresponding alkynyl product in good yield.

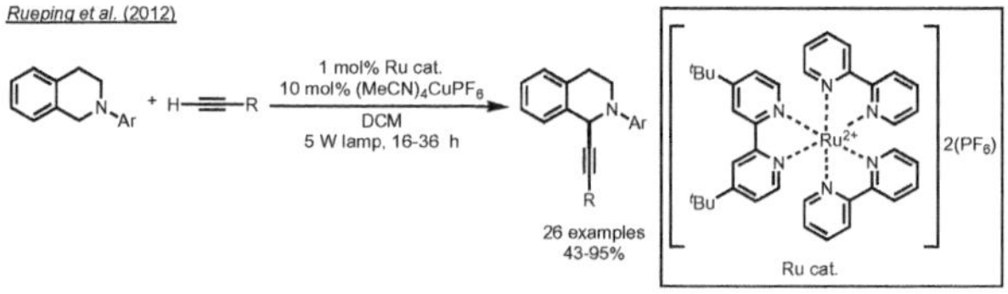

(6)

Although the abovementioned oxidative strategy is effective for the formation of CDC products using oxygen as a terminal oxidant, in some cases, a major by-product such as an amide is formed through the oxidative decomposition of the reactive iminium ion in the presence of oxygen because of the instability of the iminium ion. In this context, Stephenson developed a method to generate stable isoquinoline iminium salts with the use of bromotrichloromethane ($BrCCl_3$) as a terminal oxidant (Scheme 6) [39]. In the first step, photoexcited Ru^{II*} undergoes reductive quenching with isoquinolines to furnish radical cation intermediate **I** and Ru^I. The ruthenium catalyst is regenerated by the reduction of $BrCCl_3$ to a bromide anion and a trichloromethane radical. The formed *C*-centered radical then abstracts a hydrogen from **I**, resulting in the formation of chloroform and bromine-stabilized iminium salt **II**. After fully converting tetrahydroisoquinoline to **II**, the light is turned off, and a variety of nucleophiles are added. The reaction shows a broad scope of nucleophiles such as nitroalkanes, allyl silanes, silyl enol ethers, malonate, β-ketoesters, indoles, and acetylides.

An intramolecular C–N bond-forming CDC-type reaction catalyzed by a photoredox catalyst was also developed by Xiao and co-workers (Scheme 7) [40], involving the oxidation of the benzylic C–H bond of diamines derived from natural amino acids to form iminium ion intermediates. An intramolecular nucleophilic attack led to the formation of an iminium carbon from a secondary amine to afford the corresponding tetrahydroimidazoles in good to excellent yield with good diastereoselectivities. The authors also proposed that the addition of nitrogen to the iminium ion from its *Re* face is more favorable than on its *Si* face because of steric repulsion.

In 2013, Wu and co-workers found that secondary amines could also be used in an intermolecular CDC reaction via the combination of a ruthenium photoredox catalyst and copper (II) as a Lewis acid (Scheme 8) [41]. Under the optimized

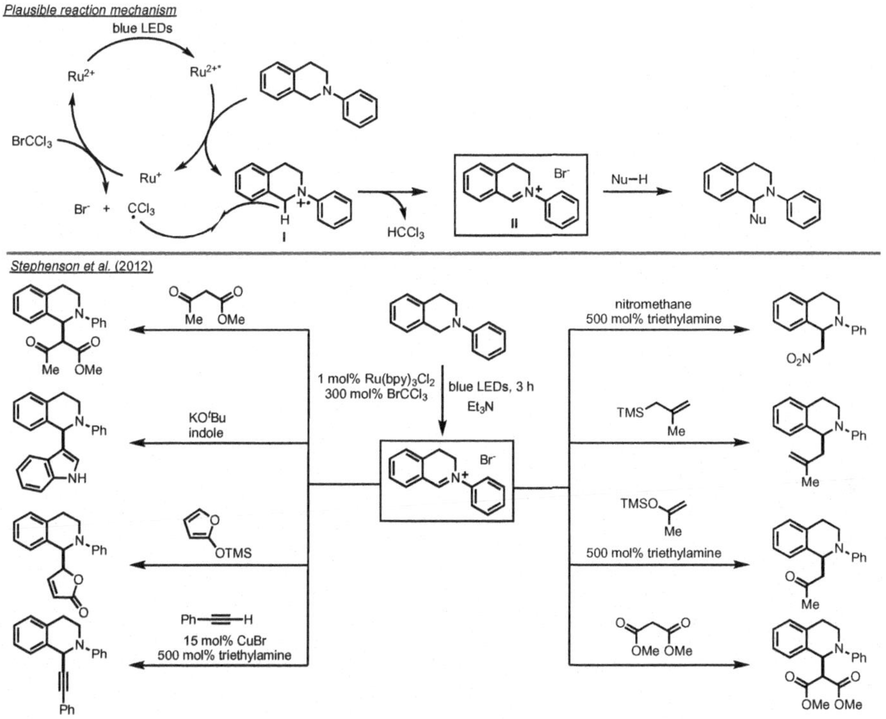

Scheme 6 Stephenson developed CDC reaction using a stable iminium salt generated by photocatalysis

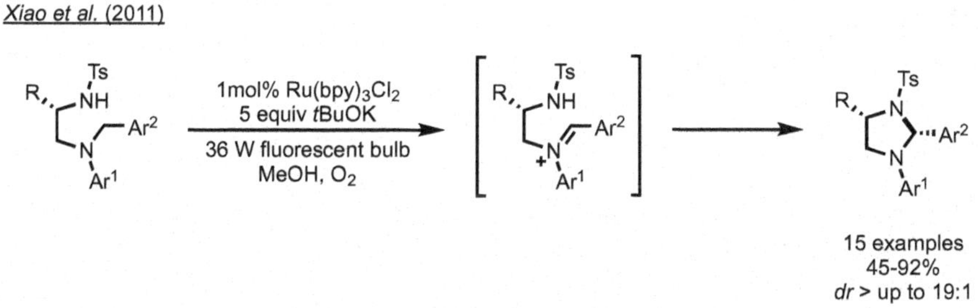

Scheme 7 Intramolecular CDC-type reaction of diamines leading to tetrahydroimidazoles

reaction conditions, a glycine-derived amino acid was oxidized to the corresponding iminium ion. Subsequently, intramolecular nucleophilic attack took place, mediated by the copper salt to afford the CDC product in good yield with poor diastereoselectivity. Their developed strategy was used to functionalize secondary amines to construct new C–C bonds with a variety of carbonyls at the α-position.

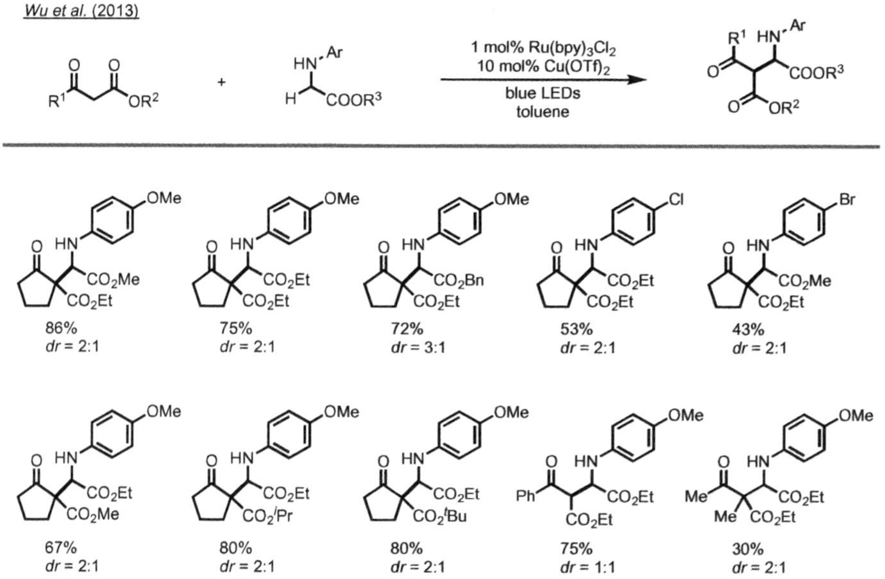

Scheme 8 CDC reaction of secondary amines with diones: a combination of a Ru photoredox catalyst with copper as a Lewis acid

Generally, amides are more stable to oxidation compared to neutral amines because the reducing ability of amides is weaker than that of amines. Therefore, the abovementioned CDC reaction strategy could not be applied to amides. Stephenson and co-workers investigated the CDC reaction of an *N*-alkyl amide with electron-rich (hetero)arenes as nucleophiles in combination with a ruthenium photoredox catalyst and persulfate as an oxidant (Scheme 9) [42]. The reaction of a variety of amides with (hetero)arenes in the presence of $Ru(bpy)_3Cl_2$ and $(NH_4)_2S_2O_8$ irradiated with a blue LED gave the corresponding CDC product in moderate to good yield. Although the developed reaction also proceeded through thermolysis of persulfate at 55 °C, the thermolysis resulted in lower yields and selectivities for this CDC reaction compared with photocatalysis. Interestingly, the mechanism put forward by the authors was proposed to proceed via Ru^{II}/Ru^{III} as a result of the strong oxidizing ability of persulfate. The reaction was initiated by the oxidative quenching of excited Ru^{II*} with persulfate to generate Ru^{III} and a sulfate radical anion [43]. The sulfate radical anion then abstracted a hydrogen from the amide to generate an amido radical. Oxidation of the amido radical led to the formation of an *N*-acyliminium intermediate by Ru^{III} (+1.27 V vs. SCE). Finally, nucleophilic trapping by electron-rich (hetero)arenes furnished the desired CDC products. In this context, He and co-workers found that benzoxazin-2-ones could also be used in the photocatalytic CDC reaction, coupled with indoles [44].

Stephenson and co-workers also applied their developed photocatalytic CDC reaction to the total synthesis of (−)−pseudovincadifformine [45], which is a vinca

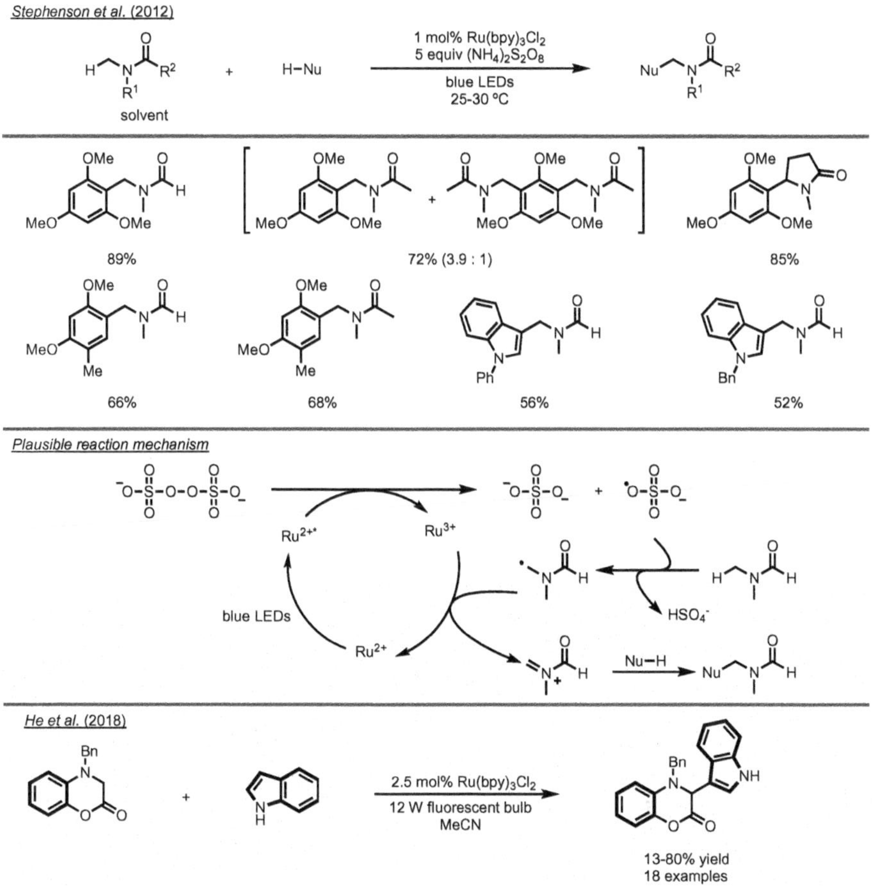

Scheme 9 CDC reaction of amides with (hetero)arenes

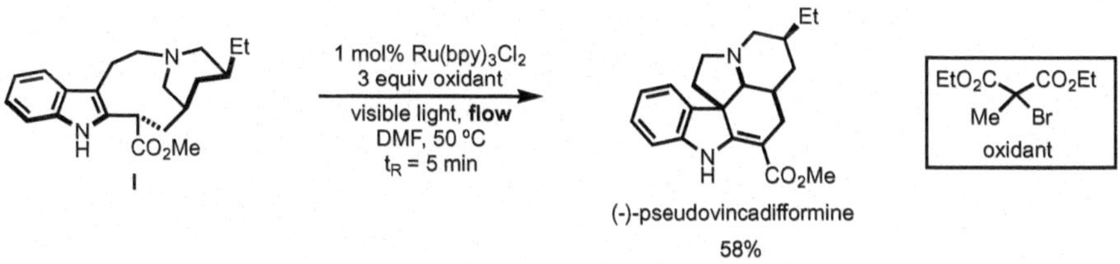

Scheme 10 Total synthesis of (−)-pseudovincadifformine using a photoredox CDC-type reaction

alkaloid (Scheme 10) [46]. Through the exposure of intermediate **I** to photoredox conditions in a flow reactor, an iminium intermediate was formed and underwent cyclization to give the desired (−)-pseudovincadifformine in 58% yield. It is noteworthy that this is a rare example of the use of a photoredox catalytic CDC reaction for the total synthesis of a natural product.

Although various reactions have been developed using a photoredox catalyst, asymmetric CDC reactions still remain an important challenge to solve, with only a few examples reported. In this context, Rovis and co-workers achieved the asymmetric CDC reaction of tetrahydroisoquinolines with aldehydes via the combination of visible-light photocatalysis with NHC catalysis (Scheme 11) [47]. This dual catalytic system enabled the addition of a broad range of aldehydes to tetrahydroisoquinolines in moderate to excellent yields with high enantioselectivities. A variety of tetrahydroisoquinolines were investigated, and the results showed that electron-rich substrates participated in the reaction to furnish the products in good yield. In the case of *N*-4-trifluoromethylphenyl tetrahydroisoquinolines, the reactions did not proceed under the conditions mainly because of the formation of radical dimerization products.

Most recently, the catalytic asymmetric cross-dehydrogenative coupling of two $C(sp^3)$–H bonds using a combination of a ruthenium-based photoredox catalyst with prolinol-based Jørgensen's catalyst was carried out by Pericàs and co-workers (Scheme 12) [48]. They developed the asymmetric catalytic CDC reaction of xanthenes with aldehydes using a visible-light-responsive photoredox catalyst. The developed reaction showed good functional group tolerance and moderate to high enantioselectivities under the conditions. They also investigated the reaction mechanism by means of several experimental studies and computational studies. The reaction was initiated by the reduction of bromotrichloroethane ($E_{1/2}$ = −0.18 V vs. SCE) through single-electron reduction from excited state Ru^{II*} ($E_{1/2}$ = −0.81 V vs. SCE) to generate a CCl_3 radical. The generated *C*-centered radical then abstracted a hydrogen from xanthene to form xanthenyl radical **I**, which was then oxidized to xanthene cation **II** by Ru^{III}. The formed cation **II** was then

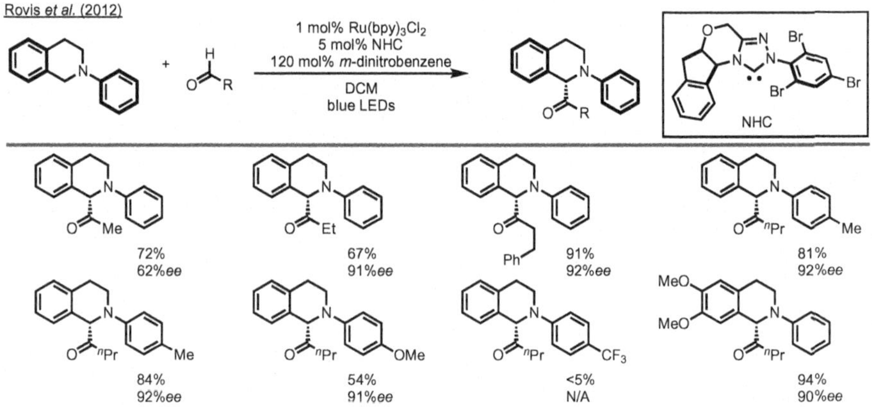

Scheme 11 Asymmetric CDC reaction of aldehydes with tetrahydroisoquinolines

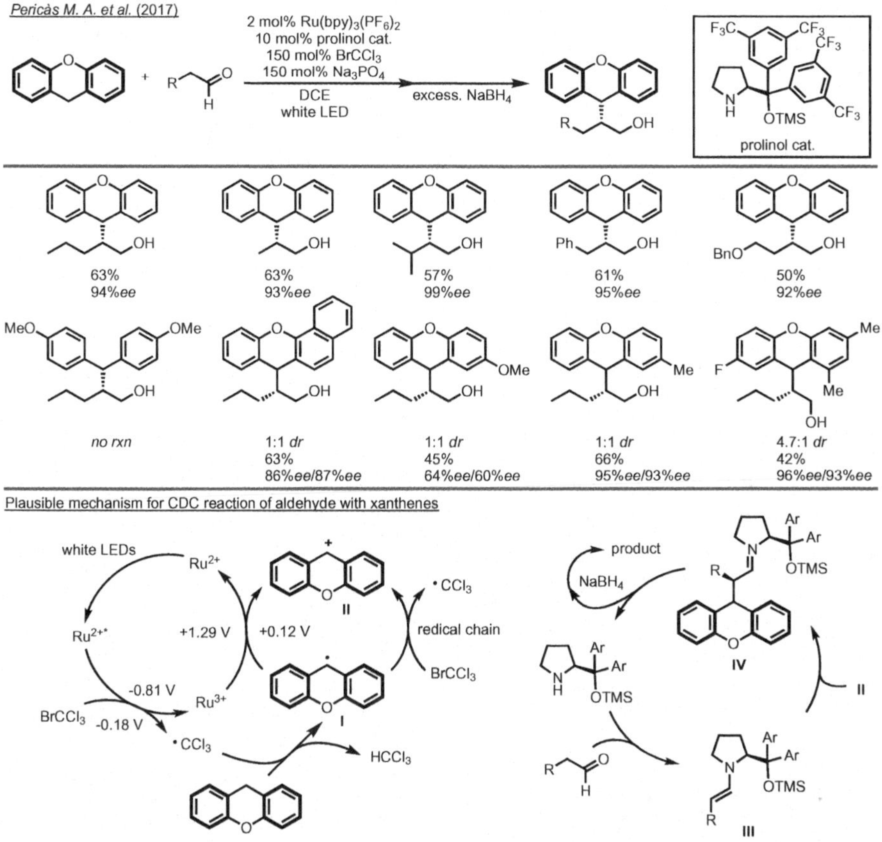

Scheme 12 Asymmetric CDC reaction of aldehydes with xanthenes

attacked by enamine intermediate **III**, generated by the reaction of the aldehyde with prolinol, to furnish **IV**.

3.2 *Iridium-Based Photoredox Catalysis*

Over the past decade, visible-light-responsive photocatalysis using the abovementioned ruthenium polypyridyl complexes was extensively used for novel chemical transformations. Moreover, iridium-based photocatalysis was also vigorously investigated [49–51]. In this section, a CDC reaction using an iridium-based photoredox catalyst is described.

As previously mentioned, Stephenson and co-workers achieved the highly effective intramolecular CDC reaction of tetrahydroisoquinolines with a ruthenium photocatalyst in 2010. In their study, the authors also described that this oxidative

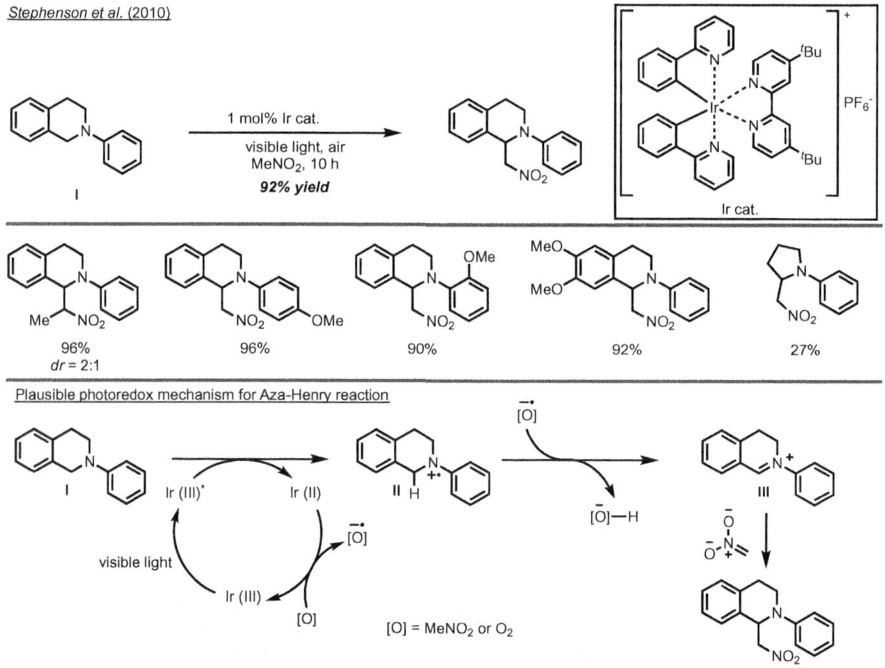

Scheme 13 Visible-light-driven CDC reaction: oxidative Aza–Henry reaction via iridium catalysis

Aza–Henry reaction also proceeded via iridium photocatalysis (Scheme 13) [31]. The developed reaction represented the first example of an iridium photocatalysis-based CDC reaction. Notably, the use of the iridium catalyst significantly accelerated the reaction to give the CDC product in an improved yield. Under the optimized conditions, the reaction was highly efficient and had good selectivity and good functional group tolerance.

In 2012, Rueping and co-workers reported the visible-light photoredox-catalyzed CDC reaction of glycine derivatives with indoles in combination with Lewis acid catalysis (Scheme 14) [52]. The reaction of various glycine

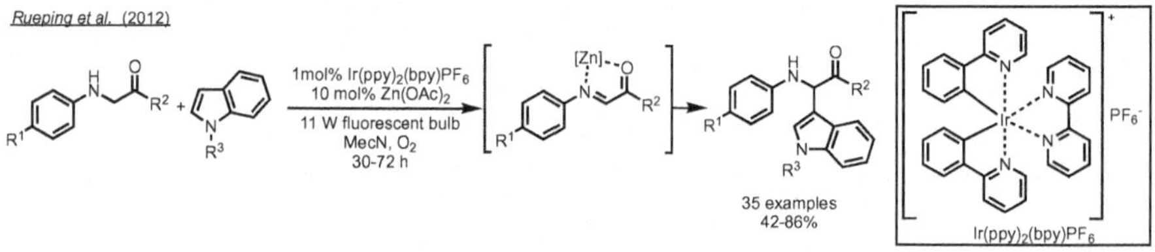

Scheme 14 CDC reaction of glycines with indoles via a combination of iridium photoredox catalysis and Lewis acid catalysis

derivatives with indoles in the presence of 1 mol% $Ir(ppy)_2(bpy)PF_6$ and 10 mol% $Zn(OAc)_2$ under visible-light irradiation conditions gave the corresponding CDC products in moderate to good yields. The reaction was also successfully extended to more complex peptides. The role of the Lewis acid was unclear, but the authors suggested that the Lewis acid catalyst influenced the formation of activated electrophiles such as imines.

In 2015, MacMillan and co-workers developed the iridium-based photocatalytic CDC reaction of cyclic and acyclic ethers with heteroarenes [53] (Scheme 15). The developed photoredox catalysis could be used for a broad range of reactions of various ethers with electron-deficient arenes, providing α-oxyarylated heteroarenes in high yield. Excited Ir^{III*} was proposed by the authors to catalyze the decomposition of sodium persulfate to the sulfate radical anion and sulfate. Hydrogen abstraction of the weakest C–H bond of the ether, e.g., the α-position, by the sulfate radical anion generated the desired electron-rich radical, which then underwent a Minisci-type addition to a protonated electron-deficient heteroarene [54, 55] and then, through the loss of a proton, furnished the radical intermediate. This formed radical intermediate then underwent a second SET with Ir^{IV} to regenerate ground state Ir^{III} and the desired CDC product.

Wu and co-workers described the dehydrogenative synthesis of indoles from various *N*-aryl enamines in the presence of a catalytic amount of an iridium photocatalyst and cobalt catalyst [56] (Scheme 16). Under LED light irradiation conditions, various *N*-aryl enamines were converted to the corresponding indole derivatives in good to excellent yield. Interestingly, the cobalt complex acted as a catalyst to trap protons and electrons from the substrates to produce H_2 gas, as

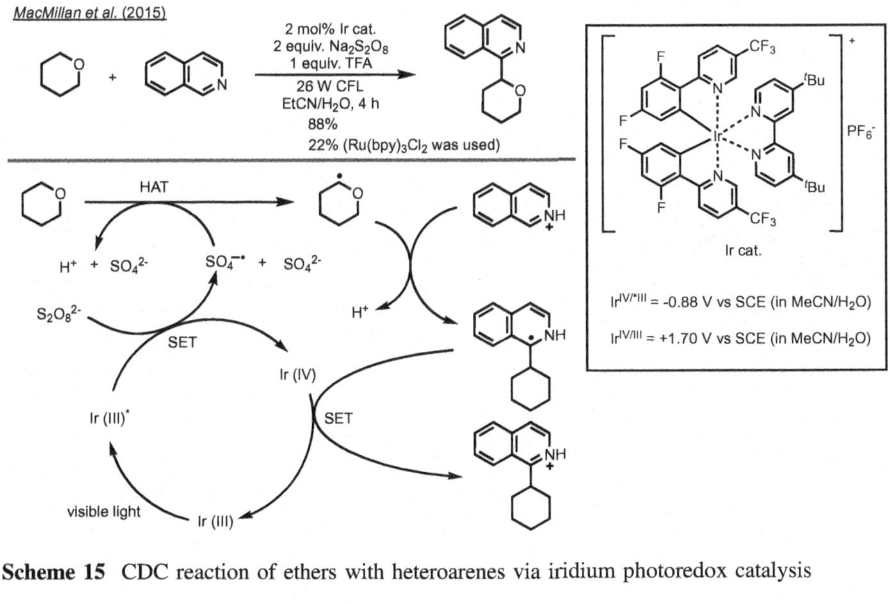

Scheme 15 CDC reaction of ethers with heteroarenes via iridium photoredox catalysis

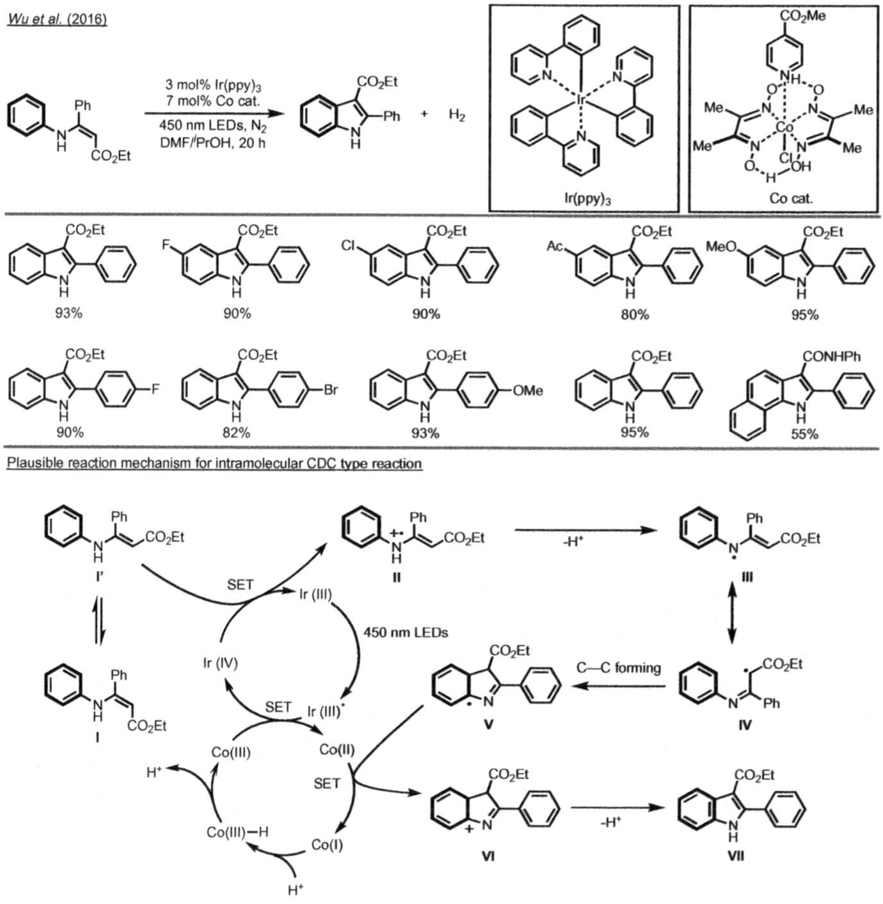

Scheme 16 Intramolecular dehydrogenative C–C bond-forming cyclization: dual catalytic system involving an iridium photocatalyst and cobalt catalyst

determined by gas chromatography, as a by-product in the reaction. A mechanistic investigation revealed that the reaction initiated the oxidative quenching of excited state Ir^{III*} ($Ir^{III*/II}$ = 0.31 V vs. SCE) with Co^{III} ($Co^{III/II}$ = −0.51 V vs. SCE), leading to the formation of Ir^{IV} and Co^{II} [57]. The formed Ir^{IV} was then reacted with enamines **I′** to abstract an electron, furnishing enamine radical cation **II** in the process and regenerating Ir^{III}. The generated **II** then released a proton to form iminium radical **III**, which existed in equilibrium with **IV**. Subsequently, intramolecular radical cyclization occurred to give **V**, which was oxidized with Co^{II} to furnish **VI**. Finally, deprotonation and tautomerization afforded the desired indoles. The reduced Co^{I} then reacted with a proton to form cobalt hydride (Co^{III}–H), which was protonated with the liberated proton to subsequently evolve H_2 and regenerate Co^{III} [58].

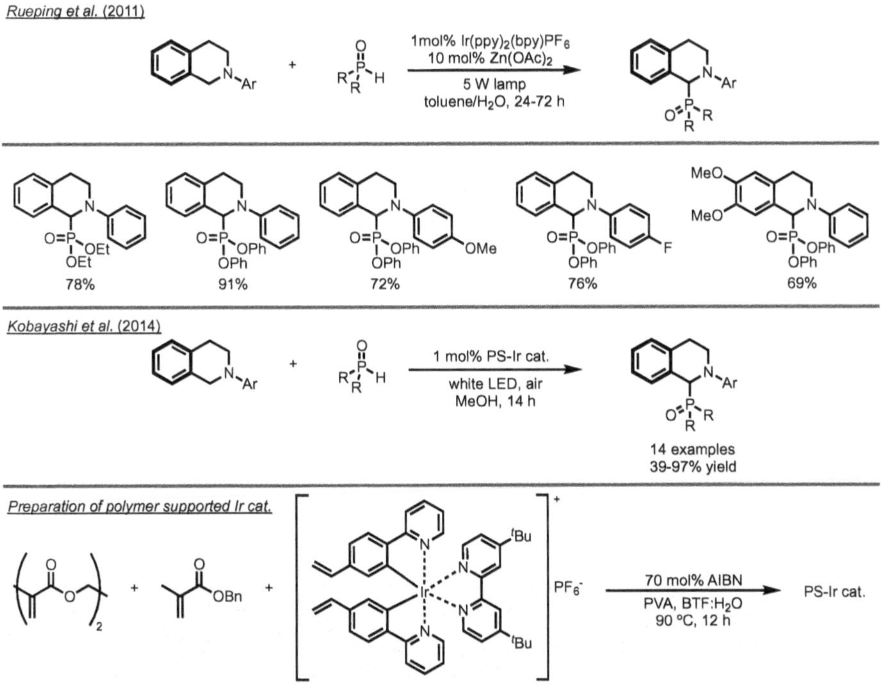

Scheme 17 Visible-light-driven CDC reaction: Aza–Henry reaction via Ru catalysis

The iminium ion generated from the photocatalysis can be trapped by a wide range of nucleophiles. In this context, phosphites are suitable coupling partners for CDC reaction involving photocatalysis. Rueping and co-workers reported the CDC reaction of tetrahydroisoquinolines with a variety of phosphites using an iridium photocatalyst (Scheme 17) [59] to synthesize various biologically interesting α-phosphoryl amines in a highly concise manner. Another method to realize C–P formation using a CDC reaction with photocatalysis was also reported by Kobayashi et al. [60]. They developed a polymer-supported iridium-based polypyridyl complex by means of a suspension polymerization method, and the developed polymer-supported catalyst was shown to be highly active and recyclable when used in a visible-light-catalyzed C–P bond-forming CDC reaction.

3.3 Cobalt-Based Photocatalysis

In 2015, Wu and co-workers discovered that in situ generated cobalt photocatalysts such as $Co(dmgH)_2Cl_2$ could act as photocatalysts in intramolecular CDC reactions to construct C–C bonds (Scheme 18) [61]. They demonstrated that versatile nucleophiles such as indoles could take part in the CDC reaction with

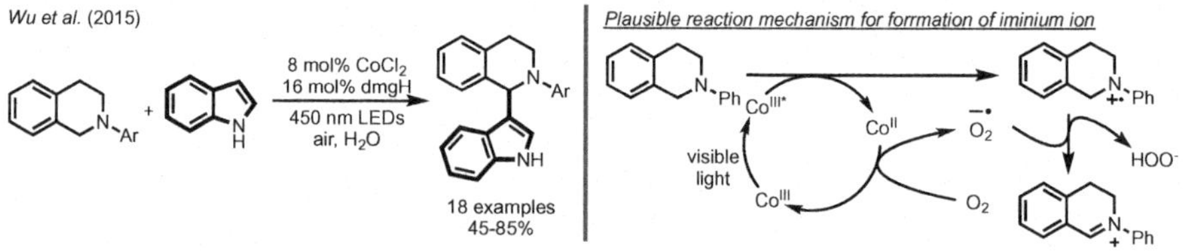

Scheme 18 Visible-light-driven CDC reaction via cobalt catalysis

tetrahydroisoquinolines under aerobic conditions. In their mechanistic investigations, a superoxide radical anion $O_2^{\cdot-}$ was found to be a crucial intermediate in the reaction, which was generated by electron transfer between the Co complex and the tetrahydroisoquinolines. On the basis of the results, the following mechanism was proposed. The reaction was initiated by the photoexcitation of preformed $Co^{III}(dmgH)_2Cl_2$ under visible-light irradiation, and this species then abstracted an electron from tetrahydroisoquinoline to generate an amine radical cation and Co^{II}. The formed Co^{II} then reacted with molecular oxygen to regenerate Co^{III} and a superoxide radical anion. After $O_2^{\cdot-}$ deprotonated the amine radical cation, single-electron oxidation furnished the iminium intermediate. Using the cobalt salt with dimethylglyoxime as a ligand to form a visible-light photosensitizer in situ, the first example of a cobaloxime-catalyzed photo-driven CDC reaction was achieved.

3.4 Platinum-Based Photocatalysis

Owing to their visible-light absorption, long excited state lifetimes, high quantum yields, and high chemical stability, platinum polypyridyl complexes have been used as photocatalysts [62]. In 2013, Wu and co-workers first reported the Pt^{II} polypyridyl complex-catalyzed oxidation of tetrahydroisoquinolines to form iminium intermediates under aerobic conditions. The formed iminium intermediates reacted with a wide range of nucleophiles to form CDC-type products (Scheme 19) [63]. In

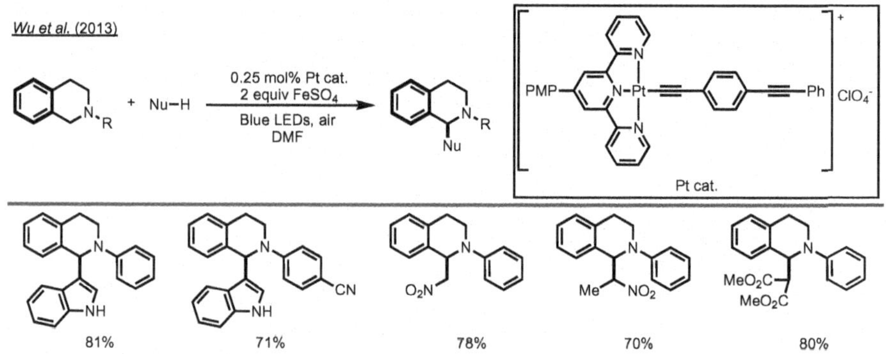

Scheme 19 Visible-light-driven CDC reaction: platinum-based photoredox catalysis

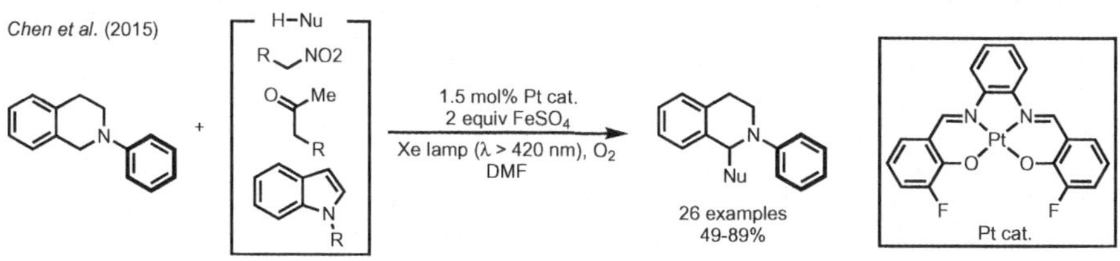

Scheme 20 Platinum Schiff base complex-catalyzed CDC reaction of tetrahydroisoquinolines

this reaction, the addition of $FeSO_4$ remarkably improved the reaction selectivity because of the trapping of a hydroperoxide free radical by an excess amount of $FeSO_4$ to suppress the undesired amide formation.

Pt^{II} Schiff base complexes can be easily synthesized from readily and commercially available chemicals. They have been found to display broad visible-light absorption and have long-lived phosphorescence emission in solution [64]. Therefore, the use of Pt complexes in oxidative transformations has long been in demand. In 2015, Chen and co-workers developed a platinum Schiff base complex-catalyzed visible-light-driven CDC reaction [65]. In the presence of a 1.5 mol% Pt Schiff base complex, two equivalents of $FeSO_4$, a variety of carbon nucleophiles, and tetrahydroisoquinolines were found to react well together to produce the desired CDC reaction products in moderate to excellent yield (Scheme 20).

3.5 *Palladium-Based Photocatalysis*

Although a tremendous number of studies regarding palladium-catalyzed cross-coupling reactions and C–H direct functionalization reactions via the thermal activation of a catalyst have been widely developed [66], the use of a palladium complex as a photoredox catalyst has not been well investigated [67, 68]. In 2011, Borisov and co-workers reported a Pd–porphyrin complex that exhibited a long triplet excited state lifetime [69], and Che envisioned that the long-lived excited state of the complex would render it a good photocatalyst for use in visible-light-induced transformations. In this context, Che and co-workers showed that the Pd–porphyrin complex is an efficient and robust catalyst for use in visible-light-driven CDC reactions (Scheme 21) [70]. The developed complex showed an excellent activity toward intra-/intermolecular CDC reactions, resulting in α-functionalized tertiary amines being synthesized in good to excellent yields under visible-light irradiation. The practical application of the developed complex was revealed in photochemical reactions performed using low catalyst loading (0.01 mol%). Mechanistic studies revealed that the palladium-based photocatalysis proceeded through singlet oxygen oxidation.

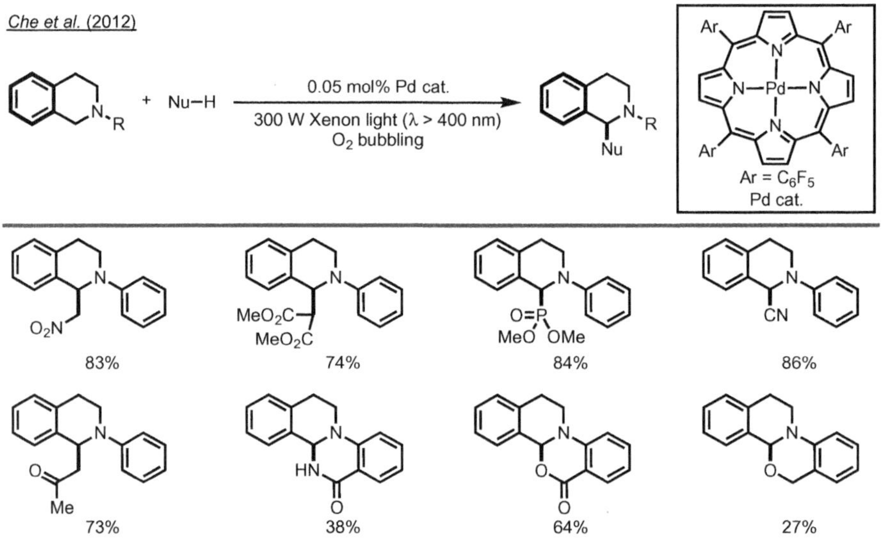

Scheme 21 Visible-light-driven CDC reaction: palladium-based photoredox catalysis

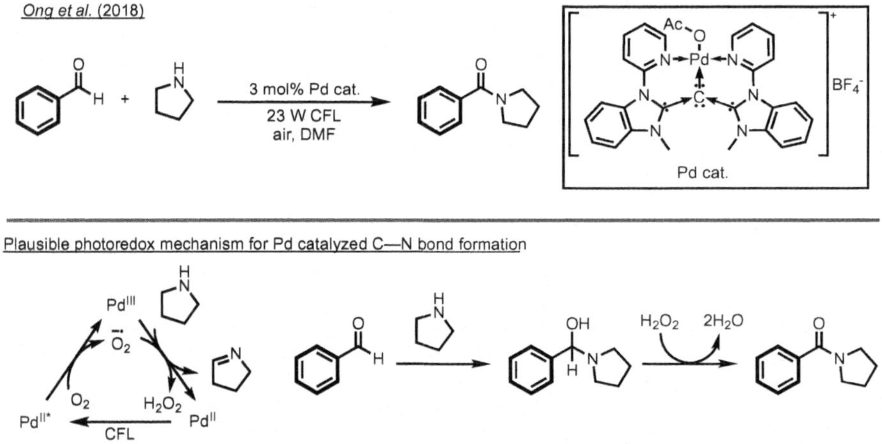

Scheme 22 Palladium-based photocatalyzed direct C–N bond-forming reaction

In 2018, Wang, Jurca, and Ong reported the Pd-photoredox-catalyzed cross-dehydrogenative coupling reaction of aldehydes with secondary amines to furnish the corresponding amides (Scheme 22) [71]. The reaction showed that the palladium bis(py-carbodicarben) complex acted as a photoredox catalyst (λ_{abs} = 499 nm; E (Pd^{III}/Pd^{II*}) = −1.31 V) under the conditions to generate a superoxide radical from molecular oxygen by oxidative quenching of the excited state of Pd^{II*} to Pd^{III}.

4 Organic-Dye-Catalyzed CDC Reactions

Visible-light photoredox catalysis has become an important tool for chemists. Recent advances in photoredox catalysis have predominantly focused on the use of traditional metal chromophores and their synthetic applications. In this class of compounds, ruthenium- and iridium-based polypyridyl complexes lead the way because of their visible-light absorption, long excited state lifetimes, tunable redox potentials, and high chemical stability. Thus, transition-metal-based photocatalysis is now broadly used in industrial and pharmaceutical processes. On the other hand, although Ru- and Ir-based photocatalysts represent superior catalytic activity, the development of catalysts and reactions based on rare metals is required to undergo a major reform because of the sustainability of these materials and their use in the future. Moreover, the platinum group metals, including Ru, Pd, Ir, and Pt, produce a broad spectrum of toxic effects on organisms. Therefore, the researcher-related synthesis of active pharmaceutical ingredients (APIs) has to comply with the permitted daily exposure (PDE) limit for platinum group metals (the PDE for metals in APIs; Pd: 100 μg/day, Pt: 100 μg/day, sum of Ir and Ru: 100 μg/day). In this context, the use of organo-photocatalysts as alternatives for well-known metal-based photocatalysts is in high demand. Over the past decade, several organo-photocatalytic CDC reactions have been reported. In this section, CDC reactions using organo-photoredox catalysts are described.

Drawing upon previous photoredox methodologies used for the CDC reaction of tetrahydroisoquinoline, König and co-workers demonstrated that Eosin Y (EY), which is a fluorescein derivative, is a suitable alternative that can be used to replace metal-based photocatalysts (Scheme 23) [72]. Moderate to excellent yields of the CDC products were observed for the Aza–Henry reaction and Mannich-type reaction when tetrahydroisoquinoline was used as a starting material. Moreover, the developed reaction was also applied to a C–P bond-forming CDC reaction, resulting in a high yield of product. The use of an Eosin Y bis-tetrabutylammonium salt (TBA–Eosin Y) as a photocatalyst under a similar set of conditions was reported shortly after by Wu and co-workers in 2012 [73]. This organo-soluble TBA–Eosin Y was used to catalyze a visible-light-driven CDC reaction with a variety of nucleophiles in moderate to good yields. The proposed mechanism involves the transfer of a single-electron from tetrahydroisoquinoline to the excited state of Eosin Y* to give an aminyl radical cation, which then loses a hydrogen atom to generate an iminium ion intermediate. The formed iminium ion is subsequently trapped by the nucleophile to form the CDC products. Electron spin resonance (ESR) measurements revealed that superoxide radical anions ($O_2^{\cdot-}$) were the active species in the transformation rather than singlet oxygen ($^1O_2$) during visible-light irradiation.

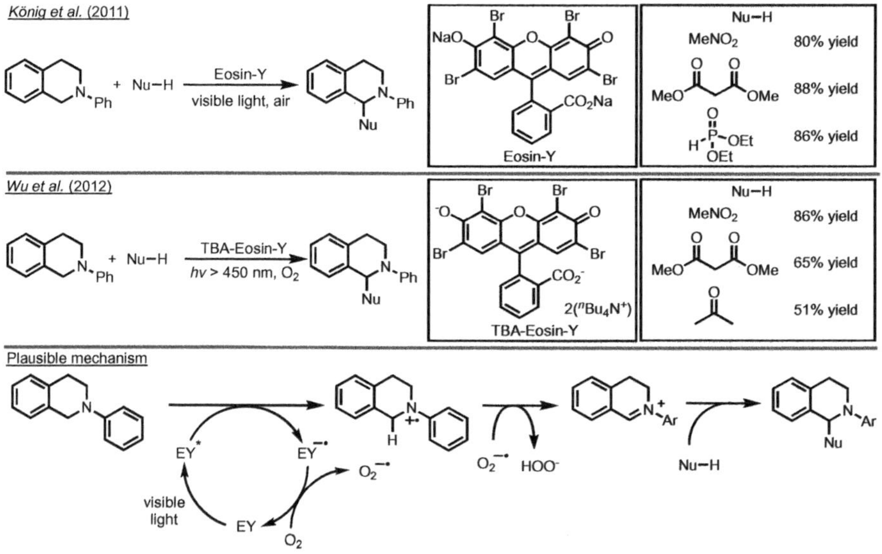

Scheme 23 Visible-light-driven CDC reaction: Aza–Henry reaction via Eosin Y catalysis

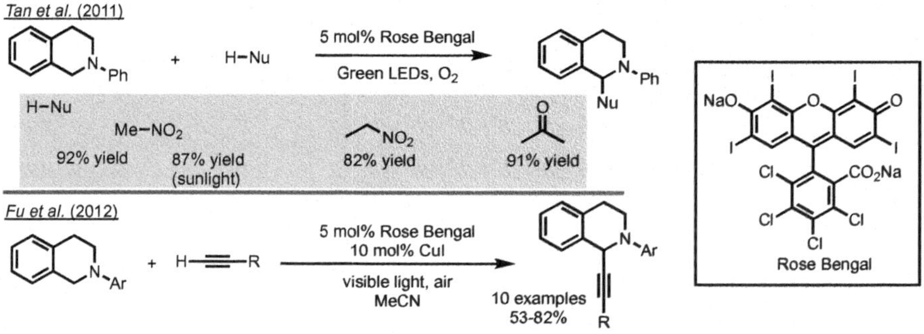

Scheme 24 Visible-light-driven CDC reaction through rose bengal catalysis

Tan and co-workers utilized rose bengal in the oxidation of tetrahydroisoquinolines to their corresponding iminium salts under aerobic oxidative conditions (Scheme 24) [74]. Owing to the absorption wavelength of rose bengal (λ_{abs} = 559 nm), the use of green LEDs was critical for the reaction to progress. Aza–Henry and Mannich-type reactions then occurred to form the CDC products in moderate to good yields under green LED irradiation conditions. Interestingly, the developed reaction can be applied using sunlight. In addition, the protocol was applicable to photo-driven CDC reactions using terminal alkynes as a coupling partner. Fu and co-workers utilized rose bengal and a household light to irradiate visible light to demonstrate alkynylation of tetrahydroisoquinolines. The reaction was found to tolerate the addition of various functional groups on the alkyne coupling partner [75]. From mechanistic investigations, the reaction mechanism was shown to proceed via the same reaction pathway to form an iminium ion intermediate, as discussed above (Scheme 23).

BODIPY derivatives are well-known organic dyes that are widely used as chemosensors, laser dyes, and photosensitizers. Recently, their applications in CDC reactions have been reported. Wu and co-workers employed simple BODIPY in the aforementioned CDC reaction of tetrahydroisoquinolines (Scheme 25) [76]. Generally, the photoexcitation of dyes results in the formation of a high-energy singlet excited state, which then undergoes intersystem crossing (ISC) to yield a triplet excited state. In this reaction, the authors employed simple BODIPY as a photocatalyst to use their singlet excited state and reported that it was successful in promoting intermolecular C–C bond-forming and C–P bond-forming CDC reactions. The key to their BODIPY design was to remove heavy atoms such as iodine and bromine from BODIPY to suppress the ISC to form a triplet state. ESR studies revealed that the singlet state BODIPY acts as a catalyst for the reaction, rather than triplet excited state and singlet oxygen. By contrast, Xue and co-workers reported that triplet excited state BODIPY catalyzed a C–N bond-forming CDC reaction [77]. Their developed protocol of direct amide synthesis from amines and aldehydes showed a broad substrate scope, requiring mild conditions and only air as an oxidant. They employed dibromo-substituted BODIPY to enhance the ISC from the singlet excited state to the triplet excited state using the heavy atom effect [78]. On the basis of their mechanistic investigations, the reaction was shown to be initiated by the photoexcitation of BODIPY to a high-energy singlet excited state, which immediately undergoes ISC to convert to the long-lived triplet excited state. Two possible mechanisms to form the oxidant were proposed. Molecular oxygen was reacted with the catalyst to form singlet oxygen ($^1O_2$) while regenerating the ground

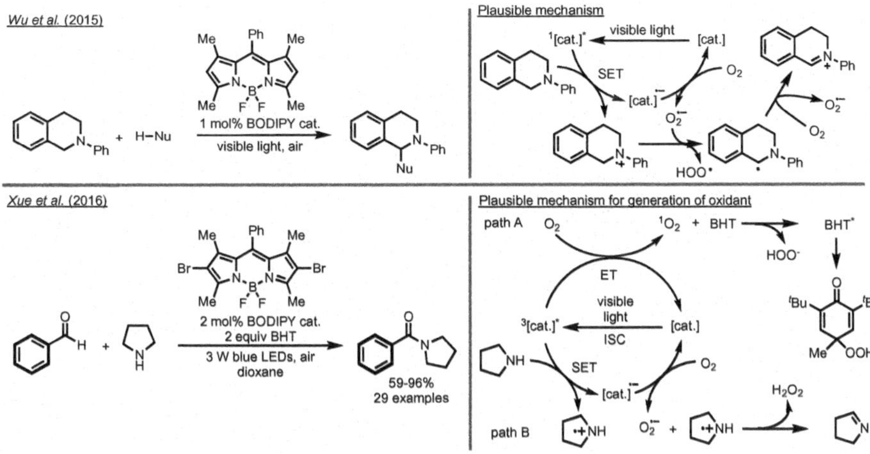

Scheme 25 BODIPY-based photocatalyzed CDC reactions

state of BODIPY. $^1O_2$ was then reacted with butylated hydroxytoluene (BHT) to give BHT–OOH (path A). On the other hand, the triplet excited state of BODIPY reacted with amines, leading to cascade electron transfer to the O_2 to furnish $O_2^{\cdot-}$. A hydrogen atom was then abstracted from the amine radical cation by $O_2^{\cdot-}$ to form HOOH (path B).

Phenazinium salts are well-known electron acceptors and also act as photocatalysts. In 2014, Leow and co-workers disclosed a visible-light-induced C–N bond-forming CDC reaction of amines with aldehydes, and amides were directly constructed under mild reaction conditions using a phenazinium salt as a catalyst (Scheme 26) [79]. They proposed that H_2O_2, generated photocatalytically under aerobic conditions, is believed to oxidize the hemiaminal intermediate, which is formed by the reaction of an amine with an aldehyde, as described above (Scheme 25).

The strong oxidative potential of acridinium salts has also been well investigated for their use in photocatalysis [80]. Lei reported that merging visible-light-responsive acridinium salts and a cobalt complex led to the catalysis of the C–N bond-forming CDC reaction of electron-rich arenes with pyrazoles (Scheme 27) [81]. Dialkyl arenes such as xylenes and monoalkyl arenes such as toluene ^{t}Bu-benzene were converted to the corresponding products in good yield with good site selectivity. Unfortunately, the reaction using *p*-nitrotoluene resulted in no reaction because of the inhibition of the nitro group, possibly indicating that the reaction proceeded via a radical process. Moreover, a variety of pyrazoles were successfully used in the reaction system to furnish the corresponding *N*-aryl pyrazoles. Mechanistic investigations revealed that the reaction was initiated by the photoexcitation of Acr^+ under blue LED irradiation conditions, leading to the formation of excited state Acr^{+*}. Arenes ($E_{1/2\ red}$ = +1.90 − 2.00 V) were oxidized by the generated Acr^{+*} ($E_{1/2\ red}$ = +2.06 V) to furnish the corresponding arene radical cation and acridinium radical $Acr^{\cdot}$. The generated arene radical cation was then trapped by the nucleophilic pyrazole to form a biaryl radical. Then, the radical adduct underwent electron transfer with Co^{II} to furnish a biaryl cation, which subsequently released a proton to produce the corresponding product. The released proton was then captured by Co^{I} in the reaction system yielding Co^{III}–H, the protonation of which resulted in the regeneration of Co^{III} and the evolution of hydrogen gas.

Anthraquinones, quinone-type dyes, are a class of compounds that show strong oxidation potential in their excited (triplet) state. Although unsubstituted

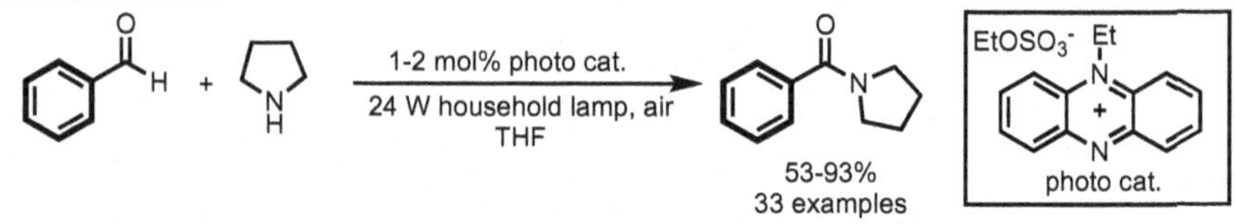

Scheme 26 Phenazinium-based photocatalyzed direct C–N bond-forming reaction

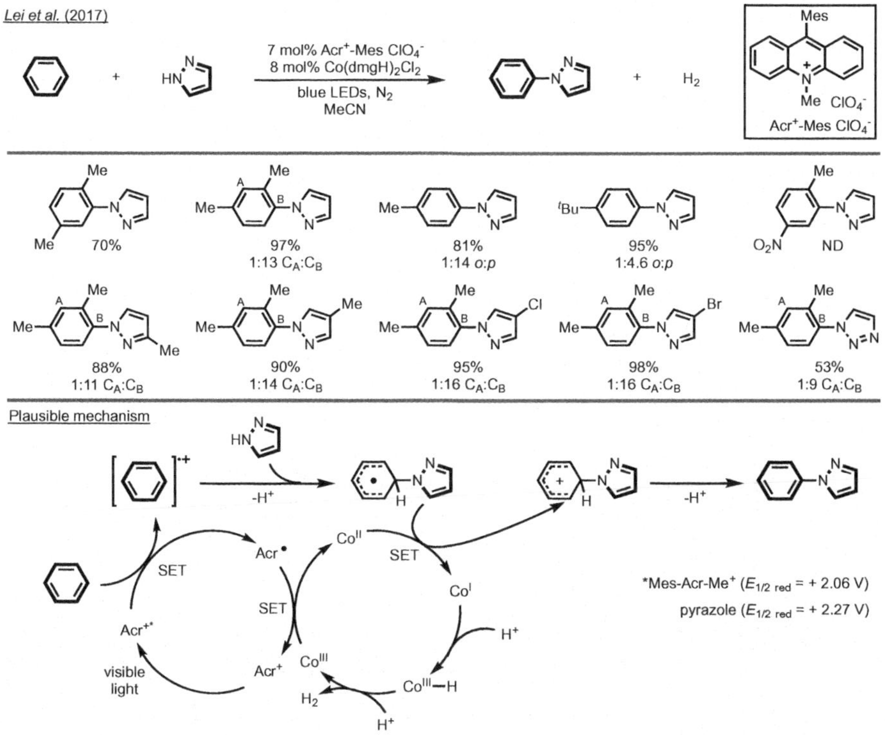

Scheme 27 C–N bond-forming CDC reaction catalyzed by an acridinium salt and cobalt complex

9,10-anthraquinone (AQN) generally requires UV light to undergo excitation, substituted AQN acts as a visible-light photocatalyst. In this context, several examples of visible-light-driven CDC reactions using AQN as a photocatalyst have been reported. Our group developed a visible-light-driven C–C bond-forming CDC reaction through AQN photocatalysis (Scheme 28) [82]. The reaction of the aforementioned tetrahydroquinolines with a variety of nucleophiles, including nitromethanes, ketones, and indoles, in the presence of a catalytic amount of 2-Cl-anthraquinone proceeded smoothly to furnish the corresponding CDC

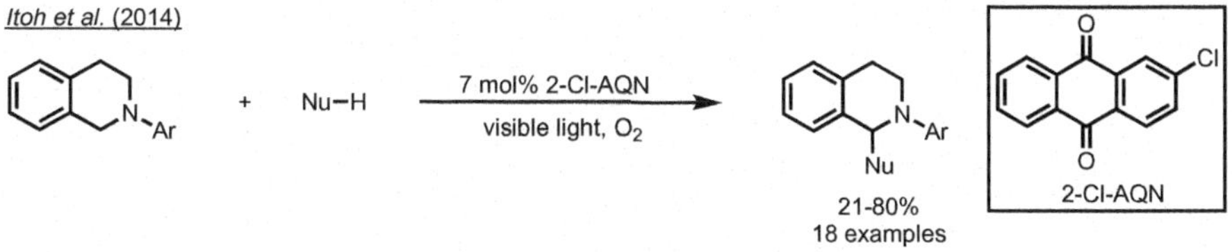

Scheme 28 Visible-light-driven CDC reaction through anthraquinone catalysis

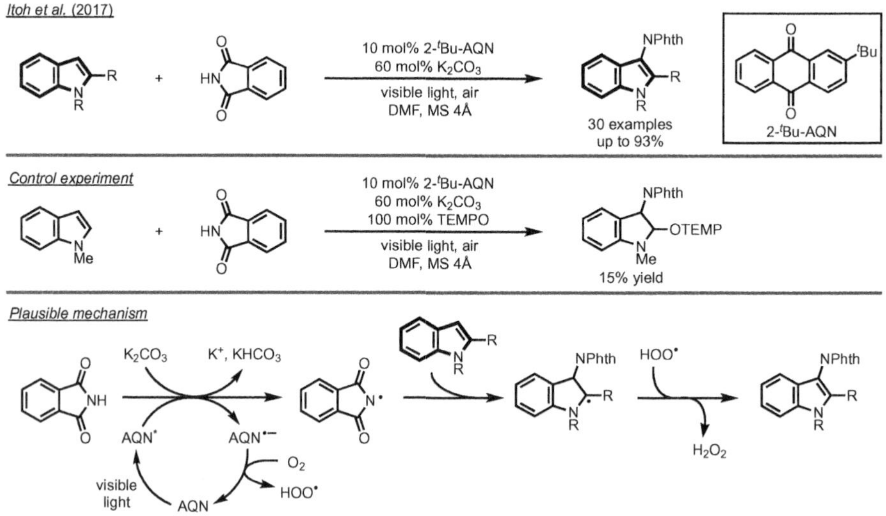

Scheme 29 C–N bond-forming CDC reaction catalyzed by an anthraquinone-based photocatalyst

products in moderate to good yield. The mechanism of the reaction proceeded via the same pathway to form an iminium ion intermediate, as discussed before.

Using AQN derivatives as photocatalysts and visible light, our group realized a mild direct amination CDC reaction using phthalimide as an amination reagent (Scheme 29) [83]. In this reaction, it was possible to use various indole derivatives to furnish the C–N bond-forming CDC products in good yield. A TEMPO adduct was observed when the reaction was performed with 100 mol% of TEMPO as a radical scavenger. The results indicated that the amination reaction proceeded via the generation of an *N*-centered radical, followed by the addition to an indole. On the basis of several experiments, the reaction was found to be initiated by the formation of an *N*-centered radical through the deprotonation of phthalimide and single-electron transfer to AQN*. The radical species reacted with the indoles, followed by one-electron oxidation and aromatization. On the other hand, $AQN^{\cdot -}$ was oxidized by molecular oxygen to AQN, which was then converted by visible light to AQN* to complete the catalytic system.

5 Iodine-Based CDC Reactions

Iodine is mainly produced in Chile (20,000 metric tons in 2017) and Japan (10,600 metric tons in 2017), with a bulk price of around \$20/kg. Iodine can adopt various oxidation states (−1, 0, +1, +3, +5, +7) in inorganic and organic derivatives. Therefore, molecular iodine is known as an inexpensive and abundant nonmetal oxidant that can be used in oxidative transformations. Moreover, recent developments in iodine chemistry have shown that a catalytic amount of iodine or iodide can be merged with a stoichiometric amount of a co-oxidant such as TBHP, *m*CPBA, or hydrogen peroxide and be used for various transformations as an alternative to traditional transformations that use transition metal complexes under thermal reaction conditions [84–86]. Using these recent developments, it is possible to use inexpensive, nonmetallic, and harmless reagents for the synthesis of fine chemicals such as pharmaceuticals and agrochemicals. Although we have discussed and shown the recent developments in visible-light-driven CDC reactions via organo-photocatalysis (Sect. 1.4), these organo-catalysts are not cheap enough to use on an industrial scale (Eosin Y: $90/25 g, rose bengal: $100/1 g, BODIPY: $380/250 mg, Mes-Acr$^+$ BF_4^-: $200/1 g; from Sigma-Aldrich in 2018). By contrast, our group have developed several transformations using catalytic amounts of iodine/iodide under aerobic or oxygen atmosphere irradiated with visible light [87–89]. In this section, the recent development in iodine-based visible-light-promoted catalytic CDC reactions will be described.

In 2013, a visible-light-promoted molecular-iodine-catalyzed oxidative C–C bond formation reaction through a CDC reaction between tetrahydroisoquinolines with carbon nucleophiles was reported by our group (Scheme 30) [90]. Although a similar reaction using a metal photoredox catalyst and organo-photocatalyst had been previously developed, as discussed above, the aerobic oxidative transformation mediated by iodine in the presence of visible light represented the first example of this class of C–C bond-forming CDC reactions. The reaction of tetrahydroisoquinolines with nucleophiles such as nitroalkanes, malonate esters, and acetone in the presence of 5 mol% molecular iodine and acetic acid proceeded smoothly via a

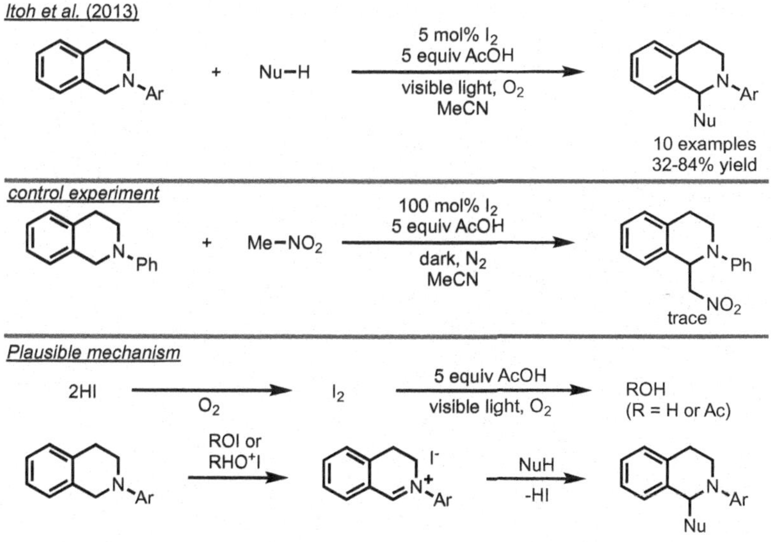

Scheme 30 Visible-light-driven CDC reaction through iodine catalysis

CDC reaction to afford the corresponding products in moderate to excellent yield. Previously, it was reported by Shirasaka and co-workers that the oxidation of tetrahydroisoquinoline with molecular iodine in the presence of base took place to give the corresponding iminium iodides [91]. However, the developed reaction using a stoichiometric amount of molecular iodine under dark conditions resulted in a trace amount of product formation; this observation revealed that the reaction requires all three components: iodine, visible light, and oxygen. From mechanistic investigations, a plausible mechanism for the developed CDC reaction was proposed. It is known that molecular iodine can be converted to the more reactive I^+ species using an oxidizing reagent. The active species in the reaction has not yet been clarified; however, hypoiodous acid or acetyl hypoiodite is generated from I_2 in situ under aerobic photooxidative conditions in the presence of H_2O or AcOH [92]. Tetrahydroisoquinoline is oxidized to an iminium ion by ROI or its protonated species (protonated by AcOH and/or HI) generated in situ from I_2. The intermediate undergoes an acid-promoted addition of the carbon nucleophile to form the desired product. Finally, HI is oxidized to molecular iodine under oxygen atmosphere.

Our group reported the practical transformation of β-keto esters to their corresponding α-malonic acid esters in the presence of molecular oxygen catalyzed by calcium iodide under visible-light irradiation [93]. Mechanistic investigations revealed that the reactive intermediate of this reaction was the carbon-centered radical, which was generated by homolytic cleavage of the C–I bond assisted by visible-light irradiation. According to this result, our group developed a photooxidative C–C bond-forming reaction through an intramolecular CDC reaction by means of the photooxidative generation of a *C*-centered radical (Scheme 31) [94]. The reaction of indole with a pendant malonate attachment in the presence of calcium iodide under visible-light irradiation gave the corresponding product in excellent yield with good selectivity. Moreover, the developed reaction showed good functional group tolerance under the optimized conditions. In the mechanistic investigations, the reaction of an iodinate starting material, which is assumed to be an intermediate for the reaction, in the presence of 20 mol% CaI_2 under air and fluorescent lamp irradiation gave the corresponding product in 81% yield. On the basis of the results, a plausible reaction mechanism for the photooxidative cyclization reaction was also proposed. We first envisioned that the reaction proceeded via a radical cyclization reaction, generated by the homolytic cleavage of the C–I bond mediated by visible-light irradiation (Scheme 31, route 1). A detailed mechanistic study revealed that another pathway may exist. In the presence of iodine with calcium hydroxide under dark conditions, the desired cyclization reaction also proceeded. Route 2 shown in Scheme 31 shows the ionic mechanism for our cyclization reaction. It is likely that the electrophilic activation of the indole mediated by molecular iodine took place, leading to a cyclization reaction from malonate to give the desired product (route 2). Furthermore, this is the first example of a metal-free direct photo-aerobic radical dehydrogenative cyclization reaction leading to the formation of polycyclic indoles.

The developed reaction was also applied for the intermolecular CDC reaction of thiophene with carbonyls (Scheme 32) [95]. The use of molecular iodine as a

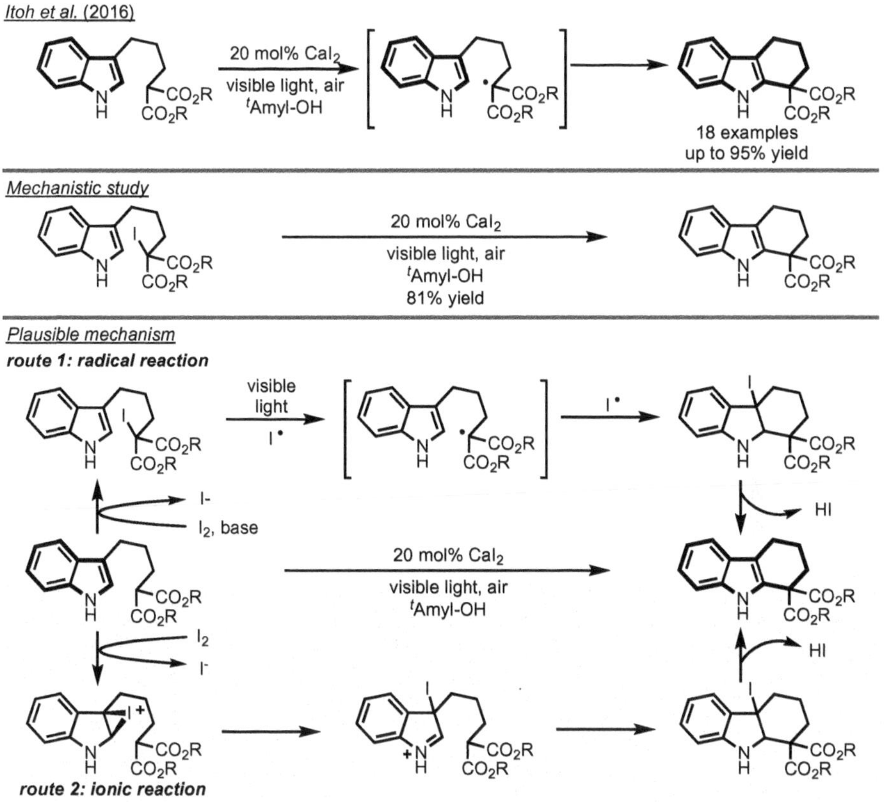

Scheme 31 Visible-light-driven intramolecular CDC-type reaction through iodine catalysis

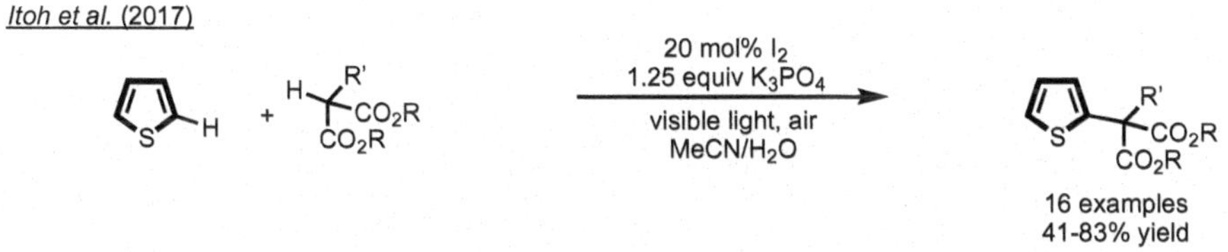

Scheme 32 Visible-light-driven intermolecular CDC-type reaction through iodine catalysis

catalyst in the presence of potassium phosphate under visible-light irradiation led to smooth C–C bond formation to give the product. Under the optimized conditions, the reaction demonstrated high efficiency, selectivity, and functional group tolerance. From mechanistic studies, the reaction mechanism was proposed to proceed via the same reaction pathway to form a *C*-centered radical as an intermediate, as discussed above (Scheme 31).

These reactions proceed through the activation of cheap and abundant molecular iodine and the formation of C–C bonds. In contrast to traditional methods that can be used to generate similar products, the advantages of this methodology are its mild reaction conditions, operational simplicity, and the low cost and abundance of the molecular iodine catalyst.

6 Conclusions and Outlook

In this chapter, representative examples for the CDC reaction via visible-light photocatalysis have been introduced. The CDC reaction is currently one of the most widely used methods to form carbon–carbon, carbon–nitrogen, and carbon–oxygen bonds in the synthesis of particularly complex molecules. As has been introduced, various photocatalytic CDC reactions have been reported. Furthermore, it has generally been found that the C–H bond can be selectively reacted, resulting in site-selective oxidation (particularly at the α-position of amines). On the other hand, only a few examples of asymmetric CDC reactions involving visible-light photocatalysis have been reported so far. It is easy to imagine that further developments in this area will be made through the development of a novel methodology that will enable asymmetric reactions to be realized. Using the protocol investigated in this chapter, the synthesis of complex molecules, including naturally occurring products and pharmaceuticals, will be the priority for the use of this reaction in the future.

References

1. Tsuji J, Takahashi H (1965) Organic syntheses by means of noble metal compounds. XII. Reaction of the cyclooctadiene-palladium chloride complex with ethyl malonate. J Am Chem Soc 87:3275–3276
2. Trost BM, Dietsche TJ (1973) New synthetic reactions. Asymmetric Induction in Allylic Alkylations 95:8200–8201
3. Mizoroki T, Mori K, Ozaki A (1971) Arylation of olefin with aryl iodide catalyzed by palladium. Bull Chem Soc Jpn 44:581–581
4. Miyaura N, Suzuki A (1979) Stereoselective synthesis of arylated (*E*)-alkenes by the reaction of alk-1-enylboranes with aryl halides in the presence of palladium catalyst. J Chem Soc Chem Commun 866–867
5. Negishi E, Baba S (1976) Novel stereoselective alkenyl–aryl coupling via nickel-catalysed reaction of alkenylanes with aryl halides. J Chem Soc Chem Commun 596–597
6. Miyaura N (2002) Cross-coupling reactions. A practical guide. Springer, Bellin
7. Nishihara Y (2013) Applied cross-coupling reactions. Lecture note in chemistry 80. Springer, Berlin
8. Yu JQ, Shi Z (2010) C-H activation in topics in current chemistry. Springer, Berlin
9. Ackermann L (2009) Modern arylation methods. Wiley, Chichester, UK
10. Dixneuf PH, Doucet H (2016) C-H bond activation and catalytic functionalization in topics in current chemistry. Springer, Berlin

11. Li CJ, Li Z (2006) Green chemistry: the development of cross-dehydrogenative coupling (CDC) for chemical synthesis. Pure Appl Chem 78:935–945
12. Li CJ (2009) Cross-dehydrogenative coupling (CDC): exploring C–C bond formations beyond functional group transformations. Acc Chem Res 42:335–344
13. Scheuermann CJ (2010) Beyond traditional cross couplings: the scope of the cross dehydrogenative coupling reaction. Chem Asian J 5:436–451
14. Shaw MH, Twilton J, Macmillan DWC (2016) Photoredox catalysis in organic chemistry. J Org Chem 81:6898–6926
15. Ghosh I, Marzo L, Das A, Shaikh R, König B (2016) Visible light mediated photoredox catalytic arylation reactions. Acc Chem Res 49:1566–1577
16. Gentry EC, Knowles RR (2016) Synthetic applications of proton-coupled electron transfer. Acc Chem Res 49:1546–1556
17. Liu Q, Wu LZ (2017) Recent advances in visible-light-driven organic reactions. Nat Sci Rev 4:359–380
18. Marzo L, Pagire SK, Reiser O, König B (2018) Visible-light photocatalysis: does it make a difference in organic synthesis? Angew Chem Int Ed 57:10034–10072
19. Buckles RE (1955) Illumination of cis- and trans-stilbenes in dilute solutions. J Am Chem Soc 77:1040–1041
20. Wood CS, Mallory FB (1964) Photochemistry of stilbenes. IV. The preparation of substituted phenanthrenes. J Org Chem 29:3373–3377
21. Mallory FB, Wood CS, Gordon JT (1964) Photochemistry of stilbenes. III. Some aspects of the mechanism of photocyclization to phenanthrenes. J Am Chem Soc 86:3094–3102
22. Carruthers W (1967) Photocyclization of some stilbene analogues. Synthesis of Dibenzo-[*a,l*] pyrene. J Chem Soc C 0:1525–1527
23. Sholz M, Mühlstädt M, Dietz F (1967) Chemie angeregter zustände. I. Mitt. Die richtung der photocyclisierung naphthalinsubstituierter äthylene. Tetrahedron 8:665–668
24. Dietz F, Scholz M (1968) Chemie angeregter zustände—IV: Die Photocyclisierung der Drei Isomeren Distyrybenzole. Tetrahedron 24:6845–6849
25. Laarhoven WH, Cuppen THJHM, Nivard RJF (1970) Photodehydrocyclizations in stilbene-like compounds II: photochemistry of distyrylbenzenes. Tetrahedron 26:1069–1083
26. Mallory FB, Mallory CW (1983) Photocyclization of stilbenes. VII. Unusual fluorine atom rearrangement in the photocyclization of 1-fluoro[5]helicenes. J Org Chem 48:526–532
27. Yamamoto K, Harada T, Okamoto Y, Chikamatsu H, Nakazaki M, Kai Y, Nakao T, Tanaka M, Harada S, Kasai N (1988) Synthesis and molecular structure of [7]circulene. J Am Chem Soc 110:3578–3584
28. Kalyanasundaram K (1982) Photophysics, photochemistry, and solar energy conversion with tris(bipyridyl)ruthenium(II) and its analogues. Coord Chem Rev 46:159–244
29. Juris A, Balzani V (1988) Ru(II) polypyridine complexes: photophysics, photochemistry, electrochemistry, and chemiluminescence. Coord Chem Rev 84:85–277
30. Prier CK, Rankic DA, Macmillan DWC (2013) Visible light photoredox catalysis with transition metal complexes: applications in organic synthesis. Chem Rev 113:5322–5363
31. Condie AG, González-Gómez JC, Stephenson CRJ (2010) Visible-light photoredox catalysis: Aza-Henry reactions via C–H functionalization. J Am Chem Soc 132:1464–1465
32. Girard SA, Knauber T, Li CJ (2014) The cross-dehydrogenative coupling of Csp3–H bonds: a versatile strategy for C–C bond formations. Angew Chem Int Ed 53:74–100
33. Lakshman M, Vuram PK (2017) Cross-dehydrogenative coupling and oxidative-amination reactions of ethers and alcohols with aromatics and heteroaromatics. Chem Sci 8:5845–5888
34. Nicholas AMP, Arnold DR (1982) Thermochemical parameters for organic radicals and radical ions. Part 1. The estimation of the pKa of radical cations based on thermochemical calculations. Can J Chem 60:2165–2179
35. Wayner DDM, Dannenberg JJ, Griller D (1986) Oxidation potentials of α-aminoalkyl radicals: bond dissociation energies for related radical cations. Chem Phys Lett 131:189–191

36. Rueping M, Vila C, Koenigs RM, Poscharny K, Fabry DC (2011) Dual catalysis: combining photoredox and Lewis base catalysis for direct Mannich reactions. Chem Commun 47:2360–2362
37. Zhao G, Yang C, Guo L, Sun H, Chen C, Xia W (2012) Visible light-induced oxidative coupling reaction: easy access to Mannich-type products. Chem Commun 48:2337–2339
38. Rueping M, Koenigs RM, Poscharny K, Fabry DC, Leonori D, Vila C (2012) Dual catalysis: combination of photocatalytic aerobic oxidation and metal catalyzed alkynylation reactions—C–C bond formation using visible light. Chem Eur J 18:5170–5174
39. Freeman DB, Furst L, Condie AG, Stephenson CRJ (2012) Functionally diverse nucleophilic trapping of iminium intermediates generated utilizing visible light. Org Lett 14:94–97
40. Xuan J, Cheng Y, An J, Lu LQ, Zhang XX, Xiao WJ (2011) Visible light-induced intramolecular cyclization reactions of diamines: a new strategy to construct tetrahydroimidazoles. Chem Commun 47:8337–8339
41. Gao XW, Meng QY, Xiang M, Chen B, Feng K, Tung CH, Wu LZ (2013) Combining visible light catalysis and transition metal catalysis for the alkylation of secondary amines. Adv Synth Catal 355:2158–2164
42. Dai C, Meschini F, Narayanam JMR, Stephenson CRJ (2012) Friedel—crafts amidoalkylation via thermolysis and oxidative photocatalysis. J Org Chem 77:4425–4431
43. Memming R (1969) Mechanism of the electrochemical reduction of persulfates and hydrogen peroxide. J Electrochem Soc 116:785–790
44. Zhang GY, Yu KX, Zhang C, Guan Z, He YH (2018) Oxidative cross-dehydrogenative-coupling reaction of 3,4-dihydro-1,4-benzoxazin-2-ones through visible-light photoredox catalysis. Eur J Org Chem 525–531
45. Beatty JW, Stephenson CRJ (2014) Synthesis of (−)-pseudotabersonine, (−)-pseudovincadifformine, and (+)-coronaridine enabled by photoredox catalysis in flow. J Am Chem Soc 136:10270–10273
46. Moudi M, Go R, Yien CYS, Nazre M (2013) Vinca alkaloids. Int J Prev Med 4:1231–1235
47. DiRocco DA, Rovis T (2012) Catalytic asymmetric α-acylation of tertiary amines mediated by a dual catalysis mode: N-heterocyclic carbene and photoredox catalysis. J Am Chem Soc 134:8094–8097
48. Larionov E, Mastandrea MM, Pericàs MA (2017) Asymmetric visible-light photoredox cross-dehydrogenative coupling of aldehydes with xanthenes. ACS Catal. 7:7008–7013
49. Flamigni L, Barbieri A, Sabatini C, Ventura B, Barigelletti F (2007) Photochemistry and photophysics of coordination compounds: iridium. Top Curr Chem 281:143–203
50. Ulbricht C, Beyer B, Friebe C, Winter A, Schubert US (2009) Recent developments in the application of phosphorescent iridium(III) complex systems. Adv Mater 21:4418–4441
51. Nagib DA, Scott ME, MacMillan DWC (2009) Enantioselective α-trifluoromethylation of aldehydes via photoredox organocatalysis. J Am Chem Soc 131:10875–10877
52. Zhu S, Rueping M (2012) Merging visible-light photoredox and Lewis acid catalysis for the functionalization and arylation of glycine derivatives and peptides. 48:11960–11962
53. Jin J, MacMillan DWC (2014) Direct α-arylation of ethers through the combination of photoredox-mediated C–H functionalization and the minisci reaction. Angew Chem Int Ed 54:1565–1569
54. Minisci F, Bernardi R, Bertini F, Galli R, Perchinummo M (1971) Nucleophilic character of alkyl radicals—VI: a new convenient selective alkylation of heteroaromatic bases. Tetrahedron 27:3575–3579
55. Minisci F, Vismara E, Fontana F, Morini G, Serravalle M, Giordano C (1987) Polar effects in free-radical reactions. Solvent and isotope effects and effects of base catalysis on the regio- and chemoselectivity of the substitution of protonated heteroaromatic bases by nucleophilic carbon-centered radicals. J Org Chem 52:730–736
56. Wu CJ, Meng QY, Lei T, Zhong JJ, Liu WQ, Zhao LM, Li ZJ, Chen B, Tung CH, Wu LZ (2016) An Oxidant-free strategy for indole synthesis via intramolecular C–C bond construction under visible light irradiation: cross-coupling hydrogen evolution reaction. ACS Catal. 6:4635–4639

57. Teegardin K, Day JI, Chan J, Weaver J (2016) Advances in photocatalysis: a microreview of visible light mediated ruthenium and iridium catalyzed organic transformations. Org Process Res Dev 20:1156–1163
58. Wakerley DW, Reisner E (2014) Development and understanding of cobaloxime activity through electrochemical molecular catalyst screening. Phys Chem Chem Phys 16:5739–5746
59. Rueping M, Zhu S, Koenings RM (2011) Photoredox catalyzed C–P bond forming reactions —visible light mediated oxidative phosphonylations of amines. Chem Commun 47:8769–8681
60. Yoo WJ, Kobayashi S (2014) Efficient visible light-mediated cross dehydrogenative coupling reactions of tertiary amines catalyzed by a polymer-immobilized iridium-based photocatalyst. Green Chem 16:2438–2442
61. Wu CJ, Zhong JJ, Meng QY, Lei T, Gao XW, Tung CH, Wu LZ (2015) Cobalt-catalyzed cross-dehydrogenative coupling reaction in water by visible light. Org Lett 17:884–887
62. Winter A, Newkome GR, Schubert US (2011) Catalytic applications of terpyridines and their transition metal complexes. ChemCatChem 3:1384–1406
63. Zhong JJ, Meng QY, Wang GX, Liu Q, Chen B, Feng K, Tung CH, Wu LZ (2013) A highly efficient and selective aerobic cross-dehydrogenative-coupling reaction photocatalyzed by a platinum(II) terpyridyl complex. Chem Eur J 19:6443–6450
64. Che CM, Kwok CC, Lai SW, Rausch AF, Finkenzeller WJ, Zhu N, Yersin H (2010) Photophysical properties and OLED applications of phosphorescent platinum(II) Schiff base complexes. Chem Eur J 16:233–247
65. Shelar DP, Li TT, Chen Y, Fu WF (2015) Platinum(II) Schiff Base complexes as photocatalysts for visible-light-induced cross-dehydrogenative coupling reactions. ChemPlusChem 80:1541–1546
66. Tsuji J (2005) Palladium reagents and catalysts: new perspectives for the 21st century. Wiley, Chichester, UK
67. Kurandina D, Parasram M, Gevergyan V (2017) Visible light-induced room-temperature heck reaction of functionalized alkyl halides with vinyl arenes/heteroarenes. Angew Chem Int Ed 56:14212–14216
68. Parasram M, Chuentragool P, Wang Y, Shi Y, Gevorgyan V (2017) General, auxiliary-enabled photoinduced Pd-catalyzed remote desaturation of aliphatic alcohols. J Am Chem Soc 139:14857–14860
69. Borisov SM, Lehner P, Klimant I (2011) Novel optical trace oxygen sensors based on platinum(II) and palladium(II) complexes with 5,10,15,20-*meso*-tetrakis-(2,3,4,5,6-pentafluorphenyl)-porphyrin covalently immobilized on silica-gel particles. Anal Chim Acta 690:108–115
70. To WP, Liu Y, Lau TC, Che CM (2013) A robust palladium(II)-porphyrin complex as catalyst for visible light—induced oxidative C–H functionalization. Chem Eur J 19:5654–5664
71. Hsu YC, Wang VCC, Au-Yeung KC, Tsai CY, Chang CC, Lin BC, Chan YT, Hsu CP, Yap GPA, Jurca T, Ong TG (2018) One-pot tandem photoredox and cross-coupling catalysis with a single palladium carbodicarbene complex. Angew Chem Int Ed 57:4622–4626
72. Hari DP, König B (2011) Eosin Y catalyzed visible light oxidative C–C and C–P bond formation. Org Lett 13:3852–3855
73. Liu Q, Li YN, Zhang HH, Chen B, Tung CH, Wu LZ (2012) Reactivity and mechanistic insight into visible-light-induced aerobic cross—dehydrogenative coupling reaction by organophotocatalysts. Chem Eur J 18:620–627
74. Pan Y, Kee CW, Chen L, Tan CH (2011) Dehydrogenative coupling reactions catalysed by rose bengal using visible light irradiation. Green Chem 13:2682–2685
75. Fu W, Guo W, Zou G, Xu C (2012) Selective trifluoromethylation and alkynylation of tetrahydroisoquinolines using visible light irradiation by Rose Bengal. J Fluorine Chem 140:88–94

76. Wang XZ, Meng QY, Zhong JJ, Gao XW, Lei T, Zhao LM, Li ZJ, Chen B, Tung CH, Wu LZ (2015) The singlet excited state of BODIPY promoted aerobic cross-dehydrogenative-coupling reactions under visible light. Chem Commun 51:11256–11259
77. Wang XF, Yu SS, Wang C, Xue D, Xiao J (2016) BODIPY catalyzed amide synthesis promoted by BHT and air under visible light. Org Biomol Chem 14:7028–7037
78. Zhang XF (2018) BODIPY photosensitizers based on PET and heavy atom effect: a comparative study on the efficient formation of excited triplet state and singlet oxygen in BODIPY dimers and monomers. J Photochem Photobiol A: Chem 355:431–443
79. Leow D (2014) Phenazinium salt-catalyzed aerobic oxidative amidation of aromatic aldehydes. Org Lett 16:5812–5815
80. Fukuzumi S, Ohkubo K (2014) Organic synthetic transformations using organic dyes as photoredox catalysts. Org Biomol Chem 12:6059–6071
81. Niu L, Yi H, Wang S, Liu T, Liu J, Lei A (2017) Photo-induced oxidant-free oxidative C-H/N–H cross-coupling between arenes and azoles. Nat Commun 8:14226–14232
82. Yamaguchi T, Nobuta T, Tada N, Miura T, Nakayama T, Uno B, Itoh A (2014) Aerobic Photooxidative carbon–carbon bond formation between tertiary amines and carbon nucleophiles using 2-chloroanthra-9,10-quinone. Synlett 25:1453–1457
83. Yamaguchi T, Yamaguchi E, Itoh A (2017) Cross-dehydrogenative C–H amination of indoles under aerobic photo-oxidative conditions. Org Lett 19:1282–1285
84. Yusubov MS, Zhdankin VV (2015) Iodine catalysis: a green alternative to transition metals in organic chemistry and technology. Resour-Efficient Technol 1:49–67
85. Yoshimura A, Zhdankin VV (2016) Advances in synthetic applications of hypervalent iodine compounds. Chem Rev 116:3328–3435
86. Breugst M, Heiden D (2018) Mechanism in iodine catalysis. Chem Eur J 24:9187–9199
87. Tada N, Shomura M, Cui L, Nobuta T, Miura T, Itoh A (2011) Catalytic oxidative cleavage of 1,3-diketones to carboxylic acids by aerobic photooxidation with iodine. Synlett 19:2896–2900
88. Tada N, Ishigami T, Cui L, Ban K, Miura T, Itoh A (2013) Calcium iodide catalyzed photooxidative oxylactonization of oxocarboxylic acids using molecular oxygen as terminal oxidant. Tetrahedron Lett 54:256–258
89. Nagasawa Y, Matsusaki Y, Hotta T, Nobuta T, Tada N, Miura T, Itoh A (2014) Aerobic photooxidative synthesis of benzimidazoles from aromatic aldehydes and diamines using catalytic amounts of magnesium iodide. Tetrahedron Lett 55:6543–6546
90. Nobuta T, Fujiya A, Yamaguchi T, Tada N, Miura T, Itoh A (2013) Molecular-iodine-catalyzed aerobic photooxidative C–C bond formation between tertiary amines and carbon nucleophiles. RSC Adv 3:10189–10192
91. Shirasaka T, Takuma Y, Shimpuku T, Imaki N (1990) Practical synthesis of 5,6,7,8-tetrahydro-4-methoxy-6-methyl—1,3-dioxolo[4,5-g]isoquinolin-5-ol. J Org Chem 55:3767–3771
92. Lengyel I, Epstein IR, Kustin K (1993) Kinetics of iodine hydrolysis. Inorg Chem 32:5880–5882
93. Kanai N, Nakayama H, Tada N, Itoh A (2010) Tandem oxidation/rearrangement of β-ketoesters to tartronic esters with molecular oxygen catalyzed by calcium iodide under visible light irradiation with fluorescent lamp. Org Lett 12:1948–1951
94. Yamaguchi E, Sudo Y, Tada N, Itoh A (2016) Rare metal-free photo-aerobic intramolecular dehydrogenative cyclization reaction towards polycyclic heteroarenes. Adv Synth Catal 358:3191–3195
95. Sudo Y, Yamaguchi E, Itoh A (2017) Cross-dehydrogenative C–H amination of indoles under aerobic photo-oxidative conditions. Org Lett 19:1282–1285

Application of Electrochemical Cross-Dehydrogenative Couplings in the Syntheses of Heterocycles

Amit R. Nath, Chin Fei Chee and Noorsaadah Abd Rahman

Abstract Heterocycles are becoming ubiquitous in the modern chemistry nowadays. However, the synthesis of heterocycles remains challenging. Issues like simplicity, selectivity, environmental friendliness, and availability of starting materials are still key factors to be considered before planning for a synthesis. In the past three decades, the transition-metal-catalyzed cross-coupling and direct oxidative cross-coupling reactions have been commonly utilized for the construction of heterocycles. Although these strategies are efficient, the use of organic halides and organometallic reagents resulted in generating undesirable chemical wastes. Recently, cross-dehydrogenative coupling (CDC) or cross-coupling with hydrogen evolution has been recognized as a powerful tool for the synthesis and functionalization of heterocycles. The benefits of using electrochemical cross-dehydrogenative coupling method over conventional cross-couplings are the avoidance of reagent waste and the mild reaction conditions. The method maximizes atom efficiency by replacing stoichiometric redox reagents with electrical current. Majority of the electrochemical cross-dehydrogenative coupling reactions can be carried out at an ambient temperature. It is chemoselective and tolerates with many functional groups. In this chapter, an overview of the electrochemical CDC strategies for carrying out carbon–carbon (C–C), carbon–nitrogen (C–N), carbon–oxygen (C–O) bond formations as well as carbon–hydrogen (C–H) functionalization is presented with examples of reaction and mechanistic insight. This chapter is organized based on the aforementioned reaction types, and majority of information retrieved between those of 2000 and 2018 will hopefully serve as a useful reference to readers.

A. R. Nath (✉) · C. F. Chee
Nanotechnology and Catalysis Research Centre, University of Malaya,
50603 Kuala Lumpur, Malaysia
e-mail: arnath@um.edu.my

C. F. Chee
e-mail: cheechinfei@um.edu.my

N. A. Rahman
Department of Chemistry, University of Malaya, 50603 Kuala Lumpur, Malaysia
e-mail: noorsaadah@um.edu.my

A. Srivastava and C. K. Jana (eds.), *Heterocycles via Cross Dehydrogenative Coupling*, https://doi.org/10.1007/978-981-13-9144-6_13

Keywords Electrochemical cross-dehydrogenative coupling · Heterocycles · Oxidative cross-coupling · Green synthesis

1 Introduction

Heterocycles are becoming ubiquitous in the modern chemistry nowadays. However, the synthesis of heterocycles remains challenging. Issues like simplicity, selectivity, environmental friendliness, and availability of starting materials are still key factors to be considered before planning for a synthesis. In the past three decades, the transition-metal-catalyzed cross-coupling reaction has been considered one of the most useful reactions for the construction of heterocycles. This reaction is still remained important for the construction of C–C, C–N, C–O, C–S, C–P, C–M bonds [1]. Many name reactions including Suzuki reaction, Heck reaction, Stille reaction, Negishi reaction, Kumada reaction, and Buchwald–Hartwig reaction have been developed through transition metal catalysis [2]. Generally, the reaction involves an organic electrophile (R^1–X) and an organometallic nucleophile (R^2–M) in the presence of a metal catalyst [M] (Scheme 1) [2]. The widely used metal catalysts are Ni and Pd and transition metal complexes from groups 8 to 10 [3]. These metal catalysts can undergo reduction–oxidation exchange easily in order to complete the catalytic cycle. They are less sensitive to oxygen and tend to react without the intervention of radical intermediates, which can eventually lead to side

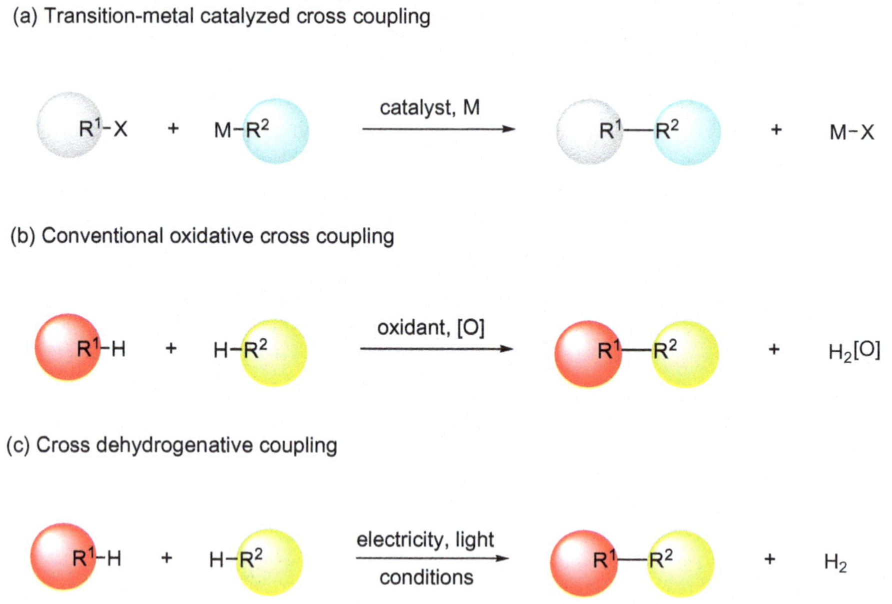

Scheme 1 General strategies for cross-coupling reactions

products such as those from homocoupling [3]. In spite of the fact that these cross-couplings are steadfast and proficient, the utilization of organic halides and organometallic reagents certainly forms unwanted by-products (Scheme 1a). In addition, these organometallic reagents and halides are not readily accessible and require being obtained through multi-step reactions [2].

Apart from transition-metal-catalyzed cross-couplings, direct oxidative cross-coupling has also been considered an efficient tool for the construction of C–C or C–X (X=N, O, S, P) bonds (Scheme 1b). In this reaction, stoichiometric amounts of oxidants, such as copper (II) and silver (I) salts, hypervalent metal salts, organic halides, pseudohalides (triflates, mesylates, and tosylates), and peroxides, are required to remove the hydrogen atoms in the bond formation processes. However, these chemical oxidants are expensive and toxic. The by-product resulted from this oxidative cross-coupling (i.e., $H_2[O]$) is unable to be reused and hence reduces the atom economy of the total transformation. The use of peroxides could also cause some safety issues in technical-scale operations [4].

Cross-dehydrogenative coupling (CDC) or cross-coupling with hydrogen evolution has been familiar as a powerful tool for the synthesis and functionalization of heterocycles [5, 6]. A huge variety of heterocycles can be synthesized from readily accessible reactants. In recent years, the development of direct cross-couplings using substrates without pre-functionalized C–H and X–H (X=N, O, S, P) bonds has received considerable attention [7, 2]. These new methods enable the C–C or C–X (X=N, O, S, P) bond formations via C–H or X–H bond activations without the necessary of coupling precursors C–X′ or C–M (X′ = halides or pseudohalides; M = metal) and oxidants. Formally, the CDC reaction occurs with a loss of an equivalent of hydrogen gas as the only by-product. The reaction is straightforward and uses readily available reagents. Therefore, direct cross-couplings are becoming highly attractive approaches in terms of efficiency, economy, and environmental impact [2].

In order to allow an oxidative R^1–H/R^2–H cross-coupling with hydrogen evolution, it is necessary to activate the C–H/X–H bond via electrochemical oxidation or photochemical oxidation [2].

In photochemical oxidation, a photocatalyst is used to absorb photonic energy. Upon absorption of photonic energy, the photocatalyst is excited to its triplet state, which can then engage in single-electron transfer (SET) events with organic substrates [8]. In electrochemical oxidation, electron transfer between the electrode and the substrate leads to the formation of a reactive intermediate, i.e., radical, radical cation, cation, or anion, depending on the electron configuration of the substrate and on the type of the redox processes (oxidation or reduction) (Scheme 2) [9].

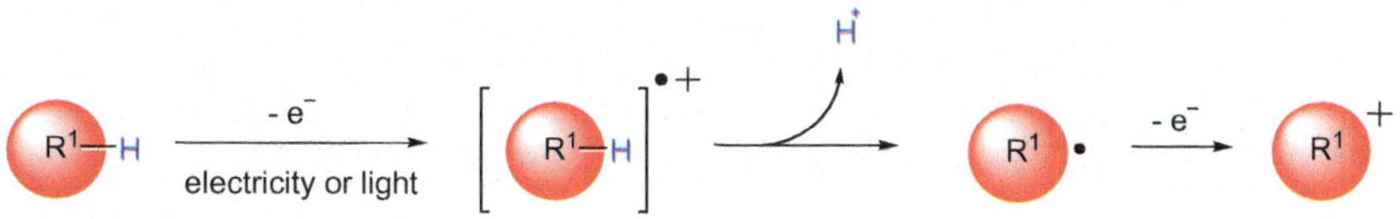

Scheme 2 Generation of radicals, radical cations, and cations

2 Electrochemical Cross-Dehydrogenative Coupling Reaction: Fundamental Principles

Electrolysis is a process in which electrical energy from a direct current (DC) supply produces a non-spontaneous chemical reaction in an electrochemical cell. Electrolysis is carried out in an electrochemical cell that comprised of an electroactive species (often referred to as substrate), electrolyte, solvent, and two electrodes (anode and cathode) or three electrodes (working, counter, and reference) (Fig. 1). The anode is connected to the positive pole of a power source (galvanostat or potentiostat), while the cathode is connected to the negative pole. In general, electrodes are made from inert materials that enable electron transfer in solution. Examples of these inert materials are graphite, glassy carbon, reticulated vitreous carbon (RVC), magnesium, platinum, and stainless steel. The term "electrolysis" is not often used in electrochemical oxidative cross-coupling reaction. It is often referred to as anodic oxidation or cathodic reduction. An anodic oxidation (working electrode) is balanced by cathodic reduction at the cathode. Electrons donated by the substrates at the anode move to the cathode and complete the circuit. The solvent molecules, protons, or other species are reduced at the cathode. In a cathodic reduction, a sacrificial anode made of oxidizable materials (e.g., Mg, Zn, or Fe) is used to prevent the undesired oxidation of reactive intermediates [10].

The electrochemical cells can be categorized into undivided cells and divided cells (Fig. 2). In undivided cells, the anode and cathode are placed in the same compartment, thus allowing for both reduction and oxidation to occur within the same compartment. The construction of undivided cells is easy. However, the desired product or starting material must be stable toward the electrodes [4]. In divided cells, the anode and cathode are physically separated through the use of a

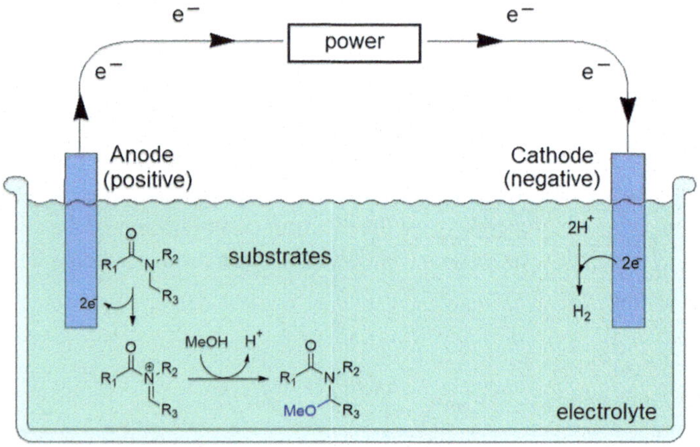

Fig. 1 Schematic diagram of a simple electrochemical cell with an illustration of an example of electrochemical cross-dehydrogenative coupling reaction

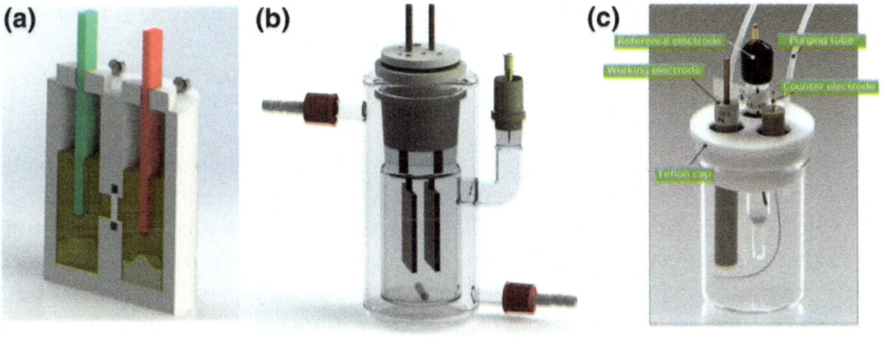

Fig. 2 **a** Cross sections of a divided cell for electrosynthesis (reproduced from Ref. [4] with permission from the American Chemical Society), **b** an undivided two-electrode cells (reproduced from Ref. [116] with permission from the Royal Society of Chemistry), and **c** an undivided three-electrode cell (reproduced from with permission from ALS Co. Ltd, Japan)

porous frit that is made of porous glass, ceramic, polymers, etc. This setup allows for the transfer of ions and enables the two half-reactions of the electrolysis to occur separately [11].

Most of the electrochemical cross-coupling reactions rely on fluorinated solvents as reaction media [12]. In particular, 1,1,1,3,3,3-hexafluoroisopropanol (HFIP) seems to be ideal and the major source of selectivity by a solvent effect [12]. When choosing a solvent, the solubility of the substrates, products, and electrolyte has to be considered. Solvents that can also be utilized in electrochemical processes include acetonitrile, dichloromethane, methanol, dimethylformamide, trifluoroethanol, trifluoroacetic acid, formic acid, water, ionic liquids, and supercritical fluids. However, formic acid is not sufficiently stable under electrolysis conditions [13]. Typically, a supporting electrolyte is added to increase the conductivity of the reaction solution. Examples of the supporting electrolytes are alkali metal perchlorates and tetraalkyl ammonium salts [11]. These electrolytes can be recycled upon recovery [14]. The chosen solvent and the supporting electrolyte should be inert toward the substrates and products [9].

In direct electrolysis, substrate molecules undergo electron transfers with the electrode surface and such electron transfers are called heterogeneous process [4]. Subsequently, a further electron transfer via other intermediates or by contact with the electrode can occur to generate a neutral or charged non-radical derivative [4]. The electron transfer with the electrode can also occur indirectly via a mediator (redox catalyst) that forms a stabilized intermediate species. This reactive species can then oxidize or reduce a substrate molecule homogeneously in an indirect electrolytic process [10]. Examples of these mediators are triarylamines, nitroxyl radicals, and ionic halides [10].

The potential of the electrode (E)—that is the difference between the potential at the electrode of interest and the selected reference electrode, such as the saturated calomel electrode (SCE)—will determine whether a specific electron transfer process is thermodynamically feasible, and is given by Eq. (1):

$$\Delta G = -nFE \tag{1}$$

where ΔG is the free energy change, F is Faraday's constant (96,485 Coulombs mol^{-1}), and n is the number of electrons involved in the overall reaction [11].

Typically, a cyclic voltammetry (CV) analysis is performed before an electrolysis experiment. Results from this analysis can provide information like which functional groups may be first oxidized or reduced, as well as the magnitude of the free energy that is needed to promote electron transfer in the studied electrochemical process [11, 15]. A high-throughput screening (HTS) system is used to obtain the optimum reaction conditions of an electrolysis experiment [4].

In general, the electrolysis experiments are conducted using either a controlled current or controlled potential operational modes. In a controlled current experiment, the current is held constant while the voltage gradually increases until the potential is reached for the electroactive species [11]. Electrolysis at controlled potential leads to selective conversion as a result of the applied potential at the working electrode since it is chosen to match that of the substrate. Unfortunately, this requires a three-electrode arrangement [4]. Substrate concentration is also important in an electrolysis experiment. A decrease in substrate concentration is accompanied by a decrease in the current and thus the rate of the reaction [11].

3 Recent Development of Electrochemical Cross-Coupling Reactions

Although the application of electric current to drive non-spontaneous organic reactions had begun 185 years ago, [16] the actual beginning of electrochemical oxidative cross-coupling reaction is considered to be the year 1848 when Kolbe reported the first anodic oxidation [17, 18]. In his pioneering work, carboxylate anions are oxidized at an anode to generate carbon-centered free radicals, which subsequently undergo homocoupling to produce the new C–C bond. Historical development of electrochemical cross-coupling reactions is shown in Fig. 3. Though reports on electro-transformations of the organic molecule were scarce during 1840–1950. However, Hickling's invention of the potentiostat in 1942 has opened a new dimension in electrosynthesis.

In 1958, Corey and co-workers used electrolysis as a key step to synthesize natural products pentacyclosqualene and β-onoceradiene [19]. Additional advances in preparative electrolysis include the use of indirect electrolysis in which a redox mediator undergoes electron transfer at an electrode to afford an electrochemically generated reagent that triggers the reaction of interest [11]. The use of inorganic mediators was found as early as 1900 when chromium salts were harnessed to facilitate the anodic synthesis of quinones [20]. Organic redox mediators such as triarylamines

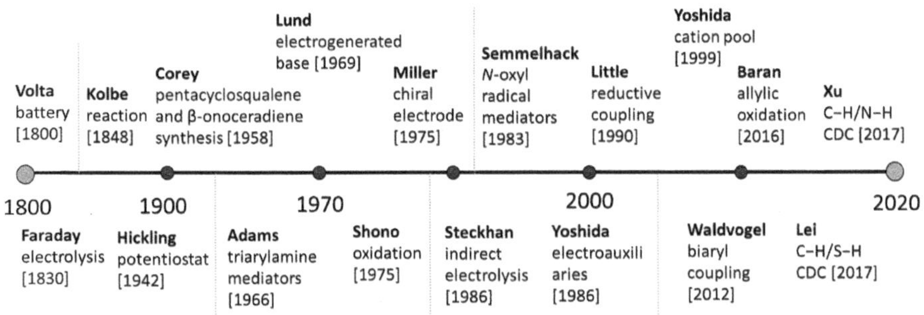

Fig. 3 Historical development of electrochemical cross-coupling reactions

and nitroxyl radicals were popularized after the 1970s [10]. In 1975, Miller invented the chiral electrode for asymmetric synthesis [21]. In the same year, Shono et al. developed the Shono oxidation of carbamates [22]. In the late 1980s, Yoshida introduced the concept of using electroauxiliaries in CDC reaction. In this concept, sulfur- and silicon-containing functional groups are incorporated into substrate molecules to lower their electrochemical potentials [23]. This facilitates either oxidation or reduction of the starting material in the presence of the desired nucleophile and reduces the likelihood of overoxidation or overreduction of the resultant product by promoting electron transfer in a more selective and predictable fashion [11]. Important contributions during the twenty-first century include Waldvogel's work on selective biaryl cross-coupling reactions [13], Yoshida's "the pool" method [24], Little's cathodic reduction for ring-forming reactions [25, 26], Baran's allylic oxidation [14], Xu's work on the syntheses of *N*-heterocycles [27, 28], Lei's dehydrogenative C–H/S–H cross-coupling [29], and among others.

4 Advantages of Electrochemical Cross-Dehydrogenative Coupling

The benefits of using electrochemical oxidative cross-coupling method over conventional cross-couplings are the avoidance of reagent waste and the mild reaction conditions. The use of electrochemistry in the context of CDC is considered a "green" and sustainable alternative to transition-metal-catalyzed and redox-based methods [11]. The method maximizes atom efficiency by replacing stoichiometric redox reagents with electrical current. Majority of the electrochemical oxidative cross-coupling reactions can be carried out at an ambient temperature. It is chemoselective and tolerates many functional groups [30]. In the industrial context, electrochemistry offers a way to design economic, scalable, and inherently safe process that can easily be switched off [31]. Therefore, electrochemical oxidative cross-coupling constitutes an appealing strategy for the construction of heterocycles under mild and environmentally friendly conditions.

5 Scope of this Chapter

5.1 Carbon–Carbon (C–C) Bond Formation

A number of review articles have been recently published that summarize the impressive advances made in the electrochemical oxidative cross-couplings. In this chapter, an overview of the electrochemical CDC strategies for carrying out carbon–carbon (C–C), carbon–nitrogen (C–N), carbon–oxygen (C–O) bond formations as well as carbon–hydrogen (C–H) functionalization is presented with examples of reaction and mechanistic insight. This chapter is organized based on the aforementioned reaction types and majority of information retrieved between those of 2000 and 2018.

5.1.1 Intermolecular C–C Bond Formation

Most of the cross-dehydrogenative coupling (CDC) reactions inevitably use stoichiometric oxidants, e.g., DDQ, TBHP, and IBX, providing undesired wastes. For example, enantioselective CDC reactions with carbonyl molecules enabled by the visible-light-promoted hydrogen transfer method also need stoichiometric amounts of nitrobenzenes as oxidants [32]. Previously, Li disclosed that electrochemical oxidation was a viable approach in CDC reactions [33]. Luo and co-workers envisioned to utilize chiral primary amine catalysts with electrochemical C–H oxidation in order to remove the use of any chemical oxidants. Aminocatalysis is found to be successful in enabling the stereoselective C–H functionalizations [34–36]. They developed a protocol based on enamine catalysis and electrochemical C–H oxidation providing stereoselective electro-coupling of simple ketones with tertiary amines by the combined use of anodic C–H oxidation and chiral primary amine catalysis. The authors identified the chiral primary amine as the optimal catalyst (derived from chiral trans-*N*, *N*-diaminocyclohexanes) and trifluoroethanol (CF_3CH_2OH) as the ideal additive. The established oxidant-free coupling reaction tolerated an array of *N*-aryl substituted tetrahydroisoquinolines and various simple ketones to provide the asymmetric oxidative coupling products in reasonable yields with good to excellent diastereo- and enantioselectivities (Scheme 3) [37].

The oxidative deprotonation of organic compounds to generate carbocations followed by their reactions with carbon nucleophiles is one of the most straightforward approaches to C–C bond formation. Generally, this type of transformation is difficult to achieve due to high reactivity of carbocations. In order to skip this problem, a two-step transformation procedure was used (shown in Scheme 4) to achieve an oxidative C–C bond formation [38–40].

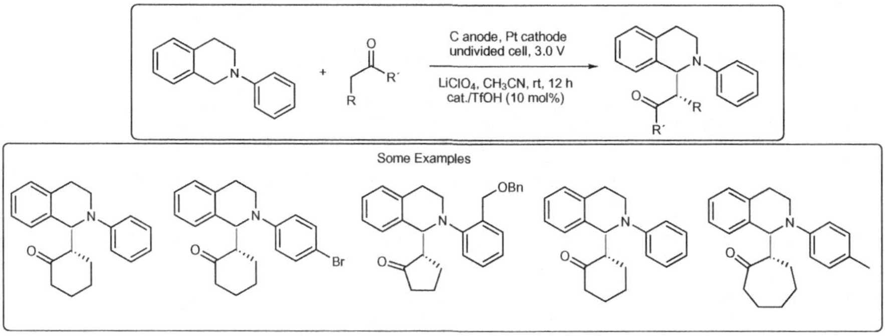

Scheme 3 Electrochemical CDC of tertiary amines with simple ketones

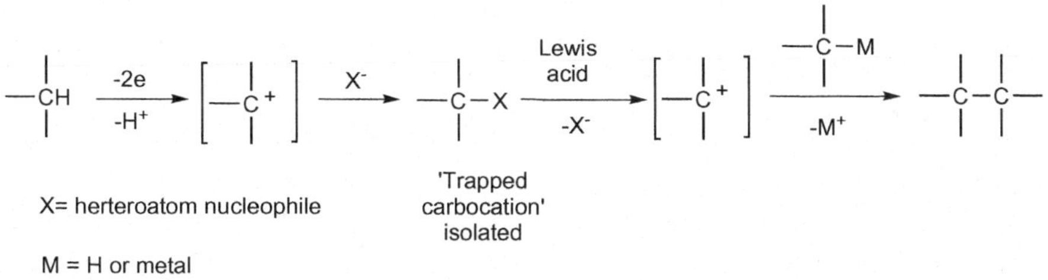

Scheme 4 Two-step procedure for carbon–carbon bond formation involving temporary trapping of an oxidatively generated carbocation

Scheme 5 Oxidative generation of a carbocation and its reaction with a carbon nucleophile in situ

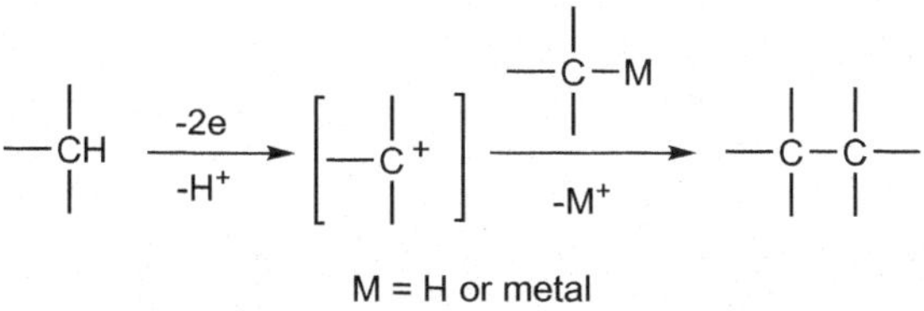

First, a carbocation is electrochemically produced in the presence of an excess quantity of a heteroatom nucleophile (X^{-}) of high oxidation potential (e.g., methanol) [38–40]. The nucleophile served as an agent to trap the carbocation intermediate. The resulting "trapped carbocation" is isolated and then used for the regeneration of the carbocation under non-oxidative conditions. When the "trapped carbocation" is treated with Lewis acid, the carbocation is generated and reacted with carbon nucleophiles to obtain desired C–C bond formation product [38–40]. However, this two-step technique is time consuming in comparison with the direct reaction procedure shown in Scheme 5.

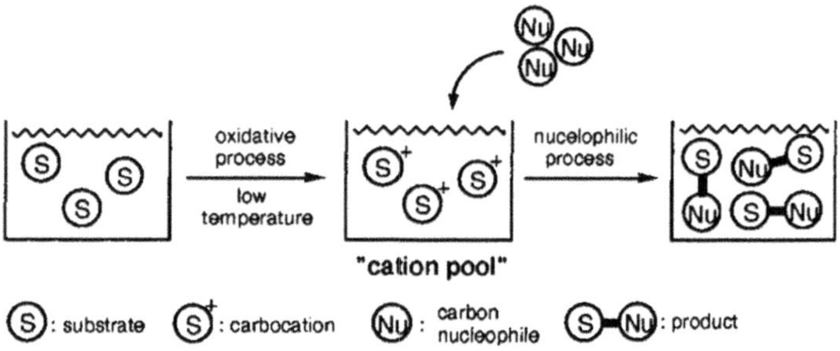

Fig. 4 "Cation pool" method for the oxidative C–C bond formation (reproduced from Yoshida et al. [41] with permission from the American Chemical Society)

To overcome this tedious process, a "cation pool" method can be used. The "cation pool" method involves generation and accumulation of cations, such as *N*-acyliminium ions, through electrolysis at low temperatures. The nucleophile is subsequently introduced to the reaction mixture under non-oxidative conditions, which allows for easily oxidized nucleophiles to be used as shown in Fig. 4.

Early work from Yoshida and co-workers involving the production of "cation pools" from carbamates using low-temperature electrolysis enabled the use of a collection of carbon nucleophiles, including allyl silanes, enol silyl ethers, enol acetates as well as aromatic and 1,3-dicarbonyl compounds (Scheme 6) [41]. The electrochemical oxidation of *N*-(methoxycarbonyl)pyrrolidine as carbamate precursor was carried out in CH_2Cl_2 using a graphite felt anode at −72 °C in the absence of allyl silanes. At this temperature, the iminium cation was collected without any decomposition. After the electrolysis was completed, allyl silanes (2 equiv.) were added to the newly generated "iminium cation pool" containing iminium cation to access the expected carbon–carbon bond formation.

5.1.2 Intramolecular C–C Bond Formation

Intramolecular annulation is one of the most straightforward methods to produce five-membered heterocycles. One typical method is 1,3-dipole cycloaddition, which usually requires the use of nitrones, azomethine ylides, azomethine imine, and azides as functionalized substrates [42, 43]. Lei and co-workers developed iodine-catalyzed oxidative annulation for the synthesis of heterocycles [44, 118, 45]. Since iodide salts have been reported to act as redox mediators under electrochemical conditions, they studied iodine-promoted oxidative annulation reactions under electrochemical conditions and applied to the synthesis of indoles (Scheme 7) [46]. In their method, platinum plate was used as anode and cathode material in the electrolysis. KI was utilized as the supporting electrolyte, and the iodide acted as a redox mediator for the substrate's oxidation. Amidyl radical was

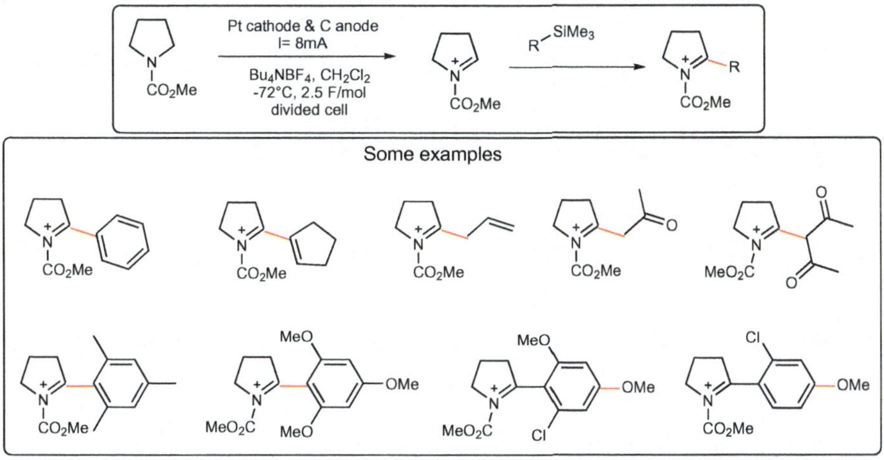

Scheme 6 C–C bond formation of carbamates using cation pool method

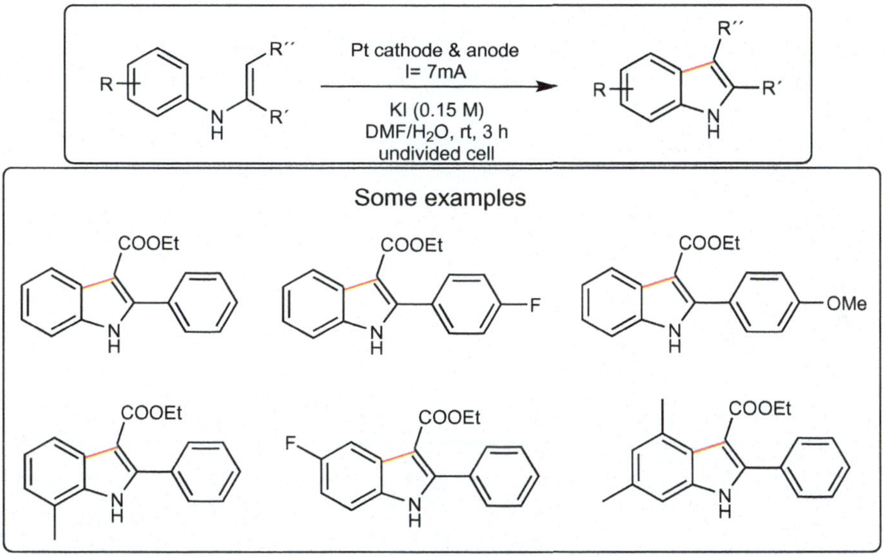

Scheme 7 Electrochemical intramolecular dehydrogenative C–C bond formation of *N*-aryl enamines mediated by iodides

anticipated to be produced from the oxidation of the substrate by the I^+ generated in situ.

Xu group reported an electrochemical CDC reaction utilizing a redox catalyst Cp_2Fe for the preparation of functionalized 3-fluorooxindoles (Scheme 8) [47]. Cp_2Fe is a low-cost organometallic catalyst for the electrocyclization of the

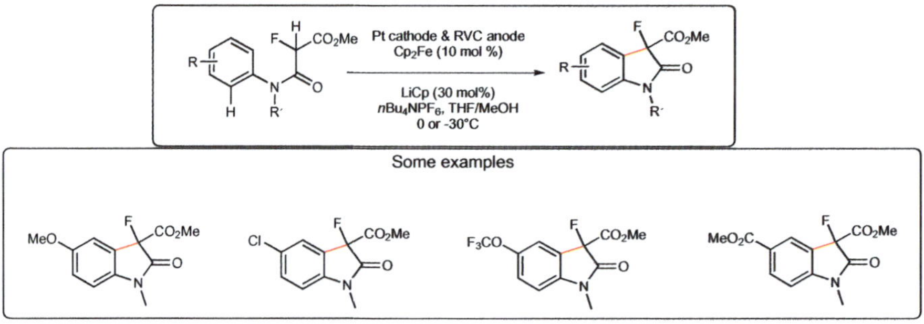

Scheme 8 Electrochemical intramolecular dehydrogenative C–C bond formation of malonate amide mediated by Cp_2Fe

malonate amides. It is also an oxidative radical reaction promotor [48, 49]. In order to efficiently generate the base and heat-sensitive fluorinated oxindoles, the oxidant and base are required to form in situ at low temperature. Monofluoroalkyl radical was anticipated to be formed from the electron transfer between deprotonated malonate amide and Cp_2Fe^+ [47].

Another example of C–C intramolecular cyclization can be found in Moeller's synthesis of alliacol A. In this example, a tandem anodic coupling–Friedel–Crafts alkylation strategy is used (Scheme 9). In the synthesis, the absolute stereochemistry of the final natural product is set relative to a methyl group that is incorporated early in the sequence using an asymmetric Michael reaction [50, 51].

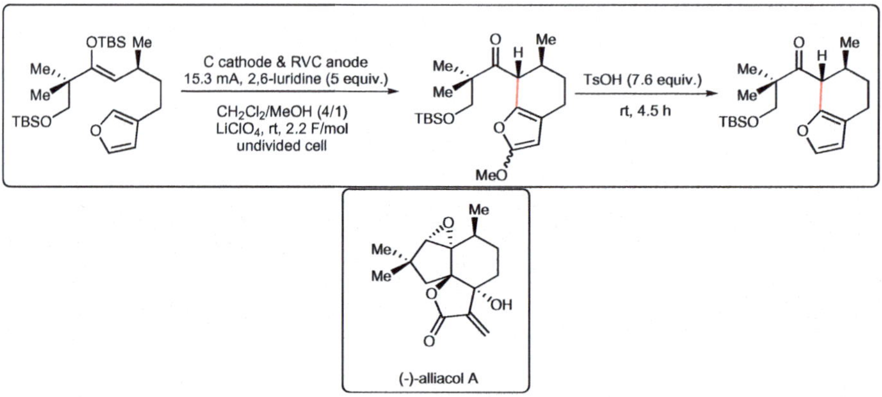

Scheme 9 Synthesis of alliacol A via anodic coupling reaction

5.2 *Carbon–Nitrogen (C–N) Bond Formation*

5.2.1 Intermolecular C–N Bond Formation

Carbon–nitrogen bond formation is important in synthetic organic chemistry, due to the high prevalence of the nitrogen-containing heterocycles in nature [52–54]. The electrochemical amination is straightforward and atom economy due to the use of nitrogen sources with N–H bonds to form C–N bond. The reaction does not require the need of starting materials with specific reactive groups [55].

It has been revealed that an amide or carbamate is oxidized on the anode to generate an iminium cation, which is trapped in situ by a nucleophile [56–58]. However, only compounds with high oxidation potentials, such as methanol and cyanide ions, have been employed as nucleophiles. Although amines have lower oxidation potentials, which should facilitate the electrochemical oxidation process, protocols utilizing amines have been employed less frequently compared to their amide and carbamate counterparts. Huang and Gong recently disclosed a method for the electrochemical oxidation of γ-lactams. Direct trapping of the generated electrophilic acyliminium ions with anilines could be accomplished despite their low oxidation potential (Scheme 10) [59].

This type of oxidation of aliphatic amines can also be extended to the synthesis of sulfonyl amidines [60]. Different tertiary amines were initially evaluated and were shown to efficiently afford the sulfonyl amidine derivatives upon reaction with sulfonyl azides (Scheme 11).

Employing Et_3N as the amine, these transformations presumably include anodic oxidation of the aliphatic amine to produce an iminium intermediate, which tautomerizes into the corresponding enamine (Scheme 12). Subsequent 1,3-dipolar cycloaddition between the generated enamine and tosyl azide furnishes

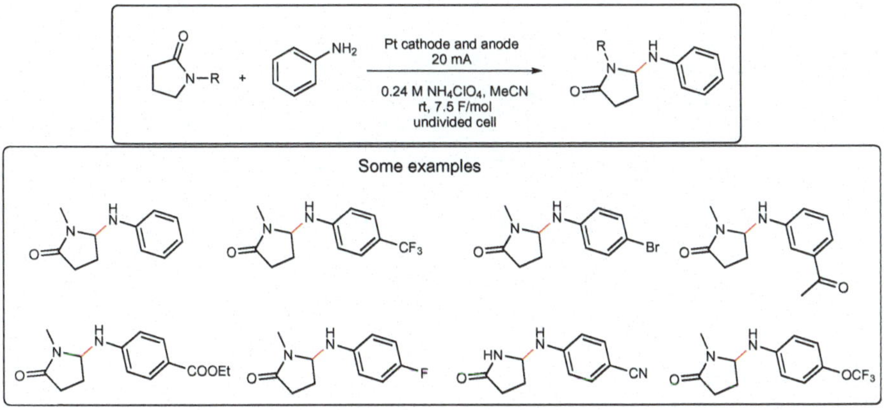

Scheme 10 Electrochemical oxidative C–H/N–H coupling between γ-lactams and anilines

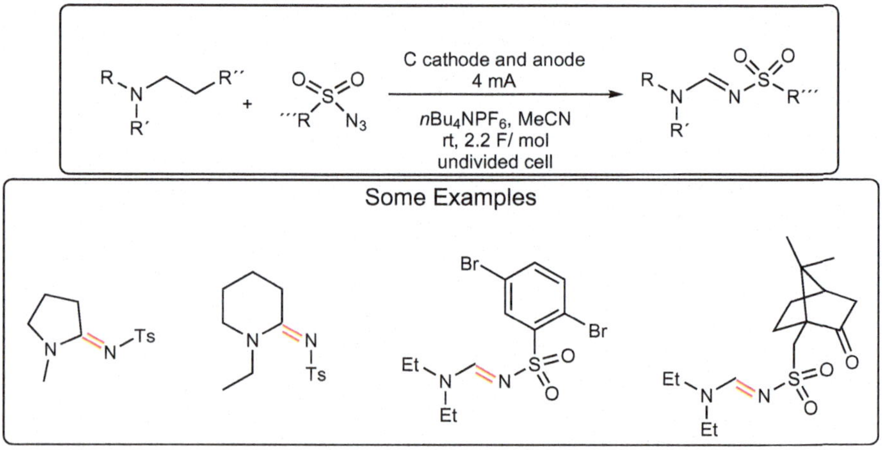

Scheme 11 Electrochemical synthesis of sulfonyl amidines from aliphatic amines and sulfonyl azides

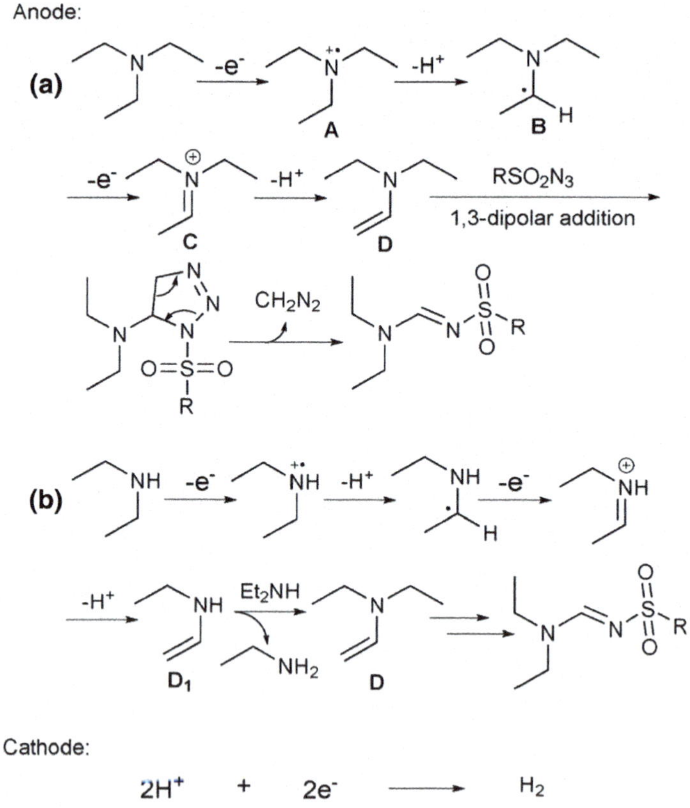

Scheme 12 Proposed mechanism for the imidation of **a** triethylamine and **b** diethylamine (reproduced from Zhang et al. [60] with permission of Royal Society of Chemistry)

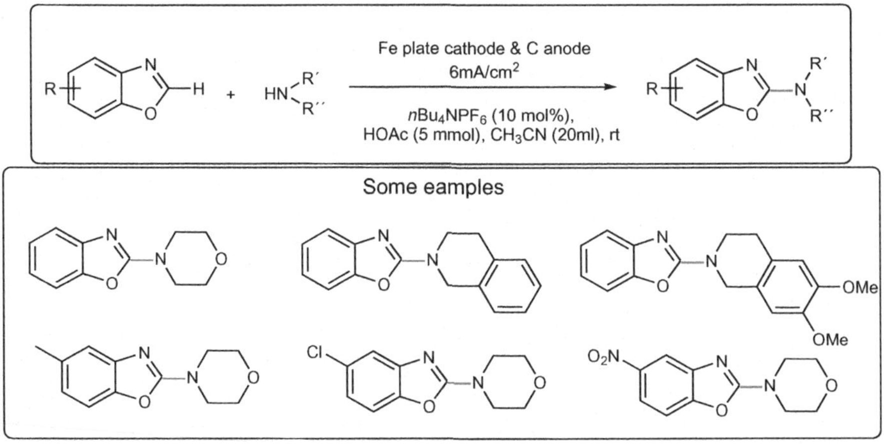

Scheme 13 Electrochemical cross-coupling between benzoxazoles and secondary amine mediated by iodides

cycloadduct. Finally, extrusion of CH_2N_2 delivers the desired amidine product. The synthesis of amidine derivatives was also accomplished using secondary and primary aliphatic amines.

In 2014, Zeng and his co-workers developed an electrochemical degenerative cross-coupling for C–N bond formation (Scheme 13) [61]. The reaction which utilized benzoxazoles and secondary amines as starting materials was carried out in an undivided cell equipped with carbon as anode and Fe as cathode. A series of 2-aminobenzoxazole adducts were synthesized in good yields using a redox catalyst tetraalkylammonium iodides. The benzoxazoles are hypothesized to undergo an acid-promoted nucleophilic addition reaction prior to the C–N bond formation. However, only secondary alkylamines are suitable in this electrochemical reaction [61].

Later, Zeng and his co-workers reported a NH_4I mediated electrochemical cross-coupling of ketones with secondary amines (Scheme 14) [62]. The reaction was carried out in an undivided cell equipped with graphite plates as electrode for both anode and cathode. In this reaction, iodine is derived from the anodic oxidation of iodide. Reaction of iodine with the ketone generated an α-iodoketone intermediate, which further reacted with secondary amine to furnish the α-aminated product. However, only secondary alkylamines are suitable in this reaction.

Recently, Ackermann and co-workers reported an electrochemical cross-coupling of benzoxazoles with secondary amine (Scheme 15) [63]. The reaction is accomplished in an undivided cell equipped with RVC as anode and Pt plate as cathode. Aryl C(sp2)–H and benzylic C(sp3)–H can also be aminated in this reaction.

In 2018, the same group reported the electrochemical amination using a cobalt catalyst (Scheme 16) [64]. An *N,O*-bidentate directing group is required for this

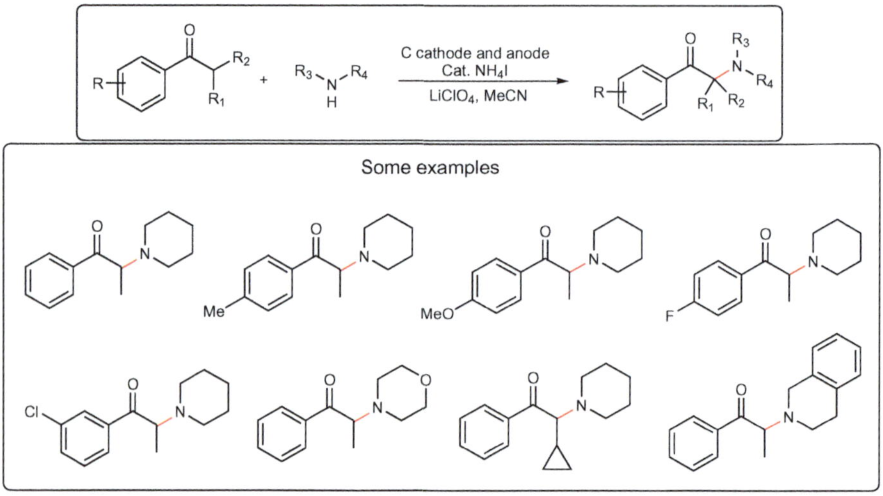

Scheme 14 Electrochemical cross-coupling of ketones with secondary amines mediated by NH_4I

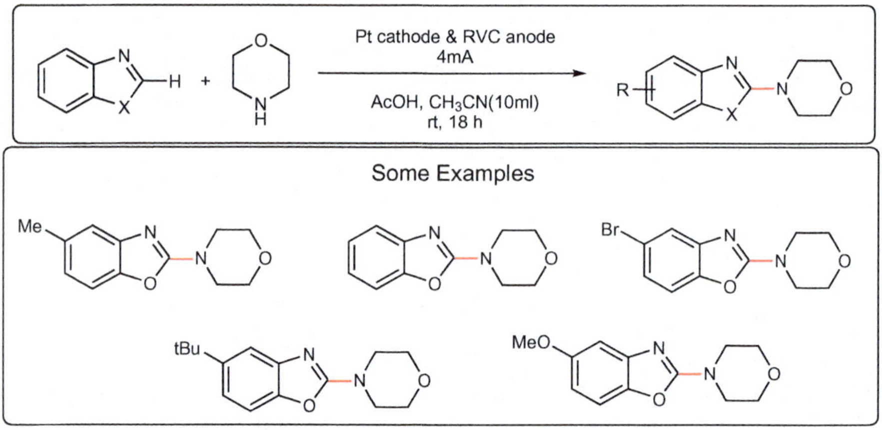

Scheme 15 Catalyst and chemical oxidant-free electrochemical cross-coupling of benzoxazole and secondary amine

CDC approach. However, not many nitrogen nucleophiles are suitable in this protocol. The Co catalyst is recycled by anodic oxidation, and hydrogen is evolved by the cathodic reduction.

In the same year, Lei group reported a similar directing group strategy in electrochemical CDC reaction (Scheme 17) [65]. In their method, *N*-(quinolin-8-yl) carboxamide was utilized as directing group. The use of undivided cell is not favorable due to the cathodic reduction of cobalt catalyst.

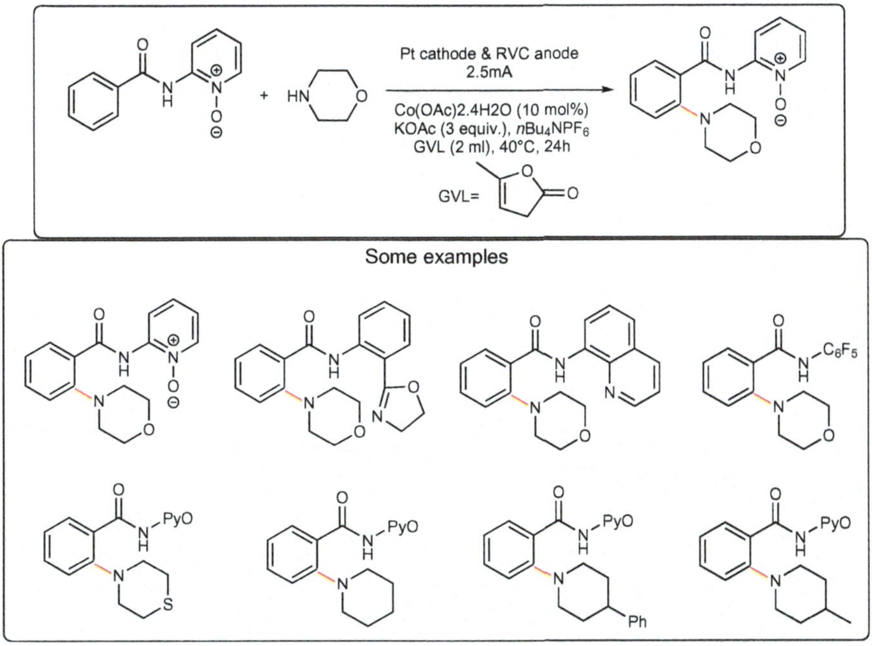

Scheme 16 Cobalt-catalyzed electrochemical C–H amination

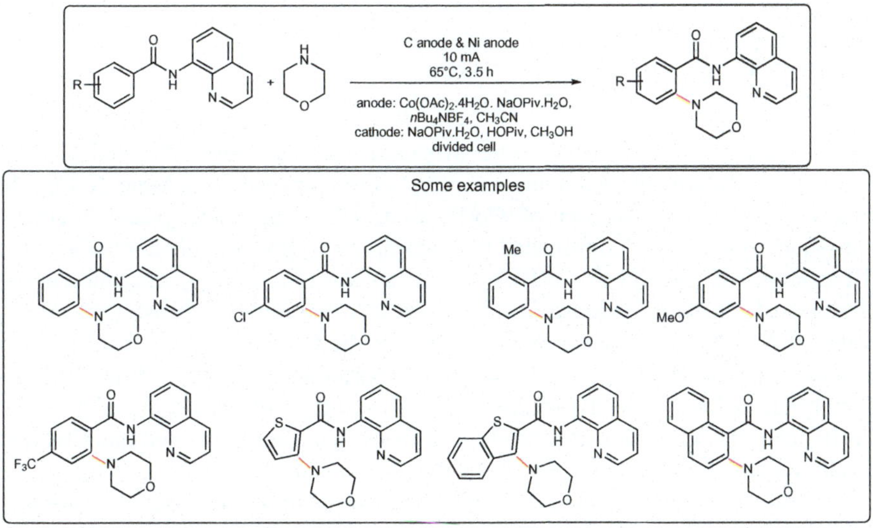

Scheme 17 Cobalt-catalyzed electrochemical amination between arene and morpholine [65]

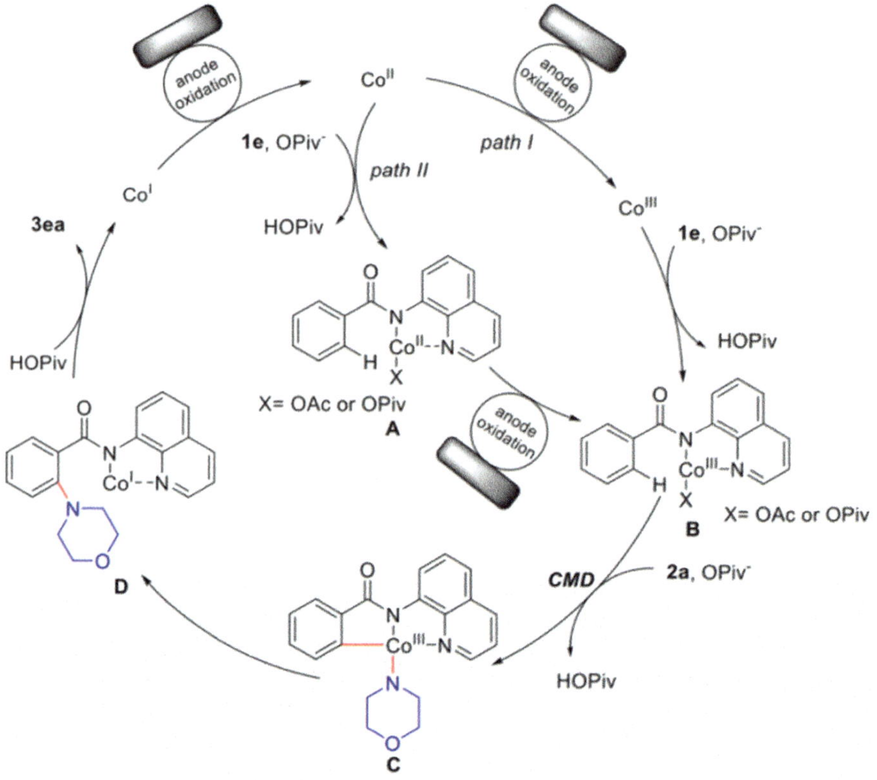

Scheme 18 Proposed mechanism for amination between arenes and morpholine (reproduced from Gao et al. [65] with permission of American Chemical Society)

Two plausible mechanisms were proposed for the cobalt-catalyzed electrochemical amination between arene and morpholine as shown in Scheme 18 [65]. In path I, oxidative conversion of Co(II) to Co(III) was taken place at anode. Subsequently Co(III) is coordinated to *N*-(quinolin-8-yl)benzamide to obtain an Co (III)-complex B. In path II, Co(II) is bonded to *N*-(quinolin-8-yl)benzamide in the presence of a base to obtain a Co(II)-complex A before electrolysis. This Co(II)-complex A is then oxidized at anode to provide the Co(III)-complex B. C–H bond was activated in the presence of a base and morpholine reacted with the Co(III)-complex B to form a Co(III)-complex C, followed by reductive elimination of the Co(III)-complex C to the expected product. Co(I) ions were converted into Co(II) at anodic oxidation and completed the catalytic cycles of the cobalt catalyst.

Lei and co-workers reported a direct electrolysis of acyclic diaryl amines without the necessary of external oxidant (Scheme 19) [66]. In this reaction, anodic oxidation of phenothiazine generated a radical cation and subsequently reacted with phenols to a range of coupling products.

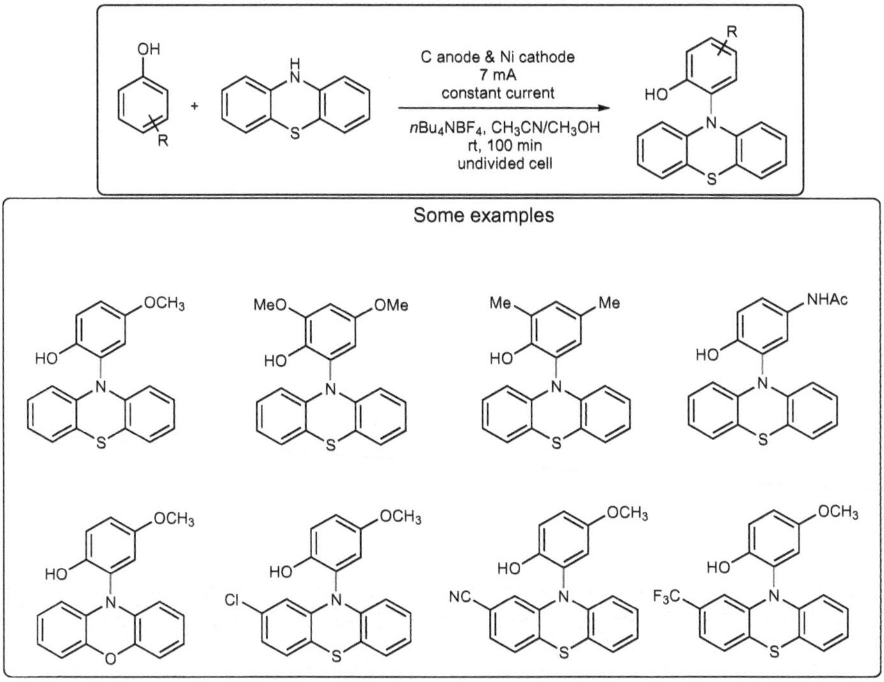

Scheme 19 Electrochemical amination between phenol and phenothiazine

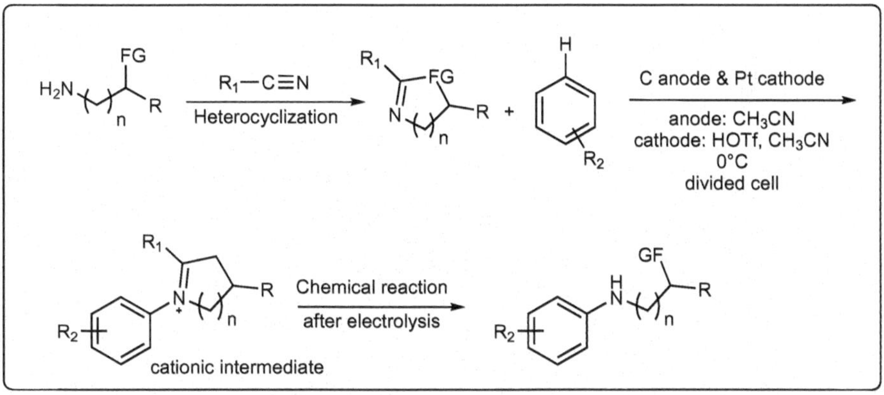

Scheme 20 Electrochemical oxidative amination between arenes and primary amines through a heterocyclization approach

In 2015, Yoshida demonstrated an electrochemical oxidative amination between aromatic compounds and primary alkylamines bearing a hydroxyl group or an amino group (Scheme 20) [67]. Prior to the electrochemical reaction, primary alkylamines were cyclized by treating with nitrile. The resulting nitrogen heterocycles and aromatic substrate were taken into an anodic chamber of a divided electrochemical cell to attain oxidative C–N cross-coupling reaction, yielding a cationic intermediate. The cationic intermediate is then chemically converted to the desired cross-coupling products under non-oxidative conditions [67]. This method avoids direct use of alkylamines in electrochemical CDC reaction because they are easily oxidized prior reaction to aromatic substrates. A variety of electro-rich and electro-neutral arenes can be used in this approach, and up to 93% yield of product can be obtained.

In 2013, Wang and his co-workers revealed an electrochemical approach to prepare α-ketoesters from acetophenones and amines to avoid the occurrence of haloform reaction (Scheme 21) [68]. The reaction was performed in an undivided cell equipped with Pt plates for both electrodes. Several aromatic and aliphatic amines along with some ammonium acetate are compatible with this reaction protocol and provide noble product yields.

A plausible mechanism for the electrochemical C–H/N–H cross-coupling mediated by nBu_4NI is illustrated in Scheme 22. It is hypothesized that anodic oxidation of iodide provides iodine that converts acetophenones to acetophenone radical. This radical readily reacts with oxygen to 2-oxo-2-phenylacetaldehyde. Subsequently, nucleophilic addition of an amine to 2-oxo-2-phenylacetaldehyde afforded the desired molecule. Alcohol can also be used in this reaction instead of amines [68].

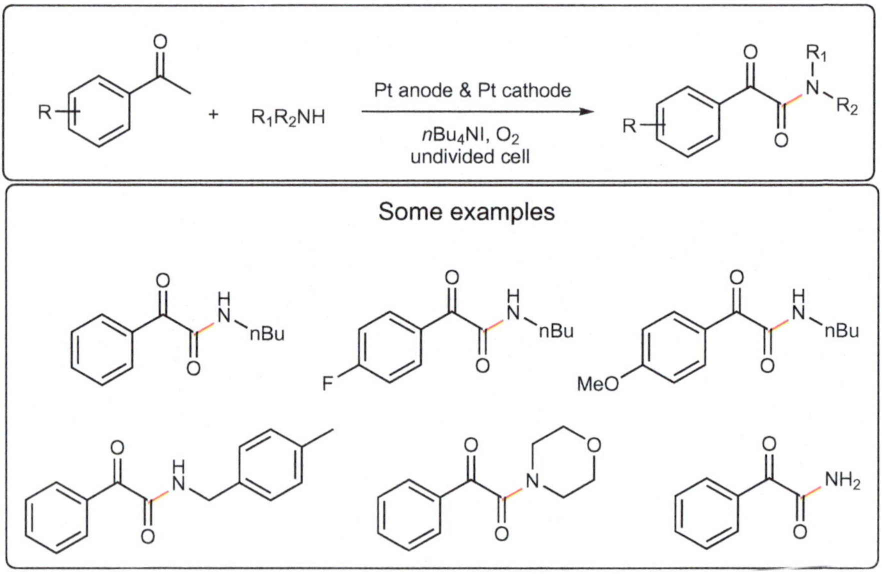

Scheme 21 Electrochemical C–H/N–H cross-coupling mediated by nBu_4NI

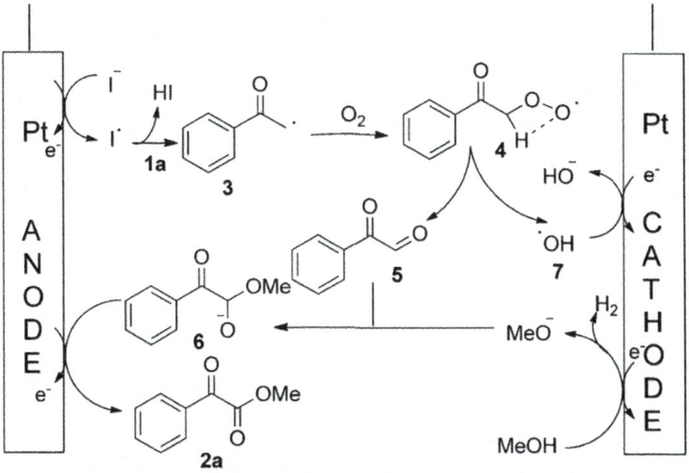

Scheme 22 Mechanism for the formation of α-ketoesters (reproduced from Zhang et al. [68] with permission of John Wiley and Sons)

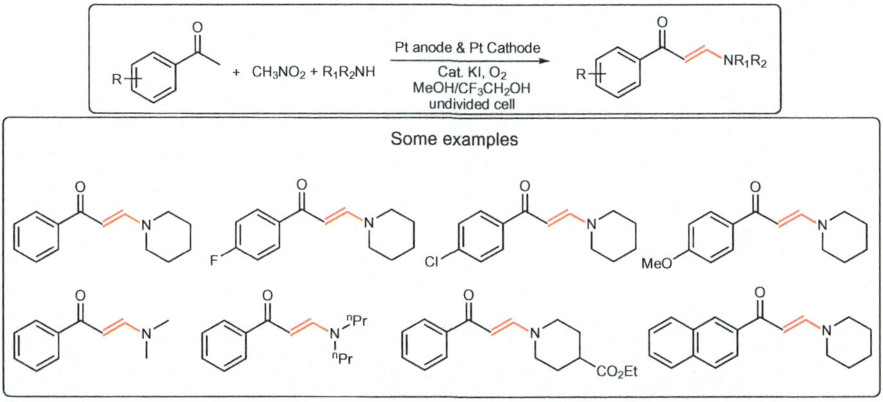

Scheme 23 Electrochemical synthesis of enaminones with nitromethane as a carbon source

In 2015, Wang revealed an efficient protocol for the preparation of enaminones using KI as a redox catalyst (Scheme 23) [69]. This electrochemical CDC method was found highly efficient in the synthesis of enaminones with the use of the diverse substrate.

According to the proposed mechanism shown in Scheme 24, anodic oxidation of tertiary amine provides an iminium ion, where tertiary amine might be formed by the reaction of secondary amine and α-iodoketone. The iminium ion may attack the nitromethane to form adduct which provides nitroalkene by the releasing piperidine molecule. Subsequently, nitroalkene undergoes 1,4-addition and β-H elimination to afford the desired products.

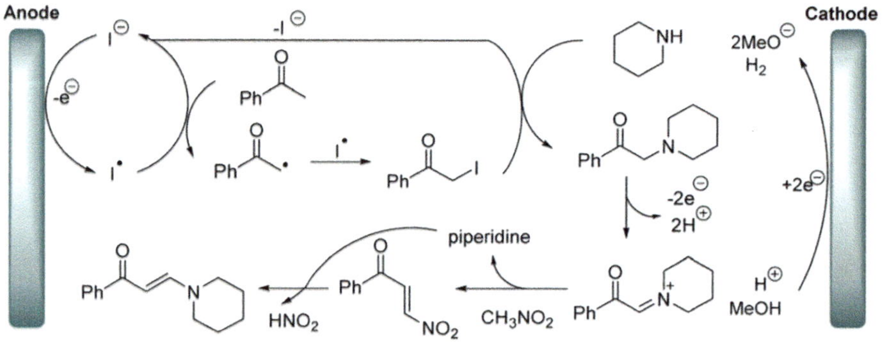

Scheme 24 Proposed mechanism for the synthesis of enaminones with nitromethane as a carbon source (reproduced from Xu et al. [69] with the permission of Royal Society of Chemistry)

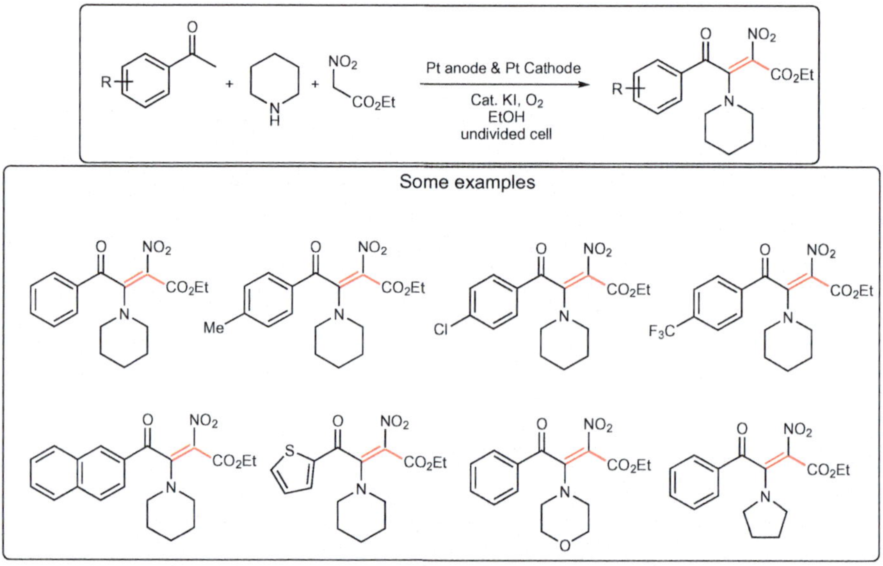

Scheme 25 Electrochemical synthesis of enaminones mediated by KI

Nitroacetate can be used as a substrate instead of nitromethane for the iodine-mediated synthesis of nitro-containing enaminones (Scheme 25) [70]. Unlikely the proposed mechanism in Scheme 24, proton was eliminated (instead of nitro group) in the final step to obtain the nitro-containing enaminones. The reaction was performed in an undivided cell equipped with Pt plate for both electrodes for optimum yields.

A recyclable redox mediator based on the iodine(I)/iodine(III) redox couple for C–N bond formation was reported by the Francke group in 2016 (Scheme 26) [71].

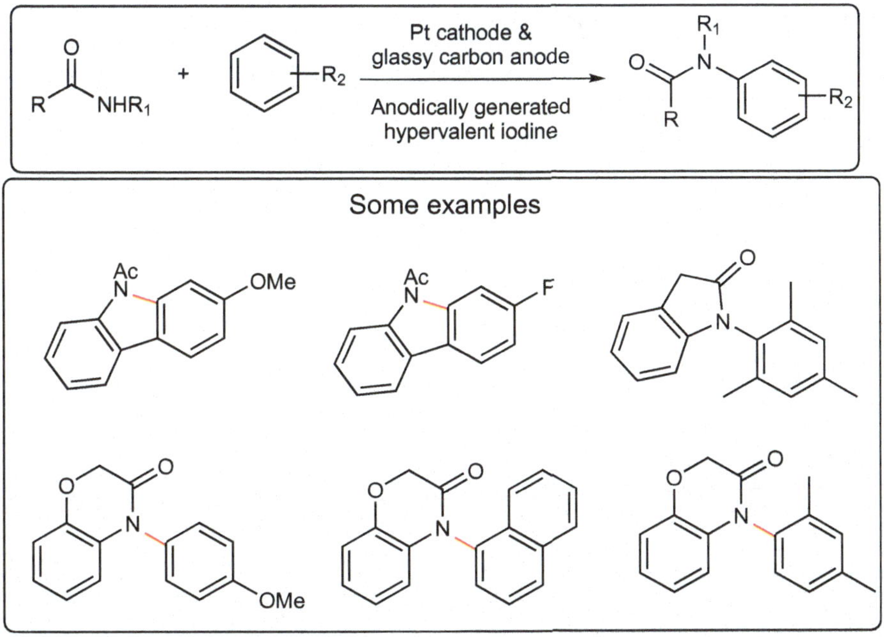

Scheme 26 Electrochemical dehydrogenative C–N bond formation

It was established that 1,1,1,3,3,3-hexafluoroisopropanol plays a vital role as a solvent in anodic oxidation for the formation of active iodine(III) intermediate and its successive chemical transformation. In addition, the supporting electrolyte can combine with redox mediator through redox-active iodophenyl moiety and an alkylammonium group that also allow simple recovery to reuse both components. Interestingly, both intermolecular and intramolecular C–N bond formations can be obtained with amides and simple arenes in this electrochemical technique.

Apart from the above achievements of C(sp2)–H direct amination, the strategy of C–H/N–H cross-coupling via a *N*-radical intermediate has also been applied to the amination of C(sp3)–H bonds. As a limited example, Lei and co-workers recently developed an electrooxidative method for the intermolecular C(sp3)–H/N–H cross-coupling under metal and oxidant-free conditions (Scheme 27) [72]. The reactive C(sp3)–H bonds of benzylic, allylic, and those adjacent to O, S, and N atoms were smoothly aminated in this protocol.

In 2002, Yudin described an electrochemical oxidation for olefin aziridination with *N*-aminophthalimide to replace the toxic metal oxidants (Scheme 28) [73]. A scope of a variety of substrate was measured for the alkenes in intermolecular electrochemical annulation reactions. The reactions were accomplished in a divided cell at a certain potential. Carboxylates were found to be vital intermediate for the completion of this conversion [74]. They utilized the stoichiometric amount of NEt_3 and HOAc into the anodic chamber instead of using supporting electrolytes.

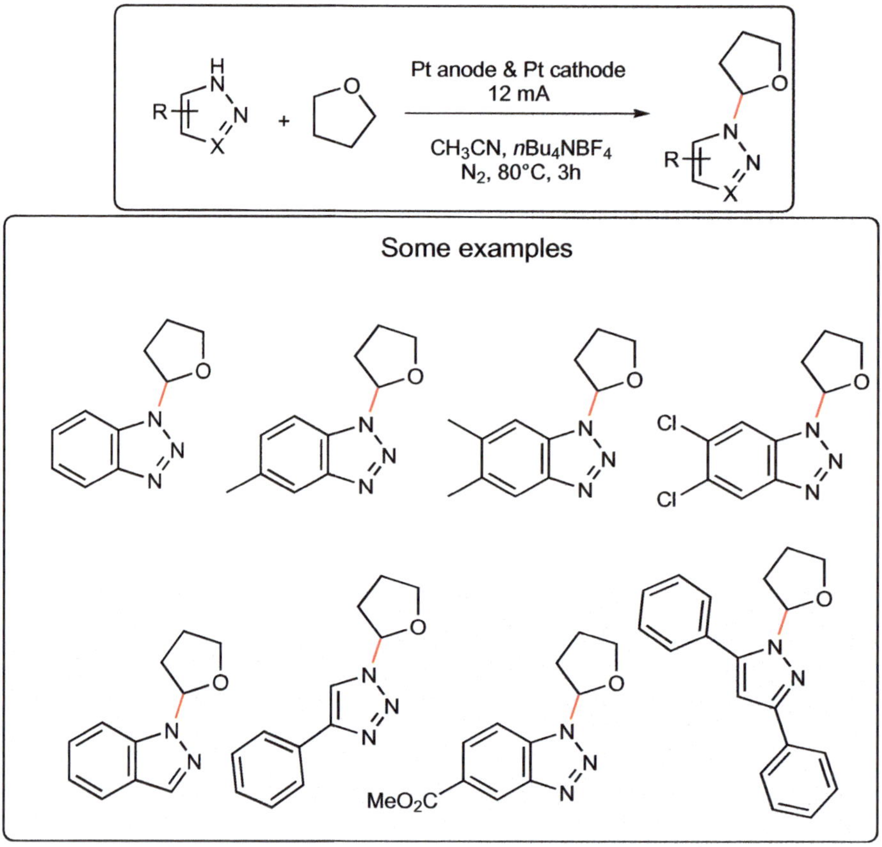

Scheme 27 Electrochemical intermolecular oxidative C–H/N–H cross-coupling

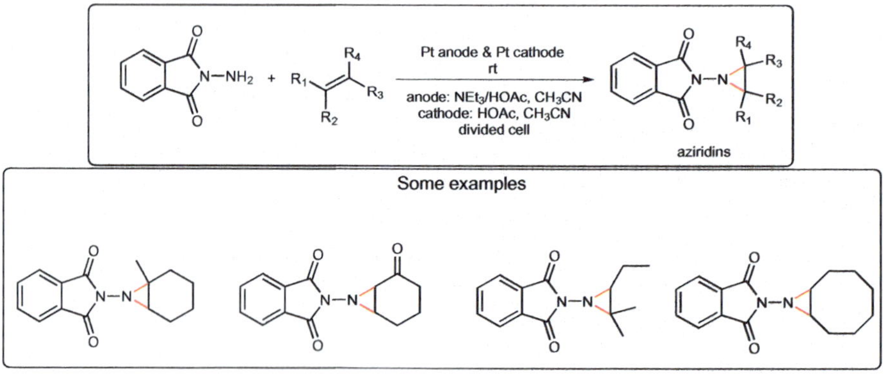

Scheme 28 Electrochemical dehydrogenative [1+2] annulation between *N*-aminophthalimide and alkene

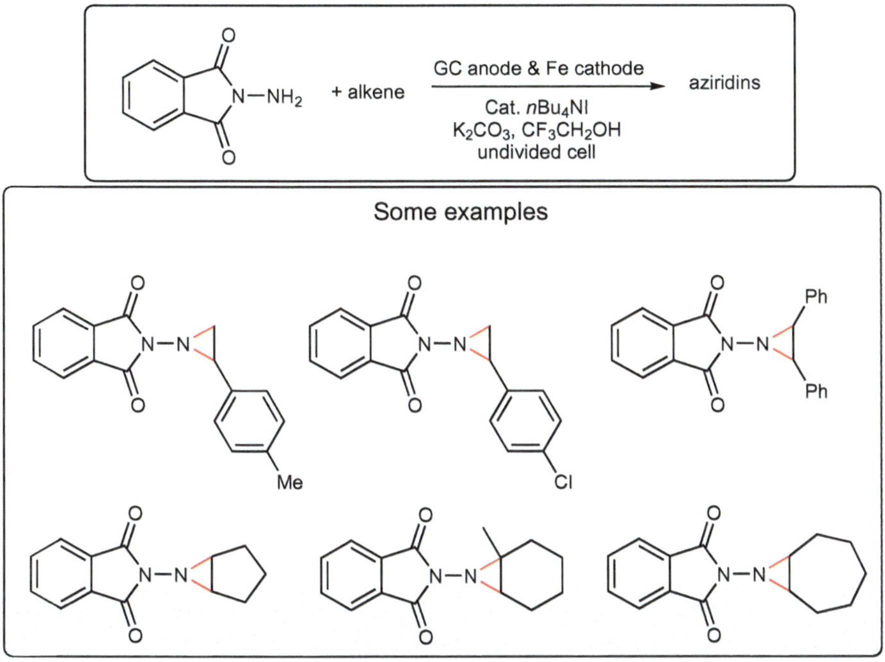

Scheme 29 Electrochemical aziridination of alkenes mediated by $n\mathrm{Bu_4NI}$

In 2015, Zeng accomplished the same transformation mediated by $n\mathrm{Bu_4NI}$ as redox catalyst with constant current flow (Scheme 29) [75]. The reaction was performed in an undivided cell equipped with GC anode and Fe plate as cathode. The formation of nitrogen radical anticipated the key step in this reaction system. Oxidative annulation can be afforded with several substituted styrenes and $\mathrm{PhtiNH_2}$ in notable yields.

5.2.2 Intramolecular C–N Bond Formation

Nitrogen-centered radicals are adaptable intermediates for the development of nitrogen-bearing molecules. However, the oxidative activation of the N–H bond has drawn great interest but the formation of amidyl radicals is still remaining as a challenge [76–78]. In the field of electronic C–N bond formation from inactivated C–H and N–H bonds, a series of achievements have been made by Xu's group. In 2016, Xu and co-workers reported an electrocatalytic intramolecular hydroamination reaction by utilizing the cheap organometallic ferrocene ([Cp2Fe]) as the redox catalyst, which obviated the requirements for a stoichiometric strong oxidant and/or expensive metal catalyst (Scheme 30) [48]. According to the controlled cyclic voltammetry experiments with allylic carbamate and [Cp2Fe] in a solvent of MeOH with or without THF, no electron transfer process occurred just in MeOH,

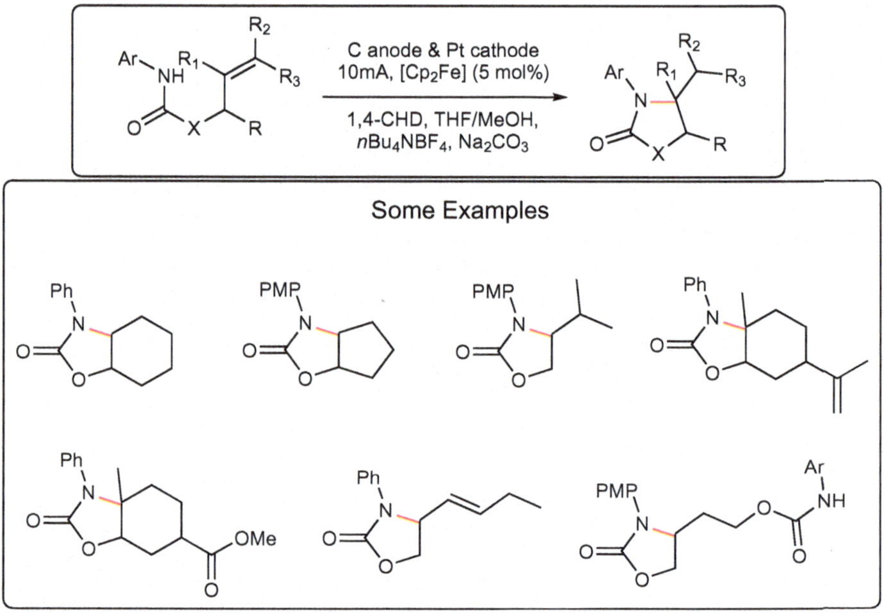

Scheme 30 Electrochemical hydroamination of alkenes

suggesting the importance of the less polar media to the reaction success. The substitution pattern and configuration of the alkene C=C bond in allylic carbamate did not have a negative impact on the cyclization efficiency.

A mechanism was proposed using cyclohex-2-en-1-yl 2-phenylacetate (1d) as a model substrate (Scheme 31). At first, an anodic transformation of Fc to Fc^+ might be taken place by employing electric current and simultaneously MeOH would be reduced to MeO^- and H_2 in cathode. Afterward, the substrate 1d would convert into

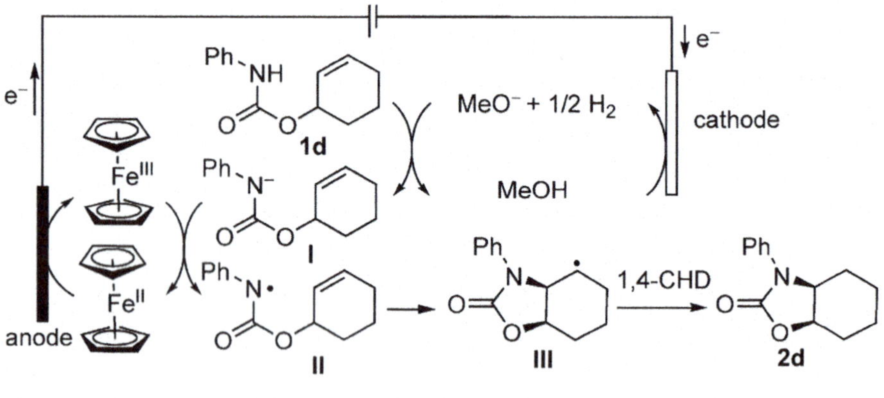

Scheme 31 Proposed mechanism for electrochemical generation of amidyl radical for hydroamidation (reproduced from Zhu et al. [48] with permission from John Wiley and Sons)

its conjugate base I by deprotonation. Subsequently, single-electron transfer would be taken place from Fc^+ to I to provide amidyl intermediate radical II and Fc^+ turns back to Fc. This electron transfer step is facilitated in less polar media because of reduced solvation of the ionic species.

In 2017, the same group reported aza-Wacker-type cyclization in electrochemistry of tri- and tetra-substituted alkenes, a metal- and reagent-free electrochemical intramolecular oxidative amination reaction, wherein the *N*-centered radical was produced by direct anodic oxidation of the amidyl N–H bond tethered to a poly-substituted alkene moiety (Scheme 32) [79].

In 2016, Xu reported an electrochemical intramolecular C–N [3+2] cyclization of aniline and alkyne moiety that can provide indoles and azaindoles (Scheme 33) [49]. The reaction was carried out in an undivided cell using Cp_2Fe as redox mediator. The formation of amidyl intermediate radical by anodic oxidation with the help of Cp_2Fe^+ was the key step for this transformation. High reaction efficiency and excellent functionality were obtained in the electrochemical approach.

In 2017, the same group utilized the above reaction in a cascade cyclization for the preparation of nitrogen-doped polyaromatic compounds (Scheme 34) [80]. Easily accessible urea-tethered dynes were utilized in this electrochemical cyclization to synthesize nitrogen-doped polyaromatic hydrocarbons (PAHs). The use of Cp_2Fe (ferrocene) as a redox catalyst allows to obtain a diverse electron-rich PAHs that even includes helicene-like structure.

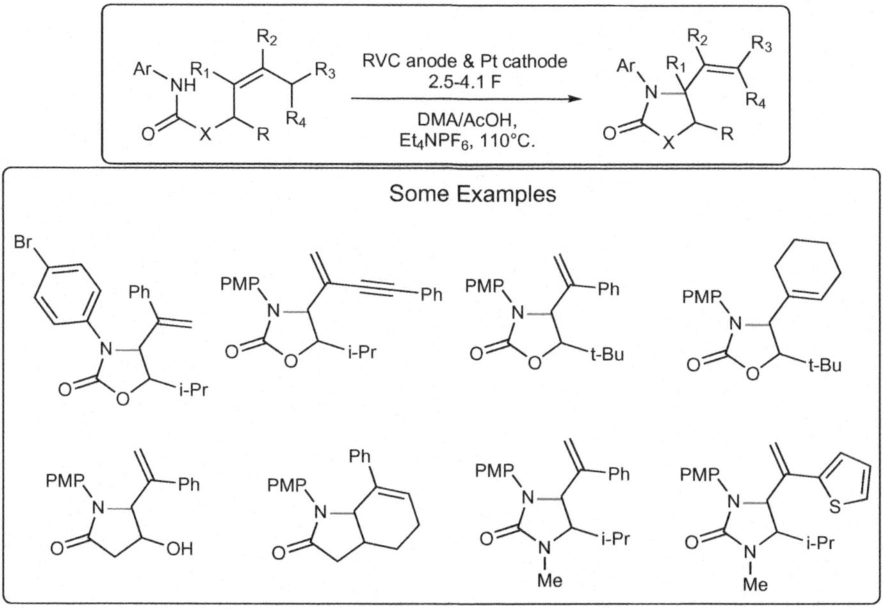

Scheme 32 Electrochemical metal-free intramolecular amination of tri-/tetra-substituted alkenes

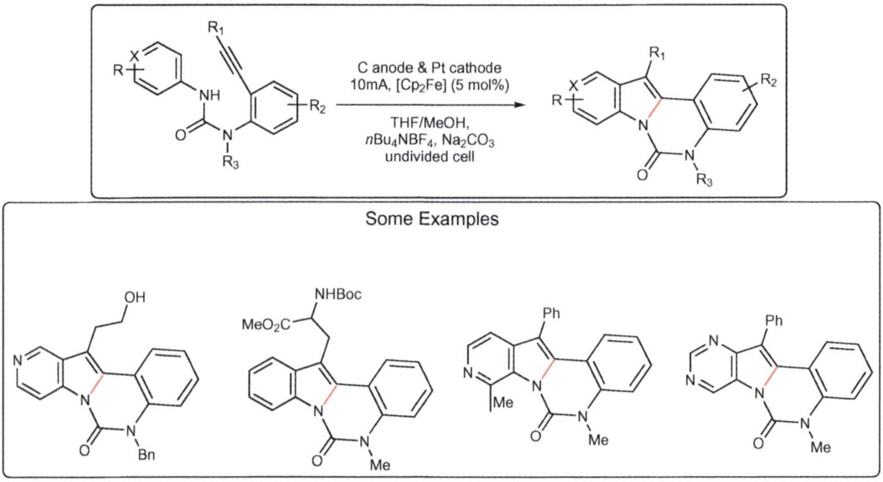

Scheme 33 Electrochemical synthesis of indoles and azaindoles

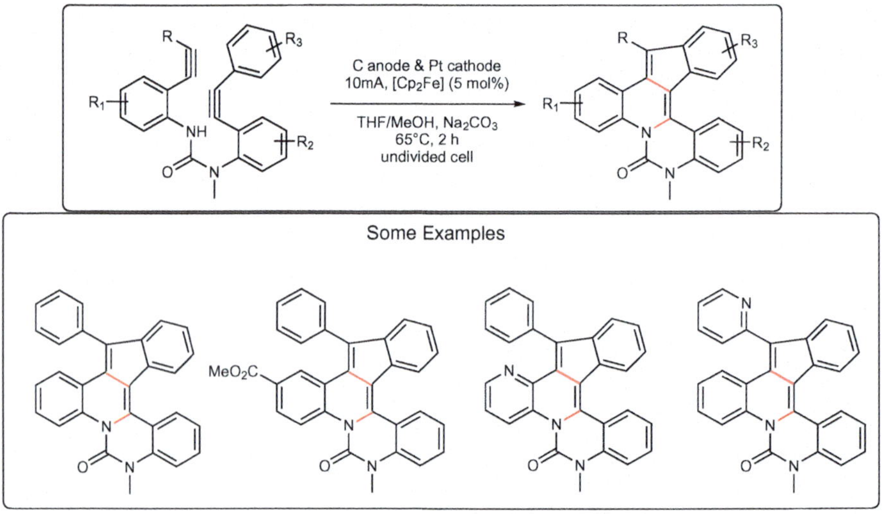

Scheme 34 Electrochemical annulation for the synthesis of nitrogen-doped polyaromatic compounds

In 2016, Xu's group reported an electrochemical approach to access the amidinyl radical intermediate via the anodic N–H bond cleavage of substrate (Scheme 35) [81], which was followed by intramolecular cyclization onto (hetero)arenes and rearomatization to give a diverse range of functionalized tetracyclic benzimidazoles or pyridoimidazoles in a high-efficiency process. Pre-prepared substrates derived from 2-(benzylamino)benzonitrile were used to proceed the electrochemical

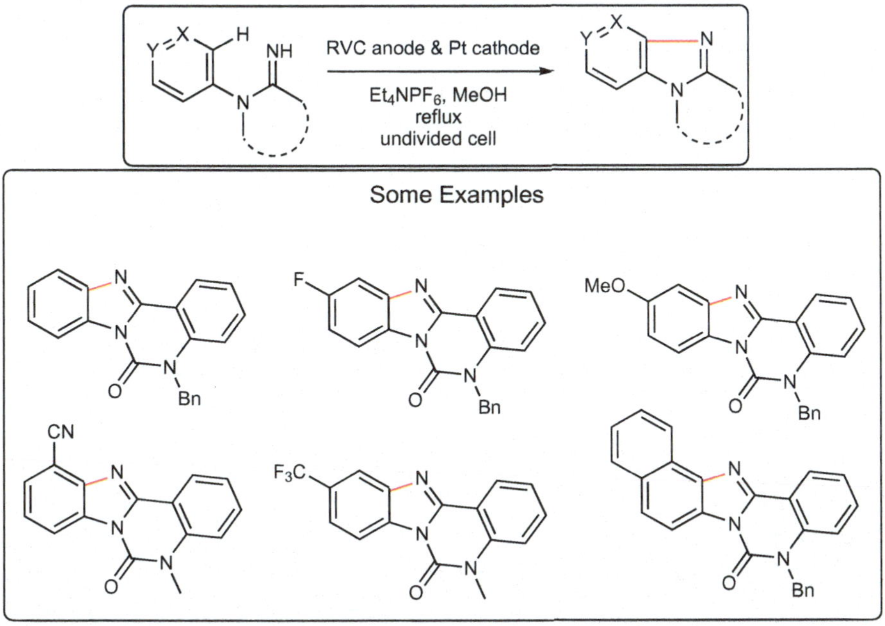

Scheme 35 Amidinyl radical formation through anodic NH bond cleavage

reaction, which required three-necked round-bottomed flask equipped with the electrodes–RVC as anode and Pt as cathode, Et_4NPF_6 was used as electrolyte in refluxing MeOH, and constant current flow was kept at 10 mA that affords benzimidazole product in high yields.

In 2018, Zang and co-workers reported electrochemical generation of *N*-acyloxy amidyl radicals via an inner-sphere electron transfer process using NaBr as the catalyst and electrolyte (Scheme 36) [82]. At first, methoxide was formed in anodic oxidation, which subsequently reacted with bromine to form a species bearing N–Br bond in situ. The key intermediate *N*-acyloxy amidyl was generated via the homolytic cleavage of N–Br bond. Afterward, lactam molecule was formed through radical addition, oxidation, and deprotonation.

In 2017, electrochemical synthesis of *N*-heteroaromatic products through anodic condensation between biaryl aldehydes and NH_3 (Scheme 37) was reported by Xu and co-workers [83]. This electrolytic reaction proceeded via C–H/N–H cross-coupling to produce functionalized polycyclic heteroaromatic compounds in a chemo- and regioselective manner. This electrochemical oxidative condensation of substrate biaryl aldehyde with NH_3 was taken into a three-necked round-bottomed flask (an undivided cell) equipped with a reticulated vitreous carbon (RVC) anode and a Pt plate cathode. A mixed solvent of hexafluoro-2-propanol (HFIP)/MeOH (5:1) was utilized at room temperature to obtain the highest yield of phenanthridines.

In the same year, Wang and co-workers demonstrated the electrochemically oxidative cyclization of 2′-aminoacetophenones for the synthesis of Isatins utilizing

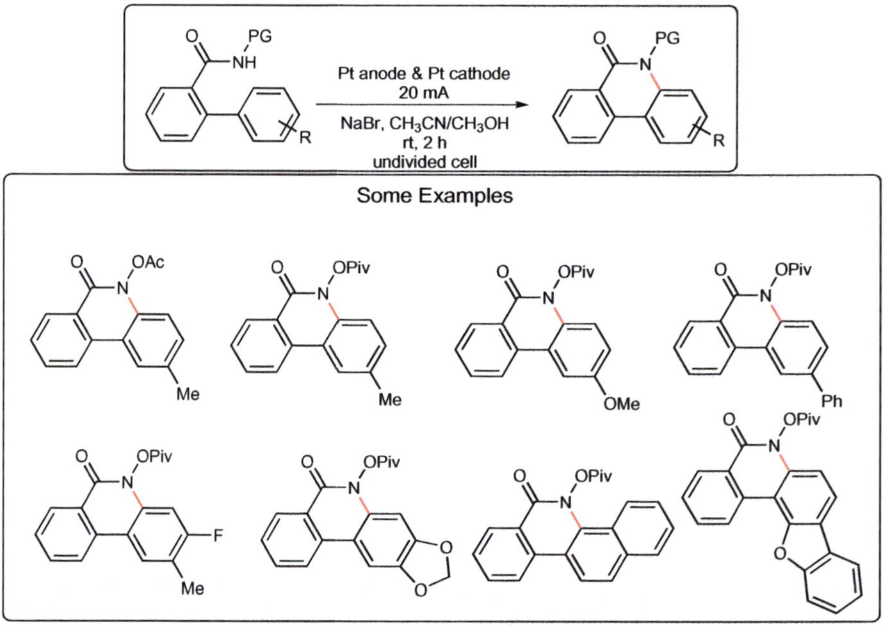

Scheme 36 Electrochemical formation of *N*-acyloxy amidyl radical for lactam preparation

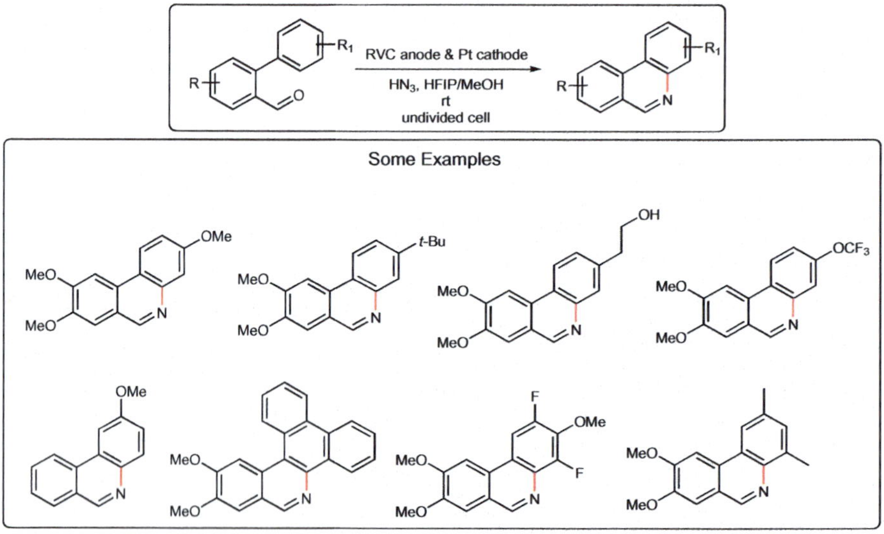

Scheme 37 Synthesis of phenanthridines through C–N bond-forming cyclization

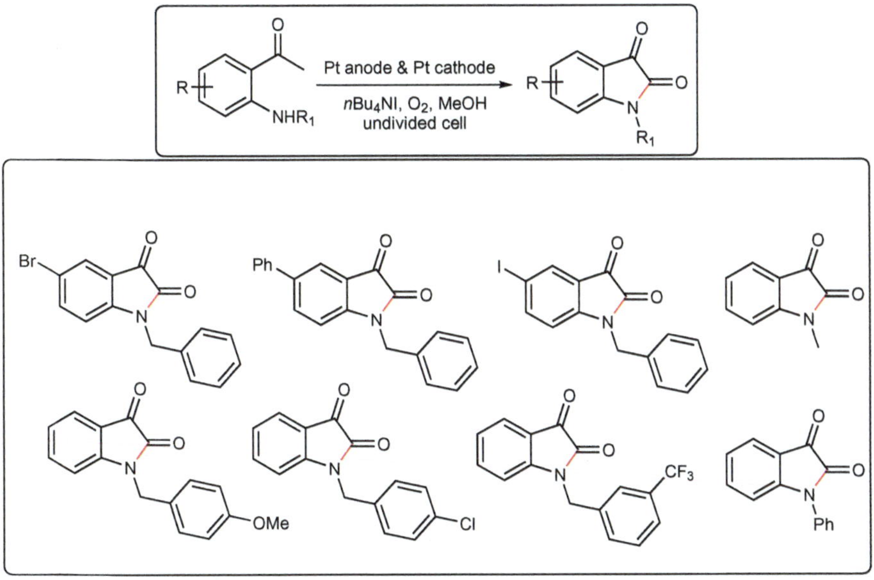

Scheme 38 Electrochemical intramolecular C–H/N–H cross-coupling for the synthesis of isatins

O_2 in the presence of n-Bu_4NI as redox catalyst (Scheme 38) [84]. The reaction was performed with the substrate 1-(2-(benzylamino)phenyl)ethanone in an undivided cell equipped with a pair of platinum plate electrodes in the presence of KI as electrolyte and MeOH as solvent under an oxygen atmosphere.

In 2016, Zeng disclosed an effective electrochemical approach for the amino-oxygenation of styrenes to prepare 3-alkoxy-substituted indolines (Scheme 39) [85]. The reaction was accomplished in an undivided cell equipped with graphite anode and cathode as electrode. Several substituted *N*-(2-vinylphenyl) sulfonamides can be afforded by this electrochemical approach.

A proposed mechanism for the formation of 3-methoxyphenylsulfonylindolines from *N*-(2-vinylphenyl) sulfonamide was shown in Scheme 40. The reaction was triggered by the anodic conversion of iodide into iodine that would react with 3-methoxyphenylsulfonylindolines (1) to generate the iodonium intermediate (3). Then, the intermediate 3 would experience an intramolecular nucleophilic attack of sulfonamide nitrogen to afford the expected 3-iodo-1-arylsulfonyl indoline (4).

In 2016, Waldvogel and his group described intramolecular dehydrogenative N–H/N–H cross-coupling to obtain 1,2-diarylpyrazolidin-3-5-diones via anodic oxidation (Scheme 41) [86]. The reaction was carried out in an undivided cell equipped with a graphite anode and a platinum cathode using HFIP as the solvent. This dehydrogenative cyclization was found excellent functional group functionality along with good product yields.

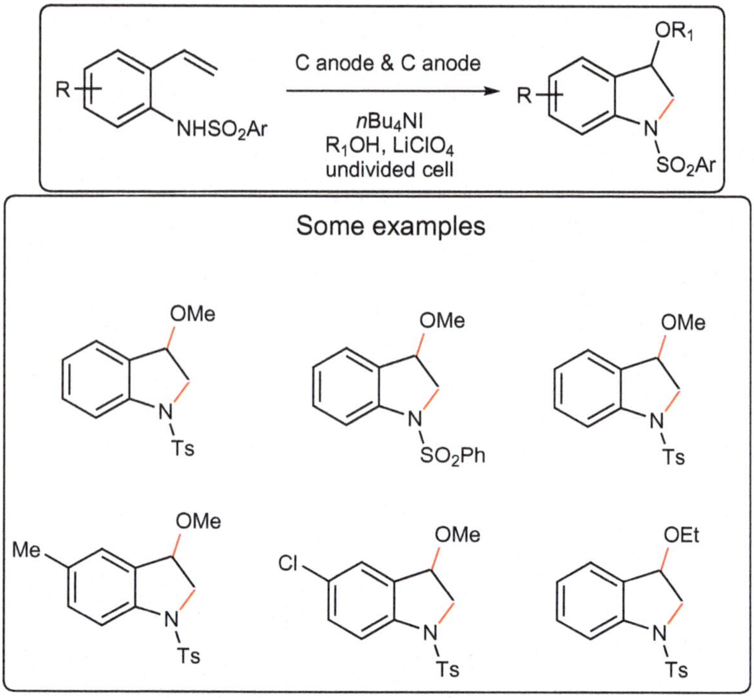

Scheme 39 Electrochemical synthesis of indolines mediated by n-Bu$_4$NI

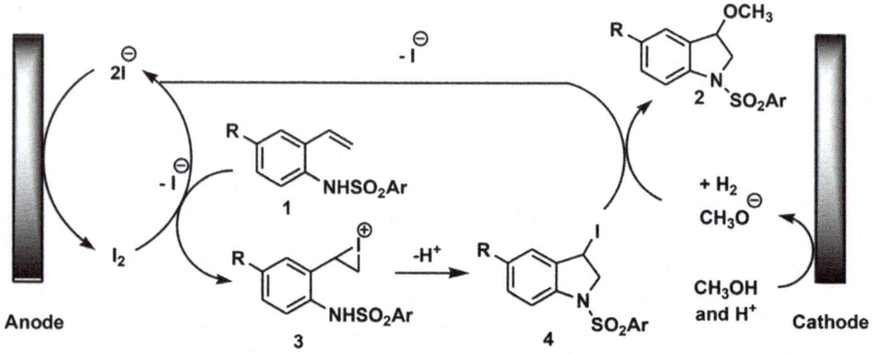

Scheme 40 Plausible mechanism for the formation of indoles mediated by nBu$_4$NI (reproduced from Liang et al. [85] with permission of Royal Society of Chemistry)

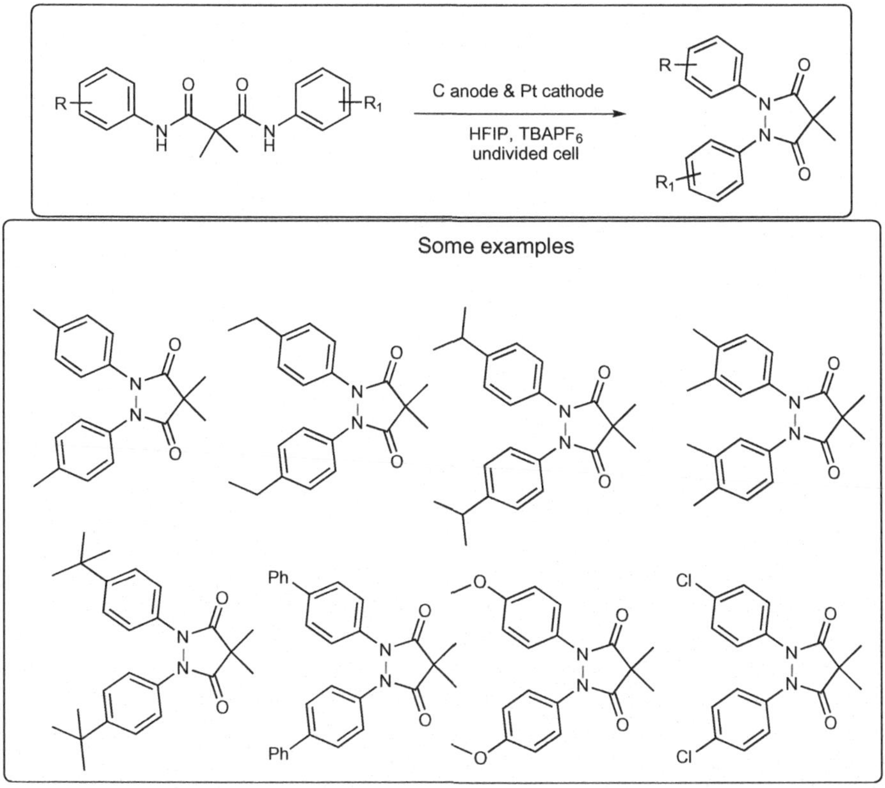

Scheme 41 Electrochemical intermolecular dehydrogenative N–N bond formation

5.3 Carbon–Oxygen (C–O) Bond Formation

5.3.1 Intermolecular C–O Bond Formation

Nitrogen-based heterocycles constitute an essential structural motif and are commonly found in natural products. In 1975, Shono and co-workers disclosed an electrochemical method for oxidation of carbamates to *N*-carbamoyl iminium ions [22, 57, 87, 88]. The reaction proceeds through the initial formation of a nitrogen-centered radical (vide infra), which is subsequently oxidized to an iminium ion intermediate that can be trapped with an alcoholic solvent molecule [89]. This allows for functionalization of the α-position adjacent to the nitrogen atom in heterocycles [90–92, 117]. Inspired by these facts, in 2013 Fuchigami and Baba reported an approach for anodic substitutions at the α-position of the nitrogen atom of imines [93]. They demonstrated anodic oxidation of imidates and imines to yield α-methoxylated or α-acetoxylated products (Scheme 42). The reaction was operated under a constant current utilizing an undivided cell equipped with Pt cathode and anode along with a mixture of supporting electrolytes Et_4NBF_4 and $(NH_4)_2SO_4$.

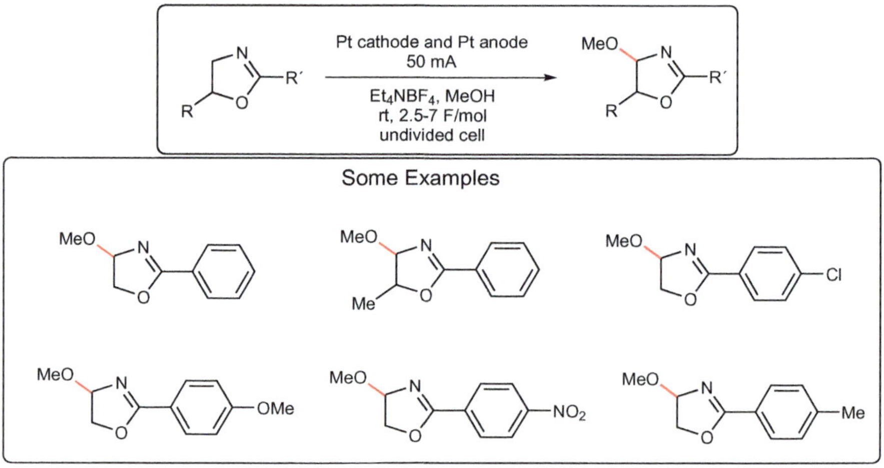

Scheme 42 Formation of intermolecular C–O bond via anodic oxidation of imides

Considering the versatility of Shono-type oxidation reactions, it is perhaps unsurprising that these reactions have found numerous synthetic applications. The groups of Moeller and Aube developed a practical two-step protocol for the late-stage functionalization of lactams by combining an intramolecular Schmidt reaction of keto azides followed by electrochemical anodic oxidation. The applicability of the Shono oxidation was exemplified across a range of ring systems. Finally, the versatility of the produced methoxyhemiaminals was illustrated by synthesizing an array of functionalized lactam products through the addition of various nucleophilic partners to the subsequent in situ generated *N*-acyliminium ions (Scheme 43) [94].

Schmalz and co-workers also utilized the Shono oxidation in the context of accessing a tricyclic dipeptide mimetic starting from L-proline. The anodic

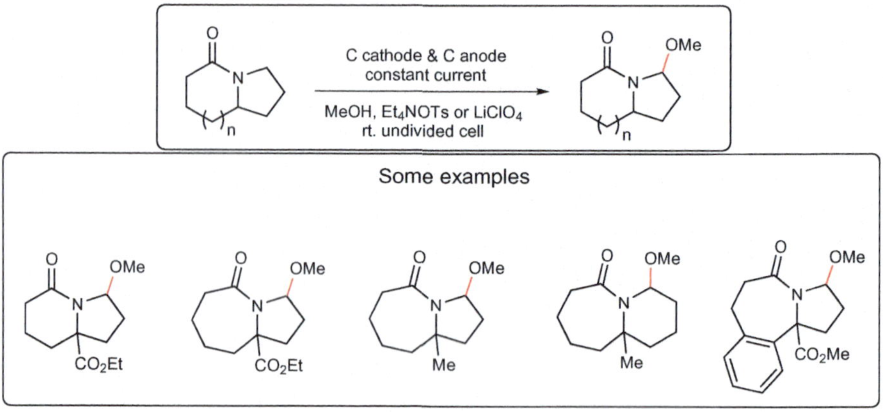

Scheme 43 Electrochemical anodic oxidation of polycyclic lactams

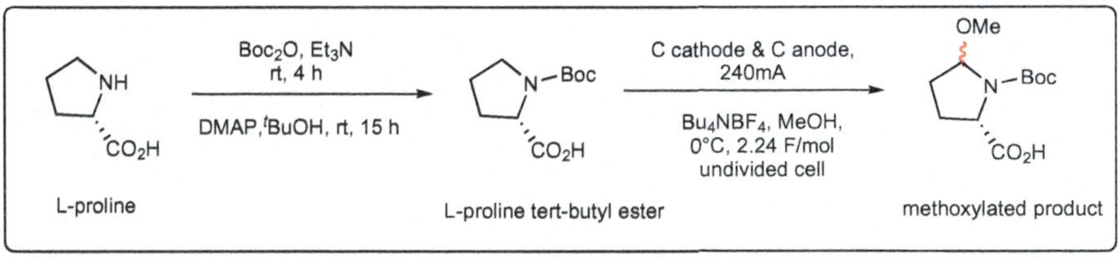

Scheme 44 Oxidative methoxylation in the preparation of a peptidomimetic

oxidation of L-proline tert-butyl ester afforded gram-scale quantities of the desired methoxylated product in virtually quantitative yield (Scheme 44). Building block was subsequently converted into diproline mimetic through peptide coupling, ring-closing metathesis, and protecting group adjustment [95].

Intermolecular [3+2] annulation is one of the most straightforward approaches to construct five-membered heterocycles. In 2017, the Lei group reported a dehydrogenative [3+2] annulation between phenol and indole derivatives for accessing benzofuroindolines using an undivided cell [96]. Here, various *N*-acetylindoles bearing different C-3 substituents were compatible with the electrochemical annulation and afforded the respective benzofuro[3,2-b]-indolines in high to excellent yields (Scheme 45).

However, applying 2-substituted *N*-acetylindoles resulted in a switch in selectivity and instead produced the corresponding benzofuro[2,3-b]indoline core (Scheme 46). CV experiments of p-methoxyphenol and various *N*-acetylindoles revealed that the oxidation potentials were quite similar, suggesting that anodic oxidation of both coupling components might occur under the electrolytic conditions.

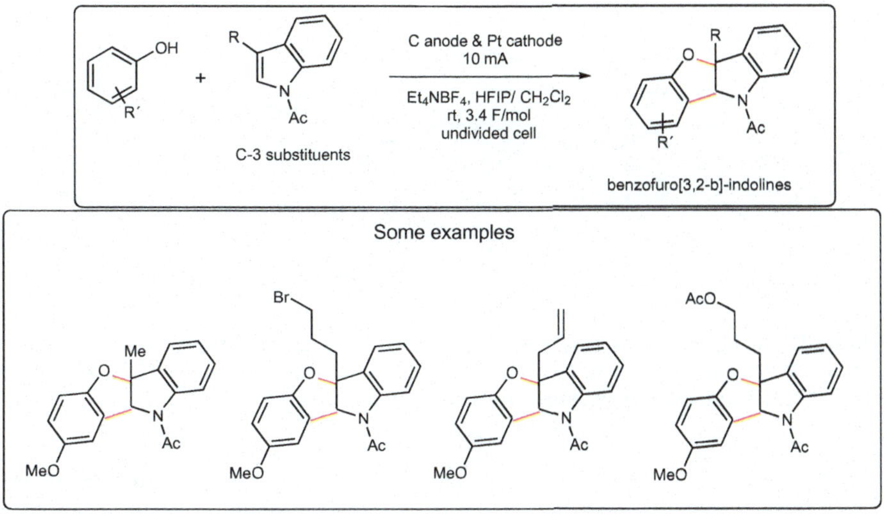

Scheme 45 Electrochemical synthesis of benzofuro[3,2-b]-indolines

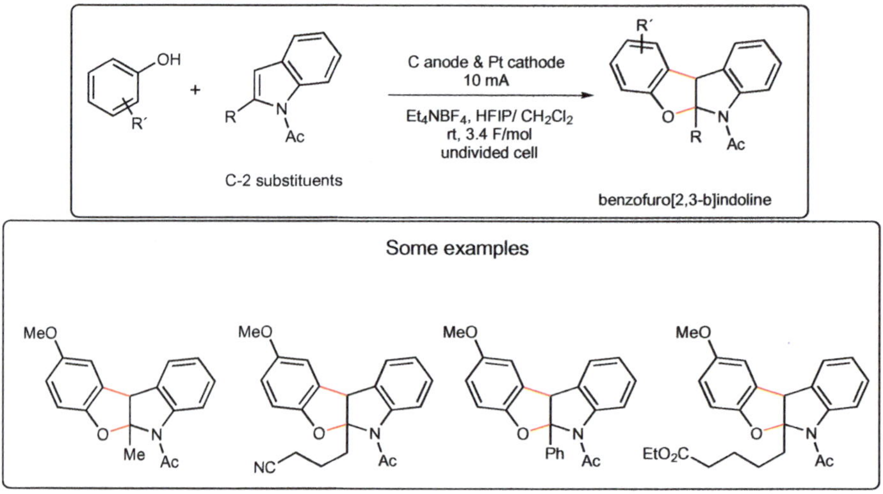

Scheme 46 Electrochemical synthesis of benzofuro[2,3-b]-indolines

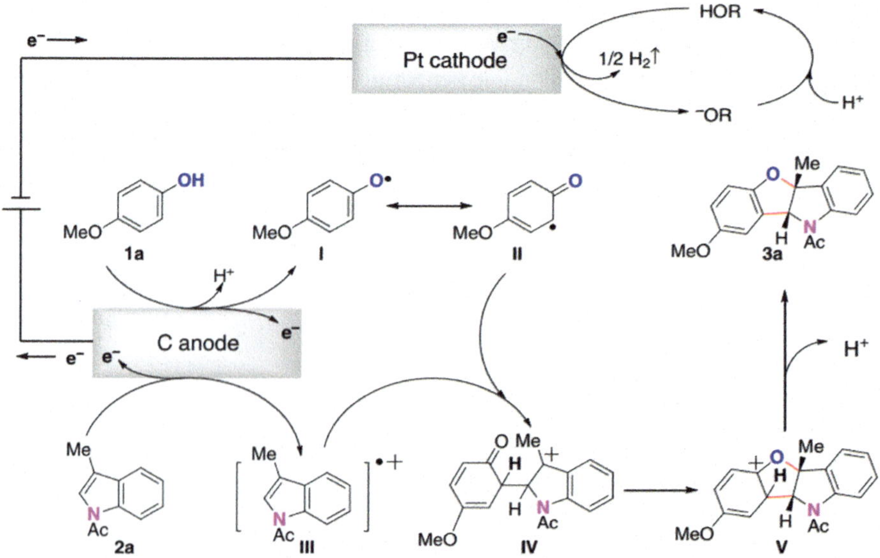

Scheme 47 A plausible reaction mechanism between p-methoxyphenol and 3-methyl-*N*-acetylindole (regenerated figure from external oxidant-free electrooxidative [3+2] annulation between phenol and indole derivatives) (reproduced from Liu et al. [96])

According to the proposed mechanism shown in Scheme 47, a phenol oxygen radical (I) was generated by the anodic oxidation of p-methoxyphenol (1a). This radical I has undergone isomerization to give carbon radical II. Simultaneously,

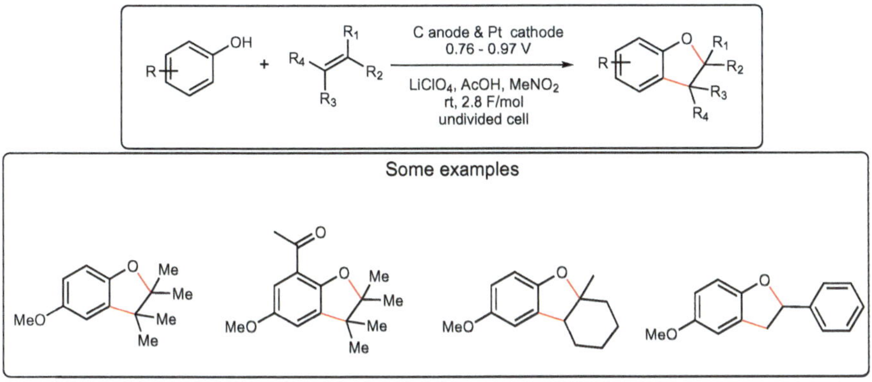

Scheme 48 Synthesis of dibenzofuran derivatives via intermolecular cycloaddition

3-methyl-*N*-acetylindole (2a) was oxidized by anode to give cation intermediate III. Direct cross-coupling of radical II with the cation intermediate III would provide cation intermediate IV. Subsequently, intramolecular cyclization and deprotonation of cation intermediate IV afforded benzofuro[3,2-b]indoline (3a). Hydrogen gas was evolved by the cathodic reduction of HFIP.

Chiba and co-workers highlighted that the anodically generated phenoxonium ions could engage in intermolecular [3+2] cycloaddition with unactivated alkenes to afford dihydrobenzofuran derivatives (Scheme 48) [97, 98]. The reaction has also been conducted using a temperature-controlled multi-phase solution consisting of cyclohexane and $LiClO_4/MeNO_2$ [99].

5.3.2 Intramolecular C–O Bond Formation

Since the prevalence of tetrahydrofuran and tetrahydropyran rings in biologically relevant natural products, Moeller group were interested in determining an approach for heterocyclic trapping groups to provide a unique pathway to heterocycles. In 2000, Moeller and co-workers published an electrochemical cyclization of enol ethers for the preparation of tetrahydrofuran and tetrahydropyran rings (Scheme 49) [100]. The reaction was carried out in an undivided cell equipped with RVC anode and Pt cathode under a constant current flow. A radical cation intermediate was anticipated to be formed by the oxidation of electron-rich alkenes in anode. This intermediate readily easily forms an unstable radical with the hydroxyl group to provide the respective tetrahydrofuran and tetrahydropyran analogs. Afterward, the generated radical has undergone the oxidation reaction in anode and subsequently reacted with MeOH to provide the final C–O annulation products. Tetrahydropyrrole and piperidine can also be obtained in this reaction by replacing the hydroxyl group of the starting materials to a protected nitrogen group [101–103].

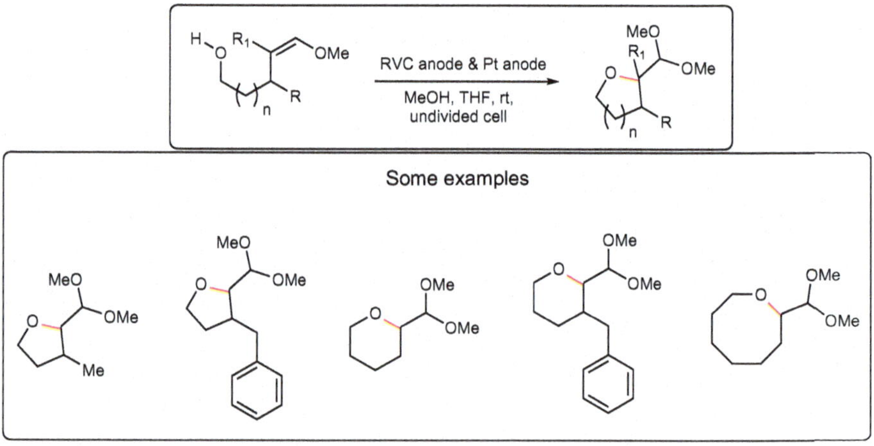

Scheme 49 Electrochemical synthesis of tetrahydrofuran and tetrahydropyran rings

In 2017, Xu disclosed an electrochemical effectual approach for the formation of lactones by oxidative annulation of carboxylic acid (Scheme 50) [104]. The reaction was carried out in an undivided cell where Pt plates were used as anode and cathode and nBu4NI as redox mediator. Both intermolecular and intramolecular products can be obtained in good yields by this reaction procedure. Acyl hypoiodite species were anticipated to be involved in the chemical conversion according to the authors' claims.

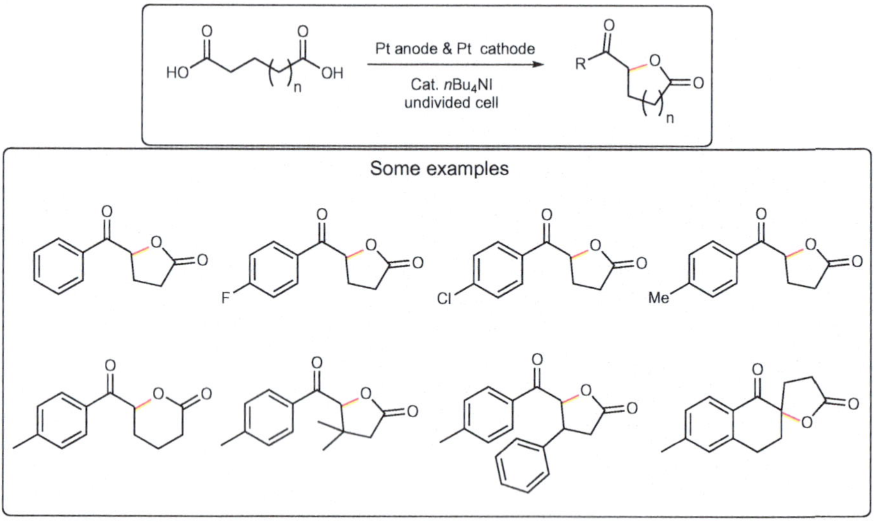

Scheme 50 Dehydrogenetive annulation of carboxylic acids to lactones

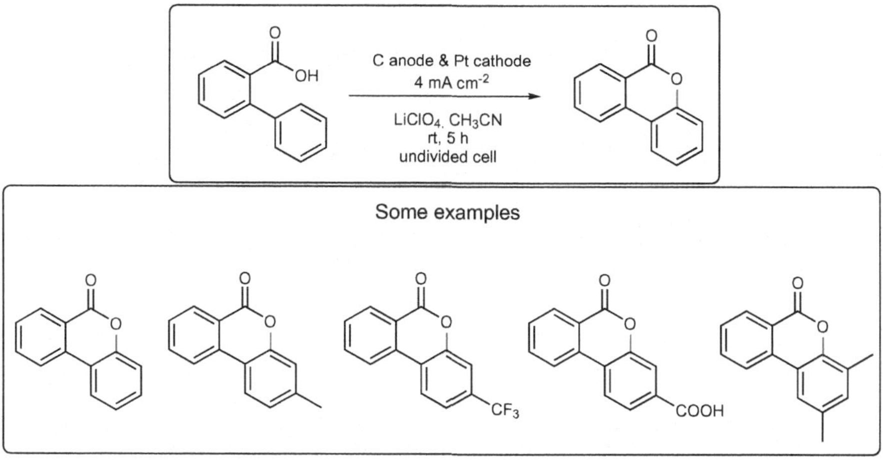

Scheme 51 Electrochemical approach of intramolecular C–O bond formation to obtain aromatic lactones

Later, the same group reported an electrochemical approach for the construction of C–O bond to afford aromatic lactones by the use of heterocycle substrates (Scheme 51). Electrochemical cyclization of 2-(hetero)arylbenzoic acids were taken into an undivided cell equipped with carbon cathode and Pt anode. This electrochemical C–O annulation was found excellent functional group compatibility and an effectual route for the preparation of biaryl lactones in good yields [105].

Okimoto and co-workers have also demonstrated that Shono-type oxidation of various amine-containing motifs carrying pendant nucleophiles can be employed to access a diverse set of heterocyclic compounds through intramolecular C–C or C–X bond formation processes [106–109]. In 2012, they demonstrated the iodine-mediated electroxidative cyclization of hydroxyamino derivatives containing a benzyl group (Scheme 52) [110]. When a hydroxyl group is tied to the substrate, the subsequent nucleophilic addition to the iminium species would provide oxazolidine. The plausible mechanism showed that KI acted as the electron transfer mediator, and the use of a small amount of NaOMe influences the product yields.

In 2013, Zeng disclosed an intramolecular dehydrogenative C(sp2)–H/O–H cross-coupling for the preparation of benzoxazoles catalyzed by sodium iodide (Scheme 53) [111]. Diverse 2-substituted benzoxazoles were accessed in a notable yield using an undivided cell equipped with GC anode and Fe plate as cathode. Unfortunately, electron-deficient imines are limited for this reaction condition.

According to the plausible mechanism (Scheme 54), hypoiodite IO_2^- could be formed from in situ molecular iodine in the basic aqueous reaction medium, and oxidation of iodine-containing intermediates could produce benzoxazoline that subsequently converted into the benzoxazole adduct [111].

In 2013, Wang and co-workers accomplished an efficient electrochemical synthesis of oxadiazoles from benzoylhydrazines mediated by KI (Scheme 55) [112].

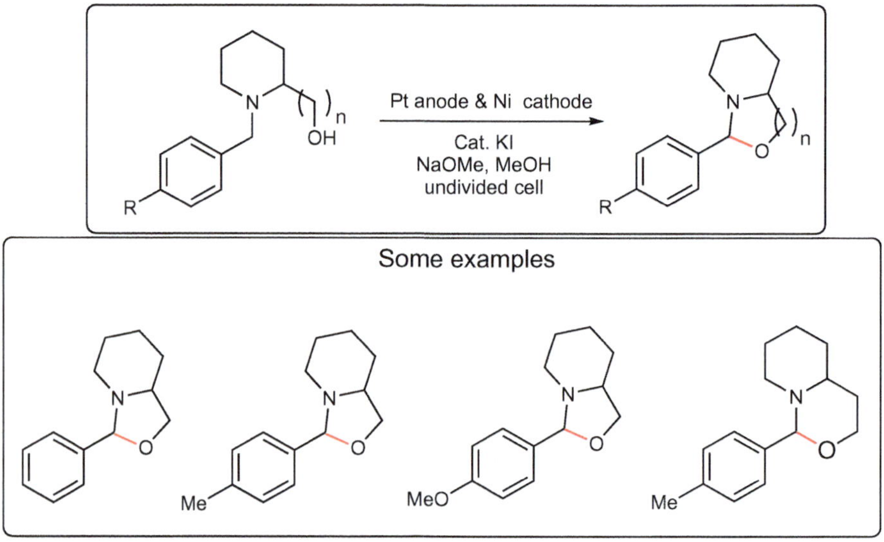

Scheme 52 Iodide-mediated intramolecular dehydrogenative C–H/O–H cross-coupling

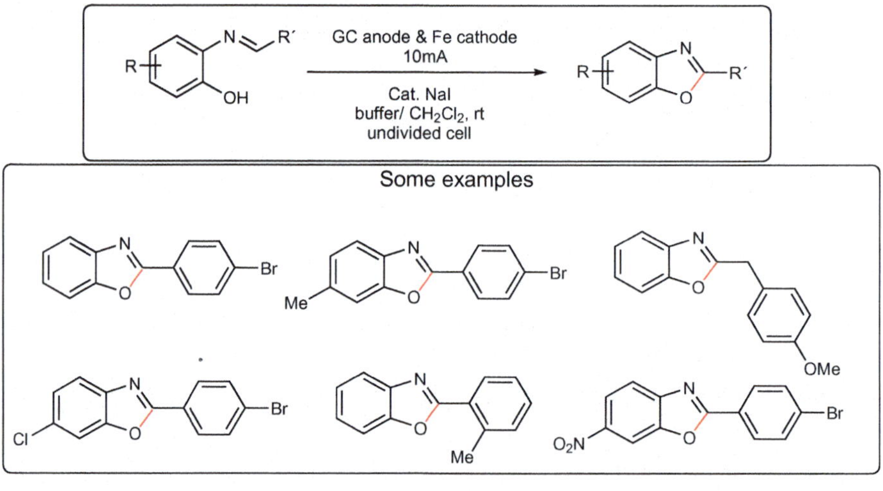

Scheme 53 Electrochemical C–O bond formation medicated by NaI

Several substituted oxadiazoles can be attained in good yields under mild reaction conditions. Gratifyingly, thiophene and furan ring were found to be tolerated in this electrochemical synthesis.

A plausible reaction pathway was proposed as in Scheme 56. Initially, an iodine radical was formed by the oxidation of an iodine anion that subsequently converted into radical 2 by accepting hydrogen atom from benzoylhydrazine. The radical 2

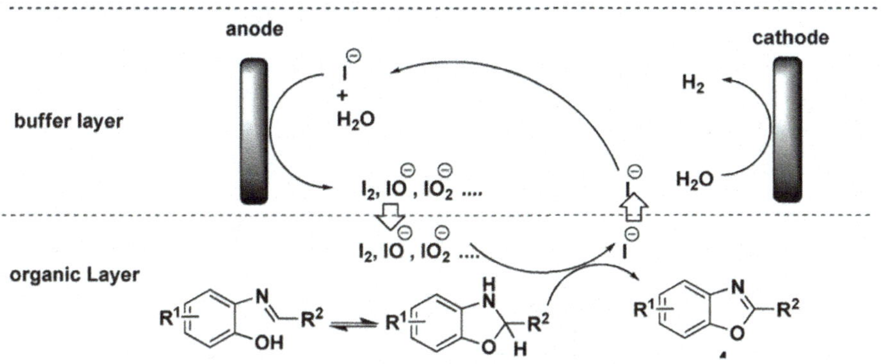

Scheme 54 A proposed mechanism of the iodide-mediated synthesis of benzoxazoles (reproduced from Li et al. [111] with the permission of John Wiley and Sons)

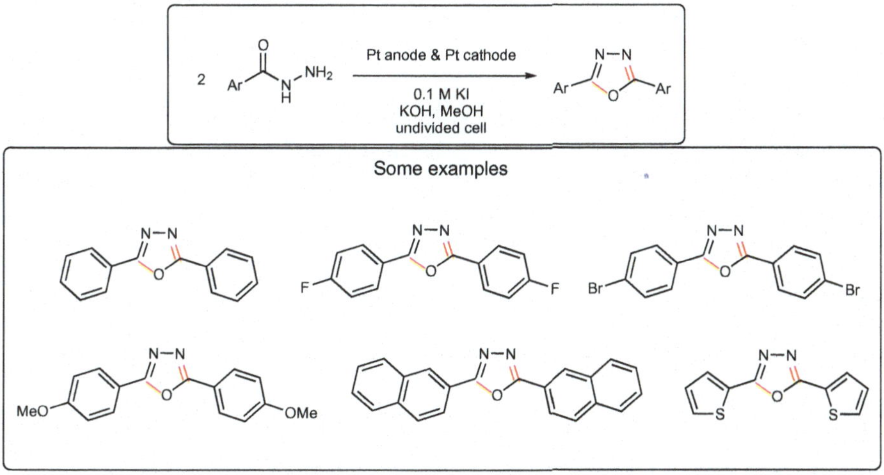

Scheme 55 KI-mediated electrochemical synthesis of oxadiazoles

readily reacted with other iodine radicals in the following steps to produce radical 3 [113]. The radical 3 might be transformed to benzaldehyde radical 4 by releasing molecular nitrogen. This radical 4 combined with a hydrogen released from solvent MeOH to provide benzaldehyde that can react with benzoylhydrazine to afford *N*-benzylidenebenzohydrazide, which converted into 5 in the alkaline medium. And the final product could be obtained by oxidation of 5.

In 2015, Hilt's group described an intramolecular cyclization of 1,4-dienols to tetrahydrofuran and pyran derivatives catalyzed by iodonium cation (Scheme 57) [114]. Regioselectivities were attained in good yields for the construction of single diastereomers, where an H-type divided cell was utilized with carbon anode and

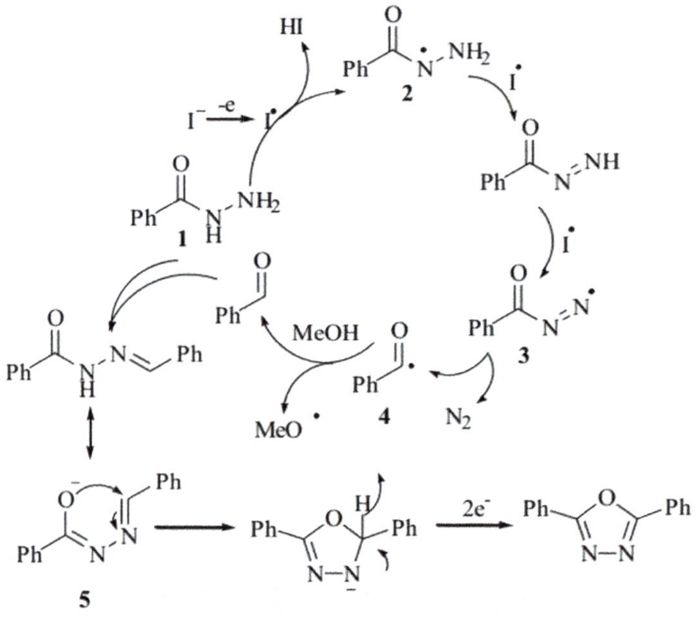

Scheme 56 Proposed mechanism for oxidation of oxadiazoles (reproduced from Li et al. [113] with the permission of Royal Society of Chemistry)

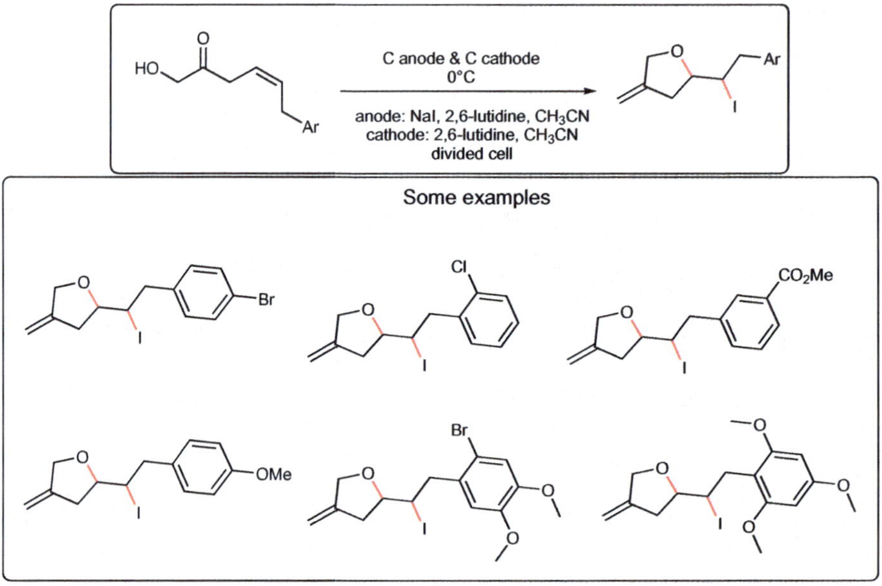

Scheme 57 Electrochemical defunctionalization of alkenes

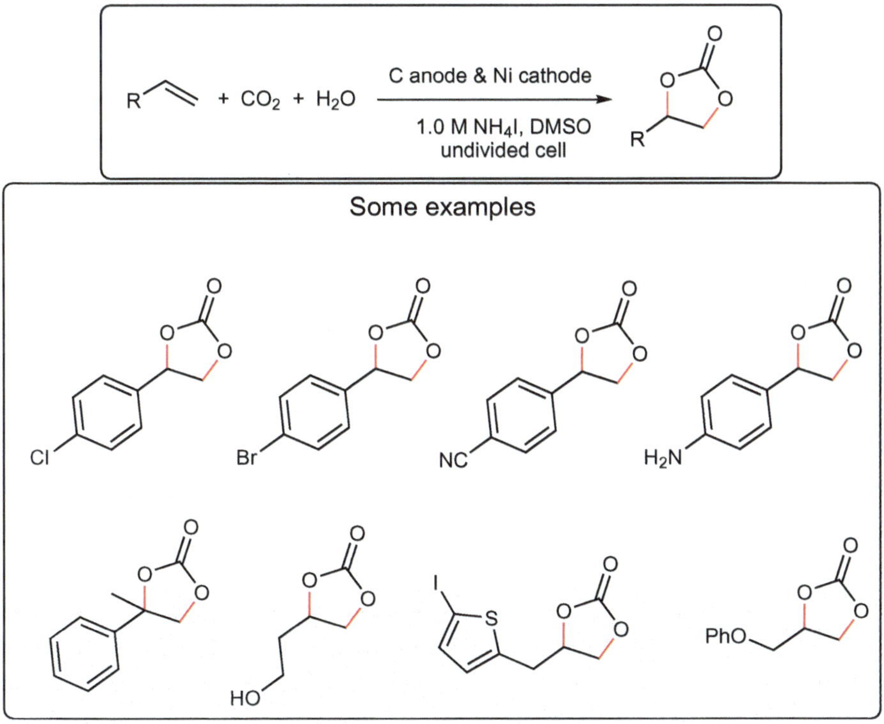

Scheme 58 Electrochemical conversion of CO_2 and olefins into cyclic carbonates mediated by NH_4I

cathode as electrodes. This reaction began with the production of an iodonium ion. Subsequently, intramolecular attack of OH provides the iodoalkoxylated products.

In 2013, Jiang disclosed the electrochemical transformation of CO_2 and olefins to cyclic carbonates mediated by NH_4I (Scheme 58) [115]. A diverse of olefins can be converted into the respective products in a notable yield using undivided cell with C anode and Ni cathode. Remarkably, aliphatic olefins can also be utilized in this methodology.

Mechanistic elucidation (Scheme 59) showed that in situ electrochemically generated I_2 and NH_3 played vital roles in the chemical transformation. The reaction of I_2 with alkenes and water produced the iodohydrin that was confirmed by GC-MS and NMR spectra. Iodohydrin might undergo the intermolecular cyclization to obtain the expected product by trapping CO_2 with the assistance of NH_3.

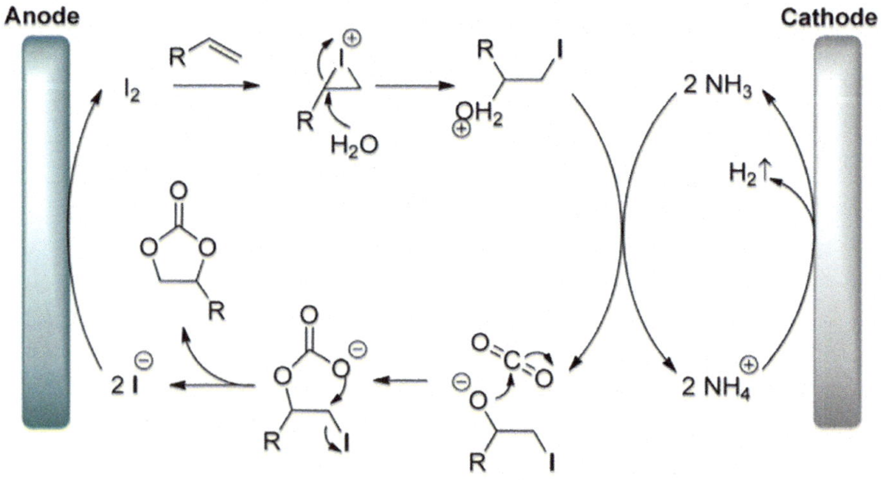

Scheme 59 Plausible mechanism of electrochemical conversion of CO_2 (reproduced from Gao et al. [115] with the permission of Elsevier)

6 Conclusion

This chapter covering fundamental aspect and highlights of electrochemical cross-dehydrogenative coupling reactions will hopefully serve as a useful reference to readers. It is environmentally friendly because hydrogen gas is produced as the only by-product. No sacrificial reagent is required. The electrochemical CDC reactions are tolerated to a wide range of functional groups because no external chemical oxidant is required [2]. Nevertheless, this synthetic strategy still needs further improvement. As Baran has pointed out the issue of reproducibility and the lack of standardized equipment, these issues will be a barrier to the widespread application of electrochemical cross-coupling reactions [10]. We believe electrochemical cross-dehydrogenative coupling will soon become a preferred method for the syntheses of heterocycles.

References

1. Meijere AD, Diederich FO (2004) Metal-catalyzed cross-coupling reactions. Wiley-VCH, Weinheim; Chichester
2. Tang S, Zeng L, Lei AW (2018) Oxidative R-1-H/R-2-H cross-coupling with hydrogen evolution. J Am Chem Soc 140:13128–13135
3. Miyaura N (2002) Cross-coupling reactions: a practical guide. Springer, Berlin
4. Gutz C, Klockner B, Waldvogel SR (2016) Electrochemical screening for electroorganic synthesis. Org Process Res Dev 20:26–32

5. Li CJ (2009) Cross-dehydrogenative coupling (CDC): exploring C–C bond formations beyond functional group transformations. Acc Chem Res 42:335–344
6. Yeung CS, Dong VM (2011) Catalytic dehydrogenative cross-coupling: forming carbon-carbon bonds by oxidizing two carbon-hydrogen bonds. Chem Rev 111:1215–1292
7. Tang S, Liu YC, Lei AW (2018) Electrochemical oxidative cross-coupling with hydrogen evolution: a green and sustainable way for bond formation. Chem 4:27–45
8. Meng QY, Zhong JJ, Liu Q, Gao XW, Zhang HH, Lei T, Li ZJ, Feng K, Chen B, Tung CH, Wu LZ (2013) A cascade cross-coupling hydrogen evolution reaction by visible light catalysis. J Am Chem Soc 135:19052–19055
9. Volke J, Liška FE (1994) Electrochemistry in organic synthesis. Springer, Berlin, Heidelberg
10. Yan M, Kawamata Y, Baran PS (2017) Synthetic organic electrochemical methods since 2000: on the verge of a renaissance. Chem Rev 117:13230–13319
11. Karkas MD (2018) Electrochemical strategies for C-H functionalization and C-N bond formation. Chem Soc Rev 47:5786–5865
12. Riehl B, Dyballa KM, Franke R, Waldvogel SR (2017) Electro-organic synthesis as a sustainable alternative for dehydrogenative cross-coupling of phenols and naphthols. Synth Stuttg 49:252–259
13. Kirste A, Elsler B, Schnakenburg G, Waldvogel SR (2012) Efficient anodic and direct phenol-arene C, C cross-coupling: the benign role of water or methanol. J Am Chem Soc 134:3571–3576
14. Horn EJ, Rosen BR, Chen Y, Tang JZ, Chen K, Eastgate MD, Baran PS (2016) Scalable and sustainable electrochemical allylic C–H oxidation. Nature 533:77–81
15. Fry AJ (1989) Synthetic organic electrochemistry. Wiley, New York
16. Lund H (2002) A century of organic electrochemistry. J Electrochem Soc 149:S21–S33
17. Kolbe H (1848) Zersetzung der Valeriansäure durch den elektrischen Strom. Justus Liebigs Ann Chem 64:339–341
18. Kolbe H (1849) Untersuchungen über die Elektrolyse organischer Verbindungen. Justus Liebigs Annalen der Chemie 69:257–294
19. Corey EJ, Sauers RR (1959) The Synthesis of Pentacyclosqualene (8,8′-Cycloönocerene) and the α- and β-Onoceradienes1. J Am Chem Soc 81:1739–1743
20. le Blanc M (1900) Über einen Versuch zur Demonstration des elektrolytischen Lösungsdruckes. Zeitschrift für Elektrochemie 7:287–290
21. Watkins BF, Behling JR, Kariv E, Miller LL (1975) Chiral electrode. J Am Chem Soc 97:3549–3550
22. Shono T, Hamaguchi H, Matsumura Y (1975) Electroorganic chemistry. XX. Anodic oxidation of carbamates. J Am Chem Soc 97:4264–4268
23. Yoshida J, Murata T, Isoe S (1986) Electrochemical oxidation of organosilicon compounds. 1. oxidative cleavage of carbon-silicon bond in allylsilanes and benzylsilanes. Tetrahedron Lett 27:3373–3376
24. Yoshida J, Shimizu A, Hayashi R (2018) Electrogenerated cationic reactive intermediates: the pool method and further advances. Chem Rev 118:4702–4730
25. Sowell CG, Wolin RL, Little RD (1990) Electroreductive cyclization reactions—stereoselection, creation of quaternary centers in bicyclic frameworks, and a formal total synthesis of quadrone. Tetrahedron Lett 31:485–488
26. Little RD, Schwaebe MK (1997) Reductive cyclizations at the cathode. Electroorg Synth: Bond Form Anode Cathode 185:1–48
27. Zhao HB, Liu ZJ, Song JS, Xu HC (2017) Reagent-Free C–H/N–H cross-coupling: regioselective synthesis of *N*-heteroaromatics from biaryl aldehydes and NH3. Angew Chem Int Ed 56:12732–12735
28. Xiong P, Xu HH, Xu HC (2017) Metal- and reagent-free intramolecular oxidative amination of tri- and tetrasubstituted alkenes. J Am Chem Soc 139:2956–2959
29. Wang P, Tang S, Huang PF, Lei AW (2017) Electrocatalytic oxidant-free dehydrogenative C–H/S–H cross-coupling. Angew Chem-Int Ed 56:3009–3013

30. Horn EJ, Rosen BR, Baran PS (2016) Synthetic organic electrochemistry: an enabling and innately sustainable method. Acs Central Science 2:302–308
31. Wiebe A, Riehl B, Lips S, Franke R, Waldvogel SR (2017) Unexpected high robustness of electrochemical cross-coupling for a broad range of current density. Sci Adv 3:eaao3920
32. Yang Q, Zhang L, Ye C, Luo S, Wu LZ, Tung CH (2017) Visible-light-promoted asymmetric cross-dehydrogenative coupling of tertiary amines to ketones by synergistic multiple catalysis. Angew Chem Int Ed 56:3694–3698
33. Baslé O, Borduas N, Dubois P, Chapuzet JM, Chan TH, Lessard J, Li CJ (2010) Aerobic and electrochemical oxidative cross-dehydrogenative-coupling (CDC) reaction in an imidazolium-based ionic liquid. Chem Eur J 16:8162–8166
34. Zhao Y-L, Wang Y, Luo Y-C, Fu X-Z, Xu P-F (2015) Asymmetric C–H functionalization involving organocatalysis. Tetrahedron Lett 56:3703–3714
35. Zhang F-L, Hong K, Li T-J, Park H, Yu J-Q (2016) Functionalization of C (sp3)–H bonds using a transient directing group. Science 351:252–256
36. Yang K, Li Q, Liu Y, Li G, Ge H (2016) Catalytic C–H arylation of aliphatic aldehydes enabled by a transient ligand. J Am Chem Soc 138:12775–12778
37. Fu N, Li L, Yang Q, Luo S (2017) Catalytic asymmetric electrochemical oxidative coupling of tertiary amines with simple ketones. Org Lett 19:2122–2125
38. Shono T, Matsumura Y, Tsubata K (1981) Electroorganic chemistry. 46. A new carbon-carbon bond forming reaction at the. alpha.-position of amines utilizing anodic oxidation as a key step. J Am Chem Soc 103:1172–1176
39. Li W, Moeller KD (1996) Conformationally restricted TRH analogs: the compatibility of a 6, 5-bicyclic lactam-based mimetic with binding to TRH-R. J Am Chem Soc 118:10106–10112
40. Mori M, Kagechika K, Sasai H, Shibasaki M (1991) New synthesis of 4-acetoxy-2-azetidinones by use of electrochemical oxidation. Tetrahedron 47:531–540
41. Yoshida J-I, Suga S, Suzuki S, Kinomura N, Yamamoto A, Fujiwara K (1999) Direct oxidative carbon–carbon bond formation using the "cation pool" method. 1. generation of iminium cation pools and their reaction with carbon nucleophiles. J Am Chem Soc 121:9546–9549
42. Pandey G, Banerjee P, Gadre SR (2006) Construction of enantiopure pyrrolidine ring system via asymmetric [3+2]-cycloaddition of azomethine ylides. Chem Rev 106:4484–4517
43. Stanley LM, Sibi MP (2008) Enantioselective copper-catalyzed 1, 3-dipolar cycloadditions. Chem Rev 108:2887–2902
44. Tang S, Liu K, Long Y, Gao X, Gao M, Lei A (2015) Iodine-catalyzed radical oxidative annulation for the construction of dihydrofurans and indolizines. Org Lett 17:2404–2407
45. Tang S, Gao X, Lei A (2016) Decarboxylative (4+1) oxidative annulation of malonate monoesters with 2-vinylpyridine derivatives. Adv Synth Catal 358:2878–2882
46. Tang S, Gao X, Lei A (2017) Electrocatalytic intramolecular oxidative annulation of *N*-aryl enamines into substituted indoles mediated by iodides. Chem Commun 53:3354–3356
47. Wu ZJ, Xu HC (2017) Synthesis of C3-fluorinated oxindoles through reagent-free cross-dehydrogenative coupling. Angew Chem 129:4812–4816
48. Zhu L, Xiong P, Mao ZY, Wang YH, Yan X, Lu X, Xu HC (2016) Electrocatalytic generation of amidyl radicals for olefin hydroamidation: use of solvent effects to enable anilide oxidation. Angew Chem Int Ed 55:2226–2229
49. Hou ZW, Mao ZY, Zhao HB, Melcamu YY, Lu X, Song J, Xu HC (2016) Electrochemical C–H/N–H functionalization for the synthesis of highly functionalized (Aza) indoles. Angew Chem Int Ed 55:9168–9172
50. Mihelcic J, Moeller KD (2003) Anodic cyclization reactions: the total synthesis of alliacol A. J Am Chem Soc 125:36–37
51. Mihelcic J, Moeller KD (2004) Oxidative cyclizations: The asymmetric synthesis of (−)-alliacol A. J Am Chem Soc 126:9106–9111
52. Collet F, Dodd RH, Dauban P (2009) Catalytic C–H amination: recent progress and future directions. Chem Commun 34:5061–5074

53. Suzuki C, Hirano K, Satoh T, Miura M (2015) Direct synthesis of N–H carbazoles via iridium (III)-catalyzed intramolecular C–H amination. Org Lett 17:1597–1600
54. Nath AR, Yehye WA (2018) Acid hydrazide: a potential reagent for the synthesis of semicarbazones. Synthesis 50:4301–4312
55. Zhao Y, Xia W (2018) Recent advances in radical-based C–N bond formation via photo-/electrochemistry. Chem Soc Rev 47:2591–2608
56. Lennartz M, Sadakane M, Steckhan E (1999) Electrochemical oxidation of (R)-4-hydroxy-2-pyrrolidone: a key building block for stereoselective *N*-acyliminium ion coupling reactions. Tetrahedron 55:14407–14420
57. Jones AM, Banks CE (2014) The Shono-type electroorganic oxidation of unfunctionalised amides. Carbon–carbon bond formation via electrogenerated *N*-acyliminium ions. Beilstein J Org Chem 10:3056–3072
58. Kabeshov MA, Musio B, Murray PR, Browne DL, Ley SV (2014) Expedient preparation of nazlinine and a small library of indole alkaloids using flow electrochemistry as an enabling technology. Org Lett 16:4618–4621
59. Gong M, Huang JM (2016) Electrochemical oxidative C-H/N–H coupling between γ-lactams and anilines. Chem Eur J 22:14293–14296
60. Zhang L, Su J-H, Wang S, Wan C, Zha Z, Du J, Wang Z (2011) Direct electrochemical imidation of aliphatic amines via anodic oxidation. Chem Commun 47:5488–5490
61. Gao W-J, Li W-C, Zeng C-C, Tian H-Y, Hu L-M, Little RD (2014) Electrochemically initiated oxidative amination of benzoxazoles using tetraalkylammonium halides as redox catalysts. J Org Chem 79:9613–9618
62. Liang S, Zeng C-C, Tian H-Y, Sun B-G, Luo X-G, Ren F-Z (2016) Electrochemically oxidative α-C–H functionalization of ketones: a cascade synthesis of α-amino ketones mediated by NH4I. J Org Chem 81:11565–11573
63. Qiu Y, Struwe J, Meyer TH, Oliveira JCACA, Ackermann L (2018) Catalyst-and reagent-free electrochemical azole C-H amination. Chem Eur J 24:12784
64. Sauermann N, Mei R, Ackermann L (2018) Electrochemical C–H amination by cobalt catalysis in a renewable solvent. Angew Chem Int Ed 57:5090–5094
65. Gao X, Wang P, Zeng L, Tang S, Lei A (2018) Cobalt (II)-catalyzed electrooxidative C–H amination of arenes with alkylamines. J Am Chem Soc 140:4195–4199
66. Tang S, Wang S, Liu Y, Cong H, Lei A (2018) Electrochemical oxidative C–H amination of phenols: access to triarylamine derivatives. Angew Chem 130:4827–4831
67. Morofuji T, Shimizu A, Yoshida J-I (2015) Heterocyclization approach for electrooxidative coupling of functional primary alkylamines with aromatics. J Am Chem Soc 137:9816–9819
68. Zhang Z, Su J, Zha Z, Wang Z (2013) Electrochemical synthesis of the aryl α-ketoesters from acetophenones mediated by KI. Chem Eur J 19:17711–17714
69. Xu K, Zhang Z, Qian P, Zha Z, Wang Z (2015) Electrosynthesis of enaminones directly from methyl ketones and amines with nitromethane as a carbon source. Chem Commun 51:11108–11111
70. Li Y, Gao H, Zhang Z, Qian P, Bi M, Zha Z, Wang Z (2016) Electrochemical synthesis of α-enaminones from aryl ketones. Chem Commun 52:8600–8603
71. Broese T, Francke R (2016) Electrosynthesis Using a Recyclable Mediator-Electrolyte System Based on Ionically Tagged Phenyl Iodide and 1,1,1,3,3,3-Hexafluoroisopropanol. Org Lett 18:5896–5899
72. Wu J, Zhou Y, Zhou Y, Chiang C-W, Lei A (2017) Electro-oxidative C (sp3)–H amination of azoles via intermolecular oxidative C (sp3)–H/N–H cross-coupling. ACS Catal 7:8320–8323
73. Siu T, Yudin AK (2002) Practical olefin aziridination with a broad substrate scope. J Am Chem Soc 124:530–531
74. Siu T, Picard CJ, Yudin AK (2005) Development of electrochemical processes for nitrene generation and transfer. J Org Chem 70:932–937

75. Chen J, Yan W-Q, Lam CM, Zeng C-C, Hu L-M, Little RD (2015) Electrocatalytic aziridination of alkenes mediated by n-Bu4NI: a radical pathway. Org Lett 17:986–989
76. Nicolaou K, Baran P, Zhong Y-L, Barluenga S, Hunt K, Kranich R, Vega J (2002) Iodine (V) reagents in organic synthesis. Part 3. New routes to heterocyclic compounds via o-iodoxybenzoic acid-mediated cyclizations: generality, scope, and mechanism. J Am Chem Soc 124:2233–2244
77. Wang Y-F, Chen H, Zhu X, Chiba S (2012) Copper-catalyzed aerobic aliphatic C–H oxygenation directed by an amidine moiety. J Am Chem Soc 134:11980–11983
78. Miller DC, Choi GJ, Orbe HS, Knowles RR (2015) Catalytic olefin hydroamidation enabled by proton-coupled electron transfer. J Am Chem Soc 137:13492–13495
79. Xiong P, Xu H-H, Xu H-C (2017) Metal-and reagent-free intramolecular oxidative amination of tri-and tetrasubstituted alkenes. J Am Chem Soc 139:2956–2959
80. Hou Z-W, Mao Z-Y, Song J, Xu H-C (2017) Electrochemical synthesis of polycyclic *N*-heteroaromatics through cascade radical cyclization of diynes. ACS Catal 7:5810–5813
81. Zhao HB, Hou ZW, Liu ZJ, Zhou ZF, Song J, Xu HC (2017) Amidinyl radical formation through anodic N–H bond cleavage and its application in aromatic C–H bond functionalization. Angew Chem Int Ed 56:587–590
82. Zhang S, Li L, Xue M, Zhang R, Xu K, Zeng C (2018) Electrochemical formation of N-acyloxy amidyl radicals and their application: regioselective intramolecular amination of sp2 and sp3 C-H bonds. Organ Lett 4:3443
83. Zhao HB, Liu ZJ, Song J, Xu HC (2017) Reagent-free C–H/N–H cross-coupling: regioselective synthesis of *N*-heteroaromatics from biaryl aldehydes and NH3. Angew Chem Int Ed 56:12732–12735
84. Qian P, Su J-H, Wang Y, Bi M, Zha Z, Wang Z (2017) Electrocatalytic C-H/N–H coupling of 2′-aminoacetophenones for the synthesis of isatins. J Org Chem 82:6434–6440
85. Liang S, Zeng C-C, Luo X-G, Ren F-Z, Tian H-Y, Sun B-G, Little RD (2016) Electrochemically catalyzed amino-oxygenation of styrenes: n-Bu 4 NI induced C–N followed by a C–O bond formation cascade for the synthesis of indolines. Green Chem 18:2222–2230
86. Gieshoff T, Schollmeyer D, Waldvogel SR (2016) Access to Pyrazolidin-3, 5-diones through Anodic N–N Bond Formation. Angew Chem Int Ed 55:9437–9440
87. Shono T, Matsumura Y, Tsubata K, Sugihara Y, Yamane S, Kanazawa T, Aoki T (1982) Electroorganic chemistry. 60. Electroorganic synthesis of enamides and enecarbamates and their utilization in organic synthesis. J Am Chem Soc 104:6697–6703
88. Shono T (1988) Synthesis of alkaloidal compounds using an electrochemical reaction as a key step. Springer, Electrochemistry III
89. Alfonso-Súarez P, Kolliopoulos AV, Smith JP, Banks CE, Jones AM (2015) An experimentalist's guide to electrosynthesis: the Shono oxidation. Tetrahedron Lett 56:6863–6867
90. Palasz PD, Utley JH, Hardstone JD (1984) Electro-organic reactions. Part 23. Regioselectivity and the stereochemistry of anodic methoxylation of *N*-acylpiperidines and *N*-acylmorpholines. J Chem Soc, Perkin Trans 2:807–813
91. Shono T, Matsumura Y, Tsubata K (1985a) Anodic oxidation of *N*-carbomethoxypyrrolidine: 2-Methoxy-*N*-carbomethoxypyrrolidine. Organic Syntheses 206–206
92. Shono T, Matsumura Y, Inoue K (1983) Electroorganic chemistry. 71. Anodic. alpha.-methoxylation of *N*-carbomethoxylated or *N*-acylated. alpha.-amino acid esters and. alpha.-amino-. beta.-lactams. J Org Chem 48:1388–1389
93. Baba D, Fuchigami T (2003) Anodic methoxylation and acetoxylation of imines and imidates. Tetrahedron Lett 44:3133–3136
94. Frankowski KJ, Liu R, Milligan GL, Moeller KD, Aubé J (2015) Practical electrochemical anodic oxidation of polycyclic lactams for late stage functionalization. Angew Chem Int Ed 54:10555–10558

95. Reuter C, Huy P, Neudörfl JM, Kühne R, Schmalz HG (2011) Exercises in pyrrolidine chemistry: gram scale synthesis of a pro–pro dipeptide mimetic with a polyproline type II helix conformation. Chem Eur J 17:12037–12044
96. Liu K, Tang S, Huang P, Lei A (2017) External oxidant-free electrooxidative [3+2] annulation between phenol and indole derivatives. Nat Commun 8:775
97. Chiba K, Fukuda M, Kim S, Kitano Y, Tada M (1999) Dihydrobenzofuran synthesis by an anodic [3+2] cycloaddition of phenols and unactivated alkenes. J Org Chem 64:7654–7656
98. Imada Y, Yamaguchi Y, Shida N, Okada Y, Chiba K (2017) Entropic electrolytes for anodic cycloadditions of unactivated alkene nucleophiles. Chem Commun 53:3960–3963
99. Chiba K, Jinno M, Kuramoto R, Tada M (1998) Stereoselective Diels-Alder reaction of electrogenerated quinones on a PTFE-fiber coated electrode in lithium perchlorate/nitromethane. Tetrahedron Lett 39:5527–5530
100. Sutterer A, Moeller KD (2000) Reversing the polarity of enol ethers: an anodic route to tetrahydrofuran and tetrahydropyran rings. J Am Chem Soc 122:5636–5637
101. Xu H-C, Moeller KD (2010) Intramolecular anodic olefin coupling reactions and the synthesis of cyclic amines. J Am Chem Soc 132:2839–2844
102. Xu H-C, Moeller KD (2008) Intramolecular anodic olefin coupling reactions: the use of a nitrogen trapping group. J Am Chem Soc 130:13542–13543
103. Xu H-C, Campbell JM, Moeller KD (2013) Cyclization reactions of anode-generated amidyl radicals. J Org Chem 79:379–391
104. Zhang S, Lian F, Xue M, Qin T, Li L, Zhang X, Xu K (2017) Electrocatalytic dehydrogenative esterification of aliphatic carboxylic acids: access to bioactive lactones. Org Lett 19:6622–6625
105. Tao XZ, Dai JJ, Zhou J, Xu J, Xu HJ (2018) Electrochemical C-O bond formation: facile access to aromatic lactones. Chem Eur J 24:6932–6935
106. Moeller KD, Sharif T (1989) Electrochemical amide oxidations in the presence of monomethoxylated phenyl rings. An unexpected relationship between the chemoselectivity of the oxidation and the location of the methoxy substituent. Tetrahedron Lett 30:1213–1216
107. Endoma MA, Butora G, Claeboe CD, Hudlicky T, Abboud KA (1997) Chemoenzymatic and electrochemical oxidations in the synthesis of octahydroisoquinolines for conversion to morphine. Relative merits of radical vs. acid-catalyzed cyclizations. Tetrahedron Lett 38:8833–8836
108. Okimoto M, Yoshida T, Hoshi M (2006) Electrooxidative cyclization of hydroquinolyl alcohols. Heterocycles 68:2563–2570
109. Okimoto M, Yoshida T, Hoshi M, Hattori K, Komata M, Numata K, Tomozawa K (2007) Electrooxidative cyclization of hydroquinolyl alcohols, hydroquinolylamines, and dimethyl aminomalonates. Aust J Chem 60:236–242
110. Okimoto M, Ohashi K, Yamamori H, Nishikawa S, Hoshi M, Yoshida T (2012) Electrooxidative cyclization of hydroxyamino compounds possessing a benzyl group. Synthesis 44:1315–1322
111. Li WC, Zeng CC, Hu LM, Tian HY, Little RD (2013) Efficient indirect electrochemical synthesis of 2-substituted benzoxazoles using sodium iodide as mediator. Adv Synth Catal 355:2884–2890
112. Ma H-Y, Zha Z-G, Zhang Z-L, Meng L, Wang Z-Y (2013) Electrosynthesis of oxadiazoles from benzoylhydrazines. Chin Chem Lett 24:780–782
113. Li X, Xu X, Zhou C (2012) Tetrabutylammonium iodide catalyzed allylic sulfonylation of α-methyl styrene derivatives with sulfonylhydrazides. Chem Commun 48:12240–12242
114. Röse P, Emge S, Yoshida J-I, Hilt G (2015) Electrochemical selenium-and iodonium-initiated cyclisation of hydroxy-functionalised 1, 4-dienes. Beilstein J Org Chem 11:174

115. Gao X, Yuan G, Chen H, Jiang H, Li Y, Qi C (2013) Efficient conversion of CO2 with olefins into cyclic carbonates via a synergistic action of I2 and base electrochemically generated in situ. Electrochem Commun 34:242–245
116. Gieshoff T, Kehl A, Schollmeyer D, Moeller KD, Waldvogel SR (2017) Electrochemical synthesis of benzoxazoles from anilides - a new approach to employ amidyl radical intermediates. Chem Commun 53:2974–2977
117. Shono T, Matsumura Y, Uchida K, Kobayashi H (1985) A new [3+3]-type annelation useful for the formation of piperidine skeletons. J Org Chem 50:3243–3245
118. Tang S, Liu K, Long Y, Qi X, Lan Y, Lei A (2015) Tuning radical reactivity using iodine in oxidative C (sp 3)–H/C (sp)–H cross-coupling: an easy way toward the synthesis of furans and indolizines. Chem Commun 51:8769–8772

Printed by Printforce, the Netherlands